GROUP THEORY IN PHYSICS

Volume III

Techniques of Physics

Editor

N. H. MARCH

Department of Theoretical Chemistry, University of Oxford, Oxford, England

Techniques of physics find wide application in biology, medicine, engineering and technology generally. This series is devoted to techniques that have found and are finding application. The aim is to clarify the principles of each technique, to emphasize and illustrate the applications, and to draw attention to new fields of possible employment.

1. D. C. Champeney: *Fourier Transforms and their Physical Applications*
2. L. B. Pendry: *Low Energy Electron Diffraction*
3. K. G. Beauchamp: *Walsh Functions and their Applications*
4. V. Cappellini, A. G. Constantinides and P. Emiliani: *Digital Filters and their Applications*
5. G. Rickayzen: *Green's Functions and Condensed Matter*
6. M. C. Hutley: *Diffraction Gratings*
7. J. F. Cornwell: *Group Theory in Physics, Vols I and II*
8. N. H. March and B. M. Deb: *The Single-Particle Density in Physics and Chemistry*
9. D. B. Pearson: *Quantum Scattering and Spectral Theory*
10. J. F. Cornwell: *Group Theory in Physics, Vol. III: Supersymmetries and Infinite-Dimensional Algebras*
11. J. M. Blackledge: *Quantitative Coherent Imaging*

GROUP THEORY IN PHYSICS

Volume III

Supersymmetries and Infinite-Dimensional Algebras

J. F. CORNWELL

Department of Physics and Astronomy
University of St Andrews, Scotland

ACADEMIC PRESS

Harcourt Brace Jovanovich, Publishers

London San Diego New York Berkeley
Boston Sydney Tokyo Toronto

ACADEMIC PRESS LIMITED
24–28 Oval Road
London NW1

US edition published by
Academic Press Inc.
San Diego, CA 92101

British Library Cataloguing in Publication Data

Cornwell, J. F.
Group theory in physics.—(Techniques of physics,
ISSN 0308-5392; 10)
Vol. 3
1. Groups, Theory of 2. Mathematical physics
I. Title II. Series
530.1′5222 QC20.7.G76

ISBN 0-12-189805-9

Typeset by Thomson Press (India) Limited, New Delhi
Printed in Great Britain by St Edmundsbury Press Ltd., Bury St Edmunds, Suffolk

Preface

As was said in the preface to the first two volumes of *Group Theory in Physics*, twenty years or so ago group theory could have been regarded merely as providing a very valuable tool for the elucidation of the symmetry aspects of physical problems, but recent developments, particularly in high-energy physics, have transformed its role, so that it now occupies a crucial and indispensable position at the centre of the stage. These developments have taken physicists increasingly deeper into the fascinating world of the pure mathematicians, and have led to an ever-growing appreciation of their achievements, the full recognition of which has been hampered to some extent by the style in which much of modern pure mathematics is presented. As with the previous two volumes, one of the main objectives of the present work is to try to overcome this communication barrier, and to present to theoretical physicists and others some important mathematical developments in a form that should be easier to comprehend and appreciate.

The previous two volumes gave a general introduction to the subject, which started right from the basic concepts and included a detailed account of the theory of Lie groups and finite-dimensional Lie algebras. The present volume is largely devoted to the extensions of these latter ideas in two different directions, first to finite-dimensional Lie *super*-algebras and then to *infinite*-dimensional Lie algebras, particularly to the Kac–Moody and Virasoro algebras. The combination of these two generalizations in a theory of infinite-dimensional Lie superalgebras is also treated, but in less detail. These ideas and their applications have generated an immense amount of interest and activity in the last fifteen years, and have given rise to the publication of a vast number of papers. Although the present account will include applications of these concepts to particle physics, including its latest manifestation in the form of string and superstring theory, the emphasis will be almost completely on the algebraic aspects of these subjects. Together with the limitations of space, this has meant that the quantum field theory aspect and the treatment of supergravity have had to be omitted.

Although examination of the contents list will show in detail the list of topics that are treated, it might be helpful to indicate here in broad outline the logical dependence of the various chapters on each other.

(i) Chapters 20–24 form a single sequence in which each chapter makes extensive use of the chapter that precedes it, the whole sequence providing an introduction to the Poincaré supersymmetric theories that have featured so prominently and extensively in the theoretical physics literature in the last fifteen years.

(ii) Chapter 25, which deals with simple Lie superalgebras, follows on from Chapter 22, and is independent of the three chapters that immediately precedes it (apart from some allusions in its Section 6). It is augmented by Appendix M, which summarizes a considerable amount of detailed data on the classical simple complex Lie superalgebras.

(iii) Chapters 26–28 form another single sequence, this time dealing with the infinite-dimensional Kac–Moody and Virasoro algebras. Apart from the superalgebra considerations of Chapter 26, Section 6, and Chapter 28, Section 4, these can be read independently of the chapters that appear before them. Detailed information on the complex affine Kac–Moody algebras is presented in Appendix N. (Much of the treatment of the Virasoro algebra in Chapter 28 does not depend on Kac–Moody algebras.)

(iv) Although the presentation of the algebraic aspects of string and superstring theory in Chapter 29 makes repeated reference to the Virasoro algebra, a knowledge of Kac–Moody algebras is only assumed in its penultimate section.

(v) The description of Clifford algebras in Appendix L is designed both to be read as a self-contained account and to provide a compendium of results that are needed elsewhere.

As in the previous two volumes, the *proofs* of theorems have been divided into three categories. First there are those that by virtue of the direct nature of their arguments assist in the appreciation of the theorems. These are included in the main text. Then there are proofs that are worth recording but which are more lengthy or indirect. These are relegated to Appendix K or O. Finally there are proofs that are just too long, or involve ideas that have not been developed in these volumes. For these, all that is given is a reference or references to works where they may be found.

I should like to thank Dr D. H. Hartley for his useful comments on the first draft of certain chapters, Mrs P. Askings for assistance with the typing, and my daughter Rebecca for producing the diagrams.

J.F. CORNWELL
St Andrews

To my wife Elizabeth and daughters
Rebecca and Jane

Contents

Contents of Volume I

Contents of Volume II

Part D

Lie Superalgebras, Lie Supergroups and their Applications

20

Introduction to Superalgebras and Supermatrices

1 The notion of grading

Much of this book is concerned with various *super* objects such as *superalgebras*, *supergroups* and *supermatrices*. The essential feature of all of these is that they are *graded*. The object of this first chapter is to introduce the concept of a graded algebraic structure, considering first the idea of a *graded vector space*. Superalgebras are obtained by enriching such a space by adding appropriate products. The simplest type is the *associative superalgebra* that will be examined in Section 2. The *Grassmann algebras* of Section 3 provide particularly important examples that will feature frequently throughout this book. Their first occurrence will be in Section 4, where they provide the matrix elements of supermatrices, which will themselves play an important role in the theory of supergroups.

The simplest example of a graded structure is provided by the integers, each of which is either *even* or *odd*. Moreover for ordinary addition it is well known that

$$\left.\begin{aligned} \text{even integer} + \text{even integer} &= \text{even integer},\\ \text{even integer} + \text{odd integer} &= \text{odd integer},\\ \text{odd integer} + \text{odd integer} &= \text{even integer}. \end{aligned}\right\} \qquad (20.1)$$

The operation of addition can be regarded as the group "product" of the additive group of integers. Denoting this product by •, (20.1) can be reexpressed as

$$\begin{aligned} \text{even} \bullet \text{even} &= \text{even}, \\ \text{even} \bullet \text{odd} &= \text{odd}, \\ \text{odd} \bullet \text{odd} &= \text{even}. \end{aligned} \tag{20.2}$$

Another example is provided by the cyclic group of order 2, Z_2, whose two elements E (the identity) and A are such that

$$EE = E, \quad EA = AE = A, \quad AA = E. \tag{20.3}$$

With E being described as being *even* and A as being *odd*, (20.3) is in agreement with the grading (20.2). For this reason a structure with the grading (20.2) is sometimes said to be "Z_2 graded".

It is possible to construct other types of grading, but in this book the word "graded" will henceforth be used to describe a structure of the type (20.2) (unless it is explicitly stated otherwise).

The first stage in applying the idea of grading to linear algebra is to define the concept of a *graded vector space.* To do this, suppose that V is a real or complex vector space of dimension $m + n$, where m and n are any two positive integers, and suppose that $a_1, a_2, \ldots, a_{m+n}$ is a basis for V. (An introductory account of vector spaces was given in Appendix B.) Then any element a of V can be written in the form

$$a = \sum_{j=1}^{m+n} \mu_j a_j,$$

where the coefficients μ_j are real or complex numbers (as appropriate). This space can be graded by arbitrarily decreeing that every element of the form

$$a = \sum_{j=1}^{m} \mu_j a_j$$

is to be regarded as being *even*, while every element of the form

$$a = \sum_{j=m+1}^{m+n} \mu_j a_j$$

is to be said to be *odd.* Thus the even elements involve only the first m basis elements and the odd elements involve only the remaining n basis elements. (Whenever a graded vector space is encountered later it will be assumed that its basis elements are ordered in this way.) Any element $a \in V$ that is either even or odd is said to be *homogeneous*, and the *degree* (or *parity*) of such elements is defined by

$$\deg a = \begin{cases} 0 & \text{if } a \text{ is even,} \\ 1 & \text{if } a \text{ is odd.} \end{cases} \tag{20.4}$$

The set of even elements of V form a subspace of V that is called the *even subspace* and which will be denoted by V_0. Similarly, the odd elements of V form the *odd subspace* V_1. Clearly V is the direct sum of V_0 and V_1, that is, $V = V_0 \oplus V_1$.

For a vector space the idea of grading only assumes its full significance when a product is defined for each pair of elements. When this product is associative the resulting structure is called an *associative superalgebra.* This type of algebra will be discussed in the next section. The slightly more complicated concept of a *Lie superalgebra*, which is a graded vector space with a certain non-associative product, will be deferred until Chapter 21.

2 Associative superalgebras

Definition *Associative superalgebra*

A graded vector space V for which for every pair of elements a and b there exists a product ab that is also in V, and for which

(i) for all $a, b, a', b' \in V$ and μ, λ, μ', λ' of the field of V (i.e. real or complex numbers as appropriate)

$$(\mu a + \mu' a')(\lambda b + \lambda' b') = \mu\lambda(ab) + \mu\lambda'(ab') + \mu'\lambda(a'b) + \mu'\lambda'(a'b'), \quad (20.5)$$

(ii) for all $a, b, c \in V$

$$(ab)c = a(bc), \quad (20.6)$$

and

(iii) the product satisfies the grading multiplication rule (20.2),

is called an *associative superalgebra.*

Example I *The matrix associative superalgebras* $M(p/q; \mathbb{C})$ *and* $M(p/q; \mathbb{R})$

Some of the most interesting and useful examples of associative superalgebras are provided by matrices, for which, with the product being taken to be ordinary matrix multiplication, the linearity and associative requirements (20.5) and (20.6) are automatically satisfied. Suppose that p and q are any two positive integers, and consider a $(p+q) \times (p+q)$ matrix $\mathbf{M}$ with complex entries and with the partitioning

$$\mathbf{M} = \begin{bmatrix} \mathbf{A} & \mathbf{B} \\ \mathbf{C} & \mathbf{D} \end{bmatrix}, \quad (20.7)$$

where $\mathbf{A}$, $\mathbf{B}$, $\mathbf{C}$ and $\mathbf{D}$ are submatrices with dimensions $p \times p$, $p \times q$, $q \times p$ and $q \times q$ respectively (cf. Appendix A, Section 1). Then

(i) $\mathbf{M}$ is said to be *even* if $\mathbf{B} = \mathbf{0}$ and $\mathbf{C} = \mathbf{0}$; that is, if

$$\mathbf{M} = \begin{bmatrix} \mathbf{A} & \mathbf{0} \\ \mathbf{0} & \mathbf{D} \end{bmatrix}; \tag{20.8a}$$

whereas

(ii) $\mathbf{M}$ is said to be *odd* if $\mathbf{A} = \mathbf{0}$ and $\mathbf{D} = \mathbf{0}$; that is, if

$$\mathbf{M} = \begin{bmatrix} \mathbf{0} & \mathbf{B} \\ \mathbf{C} & \mathbf{0} \end{bmatrix}. \tag{20.8b}$$

Thus if

$$\mathbf{N} = \begin{bmatrix} \mathbf{E} & \mathbf{0} \\ \mathbf{0} & \mathbf{H} \end{bmatrix} \tag{20.9a}$$

is another *even* matrix with the *same* partitioning, and $\mathbf{M}$ is the even matrix of (20.8a), then

$$\mathbf{MN} = \begin{bmatrix} \mathbf{AE} & \mathbf{0} \\ \mathbf{0} & \mathbf{DH} \end{bmatrix},$$

which is another *even* matrix. Similarly if $\mathbf{M}$ is the *odd* matrix of (20.8b) and $\mathbf{N}$ is the *even* matrix of (20.9a) then

$$\mathbf{MN} = \begin{bmatrix} \mathbf{0} & \mathbf{BH} \\ \mathbf{CE} & \mathbf{0} \end{bmatrix},$$

which is *odd*. Likewise if

$$\mathbf{N} = \begin{bmatrix} \mathbf{0} & \mathbf{F} \\ \mathbf{G} & \mathbf{0} \end{bmatrix} \tag{20.9b}$$

is any *odd* matrix with the same partitioning, and $\mathbf{M}$ is the *even* matrix (20.8a), then

$$\mathbf{MN} = \begin{bmatrix} \mathbf{0} & \mathbf{AF} \\ \mathbf{DG} & \mathbf{0} \end{bmatrix},$$

which is also *odd*. Finally if $\mathbf{M}$ is the *odd* matrix (20.8b) and $\mathbf{N}$ is the *odd* matrix (20.9b) then

$$\mathbf{MN} = \begin{bmatrix} \mathbf{BG} & \mathbf{0} \\ \mathbf{0} & \mathbf{CF} \end{bmatrix},$$

which is again *even*. Thus in every case the grading rules (20.2) are satisfied. As the set of all complex linear combinations of these matrices form a complex graded vector space, they constitute an associative superalgebra, which will be denoted by $M(p/q; \mathbb{C})$.

The subset of $M(p/q; \mathbb{C})$ consisting of all those matrices that have real entries will be denoted by $M(p/q; \mathbb{R})$. It too is an associative superalgebra, but its underlying vector space is real.

Definition *Commutative associative superalgebra*

An associative algebra is said to be *commutative* if

$$ba = (-1)^{(\deg a)(\deg b)} ab \tag{20.10}$$

for all homogeneous a and b of the algebra; that is, if

$$ba = \begin{cases} -ab & \text{if both } a \text{ and } b \text{ are odd,} \\ ab & \text{otherwise.} \end{cases} \tag{20.11}$$

3 Grassmann algebras

Grassmann algebras are particular examples of associative superalgebras that will play a very important part in the developments that follow. The first stage in the construction of such an algebra is to consider a set of L generators $\mathcal{E}_1, \mathcal{E}_2, \ldots, \mathcal{E}_L$, which are assumed to have products $\mathcal{E}_j\mathcal{E}_k$ such that

(i) for all j, k, $l = 1, 2, \ldots, L$,

$$(\mathcal{E}_j\mathcal{E}_k)\mathcal{E}_l = \mathcal{E}_j(\mathcal{E}_k\mathcal{E}_l); \tag{20.12}$$

(ii) for all $j, k = 1, 2, \ldots, L$,

$$\mathcal{E}_j\mathcal{E}_k = -\mathcal{E}_k\mathcal{E}_j; \tag{20.13}$$

and

(iii) each non-zero product

$$\mathcal{E}_{j_1}\mathcal{E}_{j_2}\cdots\mathcal{E}_{j_r}$$

involving r generators is linearly independent of products involving less than r generators.

It should be noted that (20.13) implies that

$$\mathcal{E}_j\mathcal{E}_j = 0 \tag{20.14}$$

for all $j = 1, 2, \ldots, L$.

This set of generators and products may be supplemented by introducing an identity, which is denoted by 1, and which is assumed to be such that

$$11 = 1 \tag{20.15}$$

and

$$1\mathcal{E}_j = \mathcal{E}_j 1 = \mathcal{E}_j \tag{20.16}$$

for all $j = 1, 2, \ldots, L$. It then follows that

$$1(\mathcal{E}_{j_1}\mathcal{E}_{j_2}\cdots\mathcal{E}_{j_r}) = (\mathcal{E}_{j_1}\mathcal{E}_{j_2}\cdots\mathcal{E}_{j_r})1 = \mathcal{E}_{j_1}\mathcal{E}_{j_2}\cdots\mathcal{E}_{j_r} \tag{20.17}$$

for any product of the generators.

(These products $\mathcal{E}_j\mathcal{E}_k$ are sometimes written in the literature as "wedge products" $\mathcal{E}_j \wedge \mathcal{E}_k$, and the resulting algebras are sometimes called "exterior algebras".)

Example I *Generators and their products for the case $L = 3$*

For $L = 3$ there are 3 generators $\mathcal{E}_1$, $\mathcal{E}_2$ and $\mathcal{E}_3$. The *only three* linearly independent non-trivial *pairs* are $\mathcal{E}_1\mathcal{E}_2$, $\mathcal{E}_2\mathcal{E}_3$ and $\mathcal{E}_1\mathcal{E}_3$, for, by (20.14), $\mathcal{E}_1\mathcal{E}_1 = \mathcal{E}_2\mathcal{E}_2 = \mathcal{E}_3\mathcal{E}_3 = 0$, and, by (20.13), $\mathcal{E}_2\mathcal{E}_1 = -\mathcal{E}_1\mathcal{E}_2$, $\mathcal{E}_3\mathcal{E}_2 = -\mathcal{E}_2\mathcal{E}_3$, and $\mathcal{E}_3\mathcal{E}_1 = -\mathcal{E}_1\mathcal{E}_3$. There is *only one* independent non-trivial *triple*, namely $\mathcal{E}_1\mathcal{E}_2\mathcal{E}_3$, for triples such as $\mathcal{E}_1\mathcal{E}_1\mathcal{E}_2$ and $\mathcal{E}_2\mathcal{E}_3\mathcal{E}_2$ that involve a repeated generator are identically zero by (20.14), while every other triple with all factors different, such as $\mathcal{E}_3\mathcal{E}_2\mathcal{E}_1$, can be recast as multiples of $\mathcal{E}_1\mathcal{E}_2\mathcal{E}_3$ by repeated use of (20.13), so that, for example, $\mathcal{E}_3\mathcal{E}_2\mathcal{E}_1 = -\mathcal{E}_1\mathcal{E}_2\mathcal{E}_3$. Moreover there are *no* non-trivial products involving 4 or more factors, because in such a product at least one generator must be repeated. (For example, for $\mathcal{E}_1\mathcal{E}_2\mathcal{E}_3\mathcal{E}_1$, applying (20.13) twice gives $\mathcal{E}_1\mathcal{E}_1\mathcal{E}_2\mathcal{E}_3$, which is zero by (20.14).) Thus, with the identity included, the only independent products of generators are

$$1,\quad \mathcal{E}_1,\quad \mathcal{E}_2,\quad \mathcal{E}_3,\quad \mathcal{E}_1\mathcal{E}_2,\quad \mathcal{E}_2\mathcal{E}_3,\quad \mathcal{E}_1\mathcal{E}_3,\quad \mathcal{E}_1\mathcal{E}_2\mathcal{E}_3. \tag{20.18}$$

There are 8 ($= 2^3$) members in this set.

Generalizing the considerations of Example I to any positive integer value of L, it is easily shown that the number of non-zero independent generators and their products (with 1 included) is 2^L, none of these products contains more than L factors, and, except for the identity 1, each one can be written in the form

$$\mathcal{E}_{j_1}\mathcal{E}_{j_2}\cdots\mathcal{E}_{j_r},$$

where

$$1 \leqslant r \leqslant L$$

and

$$1 \leqslant j_1 < j_2 < j_3 < \cdots < j_r \leqslant L.$$

As it is convenient to have a notation in which such a product is written more succinctly, let μ be an *index set* that contains $N(\mu)$ *different* integers with values between 1 and L inclusive, i.e.

$$\mu = \{ j_1, j_2, \ldots, j_{N(\mu)} \}, \tag{20.19}$$

where the integers $j_1, j_2, \ldots, j_{N(\mu)}$ are assumed to be ordered in such a way that

$$1 \leqslant j_1 < j_2 < j_3 < \cdots < j_{N(\mu)} \leqslant L, \tag{20.20}$$

and *define* $\mathcal{E}_\mu$ by

$$\mathcal{E}_\mu = \mathcal{E}_{j_1}\mathcal{E}_{j_2}\cdots\mathcal{E}_{j_{N(\mu)}}. \tag{20.21}$$

The identity 1 can be incorporated into this scheme by the definition $\mathcal{E}_\varnothing = 1$, where $\varnothing$ is the "empty" index set. Thus in this notation the independent products of generators for the $L = 3$ case of Example I are

$$\begin{aligned}
&\mathcal{E}_\varnothing = 1,\\
&\mathcal{E}_{\{1\}} = \mathcal{E}_1, \quad \mathcal{E}_{\{2\}} = \mathcal{E}_2, \quad \mathcal{E}_{\{3\}} = \mathcal{E}_3,\\
&\mathcal{E}_{\{1,2\}} = \mathcal{E}_1\mathcal{E}_2, \quad \mathcal{E}_{\{2,3\}} = \mathcal{E}_2\mathcal{E}_3, \quad \mathcal{E}_{\{1,3\}} = \mathcal{E}_1\mathcal{E}_3,\\
&\mathcal{E}_{\{1,2,3\}} = \mathcal{E}_1\mathcal{E}_2\mathcal{E}_3.
\end{aligned}$$

Clearly, if μ and μ' are two index sets with a *common* element then

$$\mathcal{E}_\mu\mathcal{E}_{\mu'} = 0. \tag{20.22}$$

For example, if $\mu = \{1, 2\}$ and $\mu' = \{1, 3\}$ then

$$\mathcal{E}_{\{1,2\}}\mathcal{E}_{\{1,3\}} = (\mathcal{E}_1\mathcal{E}_2)(\mathcal{E}_1\mathcal{E}_3) = -(\mathcal{E}_1\mathcal{E}_1)(\mathcal{E}_2\mathcal{E}_3) = 0.$$

Similarly, if μ and μ' have no element in common then

$$\mathcal{E}_\mu\mathcal{E}_{\mu'} = \pm\,\mathcal{E}_{\mu''}, \tag{20.23}$$

where μ'' is the index set consisting of all the integers of μ together with those of μ', the sign in the right-hand side of (20.23) being determined by the number of permutations required to order μ'' in accordance with (20.20). Thus, for instance, if $\mu = \{1, 3\}$ and $\mu' = \{2\}$ then $\mu'' = \{1, 2, 3\}$ and

$$\mathcal{E}_\mu\mathcal{E}_{\mu'} = (\mathcal{E}_1\mathcal{E}_3)\mathcal{E}_2 = -\,\mathcal{E}_1\mathcal{E}_2\mathcal{E}_3 = -\,\mathcal{E}_{\mu''}.$$

Finally

$$\mathcal{E}_\varnothing\mathcal{E}_\mu = \mathcal{E}_\mu\mathcal{E}_\varnothing = \mathcal{E}_\mu \tag{20.24}$$

for any index set μ.

The last stage in the construction is to set up linear combinations of all the $\mathcal{E}_\mu$ (for a fixed value of L); that is, to consider the quantities of the form

$$B = \sum_\mu B_\mu\mathcal{E}_\mu, \tag{20.25}$$

where the coefficients B_μ are either all real numbers or are all complex numbers, and the sum is over all the index sets μ. The product of two such quantities B and B' may be defined by

$$BB' = \sum_{\mu}\sum_{\mu'} B_\mu B'_{\mu'}(\mathcal{E}_\mu \mathcal{E}_{\mu'}), \tag{20.26}$$

where B is given by (20.25) and B' is assumed to be given by the similar expression

$$B' = \sum_{\mu} B'_\mu \mathcal{E}_\mu.$$

If the index set μ contains $N(\mu)$ integers then the corresponding element $\mathcal{E}_\mu$ may be said to be of *level* $N(\mu)$. Thus if $\mathcal{E}_{\mu'}$ is of level $N(\mu')$ then the product $\mathcal{E}_\mu \mathcal{E}_{\mu'}$ is either 0 or is of level $N(\mu) + N(\mu')$, and so the level of a product *never decreases*. This has two important consequences. First, $\mathcal{E}_\mu \mathcal{E}_{\mu'}$ is of level 0 if and only if $\mathcal{E}_\mu$ and $\mathcal{E}_{\mu'}$ are *both* of level 0; that is, the identity element $\mathcal{E}_\varnothing$ can only be obtained in a product $\mathcal{E}_\mu \mathcal{E}_{\mu'}$ if *both* factors $\mathcal{E}_\mu$ and $\mathcal{E}_{\mu'}$ are themselves equal to the identity $\mathcal{E}_\varnothing$. Secondly, if the L is finite and if the identity coefficient $B_\varnothing$ of B in (20.25) is 0, then there exists a positive integer r such that $B^r = 0$, and B is said to be "*nilpotent*".

The set of all linear combinations (20.25) form a *vector space* for which the 2^L quantities $\mathcal{E}_\mu$ form a basis. Of course with the coefficients B_μ of (20.25) all real this is a *real* vector space; and if the B_μ are complex, it is a *complex* vector space. Moreover (20.25) implies that the linearity condition (20.5) is satisfied, while (20.12) and (20.15) yield the associative property (20.6).

It remains only to set up a grading, which can be done by defining the *degree* of a basis element $\mathcal{E}_\mu$ of level $N(\mu)$ by

$$\deg \mathcal{E}_\mu = (-1)^{N(\mu)}, \tag{20.27}$$

so that $\mathcal{E}_\mu$ is even if $N(\mu)$ is even and $\mathcal{E}_\mu$ is odd if $N(\mu)$ is odd. In particular the identity $\mathcal{E}_\varnothing$ is even, as $N(\varnothing) = 0$. The most general even element is then of the form (20.25), provided that the sum only involves the even basis elements $\mathcal{E}_\mu$, and likewise for a general odd element the sum (20.25) can only involve the odd basis elements. Equations (20.26), (20.22), (20.23) and (20.24) then imply that the grading rule (20.2) is satisfied.

In the case in which the vector space is real, this structure is a real associative superalgebra. It is known as a *real Grassmann algebra*, and will be denoted by $\mathbb{R}B_L$. If B is a member of $\mathbb{R}B_L$ then B will be described as being a *real* Grassmann element. Similarly, if the vector space is complex then the structure is a complex associative superalgebra, which is called a *complex Grassmann algebra*, and which will be denoted by $\mathbb{C}B_L$. Both are of dimension 2^L in their respective fields. The subset of even elements of $\mathbb{R}B_L$ and the subset of odd elements of $\mathbb{R}B_L$ both form real vector spaces of dimension 2^{L-1}. They will be denoted by $\mathbb{R}B_{L0}$ and $\mathbb{R}B_{L1}$ respectively. The corresponding even and odd subspaces of $\mathbb{C}B_L$ will be denoted by $\mathbb{C}B_{L0}$ and $\mathbb{C}B_{L1}$. Both are complex vector spaces of dimension 2^{L-1}.

Equations (20.13), (20.15), (20.17) and (20.29) show that if B and B' are any two members of $\mathbb{R}B_L$ or $\mathbb{C}B_L$ then

$$BB' = \begin{cases} -B'B & \text{if } B \text{ and } B' \text{ are both odd,} \\ BB' & \text{otherwise,} \end{cases} \tag{20.28}$$

so (cf. (20.11)) $\mathbb{R}B_L$ and $\mathbb{C}B_L$ are both *commutative* associative superalgebras.

For thc original account of the formulation of these algebras see Grassmann (1911).

Definition *Complex conjugate B^* of B*

Let B be any element of $\mathbb{C}B_L$, and let its expansion of the form (20.25) be rewritten as

$$B = \sum_{\mu} (u_\mu + iv_\mu)\mathcal{E}_\mu, \tag{20.29}$$

where each u_μ and v_μ is a real number. The *complex conjugate* B^* of B is then defined to be the element of $\mathbb{C}B_L$ whose expansion is

$$B^* = \sum_{\mu} (u_\mu - iv_\mu)\mathcal{E}_\mu.$$

Thus B is real if and only if $B^* = B$.

Definition *Adjoint $B^\#$ of B*

Let B be any element of $\mathbb{C}B_L$ and let (20.29) be an expansion for B in which each u_μ and v_μ is again a real number. The *adjoint* $B^\#$ of B is then defined to be the element of $\mathbb{C}B_L$ whose expansion is

$$B^\# = \sum_{\mu} (u_\mu - iv_\mu)\mathcal{E}_\mu^\#, \tag{20.30}$$

where each $\mathcal{E}_\mu^\#$ is itself defined by

$$\mathcal{E}_\mu^\# = \begin{cases} \mathcal{E}_\mu & \text{if } \mathcal{E}_\mu \text{ is even,} \\ -i\mathcal{E}_\mu & \text{if } \mathcal{E}_\mu \text{ is odd.} \end{cases} \tag{20.31}$$

This definition implies that if B is *real*, that is, if $B \in \mathbb{R}B_L$, then

$$B^\# = \begin{cases} B & \text{if } B \text{ is also even, i.e. if } B \in \mathbb{R}B_{L0}, \\ -iB & \text{if } B \text{ is also odd, i.e. if } B \in \mathbb{R}B_{L1}. \end{cases} \tag{20.32}$$

In particular it should be noted that if B is real and odd then $B^{\#} \neq B$. That is, $B^{\#} = B$ is *not* the criterion for B to be real! For a general element $X + iY$ of $\mathbb{C}B_{L0}$, where X and Y are both members of $\mathbb{R}B_{L0}$, (20.30) and (20.31) show that

$$(X + iY)^{\#} = X - iY = (X + iY)^{*}, \tag{20.33}$$

whereas for a general element $\Theta + i\Psi$ of $\mathbb{C}B_{L1}$, where Θ and Ψ are both members of $\mathbb{R}B_{L1}$, (20.30) and (20.32) show that

$$(\Theta + i\Psi)^{\#} = -i\Theta - \Psi. \tag{20.34}$$

This adjoint operation has five useful properties:

(a) if $B \in \mathbb{C}B_{Lj}$ (for $j = 0$ or 1) then $B^{\#} \in \mathbb{C}B_{Lj}$;

(b) for any pair B and B' of $\mathbb{C}B_L$,

$$(B + B')^{\#} = B^{\#} + B'^{\#}; \tag{20.35}$$

(c) if $B \in \mathbb{C}B_L$ and b is any complex number then

$$(bB)^{\#} = b^{*}B^{\#},$$

where b^{*} here denotes the complex conjugate of b;

(d) for any pair B and B' of $\mathbb{C}B_L$,

$$(BB')^{\#} = B'^{\#}B^{\#}; \tag{20.36}$$

(e) for all B of $\mathbb{C}B_L$,

$$(B^{\#})^{\#} = B. \tag{20.37}$$

It should be noted that on the right-hand side of (20.36) the order of the factors B and B' is *reversed* relative to that of the left-hand side.

It is appropriate to mention here another adjoint operation, which has been used extensively by DeWitt (1984), and which will be denoted here by $B^{\S}$ rather than $B^{\#}$ to avoid confusion.

Definition *Adjoint $B^{\S}$ of B*

Let B be any element of $\mathbb{C}B_L$, and let (20.29) be an expansion for B in which each u_μ and v_μ is again a real number. The *adjoint* $B^{\S}$ of B is defined to be the element of $\mathbb{C}B_L$ whose expansion is

$$B^{\S} = \sum_{\mu} (u_\mu - iv_\mu)\mathcal{E}_\mu^{\S}, \tag{20.38}$$

where, if $\mathcal{E}_\mu$ is given by (20.21), $\mathcal{E}_\mu^{\S}$ is defined by

$$\mathcal{E}_\mu^{\S} = \mathcal{E}_{j_{N(\mu)}} \cdots \mathcal{E}_{j_2}\mathcal{E}_{j_1}, \tag{20.39}$$

that is, the order of generators in $\mathcal{E}_\mu^{\S}$ is *reversed*.

It is easily verified that the § operation has the same properties (a), (b), (c), (d) and (e) as the # operation, but the analogues of (20.32), (20.33), and (20.34) are much more complicated. (Indeed DeWitt (1984) defines a Grassmann element B to be "real" if $B^{\S} = B$ and to be "imaginary" if $B^{\S} = -B$. With DeWitt's definition, $\mathcal{E}_\mu$ is "real" if $\frac{1}{2}N(\mu)\{N(\mu)-1\}$ is even, and $\mathcal{E}_\mu$ is "imaginary" if $\frac{1}{2}N(\mu)\{N(\mu)-1\}$ is odd, whereas, with the definition adopted above, which will continue to be used throughout this book, *every* $\mathcal{E}_\mu$ is *real* because the expansion (20.25) for $\mathcal{E}_\mu$ involves only a single *real* coefficient B_μ (with value 1). With DeWitt's definition of reality, the set of "real" Grassman elements do *not* form an *algebra*, for if both B and B' are odd and "real" in DeWitt's sense then, by the § analogue of (20.36), $(BB')^{\S} = B'^{\S}B^{\S} = (-B')(-B) = B'B = -BB'$, so BB' is necessarily "imaginary". Indeed this argument shows that for *every* adjoint operation # possessing the property (20.36) the set of elements satisfying $B^{\#} = B$ do *not* form a closed set under multiplication, and so they *cannot* form a subalgebra of $\mathbb{C}B_L$. That is, a real subalgebra of $\mathbb{C}B_L$ can *never* be defined by means of such an adjoint operator.)

4 Supermatrices

A *supermatrix* is a special type of matrix whose elements are members of a Grassmann algebra. For convenience of exposition this will be assumed to be the complex algebra $\mathbb{C}B_L$, but this could be replaced at each stage by the corresponding real algebra $\mathbb{R}B_L$. For the rest of this section p, q, r and s will denote non-negative integers such that $p+q \geqslant 1$ and $r+s \geqslant 1$.

There are only two kinds of supermatrix, which are called "even" and "odd".

Definition *Even and odd supermatrices*

Let $\mathbf{M}$ be a $(p+q)\times(r+s)$ matrix that is partitioned in the form

$$\mathbf{M} = \begin{bmatrix} \mathbf{A} & \mathbf{B} \\ \mathbf{C} & \mathbf{D} \end{bmatrix}, \tag{20.40}$$

where $\mathbf{A}, \mathbf{B}, \mathbf{C}$ and $\mathbf{D}$ are submatrices with dimensions $p\times r$, $p\times s$, $q\times r$ and $q\times s$ respectively. Then

(i) $\mathbf{M}$ is said to be a $(p/q)\times(r/s)$ *even supermatrix* if the elements of $\mathbf{A}$ and $\mathbf{D}$ are members of the *even* Grassmann subspace $\mathbb{C}B_{L0}$ and the elements of $\mathbf{B}$ and $\mathbf{C}$ are members of the *odd* Grassmann subspace $\mathbb{C}B_{L1}$; and

(ii) $\mathbf{M}$ is said to be a $(p/q)\times(r/s)$ *odd supermatrix* if the elements of $\mathbf{A}$ and $\mathbf{D}$ are members of the *odd* Grassmann subspace $\mathbb{C}B_{L1}$ and the elements of $\mathbf{B}$ and $\mathbf{C}$ are members of the *even* Grassmann subspace $\mathbb{C}B_{L0}$.

Definition *Degree of a supermatrix*

The *degree*, $\deg \mathbf{M}$, of a supermatrix $\mathbf{M}$ is defined by

$$\deg \mathbf{M} = \begin{cases} 0 & \text{if } \mathbf{M} \text{ is even,} \\ 1 & \text{if } \mathbf{M} \text{ is odd.} \end{cases} \tag{20.41}$$

The matrix product of a $(p/q) \times (r/s)$ supermatrix $\mathbf{M}$ with a $(p'/q') \times (r'/s')$ supermatrix $\mathbf{M}'$ is defined only if $r = p'$ and $s = q'$, and is then given by the usual combination

$$(\mathbf{MM}')_{jk} = \sum_{l=1}^{r+s} M_{jl} M'_{lk} \tag{20.42}$$

(cf. (A.1)), but, as the elements M_{jl} and M'_{lk} are assumed to be members of $\mathbb{C}B_L$, each of the products $M_{jl}M'_{lk}$ is a "Grassmann" product of the type considered in the previous section. As the Grassmann product is associative (cf. (20.6)), it follows that

$$(\mathbf{MM}')\mathbf{M}'' = \mathbf{M}(\mathbf{M}'\mathbf{M}'') \tag{20.43}$$

for any three supermatrices $\mathbf{M}$, $\mathbf{M}'$ and $\mathbf{M}''$ whose partitionings are such as to make the products well defined. Moreover the grading rule (20.2) is satisfied for multiplication of supermatrices.

The most useful supermatrices are those that are either "square" (in the sense that $p = r$ *and* $q = s$) or are in the form of row or column matrices.

Definition *The sets* $\mathrm{M}(p/q; \mathbb{C}B_L)$ *and* $\mathrm{M}(p/q; \mathbb{R}B_L)$ *of square supermatrices*

The set of all $(p/q) \times (p/q)$ supermatrices with entries in $\mathbb{C}B_{L0}$ or $\mathbb{C}B_{L1}$ is denoted by $\mathrm{M}(p/q; \mathbb{C}B_L)$ (each supermatrix being necessarily either even or odd). The corresponding set with entries restricted to the corresponding real Grassmann subspaces will be denoted by $\mathrm{M}(p/q; \mathbb{R}B_L)$.

Definition *Super row matrix*

A $(1/0) \times (r/s)$ *super row matrix* is a $(1/0) \times (r/s)$ *even* supermatrix. That is, it is a $1 \times (r + s)$ matrix of the form

$$\mathbf{m} = [\mathbf{a} \quad \mathbf{b}], \tag{20.44}$$

where $\mathbf{a}$ is a $1 \times r$ row matrix whose elements are members of the *even* Grassmann subspace $\mathbb{C}B_{L0}$ and $\mathbf{b}$ is a $1 \times s$ row matrix whose elements are members of the *odd* Grassmann subspace $\mathbb{C}B_{L1}$.

Definition *Super column matrix*

A $(p/q) \times (1/0)$ *super column matrix* is a $(p/q) \times (1/0)$ *even* supermatrix. That is, it is a $(p+q) \times 1$ matrix of the form

$$\mathbf{m} = \begin{bmatrix} \mathbf{a} \\ \mathbf{c} \end{bmatrix}, \tag{20.45}$$

where $\mathbf{a}$ is a $p \times 1$ column matrix whose elements are members of the *even* Grassmann subspace $\mathbb{C}\mathrm{B}_{L0}$ and $\mathbf{c}$ is a $q \times 1$ column matrix whose elements are members of the *odd* Grassmann subspace $\mathbb{C}\mathrm{B}_{L1}$.

There is a simple connection between the supermatrices of $\mathrm{M}(p/q;\mathbb{C}\mathrm{B}_L)$ and the complex-valued matrices of $\mathrm{M}(p/q;\mathbb{C})$ that were discussed in Example I of Section 2. If every matrix element of a matrix $\mathbf{M}$ of $\mathrm{M}(p/q;\mathbb{C})$ is multiplied by the Grassmann identity 1 of $\mathbb{C}\mathrm{B}_L$ then the resulting matrix is a member of $\mathrm{M}(p/q;\mathbb{C}\mathrm{B}_L)$, and is an even supermatrix if $\mathbf{M}$ is even and is an odd supermatrix if $\mathbf{M}$ is odd. For example, if $p = q = 1$, this process produces a mapping Ψ defined by

$$\Psi\left(\begin{bmatrix} a & 0 \\ 0 & d \end{bmatrix}\right) \rightarrow \begin{bmatrix} a1 & 0 \\ 0 & d1 \end{bmatrix} \tag{20.46}$$

and

$$\Psi\left(\begin{bmatrix} 0 & b \\ c & 0 \end{bmatrix}\right) \rightarrow \begin{bmatrix} 0 & b1 \\ c1 & 0 \end{bmatrix}, \tag{20.47}$$

where a, b, c and d are complex numbers. Clearly, because of (20.15),

$$\Psi(\mathbf{M})\Psi(\mathbf{M}') = \Psi(\mathbf{M}\mathbf{M}') \tag{20.48}$$

for all $\mathbf{M}, \mathbf{M}' \in \mathrm{M}(1/1;\mathbb{C})$. The corresponding mapping from $\mathrm{M}(p/q;\mathbb{C})$ into $\mathrm{M}(p/q;\mathbb{C}\mathrm{B}_L)$ has the same property. Consequently $\mathrm{M}(p/q;\mathbb{C})$ can be identified with a subset of $\mathrm{M}(p/q;\mathbb{C}\mathrm{B}_L)$, and so all the properties that will be discussed in this section for the supermatrices of $\mathrm{M}(p/q;\mathbb{C}\mathrm{B}_L)$ can be immediately translated into corresponding properties of the matrices of $\mathrm{M}(p/q;\mathbb{C})$. Clearly there is a similar relationship between $M(p/q;\mathbb{R})$ and $\mathrm{M}(p/q;\mathbb{R}\mathrm{B}_L)$.

Definition *Unit supermatrix*

The $(p/q) \times (p/q)$ *unit supermatrix* $\mathbf{1}$ is the $(p+q) \times (p+q)$ diagonal matrix all of whose diagonal elements are equal to the identity 1 of the Grassmann algebra $\mathbb{C}\mathrm{B}_L$.

Clearly the $(p/q) \times (p/q)$ unit supermatrix is an *even* supermatrix.

Definition *Invertible supermatrix*

A supermatrix $\mathbf{M}$ of $\mathrm{M}(p/q; \mathbb{C}B_L)$ is said to be "invertible" if there exists an "inverse matrix", also in $\mathrm{M}(p/q; \mathbb{C}B_L)$, which will be denoted by $\mathbf{M}^{-1}$, such that

$$\mathbf{M}\mathbf{M}^{-1} = \mathbf{M}^{-1}\mathbf{M} = \mathbf{1}. \tag{20.49}$$

The $(1/0) \times (1/0)$ even supermatrix $\mathbf{M} = [\mathcal{E}_{\{1,2\}}]$ provides an example of a supermatrix that is *not* invertible but that has a *non-zero* determinant. This shows that the usual relationship between invertibility and the non-vanishing of the determinant does *not* generalize to supermatrices. For even supermatrices, which are the ones for which the question of invertibility is the most important, the basic results are embodied in the next theorem.

Theorem I (a) Suppose that $\mathbf{A}$ is a $(p/0) \times (p/0)$ even supermatrix (that is, $\mathbf{A}$ is a $p \times p$ matrix all of whose elements are members of $\mathbb{C}B_{L0}$). Let

$$\mathbf{A} = \sum_{\mu} \mathbf{A}_{\mu} \mathcal{E}_{\mu} \tag{20.50}$$

be the expansion corresponding to (20.25), where for each index set μ the matrix $\mathbf{A}_{\mu}$ is a $p \times p$ matrix with complex entries. Then $\mathbf{A}$ is invertible if and only if the matrix $\mathbf{A}_{\varnothing}$ corresponding to the Grassmann identity $\mathcal{E}_{\varnothing}(=1)$ is invertible.

(b) If $\mathbf{M}$ is a $(p/q) \times (p/q)$ even supermatrix, partitioned as in (20.40), then $\mathbf{M}$ is invertible if and only if its submatrices $\mathbf{A}$ and $\mathbf{D}$ are invertible.

(c) If $\mathbf{M}$ and $\mathbf{N}$ are $(p/q) \times (p/q)$ even invertible supermatrices then $\mathbf{MN}$ is also a $(p/q) \times (p/q)$ even invertible supermatrix.

Proof See Appendix K, Section 1.

Part (c) of this theorem shows that the set of supermatrices of the following definition do indeed form a group.

Definition *The group* $\mathrm{GL}(p/q; \mathbb{C}B_L)$

The group $\mathrm{GL}(p/q; \mathbb{C}B_L)$ is the set of all $(p/q) \times (p/q)$ even invertible supermatrices, the group product operation being the matrix multiplication of (20.42).

This group is a generalization of the group $\mathrm{GL}(n, \mathbb{C})$ of $n \times n$ complex-valued non-singular matrices, and will play a similar role in the development of Lie supergroups to that played by $\mathrm{GL}(n, \mathbb{C})$ in "ordinary" Lie group theory.

Although the definition (20.42) of matrix multiplication for supermatrices is the obvious and straightforward generalization of the usual definition (A.1), when the other operations on ordinary matrices (such as multiplying by a scalar, or taking the transpose) are generalized to supermatrices it is necessary to introduce extra sign factors. These are chosen in a consistent way so that the result of performing any succession of operations on a supermatrix is similar to the corresponding result on ordinary matrices. The various "super-operations" will now be considered in turn, the important properties being displayed in the form of theorems.

Definition *Scalar multiplication of a supermatrix by a Grassmann element*

Suppose that $B \in \mathbb{C}B_{L0}$ or $\mathbb{C}B_{L1}$ and that $\mathbf{M}$ is a $(p/q) \times (r/s)$ supermatrix, partitioned as in (20.40). Then $B\mathbf{M}$ and $\mathbf{M}B$ are defined as $(p/q) \times (r/s)$ supermatrices by

$$B\mathbf{M} = \begin{bmatrix} B\mathbf{1}_p & \mathbf{0} \\ \mathbf{0} & (-1)^{\deg B} B\mathbf{1}_q \end{bmatrix} \begin{bmatrix} \mathbf{A} & \mathbf{B} \\ \mathbf{C} & \mathbf{D} \end{bmatrix} \tag{20.51}$$

and

$$\mathbf{M}B = \begin{bmatrix} \mathbf{A} & \mathbf{B} \\ \mathbf{C} & \mathbf{D} \end{bmatrix} \begin{bmatrix} B\mathbf{1}_r & \mathbf{0} \\ \mathbf{0} & (-1)^{\deg B} B\mathbf{1}_s \end{bmatrix}, \tag{20.52}$$

where $\mathbf{1}_p$, $\mathbf{1}_q$, $\mathbf{1}_r$ and $\mathbf{1}_s$ are $p \times p$, $q \times q$, $r \times r$ and $s \times s$ unit matrices.

It is obvious that in general

$$B\mathbf{M} \neq \mathbf{M}B, \tag{20.53}$$

but if $\mathbf{M}, \mathbf{N} \in \mathrm{M}(p/q; \mathbb{C}B_L)$ then

$$B(\mathbf{MN}) = (B\mathbf{M})\mathbf{N}, \tag{20.54}$$

$$(\mathbf{M}B)\mathbf{N} = \mathbf{M}(B\mathbf{N}), \tag{20.55}$$

$$\mathbf{M}(\mathbf{N}B) = (\mathbf{MN})B, \tag{20.56}$$

for all B of $\mathbb{C}B_{L0}$ or $\mathbb{C}B_{L1}$ (because of the associative property (20.43)).

Definition *Supertranspose of a supermatrix*

If $\mathbf{M}$ is a $(p/q) \times (r/s)$ supermatrix, partitioned as in (20.40), then its *supertranspose* is the $(r/s) \times (p/q)$ supermatrix $\mathbf{M}^{st}$ defined by

$$\mathbf{M}^{st} = \begin{bmatrix} \tilde{\mathbf{A}} & (-1)^{\deg \mathbf{M}} \tilde{\mathbf{C}} \\ -(-1)^{\deg \mathbf{M}} \tilde{\mathbf{B}} & \tilde{\mathbf{D}} \end{bmatrix}, \tag{20.57}$$

where $\tilde{\mathbf{A}}$, $\tilde{\mathbf{B}}$, $\tilde{\mathbf{C}}$ and $\tilde{\mathbf{D}}$ are the "ordinary" transposes of $\mathbf{A}$, $\mathbf{B}$, $\mathbf{C}$ and $\mathbf{D}$

respectively. That is,

$$\mathbf{M}^{\mathrm{st}} = \begin{cases} \begin{bmatrix} \tilde{\mathbf{A}} & \tilde{\mathbf{C}} \\ -\tilde{\mathbf{B}} & \tilde{\mathbf{D}} \end{bmatrix} & \text{if } \mathbf{M} \text{ is even,} \\ \begin{bmatrix} \tilde{\mathbf{A}} & -\tilde{\mathbf{C}} \\ \tilde{\mathbf{B}} & \tilde{\mathbf{D}} \end{bmatrix} & \text{if } \mathbf{M} \text{ is odd.} \end{cases} \tag{20.58}$$

In particular, if $\mathbf{m}$ is the $(1/0) \times (r/s)$ super row matrix of (20.44) then $\mathbf{m}^{\mathrm{st}}$ is the $(r/s) \times (1/0)$ super column matrix

$$\mathbf{m}^{\mathrm{st}} = \begin{bmatrix} \tilde{\mathbf{a}} \\ -\tilde{\mathbf{b}} \end{bmatrix}. \tag{20.59}$$

Similarly, if $\mathbf{m}$ is the $(p/q) \times (1/0)$ super column matrix of (20.45) then $\mathbf{m}^{\mathrm{st}}$ is the $(1/0) \times (p/q)$ super row matrix

$$\mathbf{m}^{\mathrm{st}} = [\tilde{\mathbf{a}} \quad \tilde{\mathbf{c}}]. \tag{20.60}$$

Theorem II (a) If $\mathbf{M}$ is a $(p/q) \times (r/s)$ supermatrix, partitioned as in (20.40), then

$$(\mathbf{M}^{\mathrm{st}})^{\mathrm{st}} = \begin{bmatrix} \mathbf{A} & -\mathbf{B} \\ -\mathbf{C} & \mathbf{D} \end{bmatrix} \tag{20.61}$$

and so

$$(((\mathbf{M}^{\mathrm{st}})^{\mathrm{st}})^{\mathrm{st}})^{\mathrm{st}} = \mathbf{M}. \tag{20.62}$$

(b) If $\mathbf{M}$ is a $(p/q) \times (r/s)$ supermatrix and $\mathbf{N}$ is a $(r/s) \times (t/u)$ supermatrix then

$$(\mathbf{MN})^{\mathrm{st}} = (-1)^{(\deg \mathbf{M})(\deg \mathbf{N})} \mathbf{N}^{\mathrm{st}} \mathbf{M}^{\mathrm{st}}. \tag{20.63}$$

(c) If $B \in \mathbb{C}\mathrm{B}_{L0}$ or $\mathbb{C}\mathrm{B}_{L1}$ and $\mathbf{M}$ is a $(p/q) \times (r/s)$ supermatrix then

$$(B\mathbf{M})^{\mathrm{st}} = B(\mathbf{M})^{\mathrm{st}} \tag{20.64}$$

and

$$(\mathbf{M}B)^{\mathrm{st}} = (\mathbf{M})^{\mathrm{st}} B. \tag{20.65}$$

Proof As the proof of these statements involves no more than straightforward application of the relevant definitions, the details will be omitted.

Definition *Supertrace of a supermatrix*

If $\mathbf{M}$ is a $(p/q) \times (p/q)$ supermatrix, partitioned as in (20.40), then its *supertrace* is the Grassmann element $\operatorname{str} \mathbf{M}$ defined by

$$\operatorname{str} \mathbf{M} = \operatorname{tr} \mathbf{A} - (-1)^{\deg \mathbf{M}} \operatorname{tr} \mathbf{D}. \tag{20.66}$$

Thus

$$\operatorname{str}\mathbf{M}=\begin{cases}\operatorname{tr}\mathbf{A}-\operatorname{tr}\mathbf{D} & \text{if } \mathbf{M} \text{ is even,}\\ \operatorname{tr}\mathbf{A}+\operatorname{tr}\mathbf{D} & \text{if } \mathbf{M} \text{ is odd.}\end{cases} \tag{20.67}$$

(Here "tr" denotes the "ordinary" trace.)

Theorem III (a) If $\mathbf{M}$ and $\mathbf{N}$ are both members of $\mathrm{M}(p/q; \mathbb{C}\mathrm{B}_L)$ and are either both even or are both odd then

$$\operatorname{str}(\mathbf{M}+\mathbf{N})=\operatorname{str}\mathbf{M}+\operatorname{str}\mathbf{N}. \tag{20.68}$$

(b) If $\mathbf{M}$ is a $(p/q)\times(r/s)$ supermatrix and $\mathbf{N}$ is a $(r/s)\times(p/q)$ supermatrix then

$$\operatorname{str}(\mathbf{MN})=(-1)^{(\deg\mathbf{M})(\deg\mathbf{N})}\operatorname{str}(\mathbf{NM}). \tag{20.69}$$

(c) If $\mathbf{M}\in\mathrm{M}(p/q; \mathbb{C}\mathrm{B}_L)$ and $B\in\mathbb{C}\mathrm{B}_{L0}$ or $\mathbb{C}\mathrm{B}_{L1}$ then

$$\operatorname{str}(B\mathbf{M})=B(\operatorname{str}\mathbf{M}) \tag{20.70}$$

and

$$\operatorname{str}(\mathbf{M}B)=(\operatorname{str}\mathbf{M})B. \tag{20.71}$$

(d) If $\mathbf{M}\in\mathrm{M}(p/q; \mathbb{C}\mathrm{B}_L)$ then

$$\operatorname{str}(\mathbf{M}^{\mathrm{st}})=\operatorname{str}\mathbf{M}. \tag{20.72}$$

(e) If $\mathbf{M}\in\mathrm{M}(p/q; \mathbb{C}\mathrm{B}_L)$ and $\mathbf{S}$ is an even invertible $(p/q)\times(p/q)$ supermatrix then

$$\operatorname{str}(\mathbf{SMS}^{-1})=\operatorname{str}\mathbf{M}. \tag{20.73}$$

Proof There are no subtleties involved in the proofs of (a)–(d), which require only the application of the appropriate definitions. Part (e) follows from (b) on putting $\mathbf{N}=\mathbf{MS}^{-1}$, for $\mathbf{M}=(\mathbf{MS}^{-1})\mathbf{S}$ and $\mathbf{SMS}^{-1}=\mathbf{S}(\mathbf{MS}^{-1})$.

Definition *Superdeterminant of an even supermatrix*

If $\mathbf{M}$ is an even invertible $(p/q)\times(p/q)$ supermatrix, partitioned as in (20.40), then its *superdeterminant* is the even Grassmann element sdet $\mathbf{M}$ that is defined by

$$\operatorname{sdet}\mathbf{M}=\frac{\det(\mathbf{A}-\mathbf{BD}^{-1}\mathbf{C})}{\det\mathbf{D}}. \tag{20.74}$$

Part (b) of Theorem I shows that as $\mathbf{M}$ is assumed to be invertible both $\mathbf{A}^{-1}$ and $\mathbf{D}^{-1}$ exist. As both $\mathbf{D}$ and $\mathbf{A}-\mathbf{BD}^{-1}\mathbf{C}$ contain only members of the even Grassmann subspace $\mathbb{C}\mathrm{B}_{L0}$, no ambiguities arise in calculating the "ordinary" determinants. The superdeterminant is sometimes known as the *Berezinian*.

Theorem IV (a) If $\mathbf{M}$ and $\mathbf{N}$ are any two members of $GL(p/q; \mathbb{C}B_L)$ then

$$\text{sdet}(\mathbf{MN}) = (\text{sdet}\,\mathbf{M})(\text{sdet}\,\mathbf{N}). \tag{20.75}$$

(b) If $\mathbf{M} \in GL(p/q; \mathbb{C}B_L)$ then

$$\text{sdet}(\mathbf{M}^{st}) = \text{sdet}\,\mathbf{M}. \tag{20.76}$$

(c) If $\mathbf{M}$ is an even $(p/q) \times (p/q)$ supermatrix then $\exp \mathbf{M}$ is an even invertible $(p/q) \times (p/q)$ supermatrix and

$$\text{sdet}(\exp \mathbf{M}) = \exp(\text{str}\,\mathbf{M}). \tag{20.77}$$

(d) An expression equivalent to (20.74) is

$$\text{sdet}\,\mathbf{M} = \frac{\det \mathbf{A}}{\det(\mathbf{D} - \mathbf{C}\mathbf{A}^{-1}\mathbf{B})}. \tag{20.78}$$

Proof See Appendix K, Section 1.

Definition *Adjoint* $\mathbf{M}^{\ddagger}$ *of a supermatrix* $\mathbf{M}$

If $\mathbf{M}$ is a $(p/q) \times (r/s)$ supermatrix, partitioned as in (20.40), then its *adjoint* is the $(r/s) \times (p/q)$ supermatrix $\mathbf{M}^{\ddagger}$ defined by

$$\mathbf{M}^{\ddagger} = \begin{bmatrix} \tilde{\mathbf{A}}^{\#} & \tilde{\mathbf{C}}^{\#} \\ \tilde{\mathbf{B}}^{\#} & \tilde{\mathbf{D}}^{\#} \end{bmatrix}, \tag{20.79}$$

where, for example, $\tilde{\mathbf{A}}^{\#}$ denotes the "ordinary" transpose of the matrix $\mathbf{A}^{\#}$, this latter matrix being obtained from $\mathbf{A}$ by replacing every element of $\mathbf{A}$ by its Grassmann adjoint (as defined in (20.30)). That is, $(\tilde{\mathbf{A}}^{\#})_{jk} = (A_{kj})^{\#}$.

Theorem V (a) If $\mathbf{M}$ is a $(p/q) \times (r/s)$ supermatrix and $\mathbf{N}$ is an $(r/s) \times (t/u)$ supermatrix then

$$(\mathbf{MN})^{\ddagger} = \mathbf{N}^{\ddagger}\mathbf{M}^{\ddagger}. \tag{20.80}$$

(b) If $\mathbf{M}$ is a $(p/q) \times (r/s)$ supermatrix then

$$(\mathbf{M}^{\ddagger})^{\ddagger} = \mathbf{M}. \tag{20.81}$$

(c) If $\mathbf{M}$ is even invertible $(p/q) \times (p/q)$ supermatrix then

$$\text{sdet}(\mathbf{M}^{\ddagger}) = (\text{sdet}\,\mathbf{M})^{\#} = (\text{sdet}\,\mathbf{M})^{*}, \tag{20.82}$$

where # denotes the Grassmann adjoint operation (of (20.30)) and * denotes the Grassmann complex-conjugation operation.

Proof As only straightforward applications of the definitions are involved, no details will be given.

21

General Properties of Lie Superalgebras

1 Lie superalgebras introduced

This chapter will be devoted to presenting the basic definitions and properties of Lie superalgebras. Many of the familiar ideas of the theory of Lie algebras will reappear, the essential modification to them being that all the relevant vector spaces have to be systematically graded.

Section 2 starts with the definition of a Lie superalgebra, from which the key structural roles of the even and odd parts are easily deduced. These are illustrated by considering a number of important examples, including the Heisenberg superalgebras and the orthosymplectic Lie superalgebras. The straightforward generalizations of the Lie algebraic notions of subalgebras, direct sums, and homomorphic and isomorphic mappings form the content of Section 3. The corresponding generalization of the idea of a representation is developed in Section 4, the grading of the carrier spaces playing a crucial part. The associated concepts of the adjoint representation and the Killing form are then considered for Lie superalgebras in Section 5.

All of these fundamental considerations apply equally to all Lie superalgebras. Their detailed development for particular types of Lie superalgebra is left to subsequent chapters. In particular the "Poincaré superalgebras" will be examined in detail in Chapter 23, and the properties and classification of the "simple Lie superalgebras" will be presented in Chapter 25. Just as Lie groups can be associated with real Lie algebras, it will be shown in Chapter 22 that from real Lie superalgebras it is possible to construct "Lie supergroups".

The Lie superalgebras that will be considered in this book are essentially generalizations of Lie algebras with a Z_2 grading. Generalizations with a Z_N

grading for $N > 2$ are sometimes called "colour algebras" and have been studied by Scheunert (1979b, 1983a, b), Agrawala (1981), Green and Jarvis (1983) and Kleeman (1985).

2 Definitions and immediate consequences

Definition *Lie superalgebra*

Let $\mathscr{L}_s$ be a complex graded vector space (cf. Chapter 20, Section 1), with $\mathscr{L}_0$ and $\mathscr{L}_1$ being its even and odd subspaces, which are assumed to have dimensions m and n respectively (where $m \geqslant 0$, $n \geqslant 0$ and $m + n \geqslant 1$). Suppose that for all $a, b \in \mathscr{L}_s$ there exists a *generalized Lie product* (or *supercommutator*) $[a, b]$ with the following properties:

(i) $[a, b] \in \mathscr{L}_s$ for all $a, b \in \mathscr{L}_s$;

(ii) for all $a, b \in \mathscr{L}_s$ and any complex numbers α and β

$$[\alpha a + \beta b, c] = \alpha[a, c] + \beta[b, c]; \tag{21.1}$$

(iii) if a and b are homogeneous elements of $\mathscr{L}_s$ then $[a, b]$ is also a homogeneous element of $\mathscr{L}_s$ whose degree is $(\deg a + \deg b) \bmod 2$; that is, $[a, b]$ is odd if either a or b is odd, but $[a, b]$ is even if a and b are both even or if a and b are both odd;

(iv) for any two homogeneous elements a and b of $\mathscr{L}_s$

$$[b, a] = -(-1)^{(\deg a)(\deg b)}[a, b]; \tag{21.2}$$

(v) for any three homogeneous elements a, b and c of $\mathscr{L}_s$

$$\begin{aligned}[a, [b, c]](-1)^{(\deg a)(\deg c)} &+ [b, [c, a]](-1)^{(\deg b)(\deg a)} \\ &+ [c, [a, b]](-1)^{(\deg c)(\deg b)} = 0.\end{aligned} \tag{21.3}$$

Then $\mathscr{L}_s$ is said to be a *complex Lie superalgebra* with even dimension m and odd dimension n.

For a *real Lie superalgebra* $\mathscr{L}_s$ with even dimension m and odd dimension n the definition is exactly the same, except that $\mathscr{L}_s$ has to be a *real* graded vector space and the α and β of part (ii) must be restricted to being real.

Lie superalgebras are sometimes known alternatively as *graded Lie algebras* or *pseudo Lie algebras*. The condition (21.3) is called the *generalized Jacobi identity*.

One immediate consequence of (ii) and (iv) is that

$$[a, \beta b + \gamma c] = \beta[a, b] + \gamma[a, c] \tag{21.4}$$

for all $a, b, c \in \mathscr{L}_s$ and any two numbers β and γ (real or complex as appropriate).

Although parts (i) and (ii) are straightforward generalizations of the corresponding parts of the definitions of real and complex Lie algebras that were given in Chapter 10, Section 4, the contents of parts (iii), (iv) and (v) are more subtle and so merit close examination.

The implication of part (iii) is that as

$$\deg a = \begin{cases} 0 & \text{if } a \in \mathscr{L}_0, \\ 1 & \text{if } a \in \mathscr{L}_1, \end{cases} \tag{21.5}$$

then

$$\text{if } a \in \mathscr{L}_0 \quad \text{and} \quad b \in \mathscr{L}_0 \quad \text{then} \quad [a, b] \in \mathscr{L}_0, \tag{21.6a}$$

$$\text{if } a \in \mathscr{L}_0 \quad \text{and} \quad b \in \mathscr{L}_1 \quad \text{then} \quad [a, b] \in \mathscr{L}_1, \tag{21.6b}$$

$$\text{if } a \in \mathscr{L}_1 \quad \text{and} \quad b \in \mathscr{L}_0 \quad \text{then} \quad [a, b] \in \mathscr{L}_1, \tag{21.6c}$$

$$\text{if } a \in \mathscr{L}_1 \quad \text{and} \quad b \in \mathscr{L}_1 \quad \text{then} \quad [a, b] \in \mathscr{L}_0. \tag{21.6d}$$

That is, the generalized Lie products obey the grading rule (20.2).

Similarly, part (iv) implies that

(a) if $a \in \mathscr{L}_0$ and $b \in \mathscr{L}_0$, or if $a \in \mathscr{L}_0$ and $b \in \mathscr{L}_1$, or if $a \in \mathscr{L}_1$ and $b \in \mathscr{L}_0$, then

$$[b, a] = -[a, b], \tag{21.7}$$

and so the generalized Lie product for these cases will be written as $[a, b]_-$;

whereas

(b) if $a \in \mathscr{L}_1$ and $b \in \mathscr{L}_1$ then

$$[b, a] = [a, b], \tag{21.8}$$

and so the generalized Lie product for this case will be written as $[a, b]_+$.

The convention that will be adopted throughout the rest of this book is that $[a, b]$ will denote the generalized Lie product for arbitrary elements a and b of $\mathscr{L}_s$, but if a and b are both homogeneous *and* it is desirable that the sign in (21.7) or (21.8) be emphasized, then the generalized Lie product will be denoted by $[a, b]_-$ or $[a, b]_+$ as appropriate. (In the literature $\{a, b\}$ is sometimes employed instead of $[a, b]_+$.)

Choosing a *homogeneous basis* of $\mathscr{L}_s$ to be one in which the basis elements $a_1, a_2, \ldots, a_{m+n}$ are such that $a_1, a_2, \ldots, a_m \in \mathscr{L}_0$ and $a_{m+1}, a_{m+2}, \ldots, a_{m+n} \in \mathscr{L}_1$, the *structure constants* c^r_{pq} may be defined by

$$[a_p, a_q] = \sum_{r=1}^{m+n} c^r_{pq} a_r. \tag{21.9}$$

Then, if

$$a = \sum_{p=1}^{m+n} \alpha_p a_p$$

and

$$b = \sum_{p=1}^{m+n} \beta_p a_p$$

are any two elements of $\mathscr{L}_s$ (the coefficients $\alpha_1, \alpha_2, \ldots, \alpha_{m+n}$ and $\beta_1, \beta_2, \ldots, \beta_{m+n}$ being real if $\mathscr{L}_s$ is real and complex if $\mathscr{L}_s$ is complex),

$$[a, b] = \sum_{p,q,r=1}^{m+n} \alpha_p \beta_q c^r_{pq} a_r,$$

so that every generalized Lie product of $\mathscr{L}_s$ can be evaluated from a knowledge of the structure constants. With this choice of basis, (21.2) implies that

$$c^r_{pq} = -(-1)^{(\deg a_p)(\deg a_q)} c^r_{qp}, \tag{21.10}$$

so

$$c^r_{pq} = -c^r_{qp} \tag{21.11}$$

if either a_p or a_q is even, or if both a_p and a_q are even, *but*

$$c^r_{pq} = c^r_{qp} \tag{21.12}$$

if both a_p and a_q are odd. The relations (iii) and (v) also have implications for the structure constants, but discussion of these will be deferred until Section 5, where they will be taken up in the context of the adjoint representation of $\mathscr{L}_s$.

Part (v) of the definition involves four special cases:

(a) if $a \in \mathscr{L}_0$, $b \in \mathscr{L}_0$ and $c \in \mathscr{L}_0$ then (21.3) becomes

$$[a, [b, c]_-]_- + [b, [c, a]_-]_- + [c, [a, b]_-]_- = 0; \tag{21.13}$$

that is, the generalized Jacobi identity reduces in this case to the "ordinary" Jacobi identity;

(b) if $a \in \mathscr{L}_0$ and $b \in \mathscr{L}_0$ but $c \in \mathscr{L}_1$ then (21.3) reduces to (21.13) again;

(c) if $a \in \mathscr{L}_0$, $b \in \mathscr{L}_1$ and $c \in \mathscr{L}_1$ then (21.3) becomes

$$[a, [b, c]_+]_- + [b, [c, a]_-]_+ - [c, [a, b]_-]_+ = 0, \tag{21.14}$$

(d) if $a \in \mathscr{L}_1$, $b \in \mathscr{L}_1$ and $c \in \mathscr{L}_1$ then (21.3) becomes

$$[a, [b, c]_+]_- + [b, [c, a]_+]_- + [c, [a, b]_+]_- = 0. \tag{21.15}$$

The different roles played by the subspaces $\mathscr{L}_0$ and $\mathscr{L}_1$ are brought out clearly by the following theorem.

Theorem I (a) Provided that $m \geqslant 1$, the even subspace $\mathscr{L}_0$ is an "ordinary" Lie algebra, which is real if $\mathscr{L}_s$ is real.

(b) Provided that $m \geqslant 1$ and $n \geqslant 1$, the odd subspace $\mathscr{L}_1$ of $\mathscr{L}_s$ is a carrier space for a representation of the Lie algebra $\mathscr{L}_0$, and the matrices of the representation are real if $\mathscr{L}_s$ is a real Lie superalgebra. This representation will be referred to henceforth as *the representation of* $\mathscr{L}_0$ *on* $\mathscr{L}_1$.

Proof (a) The first statement of the set (21.6) is that $\mathscr{L}_0$ is "closed"; that is, if $a, b \in \mathscr{L}_0$ then $[a, b] \in \mathscr{L}_0$. Part (ii) of the definition of $\mathscr{L}_s$ implies that the same equation holds on $\mathscr{L}_0$. Equation (21.7) is the antisymmetry requirement of the ordinary Lie product, and, as noted above, (21.13) is the usual Jacobi identity. Thus all the conditions required for $\mathscr{L}_0$ to be a Lie algebra hold (cf. Chapter 11, Section 4). Clearly $\mathscr{L}_0$ is real if $\mathscr{L}_s$ is real.

(b) Assuming that $a_1, a_2, \ldots, a_m$ form a basis for $\mathscr{L}_0$ and $a_{m+1}, a_{m+2}, \ldots, a_{m+n}$ form a basis for $\mathscr{L}_1$, the second statement of (21.6) implies that if $a \in \mathscr{L}_0$ and $j = 1, 2, \ldots, n$ then $[a, a_{m+j}] \in \mathscr{L}_1$, and hence

$$[a, a_{m+j}]_- = \sum_{k=1}^{n} D(a)_{kj} a_{m+k}. \tag{21.16}$$

Here $D(a)_{kj}$ are a set of n^2 real numbers if $\mathscr{L}_s$ is a real Lie superalgebra and are a set of n^2 complex numbers if $\mathscr{L}_s$ is complex. Putting $c = a_{m+j}$ in (21.13) and using (21.7), it follows that

$$\left[a, \sum_{k=1}^{n} D(b)_{kj} a_{m+k}\right]_- - \left[b, \sum_{k=1}^{n} D(a)_{kj} a_{m+k}\right]_- - \sum_{k=1}^{n} D([a, b])_{kj} a_{m+k} = 0,$$

which, with a further application of (21.16), reduces to

$$\sum_{k,l=1}^{n} \{D(a)_{lk} D(b)_{kj} - D(b)_{lk} D(a)_{kj}\} a_{m+l} = \sum_{l=1}^{n} D([a, b]_-)_{lj} a_{m+l}.$$

As the basis elements $a_{m+1}, a_{m+2}, \ldots, a_{m+n}$ are linearly independent, this implies that for all $a, b \in \mathscr{L}_0$

$$\mathbf{D}([a, b]_-) = \mathbf{D}(a)\mathbf{D}(b) - \mathbf{D}(b)\mathbf{D}(a),$$

so that the matrices $\mathbf{D}(a)$ form an n-dimensional representation of $\mathscr{L}_0$. Clearly these matrices are real if $\mathscr{L}_s$ is real.

The condition $m + n \geqslant 1$ has been inserted in the definition of a Lie superalgebra in order to avoid having to consider the rather pathological case of a zero-dimensional algebra. However, the case $m \geqslant 1$ and $n = 0$ is *not* excluded, and Theorem I shows that for this case the Lie superalgebra is actually just a Lie algebra. Thus every Lie algebra can be regarded as being a special case of a Lie superalgebra. The other special case $m = 0$ and $n \geqslant 1$ is also allowed.

Theorem I shows that one way of constructing a Lie superalgebra is to start with a Lie algebra $\mathscr{L}_0$, choose a representation **D** of $\mathscr{L}_0$ with a carrier space, which may be called $\mathscr{L}_1$, and set up the vector-space direct sum $\mathscr{L}_0 \oplus \mathscr{L}_1$. With the elements of $\mathscr{L}_0$ defined to be even and those of $\mathscr{L}_1$ defined to be odd, this direct sum is a graded vector space. Also, for $a, b \in \mathscr{L}_0$ the generalized Lie product $[a, b]$ can be taken to be the Lie product of $\mathscr{L}_0$, and for $a \in \mathscr{L}_0$ and $b \in \mathscr{L}_1$ the product $[a, b]$ can be defined using (21.16). Then all the requirements on the generalized Lie product $[a, b]$ are automatically satisfied *except* those for which *both a and b* are members of $\mathscr{L}_1$. However, it is *not* an easy matter to satisfy the remaining requirements (21.8), (21.14) and (21.15) (particularly the last two), and only for some representations **D** of some Lie algebras $\mathscr{L}_0$ is this possible. Of course, to produce a *real* Lie superalgebra by this process it is necessary to start with a *real* Lie algebra and to use to *real* representation **D** of $\mathscr{L}_0$.

A Lie superalgebra $\mathscr{L}_s$ can be constructed from *any associative* superalgebra by defining the generalized Lie product by

$$[a, b] = ab - (-1)^{(\deg a)(\deg b)} ba. \tag{21.17}$$

Thus

(a) if at least one of the elements a and b is *even* then this implies that

$$[a, b]_- = ab - ba; \tag{21.18}$$

whereas

(b) if a and b are both *odd* elements then

$$[a, b]_+ = ab + ba. \tag{21.19}$$

It is easily verified that the conditions (21.7) and (21.8) as well as (21.13)–(21.15) are then satisfied. In particular, this construction applies to the *matrix associative superalgebras* $M(p/q; \mathbb{R})$ and $M(p/q; \mathbb{C})$ of Example I of Chapter 20, Section 2. Consequently it will be assumed henceforth that in any matrix realization of a Lie superalgebra the even and odd elements have the forms (20.8a) and (20.8b) respectively (for some values of p and q), and that (as in (21.17))

$$[\mathbf{M}, \mathbf{N}] = \mathbf{MN} - (-1)^{(\deg \mathbf{M})(\deg \mathbf{N})} \mathbf{NM}. \tag{21.20}$$

Thus if at least one of the matrices **M** and **N** is *even*,

$$[\mathbf{M}, \mathbf{N}]_- = \mathbf{MN} - \mathbf{NM}, \tag{21.21}$$

and if both **M** and **N** are *odd*,

$$[\mathbf{M}, \mathbf{N}]_+ = \mathbf{MN} + \mathbf{NM}. \tag{21.22}$$

Because the generalized Lie products $[\ ,\ ]_-$ and $[\ ,\ ]_+$ can sometimes be defined in this way, they are often called *commutators* and *anticommutators* respectively (and this is frequently done even in situations in which there is no underlying associative superalgebra).

Example I *Heisenberg superalgebras*

Let $b_j^\dagger$ and b_j for $j = 1, 2, \ldots, r$ be a set of fermion creation and annihilation operators, which are assumed to be such that, for $j, k = 1, 2, \ldots, r$,

$$\left.\begin{aligned} [b_j^\dagger, b_k]_+ &= \delta_{jk} 1, \\ [b_j, b_k]_+ &= 0, \\ [b_j^\dagger, b_k^\dagger]_+ &= 0, \end{aligned}\right\} \tag{21.23}$$

and, for $j = 1, 2, \ldots, r$,

$$\begin{aligned} [b_j^\dagger, 1]_- &= 0, \\ [b_j, 1]_- &= 0, \end{aligned} \tag{21.24}$$

as well, of course, as

$$[1, 1]_- = 0.$$

Here 1 is the identity operator of the space on which these creation and annihilation operators act. Defining the degrees of these operators by deg 1 $= 0$ and $\deg b_j^\dagger = \deg b_j = 1$ (for $j = 1, 2, \ldots, r$), it follows that they form a basis for a Lie superalgebra $\mathscr{L}_s$ with even dimension 1 and odd dimension $2r$.

If this algebra is extended to include a further operator b_{r+1} that is such that

$$[b_{r+1}, b_j]_+ = [b_{r+1}, b_j^\dagger]_+ = 0$$

for $j = 1, 2, \ldots, r$, and

$$[b_{r+1}, 1]_- = 0,$$

and if $\deg b_{r+1}$ is defined to have value 1, then another Lie superalgebra $\mathscr{L}'_s$ with even dimension 1 and odd dimension $2r + 1$ is obtained.

Both $\mathscr{L}_s$ and $\mathscr{L}'_s$ are known as *Heisenberg superalgebras*.

Example II *The Lie superalgebras* sl$(p/q; \mathbb{R})$, sl$(p/q; \mathbb{C})$ *and* sl$(p/q; \mathbb{H})$

The subset of all the matrices $\mathbf{M}$ of $\mathrm{M}(p/q; \mathbb{R})$ (cf. Example I of Chapter 20, Section 2) that satisfy the condition

$$\operatorname{str} \mathbf{M} = 0 \tag{21.25}$$

and for which the generalized Lie product is defined by (21.20) form a real Lie superalgebra that is denoted by sl$(p/q; \mathbb{R})$. Here p and q are positive integers, the elements of the *even* Lie algebra $\mathscr{L}_0$ are of the form

$$\mathbf{M} = \begin{bmatrix} \mathbf{A} & \mathbf{0} \\ \mathbf{0} & \mathbf{D} \end{bmatrix}, \tag{21.26}$$

where $\mathbf{A}$ and $\mathbf{D}$ are real $p \times p$ and $q \times q$ matrices respectively, the supertrace condition (21.25) implying (by (20.67)) that

$$\operatorname{tr} \mathbf{A} - \operatorname{tr} \mathbf{D} = 0, \tag{21.27}$$

while the elements of the odd part $\mathscr{L}_1$ are of the form

$$\mathbf{M} = \begin{bmatrix} \mathbf{0} & \mathbf{B} \\ \mathbf{C} & \mathbf{0} \end{bmatrix}, \tag{21.28}$$

where $\mathbf{B}$ and $\mathbf{C}$ are $p \times q$ and $q \times p$ real matrices that do not have to satisfy any constraint. As (20.69) and (21.20) together imply that

$$\operatorname{str}([\mathbf{M}, \mathbf{N}]) = 0$$

for any two matrices $\mathbf{M}$ and $\mathbf{N}$ of $\mathrm{M}(p/q; \mathbb{R})$, it is certainly true that if $\mathbf{M}, \mathbf{N} \in \mathrm{sl}(p/q; \mathbb{R})$ then $[\mathbf{M}, \mathbf{N}] \in \mathrm{sl}(p/q; \mathbb{R})$.

Clearly there are three types of element in $\mathscr{L}_0$:

(i) those for which $\operatorname{tr} \mathbf{A} = 0$ and $\mathbf{D} = \mathbf{0}$, which form a Lie subalgebra that is isomorphic to the real Lie algebra $\mathrm{sl}(p; \mathbb{R})$;

(ii) those for which $\mathbf{A} = \mathbf{0}$ and $\operatorname{tr} \mathbf{D} = 0$, which form a Lie subalgebra that is isomorphic to $\mathrm{sl}(q; \mathbb{R})$;

(iii) scalar multiples of

$$\begin{bmatrix} p^{-1}\mathbf{1}_p & \mathbf{0} \\ \mathbf{0} & q^{-1}\mathbf{1}_q \end{bmatrix}, \tag{21.29}$$

which form a one-dimensional Abelian real Lie algebra.

$\mathscr{L}_0$ is the direct sum of these three subalgebras.

Obviously the dimension m of $\mathscr{L}_0$ is given by

$$m = p^2 + q^2 - 1, \tag{21.30}$$

and the dimension n of $\mathscr{L}_1$ is given by

$$n = 2pq. \tag{21.31}$$

Similarly $\mathrm{sl}(p/q; \mathbb{C})$ is defined as the set of matrices $\mathbf{M}$ of $\mathrm{M}(p/q; \mathbb{C})$ that satisfy (21.25), and it can be regarded either as a complex Lie superalgebra whose even and odd dimensions are given by (21.30) and (21.31), or as a real Lie superalgebra with both of these dimensions being doubled. (Some other notations for $sl(p/q; \mathbb{C})$ considered a complex Lie superalgebra are $sl(p, q)$ or $sl(p, q; \mathbb{C})$ (cf. Kac 1977a) and $spl(p, q)$ (cf. Nahm and Scheunert 1976, Scheunert *et al.* 1976a, Scheunert 1979a).

Finally $sl(p/q; \mathbb{H})$ is defined as the set of matrices $\mathbf{M}$ partitioned as for $M(p/q; \mathbb{C})$, but having *quaternionic* entries in place of complex numbers, and that satisfy the constraint

$$\operatorname{str} \mathbf{M}_0 = \mathbf{0},$$

where

$$\mathbf{M} = \sum_{r=0}^{3} \mathbf{M}_r e_r,$$

e_0, e_1, e_2 and e_3 being the basic quaternions introduced in Appendix B, Section 8, and $\mathbf{M}_0$, $\mathbf{M}_1$, $\mathbf{M}_2$ and $\mathbf{M}_3$ all being members of $M(p/q; \mathbb{R})$.

Example III *The orthosymplectic Lie superalgebras* $osp(p/q; \mathbb{R})$ *and* $osp(p/q; \mathbb{C})$

Consider the set of matrices $M(p/q; \mathbb{R})$ with $p \geqslant 1$ and q positive and *even* (cf. Example I of Chapter 20, Section 2). Let $\mathbf{K}$ be the member of $M(p/q; \mathbb{R})$ that is such that

$$\mathbf{K} = \begin{bmatrix} \mathbf{G} & \mathbf{0} \\ \mathbf{0} & \mathbf{J} \end{bmatrix}, \tag{21.32}$$

where $\mathbf{G} = \mathbf{1}_p$ (the $p \times p$ unit matrix) and $\mathbf{J}$ is the $q \times q$ matrix

$$\mathbf{J} = \begin{bmatrix} \mathbf{0} & \mathbf{1}_{q/2} \\ -\mathbf{1}_{q/2} & \mathbf{0} \end{bmatrix} \tag{21.33}$$

(cf. Table 10.1). The subset of matrices $\mathbf{M}$ of $M(p/q; \mathbb{R})$ that satisfy the condition

$$\mathbf{M}^{st}\mathbf{K} + (-1)^{\deg \mathbf{M}}\mathbf{KM} = \mathbf{0} \tag{21.34}$$

and for which the generalized Lie product is defined by (21.20) form a real Lie superalgebra that is denoted by $osp(p/q; \mathbb{R})$. Closure of the algebra follows from the property (20.63) of the supertranspose, which together with (21.20) implies that

$$([\mathbf{M}, \mathbf{N}])^{st} = -[\mathbf{M}^{st}, \mathbf{N}^{st}],$$

so that if $\mathbf{M}, \mathbf{N} \in osp(p/q; \mathbb{R})$ then

$$([\mathbf{M}, \mathbf{N}])^{st}\mathbf{K} + (-\mathbf{1})^{(\deg \mathbf{M} + \deg \mathbf{N})}\mathbf{K}[\mathbf{M}, \mathbf{N}] = \mathbf{0},$$

thereby showing that $[\mathbf{M}, \mathbf{N}] \in osp(p/q; \mathbb{R})$ (as the degree of $[\mathbf{M}, \mathbf{N}]$ is $(\deg \mathbf{M} + \deg \mathbf{N}) \bmod 2$).

The *even* elements of $\mathrm{osp}(p/q;\mathbb{R})$ are of the form

$$\mathbf{M} = \begin{bmatrix} \mathbf{A} & \mathbf{0} \\ \mathbf{0} & \mathbf{D} \end{bmatrix}, \tag{21.35}$$

and therefore satisfy the constraint

$$\tilde{\mathbf{A}}\mathbf{G} + \mathbf{G}\mathbf{A} = \mathbf{0}, \tag{21.36}$$

which, with $\mathbf{G} = \mathbf{1}_p$, reduces to

$$\tilde{\mathbf{A}} + \mathbf{A} = \mathbf{0},$$

together with the constraint

$$\tilde{\mathbf{D}}\mathbf{J} + \mathbf{J}\mathbf{D} = \mathbf{0}. \tag{21.37}$$

Thus the set of $p \times p$ matrices $\mathbf{A}$ form the real *orthogonal* Lie algebra $\mathrm{so}(p)$, and the set of $q \times q$ matrices $\mathbf{D}$ form the real *symplectic* Lie algebra $\mathrm{sp}(\frac{1}{2}q, \mathbb{R})$ (cf. Table 10.1), so

$$\mathscr{L}_0 = \mathrm{so}(p) \oplus \mathrm{sp}(\tfrac{1}{2}q, \mathbb{R}).$$

It is this property that suggests the name *orthosymplectic* for $\mathrm{osp}(p/q;\mathbb{R})$.

The *odd* elements of $\mathrm{osp}(p/q;\mathbb{R})$ are of the form

$$\mathbf{M} = \begin{bmatrix} \mathbf{0} & \mathbf{B} \\ \mathbf{C} & \mathbf{0} \end{bmatrix}, \tag{21.38}$$

and therefore satisfy the constraint

$$\tilde{\mathbf{B}}\mathbf{G} - \mathbf{J}\mathbf{C} = \mathbf{0}. \tag{21.39}$$

The dimension m of the even part $\mathscr{L}_0$ of $\mathrm{osp}(p/q;\mathbb{R})$ is given by

$$m = \tfrac{1}{2}p(p-1) + \tfrac{1}{2}q(q+1), \tag{21.40}$$

and the dimension n of the odd part $\mathscr{L}_1$ is given by

$$n = pq. \tag{21.41}$$

It will be convenient, as in the analysis in Appendix M, to take for $\mathbf{G}$ either

$$\mathbf{G} = \begin{bmatrix} \mathbf{0} & \mathbf{1}_r \\ \mathbf{1}_r & \mathbf{0} \end{bmatrix} \tag{21.42}$$

if $p(=2r)$ is even, or

$$\mathbf{G} = \begin{bmatrix} \mathbf{0} & \mathbf{1}_r & \mathbf{0} \\ \mathbf{1}_r & \mathbf{0} & \mathbf{0} \\ \mathbf{0} & \mathbf{0} & \mathbf{1}_1 \end{bmatrix} \tag{21.43}$$

if $p(=2r+1)$ is odd, instead of taking $\mathbf{G} = \mathbf{1}_p$. As this merely corresponds to applying to the matrices $\mathbf{M}$ of $\mathrm{M}(p/q; \mathbb{R})$ a transformation of the type $\mathbf{M} \rightarrow \mathbf{S}\mathbf{M}\tilde{\mathbf{S}}$ with a $(p+q) \times (p+q)$ matrix $\mathbf{S}$ of the form

$$\mathbf{S} = \begin{bmatrix} \mathbf{S}_0 & \mathbf{0} \\ \mathbf{0} & \mathbf{1}_q \end{bmatrix},$$

where $\mathbf{S}_0$ is an appropriately chosen non-singular $p \times p$ matrix, the resulting matrices form a real Lie superalgebra that is isomorphic to $\mathrm{osp}(p/q; \mathbb{R})$, and which may, without confusion, be denoted by the same set of symbols.

Similarly $\mathrm{osp}(p/q; \mathbb{C})$ is defined as the set of matrices $\mathbf{M}$ of $\mathrm{M}(p/q; \mathbb{C})$ that satisfy (21.34) (with $\mathbf{G}$ being chosen to be $\mathbf{1}_p$ or taken to be given by (21.42) or (21.43)). This set can be regarded either as a complex Lie superalgebra whose even and odd dimensions are given by (21.40) and (21.41), or as a real Lie superalgebra with both of these dimensions being doubled.

Definition *Abelian Lie superalgebra*

A Lie superalgebra $\mathscr{L}_s$ is said to be *Abelian* (or *commutative*) if

$$[a, b] = 0 \tag{21.44}$$

for all $a, b \in \mathscr{L}_s$.

Of course (21.44) involves the generalized Lie product (and not just the commutator).

3 Subalgebras, direct sums and homomorphisms of Lie superalgebras

The concepts of this section are direct generalizations of those for Lie algebras, with the modifications in the definitions involving no more than replacing the Lie product by the generalized Lie product and possibly making some qualifications about grading. All the ideas apply equally to real and complex Lie superalgebras.

Definition *Subalgebra of a Lie superalgebra*

A subalgebra $\mathscr{S}$ of a Lie superalgebra $\mathscr{L}_s$ is a subset of elements of $\mathscr{L}_s$ that form a vector subspace of $\mathscr{L}_s$ (with the same field as $\mathscr{L}_s$) and that is closed with respect to the generalized Lie product of $\mathscr{L}_s$ (that is, if $a, b \in \mathscr{S}$ then $[a, b] \in \mathscr{S}$).

This definition does *not* require that the subalgebra should possess any grading properties, and so it is too general to be of much use. A more significant quantity is a "*graded* subalgebra".

Definition *Graded subalgebra of a Lie superalgebra*

A *graded subalgebra* $\mathscr{L}'_s$ of a Lie superalgebra $\mathscr{L}_s$ is a subset of elements of $\mathscr{L}_s$ that themselves form a Lie superalgebra with the same generalized Lie product and field as $\mathscr{L}_s$, and for which the even subspace $\mathscr{L}'_0$ of $\mathscr{L}'_s$ is a subspace of the even subspace $\mathscr{L}_0$ of $\mathscr{L}_s$ and the odd subspace $\mathscr{L}'_1$ of $\mathscr{L}'_s$ is a subspace of the odd subspace $\mathscr{L}_1$ of $\mathscr{L}_s$.

The possibility that one or other of the subspaces $\mathscr{L}'_0$ and $\mathscr{L}'_1$ is zero-dimensional is not excluded in this definition. $\mathscr{L}'_s$ is said to be a *proper* graded subalgebra of $\mathscr{L}_s$ if at least one element of $\mathscr{L}_s$ is not contained in $\mathscr{L}'_s$.

The following example illustrates the difference between a graded subalgebra and a subalgebra that is not graded.

Example I *Subalgebras of a Heisenberg superalgebra*

Let $\mathscr{L}_s$ be the Heisenberg superalgebra of Example I of Section 2 with $r = 1$, so its basic elements may be taken to be $a_1 = 1$, $a_2 = b_1^\dagger$ and $a_3 = b_1$, with a_1 being even and a_2 and a_3 both odd. The set of elements of the form $\mu(a_0 + a_1)$ (where μ is any complex number) form a one-dimensional vector subspace of $\mathscr{L}_s$, and, as $[a_0 + a_1, a_0 + a_1] = 0$, this subspace is a subalgebra. However, as it possesses no grading, every element being non-homogeneous, it is not a graded subalgebra of $\mathscr{L}_s$.

By contrast, the set of elements of the form μa_1 form a graded subalgebra of $\mathscr{L}_s$.

Definition *Invariant graded subalgebra of a Lie superalgebra*

A graded superalgebra $\mathscr{L}'_s$ of a Lie superalgebra $\mathscr{L}_s$ is said to be *invariant* if $[a, b] \in \mathscr{L}'_s$ for all $a \in \mathscr{L}'_s$ and $b \in \mathscr{L}_s$.

Invariant graded subalgebras are sometimes known as "graded ideals". Because of the assumed grading properties, it is true that $[a, b] \in \mathscr{L}'_s$ for all $a \in \mathscr{L}'_s$ and $b \in \mathscr{L}_s$ if and only if $[b, a] \in \mathscr{L}'_s$ for all $a \in \mathscr{L}'_s$ and $b \in \mathscr{L}_s$; that is, the asymmetry in the definition does not matter. By contrast, if the concept of invariance is extended to non-graded subalgebras, "left-invariance" does *not* imply "right-invariance" (and vice versa).

Let $\mathscr{L}'_s$ and $\mathscr{L}''_s$ be any two subspaces of a Lie superalgebra $\mathscr{L}_s$, and let $[\mathscr{L}'_s, \mathscr{L}''_s]$ denote the subset of $\mathscr{L}_s$ that consists of all linear combinations (with coefficients in the same field as that of $\mathscr{L}_s$) of elements of the form $[a', a'']$, where $a' \in \mathscr{L}'_s$ and $a'' \in \mathscr{L}''_s$.

Theorem I If $\mathscr{L}'_s$ and $\mathscr{L}''_s$ are two invariant graded subalgebras of a Lie superalgebra $\mathscr{L}_s$, then $[\mathscr{L}'_s, \mathscr{L}''_s]$ is either an invariant subalgebra of $\mathscr{L}_s$, or it is the "trivial" set $\{0\}$ consisting only of the zero element 0 of $\mathscr{L}_s$.

Proof This proceeds along exactly the same lines as the proof of the corresponding theorem for Lie algebras (Theorem III of Chapter 11, Section 2) and so will not be repeated.

It should be noted that this theorem does not guarantee that $[\mathscr{L}'_s, \mathscr{L}''_s]$ is *graded.*

The concept of a solvable Lie superalgebra is very similar to that of a solvable Lie algebra that was given in Chapter 13, Section 2.

Definition *Solvable Lie superalgebra* $\mathscr{L}_s$

Let $\mathscr{L}_s^{(0)} = \mathscr{L}_s$, and let $\mathscr{L}_s^{(k)} = [\mathscr{L}_s^{(k-1)}, \mathscr{L}_s^{(k-1)}]$ for each $k = 1, 2, 3, \ldots$. Then the Lie superalgebra $\mathscr{L}_s$ is said to be *solvable* if there exists a value k $(= 1, 2, 3, \ldots)$ for which $\mathscr{L}_s^{(k)}$ is the trivial set $\{0\}$.

Theorem II If $\mathscr{L}_s$ is a Lie superalgebra with even subspace $\mathscr{L}_0$ and odd subspace $\mathscr{L}_1$ then the sets of elements in the vector-space direct sums

$$[\mathscr{L}_1, \mathscr{L}_1] \oplus \mathscr{L}_1$$

and

$$\mathscr{L}_0 \oplus [\mathscr{L}_0, \mathscr{L}_1]$$

both form invariant graded subalgebras of $\mathscr{L}_s$.

Proof By the fourth statement of (21.6), $[\mathscr{L}_1, \mathscr{L}_1]$ is contained in $\mathscr{L}_0$, so $[\mathscr{L}_1, \mathscr{L}_1] \oplus \mathscr{L}_0$ is certainly graded. Also, if a, $b \in \mathscr{L}_1$ and $c \in \mathscr{L}_0$ (21.2) and (21.3) show that

$$[[a, b], c] = [a, [b, c]] + [b, [c, a]],$$

which implies that $[[a, b], c] \in [\mathscr{L}_1, \mathscr{L}_1]$, whereas if $a, b, c \in \mathscr{L}_1$ then $[[a, b], c] \in \mathscr{L}_1$. Similarly if $a \in \mathscr{L}_1$ and $c \in \mathscr{L}_0$ then $[a, c] \in \mathscr{L}_1$, and if $a, c \in \mathscr{L}_1$ then $[a, c] \in [\mathscr{L}_1, \mathscr{L}_1]$. Consequently $[\mathscr{L}_1, \mathscr{L}_1] \oplus \mathscr{L}_1$ is an invariant graded subalgebra of $\mathscr{L}_s$. The proof for $\mathscr{L}_0 \oplus [\mathscr{L}_0, \mathscr{L}_1]$ is similar.

Definition *Homomorphic mapping of a Lie superalgebra* $\mathscr{L}_s$ *onto a Lie superalgebra* $\mathscr{L}'_s$

Let ψ be a mapping of a Lie superalgebra $\mathscr{L}_s$ onto a Lie superalgebra $\mathscr{L}'_s$ having the same field such that

(i) for all $a, b \in \mathscr{L}_s$ and all α, β of the field,

$$\psi(\alpha a + \beta b) = \alpha\psi(a) + \beta\psi(b); \tag{21.45}$$

(ii) for all $a, b \in \mathscr{L}_s$,

$$\psi([a, b]) = [\psi(a), \psi(b)]; \tag{21.46}$$

and

(iii) ψ maps the even subspace $\mathscr{L}_0$ of $\mathscr{L}_s$ onto the even subspace $\mathscr{L}'_0$ of $\mathscr{L}'_s$ and the odd subspace $\mathscr{L}_1$ of $\mathscr{L}_s$ onto the odd subspace $\mathscr{L}'_1$ of $\mathscr{L}'_s$.

Then ψ is said to be a *homomorphic mapping* of $\mathscr{L}_s$ onto $\mathscr{L}'_s$.

Definition *Isomorphic mapping of a Lie superalgebra $\mathscr{L}_s$ onto a Lie superalgebra $\mathscr{L}'_s$*

A mapping ψ of a Lie superalgebra $\mathscr{L}_s$ onto a Lie superalgebra $\mathscr{L}'_s$ with the same field is said to be *isomorphic* if it is both homomorphic and one-to-one.

In the special case in which $\mathscr{L}'_s$ is identical to $\mathscr{L}_s$ (so that ψ is a mapping of $\mathscr{L}_s$ onto itself) an isomorphic mapping is called an *automorphism.* With the product of two automorphisms ϕ and ψ of $\mathscr{L}_s$ being defined by $(\phi\psi)(a) = \phi(\psi(a))$ (for all $a \in \mathscr{L}_s$), the set of all automorphisms form a group, which will be denoted by $\mathrm{Aut}(\mathscr{L}_s)$.

Example II *The grading automorphism*

The "grading automorphism" ψ of $\mathscr{L}_s$ is defined for all homogeneous $a \in \mathscr{L}_s$ by

$$\psi(a) = (-1)^{\deg a} a \tag{21.47}$$

and is then extended by linearity to the whole of $\mathscr{L}_s$. Thus if a is *any* element of $\mathscr{L}_s$, and a_{even} and a_{odd} are its even and odd components, so that

$$a = a_{\text{even}} + a_{\text{odd}}$$

with $a_{\text{even}} \in \mathscr{L}_0$ and $a_{\text{odd}} \in \mathscr{L}_1$, (21.47) implies that

$$\psi(a) = a_{\text{even}} - a_{\text{odd}}, \tag{21.48}$$

and consequently

$$a_{\text{even}} = \tfrac{1}{2}\{a + \psi(a)\}$$

and

$$a_{\text{odd}} = \tfrac{1}{2}\{a - \psi(a)\}.$$

Theorem III Suppose that there exists a homomorphic mapping of a Lie superalgebra $\mathscr{L}_s$ onto a Lie superalgebra $\mathscr{L}'_s$ with the same field. Then $\mathscr{L}_s$ and $\mathscr{L}'_s$ have the same dimensions for their even subspaces $\mathscr{L}_0$ and $\mathscr{L}'_0$ and the same dimensions for their odd subspaces $\mathscr{L}_1$ and $\mathscr{L}'_1$ if and only if ψ is an *isomorphic* mapping.

Proof This is again a straightforward generalization of the corresponding theorem for Lie algebras (Theorem I of Chapter 11, Section 3), so the details will be omitted.

Definition *Direct sum of two Lie superalgebras*

A Lie superalgebra $\mathscr{L}_s$ is said to be the *direct sum* of two Lie superalgebras $\mathscr{L}'_s$ and $\mathscr{L}''_s$ (all with the same field) if

(i) the vector space of $\mathscr{L}_s$ is the direct sum of the vector spaces of $\mathscr{L}'_s$ and $\mathscr{L}''_s$;

(ii) the even subspace $\mathscr{L}_0$ of $\mathscr{L}_s$ is the direct sum of the even subspaces $\mathscr{L}'_0$ and $\mathscr{L}''_0$ of $\mathscr{L}'_s$ and $\mathscr{L}''_s$, and the odd subspace $\mathscr{L}_1$ of $\mathscr{L}_s$ is the direct sum of the odd subspaces $\mathscr{L}'_1$ and $\mathscr{L}''_1$ of $\mathscr{L}'_s$ and $\mathscr{L}''_s$; and

(iii) for all $a' \in \mathscr{L}'_s$ and $a'' \in \mathscr{L}''_s$

$$[a', a''] = 0. \tag{21.49}$$

This is expressed by writing $\mathscr{L}_s = \mathscr{L}'_s \oplus \mathscr{L}''_s$.

4 Graded representations of Lie superalgebras

Although most of the ideas on representations of Lie superalgebras are straightforward generalizations of those given for Lie algebras in Chapter 11, Section 4, some new features appear, particularly in respect of the grading of representations. All the considerations of this section apply equally to real and complex Lie superalgebras.

Definition *Graded representation of a Lie superalgebra $\mathscr{L}_s$*

Suppose that for every $a \in \mathscr{L}_s$ there exists a matrix $\mathbf{\Gamma}(a)$ from the set $\mathbf{M}(d_0/d_1; \mathbb{C})$ (cf. Example I of Chapter 20, Section 2) such that

(i) for all $a, b \in \mathscr{L}_s$ and all α, β of the field of $\mathscr{L}_s$,

$$\mathbf{\Gamma}(\alpha a + \beta b) = \alpha \mathbf{\Gamma}(a) + \beta \mathbf{\Gamma}(b); \tag{21.50}$$

(ii) for all $a, b \in \mathscr{L}_s$,

$$\mathbf{\Gamma}([a, b]) = [\mathbf{\Gamma}(a), \mathbf{\Gamma}(b)]; \tag{21.51}$$

and

(iii) if $a \in \mathscr{L}_0$, the even subspace of $\mathscr{L}_s$, then $\mathbf{\Gamma}(a)$ has the form

$$\mathbf{\Gamma}(a) = \begin{bmatrix} \mathbf{\Gamma}_{00}(a) & \mathbf{0} \\ \mathbf{0} & \mathbf{\Gamma}_{11}(a) \end{bmatrix}, \tag{21.52}$$

where $\mathbf{\Gamma}_{00}(a)$ and $\mathbf{\Gamma}_{11}(a)$ are $d_0 \times d_0$ and $d_1 \times d_1$ submatrices respectively; and if $a \in \mathscr{L}_1$, the odd subspace of $\mathscr{L}_s$, then $\mathbf{\Gamma}(a)$ has the form

$$\mathbf{\Gamma}(a) = \begin{bmatrix} \mathbf{0} & \mathbf{\Gamma}_{01}(a) \\ \mathbf{\Gamma}_{10}(a) & \mathbf{0} \end{bmatrix}, \tag{21.53}$$

where $\mathbf{\Gamma}_{01}(a)$ and $\mathbf{\Gamma}_{10}(a)$ are $d_0 \times d_1$ and $d_1 \times d_0$ submatrices respectively.

Then these matrices $\mathbf{\Gamma}(a)$ are said to form a (d_0/d_1)-*dimensional graded representation* of $\mathscr{L}_s$.

Clearly the even elements of $\mathscr{L}_s$ are represented by even matrices of $\mathrm{M}(d_0/d_1;\mathbb{C})$ and the odd elements are represented by odd matrices, so that

$$\deg \mathbf{\Gamma}(a) = \deg a \tag{21.54}$$

for all homogeneous $a \in \mathscr{L}_s$. In (21.51) $[a, b]$ is the generalized Lie product of $\mathscr{L}_s$, and $[\mathbf{\Gamma}(a), \mathbf{\Gamma}(b)]$ is defined (in accordance with (21.20) and (21.54)) for all homogeneous $a, b \in \mathscr{L}_s$ by

$$[\mathbf{\Gamma}(a), \mathbf{\Gamma}(b)] = \mathbf{\Gamma}(a)\mathbf{\Gamma}(b) - (-1)^{(\deg a)(\deg b)}\mathbf{\Gamma}(b)\mathbf{\Gamma}(a). \tag{21.55}$$

In specifying a graded representation of $\mathscr{L}_s$, it is obviously sufficient to specify *only* the matrices $\mathbf{\Gamma}(a_j)$ for every basis element a_j of a homogeneous basis of $\mathscr{L}_s$. Clearly the set of matrices $\mathbf{\Gamma}(a)$ themselves form a Lie superalgebra with the same field as $\mathscr{L}_s$, and there is a homomorphic mapping of $\mathscr{L}_s$ onto this Lie superalgebra. The matrices $\mathbf{\Gamma}(a)$, $\mathbf{\Gamma}_{00}(a)$ and $\mathbf{\Gamma}_{11}(a)$ of (21.52) provide representations of the Lie algebra $\mathscr{L}_0$.

To rephrase the definition in terms of modules, let the "carrier space" V be a $(d_0 + d_1)$-dimensional inner product space that is assumed to be a graded vector space with even part V_0 having dimension d_0 and the odd part V_1 having dimensional d_1. Let $\psi_j (j = 1, 2, \ldots, d_0)$ form a basis for V_0 and let $\psi_j (j = d_0 + 1, d_0 + 2, \ldots, d_0 + d_1)$ form a basis for V_1. Then for each $a \in \mathscr{L}_s$ a linear operator $\Phi(a)$ may be defined by

$$\Phi(a)\psi_j = \sum_{k=1}^{d_0+d_1} \Gamma(a)_{kj}\psi_k, \quad j = 1, 2, \ldots, d_0 + d_1 \tag{21.56}$$

(cf. (11.9)), and this definition can be extended to the whole of V by the requirement that

$$\Phi(a)\left\{\sum_{j=1}^{d_0+d_1} b_j\psi_j\right\} = \sum_{j=1}^{d_0+d_1} b_j\{\Phi(a)\psi_j\} \tag{21.57}$$

for any set of complex numbers b_j ($j = 1, 2, \ldots, d_0 + d_1$). It then follows that the set of operators $\Phi(a)$ form a Lie superalgebra and there is a homomorphic mapping of $\mathscr{L}_s$ onto this Lie superalgebra. In this Lie superalgebra of operators the generalized Lie product is given for all homogeneous a and b of $\mathscr{L}_s$ by

$$[\Phi(a), \Phi(b)] = \Phi(a)\Phi(b) - (-1)^{(\deg a)(\deg b)}\Phi(b)\Phi(a) \tag{21.58}$$

(cf. (21.55)). Also, if the basis $\psi_1, \psi_2, \ldots$ of V is chosen to be an orthonormal set then

$$\Gamma(a)_{kj} = (\psi_k, \Phi(a)\psi_j) \tag{21.59}$$

for all $a \in \mathscr{L}_s$ and all $j, k = 1, 2, \ldots, d_0 + d_1$. Finally, (21.52), (21.53) and (21.56) imply that

$$\text{if } a\in\mathscr{L}_0 \quad \text{and} \quad \psi\in \mathrm{V}_0 \quad \text{then} \quad \Phi(a)\psi\in\mathrm{V}_0, \tag{21.60a}$$

$$\text{if } a\in\mathscr{L}_0 \quad \text{and} \quad \psi\in \mathrm{V}_1 \quad \text{then} \quad \Phi(a)\psi\in\mathrm{V}_1, \tag{21.60b}$$

$$\text{if } a\in\mathscr{L}_1 \quad \text{and} \quad \psi\in \mathrm{V}_0 \quad \text{then} \quad \Phi(a)\psi\in\mathrm{V}_1, \tag{21.60c}$$

$$\text{if } a\in\mathscr{L}_1 \quad \text{and} \quad \psi\in \mathrm{V}_1 \quad \text{then} \quad \Phi(a)\psi\in\mathrm{V}_0. \tag{21.60d}$$

It is purely a matter of convention which of these two subspaces of V is described as being "even", for their interchange merely produces a *graded* representation of $\mathscr{L}_s$ that is *equivalent* to the original one, but with $\mathbf{\Gamma}_{00}(a)\leftrightarrow\mathbf{\Gamma}_{11}(a)$ in (21.52), $\mathbf{\Gamma}_{01}(a)\leftrightarrow\mathbf{\Gamma}_{10}(a)$ in (21.53), and $d_0\leftrightarrow d_1$. Apart from this the only similarity transformations that are relevant are those that transform V_0 into V_0 and V_1 into V_1, and for these the associated matrices have the form

$$\mathbf{S} = \begin{bmatrix} \mathbf{S}_0 & \mathbf{0} \\ \mathbf{0} & \mathbf{S}_1 \end{bmatrix}, \tag{21.61}$$

where $\mathbf{S}_0$ and $\mathbf{S}_1$ are $d_0 \times d_0$ and $d_1 \times d_1$ non-singular submatrices. That is, each such $\mathbf{S}$ must be an even invertible member of $\mathrm{M}(d_0/d_1; \mathbb{C})$. Then, with

$$\mathbf{\Gamma}'(a) = \mathbf{S}^{-1}\mathbf{\Gamma}(a)\mathbf{S}$$

(for all $a\in\mathscr{L}_s$), the set of matrices $\mathbf{\Gamma}'(a)$ form a (d_0/d_1)-dimensional graded representation of $\mathscr{L}_s$ with the same carrier space V (and, indeed, with the same even and odd subspaces as $\mathbf{\Gamma}$). The graded representations $\mathbf{\Gamma}$ and $\mathbf{\Gamma}'$ are said to be *equivalent*.

Concepts involving reducibility are most neatly formulated in terms of modules.

Definition *Reducible graded representation of a Lie superalgebra* $\mathscr{L}_s$

A graded representation of $\mathscr{L}_s$ for which the module consists of the operators $\Phi(a)$ acting on the graded vector space V is said to be *reducible* if V possesses a proper subspace V′ such that $\Phi(a)$ maps every vector of V′ into V′ for every $a \in \mathscr{L}_s$.

Here it is assumed that the even part of V'_0 of V′ is a subspace of the even part V_0, and the odd part V'_1 of V′ is a subspace of the odd part V_1 of V. The subspace V′ is said to be *invariant*.

Definition *Irreducible graded representation of a Lie superalgebra* $\mathscr{L}_s$

A graded representation of a Lie superalgebra is said to be *irreducible* if it is not reducible.

Definition *Completely reducible graded representation of a Lie superalgebra* $\mathscr{L}_s$

A graded representation of $\mathscr{L}_s$ for which the module consists of the operators $\Phi(a)$ acting on the graded vector space V is said to be *completely reducible* if V can be written as a direct sum of graded subspaces, each of which is invariant, and none of which has an invariant subspace.

The generalization of Schur's Lemma for Lie superalgebras is more complicated than the corresponding result for Lie algebras (Theorem II of Chapter 11, Section 4).

Theorem I Suppose that the matrices $\mathbf{\Gamma}(a)$ form a (d_0/d_1)-dimensional *irreducible* graded representation of a Lie superalgebra $\mathscr{L}_s$, and that $\mathbf{M}$ is a matrix of $\mathrm{M}(d_0/d_1; \mathbb{C})$ such that

$$[\mathbf{M}, \mathbf{\Gamma}(a)] = \mathbf{0} \tag{21.62}$$

for all $a \in \mathscr{L}_s$. We then have the following.

(a) If $\mathbf{M}$ is *even* then $\mathbf{M}$ must have the form

$$\mathbf{M} = \kappa \mathbf{1}, \tag{21.63}$$

where κ is any constant and $\mathbf{1}$ is the $(d_0 + d_1)$-dimensional unit matrix.

(b) If $\mathbf{M}$ is *odd* and

$$d_0 = d_1 \tag{21.64}$$

and there exists a non-singular $d_0 \times d_0$ matrix $\mathbf{B}$ such that

$$\mathbf{\Gamma}_{11}(a) = \mathbf{B}^{-1} \mathbf{\Gamma}_{00}(a) \mathbf{B} \tag{21.65}$$

for all $a \in \mathscr{L}_0$, and

$$\Gamma_{01}(a) = -\mathbf{B}\Gamma_{10}(a)\mathbf{B} \tag{21.66}$$

for all $a \in \mathscr{L}_1$, then $\mathbf{M}$ must have the form

$$\mathbf{M} = \mu \begin{bmatrix} \mathbf{0} & \mathbf{B} \\ \mathbf{B}^{-1} & \mathbf{0} \end{bmatrix}, \tag{21.67}$$

where μ is any constant. However, if any of the conditions (21.64), (21.65) or (21.66) are *not* satisfied then there is *no* odd $\mathbf{M}$ of $\mathrm{M}(d_0/d_1;\mathbb{C})$ satisfying (21.62). (In (21.65) and (21.66) the submatrices $\Gamma_{00}(a)$, $\Gamma_{11}(a)$, $\Gamma_{01}(a)$ and $\Gamma_{10}(a)$ are as defined in (21.52) and (21.53).)

Proof See Appendix K, Section 2.

Example I *Graded irreducible representations of some Heisenberg superalgebras*

(a) Consider the Heisenberg superalgebra $\mathscr{L}_s$ of Example I of Section 2 with $r = 1$, taking the homogeneous basis elements to be $a_1 = 1$, $a_2 = b_1^\dagger$ and $a_3 = b_1$. A (1/1)-dimensional irreducible graded representation of $\mathscr{L}_s$ is provided by

$$\Gamma(a_1) = \left[\begin{array}{c|c} 1 & 0 \\ \hline 0 & 1 \end{array}\right], \quad \Gamma(a_2) = \left[\begin{array}{c|c} 0 & 1 \\ \hline 0 & 0 \end{array}\right], \quad \Gamma(a_3) = \left[\begin{array}{c|c} 0 & 0 \\ \hline 1 & 0 \end{array}\right] \tag{21.68}$$

(where dividing lines have been inserted to emphasize the partitioning), so that, for example, $\Gamma_{01}(a_2) = [1]$ and $\Gamma_{10}(a_2) = [0]$. Although the condition (21.64) holds, it is clear that the condition (21.66) can never be achieved, so (21.63) provide the only homogeneous matrices of $\mathrm{M}(1/1;\mathbb{C})$ that satisfy (21.62).

(b) Consider the Heisenberg superalgebra $\mathscr{L}'_s$ of Example I of Section 2 with $r = 1$, taking the homogeneous basis elements to be $a_1 = 1$, $a_2 = b_1^\dagger$, $a_3 = b_1$ and $a_4 = b_2$. A (2/2)-dimensional irreducible graded representation of $\mathscr{L}'_s$ is provided by

$$\Gamma(a_1) = \left[\begin{array}{cc|cc} 1 & 0 & 0 & 0 \\ 0 & 1 & 0 & 0 \\ \hline 0 & 0 & 1 & 0 \\ 0 & 0 & 0 & 1 \end{array}\right], \quad \Gamma(a_2) = \left[\begin{array}{cc|cc} 0 & 0 & 1 & 0 \\ 0 & 0 & 0 & 0 \\ \hline 0 & 0 & 0 & 0 \\ 0 & 1 & 0 & 0 \end{array}\right],$$

$$\Gamma(a_3) = \left[\begin{array}{cc|cc} 0 & 0 & 0 & 0 \\ 0 & 0 & 0 & 1 \\ \hline 1 & 0 & 0 & 0 \\ 0 & 0 & 0 & 0 \end{array}\right], \quad \Gamma(a_4) = \left[\begin{array}{cc|cc} 0 & 0 & 0 & \frac{1}{2} \\ 0 & 0 & -1 & 0 \\ \hline 0 & -\frac{1}{2} & 0 & 0 \\ 1 & 0 & 0 & 0 \end{array}\right]$$

(the partitioning again being emphasized by the insertion of dividing lines). In this case the conditions (21.64)–(21.66) are satisfied with

$$\mathbf{B} = \begin{bmatrix} 0 & 2^{-1/2}i \\ 2^{1/2}i & 0 \end{bmatrix},$$

so the homogeneous matrices (21.63) *and* (21.67) satisfy (21.62) in this case.

The supertranspose operation that was introduced in Chapter 20, Section 4, can be applied to the graded matrices of $M(d_0/d_1;\mathbb{C})$, thereby enabling contragredient graded representations to be defined.

Definition *Contragredient graded representation of a Lie superalgebra* $\mathscr{L}_s$

If the matrices $\Gamma(a)$ $(a \in \mathscr{L}_s)$ form a (d_0/d_1)-dimensional graded representation of $\mathscr{L}_s$ then the set of $(d_0 + d_1) \times (d_0 + d_1)$ matrices defined by

$$\Gamma^C(a) = -\Gamma(a)^{st} \tag{21.69}$$

for all $a \in \mathscr{L}_s$ also form a (d_0/d_1)-dimensional graded representation of $\mathscr{L}_s$. This representation Γ^C is known as the representation *contragredient* to Γ.

Thus, by (20.58) and (21.69), if a is *even*, so that $\Gamma(a)$ is given by (21.52), then

$$\Gamma^C(a) = \begin{bmatrix} -\tilde{\Gamma}_{00}(a) & \mathbf{0} \\ \mathbf{0} & -\tilde{\Gamma}_{11}(a) \end{bmatrix}; \tag{21.70}$$

whereas if a is *odd*, so that $\Gamma(a)$ is given by (21.53),

$$\Gamma^C(a) = \begin{bmatrix} \mathbf{0} & \tilde{\Gamma}_{10}(a) \\ -\tilde{\Gamma}_{01}(a) & \mathbf{0} \end{bmatrix}. \tag{21.71}$$

It is obvious that these matrices have the required grading property, and it is easily checked that they form a representation of $\mathscr{L}_s$.

The notion of the *direct product* of two representations of a Lie algebra that was discussed in detail in Chapter 11, Section 4, can be generalized to the situation of graded representations of a Lie superalgebra $\mathscr{L}_s$. To this end, let $\Phi'(a)$ $(a \in \mathscr{L}_s)$ and $V' = V'_0 \oplus V'_1$ be a set of linear operators and a graded vector space that together form a module for a (d'_0/d'_1)-dimensional graded representation Γ' of $\mathscr{L}_s$, and let $\Phi''(a)$ $(a \in \mathscr{L}_s)$ and $V'' = V''_0 \oplus V''_1$ be the corresponding quantities for a (d''_0/d''_1)-dimensional graded representation Γ'' of $\mathscr{L}_s$. Then, for any homogeneous $a \in \mathscr{L}_s$ and any homogeneous $\psi' \in V'$ and $\psi'' \in V''$, *define* the *direct product* operators $\Phi(a)$ by

$$\Phi(a)\{\psi' \otimes \psi''\} = \{\Phi'(a)\psi'\} \otimes \psi'' + (-1)^{(\deg a)(\deg \psi')}\psi' \otimes \{\Phi''(a)\psi''\} \tag{21.72}$$

(cf. (11.31)). Thus, if a is *even*, $\Phi(a)$ maps the subspaces $V'_0 \otimes V''_0$, $V'_0 \otimes V''_1$, $V'_1 \otimes V''_0$ and $V'_1 \otimes V''_1$ into themselves. By contrast, if a is *odd*, $\Phi(a)$ maps $\{(V'_0 \otimes V''_0) \oplus (V'_1 \otimes V''_1)\}$ into $\{(V'_0 \otimes V''_1) \oplus (V'_1 \otimes V''_0)\}$ and $\{(V'_0 \otimes V''_1) \oplus (V'_1 \otimes V''_0)\}$ into $\{(V'_0 \otimes V''_0) \oplus (V'_1 \otimes V''_1)\}$. Consequently the direct product space $V = V' \otimes V''$ possesses a natural grading, with the even part being the $(d'_0 d''_0 + d'_1 d''_1)$-dimensional subspace $\{(V'_0 \otimes V''_0) \oplus (V'_1 \otimes V''_1)\}$, and the odd part being the $(d'_0 d''_1 + d'_1 d''_0)$-dimensional subspace $\{(V'_0 \otimes V''_1) \oplus (V'_1 \otimes V''_0)\}$. It is easily checked that, with the definition (21.72), $[\Phi(a), \Phi(b)] = \Phi([a, b])$ for $a, b \in \mathscr{L}_s$, and hence these operators acting on $V = V' \otimes V''$ form a module of a graded representation of $\mathscr{L}_s$.

The construction of graded representations of direct sums of Lie superalgebras is very similar. Let $\mathscr{L}'_s$ and $\mathscr{L}''_s$ be two Lie superalgebras (with the same field) that possess graded representations of dimensions (d'_0/d'_1) and (d''_0/d''_1) respectively, for which the modules consist of operators $\Phi'(a')$ $(a' \in \mathscr{L}'_s)$ and $\Phi''(a'')$ $(a'' \in \mathscr{L}''_s)$ acting on graded vector spaces $V' = V'_0 \oplus V'_1$ and $V'' = V''_0 \oplus V''_1$ respectively. Then, for each homogeneous $a' \in \mathscr{L}'_s$, $a'' \in \mathscr{L}''_s$, $\psi' \in V'$ and $\psi'' \in V''$, linear operators corresponding to elements of $\mathscr{L}_s = \mathscr{L}'_s \oplus \mathscr{L}''_s$ can be defined by

$$\Phi(a' + a'')\{\psi' \otimes \psi''\} = \{\Phi'(a')\psi'\} \otimes \psi'' + (-1)^{(\deg a')(\deg \psi')} \psi' \otimes \{\Phi''(a'')\psi''\} \tag{21.73}$$

(cf. (11.56) and (21.72)). Again a grading can be introduced on $V = V' \otimes V''$ by taking the subspace $\{(V'_0 \otimes V''_0) \oplus (V'_1 \otimes V''_1)\}$ to be even and the subspace $\{(V'_0 \otimes V''_1) \oplus (V'_1 \otimes V''_0)\}$ to be odd. It is not difficult to check that the operators $\Phi(a)$ of (21.73) provide a graded representation of $\mathscr{L}_s = \mathscr{L}'_s \oplus \mathscr{L}''_s$ in which

$$[\Phi(a'), \Phi(a'')] = 0 \tag{21.74}$$

for all homogeneous $a' \in \mathscr{L}'_s$ and $a'' \in \mathscr{L}''_s$ (in accordance with (21.49)).

For a discussion of such concepts as induced representations see Scheunert (1985).

5 The adjoint representation and the Killing form of a Lie superalgebra

The adjoint representation and the Killing form play just as important a role for Lie superalgebras as they do for Lie algebras.

Theorem I Let $\mathscr{L}_s$ be a real or complex Lie superalgebra with even dimension m and odd dimension n, and let $a_1, a_2, \ldots, a_{m+n}$ be a homogeneous basis for $\mathscr{L}_s$ (i.e. a basis in which $a_1, a_2, \ldots, a_m$ are even and

$a_{m+1}, a_{m+2}, \ldots, a_{m+n}$ are odd). For any $a \in \mathscr{L}_s$ let $\mathbf{ad}(a)$ be the $(m+n) \times (m+n)$ matrix defined by

$$[a, a_j] = \sum_{k=1}^{m+n} \{\mathbf{ad}(a)\}_{kj} a_k \tag{21.75}$$

for $j = 1, 2, \ldots, m+n$. Then the set of matrices $\mathbf{ad}(a)$ form an (m/n)-dimensional graded representation of $\mathscr{L}_s$ called the *adjoint representation* of $\mathscr{L}_s$.

Proof It is obvious that the matrices $\mathbf{ad}(a)$ are graded in the way that is implied. The rest of the proof then proceeds as for the corresponding theorem on Lie algebras (Theorem I of Chapter 11, Section 5) (replacing Lie products by generalized Lie products, the Jacobi identity by the generalized Jacobi identity, and so on).

In terms of the corresponding module, the operators $\mathrm{ad}(a)$ may be defined for each $a \in \mathscr{L}_s$ by

$$\mathrm{ad}(a)b = [a, b] \tag{21.76}$$

for all $b \in \mathscr{L}_s$. The carrier space is then $\mathscr{L}_s$ itself, the even subspace being $\mathscr{L}_0$ and the odd subspace $\mathscr{L}_1$.

Equations (21.9) and (21.75) together imply that

$$\{\mathbf{ad}(a_p)\}_{rq} = c^r_{pq} \tag{21.77}$$

for $p, q, r = 1, 2, \ldots, m+n$, where c^r_{pq} are the structure constants of $\mathscr{L}_s$. Thus if $\mathscr{L}_s$ is a real Lie superalgebra then all the elements of $\mathbf{ad}(a)$ are real for all $a \in \mathscr{L}_s$.

Because of the connection between the set of graded matrices $\mathrm{M}(p/q; \mathbb{C})$ and the set of supermatrices $\mathrm{M}(p/q; \mathbb{C}\mathrm{B}_L)$ that was discussed in Chapter 20, Section 4, the definition given in (20.66) for the supertrace applies to members of $\mathrm{M}(p/q; \mathbb{C})$. However, as the "diagonal submatrices" $\mathbf{A}$ and $\mathbf{D}$ are both zero matrices if $\mathbf{M}$ is odd, the definition of the supertrace for $\mathbf{M} \in \mathrm{M}(p/q; \mathbb{C})$ reduces to

$$\mathrm{str}\,\mathbf{M} = \begin{cases} \mathrm{tr}\,\mathbf{A} - \mathrm{tr}\,\mathbf{D} & \text{if } \mathbf{M} \text{ is even,} \\ 0 & \text{if } \mathbf{M} \text{ is odd,} \end{cases} \tag{21.78}$$

$\mathbf{M}$ being assumed to be given by (20.8a) or (20.8b) as appropriate. Then, by (20.69) and (21.20),

$$\mathrm{str}([\mathbf{M}, \mathbf{N}]) = 0 \tag{21.79}$$

for all $\mathbf{M}$ and $\mathbf{N}$ of $\mathrm{M}(p/q; \mathbb{C})$.

Definition *Killing form of a Lie superalgebra*

The *Killing form* $B(a, b)$ corresponding to any two elements a and b of a Lie superalgebra $\mathscr{L}_s$ is defined by

$$B(a, b) = \text{str}\,\{\mathbf{ad}(a)\mathbf{ad}(b)\}, \tag{21.80}$$

where $\mathbf{ad}(a)$ denotes the matrix representing $a \in \mathscr{L}_s$ in the adjoint representation of $\mathscr{L}_s$ (as defined in (21.75)) and "str" denotes the supertrace (cf. (21.78)) of the matrix product.

If $\mathscr{L}_s$ is a real Lie superalgebra then all the matrix elements of $\mathbf{ad}(a)$ are real for all $a \in \mathscr{L}_s$, and hence in this case $B(a, b)$ is real for all $a, b \in \mathscr{L}_s$. This will not be so when $\mathscr{L}_s$ is complex.

Theorem II The Killing form of a Lie superalgebra $\mathscr{L}_s$ has the following properties:

(a) for all $a, b \in \mathscr{L}_s$,

$$B(b, a) = (-1)^{(\deg a)(\deg b)} B(a, b); \tag{21.81}$$

(b) for all $a, b \in \mathscr{L}_s$ and all α, β of the field of $\mathscr{L}_s$,

$$B(\alpha a, \beta b) = \alpha\beta B(a, b); \tag{21.82}$$

(c) for all $a, b, c \in \mathscr{L}_s$,

$$B(a, b + c) = B(a, b) + B(a, c); \tag{21.83}$$

(d) for all $a \in \mathscr{L}_0$ and $b \in \mathscr{L}_1$ (or $a \in \mathscr{L}_1$ and $b \in \mathscr{L}_0$),

$$B(a, b) = 0; \tag{21.84}$$

(e) if ψ is any automorphism of $\mathscr{L}_s$ then

$$B(\psi(a), \psi(b)) = B(a, b) \tag{21.85}$$

for all $a, b \in \mathscr{L}_s$;

(f) for all $a, b, c \in \mathscr{L}_s$,

$$B([a, b], c) = B(a, [b, c]); \tag{21.86}$$

(g) if $\mathscr{L}'_s$ is an invariant graded subalgebra of $\mathscr{L}_s$, and $B'(\ ,\ )$ denotes the Killing form of $\mathscr{L}'_s$ considered as a Lie superalgebra in its own right, then

$$B(a, b) = B'(a, b) \tag{21.87}$$

for all $a, b \in \mathscr{L}'_s$.

Proof Part (a) follows immediately from (20.69) and (21.80), while (b) and (c) are obvious. For (d), under the conditions stated, $\mathbf{ad}(a)\mathbf{ad}(b)$ is an odd member of $M(m, n; \mathbb{C})$, and hence, by (21.78), its supertrace is zero. Parts (e)–(g) are proved in the same way as in the corresponding theorem for Lie algebras (Theorem I of Chapter 13, Section 2), making, of course, the appropriate generalizations.

Because of the property (21.81), the Killing form of $\mathscr{L}_s$ is said to be *supersymmetric*, while (21.82) and (21.83) (taken with (21.81)) imply that it is *bilinear*. Similarly, because $B(\ , \ )$ satisfies (21.84), it is said to be *consistent*, and, by virtue of (21.86), it is also described as being *invariant*.

Theorem III Let $\mathscr{L}'_s$ be an invariant graded subalgebra of $\mathscr{L}_s$, and let $\mathscr{L}'^{\perp}_s$ be the "orthogonal complement of $\mathscr{L}'_s$ in $\mathscr{L}_s$ with respect to the Killing form $B(\ , \ )$ of $\mathscr{L}_s$", which is defined to be the subset of elements of $\mathscr{L}_s$ such that $a \in \mathscr{L}'^{\perp}_s$ if

$$B(a, b) = 0 \tag{21.88}$$

for all $b \in \mathscr{L}'_s$. Then $\mathscr{L}'^{\perp}_s$ is also an invariant graded subalgebra of $\mathscr{L}_s$.

Proof See Appendix K, Section 3.

As the proof involved only the properties (21.81)–(21.84) and (21.86), the same result is true for the orthogonal complement defined in terms of any other bilinear supersymmetric consistent invariant form on $\mathscr{L}_s$.

22

Superspace and Lie Supergroups

1 Grassmann variables as coordinates

The main object of this chapter is to introduce the idea of superspace and the concept of a Lie supergroup. The superspace that will be discussed here is a generalization of Euclidean space in which each point is specified by a set of elements from a real Grassmann algebra rather than just being specified by a set of real numbers. Likewise a Lie supergroup is a generalization of a Lie group in which the elements are parameterized by points of a superspace instead of points of a Euclidean space.

Section 2 is devoted to superspace. The basic ideas involved in the setting up of a superspace are introduced in Subsection 2(*a*). The formulation of the concepts of continuity, differentiability, superanalyticity and superdifferentiability for functions defined on superspace is described in Subsections 2(*b*) and 2(*c*), and the case where these functions form supermatrices is developed in Subsection 2(*d*).

For convenience, the Lie supergroup discussion of Section 3 will be confined to *linear* Lie supergroups. The presentation in Subsection 3(*a*) will be arranged in such a way that each stage in the construction of a linear Lie supergroup appears as a simple generalization of the corresponding stage in the construction of a linear Lie group as described in Chapter 3. The rather subtle relationship of linear Lie supergroups to real Lie superalgebras will be examined in Subsection 3(*b*).

2 Analysis on superspace

(a) *The superspace* $\mathbb{R}B_L^{m,n}$

One of the most important features of a linear Lie group of order m is that there exists a one-to-one correspondence between a set of elements of the group close to its identity and a set of points $\mathbf{x} = (x^1, x^2, \dots, x^m)$ of $\mathbb{R}^m$ close to the origin of $\mathbb{R}^m$, the origin of $\mathbb{R}^m$ corresponding to the identity of the group. Here two important attributes of $\mathbb{R}^m$ are involved. Firstly, $\mathbb{R}^m$ can be regarded as being m copies of the set of real numbers $\mathbb{R}$. Secondly, $\mathbb{R}^m$ possesses a *metric* (or *distance function*) d, which is defined to be such that the distance between two points $\mathbf{x} = (x^1, x^2, \dots, x^m)$ and $\mathbf{y} = (y^1, y^2, \dots, y^m)$ is

$$d(\mathbf{x}, \mathbf{y}) = \|\mathbf{x} - \mathbf{y}\|, \tag{22.1}$$

where the *norm* $\|\mathbf{x}\|$ is itself defined by

$$\|\mathbf{x}\| = \{(x^1)^2 + (x^2)^2 + \cdots + (x^m)^2\}^{1/2} \tag{22.2}$$

and is known as the "Euclidean norm". It is easily shown that this metric has the properties that

(i) $d(\mathbf{x}, \mathbf{y}) = d(\mathbf{y}, \mathbf{x})$ for all $\mathbf{x}$ and $\mathbf{y}$ of $\mathbb{R}^m$,

(ii) $d(\mathbf{x}, \mathbf{x}) = 0$ for all $\mathbf{x}$ of $\mathbb{R}^m$,

(iii) $d(\mathbf{x}, \mathbf{y}) > 0$ if $\mathbf{x} \neq \mathbf{y}$, and

(iv) If $\mathbf{x}$, $\mathbf{y}$ and $\mathbf{z}$ are any three points of $\mathbb{R}^m$ then

$$d(\mathbf{x}, \mathbf{z}) \leqslant d(\mathbf{x}, \mathbf{y}) + d(\mathbf{y}, \mathbf{z}).$$

An *open sphere* of radius r centred on the point $\mathbf{y}$ is then defined to be the set of points $\mathbf{x}$ of $\mathbb{R}^m$ such that

$$d(\mathbf{x}, \mathbf{y}) < r.$$

(In this language the set of points satisfying (3.2) form the open sphere of radius η centred on the origin.) Further, a set of points V of $\mathbb{R}^m$ is said to form an *open set* of $\mathbb{R}^m$ if, for *every* point $\mathbf{y}$ of V, there exists an open sphere centred on $\mathbf{y}$ of some radius r (which may depend on $\mathbf{y}$) that is *completely* contained in V.

As a Grassmann generalization of $\mathbb{R}^m$, consider the space $\mathbb{R}B_L^{m,n}$, which is defined to consist of m copies of the even subspace $\mathbb{R}B_{L0}$ of the real Grassmann algebra $\mathbb{R}B_L$ and n copies of the odd subspace $\mathbb{R}B_{L1}$ of $\mathbb{R}B_L$. (Here L may be taken to be either a finite positive integer or to be infinite.) The m copies of $\mathbb{R}B_{L0}$ will be denoted by $X^1, X^2, \dots, X^m$, and the n copies of $\mathbb{R}B_{L1}$ will be indicated by $\Theta^1, \Theta^2, \dots, \Theta^n$. Each real even Grassmann element X^j (for $j = 1, 2, \dots, m$) has an expansion in terms of the basis elements $\mathcal{E}_\mu$ of $\mathbb{C}B_L$ of the form

$$X^j = \sum_{\mu} X^j_{\mu} \mathcal{E}_{\mu} \tag{22.3}$$

(cf. (20.25)), where the coefficients X^j_{μ} are all real numbers, and only even index sets μ appear. Similarly, for each real odd Grassmann element Θ^k (for $k = 1, 2, \ldots, n$),

$$\Theta^k = \sum_{\mu} \Theta^k_{\mu} \mathcal{E}_{\mu}, \tag{22.4}$$

where the coefficients Θ^k_{μ} are again all real numbers but only odd index sets μ appear. It is convenient to make the notation more concise by regarding $X^1, X^2, \ldots, X^m$ as elements of an m-component quantity $\mathbf{X}$, and $\Theta^1, \Theta^2, \ldots, \Theta^n$ as elements of a n-component quantity $\mathbf{\Theta}$, and by defining $(\mathbf{X}; \mathbf{\Theta})$, a typical element of $\mathbb{R}\mathrm{B}_L^{m,n}$, by

$$(\mathbf{X}; \mathbf{\Theta}) = (X^1, X^2, \ldots, X^m, \Theta^1, \Theta^2, \ldots, \Theta^n).$$

As $\mathbb{R}\mathrm{B}_{L0}$ and $\mathbb{R}\mathrm{B}_{L1}$ are both real vector spaces of dimension 2^{L-1}, $\mathbb{R}\mathrm{B}_L^{m,n}$ is a real vector space of dimension $(m+n)2^{L-1}$. It is usual to choose L to be such that $L \geqslant n$, but Rogers (1986a) has demonstrated that there are considerable advantages in taking $L \geqslant 2n$. Apart from this restriction, the value of L is normally quite arbitrary.

To allow analysis to be performed on the space $\mathbb{R}\mathrm{B}_L^{m,n}$, it has to be provided with a metric. There are two alternative choices that are both convenient in certain respects. Each follows from a choice of a norm on $\mathbb{C}\mathrm{B}_L$. Suppose first that L is finite, and let B be any element of $\mathbb{C}\mathrm{B}_L$, whose expansion in terms of the basis elements $\mathcal{E}_{\mu}$ is

$$B = \sum_{\mu} B_{\mu} \mathcal{E}_{\mu}$$

(as in (20.25)), where B_{μ} are a set of complex numbers. One choice is the "Euclidean norm", which may be defined by generalizing (22.2) to give

$$\| B \| = \left\{ \sum_{\mu} |B_{\mu}|^2 \right\}^{1/2}. \tag{22.5}$$

The other choice is to take

$$\| B \| = \sum_{\mu} |B_{\mu}|. \tag{22.6}$$

With both choices it is true that

(i) $\| B \| \geqslant 0$ for all $B \in \mathbb{C}\mathrm{B}_L$;

(ii) $\| B \| = 0$ if and only if $B = 0$;

(iii) $\| B + B' \| \leqslant \| B \| + \| B' \|$ for all B and B' of $\mathbb{C}\mathrm{B}_L$; and

(iv) $\| cB \| = |c| \, \| B \|$ for all $B \in \mathbb{C}\mathrm{B}_L$ and any complex number c.

For $\mathbb{R}\mathrm{B}_L^{m,n}$ the Euclidean norm may be defined by

$$\| (\mathbf{X}; \mathbf{\Theta}) \| = \left\{ \sum_{j=1}^{m} \sum_{\mu} | X_{\mu}^{j} |^2 + \sum_{k=1}^{n} \sum_{\mu} | \Theta_{\mu}^{k} |^2 \right\}^{1/2}, \tag{22.7}$$

and the norm corresponding to (22.6) may be defined by

$$\| (\mathbf{X}; \mathbf{\Theta}) \| = \sum_{j=1}^{m} \| X^j \| + \sum_{k=1}^{n} \| \Theta^k \|, \tag{22.8a}$$

so that

$$\| (\mathbf{X}; \mathbf{\Theta}) \| = \sum_{j=1}^{m} \sum_{\mu} | X_{\mu}^{j} | + \sum_{k=1}^{n} \sum_{\mu} | \Theta_{\mu}^{k} |. \tag{22.8b}$$

With each of these choices of norm a metric d may be introduced (cf. (22.1)) by the prescription

$$d((\mathbf{X}; \mathbf{\Theta}), (\mathbf{X}'; \mathbf{\Theta}')) = \| (\mathbf{X}; \mathbf{\Theta}) - (\mathbf{X}'; \mathbf{\Theta}') \|. \tag{22.9}$$

It is easily checked that this has essentially the same properties as for $\mathbb{R}^m$:

(i) $d((\mathbf{X}; \mathbf{\Theta}), (\mathbf{X}'; \mathbf{\Theta}')) = d((\mathbf{X}'; \mathbf{\Theta}'), (\mathbf{X}; \mathbf{\Theta}))$ for all $(\mathbf{X}; \mathbf{\Theta})$ and $(\mathbf{X}'; \mathbf{\Theta}')$ of $\mathbb{R}\mathrm{B}_L^{m,n}$;

(ii) $d((\mathbf{X}; \mathbf{\Theta}), (\mathbf{X}; \mathbf{\Theta})) = 0$ for all $(\mathbf{X}; \mathbf{\Theta})$ of $\mathbb{R}\mathrm{B}_L^{m,n}$;

(iii) $d((\mathbf{X}; \mathbf{\Theta}), (\mathbf{X}'; \mathbf{\Theta}')) > 0$ if $(\mathbf{X}; \mathbf{\Theta}) \neq (\mathbf{X}'; \mathbf{\Theta}')$;

and

(iv) if $(\mathbf{X}; \mathbf{\Theta})$, $(\mathbf{X}'; \mathbf{\Theta}')$ and $(\mathbf{X}''; \mathbf{\Theta}'')$ are any three points of $\mathbb{R}\mathrm{B}_L^{m,n}$ then
$$d((\mathbf{X}; \mathbf{\Theta}), (\mathbf{X}''; \mathbf{\Theta}'')) \leqslant d((\mathbf{X}; \mathbf{\Theta}), (\mathbf{X}'; \mathbf{\Theta}')) + d((\mathbf{X}'; \mathbf{\Theta}'), (\mathbf{X}''; \mathbf{\Theta}'')).$$

For each metric an *open sphere* of radius r centred on the point $(\mathbf{X}'; \mathbf{\Theta}')$ is then defined to be the set of points $(\mathbf{X}; \mathbf{\Theta})$ of $\mathbb{R}\mathrm{B}_L^{m,n}$ such that

$$d((\mathbf{X}; \mathbf{\Theta}), (\mathbf{X}'; \mathbf{\Theta}')) < r,$$

and a set of points U of $\mathrm{RB}_L^{m,n}$ is said to form an *open set* of $\mathbb{R}\mathrm{B}_L^{m,n}$ if for *every* point $(\mathbf{X}'; \mathbf{\Theta}')$ of U there exists an open sphere centred on $(\mathbf{X}'; \mathbf{\Theta}')$ of some radius r (which may depend on $(\mathbf{X}'; \mathbf{\Theta}')$) that is *completely* contained in U.

Although the open spheres defined with respect to these metrics clearly do not coincide, it is nevertheless true that if U is an open set with respect to one metric then it is an open set with respect to the other, and so they define equivalent topologies on $\mathbb{R}B_L^{m,n}$. (More precisely, as each collection of open sets introduced here constitutes a topology in the sense of the definition of a topological space that is given in Appendix J, Section 2, and as these two collections coincide, both metrics produce the same topology.) Consequently, as far as topological properties are concerned, both choices are equivalent. However, although the Euclidean norm may appear a more natural choice, Rogers (1980) has pointed out that with the choice (22.6)

(i) $\|\mathcal{E}_\varnothing\| = 1$, where $\mathcal{E}_\varnothing$ $(= 1)$ is the identity of $\mathbb{C}B_L$;

(ii) for all B and B' of $\mathbb{C}B_L$,

$$\|BB'\| \leqslant \|B\| \, \|B'\|;$$

and

(iii) $\mathbb{C}B_L$ is "complete" in the sense that every Cauchy sequence of elements of $\mathbb{C}B_L$ converges to an element of $\mathbb{C}B_L$ (a Cauchy sequence being as defined in Appendix B, Section 3).

These are the requirements for an algebra possessing a norm to be a *Banach algebra*, and, as Rogers (1980) has demonstrated, there are a number of advantages in being able to take $\mathbb{C}B_L$ to be a Banach algebra. Consequently the norm (22.6) will be used henceforth.

For the case in which L is infinite the discussion is similar, but the right-hand sides of (22.5)–(22.8a, b) all involve infinite series, and so the corresponding norms and metrics are meaningful only for those elements of $\mathbb{C}B_\infty$ or $\mathbb{R}B_\infty^{m,n}$ for which these series converge. Rogers (1980) has shown that with the norm (22.6) and with this restriction the Grassmann algebra $\mathbb{C}B_\infty$ is again a Banach algebra.

Just as an analytic manifold of dimension m is obtained by patching together subspaces each of which is homeomorphic to some open set of $\mathbb{R}^m$ (cf. Appendix J, Section 3), a *supermanifold* of even dimension m and odd dimension n may be set up as a topological space whose subspaces are each homeomorphic to some open set of $\mathbb{R}B_L^{m,n}$. For details of this construction see Rogers (1980), Boyer and Gitler (1980, 1984) and Jadczyk and Pilch (1981). This results in a precise formulation of the *superspace* that has been the subject of many papers in the theoretical physics literature since the pioneering work of Volkov and Akulov (1973) and Salam and Strathdee (1974a). An alternative extension of the idea of an analytic manifolds that also involves Grassmann algebras has been developed by Berezin and Leites (1975), Kostant (1977) and Leites (1980c). The resulting structures, which incorporate sheaves defined on analytic manifolds, have been called "graded manifolds" by Kostant (1977), and it is convenient to use this name to distinguish these structures from those built on the $\mathbb{R}B_L^{m,n}$ spaces, for which the term "supermanifold" will be reserved. For discussions of the relationships between supermanifolds and graded manifolds see for example Rogers (1980, 1986a), Batchelor (1980, 1984), Rothstein (1986) and Sanchez (1988). The present book deals solely with the $\mathbb{R}B_L^{m,n}$ formulation.

(b) *Differentiable functions on* $\mathbb{R}B_L^{m,n}$

Two types of "Grassmann-valued" functions will now be discussed. One is defined on an open set of $\mathbb{R}^m$, the other on an open set of $\mathbb{R}B_L^{m,n}$. Although the latter is more important in applications to Lie supergroups, the former will be considered first as it is the more straightforward.

A Grassmann-valued function $\mathscr{F}$ can be defined on an open set V of $\mathbb{R}^m$ by assigning to each element $\mathbf{x} = (x^1, x^2, \ldots, x^m)$ of V an element $\mathscr{F}(x)$ of the Grassmann algebra $\mathbb{C}B_L$. Such a function can be expanded in the form (cf. (20.25)).

$$\mathscr{F}(\mathbf{x}) = \sum_{\mu} \mathscr{F}_{\mu}(\mathbf{x})\mathcal{E}_{\mu}, \tag{22.10}$$

where the $\mathscr{F}_{\mu}(\mathbf{x})$ are all complex-valued functions of $\mathbf{x}$ in $\mathbf{V}$ and the sum is over all index sets μ of the type (20.22). $\mathscr{F}(\mathbf{x})$ is said to be *even* if only the even basis elements $\mathcal{E}_{\mu}$ of $\mathbb{C}B_L$ appear in this expansion, and to be *odd* if only odd $\mathcal{E}_{\mu}$ appear, so every Grassmann-valued function $\mathscr{F}(\mathbf{x})$ can be written as the sum of an even and an odd function.

Definition *Continuity of a Grassmann-valued function defined on an open set of* $\mathbb{R}^m$

The function $\mathscr{F}(\mathbf{x})$ defined on the open set V of $\mathbb{R}^m$ is said to be *continuous* at a point $\mathbf{x}'$ of V if $\mathscr{F}(\mathbf{x}) \rightarrow \mathscr{F}(\mathbf{x}')$ as $\mathbf{x} \rightarrow \mathbf{x}'$. This can be expressed more precisely in terms of the metrics for $\mathbb{C}B_L$ and $\mathbb{R}^m$ that were introduced above by requiring that for any real number $\varepsilon > 0$ there exists a real number $\delta > 0$ such that

$$d(\mathscr{F}(\mathbf{x}), \mathscr{F}(\mathbf{x}')) < \varepsilon$$

for all $\mathbf{x} \in \mathrm{V}$ for which $d(\mathbf{x}, \mathbf{x}') < \delta$.

It is easily shown that $\mathscr{F}(\mathbf{x})$ is continuous at $\mathbf{x}'$ if and only all the complex-valued component functions $\mathscr{F}_{\mu}(\mathbf{x})$ of (22.10) are continuous at $\mathbf{x}'$.

Definition *Differentiability of a Grassmann-valued function defined on an open set of* $\mathbb{R}^m$

The function $\mathscr{F}(\mathbf{x})$ defined on the open set V of $\mathbb{R}^m$ is said to be *differentiable* with respect to x^j (for some j in the set $\{1, 2, \ldots, m\}$) at a point $\mathbf{x}'$ of V and is said to have partial derivative $\partial\mathscr{F}(\mathbf{x})/\partial x^j|_{\mathbf{x}=\mathbf{x}'}$ if for any real number $\varepsilon > 0$ there exists a real number $\delta > 0$ such that

$$d\left(\left.\frac{\partial\mathscr{F}(\mathbf{x})}{\partial x^j}\right|_{\mathbf{x}=\mathbf{x}'}, \frac{\mathscr{F}(\mathbf{x}) - \mathscr{F}(\mathbf{x}')}{x^j - x'^j}\right) < \varepsilon \tag{22.11}$$

for all $\mathbf{x}\in$V for which $d(\mathbf{x}, \mathbf{x}') < \delta$. If this partial derivative exists for all $\mathbf{x}'\in$V it defines a Grassmann-valued function $\partial\mathscr{F}(\mathbf{x})/\partial x^j$ in V.

It is obvious that $\partial\mathscr{F}(\mathbf{x})/\partial x^j$ exists at $\mathbf{x}'$ if and only if the partial derivatives $\partial\mathscr{F}_\mu(\mathbf{x})/\partial x^j$ exist at $\mathbf{x} = \mathbf{x}'$ for all the complex-valued component functions $\mathscr{F}_\mu(\mathbf{x})$ of (22.10). Then

$$\frac{\partial\mathscr{F}(\mathbf{x})}{\partial x^j} = \sum_\mu \frac{\partial\mathscr{F}_\mu(\mathbf{x})}{\partial x^j}\,\mathcal{E}_\mu \tag{22.12}$$

for all $\mathbf{x}\in$V.

Clearly, higher derivatives can be obtained by repeating this process.

Definition *The sets* C^p(V) *and* C^∞(V)

Suppose that the Grassmann-valued function $\mathscr{F}(\mathbf{x})$ is defined on the open set V of $\mathbb{R}^m$. If $\mathscr{F}(\mathbf{x})$ is continuous on V then $\mathscr{F}(\mathbf{x})$ is said to be a member of C^0(V). If p is a finite positive integer and all derivatives of $\mathscr{F}(\mathbf{x})$ of order p exist on V then $\mathscr{F}(\mathbf{x})$ is said to be a member of the set C^p(V). If $\mathscr{F}(\mathbf{x})\in C^p$(V) for every positive integer p then $\mathscr{F}(\mathbf{x})$ is said to be a member of the set C^∞(V).

Most of these ideas can be generalized immediately for a Grassmann-valued function defined on an open set of $\mathbb{R}B_L^{m,n}$ rather than $\mathbb{R}^m$. Such a function may be defined by assigning to each point $(\mathbf{X}; \mathbf{\Theta}) = (X^1, X^2, \ldots, X^m, \Theta^1, \Theta^2, \ldots, \Theta^n)$ of an open set U of $\mathbb{R}B_L^{m,n}$ an element $F(\mathbf{X}; \mathbf{\Theta})$ of the Grassmann algebra $\mathbb{C}B_L$. The analogue of (22.10) is

$$F(\mathbf{X}; \mathbf{\Theta}) = \sum_\mu F_\mu(\mathbf{X}; \mathbf{\Theta})\mathcal{E}_\mu, \tag{22.13}$$

where each of the $F_\mu(\mathbf{X}; \mathbf{\Theta})$ is a complex-valued function of $(\mathbf{X}; \mathbf{\Theta})$ in U and the sum is over all index sets μ of the type (20.22). Again $F(\mathbf{X}; \mathbf{\Theta})$ is said to be *even* if only the even basis elements $\mathcal{E}_\mu$ of $\mathbb{C}B_L$ appear in this expansion, and to be *odd* if only odd $\mathcal{E}_\mu$ appear, so every Grassmann-valued function $F(\mathbf{X}; \mathbf{\Theta})$ on U can again be written as the sum of an even and an odd function.

Definition *Continuity of a Grassmann-valued function defined on an open set of* $\mathbb{R}B_L^{m,n}$

The function $F(\mathbf{X};\mathbf{\Theta})$ defined on the open set U of $\mathbb{R}B_L^{m,n}$ is said to be *continuous* at a point $(\mathbf{X}';\mathbf{\Theta}')$ of U if $F(\mathbf{X};\mathbf{\Theta})\rightarrow F(\mathbf{X}';\mathbf{\Theta}')$ as $(\mathbf{X};\mathbf{\Theta})\rightarrow(\mathbf{X}';\mathbf{\Theta}')$. This can be expressed more precisely in terms of the metrics for $\mathbb{C}B_L$ and $\mathbb{R}B_L^{m,n}$ that were introduced above by requiring that for any real number $\varepsilon>0$ there exists a real number $\delta>0$ such that

$$d(F(\mathbf{X};\mathbf{\Theta}),\ F(\mathbf{X}';\mathbf{\Theta}'))<\varepsilon$$

for all $(\mathbf{X};\mathbf{\Theta})\in\mathrm{U}$ for which $d((\mathbf{X};\mathbf{\Theta}),\ (\mathbf{X}';\mathbf{\Theta}'))<\delta$.

The concept of differentiability for a function defined on $\mathbb{R}B_L^{m,n}$ is more subtle, and requires careful definition and discussion. The difficulty is that, whereas (22.11) involves dividing the Grassmann-valued quantity $F(\mathbf{x}')-F(\mathbf{x})$ by the real number $x^j-x'^j$, division of the corresponding quantity $F(\mathbf{X}';\mathbf{\Theta}')-F(\mathbf{X};\mathbf{\Theta})$ by an element $(\mathbf{X}';\mathbf{\Theta}')-(\mathbf{X};\mathbf{\Theta})$ of $\mathbb{R}B_L^{m,n}$ is not defined. The following definition was first given by Rogers (1980).

Definition *Differentiability of a Grassmann-valued function defined on an open set of* $\mathbb{R}B_L^{m,n}$

Let $F(\mathbf{X};\mathbf{\Theta})$ be a *continuous* function that takes values in $\mathbb{C}B_L$ and is defined on an open set U of $\mathbb{R}B_L^{m,n}$. Suppose that there exist m functions $\partial F(\mathbf{X};\mathbf{\Theta})/\partial X^j$ (for $j=1,2,\ldots,m$) and n functions $\partial F(\mathbf{X};\mathbf{\Theta})/\partial\Theta^k$ (for $k=1,2,\ldots,n$) that all have values in $\mathbb{C}B_L$, are defined for all $(\mathbf{X};\mathbf{\Theta})$ in U and are such that

$$F(\mathbf{X}+\mathbf{Y};\mathbf{\Theta}+\mathbf{\Psi})=F(\mathbf{X};\mathbf{\Theta})+\sum_{j=1}^{m}Y^j\frac{\partial F(\mathbf{X};\mathbf{\Theta})}{\partial X^j}$$
$$+\sum_{k=1}^{n}\Psi^k\frac{\partial F(\mathbf{X};\mathbf{\Theta})}{\partial\Theta^k}+\|(\mathbf{Y};\mathbf{\Psi})\|\,\eta(\mathbf{Y};\mathbf{\Psi}),\qquad(22.14)$$

where $(\mathbf{X};\mathbf{\Theta})$ and $(\mathbf{X}+\mathbf{Y};\mathbf{\Theta}+\mathbf{\Psi})$ are both points in U, and $\eta(\mathbf{Y};\mathbf{\Psi})$ is a function defined on $\mathbb{R}B_L^{m,n}$ with values in $\mathbb{C}B_L$ that is such that

$$\|\eta(\mathbf{Y};\mathbf{\Psi})\|\rightarrow 0\quad\text{as }\|(\mathbf{Y};\mathbf{\Psi})\|\rightarrow 0.\qquad(22.15)$$

Then the function $F(X;\Theta)$ is said to be *differentiable* in U and the quantities $\partial F(\mathbf{X};\mathbf{\Theta})/\partial X^j$ and $\partial F(\mathbf{X};\mathbf{\Theta})/\partial\Theta^k$ are described as being its partial derivatives.

The even partial derivatives $\partial F(\mathbf{X};\mathbf{\Theta})/\partial X^j$ are *completely* defined by this prescription. So too are the odd derivatives $\partial F(\mathbf{X};\mathbf{\Theta})/\partial\Theta^k$, *except* that if L is finite the component of each odd derivative $\partial F(\mathbf{X};\mathbf{\Theta})/\partial\Theta^k$ that is proportional to the "highest" basis element $\mathcal{E}_{\{1,2,\ldots,L\}}$ of $\mathbb{C}B_L$ is completely *arbitrary*. This is so because if $\partial F(\mathbf{X};\mathbf{\Theta})/\partial\Theta^k$ satisfies (22.14) then so too does $\{\partial F(\mathbf{X};\mathbf{\Theta})/\partial\Theta^k + \lambda^k(\mathbf{X};\mathbf{\Theta})\mathcal{E}_{\{1,2,\ldots,L\}}\}$, where $\lambda^k(\mathbf{X};\mathbf{\Theta})$ is an arbitrary complex-valued function of $(\mathbf{X};\mathbf{\Theta})$ in U—the second term always giving *zero* contribution to the right-hand side of (22.14) because $\Psi^k\mathcal{E}_{\{1,2,\ldots,L\}}=0$ (as Ψ^k is an odd element of $\mathbb{C}B_L$). On one hand it is clear that this arbitrariness cannot matter very much, because L can be chosen to be arbitrarily large, and so the undetermined term proportional to $\mathcal{E}_{\{1,2,\ldots,L\}}$ can, in a sense, be made to recede from view. Moreover, in certain applications in connection with linear Lie supergroups the undetermined term simply disappears. On the other hand this arbitrariness spoils such results as the graded Leibnitz formulae for the derivatives of the product of two Grassmann-valued functions. Fortunately Rogers (1986a) has given a procedure for uniquely specifying the $\mathcal{E}_{\{1,2,\ldots,L\}}$ terms of the derivatives $\partial F(\mathbf{X};\mathbf{\Theta})/\partial\Theta^k$ in such a way that the graded Leibnitz formula is always valid. This prescription will be described in the next subsection.

Second derivatives can be defined by replacing $F(\mathbf{X};\mathbf{\Theta})$ in (22.14) by each of the first derivatives $\partial F(\mathbf{X};\mathbf{\Theta})/\partial X^j$ and $\partial F(\mathbf{X};\mathbf{\Theta})/\partial\Theta^k$ in turn, and higher derivatives can be obtained by repeating this process. For higher-order odd derivatives the arbitrariness of the $\mathcal{E}_{\{1,2,\ldots,L\}}$ term of the first derivative spreads to other terms. As an example, suppose that $L=4$, $n=2$ and that $\partial F(\mathbf{X};\mathbf{\Theta})/\partial\Theta^1=\lambda^1(\mathbf{X};\mathbf{\Theta})\mathcal{E}_{\{1,2,3,4\}}$, where $\lambda^1(\mathbf{X};\mathbf{\Theta})$ is an arbitrary complex-valued function of $(\mathbf{X};\mathbf{\Theta})$. One subset of such functions is provided by $\lambda^1(\mathbf{X};\mathbf{\Theta})\mathcal{E}_{\{1,2,3,4\}}=\gamma\mathcal{E}_{\{1,2,3\}}\Theta^2$, where γ is an arbitrary complex number. Clearly $\partial\{\gamma\mathcal{E}_{\{1,2,3\}}\Theta^2\}/\partial\Theta^2=\gamma\mathcal{E}_{\{1,2,3\}}+\lambda'(\mathbf{X};\mathbf{\Theta})\mathcal{E}_{\{1,2,3,4\}}$, where $\lambda'(\mathbf{X};\mathbf{\Theta})$ is another arbitrary complex-valued function of $(\mathbf{X};\mathbf{\Theta})$. Repeating with $\gamma\mathcal{E}_{\{1,2,3\}}\Theta^2$ replaced by $\gamma\mathcal{E}_{\{1,2,4\}}\Theta^2$, $\gamma\mathcal{E}_{\{1,3,4\}}\Theta^2$ and $\gamma\mathcal{E}_{\{2,3,4\}}\Theta^2$ in turn, it is clear that the second derivative of $F(\mathbf{X};\mathbf{\Theta})$ with respect to Θ^1 and Θ^2 contains arbitrary terms proportional to $\mathcal{E}_{\{1,2,3,4\}}$, *and* to $\mathcal{E}_{\{1,2,3\}}$, $\mathcal{E}_{\{1,2,4\}}$, $\mathcal{E}_{\{1,3,4\}}$ and $\mathcal{E}_{\{2,3,4\}}$. The argument can be repeated to show that for the higher odd derivatives of order p the coefficients of $\mathcal{E}_\mu$ for all the index sets μ containing $L-p+1$ integers are arbitrary complex-valued functions of $(\mathbf{X};\mathbf{\Theta})$. Rogers' (1986a) prescription automatically removes all the arbitrariness in all these higher derivatives as well.

One immediate consequence of the definition (22.14) is that if $F(\mathbf{X};\mathbf{\Theta})$ is an even function then its derivatives $\partial F(\mathbf{X};\mathbf{\Theta})/\partial X^j$ are all even and its derivatives $\partial F(\mathbf{X};\mathbf{\Theta})/\partial\Theta^k$ are all odd, whereas if $F(\mathbf{X};\mathbf{\Theta})$ is an odd function then the $\partial F(\mathbf{X};\mathbf{\Theta})/\partial X^j$ are all odd and the $\partial F(\mathbf{X};\mathbf{\Theta})/\partial\Theta^k$ are all even.

Definition *The sets* $\mathrm{G}^p(\mathrm{U})$ *and* $\mathrm{G}^\infty(\mathrm{U})$

Suppose that the Grassmann-valued function $F(\mathbf{X};\mathbf{\Theta})$ is defined on the open set U of $\mathbb{R}B_L^{m,n}$. If $F(\mathbf{X};\mathbf{\Theta})$ is continuous on U then $F(\mathbf{X};\mathbf{\Theta})$ is said to be a member of $\mathrm{G}^0(\mathrm{U})$. If p is a finite positive integer and all derivatives of $F(\mathbf{X};\mathbf{\Theta})$

of order p exist on U, then $F(\mathbf{X};\mathbf{\Theta})$ is said to be a member of the set $G^p(U)$. If $F(\mathbf{X};\mathbf{\Theta})\in G^p(U)$ for every positive integer p then $F(\mathbf{X};\mathbf{\Theta})$ is described as being a member of the set $G^\infty(U)$.

By expanding $F(\mathbf{X};\mathbf{\Theta})$, X^j and Θ^k in terms of their real or complex components $F_\mu(\mathbf{X};\mathbf{\Theta})$, X^j_μ and Θ^k_μ using (22.13), (22.3) and (22.4) respectively, it is clear that (22.14) implies that if the partial derivatives $\partial F(\mathbf{X};\mathbf{\Theta})/\partial X^j$ and $\partial F(\mathbf{X};\mathbf{\Theta})/\partial\Theta^k$ exist then so too do the complex-valued derivatives $\partial F_\mu(\mathbf{X};\mathbf{\Theta})/\partial X^j_{\mu'}$ and $\partial F_\mu(\mathbf{X};\mathbf{\Theta})/\partial\Theta^k_{\mu'}$ for all index sets μ and μ' of the type (20.22). As each of the $m+n$ partial derivatives $\partial F(\mathbf{X};\mathbf{\Theta})/\partial X^j$ and $\partial F(\mathbf{X};\mathbf{\Theta})/\partial\Theta^k$ when expressed in terms of expansions of the form (22.13) involve 2^L-complex-valued component functions, the whole set of these "Grassmann derivatives" depend on $(m+n)2^L$ such functions. However, there are $(m+n)2^{2L-1}$ complex-valued functions of the form $\partial F_\mu(\mathbf{X};\mathbf{\Theta})/\partial X^j_{\mu'}$ and $\partial F_\mu(\mathbf{X};\mathbf{\Theta})/\partial\Theta^k_{\mu'}$. Consequently if $F(\mathbf{X};\mathbf{\Theta})\in G^1(U)$ then these latter functions must satisfy a set of constraints, which in turn imply that there are restrictions on the form of $F(\mathbf{X};\mathbf{\Theta})$. By continuing such an analysis, it becomes apparent that if $F(\mathbf{X};\mathbf{\Theta})\in G^\infty(U)$ then the form of $F(X;\Theta)$ must be quite severely restricted, as Theorem II of the next subsection will show.

By means of the expansions (22.3) and (22.4), a function $F(\mathbf{X};\mathbf{\Theta})$ of $(\mathbf{X};\mathbf{\Theta})$ can also be considered to be a function of the $(m+n)2^{L-1}$ real variables X^j_μ (for $j=1,2,\dots,m$ and all even index sets μ) and Θ^k_μ (for $k=1,2,\dots,n$ and all odd index sets μ), as can its component complex-valued functions $F_{\mu'}(\mathbf{X};\mathbf{\Theta})$ of (22.13). The derivatives $\partial F(\mathbf{X};\mathbf{\Theta})/\partial X^j_\mu$ can be defined by

$$\frac{\partial F(\mathbf{X};\mathbf{\Theta})}{\partial X^j_\mu}=\sum_{\mu'}\frac{\partial F_{\mu'}(\mathbf{X};\mathbf{\Theta})}{\partial X^j_\mu}\mathcal{E}_{\mu'},$$

and so it follows immediately from (22.14) (on putting $X^j=X^j_\mu\mathcal{E}_\mu$ with no summation implied) that

$$\frac{\partial F(\mathbf{X};\mathbf{\Theta})}{\partial X^j_\mu}=\mathcal{E}_\mu\frac{\partial F(\mathbf{X};\mathbf{\Theta})}{\partial X^j}. \tag{22.16}$$

Similarly the derivatives $\partial F(\mathbf{X};\mathbf{\Theta})/\partial\Theta^k_\mu$ can be defined by

$$\frac{\partial F(\mathbf{X};\mathbf{\Theta})}{\partial\Theta^k_\mu}=\sum_{\mu'}\frac{\partial F_{\mu'}(\mathbf{X};\mathbf{\Theta})}{\partial\Theta^k_\mu}\mathcal{E}_{\mu'},$$

and so it follows immediately from (22.14) that

$$\frac{\partial F(\mathbf{X};\mathbf{\Theta})}{\partial\Theta^k_\mu}=\mathcal{E}_\mu\frac{\partial F(\mathbf{X};\mathbf{\Theta})}{\partial\Theta^k}. \tag{22.17}$$

As the $\mathcal{E}_\mu$ appearing in (22.17) are necessarily *odd*, the arbitrary term

$\lambda^k(\mathbf{X};\mathbf{\Theta})\mathcal{E}_{\{1,2,\ldots,L\}}$ in $\partial F(\mathbf{X};\mathbf{\Theta})/\partial\Theta^k$ makes *zero* contribution to $\partial F(\mathbf{X};\mathbf{\Theta})/\partial\Theta^k_\mu$, so $\partial F(\mathbf{X};\mathbf{\Theta})/\partial\Theta^k_\mu$ is *uniquely* determined by (22.17).

Among the $\mathbb{C}B_L$ valued functions defined on $\mathbb{R}B_L^{m,n}$ are those involving products of the odd Grassmann variables $\Theta^1,\Theta^2,\ldots,\Theta^n$. Because $\Theta^k\Theta^k=0$ for each k in the set $\{1,2,\ldots,n\}$, no such product can involve any particular Θ^k more than once. It is convenient to let Λ denote a subset of the integers $\{1,2,\ldots,n\}$, with no integer appearing more than once in Λ. Suppose that Λ contains $N(\Lambda)$ members, where necessarily $0\leqslant N(\Lambda)\leqslant n$, and that, if $1\leqslant N(\Lambda)$, these are $k_1,k_2,\ldots,k_{N(\Lambda)}$, so that

$$\Lambda=\{k_1,k_2,\ldots,k_{N(\Lambda)}\}. \tag{22.18}$$

These will be assumed to be ordered so that

$$1\leqslant k_1<k_2<\cdots<k_{N(\Lambda)}\leqslant n. \tag{22.19}$$

The corresponding product of odd Grassmann variables Θ^k will be denoted by Θ^Λ, so that if $1\leqslant N(\Lambda)$,

$$\Theta^\Lambda=\Theta^{k_1}\Theta^{k_2}\ldots\Theta^{k_{N(\Lambda)}}. \tag{22.20}$$

Thus, for example, with $\Lambda=\{1,2,4\}$, $N(\Lambda)=3$ and $\Theta^\Lambda=\Theta^1\Theta^2\Theta^4$. If $\Lambda=\varnothing$, the empty set, the corresponding definition is

$$\Theta^\varnothing=\mathcal{E}_\varnothing. \tag{22.21}$$

Clearly (22.14) implies immediately that

$$\partial\Theta^\Lambda/\partial X^j=0 \tag{22.22}$$

for $j=1,2,\ldots,m$. For the odd derivatives $\partial\Theta^\Lambda/\partial\Theta^k$ the situation is more complicated. Let $\Lambda-\{k\}$ be the subset of $\{1,2,\ldots,n\}$ that for $k\in\Lambda$ is defined to be the set obtained from Λ by removing k, while for $k\notin\Lambda$ the set $\Lambda-\{k\}$ is defined to be the empty set $\varnothing$. If $k\in\Lambda$, let $p(k)$ be the position of k in the set Λ when Λ is ordered according to the rule (22.19). Define $\Theta^{\Lambda/k}$ by

$$\Theta^{\Lambda/k}=\begin{cases}(-1)^{p(k)-1}\Theta^{\Lambda-\{k\}} & \text{if } k\in\Lambda,\\ 0 & \text{if } k\notin\Lambda.\end{cases} \tag{22.23}$$

Then (22.14) gives

$$\frac{\partial\Theta^\Lambda}{\partial\Theta^k}=\Theta^{\Lambda/k}+\lambda^k(X;\Theta)\mathcal{E}_{\{1,2,\ldots,L\}},$$

where $\lambda^k(\mathbf{X};\mathbf{\Theta})$ is an arbitrary complex-valued function of $(\mathbf{X};\mathbf{\Theta})$. For example, if $\Lambda=\{1,3,4\}$ and $k=3$ then $\Lambda-\{k\}=\{1,3,4\}-\{3\}=\{1,4\}$ and $p(k)=p(3)=2$, so $\Theta^{\Lambda/k}=-\Theta^{\{1,4\}}=-\Theta^1\Theta^4$, and hence

$$\frac{\partial\Theta^{\{1,3,4\}}}{\partial\Theta^3}=-\Theta^1\Theta^4+\lambda^3(\mathbf{X};\mathbf{\Theta})\mathcal{E}_{\{1,2,\ldots,L\}};$$

whereas if $\Lambda = \{1, 3, 4\}$ and $k = 2$ then $k \notin \Lambda$, so $\Theta^{\Lambda/k} = 0$, and hence

$$\frac{\partial \Theta^{\{1,3,4\}}}{\partial \Theta^2} = \lambda^2(\mathbf{X}; \boldsymbol{\Theta})\mathcal{E}_{\{1,2,\ldots,L\}},$$

$\lambda^2(\mathbf{X}; \boldsymbol{\Theta})$ and $\lambda^3(\mathbf{X}; \boldsymbol{\Theta})$ being arbitrary complex-valued functions of $(\mathbf{X}; \boldsymbol{\Theta})$.

(c) *Superanalytic and superdifferentiable functions on* $\mathbb{R}\mathrm{B}_L^{m,n}$

It is convenient to introduce for $\mathbb{R}\mathrm{B}_L$ two projection operators ε and s by the prescription that if $B \in \mathbb{R}\mathrm{B}_L$ has the expansion

$$B = \sum_{\mu} B_\mu \mathcal{E}_\mu$$

(cf. (20.25)) then

$$\varepsilon(B) = B_\varnothing \tag{22.24}$$

and

$$s(B) = B - B_\varnothing \mathcal{E}_\varnothing. \tag{22.25}$$

DeWitt (1984) has called the quantity $\varepsilon(B)$ (which is a real number) the *body* of B and the quantity $s(B)$ (which is a member of $\mathbb{R}\mathrm{B}_L$) the *soul* of B. If L is finite, there exists some integer r such that $s(B)^r = 0$. Clearly if $B \in \mathbb{R}\mathrm{B}_{L1}$ then $B_\varnothing = 0$ and so $\varepsilon(B) = 0$ and $s(B) = B$. Similar operators may be introduced on $\mathbb{R}\mathrm{B}_L^{m,n}$, their definitions (for which the same symbols may be used without confusion) being

$$\varepsilon(\mathbf{X}; \boldsymbol{\Theta}) = \varepsilon(X^1, X^2, \ldots, X^m, \Theta^1, \Theta^2, \ldots, \Theta^n) = (\varepsilon(X^1), \varepsilon(X^2), \ldots, \varepsilon(X^m)), \tag{22.26}$$

which is a member of $\mathbb{R}^m$, and

$$\begin{aligned} s(\mathbf{X}; \boldsymbol{\Theta}) &= s(X^1, X^2, \ldots, X^m, \Theta^1, \Theta^2, \ldots, \Theta^n) \\ &= (s(X^1), s(X^2), \ldots, s(X^m), s(\Theta^1), s(\Theta^2), \ldots, s(\Theta^n)) \end{aligned} \tag{22.27}$$

which is a member of $\mathbb{R}\mathrm{B}_L^{m,n}$. Thus, in terms of the expansion (22.3), (22.24) implies that

$$\varepsilon(\mathbf{X}; \boldsymbol{\Theta}) = \varepsilon(X^1, X^2, \ldots, X^m, \Theta^1, \Theta^2, \ldots, \Theta^n) = (X^1_\varnothing, X^2_\varnothing, \ldots, X^m_\varnothing), \tag{22.28}$$

and, as $s(\Theta^k) = \Theta^k$ for $k = 1, 2, \ldots, n$, (22.25) gives

$$\begin{aligned} s(\mathbf{X}; \boldsymbol{\Theta}) &= s(X^1, X^2, \ldots, X^m, \Theta^1, \Theta^2, \ldots, \Theta^n) \\ &= (X^1 - X^1_\varnothing \mathcal{E}_\varnothing, X^2 - X^2_\varnothing \mathcal{E}_\varnothing, \ldots, X^m - X^m_\varnothing \mathcal{E}_\varnothing, \Theta^1, \Theta^2, \ldots, \Theta^n). \end{aligned} \tag{22.29}$$

If the set of elements $(\mathbf{X};\mathbf{\Theta})$ form an open set U of $\mathbb{R}B_L^{m,n}$ then the corresponding set of "bodies" $\varepsilon(\mathbf{X};\mathbf{\Theta})$ form an open set of $\mathbb{R}^m$, which may be denoted by $\varepsilon(\mathrm{U})$.

Definition *Continuation of a C^∞ function defined on $\mathbb{R}^m$*

Suppose that L is finite and $L \geqslant n$. Let U be an open set of $\mathbb{R}B_L^{m,n}$, let $\varepsilon(\mathrm{U})$ be the corresponding open set of $\mathbb{R}^m$, and let U_0 be the open set of $\mathbb{R}B_L^{m,0}$ that is defined to be such that if $(\mathbf{X};\mathbf{\Theta})\in\mathrm{U}$ then $\mathbf{X}\in\mathrm{U}_0$. Let $\mathscr{F}(X^1_\varnothing, X^2_\varnothing,\dots,X^m_\varnothing)$ be any $C^\infty(\varepsilon(\mathrm{U}))$ function that is defined on $\varepsilon(\mathrm{U})$ and takes values in $\mathbb{C}B_L$. The *continuation* of $\mathscr{F}(X^1_\varnothing, X^2_\varnothing,\dots,X^m_\varnothing)$ to U_0 is defined to be the $\mathbb{C}B_L$ valued function $Z(\mathscr{F})$ that is specified for each $\mathbf{X}$ of U_0 by

$$Z(\mathscr{F})(\mathbf{X}) = \sum_{j_1,j_2,\dots,j_m=0}^{L} \frac{1}{j_1!\,j_2!\cdots j_m!} \times \frac{\partial^{j_1}}{\partial X^{j_1}_\varnothing}\frac{\partial^{j_2}}{\partial X^{j_2}_\varnothing}\cdots\frac{\partial^{j_m}}{\partial X^{j_m}_\varnothing}\mathscr{F}(X^1_\varnothing, X^2_\varnothing,\dots,X^m_\varnothing) \times s(X^1)^{j_1}s(X^2)^{j_2}\cdots s(X^m)^{j_m}. \tag{22.30}$$

In (22.30) $s(X^j)$ denotes the "soul" of X^j (cf. (22.25)). As each X^j is even, each $s(X^j)$ factor is also even, so if $\mathscr{F}$ is even then $Z(\mathscr{F})$ also is even, and if $\mathscr{F}$ is odd then $Z(\mathscr{F})$ is also odd. The name "continuation" reflects the fact that there is an obvious one-to-one correspondence between a set $(X^1_\varnothing, X^2_\varnothing,\dots,X^m_\varnothing)$ of $\mathbb{R}^m$ and the set $(X^1_\varnothing\mathcal{E}_\varnothing, X^2_\varnothing\mathcal{E}_\varnothing,\dots,X^m_\varnothing\mathcal{E}_\varnothing)$ of $\mathbb{R}B_L^{m,0}$, and (22.30) implies that

$$Z(\mathscr{F})(X^1_\varnothing\mathcal{E}_\varnothing, X^2_\varnothing\mathcal{E}_\varnothing,\dots,X^m_\varnothing\mathcal{E}_\varnothing) = \mathscr{F}(X^1_\varnothing, X^2_\varnothing,\dots,X^m_\varnothing),$$

so $Z(\mathscr{F})$ and $\mathscr{F}$ can be thought of as agreeing on that part of their domains that are in one-to-one correspondence. The function $Z(\mathscr{F})$ can be considered in three ways: firstly, as a function defined on U_0; secondly, as a function defined on the subspace U′ of points $(\mathbf{X};\mathbf{0})$ of $\mathbb{R}B_L^{m,n}$ that is such that $(\mathbf{X};\mathbf{0})\in\mathrm{U}'$ if and only if $(\mathbf{X};\mathbf{\Theta})\in\mathrm{U}$; and, thirdly, as a Θ-independent function defined on the whole of U. In this third sense it is a $G^\infty(\mathrm{U})$ function, as the following theorem shows.

Theorem I Suppose that L is finite and that $L \geqslant n$. If U is an open set of $\mathbb{R}B_L^{m,n}$ and $\mathscr{F}(X^1_\varnothing, X^2_\varnothing,\dots,X^m_\varnothing)$ is a $C^\infty(\varepsilon(\mathrm{U}))$ function defined on $\varepsilon(\mathrm{U})$ with values in $\mathbb{C}B_L$ and continuation $Z(\mathscr{F})(\mathbf{X})$ then the $\mathbb{C}B_L$ valued function $F(\mathbf{X};\mathbf{\Theta})$ defined on U by $F(\mathbf{X};\mathbf{\Theta}) = Z(\mathscr{F})(\mathbf{X})$ is a $G^\infty(\mathrm{U})$ function. Moreover

$$\frac{\partial F(\mathbf{X};\mathbf{\Theta})}{\partial X^j} = Z\left(\frac{\partial \mathscr{F}}{\partial X^j_\varnothing}\right)(\mathbf{X}) \tag{22.31}$$

for $j = 1, 2, \ldots, m$, and

$$\frac{\partial F(\mathbf{X};\boldsymbol{\Theta})}{\partial \Theta^k} = \lambda^k(\mathbf{X};\boldsymbol{\Theta})\mathcal{E}_{\{1,2,\ldots,L\}} \tag{22.32}$$

for $k = 1, 2, \ldots, n$, where $\lambda^k(\mathbf{X};\boldsymbol{\Theta})$ is an arbitrary complex-valued function of $(\mathbf{X};\boldsymbol{\Theta})$. That is, apart from this arbitrary $\mathcal{E}_{\{1,2,\ldots,L\}}$ term, the $\partial/\partial\Theta^k$ derivative of $F(\mathbf{X};\boldsymbol{\Theta})$ $(= Z(\mathscr{F})(\mathbf{X}))$ is *zero*.

Proof See Rogers (1980).

The next theorem is the key theorem in that it provides a complete characterization of the $G^\infty(U)$ functions.

Theorem II Suppose that L is finite and that $L \geqslant n$. Let U be an open set of $\mathbb{R}B_L^{m,n}$. Then we have the following.

(a) *Every* $G^\infty(U)$ function $F(\mathbf{X};\boldsymbol{\Theta})$ can be written in the form

$$F(\mathbf{X};\boldsymbol{\Theta}) = \sum_{\Lambda} F^\Lambda(\mathbf{X})\Theta^\Lambda, \tag{22.33}$$

where Θ^Λ is a product of Θ^k factors, as defined in (22.20), the sum being over all sets Λ of the type (22.18) that satisfy (22.19), and where each $F^\Lambda(\mathbf{X})$ is a $\mathbb{C}B_L$ valued function defined on U that is a member of $G^\infty(U)$ but is independent of the odd variables $\Theta^1, \Theta^2, \ldots, \Theta^L$. Moreover each $F^\Lambda(\mathbf{X})$ is the continuation of a unique $C^\infty(\varepsilon(U))$ function $\mathscr{F}^\Lambda(X^1_\varnothing, X^2_\varnothing, \ldots, X^m_\varnothing)$; that is, for each set Λ of the type (22.18),

$$F^\Lambda(\mathbf{X}) = Z(\mathscr{F}^\Lambda)(\mathbf{X}), \tag{22.34}$$

so that

$$F(\mathbf{X};\boldsymbol{\Theta}) = \sum_{\Lambda} Z(\mathscr{F}^\Lambda)(\mathbf{X})\Theta^\Lambda. \tag{22.35}$$

(b) The even derivatives are given for $j = 1, 2, \ldots, m$ by

$$\frac{\partial F(\mathbf{X};\boldsymbol{\Theta})}{\partial X^j} = \sum_{\Lambda} \frac{\partial F^\Lambda(\mathbf{X})}{\partial X^j}\Theta^\Lambda, \tag{22.36}$$

and hence

$$\frac{\partial F(\mathbf{X};\boldsymbol{\Theta})}{\partial X^j} = \sum_{\Lambda} \left\{ Z\left(\frac{\partial \mathscr{F}^\Lambda}{\partial X^j_\varnothing}\right)\right\}(\mathbf{X})\Theta^\Lambda. \tag{22.37}$$

(c) If $F(\mathbf{X};\boldsymbol{\Theta})$ is homogeneous, that is if $F(\mathbf{X};\boldsymbol{\Theta})$ is either entirely even or entirely odd, then its odd derivatives are given for $k = 1, 2, \ldots, n$ by

$$\frac{\partial F(\mathbf{X};\boldsymbol{\Theta})}{\partial \Theta^k} = \sum_{\Lambda} (-1)^{\deg F^\Lambda} F^\Lambda(\mathbf{X})\Theta^{\Lambda/k} + \lambda^k(\mathbf{X};\boldsymbol{\Theta})\mathcal{E}_{\{1,2,\ldots,L\}}, \tag{22.38}$$

and hence

$$\frac{\partial F(\mathbf{X};\mathbf{\Theta})}{\partial \Theta^k} = \sum_{\Lambda} (-1)^{\deg \mathscr{F}^{\Lambda}} \{Z(\mathscr{F}^{\Lambda})\}(\mathbf{X})\Theta^{\Lambda/k} + \lambda^k(\mathbf{X};\mathbf{\Theta})\mathcal{E}_{\{1,2,\ldots,L\}}, \qquad (22.39)$$

where $\lambda^k(\mathbf{X};\mathbf{\Theta})$ is an arbitrary complex-valued function of $(\mathbf{X};\mathbf{\Theta})$ and $\Theta^{\Lambda/k}$ is defined in (22.23).

Proof See Rogers (1980).

The most important observation concerning this theorem is that the expansions (22.33) and (22.35) each contain only a *finite* number of terms, because there are only 2^n different products Θ^{Λ}. For example, if $n = 1$, (22.33) becomes

$$F(\mathbf{X};\mathbf{\Theta}) = F^{\varnothing}(\mathbf{X}) + F^{\{1\}}(\mathbf{X})\Theta^1,$$

and so, by (22.38),

$$\frac{\partial F(\mathbf{X};\mathbf{\Theta})}{\partial \Theta^1} = (-1)^{\deg F^{\{1\}}} F^{\{1\}}(\mathbf{X}) + \lambda^1(\mathbf{X};\mathbf{\Theta})\mathcal{E}_{\{1,2,\ldots,L\}},$$

where $\lambda^1(\mathbf{X};\mathbf{\Theta})$ is an arbitrary complex-valued function of $(\mathbf{X};\mathbf{\Theta})$. Similarly, if $n = 2$, (22.33) becomes

$$F(\mathbf{X};\mathbf{\Theta}) = F^{\varnothing}(\mathbf{X}) + F^{\{1\}}(\mathbf{X})\Theta^1 + F^{\{2\}}(\mathbf{X})\Theta^2 + F^{\{1,2\}}(\mathbf{X})\Theta^1\Theta^2;$$

and so, for example, for $k = 2$, (22.38) gives

$$\frac{\partial F(\mathbf{X};\mathbf{\Theta})}{\partial \Theta^2} = (-1)^{\deg F^{\{2\}}} F^{\{2\}}(\mathbf{X}) - (-1)^{\deg F^{\{1,2\}}} F^{\{1,2\}}(\mathbf{X})\Theta^1 + \lambda^2(\mathbf{X};\mathbf{\Theta})\mathcal{E}_{\{1,2,\ldots,L\}},$$

where $\lambda^2(\mathbf{X};\mathbf{\Theta})$ is another arbitrary complex-valued function of $(\mathbf{X};\mathbf{\Theta})$.

The $\mathbb{C}\mathrm{B}_L$ valued functions $F^{\Lambda}(\mathbf{X})$ of (22.33) should not be confused with the complex-valued functions $F_{\mu}(\mathbf{X};\mathbf{\Theta})$ of (22.13). Indeed each $F^{\Lambda}(\mathbf{X})$ has an expansion of the form (22.13); that is,

$$F^{\Lambda}(\mathbf{X}) = \sum_{\mu} F^{\Lambda}_{\mu}(\mathbf{X})\mathcal{E}_{\mu}, \qquad (22.40)$$

where the components $F^{\Lambda}_{\mu}(\mathbf{X})$ are all complex-valued. Similarly each $\mathscr{F}^{\Lambda}(\mathbf{x})$ has an expansion of the form (22.10); that is,

$$\mathscr{F}^{\Lambda}(\mathbf{x}) = \sum_{\mu} \mathscr{F}^{\Lambda}_{\mu}(\mathbf{x})\mathcal{E}_{\mu}, \qquad (22.41)$$

where the components $\mathscr{F}^{\Lambda}_{\mu}(\mathbf{x})$ are again all complex-valued. Moreover the definition (22.30) implies that

$$Z(\mathscr{F}^{\Lambda})(\mathbf{X}) = \sum_{\mu} Z(\mathscr{F}_{\mu}^{\Lambda}\mathcal{E}_{\mu})(\mathbf{X}), \tag{22.42}$$

and the corresponding expansion (22.35) then becomes

$$F(\mathbf{X};\mathbf{\Theta}) = \sum_{\Lambda}\sum_{\mu} (\mathscr{F}_{\mu}^{\Lambda}\mathcal{E}_{\mu})(\mathbf{X})\Theta^{\Lambda}. \tag{22.43}$$

In general, a number of the terms in (22.33), (22.35) and (22.43) are *redundant*, for if $N(\Lambda)$ is the number of elements in the set Λ and $N(\mu)$ is the number of elements in the set μ then

$$Z(\mathscr{F}_{\mu}^{\Lambda}\mathcal{E}_{\mu})(\mathbf{X})\Theta^{\Lambda} = 0 \tag{22.44}$$

if

$$N(\Lambda) + N(\mu) > L. \tag{22.45}$$

(This is because $\mathscr{F}_{\mu}^{\Lambda}(\mathbf{x})\mathcal{E}_{\mu}$ consists of a product of $N(\mu)$ generators $\mathcal{E}_1, \mathcal{E}_2, \ldots$ of $\mathbb{C}\mathbf{B}_L$, and so, by (22.30), $Z(\mathscr{F}_{\mu}^{\Lambda}\mathcal{E}_{\mu})(\mathbf{X})$ is a sum of terms each of which contains a product of such generators involving not less than $N(\mu)$ factors, while Θ^{Λ} is a linear combination of similar products each of which contain not less that $N(\Lambda)$ factors of this type. Thus if (22.45) holds, at least one $\mathcal{E}_j$ must appear more than once in each product of $\mathbb{C}\mathbf{B}_L$ generators, and so each such product is identically zero.)

Particularly dangerous for first derivatives are the $Z(\mathscr{F}_{\mu}^{\Lambda}\mathcal{E}_{\mu})(\mathbf{X})\Theta^{\Lambda}$ terms in (22.43) for which

$$N(\Lambda) + N(\mu) = L + 1. \tag{22.46}$$

Although such a term provides a *zero* contribution to $F(\mathbf{X};\mathbf{\Theta})$, its contribution to the first term of the expression given in (22.38) for $\partial F(\mathbf{X};\mathbf{\Theta})/\partial\Theta^k$ is *non-zero*, as $Z(\mathscr{F}_{\mu}^{\Lambda}\mathcal{E}_{\mu})(\mathbf{X})\Theta^{\Lambda/k}$ contains a term proportional to $\mathcal{E}_{\{1,2,\ldots,L\}}$. Thus if one varies a function $\mathscr{F}_{\mu}^{\Lambda}(\mathbf{x})$ for which (22.44) holds, there is no change in $F(\mathbf{X};\mathbf{\Theta})$, but there is a change in the first term of (22.38) for $\partial F(\mathbf{X};\mathbf{\Theta})/\partial\Theta^k$. Although this can be absorbed into the arbitrary $\lambda^k(\mathbf{X};\mathbf{\Theta})\mathcal{E}_{\{1,2,\ldots,L\}}$ term, so that no contradictions ensue, the moral of this analysis is that the existence of the arbitrary term $\lambda^k(\mathbf{X};\mathbf{\Theta})\mathcal{E}_{\{1,2,\ldots,L\}}$ and the arbitrariness of the $Z(\mathscr{F}_{\mu}^{\Lambda}\mathcal{E}_{\mu})(\mathbf{X})\Theta^{\Lambda}$ terms for which (22.45) holds are *inseparably* related.

Although the $Z(\mathscr{F}_{\mu}^{\Lambda}\mathcal{E}_{\mu})(\mathbf{X})\Theta^{\Lambda}$ terms for which $N(\Lambda) + N(\mu) > L + 1$ can cause no problems for first derivatives, since they give zero contributions both to $F(\mathbf{X};\mathbf{\Theta})$ and to $\partial F(\mathbf{X};\mathbf{\Theta})/\partial\Theta^k$, there will a non-zero contribution to the odd derivatives of order p from those terms for which $N(\Lambda) + N(\mu) = L + p$ (where p is a positive integer such that $p > 1$). Consequently every redundant term satisfying (22.45) makes an appearance in this way in at least one odd derivative.

However, for every set Λ and set μ for which

$$N(\Lambda) + N(\mu) \leqslant L \tag{22.47}$$

not only can there be *no* hidden contributions to the odd derivatives, but also the functions $\mathscr{F}_\mu^\Lambda(X_\varnothing^1, X_\varnothing^2, \ldots, X_\varnothing^m)$ are *uniquely* determined for each $F(\mathbf{X}; \mathbf{\Theta})$, and conversely, *all* possible $G^\infty(U)$ functions $F(\mathbf{X}; \mathbf{\Theta})$ can be obtained by using (22.43) with all possible functions $\mathscr{F}_\mu^\Lambda(X_\varnothing^1, X_\varnothing^2, \ldots, X_\varnothing^m)$ that are differentiable to all orders in $\varepsilon(U)$ and that satisfy (22.47).

Rogers' (1986a) prescription for dealing with the arbitrariness in the $\mathcal{E}_{\{1,2,\ldots,L\}}$ term of the odd Grassmann derivatives is to restrict attention to a subset of the $G^\infty(U)$ functions that will be called the "superdifferentiable functions on U" (and which Rogers (1986a) refers to as $GH^\infty(U)$ functions). (For other discussions of this problem see Boyer and Gitler (1980, 1984).)

Definition *The superdifferentiable functions and their derivatives on* $\mathbb{R}\mathrm{B}_L^{m,n}$

(i) Choose $L \geqslant 2n$, let

$$L' = \begin{cases} \tfrac{1}{2}L & \text{if } L \text{ is even} \\ \tfrac{1}{2}(L+1) & \text{if } L \text{ is odd,} \end{cases} \tag{22.48}$$

and consider $\mathbb{C}\mathrm{B}_{L'}$ as a subspace of $\mathbb{C}\mathrm{B}_L$.

(ii) Let U be an open set of $\mathbb{R}\mathrm{B}_L^{m,n}$. Restrict the functions $\mathscr{F}^\Lambda(X_\varnothing^1, X_\varnothing^2, \ldots, X_\varnothing^m)$ to take values in the subspace $\mathbb{C}\mathrm{B}_{L'}$ alone, rather than in the whole of $\mathbb{C}\mathrm{B}_L$. Then each $\mathscr{F}^\Lambda(X_\varnothing^1, X_\varnothing^2, \ldots, X_\varnothing^m)$ has the form (22.41), where each component function $\mathscr{F}_\mu^\Lambda(X_\varnothing^1, X_\varnothing^2, \ldots, X_\varnothing^m)$ corresponds to a set μ that is a subset of $\{1, 2, \ldots, L'\}$ alone. The set of functions $F(\mathbf{X}; \mathbf{\Theta})$ that arise in this way using the expansion (22.35) (or equivalently (22.43)) will be called the *superdifferentiable functions on* U.

(iii) Define $\partial F(\mathbf{X}; \mathbf{\Theta})/\partial \Theta^k$ for $k = 1, 2, \ldots, m$ by

$$\frac{\partial F(\mathbf{X}; \mathbf{\Theta})}{\partial \Theta^k} = \sum_\Lambda (-1)^{\deg \mathscr{F}^\Lambda} \{Z(\mathscr{F}^\Lambda)\}(\mathbf{X}) \Theta^{\Lambda/k}, \tag{22.49}$$

where $\Theta^{\Lambda/k}$ is defined in (22.23).

The restriction (ii) implies that $N(\mu) \leqslant L'$, and hence, by (i), $N(\mu) \leqslant \frac{1}{2}L$. Thus, since $N(\Lambda) \leqslant n \leqslant \frac{1}{2}L$, it follows that $N(\Lambda) + N(\mu) \leqslant L$, and so *all* terms of the form $Z(\mathscr{F}_\mu^\Lambda \mathcal{E}_\mu)(\mathbf{X}; \mathbf{0})\Theta^\Lambda$ that appear in (22.43) satisfy (22.47). Consequently *none* of the redundancy problems just discussed above arise, that is, for each superdifferentiable function $F(\mathbf{X}; \mathbf{\Theta})$ on U, *all* of the component functions $\mathscr{F}_\mu^\Lambda(X_\varnothing^1, X_\varnothing^2, \ldots, X_\varnothing^m)$ are *uniquely* determined. The definition (22.49) of the odd Grassmann derivatives corresponds to taking all the arbitrary complex-valued functions $\lambda^k(\mathbf{X};\mathbf{\Theta})$ to be zero. As the functions $\mathscr{F}^{\Lambda}(X^1_{\varnothing}, X^2_{\varnothing}, \ldots, X^m_{\varnothing})$ are now all uniquely determined for each Λ, (22.49) gives a *unique* value for each odd derivative $\partial F(\mathbf{X};\mathbf{\Theta})/\partial\Theta^k$ of every superdifferentiable function $F(\mathbf{X};\mathbf{\Theta})$, and hence for every higher odd derivative.

Clearly if $F(\mathbf{X};\mathbf{\Theta})$ and $G(\mathbf{X};\mathbf{\Theta})$ are any two superdifferentiable functions on U, then $F(\mathbf{X};\mathbf{\Theta}) + G(\mathbf{X};\mathbf{\Theta})$ and $F(\mathbf{X};\mathbf{\Theta})G(\mathbf{X};\mathbf{\Theta})$ are also superdifferentiable functions on U. So too is $\lambda F(\mathbf{X};\mathbf{\Theta})$ for any complex number λ.

Theorem III If $F(\mathbf{X};\mathbf{\Theta})$ and $G(\mathbf{X};\mathbf{\Theta})$ are any two superdifferentiable functions on U, then

(a) for $j = 1, 2, \ldots, m$

$$\frac{\partial\{F(\mathbf{X};\mathbf{\Theta}) + G(\mathbf{X};\mathbf{\Theta})\}}{\partial X^j} = \frac{\partial F(\mathbf{X};\mathbf{\Theta})}{\partial X^j} + \frac{\partial G(\mathbf{X};\mathbf{\Theta})}{\partial X^j}, \tag{22.50}$$

and for $k = 1, 2, \ldots, n$

$$\frac{\partial\{F(\mathbf{X};\mathbf{\Theta}) + G(\mathbf{X};\mathbf{\Theta})\}}{\partial \Theta^k} = \frac{\partial F(\mathbf{X};\mathbf{\Theta})}{\partial \Theta^k} + \frac{\partial G(\mathbf{X};\mathbf{\Theta})}{\partial \Theta^k}; \tag{22.51}$$

(b) for $j = 1, 2, \ldots, m$

$$\frac{\partial\{F(\mathbf{X};\mathbf{\Theta})G(\mathbf{X};\mathbf{\Theta})\}}{\partial X^j} = \frac{\partial F(\mathbf{X};\mathbf{\Theta})}{\partial X^j}G(\mathbf{X};\mathbf{\Theta}) + F(\mathbf{X};\mathbf{\Theta})\frac{\partial G(\mathbf{X};\mathbf{\Theta})}{\partial X^j}, \tag{22.52}$$

and, if $F(\mathbf{X};\mathbf{\Theta})$ is homogeneous, for $k = 1, 2, \ldots, n$

$$\frac{\partial\{F(\mathbf{X};\mathbf{\Theta})G(\mathbf{X};\mathbf{\Theta})\}}{\partial \Theta^k} = \frac{\partial F(\mathbf{X};\mathbf{\Theta})}{\partial \Theta^k}G(\mathbf{X};\mathbf{\Theta}) + (-1)^{\deg F} F(\mathbf{X};\mathbf{\Theta})\frac{\partial G(\mathbf{X};\mathbf{\Theta})}{\partial \Theta^k}; \tag{22.53}$$

and

(c) for any complex number λ and for $j = 1, 2, \ldots, m$

$$\frac{\partial\{\lambda F(\mathbf{X};\mathbf{\Theta})\}}{\partial X^j} = \lambda\frac{\partial F(\mathbf{X};\mathbf{\Theta})}{\partial X^j}, \tag{22.54}$$

while for $k = 1, 2, \ldots, n$

$$\frac{\partial\{\lambda F(\mathbf{X};\mathbf{\Theta})\}}{\partial \Theta^k} = \lambda\frac{\partial F(\mathbf{X};\mathbf{\Theta})}{\partial \Theta^k}. \tag{22.55}$$

The properties (22.52) and (22.53) are known as the *graded Leibnitz formulae*.

This theorem is also true for $G^\infty(U)$ functions, *except* for the $\mathcal{E}_{\{1,2,\ldots,L\}}$ terms of (22.51), (22.53) and (22.55).

Proof The proofs of parts (a) and (c) and the first part of (b) are obvious. Attention here will be concentrated on the proof of the odd part (22.53) of (b), and the point where the superdifferentiability properties are required will be identified.

By the expansion (22.43) for $F(\mathbf{X};\boldsymbol{\Theta})$ and the corresponding expression for $G(\mathbf{X};\boldsymbol{\Theta})$,

$$G(\mathbf{X};\boldsymbol{\Theta}) = \sum_{\Lambda'}\sum_{\mu'} Z(\mathscr{G}_{\mu'}^{\Lambda'}\mathcal{E}_{\mu'})(\mathbf{X})\Theta^{\Lambda'},$$

it follows that

$$F(\mathbf{X};\boldsymbol{\Theta})G(\mathbf{X};\boldsymbol{\Theta}) = \sum_{\Lambda,\Lambda'}\sum_{\mu,\mu'} Z(\mathscr{F}_{\mu}^{\Lambda}\mathcal{E}_{\varnothing})(\mathbf{X})Z(\mathscr{G}_{\mu'}^{\Lambda'}\mathcal{E}_{\varnothing})(\mathbf{X})\mathcal{E}_{\mu}\Theta^{\Lambda}\mathcal{E}_{\mu'}\Theta^{\Lambda'}. \tag{22.56}$$

But, by (22.49),

$$\frac{\partial\{\mathcal{E}_{\mu}\Theta^{\Lambda}\mathcal{E}_{\mu'}\Theta^{\Lambda'}\}}{\partial\Theta^{k}} = (-)^{N(\mu)}\mathcal{E}_{\mu}\Theta^{\Lambda/k}\mathcal{E}_{\mu'}\Theta^{\Lambda'} + (-1)^{N(\mu)+N(\mu')+N(\Lambda)}\mathcal{E}_{\mu}\Theta^{\Lambda}\mathcal{E}_{\mu'}\Theta^{\Lambda'/k},$$

and since

$$(-1)^{\deg F(\mathbf{X};\boldsymbol{\Theta})} = (-1)^{\{N(\mu)+N(\Lambda)\}}$$

and, with $F(\mathbf{X};\boldsymbol{\Theta})$ assumed to be homogeneous,

$$(-1)^{N(\mu)} = (-1)^{\deg \mathscr{F}^{\Lambda}},$$

with a similar result holding for $(-1)^{N(\mu')}$, the result (22.53) follows immediately.

At first sight, it may appear that the only assumption that has been made is that the odd derivatives are given by (22.49); that is, that the arbitrary complex-valued functions $\lambda^k(\mathbf{X};\boldsymbol{\Theta})$ have been set to be zero, and so it might appear that with this assumption (22.53) is true for *any* $G^\infty(U)$ function. To see that this is not so, consider the example with $L=2$, $n=1$, $F(\mathbf{X};\boldsymbol{\Theta}) = \mathcal{E}_{\{1\}}\Theta^1$ and $G(\mathbf{X};\boldsymbol{\Theta}) = \mathcal{E}_{\{2\}}$. In this case $F(\mathbf{X};\boldsymbol{\Theta})G(\mathbf{X};\boldsymbol{\Theta}) = -\mathcal{E}_{\{1\}}\mathcal{E}_{\{2\}}\Theta^1 = 0$, and so $\partial\{F(\mathbf{X};\boldsymbol{\Theta})G(\mathbf{X};\boldsymbol{\Theta})\}/\partial\Theta^1 = 0$. As $F(\mathbf{X};\boldsymbol{\Theta})$ is even, the right-hand side of (22.53) is

$$\{\partial F(\mathbf{X};\boldsymbol{\Theta})/\partial\Theta'\}G(\mathbf{X};\boldsymbol{\Theta}) + F(\mathbf{X};\boldsymbol{\Theta})\{\partial G(\mathbf{X};\boldsymbol{\Theta})/\partial\Theta'\} = -\mathcal{E}_{\{1\}}\mathcal{E}_{\{2\}},$$

so (22.53) is not satisfied in this case. The discrepancy appears here because the two alternative but equivalent expressions $-\mathcal{E}_{\{1\}}\mathcal{E}_{\{2\}}\Theta^1$ and 0 for $F(\mathbf{X};\boldsymbol{\Theta})G(\mathbf{X};\boldsymbol{\Theta})$ give different derivatives. More generally, although (22.56) is valid for superdifferentiable functions, for a general $G^\infty(U)$ functions extra arbitrary $\mathcal{E}_{\mu}\Theta^{\Lambda}$ terms with $N(\Lambda)+N(\mu) > L$ can be added to the right-hand side of (22.56) without affecting the values of $F(\mathbf{X};\boldsymbol{\Theta})G(\mathbf{X};\boldsymbol{\Theta})$—but these do have an effect on their odd derivatives.

The next theorem deals with higher derivatives of superdifferentiable functions.

Theorem IV If $F(\mathbf{X};\mathbf{\Theta})$ is any superdifferentiable function defined on an open set U of $\mathbb{R}\mathrm{B}_L^{m,n}$ then

(a) for $j, j' = 1, 2, \ldots, m$

$$\frac{\partial}{\partial X^j}\frac{\partial}{\partial X^{j'}}F(\mathbf{X};\mathbf{\Theta}) = \frac{\partial}{\partial X^{j'}}\frac{\partial}{\partial X^j}F(\mathbf{X};\mathbf{\Theta});$$

(b) for $j = 1, 2, \ldots, m$ and $k = 1, 2, \ldots, n$

$$\frac{\partial}{\partial X^j}\frac{\partial}{\partial \Theta^k}F(\mathbf{X};\mathbf{\Theta}) = \frac{\partial}{\partial \Theta^k}\frac{\partial}{\partial X^j}F(\mathbf{X};\mathbf{\Theta}); \tag{22.57}$$

(c) for $k = 1, 2, \ldots, n$

$$\frac{\partial}{\partial \Theta^k}\frac{\partial}{\partial \Theta^k}F(\mathbf{X};\mathbf{\Theta}) = 0, \tag{22.58}$$

and for $k, k' = 1, 2, \ldots, n$ with $k \neq k'$

$$\frac{\partial}{\partial \Theta^k}\frac{\partial}{\partial \Theta^{k'}}F(\mathbf{X};\mathbf{\Theta}) = -\frac{\partial}{\partial \Theta^{k'}}\frac{\partial}{\partial \Theta^k}F(\mathbf{X};\mathbf{\Theta}); \tag{22.59}$$

(d) with the expansion (22.33) for $F(\mathbf{X};\mathbf{\Theta})$

$$\frac{\partial}{\partial \Theta^n}\frac{\partial}{\partial \Theta^{n-1}}\cdots\frac{\partial}{\partial \Theta^2}\frac{\partial}{\partial \Theta^1}F(\mathbf{X};\mathbf{\Theta}) = (-1)^{n \deg F^{\{1,2,\ldots,n-1,n\}}}F^{\{1,2,\ldots,n-1,n\}}(\mathbf{X}). \tag{22.60}$$

Proof These results all follow immediately from (22.37) and (22.49).

Definition *Superanalytic function defined on* $\mathbb{R}\mathrm{B}_L^{m,n}$

A superdifferentiable function $F(\mathbf{X};\mathbf{\Theta})$ defined on an open set U of $\mathbb{R}\mathrm{B}_L^{m,n}$ may be said to be *superanalytic* on U if in the expansion (22.13) all the complex-valued component functions $F_\mu(\mathbf{X};\mathbf{\Theta})$ are *analytic* functions of all of the *even* real components X_μ^j that appear in (22.3) for each $j = 1, 2, \ldots, m$.

As Theorem II shows that every superanalytic function $F(\mathbf{X};\mathbf{\Theta})$ is a *polynomial* in the *odd* Grassmann variables $\Theta^1, \Theta^2, \ldots, \Theta^n$, and hence is a *polynomial* in all of the real components Θ_μ^k of these, the complex-valued component functions $F_\mu(\mathbf{X};\mathbf{\Theta})$ are *automatically* analytic functions of all of the odd real components Θ_μ^k.

The "Berezin integral" (cf. Berezin 1979) will now be discussed briefly. Let $F(\mathbf{X};\boldsymbol{\Theta})$ be a superdifferentiable function defined on some open set U of $\mathbb{R}\mathrm{B}_L^{m,n}$ whose expansion of the form (22.33) is

$$F(\mathbf{X};\boldsymbol{\Theta}) = \sum_{\Lambda} F^{\Lambda}(\mathbf{X})\Theta^{\Lambda}.$$

Then its *Berezin integral* over the odd Grassmann variables $\Theta^1, \Theta^2, \ldots, \Theta^n$ is defined by

$$\int \mathrm{d}^n\Theta\, F(\mathbf{X};\boldsymbol{\Theta}) = F^{\{1,2,\ldots,n-1,n\}}(\mathbf{X}), \tag{22.61}$$

where, just as in (22.60), only the component function $F^{\{1,2,\ldots,n-1,n\}}(\mathbf{X})$ associated with the *highest* power $\Theta^1\Theta^2\cdots\Theta^{n-1}\Theta^n$ appears. Although (22.61) specifies a well-defined operation, it cannot be assumed that the operation possesses the usual properties of an integral. For example, (22.61) makes no mention of the "limits" of the integration, the same result being obtained for *any* open set of $\mathbb{R}\mathrm{B}_L^{m,n}$ that contains a point of $\mathbb{R}\mathrm{B}_L^{m,n}$ with even coordinates $X^1, X^2, \ldots, X^m$. For a discussion of some of the problems of the "Berezin integral" see Rogers (1984a, b, 1985a, b, 1986b, 1987a, b). (For Lie supergroups the "Berezin integral" is not related in any way to the left- and right-invariant Haar integrals (Williams and Cornwell 1984).)

(*d*) *Differentiation of supermatrices*

The ideas of the previous subsections can be applied without difficulty to supermatrices. Let $\mathbf{M}$ be a $(p/q) \times (r/s)$ supermatrix that is either even or odd and is partitioned in the form (20.40); that is,

$$\mathbf{M} = \begin{bmatrix} \mathbf{A} & \mathbf{B} \\ \mathbf{C} & \mathbf{D} \end{bmatrix}. \tag{22.62}$$

Its norm $\|\mathbf{M}\|$ may be defined in terms of the norms $\|M_{jk}\|$ of its matrix elements M_{jk} by

$$\|\mathbf{M}\| = \sum_{j=1}^{p+q} \sum_{k=1}^{r+s} \|M_{jk}\|, \tag{22.63}$$

where $\|M_{jk}\|$ is defined by (22.6) as $M_{jk} \in \mathbb{C}\mathrm{B}_L$. The supermatrix metric is then defined by analogy with (22.1) and (22.9) to be given by

$$d(\mathbf{M}, \mathbf{M}') = \|\mathbf{M} - \mathbf{M}'\|, \tag{22.64}$$

and possesses all the usual properties of a distance function.

The next stage is to consider supermatrices that are functions of the variables $(\mathbf{X};\mathbf{\Theta})$ of $\mathbb{R}\mathrm{B}_L^{m,n}$. For these, differentiability can be defined by adapting the corresponding definition that was given earlier for a function $F(\mathbf{X};\mathbf{\Theta})$.

Definition *Differentiability of a supermatrix defined on an open set of* $\mathbb{R}\mathrm{B}_L^{m,n}$

Let $\mathbf{M}(\mathbf{X};\mathbf{\Theta})$ be a *supermatrix* whose elements $M_{jk}(\mathbf{X};\mathbf{\Theta})$ are *superdifferentiable* functions of $(X;\Theta)$ that take values in $\mathbb{C}\mathrm{B}_L$ and are defined on an open set U of $\mathbb{R}\mathrm{B}_L^{m,n}$. Suppose that there exist m supermatrices $\partial\mathbf{M}(\mathbf{X};\mathbf{\Theta})/\partial X^j$ (for $j=1,2,\ldots,m$) and n supermatrices $\partial\mathbf{M}(\mathbf{X};\mathbf{\Theta})/\partial\Theta^k$ (for $k=1,2,\ldots,n$) whose elements all have values in $\mathbb{C}\mathrm{B}_L$ and are defined for all $(\mathbf{X};\mathbf{\Theta})$ in U and that are such that

$$\mathbf{M}(\mathbf{X}+\mathbf{Y};\mathbf{\Theta}+\mathbf{\Psi})=\mathbf{M}(\mathbf{X};\mathbf{\Theta})+\sum_{j=1}^{m}Y^j\frac{\partial\mathbf{F}(\mathbf{X};\mathbf{\Theta})}{\partial X^j}$$

$$+\sum_{k=1}^{n}\Psi^k\frac{\partial\mathbf{M}(\mathbf{X};\mathbf{\Theta})}{\partial\Theta^k}+\|(\mathbf{Y};\mathbf{\Psi})\|\,\boldsymbol{\eta}(\mathbf{Y};\mathbf{\Psi}).\qquad(22.65)$$

Here $(\mathbf{X};\mathbf{\Theta})$ and $(\mathbf{X}+\mathbf{Y};\mathbf{\Theta}+\mathbf{\Psi})$ are both points in U and $\boldsymbol{\eta}(\mathbf{Y};\mathbf{\Psi})$ is a supermatrix defined on $\mathbb{R}\mathrm{B}_L^{m,n}$ with values in $\mathbb{C}\mathrm{B}_L$ that is such that

$$\|\boldsymbol{\eta}(\mathbf{Y};\mathbf{\Psi})\|\to 0\quad\text{as }\|(\mathbf{Y};\mathbf{\Psi})\|\to 0.\qquad(22.66)$$

Then the supermatrix $\mathbf{M}(\mathbf{X};\mathbf{\Theta})$ is said to be *differentiable* in U and the quantities $\partial\mathbf{M}(\mathbf{X};\mathbf{\Theta})/\partial X^j$ and $\partial\mathbf{M}(\mathbf{X};\mathbf{\Theta})/\partial\Theta^k$ are described as being its partial derivatives.

Theorem V If $\mathbf{M}(\mathbf{X};\mathbf{\Theta})$ is an even or odd $(p/q)\times(r/s)$ supermatrix with superdifferentiable matrix elements defined on an open set U of $\mathbb{R}\mathrm{B}_L^{m,n}$ whose partitioning is

$$\mathbf{M}(\mathbf{X};\mathbf{\Theta})=\begin{bmatrix}\mathbf{A}(\mathbf{X};\mathbf{\Theta}) & \mathbf{B}(\mathbf{X};\mathbf{\Theta})\\ \mathbf{C}(\mathbf{X};\mathbf{\Theta}) & \mathbf{D}(\mathbf{X};\mathbf{\Theta})\end{bmatrix}\qquad(22.67)$$

then, for $j=1,2,\ldots,m$,

$$\frac{\partial\mathbf{M}(\mathbf{X};\mathbf{\Theta})}{\partial X^j}=\begin{bmatrix}\dfrac{\partial\mathbf{A}(\mathbf{X};\mathbf{\Theta})}{\partial X^j} & \dfrac{\partial\mathbf{B}(\mathbf{X};\mathbf{\Theta})}{\partial X^j}\\[2ex] \dfrac{\partial\mathbf{C}(\mathbf{X};\mathbf{\Theta})}{\partial X^j} & \dfrac{\partial\mathbf{D}(\mathbf{X};\mathbf{\Theta})}{\partial X^j}\end{bmatrix},\qquad(22.68)$$

and, for $k = 1, 2, \ldots, n$,

$$\frac{\partial \mathbf{M}(\mathbf{X};\boldsymbol{\Theta})}{\partial \Theta^k} = \begin{bmatrix} \dfrac{\partial \mathbf{A}(\mathbf{X};\boldsymbol{\Theta})}{\partial \Theta^k} & \dfrac{\partial \mathbf{B}(\mathbf{X};\boldsymbol{\Theta})}{\partial \Theta^k} \\ -\dfrac{\partial \mathbf{C}(\mathbf{X};\boldsymbol{\Theta})}{\partial \Theta^k} & -\dfrac{\partial \mathbf{D}(\mathbf{X};\boldsymbol{\Theta})}{\partial \Theta^k} \end{bmatrix}. \tag{22.69}$$

Here $\partial \mathbf{A}(\mathbf{X};\boldsymbol{\Theta})/\partial X^j$ is the submatrix whose elements are the $\partial/\partial X^j$ derivatives of the submatrix $\mathbf{A}(\mathbf{X};\boldsymbol{\Theta})$ considered as a matrix in its own right, and so on.

Proof These results follow immediately from (22.67) on using (20.51).

It should be noted that the right-hand side of (22.70) contains a couple of minus signs, which originate from the minus signs that appear when the scalar multiplication of $\partial \mathbf{M}(\mathbf{X};\boldsymbol{\Theta})/\partial \Theta^k$ by the odd Grassmann element Ψ^k in $\Psi^k\, \partial \mathbf{M}(\mathbf{X};\boldsymbol{\Theta})/\partial \Theta^k$ is calculated using (20.51). Clearly if $\mathbf{M}(\mathbf{X};\boldsymbol{\Theta})$ is an even supermatrix then $\partial \mathbf{M}(\mathbf{X};\boldsymbol{\Theta})/\partial X^j$ is an even supermatrix and $\partial \mathbf{M}(\mathbf{X};\boldsymbol{\Theta})/\partial \Theta^k$ is an odd supermatrix, whereas if $\mathbf{M}(\mathbf{X};\boldsymbol{\Theta})$ is an odd supermatrix then $\partial \mathbf{M}(\mathbf{X};\boldsymbol{\Theta})/\partial X^j$ is an odd supermatrix and $\partial \mathbf{M}(\mathbf{X};\boldsymbol{\Theta})/\partial \Theta^k$ is an even supermatrix.

By means of the expansions (22.3) and (22.4), the matrix $\mathbf{M}(\mathbf{X};\boldsymbol{\Theta})$ can also be considered to be a function of the $(m+n)2^{L-1}$ real variables X^j_μ (for $j = 1, 2, \ldots, m$ and all even index sets μ) and Θ^k_μ (for $k = 1, 2, \ldots, n$ and all odd index sets μ). For each matrix element $M_{j'k'}(\mathbf{X};\boldsymbol{\Theta})$ of $\mathbf{M}(\mathbf{X};\boldsymbol{\Theta})$ let $M_{j'k'\mu'}(\mathbf{X};\boldsymbol{\Theta})$ be the complex-valued component functions that appear in an expansion of the form (22.13), so that

$$M_{j'k'}(\mathbf{X};\boldsymbol{\Theta}) = \sum_{\mu'} M_{j'k'\mu'}(\mathbf{X};\boldsymbol{\Theta})\mathcal{E}_{\mu'}.$$

The derivatives $\partial M_{j'k'}(\mathbf{X};\boldsymbol{\Theta})/\partial X^j_\mu$ of the matrix elements $M_{j'k'}(\mathbf{X};\boldsymbol{\Theta})$ can be defined by

$$\frac{\partial M_{j'k'}(\mathbf{X};\boldsymbol{\Theta})}{\partial X^j_\mu} = \sum_{\mu'} \frac{\partial M_{j'k'\mu'}(\mathbf{X};\boldsymbol{\Theta})}{\partial X^j_\mu}\mathcal{E}_{\mu'}.$$

Let $\partial \mathbf{M}(\mathbf{X};\boldsymbol{\Theta})/\partial X^j_\mu$ be the supermatrix with elements $\partial M_{j'k'}(\mathbf{X};\boldsymbol{\Theta})/\partial X^j_\mu$. Then, as in the derivation of (22.16), it follows that

$$\frac{\partial \mathbf{M}(\mathbf{X};\boldsymbol{\Theta})}{\partial X^j_\mu} = \mathcal{E}_\mu \frac{\partial \mathbf{M}(\mathbf{X};\boldsymbol{\Theta})}{\partial X^j}. \tag{22.70}$$

With $\partial \mathbf{M}(\mathbf{X};\boldsymbol{\Theta})/\partial \Theta^k_\mu$ defined similarly, it also follows that

$$\frac{\partial \mathbf{M}(\mathbf{X};\boldsymbol{\Theta})}{\partial \Theta^k_\mu} = \mathcal{E}_\mu \frac{\partial \mathbf{M}(\mathbf{X};\boldsymbol{\Theta})}{\partial \Theta^k}. \tag{22.71}$$

3 Linear Lie supergroups

(a) *The definition of a Lie supergroup and its associated "super" Lie algebra*

Several non-equivalent definitions of the concept of a Lie supergroup have appeared in the literature. The one that follows not only has the merit of being a straightforward and natural generalization of the concept of a Lie group, but is the one that is widely used in a number of theoretical physics applications.

Definition *Linear Lie supergroup of even dimension m and odd dimension n*

Let $\mathscr{G}_s(\mathbb{C}\mathbf{B}_L)$ be a set of *even* $(p/q) \times (p/q)$ supermatrices $\mathbf{G}$ with entries in $\mathbb{C}\mathbf{B}_L$ that has a grading partitioning of the form (20.40), that is,

$$\mathbf{G} = \begin{bmatrix} \mathbf{P} & \mathbf{Q} \\ \mathbf{R} & \mathbf{S} \end{bmatrix}, \tag{22.72}$$

where $\mathbf{P}$ is a $p \times p$ matrix with elements in $\mathbb{C}\mathbf{B}_{L0}$, $\mathbf{Q}$ is a $p \times q$ matrix with elements in $\mathbb{C}\mathbf{B}_{L1}$, $\mathbf{R}$ is a $q \times p$ matrix with elements in $\mathbb{C}\mathbf{B}_{L1}$ and $\mathbf{S}$ is a $q \times q$ matrix with elements in $\mathbb{C}\mathbf{B}_{L0}$. This set is said to form a *linear Lie supergroup of even dimension m and odd dimension n* if it satisfies the following conditions (A)–(D):

(A) *$\mathscr{G}_s(\mathbb{C}\mathbf{B}_L)$ must be a group with supermatrix multiplication (as defined in Chapter 20, Section 4) providing the group multiplication operation and $\mathbf{1}_{p+q}$ being the identity of the group.*

(B) *With the metric $d(\ ,\ )$ for $\mathscr{G}_s(\mathbb{C}\mathbf{B}_L)$ defined by (22.64), there must exist a $\delta > 0$ such that every element $\mathbf{G}$ of $\mathscr{G}_s(\mathbb{C}\mathbf{B}_L)$ lying in the sphere $\mathbf{M}_\delta$ of radius δ centred on the identity $\mathbf{1}_{p+q}$ of $\mathscr{G}_s(\mathbb{C}\mathbf{B}_L)$ can be parametrized by an element $(\mathbf{X};\boldsymbol{\Theta}) = (X^1, X^2, \ldots, X^m, \Theta^1, \Theta^2, \ldots, \Theta^n)$ of $\mathbb{R}\mathbf{B}_L^{m,n}$ (no two sets of points corresponding to the same element $\mathbf{G}$* of *$\mathscr{G}_s(\mathbb{C}\mathbf{B}_L)$), the identity being parametrized by $(\mathbf{X};\boldsymbol{\Theta}) = (\mathbf{0},\mathbf{0})$, that is, by*

$$X^1 = X^2 = \cdots = X^m = \Theta^1 = \Theta^2 = \cdots = \Theta^n = 0.$$

Here the sphere $\mathbf{M}_\delta$ consists of all $\mathbf{G}$ of $\mathscr{G}_s(\mathbb{C}\mathbf{B}_L)$ such that

$$d(\mathbf{M}, \mathbf{1}_{p+q}) < \delta.$$

With this assumption, every element of $\mathscr{G}_s(\mathbb{C}\mathbf{B}_L)$ lying in $\mathbf{M}_\delta$ corresponds to one and only one point of $\mathbb{R}\mathbf{B}_L^{m,n}$, the identity corresponding to the origin $(\mathbf{0},\mathbf{0})$ of $\mathbb{R}\mathbf{B}_L^{m,n}$. Moreover, no point in $\mathbb{R}\mathbf{B}_L^{m,n}$ corresponds to more than one element $\mathbf{G}$ in $\mathbf{M}_\delta$.

(C) *There must exist an* $\eta > 0$ *such that every point* $(\mathbf{X};\boldsymbol{\Theta})$ *of the sphere* S_η *of radius* η *centred on the origin* $(\mathbf{0},\mathbf{0})$ *of* $\mathbb{R}\mathbf{B}_L^{m,n}$ *corresponds to some element* $\mathbf{G}$ *in* M_δ.

Let R_η be the set of elements $\mathbf{G}$ of $\mathscr{G}_s(\mathbb{C}\mathbf{B}_L)$ so obtained. Then R_η is a subset of M_δ, and there is a one-to-one correspondence between the elements of $\mathscr{G}_s(\mathbb{C}\mathbf{B}_L)$ in R_η and the points of $\mathbb{R}\mathbf{B}_L^{m,n}$ lying in the sphere S_η. Let $\mathbf{G}(\mathbf{X};\boldsymbol{\Theta})$ denote the element of R_η that corresponds to the point $(\mathbf{X};\boldsymbol{\Theta})$ of S_η.

(D) *Each of the matrix elements of* $\mathbf{G}(\mathbf{X};\boldsymbol{\Theta})$ *must be a superanalytic function of* $(\mathbf{X};\boldsymbol{\Theta})$ *on* S_η.

Theorem I If L is finite then the linear Lie supergroup $\mathscr{G}_s(\mathbb{C}\mathbf{B}_L)$ with m even dimensions and n odd dimensions is an ordinary linear Lie group of order $(m+n)2^{L-1}$, the basis elements of its corresponding real Lie algebra $\mathscr{L}_s(\mathbb{C}\mathbf{B}_L)$ being the $m2^{L-1}$ even $(p/q)\times(p/q)$ supermatrices

$$\mathbf{M}_\mu^j = \left.\frac{\partial \mathbf{G}(\mathbf{X};\boldsymbol{\Theta})}{\partial X_\mu^j}\right|_{\mathbf{X}=\mathbf{0},\boldsymbol{\Theta}=\mathbf{0}} \tag{22.73}$$

(for $j=1,2,\ldots,m$ and all even index sets μ), together with the $n2^{L-1}$ even $(p/q)\times(p/q)$ supermatrices

$$\mathbf{N}_\mu^k = \left.\frac{\partial \mathbf{G}(\mathbf{X};\boldsymbol{\Theta})}{\partial \Theta_\mu^k}\right|_{\mathbf{X}=\mathbf{0},\boldsymbol{\Theta}=\mathbf{0}} \tag{22.74}$$

(for $k=1,2,\ldots,m$ and all odd index sets μ).

Proof As each of the m even components X^j of $(\mathbf{X};\boldsymbol{\Theta})$ has an expansion (22.3) in terms of 2^{L-1} real components X_μ^j, and each of the n odd components Θ^k of $(\mathbf{X};\boldsymbol{\Theta})$ has an expansion (22.4) in terms of 2^{L-1} real components Θ_μ^k, then, as noted in Subsection 2(*a*), $\mathbb{R}\mathbf{B}_L^{m,n}$ can be considered to be a real vector space of dimension $(m+n)2^{L-1}$. Consequently each element $\mathbf{G}$ of $\mathscr{G}_s(\mathbb{C}\mathbf{B}_L)$ near the identity of $\mathscr{G}_s(\mathbb{C}\mathbf{B}_L)$ can be parametrized by $(m+n)2^{L-1}$ real parameters, with the identity element corresponding to all of these parameters taking the value 0.

Let $G(\mathbf{X};\boldsymbol{\Theta})_{rs}$ be the r,s element of $\mathbf{G}(\mathbf{X};\boldsymbol{\Theta})$ and let

$$G(\mathbf{X};\boldsymbol{\Theta})_{rs} = \sum_\mu G(\mathbf{X};\boldsymbol{\Theta})_{rs\mu}\,\mathcal{E}_\mu$$

be its expansion (cf. (22.13)), in which the $G(\mathbf{X};\boldsymbol{\Theta})_{rs\mu}$ are complex-valued functions of $(\mathbf{X};\boldsymbol{\Theta})$, and hence of all the $(m+n)2^{L-1}$ real parameters $X_{\mu'}^j$ and $\Theta_{\mu'}^k$ of $(\mathbf{X};\boldsymbol{\Theta})$. The assumption that $G(\mathbf{X};\boldsymbol{\Theta})_{rs}$ is superanalytic near the origin implies that all the components $G_{rs\mu}$ are analytic functions of these real parameters $X_{\mu'}^j$ and $\Theta_{\mu'}^k$. Consequently all the conditions given in Chapter 3,

Section 2 for $\mathscr{G}_s(\mathbb{C}\mathbf{B}_L)$ to be a linear Lie group of order $(m+n)2^{L-1}$ are satisfied, and (22.73) and (22.74) are the basis elements of its real Lie algebra $\mathscr{L}_s(\mathbb{C}\mathbf{B}_L)$.

This theorem shows that a general element $\mathbf{M}$ of $\mathscr{L}_s(\mathbb{C}\mathbf{B}_L)$ is an *even* $(p/q)\times(p/q)$ supermatrix with the form

$$\mathbf{M}=\sum_{j=1}^{m}\sum_{\text{even }\mu}X^j_\mu\mathbf{M}^\mu_j+\sum_{k=1}^{n}\sum_{\text{odd }\mu}\Theta^k_\mu\mathbf{N}^\mu_k, \tag{22.75}$$

where X^j_μ and Θ^k_μ are *real* parameters. Let $\mathbf{M}^1,\mathbf{M}^2,\dots,\mathbf{M}^m$ be the m *even* $(p/q)\times(p/q)$ supermatrices are defined by

$$\mathbf{M}^j=\left.\frac{\partial\mathbf{M}(\mathbf{X};\boldsymbol{\Theta})}{\partial X^j}\right|_{\mathbf{X}=\mathbf{0},\boldsymbol{\Theta}=\mathbf{0}} \tag{22.76}$$

(for $j=1,2,\dots,m$) and let $\mathbf{N}^1,\mathbf{N}^2,\dots,\mathbf{N}^m$ be the n *odd* $(p/q)\times(p/q)$ supermatrices that are defined by

$$\mathbf{N}^k=\left.\frac{\partial\mathbf{M}(\mathbf{X};\boldsymbol{\Theta})}{\partial\Theta^k}\right|_{\mathbf{X}=\mathbf{0},\boldsymbol{\Theta}=\mathbf{0}} \tag{22.77}$$

(for $k=1,2,\dots,n$). Then, by (22.70),

$$\mathbf{M}^j_\mu=\mathcal{E}_\mu\mathbf{M}^j,$$

and, by (22.71),

$$\mathbf{N}^k_\mu=\mathcal{E}_\mu\mathbf{N}^k.$$

Taking the real parameters X^j_μ and Θ^k_μ of (22.75) to be the components of the coordinates of a point $(\mathbf{X};\boldsymbol{\Theta})=(X^1,X^2,\dots,X^m,\ \Theta^1,\Theta^2,\dots,\Theta^n)$ of $\mathbb{R}^{m,n}_L$ by using them in the expansions (22.3) and (22.4), the general element $\mathbf{M}$ of $\mathscr{L}_s(\mathbb{C}\mathbf{B}_L)$ of (22.74) can be rewritten more succinctly in terms of the m real even Grassmann variables $X^1,X^2,\dots,X^m$ and the m real odd Grassmann variables $\Theta^1,\Theta^2,\dots,\Theta^n$ as

$$\mathbf{M}=\sum_{j=1}^{m}X^j\mathbf{M}^j+\sum_{k=1}^{n}\Theta^k\mathbf{N}^k. \tag{22.78}$$

As every element of a Lie group near its identity can be obtained by exponentiating an appropriate element of its real Lie algebra, a very convenient parametrization of the elements of $\mathscr{G}_s(\mathbb{C}\mathbf{B}_L)$ near its identity is given by

$$\mathbf{G}(\mathbf{X};\boldsymbol{\Theta})=\exp\mathbf{M}=\exp\left\{\sum_{j=1}^{m}X^j\mathbf{M}^j+\sum_{k=1}^{n}\Theta^k\mathbf{N}^k\right\}. \tag{22.79}$$

This parametrization is consistent with (22.76) and (22.77). As every element of $\mathscr{L}_s(\mathbb{C}B_L)$ and every element of $\mathscr{G}_s(\mathbb{C}B_L)$ near its identity can be expressed in terms of the m supermatrices $\mathbf{M}^j$ and the n supermatrices $\mathbf{N}^k$, it is appropriate to refer to $\mathbf{M}^j$ and $\mathbf{N}^k$ as the *supergenerators* of $\mathscr{L}_s(\mathbb{C}B_L)$ and of $\mathscr{G}_s(\mathbb{C}B_L)$. Because the Lie supergroup $\mathscr{G}_s(\mathbb{C}B_L)$ is a Lie group with an additional Grassmann structure, it is often known alternatively as a *super Lie group.* Likewise its real Lie algebra $\mathscr{L}_s(\mathbb{C}B_L)$ may be called a "*super*" *Lie algebra*, but it must be emphasized that $\mathscr{L}_s(\mathbb{C}B_L)$ is a Lie algebra and is *not* a Lie superalgebra.

Example I *The Lie supergroups* U(p/q) *and* SU(p/q)

The Lie supergroup U(p/q) is defined to be the set of $(p/q)\times(p/q)$ even supermatrices $\mathbf{G}$ that satisfy the condition

$$\mathbf{G}^{\ddagger}\mathbf{G} = \mathbf{1}_{p+q}, \tag{22.80}$$

where $\mathbf{G}^{\ddagger}$ is defined in (20.79). The Lie supergroup SU(p/q) is defined to be the subgroup of U(p/q) consisting of those matrices $\mathbf{G}$ of U(p/q) that also satisfy the condition

$$\text{sdet}\,\mathbf{G} = 1, \tag{22.81}$$

where "sdet" denotes the superdeterminant of (20.74).

The parametrization of U(p/q) and SU(p/q) in terms of elements of $\mathbb{R}B_{L0}$ and $\mathbb{R}B_{L1}$ will now be investigated. Consider first U(p/q). Let $\mathbf{M}$ be a member of the super Lie algebra $\mathscr{L}_s(\mathbb{C}B_L)$ of $\mathscr{G}_s(\mathbb{C}B_L) = \text{U}(p/q)$, so that if t is a real parameter taking value from $-\infty$ to ∞ then the set of supermatrices $\exp(t\mathbf{M})$ form a one-parameter subgroup of U(p/q). The condition (22.80) implies that

$$\mathbf{M}^{\ddagger} + \mathbf{M} = \mathbf{0}, \tag{22.82}$$

and so, if $\mathbf{M}$ has the grading partitioning

$$\mathbf{M} = \begin{bmatrix} \mathbf{A} & \mathbf{B} \\ \mathbf{C} & \mathbf{D} \end{bmatrix}, \tag{22.83}$$

where $\mathbf{A}$ is a $p\times p$ matrix with elements in $\mathbb{C}B_{L0}$, $\mathbf{B}$ is a $p\times q$ matrix with elements in $\mathbb{C}B_{L1}$, $\mathbf{C}$ is a $q\times p$ matrix with elements in $\mathbb{C}B_{L1}$ and $\mathbf{D}$ is a $q\times q$ matrix with elements in $\mathbb{C}B_{L0}$, then the condition (20.79) gives

$$\tilde{\mathbf{A}}^{\#} = -\mathbf{A}, \quad \tilde{\mathbf{D}}^{\#} = -\mathbf{D}, \quad \mathbf{C} = -\tilde{\mathbf{B}}^{\#}, \tag{22.84}$$

Here, for example, $(\tilde{\mathbf{A}}^{\#})_{jk} = (A_{kj})^{\#}$, the # indicating the Grassmann adjoint defined in (20.30) and (20.31). For each of the p diagonal elements A_{jj} of $\mathbf{A}$, as A_{jj} is necessarily an even element of $\mathbb{C}B_L$, (22.84) and (20.33) imply that $A_{jj} = iY_{jj}$, where $Y_{jj}\in\mathbb{R}B_{L0}$. Similarly (22.84) implies that the $\frac{1}{2}p(p-1)$ upper off-diagonal elements of $\mathbf{A}$ are arbitrary members of $\mathbb{C}B_{L0}$, and the lower off-diagonal elements are completely determined by them. Thus if $A_{jk} = X_{jk} + iY_{jk}$, where X_{jk}, $Y_{jk} \in \mathbb{R}B_{L0}$ (for $1 \leqslant j < k \leqslant p$), then, by (20.33), $A_{kj} = -X_{jk} + iY_{jk}$. Consequently the matrices $\mathbf{A}$ are specified by p^2 members of $\mathbb{R}B_{L0}$. Similarly the matrices $\mathbf{D}$ are specified by q^2 members of $\mathbb{R}B_{L0}$. Finally, (22.84) shows that each of the pq elements of $\mathbf{B}$ is an arbitrary element of $\mathbb{C}B_{L1}$, and so can be written in the form $B_{jk} = \Theta_{jk} + i\Psi_{jk}$, where Θ_{jk}, $\Psi_{jk} \in \mathbb{R}B_{L1}$, and that the elements of $\mathbf{C}$ are then given by $C_{kj} = i\Theta_{jk} + \Psi_{jk}$ (where (20.34) has been invoked). Thus $\mathbf{B}$ and $\mathbf{C}$ are together determined by $2pq$ members of $\mathbb{R}B_{L1}$. Putting these results together, it is clear that U(p/q) is a Lie supergroup with $m = p^2 + q^2$ even dimensions and $n = 2pq$ odd dimensions.

For SU(p/q) the analysis is the same except that the additional condition (22.81) implies, by (20.77), that

$$\operatorname{str} \mathbf{M} = 0, \qquad (22.85)$$

"str" indicating the supertrace, so, as $\mathbf{M}$ is even, (20.67) implies that

$$\operatorname{tr} \mathbf{A} = \operatorname{tr} \mathbf{D}. \qquad (22.86)$$

This condition reduces the number of independent members of $\mathbb{R}B_{L0}$ required to specify the diagonal elements of $\mathbf{A}$ and $\mathbf{D}$ by one, showing that SU(p/q) is a Lie supergroup with $m = p^2 + q^2 - 1$ even dimensions and $n = 2pq$ odd dimensions.

It should be noted that U(p/q) and SU(p/q) are both *non-compact* Lie groups (for $q > 0$). To see this, consider the one-parameter subgroup $\exp(t\mathbf{M})$ of U(p/q), where t is a real parameter taking all values from $-\infty$ to ∞, and where $\mathbf{M}$ has all elements zero except for its first diagonal element, which is assumed to be $i\mathcal{E}_{\{1,2\}}$. Then the first diagonal element of $\exp(t\mathbf{M})$ is $\exp(it\mathcal{E}_{\{1,2\}})$, which, since $\mathcal{E}_{\{1,2\}}^2 = 0$, is equal to $\mathcal{E}_{\varnothing} + it\mathcal{E}_{\{1,2\}}$. As no two values of t give the same value for this supermatrix element, this one-parameter subgroup is non-compact, so U(p/q) must also be non-compact. The corresponding result for SU(p/q) can be demonstrated in a similar way.

Example II *Supergenerators of the Lie supergroups* SU(2/1)

Example I above shows that the most general element $\mathbf{M}$ of the "super" Lie algebra can be written in the (2/1) × (2/1) supermatrix form

$$\mathbf{M} = \begin{bmatrix} iX^1 & X^3 + iX^4 & \Theta^1 + i\Theta^2 \\ -X^3 + iX^4 & iX^2 & \Theta^3 + i\Theta^4 \\ i\Theta^1 + \Theta^2 & i\Theta^3 + \Theta^4 & -iX^1 - iX^2 \end{bmatrix}. \qquad (22.87)$$

The supergenerators (22.76) and (22.77) are then

$$\mathbf{M}^1 = \begin{bmatrix} i\mathcal{E}_\varnothing & 0 & 0 \\ 0 & 0 & 0 \\ 0 & 0 & -i\mathcal{E}_\varnothing \end{bmatrix}, \quad \mathbf{M}^2 = \begin{bmatrix} 0 & 0 & 0 \\ 0 & i\mathcal{E}_\varnothing & 0 \\ 0 & 0 & -i\mathcal{E}_\varnothing \end{bmatrix},$$

$$\mathbf{M}^3 = \begin{bmatrix} 0 & \mathcal{E}_\varnothing & 0 \\ -\mathcal{E}_\varnothing & 0 & 0 \\ 0 & 0 & 0 \end{bmatrix}, \quad \mathbf{M}^4 = \begin{bmatrix} 0 & i\mathcal{E}_\varnothing & 0 \\ i\mathcal{E}_\varnothing & 0 & 0 \\ 0 & 0 & 0 \end{bmatrix},$$

$$\mathbf{N}^1 = \begin{bmatrix} 0 & 0 & \mathcal{E}_\varnothing \\ 0 & 0 & 0 \\ -i\mathcal{E}_\varnothing & 0 & 0 \end{bmatrix}, \quad \mathbf{N}^2 = \begin{bmatrix} 0 & 0 & i\mathcal{E}_\varnothing \\ 0 & 0 & 0 \\ -\mathcal{E}_\varnothing & 0 & 0 \end{bmatrix},$$

$$\mathbf{N}^3 = \begin{bmatrix} 0 & 0 & 0 \\ 0 & 0 & \mathcal{E}_\varnothing \\ 0 & -i\mathcal{E}_\varnothing & 0 \end{bmatrix}, \quad \mathbf{N}^4 = \begin{bmatrix} 0 & 0 & 0 \\ 0 & 0 & i\mathcal{E}_\varnothing \\ 0 & -\mathcal{E}_\varnothing & 0 \end{bmatrix},$$

where (22.36), (22.49), (22.68) and (22.69) have been used. (The "extra" minus signs in the $\mathbf{C}$ submatrices of $\mathbf{N}^1, \mathbf{N}^2, \mathbf{N}^3$, and $\mathbf{N}^4$ come from the "extra" minus signs of (22.69)).

Writing the four even supergenerators in the form $\mathbf{M}^j = \mathcal{E}_\varnothing \mathbf{m}^j$ and the four odd supergenerators in the form $\mathbf{N}^k = \mathcal{E}_\varnothing \mathbf{n}^k$, where the $\mathbf{m}^j$ and $\mathbf{n}^k$ are supermatrices whose entries are complex numbers, it is clear that

$$\tilde{\mathbf{m}}^{j\#} + \mathbf{m}^j = \mathbf{0} \tag{22.88}$$

and

$$\tilde{\mathbf{n}}^k + i\mathbf{n}^{k\#} = \mathbf{0}. \tag{22.89}$$

It will be seen in Example III of Chapter 24, Section 5 that these are the defining relations of the even and odd basis elements of the real Lie superalgebra su(2/1), as they are special cases of (24.131).

In much the same way that the concept of a linear Lie group can be extended to Lie groups that are not linear (cf. Appendix J), the linearity condition in the definition given above for a Lie supergroup can be relaxed. A careful formulation of these more general Lie supergroups in terms of G^∞ manifolds has been given by Rogers (1981).

It is useful to generalize to the case of a "super" Lie algebra $\mathscr{L}_s(\mathbb{C}\mathbf{B}_L)$ the set of operators $P(\mathbf{a})$ defined in (10.21) for the real Lie algebra so(3). To this end, let $\mathscr{G}_s(\mathbb{C}\mathbf{B}_L)$ be a set of *real* even $(p/q) \times (p/q)$ supermatrices $\mathbf{G}(X;\mathbf{\Theta})$, so that the elements of $\mathbf{G}(X;\mathbf{\Theta})$ take values in $\mathbb{R}B_L$. (As above, $(\mathbf{X};\mathbf{\Theta}) = (X^1, X^2, \ldots, X^m, \Theta^1, \Theta^2, \ldots, \Theta^n)$ is a point of $\mathbb{R}\mathbf{B}_L^{m,n}$.) Let $(\mathbf{Y};\mathbf{\Psi}) = (Y^1, Y^2, \ldots, Y^p, \Psi^1, \Psi^2, \ldots, \Psi^q)$ denote a point of $\mathbb{R}\mathbf{B}_L^{p,q}$, let $F(\mathbf{Y};\mathbf{\Psi})$ be any superanalytic function defined on

$\mathbb{R}B_L^{p,q}$ and define the operator $P(\mathbf{G}(\mathbf{X};\boldsymbol{\Theta}))$ by analogy with (1.17) by

$$P(\mathbf{G}(\mathbf{X};\boldsymbol{\Theta}))F(\mathbf{Y};\boldsymbol{\Psi}) = F(\mathbf{Y}';\boldsymbol{\Psi}'), \tag{22.90}$$

where

$$\begin{bmatrix} Y'^1 \\ Y'^2 \\ \vdots \\ Y'^p \\ \Psi'^1 \\ \Psi'^2 \\ \vdots \\ \Psi'^q \end{bmatrix} = \mathbf{G}(\mathbf{X};\boldsymbol{\Theta})^{-1} \begin{bmatrix} Y^1 \\ Y^2 \\ \vdots \\ Y^p \\ \Psi^1 \\ \Psi^2 \\ \vdots \\ \Psi^q \end{bmatrix} \tag{22.91}$$

By analogy with (22.14), define the operators $P(\mathbf{M}^j)$ (for $j = 1, 2, \ldots, m$) and $P(\mathbf{N}^k)$ (for $k = 1, 2, \ldots, n$) that each acts on $F(\mathbf{Y};\boldsymbol{\Psi})$ by the prescription

$$P(\mathbf{G}(\mathbf{X}+\mathbf{Z};\boldsymbol{\Theta}+\boldsymbol{\Phi})) = P(\mathbf{G}(\mathbf{X};\boldsymbol{\Theta})) + \sum_{j=1}^{m} Z^j P(\mathbf{M}^j) + \sum_{k=1}^{n} \Phi^k P(\mathbf{N}^k), \tag{22.92}$$

where $(\mathbf{Z};\boldsymbol{\Phi})$ is any element of $\mathbb{R}B_L^{m,n}$ close to the zero of $\mathbb{R}B_L^{m,n}$ and second-order terms in $(\mathbf{Z};\boldsymbol{\Phi})$ are neglected. After some algebra, it then follows that

$$P(\mathbf{M}^j) = -(Y^1, Y^2, \ldots, Y^p, \Psi^1, \Psi^2, \ldots, \Psi^q)(\mathbf{M}^j)^{\mathrm{st}} \begin{bmatrix} \partial/\partial Y^1 \\ \partial/\partial Y^2 \\ \vdots \\ \partial/\partial Y^p \\ \partial/\partial \Psi^1 \\ \partial/\partial \Psi^2 \\ \vdots \\ \partial/\partial \Psi^q \end{bmatrix}, \tag{22.93}$$

and

$$P(\mathbf{N}^k) = -(Y^1, Y^2, \ldots, Y^p, \Psi^1, \Psi^2, \ldots, \Psi^q)(\mathbf{N}^k)^{\mathrm{st}} \begin{bmatrix} \partial/\partial Y^1 \\ \partial/\partial Y^2 \\ \vdots \\ \partial/\partial Y^p \\ \partial/\partial \Psi^1 \\ \partial/\partial \Psi^2 \\ \vdots \\ \partial/\partial \Psi^q \end{bmatrix}, \tag{22.94}$$

where $(\mathbf{M}^j)^{st}$ and $(\mathbf{N}^k)^{st}$ are the supertransposes of $\mathbf{M}^j$ and $\mathbf{N}^k$ as defined by (20.57), and $\partial/\partial Y^j$ and $\partial/\partial \Psi^k$ are the Grassmann differential operators (see (22.14)) for functions defined on the space $\mathbb{R}\mathbf{B}_L^{p,q}$. The corresponding operators for the basis elements of the Lie algebra $\mathscr{L}_s(\mathbb{C}\mathbf{B}_L)$ are then given by

$$P(\mathbf{M}_\mu^j) = \varepsilon_\mu P(\mathbf{M}^j) \tag{22.95}$$

and

$$P(\mathbf{N}_\mu^k) = \varepsilon_\mu P(\mathbf{N}^k). \tag{22.96}$$

(b) *The relationship between Lie superalgebras and Lie supergroups*

Theorem II Let $\mathscr{L}_s$ be a real Lie superalgebra of even dimension m and odd dimension n with even basis elements $a_1, a_2, \ldots, a_m$ and odd basis elements $a_{m+1}, a_{m+2}, \ldots, a_{m+n}$, and let $\mathbf{\Gamma}$ be a faithful graded representation of $\mathscr{L}_s$. For $j = 1, 2, \ldots, m$ and for every even basis element ε_μ of $\mathbb{C}\mathbf{B}_L$ define the even supermatrix $\mathbf{M}_\mu^j$ by

$$\mathbf{M}_\mu^j = \varepsilon_\mu \mathbf{\Gamma}(a_j), \tag{22.97}$$

and for $k = 1, 2, \ldots, n$ and for every odd basis element ε_μ of $\mathbb{C}\mathbf{B}_L$ define the even supermatrix $\mathbf{N}_\mu^k$ by

$$\mathbf{N}_\mu^k = \varepsilon_\mu \mathbf{\Gamma}(a_{m+k}). \tag{22.98}$$

Then the set of $(m+n)2^{L-1}$ supermatrices defined in (22.90) and (22.91) form the basis of a real Lie algebra of dimension $(m+n)2^{L-1}$, for which the Lie product is just the matrix commutator with supermatrix multiplication. This real Lie algebra will be denoted by $\mathscr{L}_s(\mathbb{C}\mathbf{B}_L)$.

Proof All that has to be shown is that the commutator of any pair of supermatrices of the sets (22.90) and (22.91) is a real linear combination of the supermatrices of (22.90) and (22.91). For example, if $\mathbf{N}_\mu^k$ is given by (22.91) and $\mathbf{N}_\mu^k = \varepsilon_{\mu'} \mathbf{\Gamma}(a_{m+k'})$, with $\varepsilon_{\mu'}$ odd, then

$$[\mathbf{N}_\mu^k, \mathbf{N}_{\mu'}^{k'}] = \varepsilon_\mu \mathbf{\Gamma}(a_{m+k}) \varepsilon_{\mu'} \mathbf{\Gamma}(a_{m+k'}) - \varepsilon_{\mu'} \mathbf{\Gamma}(a_{m+k'}) \varepsilon_\mu \mathbf{\Gamma}(a_{m+k}).$$

But, by (20.52) and (21.53), $\mathbf{\Gamma}(a_{m+k})\varepsilon_{\mu'} = -\varepsilon_{\mu'}\mathbf{\Gamma}(a_{m+k})$ and $\mathbf{\Gamma}(a_{m+k'})\varepsilon_\mu = -\varepsilon_\mu \mathbf{\Gamma}(a_{m+k'})$, and, since $\varepsilon_{\mu'}\varepsilon_\mu = -\varepsilon_\mu \varepsilon_{\mu'}$ (because both ε_μ and $\varepsilon_{\mu'}$ are odd), it follows that

$$[\mathbf{N}_\mu^k, \mathbf{M}_{\mu'}^{k'}] = -\varepsilon_\mu \varepsilon_{\mu'} \{\mathbf{\Gamma}(a_{m+k})\mathbf{\Gamma}(a_{m+k'}) + \mathbf{\Gamma}(a_{m+k'})\mathbf{\Gamma}(a_{m+k})\},$$

where the *anti*commutator appears naturally on the right-hand side. As this anticommutator is a real linear combination of the even matrices $\mathbf{\Gamma}(a_j)$ (where $j = 1, 2, \ldots, m$), the desired result is obtained. The proofs for the other types of pairs are similar.

Theorem III If $\mathscr{L}_s(\mathbb{C}B_L)$ is the real Lie algebra whose basis elements are defined by (22.90) and (22.91) then every linear Lie group whose real Lie algebra is $\mathscr{L}_s(\mathbb{C}B_L)$ is a linear Lie supergroup of even dimension m and odd dimension n.

Proof See Rogers (1981), whose proof is valid in the non-linear case as well.

As every real Lie algebra corresponds to at least one linear group (cf. Theorem IV of Chapter 10, Section 5(*b*)), Theorems II and III taken together show that *to every real Lie superalgebra $\mathscr{L}_s$ there corresponds at least one linear Lie supergroup $\mathscr{G}_s(\mathbb{C}B_L)$*.

Moreover, the considerations of Chapter 11, Section 7, imply that to a given Lie superalgebra there is an associated family of Lie supergroups, which are all isomorphic to factor groups of a "universal covering Lie supergroup" $\tilde{\mathscr{G}}_s(\mathbb{C}B_L)$ with respect to its discrete central invariant subgroups $\mathscr{K}$.

The most general element of the real Lie algebra $\mathscr{L}_s(\mathbb{C}B_L)$ defined in Theorem II has the form

$$\mathbf{M} = \sum_{j=1}^{m} \sum_{\text{even } \mu} X^j_\mu \mathbf{M}^\mu_j + \sum_{k=1}^{n} \sum_{\text{odd } \mu} \Theta^k_\mu \mathbf{N}^\mu_k, \tag{22.99}$$

where X^j_μ and Θ^k_μ are *real* parameters. Again this can be expressed more concisely by taking the real parameters X^j_μ and Θ^k_μ of (22.75) as the components of the coordinates of a point $(\mathbf{X};\mathbf{\Theta}) = (X^1, X^2, \ldots, X^m, \Theta^1, \Theta^2, \ldots, \Theta^n)$ of $\mathbb{R}B_L^{m,n}$ by using them in the expansions (22.3) and (22.4). The resulting expression for the general element $\mathbf{M}$ of $\mathscr{L}_s(\mathbb{C}B_L)$ is then

$$\mathbf{M} = \sum_{j=1}^{m} X^j \mathbf{M}^j + \sum_{k=1}^{n} \Theta^k \mathbf{N}^k, \tag{22.100}$$

where $\mathbf{M}^j$ and $\mathbf{N}^k$ are *now* defined by

$$\mathbf{M}^j = \mathcal{E}_\varnothing \mathbf{\Gamma}(a_j) \tag{22.101}$$

and

$$\mathbf{N}^k = \mathcal{E}_\varnothing \mathbf{\Gamma}(a_{k+m}). \tag{22.102}$$

(This follows since $\mathbf{M}^j_\mu = \mathcal{E}_\mu \mathbf{M}^j$ and $\mathbf{N}^k_\mu = \mathcal{E}_\mu \mathbf{N}^k$.) Because of the form of (22.93), Berezin (1977) has called $\mathscr{L}_s(\mathbb{C}B_L)$ the *Grassmann manifold* of $\mathscr{L}_s$.

If $\mathscr{G}_s(\mathbb{C}B_L)$ is a Lie supergroup whose real Lie algebra is $\mathscr{L}_s(\mathbb{C}B_L)$, its elements near its identity can be parametrized by

$$\mathbf{G}(\mathbf{X};\mathbf{\Theta}) = \exp \mathbf{M} = \exp\left\{ \sum_{j=1}^{m} X^j \mathbf{M}^j + \sum_{k=1}^{n} \Theta^k \mathbf{N}^k \right\}. \tag{22.103}$$

Applying (22.76) and (22.77) to (22.103) shows that the supermatrices $\mathbf{M}^j$ and $\mathbf{N}^k$ as defined in (22.101) and (22.102) are the supergenerators of $\mathscr{G}_s(\mathbb{C}\mathbf{B}_L)$ (which justifies the use here of the same notation as that employed in the previous subsection).

The next theorem shows that the real Lie algebra $\mathscr{L}_s(\mathbb{C}\mathbf{B}_L)$ has an interesting structure.

Theorem IV Let $\mathscr{L}_s$ be a real Lie superalgebra of even dimension m and odd dimension n whose even part is the real Lie algebra $\mathscr{L}_0$, and let $a_1, a_2, \ldots, a_m$ be the basis elements of $\mathscr{L}_0$. Let $\mathscr{L}_s(\mathbb{C}\mathbf{B}_L)$ be the real Lie algebra whose basis elements $\mathbf{M}^j_\mu$ and $\mathbf{N}^k_\mu$ are defined by (22.97) and (22.98). Then we have the following.

(a) The subset of elements $\varepsilon_\varnothing\boldsymbol{\Gamma}(a_j)$ of $\mathscr{L}_s(\mathbb{C}B_L)$ for $j=1,2,\ldots,m$ form a real Lie algebra that is *isomorphic to* $\mathscr{L}_0$.

(b) For each value of the integer N such that $0 \leqslant N \leqslant L$ the subset of $\mathscr{L}_s(\mathbb{C}\mathbf{B}_L)$ consisting of all basis elements $\mathbf{M}^j_\mu$ and $\mathbf{N}^k_\mu$ of (22.97) and (22.98) for which the Grassmann basis elements ε_μ have level greater or equal to N forms an *invariant subalgebra* of $\mathscr{L}_s(\mathbb{C}\mathbf{B}_L)$. If $N \geqslant 1$, this Lie subalgebra is a *proper* invariant subalgebra of $\mathscr{L}_s(\mathbb{C}\mathbf{B}_L)$; and if $N > \frac{1}{2}L$, this subalgebra is also *Abelian*.

(c) The Lie algebra $\mathscr{L}_s(\mathbb{C}\mathbf{B}_L)$ is *not semi-simple*.

Proof The assertion (a) is obvious. The first statement of part (b) follows because if ε_μ has level $N(\mu)$ and $\varepsilon_{\mu'}$ has level $N(\mu')$ then $\varepsilon_\mu\varepsilon_{\mu'}$ is either zero or has level $N(\mu)+N(\mu')$. Thus if $N \leqslant N(\mu)$ and $N \leqslant N(\mu')$, so that $N \leqslant N(\mu)+N(\mu')$ as well, the subset of basis elements $\mathbf{M}^j_\mu$ and $\mathbf{N}^k_{\mu'}$ of (22.97) and (22.98) defined in part (b) form a closed set under the commutator Lie product, and hence they form the basis of a Lie subalgebra of $\mathscr{L}_s(\mathbb{C}\mathbf{B}_L)$. Clearly this is an invariant subalgebra of $\mathscr{L}_s(\mathbb{C}\mathbf{B}_L)$. If $N \geqslant 1$, the set of elements $\varepsilon_\varnothing\boldsymbol{\Gamma}(a_j)$ of part (a) are not included in this subalgebra, so the invariant subalgebra is proper. Finally, if $N(\mu) > \frac{1}{2}L$ and $N(\mu') > \frac{1}{2}L$ then the index sets μ and μ' must have at least one element in common, so that $\varepsilon_\mu\varepsilon_{\mu'}=0$, and hence the invariant subalgebra must be Abelian. $\mathscr{L}_s(\mathbb{C}\mathbf{B}_L)$ therefore does not satisfy the definition of "semi-simplicity" given in Chapter 13, Section 2.

Although Theorems II and III provide a procedure by which every real Lie superalgebra $\mathscr{L}_s$ gives rise to a real Lie algebra $\mathscr{L}_s(\mathbb{C}\mathbf{B}_L)$ and thence to at least one Lie supergroup $\mathscr{G}_s(\mathbb{C}\mathbf{B}_L)$, there is no procedure that holds in the reverse direction. This is illustrated by the following example (which was first produced by Rogers (1981)).

Example III *A Lie supergroup with no corresponding Lie superalgebra*

Consider the Lie supergroup with $m = n = 1$ that consists of the even $(2/1) \times (2/1)$ supermatrices

$$\mathbf{G}(X^1, \Theta^1) = \begin{bmatrix} \mathcal{E}_{\varnothing} & X^1 & \frac{1}{2}\mathcal{E}_{\{1,2\}}\Theta^1 \\ 0 & \mathcal{E}_{\varnothing} & 0 \\ 0 & \Theta^1 & \mathcal{E}_{\varnothing} \end{bmatrix}.$$

Then, by (22.76) and (22.77),

$$\mathbf{M}^1 = \frac{\partial \mathbf{G}(X^1, \Theta^1)}{\partial X^1} = \begin{bmatrix} 0 & \mathcal{E}_{\varnothing} & 0 \\ 0 & 0 & 0 \\ 0 & 0 & 0 \end{bmatrix}$$

and

$$\mathbf{N}^1 = \frac{\partial \mathbf{G}(X^1, \Theta^1)}{\partial \Theta^1} = \begin{bmatrix} 0 & 0 & \frac{1}{2}\mathcal{E}_{\{1,2\}} \\ 0 & 0 & 0 \\ 0 & -\mathcal{E}_{\varnothing} & 0 \end{bmatrix}.$$

As $[\mathbf{M}^1, \mathbf{M}^1]_- = \mathbf{0}$, $[\mathbf{M}^1, \mathbf{N}^1]_- = \mathbf{0}$ and $[\mathbf{N}^1, \mathbf{N}^1]_+ = -\mathcal{E}_{\{1,2\}}\mathbf{M}^1$, the supergenerators $\mathbf{M}^1$ and $\mathbf{N}^1$ do *not* form a Lie superalgebra, although, in accordance with Theorem I, the supermatrices $\mathcal{E}_\mu \mathbf{M}^1$ (with $\mathcal{E}_\mu$ even) and $\mathcal{E}_\mu \mathbf{N}^1$ (with $\mathcal{E}_\mu$ odd) do form a real Lie algebra.

23

The Poincaré Superalgebras and Supergroups

1 Introduction

As the complete Poincaré group of D space–time dimensions is the group of all transformations that leave invariant the metric of D-dimensional Minkowski space–time, it is not surprising that most of the enormous literature on supersymmetry in theoretical physics journals is concerned with the Poincaré superalgebras and supergroups. Indeed this is so much so that the adjective "supersymmetric" is very frequently used on its own and without further qualification to indicate invariance with respect to a Poincaré superalgebra or supergroup.

This chapter will be devoted to the construction of the Poincaré superalgebras and supergroups for all non-trivial values of D, and to the study of their graded irreducible representations. The results will be applied in the next chapter to the setting of supersymmetric models for particles, including supersymmetric gauge theories.

The discussion will start in Section 2 with an examination of the most straightforward case, namely the "minimal" superalgebra that contains the Poincaré Lie algebra of 4 space–time dimensions. For brevity this is often referred to as the "$N = 1$, $D = 4$ Poincaré superalgebra". In Section 2(a) this is constructed in the general "4-component formalism", and the relation of this to the "2-component formalism" is described in Section 2(b). The construction of the corresponding Lie supergroup and its action on superspace form the subject of Sections 2(c) and 2(d). In Section 3 these ideas are developed to produce the "N-extended" Poincaré superalgebras and supergroups for 4-dimensional space–times, and finally in Section 4 all of these constructions are reexamined for space–times of arbitrary dimension D.

The representations of all these Poincaré superalgebras are investigated in the remaining sections, starting in Section 5 with the $N=1$, $D=4$ case. This is generalized first to the N-extended $D=4$ situation in Section 6, and then to arbitrary space–time dimensions in Section 7.

The Poincaré superalgebras are not the only Lie superalgebras that are associated with space–time transformations, since extensions of the conformal, de Sitter and anti-de Sitter algebras also exist. These are described in Chapter 25, Section 6.

2 The $N=1$, $D=4$ Poincaré superalgebra and supergroup

(a) *The Lie superalgebra extension of the Poincaré algebra*

This subsection will be devoted to setting up the smallest Lie superalgebra that contains the Poincaré algebra for 4-dimensional space–time as its even part. This Poincaré algebra was discussed in Chapter 17, and the conventions and notation employed in that earlier treatment will again be used here, with the one exception that the components of the contravariant and covariant Minkowski space–time metric tensors will henceforth be denoted by $\eta^{\alpha\beta}$ and $\eta_{\alpha\beta}$ respectively. (The object of this change is to allow in a later chapter for differences to be emphasized between flat and curved space–times, so that from now on $g^{\alpha\beta}$ and $g_{\alpha\beta}$ will denote the components of the contravariant and covariant metric tensors of a *curved* space–time.) As many different combinations of conventions appear in the literature, it is necessary to devote a fair amount of space to presenting in detail the choices that will be adopted in the rest of this book.

With the notational change just mentioned, (17.3) and (17.99) become

$$\eta^{\alpha\beta} = \eta_{\alpha\beta} = \begin{cases} -1 & \text{if } \alpha = \beta = 1,2,3, \\ 1 & \text{if } \alpha = \beta = 4, \\ 0 & \text{if } \alpha \neq \beta, \end{cases} \tag{23.1}$$

and the four contravariant 4×4 Dirac matrices γ^α satisfy the conditions

$$\gamma^\alpha \gamma^\beta + \gamma^\beta \gamma^\beta = 2\eta^{\alpha\beta} \mathbf{1}_4 \tag{23.2}$$

for $\alpha, \beta = 1, \ldots, 4$ (cf. (17.98)), while their associated covariant Dirac matrices γ_α, defined by

$$\gamma_\alpha = \sum_{\beta=1}^{4} \eta_{\alpha\beta} \gamma^\beta, \tag{23.3}$$

satisfy the covariant conditions

$$\gamma_\alpha \gamma_\beta + \gamma_\beta \gamma_\alpha = 2\eta_{\alpha\beta}\mathbf{1}_4 \tag{23.4}$$

for α, $\beta = 1,\ldots,4$ (cf. (17.111)). Of course (23.3) implies that

$$\gamma^\alpha = \sum_{\beta=1}^{4} \eta^{\alpha\beta}\gamma_\beta \tag{23.5}$$

(for $p = 1,\ldots,4$). The matrices γ^5 and γ_5, are also required; they are defined by

$$\gamma^5 = i\gamma^1\gamma^2\gamma^3\gamma^4 \tag{23.6}$$

and

$$\gamma_5 = i\gamma_1\gamma_2\gamma_3\gamma_4 \tag{23.7}$$

(as in (17.102) and (17.111)). It is easily shown that

$$\gamma_\alpha\gamma_5 + \gamma_5\gamma_\alpha = \mathbf{0} \tag{23.8}$$

and

$$\gamma^\alpha\gamma^5 + \gamma^5\gamma^\alpha = \mathbf{0} \tag{23.9}$$

(for $\alpha = 1,\ldots,4$), and that

$$(\gamma^5)^2 = \mathbf{1}_4, \quad (\gamma_5)^2 = \mathbf{1}_4. \tag{23.10}$$

If γ'_p (for $p = 1,\ldots,5$) are a set of five 4×4 matrices defined in terms of the γ_p by

$$\gamma'_p = \mathbf{S}\gamma_p\mathbf{S}^{-1}, \tag{23.11}$$

where $\mathbf{S}$ is any non-singular 4×4 matrix (cf. (17.108)), then these matrices also satisfy (23.4), (23.7), (23.8) and (23.10). That is, the Dirac matrices are only unique up to a similarity transformation. It is obvious that if the matrices γ'^α are defined in terms of the γ'_α by

$$\gamma'^\alpha = \sum_{\beta=1}^{4} \eta^{\alpha\beta}\gamma'_\beta \tag{23.12}$$

for $\alpha = 1,\ldots,4$, if γ'^5 is defined by $\gamma'^5 = i\gamma'^1\gamma'^2\gamma'^3\gamma'^4$, and the γ^α and γ_α for $\alpha = 1,\ldots,4$ are related by (23.5), then

$$\gamma'^p = \mathbf{S}\gamma^p\mathbf{S}^{-1} \tag{23.13}$$

(for $p = 1,\ldots,5$). The behaviour of quantities under similarity transformations of the form (23.11) will play a significant role in the analysis that follows.

The *Majorana* representation provides one very useful choice of the Dirac matrices. The matrices of this representation are indicated by attaching a superscript or subscript "M," and may be taken to be

$$\left.\begin{aligned}
\gamma_1^{\mathrm{M}} &= -\gamma_{\mathrm{M}}^1 = \begin{bmatrix} -i\boldsymbol{\sigma}_3 & \mathbf{0} \\ \mathbf{0} & -i\boldsymbol{\sigma}_3 \end{bmatrix}, \\
\gamma_2^{\mathrm{M}} &= -\gamma_{\mathrm{M}}^2 = \begin{bmatrix} \mathbf{0} & \boldsymbol{\sigma}_2 \\ -\boldsymbol{\sigma}_2 & \mathbf{0} \end{bmatrix}, \\
\gamma_3^{\mathrm{M}} &= -\gamma_{\mathrm{M}}^3 = \begin{bmatrix} i\boldsymbol{\sigma}_1 & \mathbf{0} \\ \mathbf{0} & i\boldsymbol{\sigma}_1 \end{bmatrix}, \\
\gamma_4^{\mathrm{M}} &= \gamma_{\mathrm{M}}^4 = \begin{bmatrix} \mathbf{0} & -\boldsymbol{\sigma}_2 \\ -\boldsymbol{\sigma}_2 & \mathbf{0} \end{bmatrix}, \\
\gamma_5^{\mathrm{M}} &= -\gamma_{\mathrm{M}}^5 = \begin{bmatrix} -\boldsymbol{\sigma}_2 & \mathbf{0} \\ \mathbf{0} & \boldsymbol{\sigma}_2 \end{bmatrix}.
\end{aligned}\right\} \tag{23.14}$$

In (23.14) $\boldsymbol{\sigma}_1, \boldsymbol{\sigma}_2$ and $\boldsymbol{\sigma}_3$ are the usual set of three 2×2 Pauli spin matrices, i.e.

$$\boldsymbol{\sigma}_1 = \begin{bmatrix} 0 & 1 \\ 1 & 0 \end{bmatrix}, \quad \boldsymbol{\sigma}_2 = \begin{bmatrix} 0 & -i \\ i & 0 \end{bmatrix}, \quad \boldsymbol{\sigma}_3 = \begin{bmatrix} 1 & 0 \\ 0 & -1 \end{bmatrix} \tag{23.15}$$

(cf. (3.24)), to which it is convenient to add a fourth by defining

$$\boldsymbol{\sigma}_4 = \mathbf{1}_2. \tag{23.16}$$

It should be noted that all the matrices of the Majorana representation are purely *imaginary*. Of course any similarity transformation with a *real orthogonal* matrix applied to the above Majorana representation will produce an equivalent representation of the Dirac matrices with the *same* reality properties, which could equally well be called a Majorana representation. Nevertheless it is convenient to refer to the representation displayed above in (23.14) as *the* Majorana representation.

Also essential for the analysis that follows is the 4×4 "charge conjugation matrix", whose precise expression varies from representation to representation. The general charge conjugation matrix, denoted by $\mathbf{C}$, has the property that

$$\mathbf{C}^{-1}\gamma_p\mathbf{C} = -\tilde{\gamma}_p \tag{23.17}$$

for $p = 1, \ldots, 5$ (cf. (17.152)), and also

$$\tilde{\mathbf{C}} = -\mathbf{C} \tag{23.18}$$

and

$$\mathbf{C}^\dagger\mathbf{C} = \mathbf{C}\mathbf{C}^\dagger = \mathbf{1}_4 \tag{23.19}$$

(cf. (17.153) and (17.154)). These constraints determine $\mathbf{C}$ in any representation up to an arbitrary factor of the form $e^{i\mu}$, where μ is real. As in (17.167), in the Majorana representation, where the charge conjugation matrix will be denoted by $\mathbf{C}^{\mathrm{M}}$, the choice will be made that

$$\mathbf{C}^{\mathrm{M}} = -\boldsymbol{\gamma}_4^{\mathrm{M}}. \tag{23.20}$$

As $(\boldsymbol{\gamma}_4^{\mathrm{M}})^2 = \mathbf{1}_4$, this implies that

$$-\boldsymbol{\gamma}_4^{\mathrm{M}}\mathbf{C}^{\mathrm{M}} = \mathbf{1}_4. \tag{23.21}$$

With this choice, $\mathbf{C}^{\mathrm{M}}$ is also purely imaginary. Under the similarity transformation (23.11), if $\mathbf{C}'$ is the charge conjugation matrix of the transformed representation, so that

$$\mathbf{C}'^{-1}\boldsymbol{\gamma}'_p\mathbf{C}' = -\tilde{\boldsymbol{\gamma}}'_p \tag{23.22}$$

for $p = 1, \ldots, 5$ (cf. (23.17)), the corresponding transformation law for the charge conjugation matrices is

$$\mathbf{C}' = \mathbf{S}\mathbf{C}\tilde{\mathbf{S}} \tag{23.23}$$

(cf (17.172), which, it should be noted, is *not* itself of the form of a similarity transformation.

Any representation of the Dirac matrices in which the matrix $\boldsymbol{\gamma}_5$ is *diagonal* (and so, by (23.10), has diagonal entries ± 1) is known as a *chiral* representation. The particular choice of chiral representation made in (17.118), where the Dirac matrices were denoted by $\boldsymbol{\gamma}_p^{\mathrm{C}}$ and $\boldsymbol{\gamma}_{\mathrm{C}}^{p}$, was convenient for establishing the connection between the Dirac matrices and the irreducible representations $\boldsymbol{\Gamma}^{0,1/2}$ and $\boldsymbol{\Gamma}^{1/2,0}$ of the homogeneous Lorentz algebra in their "standard" forms of (17.21) and (17.22), but a slightly different choice produces neater expressions when used in the Poincaré superalgebra. This is obtained from the chiral representation of (17.118) by applying a similarity transformation with

$$\mathbf{S} = \begin{bmatrix} \mathbf{1}_2 & \mathbf{0} \\ \mathbf{0} & \boldsymbol{\sigma}_2 \end{bmatrix}. \tag{23.24}$$

Calling this representation the *modified chiral* representation, and indicating it by attaching a superscript or subscript C′ as appropriate, it follows from (17.118) that

$$\left.\begin{aligned} \boldsymbol{\gamma}_p^{\mathrm{C}'} &= -\boldsymbol{\gamma}_{\mathrm{C}'}^{p} = \begin{bmatrix} \mathbf{0} & -\boldsymbol{\sigma}_p\boldsymbol{\sigma}_2 \\ \boldsymbol{\sigma}_2\boldsymbol{\sigma}_p & \mathbf{0} \end{bmatrix} \quad \text{for } p = 1, 2, 3, \\ \boldsymbol{\gamma}_4^{\mathrm{C}'} &= \boldsymbol{\gamma}_{\mathrm{C}'}^{4} = \begin{bmatrix} \mathbf{0} & -\boldsymbol{\sigma}_2 \\ -\boldsymbol{\sigma}_2 & \mathbf{0} \end{bmatrix}, \\ \boldsymbol{\gamma}_5^{\mathrm{C}'} &= -\boldsymbol{\gamma}_{\mathrm{C}'}^{5} = \begin{bmatrix} \mathbf{1}_2 & \mathbf{0} \\ \mathbf{0} & -\mathbf{1}_2 \end{bmatrix}, \end{aligned}\right\} \tag{23.25}$$

the corresponding charge conjugation matrix deduced from (17.172) using (23.23) and (23.24) being

$$\mathbf{C}^{C'} = \begin{bmatrix} \boldsymbol{\sigma}_2 & 0 \\ 0 & \boldsymbol{\sigma}_2 \end{bmatrix}. \tag{23.26}$$

These imply the neat and useful result that

$$-\boldsymbol{\gamma}_4^{C'}\mathbf{C}^{C'} = \begin{bmatrix} \mathbf{0} & \mathbf{1} \\ \mathbf{1} & \mathbf{0} \end{bmatrix}. \tag{23.27}$$

Each representation of the Dirac matrices gives rise to a 4-dimensional "spinor" representation $\boldsymbol{\Gamma}^{\text{spin}}$ of the homogeneous Lorentz algebra by the prescription (17.134), that is, by

$$\boldsymbol{\Gamma}^{\text{spin}}(L_{\alpha\beta}) = \begin{cases} -\frac{1}{2}\boldsymbol{\gamma}_\alpha\boldsymbol{\gamma}_\beta & \text{if } \alpha \neq \beta, \\ \mathbf{0} & \text{if } \alpha = \beta, \end{cases} \tag{23.28}$$

for α, $\beta = 1, \ldots, 4$. Here the basis elements $L_{\alpha\beta}$ of the homogeneous Lorentz algebra are assumed to be such that

$$L_{\alpha\beta} = -L_{\beta\alpha} \tag{23.29}$$

(cf. (17.11)), so that $L_{11} = L_{22} = L_{33} = L_{44} = 0$, and are assumed to satisfy the commutation relations

$$[L_{\alpha\beta}, L_{\lambda\mu}]_- = \eta_{\alpha\lambda}L_{\beta\mu} - \eta_{\alpha\mu}L_{\beta\lambda} - \eta_{\beta\lambda}L_{\alpha\mu} + \eta_{\beta\mu}L_{\alpha\lambda} \tag{23.30}$$

(cf. (17.13)). Of course the different spinor representations $\boldsymbol{\Gamma}^{\text{spin}}$ corresponding to different representations of the Dirac matrices are equivalent. In particular, if $\boldsymbol{\Gamma}_{C'}^{\text{spin}}$ denotes the 4-dimensional spinor representation of the homogeneous Lorentz algebra corresponding to the modified chiral representation of the Dirac matrices then (17.137) and (17.138) can be rewritten as

$$\boldsymbol{\Gamma}_{C'}^{\text{spin}}(L_{\alpha\beta}) = \begin{bmatrix} \boldsymbol{\Gamma}_L(L_{\alpha\beta}) & \mathbf{0} \\ \mathbf{0} & \boldsymbol{\Gamma}_R(L_{\alpha\beta}) \end{bmatrix}, \tag{23.31}$$

where $\boldsymbol{\Gamma}_L(L_{\alpha\beta})$ and $\boldsymbol{\Gamma}_R(L_{\alpha\beta})$ are the 2-dimensional irreducible representations of the homogeneous Lorentz algebra corresponding to $j = 0, j' = \frac{1}{2}$ and to $j = \frac{1}{2}$, $j' = 0$ respectively. To be precise, because the modified chiral representation of (23.25) has been obtained from the chiral representation of (17.118) using the similarity transformation matrix (23.24),

$$\boldsymbol{\Gamma}_L(L_{\alpha\beta}) = \boldsymbol{\Gamma}^{0,1/2}(L_{\alpha\beta}) \tag{23.32}$$

but

$$\boldsymbol{\Gamma}_R(L_{\alpha\beta}) = \boldsymbol{\sigma}_2\boldsymbol{\Gamma}^{1/2,0}(L_{\alpha\beta})(\boldsymbol{\sigma}_2)^{-1}. \tag{23.33}$$

Then, by (23.28) and (23.25), for $\alpha > \beta$

$$\Gamma_{\mathrm{L}}(L_{\alpha\beta}) = \tfrac{1}{2}\boldsymbol{\sigma}_\alpha \boldsymbol{\sigma}_\beta \tag{23.34}$$

and

$$\Gamma_{\mathrm{R}}(L_{\alpha\beta}) = \tfrac{1}{2}(\boldsymbol{\sigma}_\alpha \boldsymbol{\sigma}_\beta)^* \tag{23.35}$$

(the matrices for $\alpha < \beta$ being given by $\Gamma_{\mathrm{L}}(L_{\alpha\beta}) = -\Gamma_{\mathrm{L}}(L_{\beta\alpha})$ and $\Gamma_{\mathrm{R}}(L_{\alpha\beta}) = -\Gamma_{\mathrm{R}}(L_{\beta\alpha})$). In deriving (23.35), use has been made of the Pauli spin matrix reality properties:

$$\boldsymbol{\sigma}_2 \boldsymbol{\sigma}_\alpha (\boldsymbol{\sigma}_2)^{-1} = \begin{cases} -\boldsymbol{\sigma}_\alpha^* & \text{if } \alpha = 1, 2, 3, \\ \boldsymbol{\sigma}_\alpha^* & \text{if } \alpha = 4. \end{cases} \tag{23.36}$$

These two representations of the homogeneous Lorentz algebra may be referred to as the *left-handed* and *right-handed* representations because, as noted in Chapter 17, Section 5, they correspond to the left-handed and right-handed parts of a 4-component spinor. They are connected by the relations

$$\Gamma_{\mathrm{R}}(L_{\alpha\beta}) = \Gamma_{\mathrm{L}}(L_{\alpha\beta})^* \tag{23.37}$$

and

$$\Gamma_{\mathrm{R}}(L_{\alpha\beta}) = -\boldsymbol{\sigma}_2 \Gamma_{\mathrm{L}}(L_{\alpha\beta})^\dagger (\boldsymbol{\sigma}_2)^{-1}, \tag{23.38}$$

for $\alpha, \beta = 1, \ldots, 4$.

The commutation relations of the Poincaré algebra of 4-dimensional space–time are

$$[M_{\alpha\beta}, M_{\lambda\mu}]_- = \frac{\hbar}{i}(\eta_{\alpha\lambda} M_{\beta\mu} - \eta_{\alpha\mu} M_{\beta\lambda} - \eta_{\beta\lambda} M_{\alpha\mu} + \eta_{\beta\mu} M_{\alpha\lambda}), \tag{23.39}$$

$$[M_{\alpha\beta}, P_\mu]_- = \frac{\hbar}{i}(\eta_{\alpha\mu} P_\beta - \eta_{\beta\mu} P_\alpha), \tag{23.40}$$

$$[P_\lambda, P_\mu]_- = 0, \tag{23.41}$$

where the P_μ (for $\mu = 1, \ldots, 4$) are the operators corresponding to covariant linear 4-momentum and the $M_{\alpha\beta}$ are generalized angular momentum operators that are related to the $L_{\alpha\beta}$ of (23.30) by

$$M_{\alpha\beta} = \frac{\hbar}{i} L_{\alpha\beta} \tag{23.42}$$

for $\alpha, \beta = 1, \ldots, 4$. Expressed in this form, the Poincaré algebra is a *complex* Lie algebra. The *real* Lie algebra that has the proper orthochronous Poincaré group and its covering group as its Lie groups has basis elements $(i/\hbar)M_{\alpha\beta}$ and iP_μ. When acting on a Hilbert space the operators P_μ are self-adjoint.

The smallest Lie superalgebra that contains the Poincaré algebra as its even part has four odd basis elements, which may be denoted Q_1, Q_2, Q_3 and Q_4, and which may be chosen so that

$$[M_{\alpha\beta}, Q_a]_- = \frac{1}{2}\frac{\hbar}{i}\sum_{b=1}^{4}(\gamma_\alpha\gamma_\beta)_{ab}Q_b \tag{23.43}$$

(for a, α, $\beta = 1,\dots,4$, with $\alpha \neq \beta$),

$$[P_\mu, Q_a]_- = 0 \tag{23.44}$$

(for $a, \mu = 1,\dots,4$) and

$$[Q_a, Q_b]_+ = -2\sum_{\mu=1}^{4}(\gamma^\mu \mathbf{C})_{ab}P_\mu \tag{23.45}$$

(for $a, b = 1,\dots,4$). These odd basis elements Q_a will be called *supertranslation generators*, but a variety of other names, such as "supercharges", "supersymmetry generators" and "spinor charges", also appear in the literature.

These commutation and anticommutation relations have been expressed in terms of an arbitrary representation of the Dirac matrices γ_μ and the corresponding charge conjugation matrix $\mathbf{C}$. If a similarity transformation of the form (23.11) is applied to the covariant Dirac matrices γ_μ, and the corresponding odd elements Q'_a are *defined* by

$$Q'_a = \sum_{b=1}^{4} S_{ab}Q_b, \tag{23.46}$$

then it is easily shown using (23.12) and (23.23) that when expressed in terms of the quantities γ'_μ, $\mathbf{C}'$ and Q'_a the equations (23.43)–(23.45) have exactly the same form. That is, they are *invariant* under similarity transformations. In particular they are equally valid in the Majorana and modified chiral representations, for which it is appropriate to write the odd basis elements as Q_a^{M} and $Q_a^{\mathrm{C}'}$ (for $a = 1,\dots,4$).

The derivation of the commutation and anticommutation relations (23.43)–(23.45) is as follows. As shown in Chapter 21, Section 2, the odd basis elements of a Lie superalgebra have to form the basis of a representation of the even part of the superalgebra. This representation, which may again be denoted by $\mathbf{D}$, can be chosen so that

$$\mathbf{D}(M_{\alpha\beta}) = -\tilde{\mathbf{\Gamma}}^{\mathrm{spin}}(M_{\alpha\beta}) \tag{23.47}$$

for $\alpha, \beta = 1,\dots,4$, and

$$\mathbf{D}(P_\mu) = \mathbf{0} \tag{23.48}$$

for $\mu = 1,\dots,4$, where $\mathbf{\Gamma}^{\mathrm{spin}}$ is the 4-dimensional spinor representation of the homogeneous Lorentz algebra defined by (23.28), to which the representation

$-\tilde{\Gamma}^{spin}$ is equivalent. Then, by (23.28) and (23.42),

$$\mathbf{D}(M_{\alpha\beta})_{ba} = \frac{1}{2}\frac{\hbar}{i}(\gamma_\alpha\gamma_\beta)_{ab}$$

for $\alpha \neq \beta$. It then remains only to determine an expression for the anticommutators $[Q_a, Q_b]_+$, which must lie in the Poincaré algebra, subject to the Jacobi identity constraints (21.14) and (21.15). However, the constraint (21.14) for the case $a = P_\mu$, $b = Q_a$ and $c = Q_b$, when taken with (23.44), implies that

$$[P_\mu, [Q_a, Q_b]_+]_- = 0,$$

and application of (23.40) and (23.41) then shows that $[Q_a, Q_b]_+$ must be a linear combination of the linear momentum basis elements P_1, P_2, P_3, and P_4 *alone*, that is,

$$[Q_a, Q_b]_+ = \sum_{\mu=1}^{4} (\mathbf{F}^\mu)_{ab} P_\mu \tag{23.49}$$

for $a, b = 1, \ldots, 4$, where $\mathbf{F}^1$, $\mathbf{F}^2$, $\mathbf{F}^3$ and $\mathbf{F}^4$ are four symmetric 4×4 matrices. The constraint (21.15) is trivially satisfied by virtue of (23.44), so only (21.14) for the case $a = M_{\alpha\beta}$, $b = Q_a$ and $c = Q_b$ needs to be investigated. By (23.49), (23.43) and (23.40), this can be reduced after some algebra to the condition that

$$2\sum_{\lambda=1}^{4} \mathbf{F}^\lambda(\eta_{\alpha\lambda}\delta^\mu_\beta - \eta_{\beta\lambda}\delta^\mu_\alpha) = \mathbf{F}^\mu(\tilde{\gamma}_\beta\tilde{\gamma}_\alpha) + (\gamma_\alpha\gamma_\beta)\mathbf{F}^\mu$$

for $\alpha \neq \beta$. But, by (23.17),

$$\tilde{\gamma}_\beta\tilde{\gamma}_\alpha = \mathbf{C}^{-1}\gamma_\beta\gamma_\alpha\mathbf{C},$$

so this condition reduces to

$$2\sum_{\lambda=1}^{4} (\mathbf{F}^\lambda\mathbf{C}^{-1})(\eta_{\alpha\lambda}\delta^\mu_\beta - \eta_{\beta\lambda}\delta^\mu_\alpha) = (\mathbf{F}^\mu\mathbf{C}^{-1})(\gamma_\beta\gamma_\alpha) + (\gamma_\alpha\gamma_\beta)(\mathbf{F}^\mu\mathbf{C}^{-1})$$

for $\alpha \neq \beta$. This is satisfied with

$$\mathbf{F}^\mu\mathbf{C}^{-1} = \kappa\gamma^\mu \tag{23.50}$$

for $\mu = 1, \ldots, 4$, where κ is *any* constant, the resulting matrices $\mathbf{F}^\mu$ being symmetric as required. Thus, by (23.49),

$$[Q_a, Q_b]_+ = \kappa\sum_{\mu=1}^{4} (\gamma^\mu\mathbf{C})_{ab} P_\mu. \tag{23.51}$$

All that remains is to make an appropriate choice for κ. To this end, consider (23.51) in the Majorana representation, in which the matrices γ^μ_{M} and $\mathbf{C}^{\mathrm{M}}$ are purely imaginary, so that their products $\gamma^\mu_{\mathrm{M}}\mathbf{C}^{\mathrm{M}}$ are purely real. As the operators P_μ can be taken to be self-adjoint when acting in a Hilbert space, so too can the

Majorana operators Q_a^{M}, that is,

$$(Q_a^{\mathrm{M}})^\dagger = Q_a^{\mathrm{M}} \tag{23.52}$$

for $a = 1, \ldots, 4$. Then, for any ψ of this Hilbert space,

$$(\psi, [Q_a^{\mathrm{M}}, Q_a^{\mathrm{M}}]_+ \psi) = 2(Q_a^{\mathrm{M}}\psi, Q_a^{\mathrm{M}}\psi) \geqslant 0.$$

But, by (23.51), (23.14) and (23.20),

$$\sum_{a=1}^{4} (\psi, [Q_a, Q_a]_+ \psi) = \kappa \sum_{\mu=1}^{4} \mathrm{tr}(\gamma_{\mathrm{M}}^{\mu}\mathbf{C}^{\mathrm{M}})(\psi, P_\mu \psi) = -4\kappa(\psi, P_4\psi).$$

With the assumption the energy operator P_4 is a positive operator, it follows that $(\psi, P_4\psi) \geqslant 0$, so κ must be *real* and *negative*. Which negative value of κ is to be used is purely a matter of convention. As $\kappa = -2$ is probably the most widely used choice (cf. Fayet and Ferrara 1977, Sohnius 1985, West 1986a, b), this is the choice that will be adopted here, but $\kappa = -1$ is quite commonly used (for example, in the early work of Salam and Strathdee 1974b, 1978).

The relations (23.39)–(23.45) define a *complex* Lie superalgebra. A corresponding *real* Lie superalgebra is obtained by using the Majorana representation for the Dirac matrices, in which the products $\gamma_\alpha\gamma_\beta$ and $\gamma^\mu\mathbf{C}$ are all real, and taking the basis elements to be $iM_{\alpha\beta}$ (for $\alpha, \beta = 1, \ldots, 4$), iP_μ (for $\mu = 1, \ldots, 4$), and $e^{-i\pi/4}Q_a^{\mathrm{M}}$ (for $a = 1, \ldots, 4$).

The generalization of (23.52) that is valid in *any* representation of the Dirac matrices is

$$\sum_{b=1}^{4} (\gamma_4\mathbf{C})_{ab} Q_b^\dagger = -Q_a \tag{23.53}$$

(for $a = 1, \ldots, 4$). (This follows because (23.53) is form-invariant under similarity transformations of the Dirac matrices, and because use of (23.21) reduces it to (23.52) in the special case of the Majorana representation.)

More elaborate Lie superalgebras with even parts that contain the Poincaré Lie algebra of 4-dimensional space–time will be considered in Section 3. In each case it will be seen that the odd part involves N copies of the odd part of the above Lie superalgebra, where $N \geqslant 2$. The extensions to D-dimensional space–times for $D \neq 4$ will be examined in Section 4. The Lie superalgebra with commutation and anticommutation relations (23.39)–(23.45) will henceforth be called the "$N = 1$, $D = 4$ Poincaré superalgebra" or the "unextended $D = 4$ Poincaré superalgebra" to distinguish it from these generalizations.

(b) *The two-component formulation for the Poincaré superalgebra*

As the commutation and anticommutation relations (23.39)–(23.45) of the Poincaré superalgebra can be reduced to equations involving only 2×2

matrices when any chiral representation of the Dirac matrices is employed, the resulting expressions are said to constitute a *two-component* formulation. They take particularly simple forms if the modified chiral representation of (23.25) and (23.26) is used. As no other chiral representation will be invoked in this connection in this book, these expressions will be called henceforth *the* two-component formulation.

The first stage is to *define* the elements Q_A and $\bar{Q}_A$ (for $A = 1, 2$) by

$$Q_1 = Q_1^{C'}, \quad Q_2 = Q_2^{C'}, \quad \bar{Q}_1 = Q_3^{C'}, \quad \bar{Q}_2 = Q_4^{C'}, \tag{23.54}$$

where the superscript C′ on the $Q_a^{C'}$ just indicates that they are the odd basis elements Q_a corresponding to the modified chiral representation. Thus Q_1, Q_2, $\bar{Q}_1$ and $\bar{Q}_2$ may themselves be taken to be the odd basis elements of the Poincaré superalgebra. (No confusion should arise between the Q_A (for $A = 1, 2$) as defined here and the Q_a of a general representation of the Dirac matrices, for the former pair will always carry a capital letter subscript whereas such a subscript will never be used for the latter set).

The first simplification, which follows immediately from (23.53) and (23.27), is that

$$(Q_A)^\dagger = \bar{Q}_A, \quad (\bar{Q}_A)^\dagger = Q_A \tag{23.55}$$

(for $A = 1, 2$), where † continues to indicate the adjoint operation. Moreover (23.43) become

$$[M_{\alpha\beta}, Q_A]_- = -\sum_{B=1}^{2} \Gamma_L(M_{\alpha\beta})_{AB} Q_B \tag{23.56}$$

and

$$[M_{\alpha\beta}, \bar{Q}_A]_- = -\sum_{B=1}^{2} \Gamma_R(M_{\alpha\beta})_{AB} \bar{Q}_B \tag{23.57}$$

for $A = 1, 2$, where $\mathbf{\Gamma}_L$ and $\mathbf{\Gamma}_R$ are defined by (23.34) and (23.35), so that, for $\alpha > \beta$,

$$\mathbf{\Gamma}_L(M_{\alpha\beta}) = \frac{1}{2}\frac{\hbar}{i}\boldsymbol{\sigma}_\alpha \boldsymbol{\sigma}_\beta \tag{23.58}$$

and

$$\mathbf{\Gamma}_R(M_{\alpha\beta}) = \frac{1}{2}\frac{\hbar}{i}(\boldsymbol{\sigma}_\alpha \boldsymbol{\sigma}_\beta)^* \tag{23.59}$$

(the matrices for $\alpha < \beta$ being given by $\mathbf{\Gamma}_L(M_{\alpha\beta}) = -\mathbf{\Gamma}_L(M_{\beta\alpha})$ and $\mathbf{\Gamma}_R(M_{\alpha\beta}) = -\mathbf{\Gamma}_R(M_{\beta\alpha})$). Of course (23.44) simply imply that

$$[P_\mu, Q_A]_- = 0 \tag{23.60}$$

and

$$[P_\mu, \bar{Q}_A]_- = 0 \tag{23.61}$$

(for $\mu = 1, \ldots, 4$ and $A = 1, 2$). Moreover, as (23.5), (23.25), (23.26) and (23.36) imply that

$$\boldsymbol{\gamma}^{p}_{\mathrm{C}'}\mathbf{C}^{\mathrm{C}'} = \begin{bmatrix} \mathbf{0} & \boldsymbol{\sigma}_p \\ \boldsymbol{\sigma}^*_p & \mathbf{0} \end{bmatrix} \tag{23.62}$$

for $p = 1, 2, 3$, and

$$\boldsymbol{\gamma}^{4}_{\mathrm{C}'}\mathbf{C}^{\mathrm{C}'} = \begin{bmatrix} \mathbf{0} & -\mathbf{1}_2 \\ -\mathbf{1}_2 & \mathbf{0} \end{bmatrix}, \tag{23.63}$$

it follows that, for A, $B = 1, 2$,

$$[Q_A, Q_B]_+ = 0, \tag{23.64}$$

$$[\bar{Q}_A, \bar{Q}_B]_+ = 0, \tag{23.65}$$

$$[Q_A, \bar{Q}_B]_+ = 2 \sum_{\mu=1}^{4} (\boldsymbol{\sigma}_\mu)_{AB} P^\mu, \tag{23.66}$$

where

$$P^\mu = \sum_{\lambda=1}^{4} \eta^{\mu\lambda} P_\lambda. \tag{23.67}$$

Just for completeness, the connection of this formulation with that of the "dotted" and "undotted" or "barred" and "unbarred" spinors will be outlined briefly. From the 2-dimensional irreducible representation $\boldsymbol{\Gamma}_{\mathrm{L}}$ of the homogeneous Lorentz algebra three other 2-dimensional representations can be constructed, namely $\boldsymbol{\Gamma}^*_{\mathrm{L}}$, $-\tilde{\boldsymbol{\Gamma}}_{\mathrm{L}}$ and $-\tilde{\boldsymbol{\Gamma}}^*_{\mathrm{L}}$. These are not independent, as

$$\boldsymbol{\Gamma}^*_{\mathrm{L}}(a) = \boldsymbol{\Gamma}_{\mathrm{R}}(a),$$

$$-\tilde{\boldsymbol{\Gamma}}_{\mathrm{L}}(a) = \boldsymbol{\sigma}_2^{-1}\boldsymbol{\Gamma}_{\mathrm{L}}(a)\boldsymbol{\sigma}_2, \tag{23.68}$$

$$-\tilde{\boldsymbol{\Gamma}}^*_{\mathrm{L}}(a) = \boldsymbol{\sigma}_2^{-1}\boldsymbol{\Gamma}_{\mathrm{R}}(a)\boldsymbol{\sigma}_2, \tag{23.69}$$

for all elements a of the homogeneous Lorentz algebra. A pair of quantities that form a basis for $\boldsymbol{\Gamma}_{\mathrm{L}}$ may (as above) be denoted by u_A (for $A = 1, 2$), a pair that form a basis for $\boldsymbol{\sigma}_2^{-1}\boldsymbol{\Gamma}_{\mathrm{L}}\boldsymbol{\sigma}_2$ may be denoted by u^A (for $A = 1, 2$), a pair forming a basis for $\boldsymbol{\sigma}_2^{-1}\boldsymbol{\Gamma}_{\mathrm{R}}\boldsymbol{\sigma}_2$ may be denoted by $\bar{u}^A$ (for $A = 1, 2$), and a pair forming a basis for $\boldsymbol{\Gamma}_{\mathrm{R}}$ may (as above) be denoted by $\bar{u}_A$ (for $A = 1, 2$). (In many papers in the literature the "barred" spinors $\bar{u}^A$ and $\bar{u}_A$ are distinguished from the "unbarred" spinors u_A and u^A by *alternatively* (cf. van der Waerden 1974) or *additionally* (cf. Ferrara *et al.* 1974) putting a dot over each index A relating to a "barred" spinor, but for typographical convenience neither of these "dotted" notations will be adopted here.) As u_A and u^A (for $A = 1, 2$) form bases for equivalent representations that are related by a similarity transformation with a 2×2 matrix $\kappa\boldsymbol{\sigma}_2$, where κ is some complex number, it follows that

$$u^A = \sum_{B=1}^{2} (\kappa \boldsymbol{\sigma}_2)^{BA} u_B \tag{23.70}$$

and

$$u_A = \sum_{B=1}^{2} ((\kappa \boldsymbol{\sigma}_2)^{-1})_{BA} u^B. \tag{23.71}$$

A convenient choice for κ is $\kappa = -i$, so that $\kappa\boldsymbol{\sigma}_2 = -\tilde{\boldsymbol{\varepsilon}}$, where $\boldsymbol{\varepsilon}$ is the 2×2 antisymmetric matrix defined by

$$\boldsymbol{\varepsilon} = \begin{bmatrix} 0 & 1 \\ -1 & 0 \end{bmatrix}. \tag{23.72}$$

That is,

$$\boldsymbol{\varepsilon} = i\boldsymbol{\sigma}_2. \tag{23.73}$$

Then (23.70) and (23.71) become

$$u^A = \sum_{B=1}^{2} \varepsilon^{AB} u_B \tag{23.74}$$

and

$$u_A = -\sum_{B=1}^{2} \varepsilon_{AB} u^B, \tag{23.75}$$

where $\varepsilon_{12} = \varepsilon^{12} = 1$, $\varepsilon_{21} = \varepsilon^{21} = -1$ and $\varepsilon_{11} = \varepsilon^{11} = \varepsilon_{22} = \varepsilon^{22} = 0$.

Similarly

$$\bar{u}^A = \sum_{B=1}^{2} (\mu \boldsymbol{\sigma}_2)^{BA} \bar{u}_B \tag{23.76}$$

and

$$\bar{u}_A = \sum_{B=1}^{2} ((\mu \boldsymbol{\sigma}_2)^{-1})_{BA} \bar{u}^B. \tag{23.77}$$

where μ is also some complex number. The simplest convention is to take $\mu = \kappa = -i$ (although some authors take $\mu = -\kappa = i$), so that (23.76) and (23.77) become

$$\bar{u}^A = \sum_{B=1}^{2} \varepsilon^{AB} \bar{u}_B \tag{23.78}$$

and

$$\bar{u}_A = -\sum_{B=1}^{2} \varepsilon_{AB} \bar{u}^B. \tag{23.79}$$

In particular these ideas apply to the supertranslation generators, and permit Q^A and $\bar{Q}^A$ to be defined in terms of the Q_A and $\bar{Q}_A$ introduced in (23.54) (for

$A = 1, 2$), but, although such elements are sometimes used in the literature, they will not be employed in this book.

(c) *The $N = 1$, $D = 4$ Poincaré supergroup*

The Lie supergroup associated with the $N = 1$, $D = 4$ Poincaré superalgebra that is of greatest interest is the one that is simply connected. To simplify the terminology, this will be referred to henceforth as *the $N = 1$, $D = 4$ Poincaré supergroup*. Moreover, to save space (and unless otherwise stated explicitly), the term "Poincaré group" will henceforth be used as an abbreviation for the covering group of the proper orthochronous Poincaré group.

The first stage in its construction along the lines described in Chapter 22, Section 3(*b*) is to set up a *faithful* finite-dimensional graded representation of the *real* $N = 1$, $D = 4$ Poincaré superalgebra. As noted in Subsection 2(*a*) above, the basis elements of this real Poincaré superalgebra may be taken to be $iM_{\alpha\beta}$ (for α, $\beta = 1, \ldots, 4$), iP_μ (for $\mu = 1, \ldots, 4$) and $e^{-i\pi/4}Q_a^{\mathrm{M}}$ (for $a = 1, \ldots, 4$), where the odd basis elements correspond to the *Majorana* representation and where the *Majorana* γ_μ^{M} representation is used for the Dirac matrices. In the development that follows the labels "M" indicating the Majorana representation will be omitted, because it will become apparent that, although the theory has to be set up initially in the Majorana representation, it is in fact valid in *all* representations of the Dirac matrices, that is, it is invariant under similarity transformations of the Dirac matrices, *provided* that the odd Grassmann parameters that are involved are made to transform appropriately.

Suppose that this faithful representation of the Poincaré superalgebra has even dimension d_0 and odd dimension d_1, and is graded as in (21.52) and (21.53). Thus, for every element a of the *even* part of the real Poincaré superalgebra, that is, for every a of the real Poincaré algebra,

$$\boldsymbol{\Gamma}(a) = \begin{bmatrix} \boldsymbol{\Gamma}_{00}(a) & \mathbf{0} \\ \mathbf{0} & \boldsymbol{\Gamma}_{11}(a) \end{bmatrix}, \tag{23.80}$$

where $\boldsymbol{\Gamma}_{00}$ and $\boldsymbol{\Gamma}_{11}$ are finite-dimensional representations of the Poincaré algebra of dimensions d_0 and d_1 respectively. As at least one of these has to be faithful, it will be assumed that $\boldsymbol{\Gamma}_{00}$ is faithful. Similarly, for every element a of the *odd* part of the real Poincaré superalgebra,

$$\boldsymbol{\Gamma}(a) = \begin{bmatrix} \mathbf{0} & \boldsymbol{\Gamma}_{01}(a) \\ \boldsymbol{\Gamma}_{10}(a) & \mathbf{0} \end{bmatrix}, \tag{23.81}$$

where $\boldsymbol{\Gamma}_{01}(a)$ and $\boldsymbol{\Gamma}_{10}(a)$ are matrices of dimensions $d_0 \times d_1$ and $d_1 \times d_0$ respectively.

The faithful representation Γ_{00} may be taken to be a 5-dimensional representation of the Poincaré algebra constructed along the lines described in Chapter 17, Section 5, but it is convenient to make a minor modification in which the representation of (17.187) is replaced by a slightly different but equivalent representation. In this process each element $[\Lambda|\mathbf{t}]$ of the proper orthochronous Poincaré group, whose homogeneous part is the proper orthochronous Lorentz transformation Λ and whose space–time translational part is $\mathbf{t}$, is assigned the 5×5 matrix $\Gamma_{00}([\Lambda|\mathbf{t}])$, where

$$\Gamma_{00}([\Lambda|\mathbf{t}]) = \begin{bmatrix} \Lambda & \mathbf{t}/\lambda \\ \mathbf{0} & \mathbf{1}_1 \end{bmatrix} \tag{23.82}$$

(cf. (17.88)). Here λ is a real constant with the dimensions of length, which is introduced to make the representation dimensionless. (The succeeding formulae simplify if units are adopted in which $\lambda = 1$ and $\hbar = 1$, but for completeness λ and $\hbar$ will be retained explicitly.) As the homogeneous part is exactly as in (17.187), the generators $L_{\alpha\beta}$ of the Lorentz algebra have the representation presented in (17.188), and so, by (23.42), the corresponding 5-dimensional representation Γ_{00} of the quantities $M_{\alpha\beta}$ is given by

$$\Gamma_{00}(M_{\alpha\beta}) = \begin{bmatrix} \mathbf{M}_{\alpha\beta} & \mathbf{0} \\ \mathbf{0} & \mathbf{0} \end{bmatrix}, \tag{23.83}$$

where $\mathbf{M}_{\alpha\beta}$ are a set of 4×4 matrices such that

$$(\mathbf{M}_{\alpha\beta})_{\lambda\mu} = \frac{\hbar}{i}(-\delta_{\alpha\lambda}\eta_{\beta\mu} + \delta_{\beta\lambda}\eta_{\alpha\mu}) \tag{23.84}$$

for $\alpha, \beta, \lambda, \mu = 1, \ldots, 4$ (cf. (17.12)). The pure space–time translations are generated by elements of the Poincaré algebra that are proportional to the linear momentum operators P_μ (for $\mu = 1, \ldots, 4$). Assuming that the elements of the Poincaré algebra are dimensionless, the part corresponding to the P_μ has the form

$$\sum_{\mu=1}^{4} \frac{t^\mu}{\hbar} i P_\mu,$$

where t^1, t^2, t^3 and t^4 have the dimensions of length. Defining $\Gamma_{00}(P_\mu)$ by

$$\Gamma_{00}([\mathbf{1}|\mathbf{t}]) = \exp\left\{ \sum_{\mu=1}^{4} \frac{t^\mu}{\hbar} i \Gamma_{00}(P_\mu) \right\},$$

where $\mathbf{t} = (t^1, t^2, t^3, t^4)$, it follows that

$$\Gamma_{00}(P_\mu) = \begin{bmatrix} \mathbf{0} & -\dfrac{i\hbar}{\lambda}\boldsymbol{\delta}_\mu \\ \mathbf{0} & 0 \end{bmatrix}, \tag{23.85}$$

where $\boldsymbol{\delta}_1, \boldsymbol{\delta}_2, \boldsymbol{\delta}_3$ and $\boldsymbol{\delta}_4$ are a set of four 4×1 matrices such that

$$(\boldsymbol{\delta}_\mu)_{\alpha 1} = \delta_{\alpha\mu}. \tag{23.86}$$

(for $\alpha, \mu = 1, \ldots, 4$). With this choice, both $\boldsymbol{\Gamma}_{00}(M_{\alpha\beta})$ and $\boldsymbol{\Gamma}_{00}(P_\mu)$ have the same dimensions as the operators $M_{\alpha\beta}$ and P_μ to which they correspond.

The other representation $\boldsymbol{\Gamma}_{11}$ of the Poincaré algebra that appears in (23.80) may be taken to be 4-dimensional and to coincide with the spinor representation $\boldsymbol{\Gamma}^{\text{spin}}$ of (23.28) on the homogeneous part, so that

$$\boldsymbol{\Gamma}_{11}(M_{\alpha\beta}) = \boldsymbol{\Gamma}^{\text{spin}}(M_{\alpha\beta}) = -\frac{1}{2}\frac{\hbar}{i}\boldsymbol{\gamma}_\alpha\boldsymbol{\gamma}_\beta \tag{23.87}$$

(for $\alpha, \beta = 1, \ldots, 4$, with $\alpha \neq \beta$), and

$$\boldsymbol{\Gamma}_{11}(P_\mu) = \mathbf{0} \tag{23.88}$$

(for $\mu = 1, \ldots, 4$).

For the supertranslation generators Q_a the 5×4 matrices $\boldsymbol{\Gamma}_{01}(Q_a)$ of (23.80) may be taken to be

$$\boldsymbol{\Gamma}_{01}(Q_a) = \begin{bmatrix} -\left(\dfrac{2\hbar}{\lambda}\right)^{1/2} \mathrm{e}^{i\pi/4}\mathbf{U}_a \\ \mathbf{0} \end{bmatrix} \tag{23.89}$$

(for $a = 1, \ldots, 4$), where $\mathbf{U}_1, \mathbf{U}_2, \mathbf{U}_3$ and $\mathbf{U}_4$ are a set of four 4×4 matrices that are defined by

$$(\mathbf{U}_a)_{\alpha b} = (\boldsymbol{\gamma}^\alpha \mathbf{C})_{ab} \tag{23.90}$$

(for $a, b, \alpha = 1, \ldots, 4$), and the zero matrix in (23.89) has one row and four columns. Finally the 4×5 matrices $\boldsymbol{\Gamma}_{10}(Q_a)$ of (23.80) may be taken to be

$$\boldsymbol{\Gamma}_{10}(Q_a) = \begin{bmatrix} \mathbf{0} & -\left(\dfrac{2\hbar}{\lambda}\right)^{1/2} \mathrm{e}^{i\pi/4}\boldsymbol{\delta}_a \end{bmatrix} \tag{23.91}$$

(for $a = 1, \ldots, 4$), where the zero matrix has four rows and one column and the 4×1 matrices $\boldsymbol{\delta}_a$ are as defined in (23.86). It is easily checked that these matrices satisfy the commutation and anticommutation relations of the $N = 1$, $D = 4$ Poincaré superalgebra.

The matrix $\boldsymbol{\Gamma}(a)$ representing a general element a of the real Poincaré Lie superalgebra can be expressed as

$$\boldsymbol{\Gamma}(a) = \frac{i}{\hbar}\left\{ \sum_{\alpha,\beta=1;\,\alpha<\beta}^{4} \omega^{\alpha\beta}\boldsymbol{\Gamma}(M_{\alpha\beta}) + \sum_{\mu=1}^{4} t^\mu \boldsymbol{\Gamma}(P_\mu) \right\} + \hbar^{-1/2}\mathrm{e}^{-i\pi/4}\sum_{a=1}^{4} \psi^a \boldsymbol{\Gamma}(Q_a), \tag{23.92}$$

where $\omega^{\alpha\beta}$ (for $\alpha, \beta = 1, \ldots, 4$, with $\alpha < \beta$) are six real dimensionless parameters, t^{μ} (for $\mu = 1, \ldots, 4$) are real parameters with the dimensions of length, and ψ^{a} (for $a = 1, \ldots, 4$) are parameters that for convenience may be taken to have the dimensions of $(\text{length})^{1/2}$ and that are real if $Q_a = Q_a^{\mathrm{M}}$, where Q_a^{M} are the "Majorana" odd basis elements. By (23.92), the corresponding "super" Lie algebra consists of the set of supermatrices of the form

$$\mathbf{M} = \frac{i}{\hbar}\left\{ \sum_{\alpha,\beta=1;\alpha<\beta}^{4} \Omega^{\alpha\beta}\boldsymbol{\Gamma}(M_{\alpha\beta}) + \sum_{\mu=1}^{4} T^{\mu}\boldsymbol{\Gamma}(P_{\mu})\right\} + \hbar^{1/2} e^{-i\pi/4} \sum_{a=1}^{4} \Psi^{a}\boldsymbol{\Gamma}(Q_a), \tag{23.93}$$

where $\Omega^{\alpha\beta}$ (for $\alpha, \beta = 1, \ldots, 4$, with $\alpha < \beta$) are six dimensionless copies of the even part $\mathbb{R}B_{L0}$ of the real Grassmann algebra $\mathbb{R}B_L$, T^{μ} (for $\mu = 1, \ldots, 4$) are another four copies of $\mathbb{R}B_{L0}$ (but with the dimensions of length), and if $Q_a = Q_a^{\mathrm{M}}$ then the Ψ^{a} (for $a = 1, \ldots, 4$) are four copies of the odd part $\mathbb{R}B_{L1}$ of $\mathbb{R}B_L$ (with the dimensions of $(\text{length})^{1/2}$), so that

$$(\Psi^{\mathrm{M}a})^{*} = \Psi^{\mathrm{M}a} \tag{23.94}$$

(for $a = 1, \ldots, 4$), the $*$ indicating the Grassmann complex conjugation operation of Chapter 20, Section 3.

Now consider the effect of a similarity transformation of the form (23.11) on the covariant Dirac matrices γ_{μ}, with the corresponding transformation laws for the charge conjugation matrix and the supertranslation generators being given by (23.23) and (23.46) respectively. As noted in Subsection 2(*a*), the commutation and anticommutation relations of the Poincaré superalgebra retain the same form under these transformations. In order that the last term of (23.93) be invariant under such a transformation, so that the element of the "super" Lie algebra specified by it is unchanged, it is necessary to assume that the odd Grassmann parameters Ψ^{a} undergo a corresponding transformation of the form

$$\Psi'^{a} = \sum_{b=1}^{4} (\tilde{\mathbf{S}}^{-1})_{ab}\Psi^{b} \tag{23.95}$$

(for $a = 1, \ldots, 4$). As it is more convenient to have a parametrization in which the parameters transform in the *same* way as the generators, rather than in this inverse fashion, and in order to neatly absorb the awkward factor $e^{-i\pi/4}$ of (23.93), it is useful to introduce four parameters $\zeta^1, \zeta^2, \zeta^3$ and ζ^4 (which are members of the odd Grassmann subspace $\mathbb{C}B_{L1}$ and have the dimensions of $(\text{length})^{1/2}$) by defining

$$\Psi^{a} = e^{i\pi/4} \sum_{b=1}^{4} \zeta^{b\#}(i\gamma_4)_{ba}, \tag{23.96}$$

where the # is the Grassmann adjoint operation of (20.30) and (20.31). Then, substituting into (23.93), the elements of the "super" Lie algebra may be reexpressed in the form

$$\mathbf{M} = \frac{i}{\hbar}\left\{ \sum_{\alpha,\beta=1;\alpha<\beta}^{4} \Omega^{\alpha\beta}\boldsymbol{\Gamma}(M_{\alpha\beta}) + \sum_{\mu=1}^{4} T^{\mu}\boldsymbol{\Gamma}(P_{\mu}) + \hbar^{1/2}\sum_{a,b=1}^{4} \zeta^{a\#}(\gamma_4)_{ab}\boldsymbol{\Gamma}(Q_b)\right\}, \tag{23.97}$$

With the ζ^a being assumed to transform under similarity transformations in the *same* way as the Q_a, so that

$$\zeta'^a = \sum_{b=1}^{4} S_{ab}\zeta^b, \tag{23.98}$$

it follows that the last term of (23.97) is indeed invariant under similarity transformations involving unitary matrices $\mathbf{S}$ (these being the only ones that will be considered (cf. Appendix L, Section 2)). Moreover (23.94) and (23.96) when taken with (20.30) and (20.31) imply that in the Majorana representation (in which γ_4^{M} is purely imaginary)

$$(\zeta^{\mathrm{M}a})^{\#} = \zeta^{\mathrm{M}a} \tag{23.99}$$

(for $a = 1, \ldots, 4$). It should be noted that it is the Grassmann *adjoint* operation rather than the Grassmann complex conjugation operation that appears here! As the ζ^a and Q^a have the same transformation behaviour under similarity transformations, and as (23.53) is a consequence of (23.52), it follows from (23.99) that in *any* representation of the Dirac matrices

$$\gamma_4 \mathbf{C}\boldsymbol{\zeta}^{\#} = -\boldsymbol{\zeta}, \tag{23.100}$$

where $\boldsymbol{\zeta}$ is the 4×1 column matrix with entries $\zeta^1, \zeta^2, \zeta^3$, and ζ^4.

These considerations show that the odd Grassmann parameters ζ^a should carry an additional label indicating the representation of the Dirac matrices to which they belong. This has already been done for the Majorana representation, in which they have been labelled by $\zeta^{\mathrm{M}a}$. In the modified chiral representation (23.25) they may be denoted by $\zeta^{\mathrm{C}'a}$ (for $a = 1, \ldots, 4$). With ζ_A and $\bar{\zeta}_A$ defined by

$$\zeta_1 = \zeta^{\mathrm{C}'1}, \quad \zeta_2 = \zeta^{\mathrm{C}'2}, \quad \bar{\zeta}_1 = \zeta^{\mathrm{C}'3}, \quad \bar{\zeta}_2 = \zeta^{\mathrm{C}'4}, \tag{23.101}$$

(cf. (23.54)), it follows from (23.27) and (23.100) that

$$(\zeta_A)^{\#} = \bar{\zeta}_A \tag{23.102}$$

(for $A = 1, 2$), and (23.97) can be rewritten as

$$\mathbf{M} = \frac{i}{\hbar}\left\{ \sum_{\alpha,\beta=1;\alpha<\beta}^{4} \Omega^{\alpha\beta}\mathbf{\Gamma}(M_{\alpha\beta}) + \sum_{\mu=1}^{4} T^{\mu}\mathbf{\Gamma}(P_{\mu}) \right\}$$

$$- \hbar^{-1/2} \sum_{A,B=1}^{2} \{\zeta_A \varepsilon_{AB}\mathbf{\Gamma}(Q_B) + \bar{\zeta}_A \varepsilon_{AB}\mathbf{\Gamma}(\bar{Q}_B)\}, \qquad (23.103)$$

where $\boldsymbol{\varepsilon}$ $(= i\boldsymbol{\sigma}_2)$ is the 2×2 antisymmetric matrix of (23.72). It must be emphasized that (23.93), (23.97) and (23.103) are *all* parametrizations of the elements $\mathbf{M}$ of the *real* "super" Lie algebra corresponding to the unextended $D = 4$ Poincaré superalgebra.

The elements of the $N = 1$, $D = 4$ Poincaré *supergroup* near the identity are supermatrices of the form $\exp \mathbf{M}$. For the elements of this supergroup with $T^{\mu} = 0$ (for $\mu = 1, \ldots, 4$) and $\zeta^{a} = 0$ (for $a = 1, \ldots, 4$)

$$\exp \mathbf{M} = \exp\left\{ \frac{i}{\hbar} \sum_{\alpha,\beta=1;\alpha<\beta}^{4} \Omega^{\alpha\beta}\mathbf{\Gamma}(M_{\alpha\beta}) \right\} \qquad (23.104)$$

$$= \begin{bmatrix} \exp\left\{ \frac{i}{\hbar} \sum_{\alpha,\beta=1;\alpha<\beta}^{4} \Omega^{\alpha\beta}\mathbf{M}_{\alpha\beta} \right\} & \mathbf{0} & \mathbf{0} \\ \mathbf{0} & \mathbf{1}\mathcal{E}_{\varnothing} & \mathbf{0} \\ \mathbf{0} & \mathbf{0} & \exp\left\{ \frac{i}{\hbar} \sum_{\alpha,\beta=1,\alpha<\beta}^{4} \mathbf{\Omega}^{\alpha\beta}\mathbf{\Gamma}^{\text{spin}}(M_{\alpha\beta}) \right\} \end{bmatrix}$$

$$= \begin{bmatrix} \mathbf{\Lambda}(\Omega^{12}, \ldots) & \mathbf{0} & \mathbf{0} \\ \mathbf{0} & \mathbf{1}\mathcal{E}_{\varnothing} & \mathbf{0} \\ \mathbf{0} & \mathbf{0} & \mathbf{\Gamma}^{\text{spin}}(\mathbf{\Lambda}(\Omega^{12}, \ldots)) \end{bmatrix}, \qquad (23.105)$$

where $\mathbf{\Lambda}(\Omega^{12}, \ldots)$ and $\mathbf{\Gamma}^{\text{spin}}(\mathbf{\Lambda}(\Omega^{12}, \ldots))$ are B_L-valued 4×4 matrices obtained from the real-valued Lorentz matrix $\mathbf{\Lambda}(\omega^{12}, \ldots)$ and its spinor representation $\mathbf{\Gamma}^{\text{spin}}(\mathbf{\Lambda}(\omega^{12}, \ldots))$ of (23.28) by replacing each real parameter $\omega^{\alpha\beta}$ by the corresponding real even Grassmann element $\Omega^{\alpha\beta}$.

Strictly speaking, the matrices $\mathbf{\Gamma}^{\text{spin}}(\mathbf{\Lambda}(\omega^{12}, \ldots))$ form a double-valued representation of the proper orthochronous Lorentz group, but on the covering group of this group this representation is single-valued. As in previous situations where this difficulty has been encountered (cf. Chapters 12 and 17), the elements of a covering group will be denoted by writing them in square brackets, so that $[\mathbf{\Lambda}|\mathbf{t}]$ is a covering group element corresponding to the Poincaré transformation $[\mathbf{\Lambda}|\mathbf{t}]$. However, to avoid an over-cumbersome notation, $\mathbf{\Gamma}^{\text{spin}}([\mathbf{\Lambda}(\omega^{12}, \ldots)|\mathbf{0}])$ will continue to be denoted by $\mathbf{\Gamma}^{\text{spin}}(\mathbf{\Lambda}(\omega^{12}, \ldots))$.

Continuing with this convention, the Poincaré supergroup supermatrix (23.105) will henceforth be denoted by $\boldsymbol{\Gamma}([\boldsymbol{\Lambda}(\Omega^{12},\ldots)|\mathbf{0}|\mathbf{0}])$, and may be thought of as corresponding to an "abstract" Poincaré supergroup (covering group) element $[\boldsymbol{\Lambda}(\Omega^{12},\ldots)|\mathbf{0}|\mathbf{0}]$. These will often be written more briefly as $\boldsymbol{\Gamma}([\boldsymbol{\Lambda}|\mathbf{0}|\mathbf{0}])$ and $[\boldsymbol{\Lambda}|\mathbf{0}|\mathbf{0}]$ respectively, in which case (23.105) becomes

$$\boldsymbol{\Gamma}([\boldsymbol{\Lambda}|\mathbf{0}|\mathbf{0}])=\begin{bmatrix}\boldsymbol{\Lambda} & \mathbf{0} & \mathbf{0}\\ \mathbf{0} & \mathbf{1}\mathcal{E}_{\varnothing} & \mathbf{0}\\ \mathbf{0} & \mathbf{0} & \boldsymbol{\Gamma}^{\mathrm{spin}}(\boldsymbol{\Lambda})\end{bmatrix}. \tag{23.106}$$

Similarly, the elements of the Poincaré supergroup for which $\Omega^{\alpha\beta}=0$ for $\alpha,\beta=1,\ldots,4$ and $\zeta^a=0$ for $a=1,\ldots,4$ may be denoted by $[\mathbf{1}|\mathbf{T}|\mathbf{0}]$, where $\mathbf{T}=(T^1,T^2,T^3,T^4)$. (It will be convenient to continue to use the convention established in Chapter 1, Section 2, in which vectors are regarded as column matrices in matrix expressions unless otherwise indicated, but will be displayed in the text as row matrices). The corresponding supermatrix $\boldsymbol{\Gamma}([\mathbf{1}|\mathbf{T}|\mathbf{0}])$ is then given by

$$\boldsymbol{\Gamma}([\mathbf{1}|\mathbf{T}|\mathbf{0}])=\exp\left\{\frac{i}{\hbar}\sum_{\mu=1}^{4}T^{\mu}\boldsymbol{\Gamma}(P_{\mu})\right\} \tag{23.107}$$

$$=\begin{bmatrix}\mathcal{E}_{\varnothing}\mathbf{1}_4 & \sum_{\mu=1}^{4}\dfrac{T^{\mu}\boldsymbol{\delta}_{\mu}}{\lambda} & \mathbf{0}\\ \mathbf{0} & \mathcal{E}_{\varnothing}\mathbf{1}_1 & \mathbf{0}\\ \mathbf{0} & \mathbf{0} & \mathcal{E}_{\varnothing}\mathbf{1}_4\end{bmatrix}. \tag{23.108}$$

Finally the elements of the Poincaré supergroup for which $\Omega^{\alpha\beta}=0$ for $\alpha,\beta=1,\ldots,4$ and $T^{\mu}=0$ for $\mu=1,\ldots,4$ may be denoted by $[\mathbf{1}|\mathbf{0}|\boldsymbol{\zeta}]$, where $\boldsymbol{\zeta}=(\zeta^1,\zeta^2,\zeta^3,\zeta^4)$. The corresponding supermatrix $\boldsymbol{\Gamma}([\mathbf{1}|\mathbf{0}|\boldsymbol{\zeta}])$ is then (on using (20.51), (23.81), (23.89), (23.90) and (23.91)) given by

$$\boldsymbol{\Gamma}([\mathbf{1}|\mathbf{0}|\boldsymbol{\zeta}])=\exp\left\{\frac{i}{\hbar^{1/2}}\sum_{a,b=1}^{4}\zeta^{a\#}(\gamma_4)_{ab}\boldsymbol{\Gamma}(Q_b)\right\} \tag{23.109}$$

$$=\exp\left\{i\left(\frac{2}{\lambda}\right)^{1/2}\begin{bmatrix}\mathbf{0} & \mathbf{0} & -\sum_{a=1}^{4}\dfrac{\Psi^{a}\mathbf{U}_a}{\lambda}\\ \mathbf{0} & \mathbf{0} & \mathbf{0}\\ \mathbf{0} & \sum_{a=1}^{4}\dfrac{\Psi^{a}\boldsymbol{\delta}_a}{\lambda} & \mathbf{0}\end{bmatrix}\right\} \tag{23.110}$$

(on temporarily reverting to the notation of (23.96)). In the expansion of the exponential of (23.110) the second-order term is zero because the only submatrix that is not obviously zero has matrix elements of the form

$$\sum_{a,b=1}^{4} \Psi^a \Psi^b (\mathbf{U}_a \boldsymbol{\delta}_b)_{\mu 1} = \sum_{a,b=1}^{4} \Psi^a \Psi^b (\gamma^\mu \mathbf{C})_{ab}, \tag{23.111}$$

which are themselves zero, as $(\gamma^\mu \mathbf{C})_{ab}$ is symmetric with respect to the interchange of a and b whereas $\Psi^a \Psi^b$ is antisymmetric (as $\Psi^a, \Psi^b \in \mathbf{B}_{L1}$). Consequently all higher-order terms are also zero, and so

$$\boldsymbol{\Gamma}([\mathbf{1}|\mathbf{0}|\boldsymbol{\zeta}]) = \mathcal{E}_\varnothing \mathbf{1}_9 + \left(\frac{i}{\hbar^{1/2}}\right) \sum_{a,b=1}^{4} \zeta^{a\#}(\gamma_4)_{ab} \boldsymbol{\Gamma}(Q_b). \tag{23.112}$$

A general element $[\boldsymbol{\Lambda}|\mathbf{T}|\boldsymbol{\zeta}]$ of the $N = 1, D = 4$ Poincaré supergroup may be defined as that abstract group element whose supermatrix $\boldsymbol{\Gamma}([\boldsymbol{\Lambda}|\mathbf{T}|\boldsymbol{\zeta}])$ is *defined* by

$$\boldsymbol{\Gamma}([\boldsymbol{\Lambda}|\mathbf{T}|\boldsymbol{\zeta}]) = \boldsymbol{\Gamma}([\mathbf{1}|\mathbf{0}|\boldsymbol{\zeta}])\boldsymbol{\Gamma}([\mathbf{1}|\mathbf{T}|\mathbf{0}])\boldsymbol{\Gamma}([\boldsymbol{\Lambda}|\mathbf{0}|\mathbf{0}]). \tag{23.113}$$

Thus one can write

$$[\boldsymbol{\Lambda}|\mathbf{T}|\boldsymbol{\zeta}] = [\mathbf{1}|\mathbf{0}|\boldsymbol{\zeta}][\mathbf{1}|\mathbf{T}|\mathbf{0}][\boldsymbol{\Lambda}|\mathbf{0}|\mathbf{0}], \tag{23.114}$$

which is a natural generalization of the corresponding Poincaré group result

$$[\boldsymbol{\Lambda}|\mathbf{t}] = [\mathbf{1}|\mathbf{t}][\boldsymbol{\Lambda}|\mathbf{0}], \tag{23.115}$$

which is a special case of (17.182). *Defining* the product of two Poincaré supergroup elements $[\boldsymbol{\Lambda}|\mathbf{T}|\boldsymbol{\zeta}]$ and $[\boldsymbol{\Lambda}'|\mathbf{T}'|\boldsymbol{\zeta}']$ by

$$\boldsymbol{\Gamma}([\boldsymbol{\Lambda}|\mathbf{T}|\boldsymbol{\zeta}][\boldsymbol{\Lambda}'|\mathbf{T}'|\boldsymbol{\zeta}']) = \boldsymbol{\Gamma}([\boldsymbol{\Lambda}|\mathbf{T}|\boldsymbol{\zeta}])\boldsymbol{\Gamma}([\boldsymbol{\Lambda}'|\mathbf{T}'|\boldsymbol{\zeta}']), \tag{23.116}$$

it follows after some algebra (including the use of the relation

$$\gamma_4 \tilde{\boldsymbol{\Gamma}}^{\text{spin}}(\boldsymbol{\Lambda}^{-1})^* (\gamma_4)^{-1} = \boldsymbol{\Gamma}^{\text{spin}}(\boldsymbol{\Lambda}) \tag{23.117}$$

and (23.100)) that

$$[\boldsymbol{\Lambda}|\mathbf{T}|\boldsymbol{\zeta}][\boldsymbol{\Lambda}'|\mathbf{T}'|\boldsymbol{\zeta}'] = [\boldsymbol{\Lambda}\boldsymbol{\Lambda}'|\boldsymbol{\Lambda}\mathbf{T}' + \mathbf{T} + \boldsymbol{\tau}|\boldsymbol{\zeta} + \boldsymbol{\Gamma}^{\text{spin}}(\boldsymbol{\Lambda})\boldsymbol{\zeta}'], \tag{23.118}$$

where

$$\tau^\mu = i(\tilde{\boldsymbol{\zeta}})^{\#} \gamma_4 \gamma^\mu \boldsymbol{\Gamma}^{\text{spin}}(\boldsymbol{\Lambda})\boldsymbol{\zeta}' \tag{23.119}$$

(for $\mu = 1, \ldots, 4$), which generalizes (17.182). One immediate consequence of (23.119) is that

$$[\boldsymbol{\Lambda}|\mathbf{T}|\boldsymbol{\zeta}]^{-1} = [\boldsymbol{\Lambda}^{-1}| - \boldsymbol{\Lambda}^{-1}\mathbf{T}| - \boldsymbol{\Gamma}^{\text{spin}}(\boldsymbol{\Lambda}^{-1})\boldsymbol{\zeta}]. \tag{23.120}$$

(d) *Action of the $N = 1$, $D = 4$ Poincaré supergroup on superspace*

The action of the Poincaré supergroup transformations on superspace is most easily obtained by reformulating the Poincaré group action on Minkowski space–time and then generalizing this reformulation (Salam and Strathdee 1974a). As noted in (17.181), the Poincaré group transformation law is

$$\mathbf{r}' = \Lambda\mathbf{r} + \mathbf{t}, \tag{23.121}$$

and the product of two Poincaré transformations $[\Lambda|\mathbf{t}]$ and $[\Lambda'|\mathbf{t}']$ is given by

$$[\Lambda|\mathbf{t}][\Lambda'|\mathbf{t}'] = [\Lambda\Lambda'|\Lambda\mathbf{t}' + \mathbf{t}] \tag{23.122}$$

(cf. (17.182)). In particular this implies that

$$[\mathbf{1}|\mathbf{r}][\Lambda|\mathbf{0}] = [\Lambda|\mathbf{r}]. \tag{23.123}$$

Now consider the set of *left cosets* of the proper orthochronous Poincaré group with respect to its subgroup of proper orthochronous homogeneous Lorentz transformations. (See Chapter 2, Section 4 for a discussion of left cosets.) The equation (23.123) shows that all the elements in such a left coset formed from the Poincaré group element $[\mathbf{1}|\mathbf{r}]$ have exactly the *same* translational part $\mathbf{r}$, it being their rotational parts that differ, so the coset may be specified by $\mathbf{r}$. Acting with any Poincaré group element $[\Lambda|\mathbf{t}]$ on the coset representative $[\mathbf{1}|\mathbf{r}]$ using (23.122) then produces $[\Lambda|\mathbf{t}][\mathbf{1}|\mathbf{r}] = [\Lambda|\Lambda\mathbf{r} + \mathbf{t}]$, whose translational part is $\Lambda\mathbf{r} + \mathbf{t}$. Thus the action of $[\Lambda|\mathbf{t}]$ on the coset specified by $\mathbf{r}$ gives the coset specified by $\Lambda\mathbf{r} + \mathbf{t}$. Comparison with (23.121) shows that this is exactly the same as the action of $[\Lambda|\mathbf{t}]$ on the point $\mathbf{r}$ of Minkowski space–time. Thus there is a *one-to-one correspondence* between the points of Minkowski space–time and the set of left cosets of the proper orthochronous Poincaré group with respect to its subgroup of proper orthochronous homogeneous Lorentz transformations, and the Poincaré group elements act in the *same* way on both.

The Poincaré *supergroup* generalization starts with the observation that (23.118) includes as a special case the relation

$$[\mathbf{1}|\mathbf{X}|\mathbf{\Theta}][\Lambda|\mathbf{0}|\mathbf{0}] = [\Lambda|\mathbf{X}|\mathbf{\Theta}]. \tag{23.124}$$

Considering the set of *left cosets* of the proper orthochronous Poincaré *supergroup* with respect to its subgroup of proper orthochronous homogeneous Lorentz transformations $[\Lambda|\mathbf{0}|\mathbf{0}]$, (23.124) shows that all the elements in such a left coset formed from the Poincaré group element $[\mathbf{1}|\mathbf{X}|\mathbf{\Theta}]$ have exactly the *same* translational parts $\mathbf{X}$ and $\mathbf{\Theta}$, so the coset may be specified by $\mathbf{X}$ and $\mathbf{\Theta}$. Acting with any Poincaré supergroup element $[\Lambda|\mathbf{T}|\boldsymbol{\zeta}]$ on the coset representative $[\mathbf{1}|\mathbf{X}|\mathbf{\Theta}]$ using (23.118) then produces

$$[\boldsymbol{\Lambda}|\mathbf{T}|\boldsymbol{\zeta}][\mathbf{1}|\mathbf{X}|\boldsymbol{\Theta}] = [\boldsymbol{\Lambda}|\boldsymbol{\Lambda}\mathbf{X} + \mathbf{T} + \boldsymbol{\tau}|\boldsymbol{\zeta} + \boldsymbol{\Gamma}^{\text{spin}}(\boldsymbol{\Lambda})\boldsymbol{\Theta}], \tag{23.125}$$

where

$$\tau^{\mu} = i(\tilde{\boldsymbol{\zeta}})^{\#}\gamma_4\gamma^{\mu}\boldsymbol{\Gamma}^{\text{spin}}(\boldsymbol{\Lambda})\boldsymbol{\Theta}. \tag{23.126}$$

Thus the action of $[\boldsymbol{\Lambda}|\mathbf{T}|\boldsymbol{\zeta}]$ on the coset specified by $\mathbf{X}$ and $\boldsymbol{\Theta}$ gives the coset specified by $\boldsymbol{\Lambda}\mathbf{X} + \mathbf{T} + \boldsymbol{\tau}$ and $\boldsymbol{\zeta} + \boldsymbol{\Gamma}^{\text{spin}}(\boldsymbol{\Lambda})\boldsymbol{\Theta}$. This suggests the following definition.

Definition *Action of the proper orthochronous Poincaré supergroup in superspace*

The action of a proper orthochronous Poincaré supergroup element $[\boldsymbol{\Lambda}|\mathbf{T}|\boldsymbol{\zeta}]$ causes the set of Grassmann-valued coordinates $(\mathbf{X}, \boldsymbol{\Theta})$ of a point to be transformed into another set of Grassmann-valued coordinates $(\mathbf{X}', \boldsymbol{\Theta}')$, where

$$X'^{\alpha} = \sum_{\beta=1}^{4} \Lambda_{\alpha\beta}X^{\beta} + T^{\alpha} + i(\tilde{\boldsymbol{\zeta}})^{\#}\gamma_4\gamma^{\alpha}\boldsymbol{\Gamma}^{\text{spin}}(\boldsymbol{\Lambda})\boldsymbol{\Theta} \tag{23.127}$$

(for $\alpha = 1, \ldots, 4$) and

$$\boldsymbol{\Theta}' = \boldsymbol{\zeta} + \boldsymbol{\Gamma}^{\text{spin}}(\boldsymbol{\Lambda})\boldsymbol{\Theta}. \tag{23.128}$$

It is sometimes convenient to indicate the action of (23.127) and (23.128) by the notation

$$(\mathbf{X}', \boldsymbol{\Theta}') = [\boldsymbol{\Lambda}|\mathbf{T}|\boldsymbol{\zeta}](\mathbf{X}, \boldsymbol{\Theta}). \tag{23.129}$$

Although it is inherent in the line of argument that led to this definition that if $(\mathbf{X}', \boldsymbol{\Theta}') = [\boldsymbol{\Lambda}_2|\mathbf{T}_2|\boldsymbol{\zeta}_2](\mathbf{X}, \boldsymbol{\Theta})$ and if $(\mathbf{X}'', \boldsymbol{\Theta}'') = [\boldsymbol{\Lambda}_1|\mathbf{T}_1|\boldsymbol{\zeta}_1](\mathbf{X}', \boldsymbol{\Theta}')$ then $(\mathbf{X}'', \boldsymbol{\Theta}'') = ([\boldsymbol{\Lambda}_1|\mathbf{T}_1|\boldsymbol{\zeta}_1]\ [\boldsymbol{\Lambda}_2|\mathbf{T}_2|\boldsymbol{\zeta}_2])(\mathbf{X}, \boldsymbol{\Theta})$, it is nevertheless satisfying to check that this is true directly by using the explicit expressions of (23.127) and (23.128). Such a demonstration depends on the identity

$$\sum_{\beta=1}^{4} \Lambda_{\alpha\beta}\gamma^{\beta} = \boldsymbol{\Gamma}^{\text{spin}}(\boldsymbol{\Lambda}^{-1})\gamma^{\alpha}\boldsymbol{\Gamma}^{\text{spin}}(\boldsymbol{\Lambda}) \tag{23.130}$$

(for $\alpha = 1, \ldots, 4$) (cf. (17.130)) and (23.117).

It is implicit from thc above argument that $\mathbf{X}$ and $\mathbf{X}'$ are the same type of quantity as $\mathbf{T}$, so the components of $\mathbf{X}$ and $\mathbf{X}'$ must be members of the real even Grassmann space $\mathbb{R}\mathbf{B}_{L0}$ (and have the dimensions of length). Thus

$$(X^{\mu})^{*} = X^{\mu} \tag{23.131}$$

(for $\mu = 1, \ldots, 4$), $*$ being the Grassmann complex conjugation operation, though, as each X^{μ} is even, (23.131) could equivalently be expressed as

$$(X^{\mu})^{\#} = X^{\mu} \tag{23.132}$$

(for $\mu = 1, \ldots, 4$), where # is the Grassmann adjoint operation of (20.30) and (20.31). (Of course the components of $\mathbf{X}'$ possess the same properties.)

Similarly $\mathbf{\Theta}$ and $\mathbf{\Theta}'$ are the same type of quantity as ζ, so the components of $\mathbf{\Theta}$ and $\mathbf{\Theta}'$ must be members of the odd Grassmann space $\mathbb{C}\mathrm{B}_{L1}$ (and must have the dimensions of $(\text{length})^{1/2}$). Moreover, by analogy with (23.100),

$$\gamma_4 \mathbf{C}\mathbf{\Theta}^{\#} = -\mathbf{\Theta}, \tag{23.133}$$

and it is natural to assume that $\mathbf{\Theta}$ transforms in the same way as ζ under similarity transformations of the Dirac matrices. Thus the odd Grassmann coordinates Θ^a should carry an additional label indicating the representation of the Dirac matrices to which they belong. In the Majorana representation, in which they may be denoted by $\Theta^{\mathrm{M}a}$, (23.133) reduces to

$$(\Theta^{\mathrm{M}a})^{\#} = \Theta^{\mathrm{M}a} \tag{23.134}$$

(for $a = 1, \ldots, 4$) (cf. (23.99)). Similarly, in the modified chiral representation of (23.25), in which they may be denoted by $\Theta^{\mathrm{C}'a}$, the odd Grassmann variables Θ_A and $\bar{\Theta}_A$ may be chosen to be such that

$$\Theta_1 = \Theta_1^{\mathrm{C}'}, \quad \Theta_2 = \Theta_2^{\mathrm{C}'}, \quad \bar{\Theta}_1 = \Theta_3^{\mathrm{C}'}, \quad \bar{\Theta}_2 = \Theta_4^{\mathrm{C}'} \tag{23.135}$$

(cf. (23.101)). Then

$$(\bar{\Theta}_A)^{\#} = \Theta_A, \quad (\Theta_A)^{\#} = \bar{\Theta}_A \tag{23.136}$$

(for $A = 1, 2$) (cf. (23.102)).

As the quantities $\Theta^{\mathrm{M}a}$ are odd, the condition (23.134) can be rewritten as

$$(\mathrm{e}^{i\pi/4}\Theta^{\mathrm{M}a})^* = \mathrm{e}^{i\pi/4}\Theta^{\mathrm{M}a} \tag{23.137}$$

(for $a = 1, \ldots, 4$), indicating that it is the variables $\mathrm{e}^{i\pi/4}\Theta^{\mathrm{M}a}$ that are members of the real odd Grassmann space $\mathbb{R}\mathrm{B}_{L1}$. Consequently the superspace strictly speaking is not quite $\mathbb{R}\mathrm{B}_L^{4,4}$, but, as the introduction of the phase factor $\mathrm{e}^{i\pi/4}$ of (23.137) does not affect the essential points of the analysis given in Chapter 22, this minor distinction will be ignored.

For the homogeneous Lorentz supergroup transformation $[\mathbf{\Lambda}|\mathbf{0}|\mathbf{0}]$ (23.127) and (23.128) reduce to

$$\mathbf{X}' = \mathbf{\Lambda}\mathbf{X} \tag{23.138}$$

and

$$\mathbf{\Theta}' = \mathbf{\Gamma}^{\mathrm{spin}}(\mathbf{\Lambda})\mathbf{\Theta}. \tag{23.139}$$

Similarly, for the Poincaré supergroup translation $[\mathbf{1}|\mathbf{T}|\mathbf{0}]$, (23.127) and (23.128) give

$$\mathbf{X}' = \mathbf{X} + \mathbf{T} \tag{23.140}$$

and

$$\mathbf{\Theta}' = \mathbf{\Theta}. \tag{23.141}$$

Finally, for the remaining special transformations of the form $[\mathbf{1}|\mathbf{0}|\boldsymbol{\zeta}]$, (23.127) gives

$$X'^{\mu} = X^{\mu} + i(\tilde{\boldsymbol{\zeta}})^{\#}\boldsymbol{\gamma}_4\boldsymbol{\gamma}^{\mu}\boldsymbol{\Theta} \tag{23.142}$$

(for $\mu = 1, \ldots, 4$), and (23.128) reduces to

$$\boldsymbol{\Theta}' = \boldsymbol{\Theta} + \boldsymbol{\zeta}. \tag{23.143}$$

It is the form of (23.143) that suggests the designation *supertranslations* for the Poincaré supergroup transformations $[\mathbf{1}|\mathbf{0}|\boldsymbol{\zeta}]$.

The "scalar" transformation operators $U([\boldsymbol{\Lambda}|\mathbf{T}|\boldsymbol{\zeta}])$ for scalar superfields are defined by the prescription that

$$U([\boldsymbol{\Lambda}|\mathbf{T}|\boldsymbol{\zeta}])\Phi_s(\mathbf{X};\boldsymbol{\Theta})U([\boldsymbol{\Lambda}|\mathbf{T}|\boldsymbol{\zeta}])^{-1} = \Phi_s([\boldsymbol{\Lambda}|\mathbf{T}|\boldsymbol{\zeta}](\mathbf{X};\boldsymbol{\Theta})) \tag{23.144}$$

for every Poincaré supergroup transformation $[\boldsymbol{\Lambda}|\mathbf{T}|\boldsymbol{\zeta}]$. Here the scalar superfield $\Phi_s(\mathbf{X};\boldsymbol{\Theta})$ is an operator-valued superanalytic function defined on superspace.

In particular, considering the special case of pure supertranslations, and introducing the more concise notation

$$\delta_{\zeta}\Phi_s(\mathbf{X};\boldsymbol{\Theta}) = \left[U\left(i\hbar^{-1/2}\sum_{a,b=1}^{4}(\zeta^a)^{\#}(\boldsymbol{\gamma}_4)_{ab}Q_b\right), \Phi_s(\mathbf{X};\boldsymbol{\Theta})\right]_{-}, \tag{23.145}$$

for the corresponding operator of the general odd element of the associated "super" Lie algebra, it is easily shown that

$$\delta_{\zeta} = i\sum_{\mu=1}^{4}(\tilde{\boldsymbol{\zeta}})^{\#}\boldsymbol{\gamma}_4\boldsymbol{\gamma}^{\mu}\boldsymbol{\Theta}\frac{\partial}{\partial X^{\mu}} + \sum_{a=1}^{4}\zeta^a\frac{\partial}{\partial\Theta^a}, \tag{23.146}$$

which could be rewritten as

$$\delta_{\zeta} = i\sum_{\mu=1}^{4}(\tilde{\boldsymbol{\zeta}})^{\#}\boldsymbol{\gamma}_4\boldsymbol{\gamma}^{\mu}\boldsymbol{\Theta}\frac{\partial}{\partial X^{\mu}} - \sum_{a=1}^{4}((\tilde{\boldsymbol{\zeta}})^{\#}\boldsymbol{\gamma}_4\mathbf{C})_{1a}\frac{\partial}{\partial\Theta^a}. \tag{23.147}$$

(In (23.146) and (23.147) $\partial/\partial X^{\mu}$ and $\partial/\partial\Theta^a$ are Grassmann derivatives.) This implies that

$$[U(Q_a), \Phi_s(\mathbf{X};\boldsymbol{\Theta})]_{-} = \hbar^{-1/2}\left\{\sum_{\mu=1}^{4}(\boldsymbol{\gamma}^{\mu}\boldsymbol{\Theta})_a\frac{\partial}{\partial X^{\mu}} + i\sum_{b=1}^{4}C_{ab}\frac{\partial}{\partial\Theta^b}\right\}\Phi_s(\mathbf{X};\boldsymbol{\Theta}) \tag{23.148}$$

(for $a = 1, \ldots, 4$). In terms of the variables of the 2-component formulation, (23.146) reduces to

$$\delta_{\zeta} = i\sum_{\mu=1}^{4}\sum_{A,B=1}^{2}\{\bar{\zeta}_A(\boldsymbol{\sigma}_{\mu})_{AB}\Theta_B - \bar{\Theta}_A(\boldsymbol{\sigma}_{\mu})^*_{AB}\zeta_B\}\frac{\partial}{\partial X^{\mu}} + \sum_{A=1}^{2}\left\{\bar{\zeta}_A\frac{\partial}{\partial\bar{\Theta}_A} + \zeta_A\frac{\partial}{\partial\Theta_A}\right\}. \tag{23.149}$$

Similarly, on considering (23.144) for the homogeneous Lorentz transformations $[\mathbf{\Lambda}|\mathbf{0}|\mathbf{0}]$ and using (23.138) and (23.139), it follows that

$$[U(M_{\alpha\beta}), \Phi_s(\mathbf{X};\mathbf{\Theta})]_- = \left\{\frac{\hbar}{i}\left(X_\alpha\frac{\partial}{\partial X^\beta} - X_\beta\frac{\partial}{\partial X^\alpha}\right) - \frac{1}{2}\frac{\hbar}{i}\sum_{a,b=1}^{4}\Theta^a(\gamma_\alpha\gamma_\beta)_{ba}\frac{\partial}{\partial\Theta^b}\right\}\Phi_s(\mathbf{X};\mathbf{\Theta}), \quad (23.150)$$

where

$$X_\alpha = \sum_{\beta=1}^{4}\eta_{\alpha\beta}X^\beta. \quad (23.151)$$

Likewise, by (23.140) and (23.141),

$$[U(P_\mu), \Phi_s(\mathbf{X};\mathbf{\Theta})]_- = \frac{\hbar}{i}\frac{\partial\Phi_s(\mathbf{X};\mathbf{\Theta})}{\partial X^\mu} \quad (23.152)$$

(for $\mu = 1,\ldots,4$).

The signs of the operators (23.150) and (23.152) require some comments, for the first term of the right-hand side of (23.150) and the whole of the right-hand side of (23.152) are, with the even Grassmann derivatives replaced by the corresponding ordinary derivatives, just the usual differential operators for the basis elements of the Poincaré algebra, *except* that the signs are reversed. This change of sign has nothing to do with supersymmetry, but is a consequence that (23.150) and (23.152) are supposed to act on *operator-valued* scalar fields rather than *complex-valued* scalar fields. To see the effect of this distinction most clearly (and to exclude all supersymmetry considerations), consider the ordinary Poincaré transformations on Minkowski space–time, a typical one being denoted by $[\mathbf{\Lambda}|\mathbf{t}]$. For an *operator-valued* scalar field $\Phi(\mathbf{x})$ it is assumed that there exists a set of operators $U([\mathbf{\Lambda}|\mathbf{t}])$ that are such that

$$U([\mathbf{\Lambda}|\mathbf{t}])\Phi(\mathbf{x})U([\mathbf{\Lambda}|\mathbf{t}])^{-1} = \Phi([\mathbf{\Lambda}|\mathbf{t}]\mathbf{x}) \quad (23.153)$$

for every Poincaré transformation $[\mathbf{\Lambda}|\mathbf{t}]$ ((23.144) being the supergroup generalization of this). By contrast, every *complex-valued* scalar field can be described in each coordinate system by an appropriate complex-valued function, and the effect of a Poincaré transformation $[\mathbf{\Lambda}|\mathbf{t}]$ on a system with coordinates $\mathbf{x}$ in which the appropriate function is $F(\mathbf{x})$ is given by an operator $P([\mathbf{\Lambda}|\mathbf{t}])$ defined by

$$P([\mathbf{\Lambda}|\mathbf{t}])F(\mathbf{x}) = F([\mathbf{\Lambda}|\mathbf{t}]^{-1}\mathbf{x}), \quad (23.154)$$

this being the Poincaré generalization of (1.17). (For a detailed discussion of the interpretation of this see Chapter 1, Section 3(*c*)). Clearly (23.153) contains the expression $[\mathbf{\Lambda}|\mathbf{t}]\mathbf{x}$, but in (23.154) this is replaced by $[\mathbf{\Lambda}|\mathbf{t}]^{-1}\mathbf{x}$, in which the inverse appears instead. In particular, for pure space–time translations, $U([\mathbf{1}|\mathbf{t}])\Phi(\mathbf{x})U([\mathbf{1}|\mathbf{t}])^{-1} = \Phi(\mathbf{x}+\mathbf{t})$, whereas $P([\mathbf{1}|\mathbf{t}])F(\mathbf{x}) = F(\mathbf{x}-\mathbf{t})$. As the pure translations are generated by the 4-momentum operators P_μ (for $\mu = 1, \ldots, 4$), the corresponding operators $U(P_\mu)$ and $P(P_\mu)$ may be defined by

$$U([\mathbf{1}|\mathbf{t}]) = \exp\left\{\sum_{\mu=1}^{4} \frac{t^\mu}{\hbar} iU(P_\mu)\right\}$$

and

$$P([\mathbf{1}|\mathbf{t}]) = \exp\left\{\sum_{\mu=1}^{4} \frac{t^\mu}{\hbar} iP(P_\mu)\right\}$$

respectively, so that consideration of the situation for small t^r gives

$$[U(P_\mu), \Phi(\mathbf{x})]_- = \frac{\hbar}{i}\frac{\partial\Phi(\mathbf{x})}{\partial x^\mu} \tag{23.155}$$

but

$$P(P_\mu)F(\mathbf{x}) = -\frac{\hbar}{i}\frac{\partial F(\mathbf{x})}{\partial x^\mu}.$$

Similarly, for the generators of homogeneous Lorentz transformations,

$$[U(M_{\alpha\beta}), \Phi(\mathbf{x})]_- = -\frac{\hbar}{i}\left(x_\alpha\frac{\partial}{\partial x^\beta} - x_\beta\frac{\partial}{\partial x^\alpha}\right)\Phi(\mathbf{x})$$

but

$$P(M_{\alpha\beta})F(\mathbf{x}) = \frac{\hbar}{i}\left(x_\alpha\frac{\partial}{\partial x^\beta} - x_\beta\frac{\partial}{\partial x^\alpha}\right)F(\mathbf{x}).$$

However, whereas

$$\begin{aligned}
&[P(M_{\alpha\beta}), P(M_{\lambda\mu})]_- F(\mathbf{x}) \\
&= \left(\frac{\hbar}{i}\right)^2\left[\left(x_\alpha\frac{\partial}{\partial x^\beta} - x_\beta\frac{\partial}{\partial x^\alpha}\right), \left(x_\lambda\frac{\partial}{\partial x^\mu} - x_\mu\frac{\partial}{\partial x^\lambda}\right)\right]_- F(\mathbf{x}) \\
&= P([M_{\alpha\beta}, M_{\lambda\mu}]_-)F(\mathbf{x}),
\end{aligned}$$

so that

$$[P(M_{\alpha\beta}), P(M_{\lambda\mu})]_- = P([M_{\alpha\beta}, M_{\lambda\mu}]_-),$$

implying, as expected, that the operators $P(M_{\alpha\beta})$ form a Lie algebra that is isomorphic to the homogeneous Lorentz algebra, for the operators $U(M_{\alpha\beta})$ this is *not* true. For these operators the Jacobi identity gives

$$\begin{aligned}
&[[U(M_{\alpha\beta}), U(M_{\lambda\mu})]_-, \Phi(\mathbf{x})]_- \\
&= [U(M_{\alpha\beta}), [U(M_{\lambda\mu}), \Phi(\mathbf{x})]_-]_- - [U(M_{\lambda\mu}), [U(M_{\alpha\beta}), \Phi(\mathbf{x})]_-]_-,
\end{aligned}$$

which, with a casual double application of the explicit differential expression given above, reduces to

$$\left(\frac{\hbar}{i}\right)^2\left[\left(x_\alpha\frac{\partial}{\partial x^\beta} - x_\beta\frac{\partial}{\partial x^\alpha}\right), \left(x_\lambda\frac{\partial}{\partial x^\mu} - x_\mu\frac{\partial}{\partial x^\lambda}\right)\right]_- \Phi(\mathbf{x}),$$

which is equal to

$$-[U([M_{\alpha\beta}, M_{\lambda\mu}]_-), \Phi(\mathbf{x})]_-.$$

That is, this line of argument appears to suggest that

$$[U(M_{\alpha\beta}), U(M_{\lambda\mu})]_- = -U([M_{\alpha\beta}, M_{\lambda\mu}]_-),$$

and so

$$[U(M_{\alpha\beta}), U(M_{\lambda\mu})]_- \neq U([M_{\alpha\beta}, M_{\lambda\mu}]_-)$$

in general.

At this point one must ask why it is that, although the operators $U(M_{\alpha\beta})$ generate a Lie group of operators $U([\mathbf{\Lambda}|\mathbf{0}])$ that is certainly isomorphic to the homogeneous Lorentz group, the operators $U(M_{\alpha\beta})$ themselves do *not* appear to form a Lie algebra that is isomorphic to the homogeneous Lorentz algebra. The resolution of this apparent paradox is rather subtle. It lies in the fact that if $\Phi(\mathbf{x})$ is an operator-valued scalar field then the quantities $[U(M_{\alpha\beta}), \Phi(\mathbf{x})]_-$ need *not* themselves be operator-valued scalar fields, even though $U([\mathbf{\Lambda}|\mathbf{t}])\Phi(\mathbf{x})U([\mathbf{\Lambda}|\mathbf{t}])^{-1}$ is an operator-valued scalar field for every Poincaré group transformation $[\mathbf{\Lambda}|\mathbf{t}]$. It then follows that there is *no* reason for $[U(M_{\lambda\mu}), [U(M_{\alpha\beta}), \Phi(\mathbf{x})]_-]_-$ to be equated to

$$-(\hbar/i)(x_\lambda\partial/\partial x^\mu - x_\mu\partial/\partial x^\lambda)[U(M_{pq}), \Phi(\mathbf{x})]_-,$$

so that

$$[U(M_{\lambda\mu}), [U(M_{\alpha\beta}), \Phi(\mathbf{x})]_-]_- \neq \left(-\frac{\hbar}{i}\right)^2\left(x_\lambda\frac{\partial}{\partial x^\mu} - x_\mu\frac{\partial}{\partial x^\lambda}\right)\left(x_\alpha\frac{\partial}{\partial x^\beta} - x_\beta\frac{\partial}{\partial x^\alpha}\right)\Phi(\mathbf{x}),$$

which is contrary to the assumption that was implicitly made in the above argument. Put another way, the correct form for the action of $U(M_{\alpha\beta})$ on an operator-valued field that is derived from an operator-valued scalar field but does not itself transform as a scalar is not that quoted above but is necessarily more complicated.

It remains to demonstrate by means of an example that if $\Phi(\mathbf{x})$ is an operator-valued scalar field then the quantities $[U(M_{\alpha\beta}), \Phi(\mathbf{x})]_-$ need *not* themselves be operator-valued scalar fields. To this end, consider the unitary irreducible representation of the Poincaré group that corresponds to a non-zero-mass particle with zero spin (cf. Chapter 17, Section 8). Simplifying the notation by writing $U([\Lambda|\mathbf{t}]) = \Phi_{\hat{p},0}([\Lambda|\mathbf{t}])$ and $|p\rangle = \phi_{p,0}$, (17.220) can be rewritten as

$$U([\Lambda|\mathbf{t}])|p\rangle = \exp\left\{\frac{i}{\hbar}\sum_{\mu=1}^{4} (\Lambda \mathbf{p})_{\mu} t^{\mu}\right\}|\Lambda p\rangle.$$

Defining the creation operators $a_p^{\dagger}$ by $a_p^{\dagger}|0\rangle = |p\rangle$, where $|0\rangle$ denotes the vacuum state, and assuming that $U([\Lambda|\mathbf{t}])|0\rangle = |0\rangle$ for all $[\Lambda|\mathbf{t}]$, it follows that

$$U([\Lambda|\mathbf{t}])a_p^{\dagger}U([\Lambda|\mathbf{t}])^{-1} = \left(\frac{\Lambda p^4}{p^4}\right)^{1/2} \exp\left\{\frac{i}{\hbar}\sum_{\mu=1}^{4} (\Lambda p)_{\mu} t^{\mu}\right\} a_{\Lambda p}^{\dagger}.$$

It is then easily checked that the operator-valued field defined by

$$\Phi(\mathbf{x}) = \iiint \left(\frac{\hat{p}^4}{p^4}\right)^{1/2} \exp\left\{\frac{i}{\hbar}\sum_{\mu=1}^{4} p_{\mu} x^{\mu}\right\} a_p^{\dagger}$$

(where the integral is over all 4-momenta p such that $(p)^2 = (\hat{p})^2$) transforms as a scalar under all Poincaré transformations, but $[U(M_{\alpha\beta}), \Phi(\mathbf{x})]_-$ does *not* transform in this way. (For the general construction of operator-valued fields along these lines see Weinberg (1964a, b, 1969).)

These sign considerations for scalar field operators associated with the Poincaré algebra apply in exactly the same way to scalar superfield operators associated with the Poincaré superalgebra.

The scalar superfield will be investigated further in Chapter 24, Section 3(*a*).

3 Extended Poincaré superalgebras and Poincaré supergroups for $D = 4$

Although the $N = 1$, $D = 4$ Poincaré superalgebra discussed in Section 2(*a*) is the smallest Lie superalgebra that contains the $D = 4$ Poincaré Lie algebra as its even part, it is of great interest to ask what is the most general Lie superalgebra $\mathscr{L}_s$ that contains the $D = 4$ Poincaré Lie algebra in its even part and that is physically acceptable. This question was investigated systematically by Haag *et al.* (1975), thereby extending the partial results obtained previously by Wess and Zumino (1974a, b), Salam and Strathdee (1974a, b) and Dondi and Sohnius (1974). The first part of this section is devoted to presenting their results, with the main points of the arguments involved being developed in some detail.

Haag *et al.* (1975) assumed that for a real Lie superalgebra $\mathscr{L}_s$ to be "physically acceptable" its generators must be operators acting on the Hilbert space of physical states that must commute with the S-matrix and must act additively on the states of several incoming particles, that the S-matrix has to be non-trivial, and that each mass multiplet must contain only a finite number of particles. These are exactly the same conditions that had been considered previously by Coleman and Mandula (1967), although they made the further and more restrictive requirement that the generators should form a Lie algebra. They showed that with these assumptions the Lie algebra has to be the direct sum of the Poincaré Lie algebra and a real Lie algebra whose Lie group is compact. This latter algebra may be called an "internal Lie algebra". As the even part $\mathscr{L}_0$ of the Lie superalgebra $\mathscr{L}_s$ is itself a Lie algebra, the results of Coleman and Mandula must apply to $\mathscr{L}_0$. Consequently $\mathscr{L}_0$ must be the direct sum of the $D = 4$ Poincaré Lie algebra and an internal Lie algebra $\mathscr{L}_{\text{int}}$. By Theorem III of Chapter 15, Section 6, $\mathscr{L}_{\text{int}}$ has to be an Abelian Lie algebra or compact semi-simple Lie algebra, or else is the direct sum of an Abelian Lie algebra and a compact semi-simple Lie algebra.

As noted in Theorem I of Chapter 21, Section 2, the basis elements of the odd part $\mathscr{L}_1$ of $\mathscr{L}_s$ form the carrier space of some representation of $\mathscr{L}_0$. In particular they must form the carrier space of some representation of the $D = 4$ homogeneous Lorentz algebra. Haag *et al.* (1975) have shown that the only allowed possibility is that this representation is equivalent to the direct sum of N copies of the 4-dimensional spinor representation Γ^{spin} that was introduced in Section 2(*a*), where N is some positive integer. For convenience it may be assumed that this representation is actually the direct sum of N copies of the representation $-\tilde{\Gamma}^{\text{spin}}$ (which is equivalent to Γ^{spin}), thereby providing an immediate generalization of (23.47). Denoting the $4N$ odd basis elements by Q_a^I (for $I = 1, 2, \ldots, N$ and $a = 1, \ldots, 4$), this implies that the generalization of (23.43) is

$$[M_{\alpha\beta}, Q_a^I]_- = \frac{1}{2}\frac{\hbar}{i}\sum_{b=1}^{4}(\gamma_\alpha\gamma_\beta)_{ab}Q_b^I \tag{23.156}$$

(for $I = 1, 2, \ldots, N$ and $a, \alpha, \beta = 1, \ldots, 4$, with $\alpha \neq \beta$), and the generalization of (23.44) is

$$[P_\mu, Q_a^I]_- = 0 \tag{23.157}$$

(for $I = 1, 2, \ldots, N$ and $a, \mu = 1, \ldots, 4$).

Just as in the simpler case considered in Section 2(*a*), the choice of odd basis elements is tied to the choice of the Dirac matrices. In particular, if the Majorana representation is used, the odd basis elements may be denoted by Q_a^{MI} (for $I = 1, 2, \ldots, N$ and $a = 1, \ldots, 4$), in which case (23.156) becomes

$$[M_{\alpha\beta}, Q_a^{MI}]_- = \frac{1}{2}\frac{\hbar}{i}\sum_{b=1}^{4}(\gamma_\alpha^M\gamma_\beta^M)_{ab}Q_b^{MI}. \tag{23.158}$$

Similarly, if the modified chiral representation of (23.25) is employed, the odd basis elements may be denoted by $Q_a^{C'I}$ (for $I = 1, 2, \ldots, N$ and $a = 1, \ldots, 4$), and (23.156) then reads

$$[M_{\alpha\beta}, Q_a^{C'I}]_- = \frac{1}{2}\frac{\hbar}{i}\sum_{b=1}^{4} (\gamma_\alpha^{C'}\gamma_\beta^{C'})_{ab} Q_b^{C'I}. \tag{23.159}$$

It is easiest to proceed to the next stage in the argument by using the modified chiral representation in which the 4-dimensional representation $\boldsymbol{\Gamma}^{\text{spin}}$ of the homogeneous Lorentz algebra reduces to the direct sum of the 2-dimensional "left-handed" irreducible representation $\boldsymbol{\Gamma}_{\mathrm{L}}$ and the 2-dimensional "right-handed" irreducible representation $\boldsymbol{\Gamma}_{\mathrm{R}}$. Each of these must appear N times when the representation of $\mathscr{L}_0$ on $\mathscr{L}_1$ is reduced on the homogeneous Lorentz part of $\mathscr{L}_0$. As the part of $\mathscr{L}_0$ spanned by the four 4-momentum basis elements P_μ ($\mu = 1, \ldots, 4$) is assumed to be represented trivially in the representation of $\mathscr{L}_0$ on $\mathscr{L}_1$ (cf. (23.157)), this part of $\mathscr{L}_0$ may be omitted from the considerations. Because the remaining part of $\mathscr{L}_0$ is the direct sum of the homogeneous Lorentz algebra and $\mathscr{L}_{\text{int}}$, it follows that on this part the representation of $\mathscr{L}_0$ on $\mathscr{L}_1$ must have the form

$$(\boldsymbol{\Gamma}_{\mathrm{L}} \otimes \mathbf{L}) \oplus (\boldsymbol{\Gamma}_{\mathrm{R}} \otimes \mathbf{R}),$$

where $\mathbf{L}$ and $\mathbf{R}$ are two N-dimensional representations of $\mathscr{L}_{\text{int}}$. Writing

$$Q_1^I = Q_1^{C'I}, \quad Q_2^I = Q_2^{C'I}, \quad \bar{Q}_1^I = Q_3^{C'I}, \quad \bar{Q}_2^I = Q_4^{C'I}, \tag{23.160}$$

(cf. (23.54)), it follows that the basis elements Q_A^I and $\bar{Q}_A^I$ may be chosen so that

$$[B, Q_A^I]_- = \sum_{J=1}^{N} L(B)_{JI} Q_A^J \tag{23.161}$$

and

$$[B, \bar{Q}_A^I]_- = \sum_{J=1}^{N} R(B)_{JI} \bar{Q}_A^J \tag{23.162}$$

(for all $B \in \mathscr{L}_{\text{int}}$, $A = 1, 2$ and $I = 1, 2, \ldots, N$). Of course (23.159) and (23.160) imply that

$$[M_{\alpha\beta}, Q_A^I]_- = -\sum_{B=1}^{2} \Gamma_{\mathrm{L}}(M_{\alpha\beta})_{AB} Q_B^I \tag{23.163}$$

and

$$[M_{\alpha\beta}, \bar{Q}_A^I]_- = -\sum_{B=1}^{2} \Gamma_{\mathrm{R}}(M_{\alpha\beta})_{AB} \bar{Q}_B^I \tag{23.164}$$

(for $\alpha, \beta = 1, \ldots, 4$; $A = 1, 2$ and $I = 1, 2, \ldots, N$), where $\boldsymbol{\Gamma}_{\mathrm{L}}$ and $\boldsymbol{\Gamma}_{\mathrm{R}}$ are defined by (23.34) and (23.35) (cf. (23.56) and (23.57)). As $\mathscr{L}_{\text{int}}$ is assumed to be the Lie algebra of a *compact* Lie group, the representations $\mathbf{L}$ and $\mathbf{R}$ may be taken to be made up of *anti-Hermitian* matrices, so that on exponentiation they would produce a unitary representation of the group.

By (23.160) and (23.25), equations (23.161) and (23.162) can be combined to read

$$[B, Q_a^{C'I}]_- = \frac{1}{2}\sum_{J=1}^{N}\sum_{b=1}^{4}\{\delta_{ab}(\mathbf{L}(B)+\mathbf{R}(B))_{JI} + (\gamma_5^C)_{ab}(\mathbf{L}(B)-\mathbf{R}(B))_{JI}\}Q_b^{C'J}, \tag{23.165}$$

in which form it is invariant under similarity transformations of the Dirac matrices, so that in *any* representation of the Dirac matrices

$$[B, Q_a^I]_- = \frac{1}{2}\sum_{J=1}^{N}\sum_{b=1}^{4}\{\delta_{ab}(\mathbf{L}(B)+\mathbf{R}(B))_{JI} + (\gamma_5)_{ab}(\mathbf{L}(B)-\mathbf{R}(B))_{JI}\}Q_b^J \tag{23.166}$$

(for all $B \in \mathscr{L}_{\text{int}}$, $a = 1, \ldots, 4$ and $I = 1, 2, \ldots, N$).

The adjoint condition that generalizes (23.55) is

$$(Q_A^I)^\dagger = \bar{Q}_A^I, \quad (\bar{Q}_A^I)^\dagger = Q_A^I \tag{23.167}$$

(for $A = 1, 2$ and $I = 1, 2, \ldots, N$). Assuming that the internal symmetry operators B are such that

$$B^\dagger = -B$$

(for all $B \in \mathscr{L}_{\text{int}}$), which is consistent with them forming a *real* Lie algebra, on taking the adjoint of (23.161) and comparing it with (23.162) it becomes clear that

$$\mathbf{R}(B) = \mathbf{L}(B)^* \tag{23.168}$$

(for all $B \in \mathscr{L}_{\text{int}}$). Consequently, defining the anti-Hermitian representation $\mathbf{\Gamma}^{\text{int}}$ of $\mathscr{L}_{\text{int}}$ by

$$\mathbf{\Gamma}^{\text{int}}(B) = \mathbf{L}(B) \tag{23.169}$$

(for all $B \in \mathscr{L}_{\text{int}}$), (23.166) can be rewritten as

$$[B, Q_a^I]_- = \sum_{J=1}^{N}\sum_{b=1}^{4}\{\delta_{ab}\,\mathrm{Re}\,\mathbf{\Gamma}^{\text{int}}(B)_{JI} + (\gamma_5)_{ab}\,\mathrm{Im}\,\mathbf{\Gamma}^{\text{int}}(B)_{JI}\}Q_b^J \tag{23.170}$$

(for all $B \in \mathscr{L}_{\text{int}}$, $a = 1, \ldots, 4$ and $I = 1, 2, \ldots, N$). In particular, in the Majorana representation, (23.170) becomes

$$[B, Q_a^{MI}]_- = \sum_{J=1}^{N}\sum_{b=1}^{4}\{\delta_{ab}\,\mathrm{Re}\,\mathbf{\Gamma}^{\text{int}}(B)_{JI} + (\gamma_5^M)_{ab}\,\mathrm{Im}\,\mathbf{\Gamma}^{\text{int}}(B)_{JI}\}Q_b^{MJ}, \tag{23.171}$$

(and as $(\gamma_5^M)^* = -\gamma_5^M$ (cf. (23.14)), it follows that in the Majorana representation *all* the coefficients appearing on the right-hand side of (23.171) are *real*.

It remains to deduce the form of the anticommutators $[Q^I_a, Q^J_b]_+$, which must lie in the even part $\mathscr{L}_0$. Application of the Jacobi identity constraint (21.14) for the case $a = P_\mu$, $b = Q^I_a$, $c = Q^J_b$, when taken with (23.157), implies that

$$[P_\mu, [Q^I_a, Q^J_b]_+]_- = 0,$$

and then (23.40) shows that $[Q^I_a, Q^J_b]_+$ cannot contain any elements of the homogeneous Lorentz algebra. That is,

$$[Q^I_a, Q^J_b]_+ = \sum_{\mu=1}^{4} (\mathbf{F}^\mu)^{IJ}_{ab} P_\mu + \sum_j (\mathbf{G}^j)^{IJ}_{ab} A_j, \qquad (23.172)$$

where $A_1, A_2, \ldots$ are the basis elements of $\mathscr{L}_{\text{int}}$. The Jacobi identity constraints (21.15) are satisfied on using (23.157) if it assumed that every element of $\mathscr{L}_{\text{int}}$ appearing on the right-hand side of (23.172) commutes with every odd operator Q^I_a. Moreover, by a similar argument to that presented in Section 2(*a*), Haag *et al.* (1975) have shown that the form of the right-hand side of (23.172) can be simplified to give

$$[Q^I_a, Q^J_b]_+ = -2\delta^{IJ} \sum_{\mu=1}^{4} (\gamma^\mu \mathbf{C})_{ab} P_\mu + i(\mathbf{C})_{ab} U^{IJ} + (\gamma_5 \mathbf{C})_{ab} V^{IJ}, \qquad (23.173)$$

where the U^{IJ} and V^{IJ} are members of $\mathscr{L}_{\text{int}}$. From the (21.15) Jacobi identity condition just mentioned

$$[U^{IJ}, Q^K_a]_- = [V^{IJ}, Q^K_a]_- = 0 \qquad (23.174)$$

(for $I, J, K = 1, 2, \ldots, N$ and $a = 1, \ldots, 4$). As the U^{IJ} and V^{IJ} are members of $\mathscr{L}_{\text{int}}$, and as $\mathscr{L}_0$ is the direct sum of the Poincaré Lie algebra and $\mathscr{L}_{\text{int}}$,

$$[U^{IJ}, M_{\alpha\beta}]_- = [V^{IJ}, M_{\alpha\beta}]_- = 0 \qquad (23.175)$$

(for $I, J = 1, 2, \ldots, N$ and $\alpha, \beta = 1, \ldots, 4$) and

$$[U^{IJ}, P_\mu]_- = [V^{IJ}, P_\mu]_- = 0 \qquad (23.176)$$

(for $I, J = 1, 2, \ldots, N$ and $\mu = 1, \ldots, 4$). Also, from the Jacobi identity condition (21.14) with $a = U^{IJ}$, $b = Q^K_a$, $c = Q^L_b$, (23.174) implies that

$$[U^{IJ}, [Q^K_a, Q^L_b]_+]_- = 0,$$

and hence, by (23.173),

$$[U^{IJ}, U^{KL}]_- = [U^{IJ}, V^{KL}]_- = 0 \qquad (23.177)$$

(for $I, J, K, L = 1, 2, \ldots, N$). Similarly, considering the same Jacobi identity with U^{IJ} replaced by V^{IJ},

$$[V^{IJ}, V^{KL}]_- = 0 \tag{23.178}$$

(for $I, J, K, L = 1, 2, \ldots, N$). Moreover consideration of (21.14) with $a = B(\in \mathscr{L}_{\text{int}})$, $b = Q_a^K$, $c = Q_b^L$ shows that $[B, U^{IJ}]_-$ and $[B, V^{IJ}]_-$ must be linear combinations of the U^{KL} and V^{KL}. Thus the U^{IJ} and V^{IJ} together form an Abelian invariant subalgebra of $\mathscr{L}_{\text{int}}$, and hence they must lie in the Abelian part of $\mathscr{L}_{\text{int}}$, and so they must commute with all the elements of the semi-simple part of $\mathscr{L}_{\text{int}}$, and consequently they must commute with *all* the elements of $\mathscr{L}_{\text{int}}$. That is,

$$[U^{IJ}, B]_- = [V^{IJ}, B]_- = 0 \tag{23.179}$$

(for $I, J = 1, 2, \ldots, N$ and for all $B \in \mathscr{L}_{\text{int}}$). Putting (23.174)–(23.179) together, it is clear that each U^{IJ} and V^{IJ} commutes with every element of the Lie superalgebra $\mathscr{L}_s$. For this reason the U^{IJ} and V^{IJ} are called the *central charges* of $\mathscr{L}_s$.

As the right-hand side of (23.173) has to be symmetric with respect to the simultaneous interchanges $I \leftrightarrow J$ and $a \leftrightarrow b$, and as $\mathbf{C}$ is assumed to be antisymmetric (cf. (23.18)) and so $\gamma_5 \mathbf{C}$ is also antisymmetric, it follows that

$$U^{IJ} = -U^{JI} \tag{23.180}$$

and

$$V^{IJ} = -V^{JI}. \tag{23.181}$$

(for $I, J = 1, 2, \ldots, N$). In particular this implies that

$$U^{II} = V^{II} = 0$$

(for $I = 1, 2, \ldots, N$). Thus there are at most $\frac{1}{2}N(N-1)$ independent basis elements U^{IJ} and at most $\frac{1}{2}N(N-1)$ independent basis elements V^{IJ}. Clearly these central charges *cannot* appear when $N = 1$.

Putting (23.173) into the Majorana representation, where it reads

$$[Q_a^{MI}, Q_b^{MJ}]_+ = -2\delta^{IJ} \sum_{\mu=1}^{4} (\gamma_{\mathrm{M}}^{\mu} \mathbf{C}^{\mathrm{M}})_{ab} P_\mu + i(\mathbf{C}^{\mathrm{M}})_{ab} U^{IJ} + (\gamma_5^{\mathrm{M}} \mathbf{C}^{\mathrm{M}})_{ab} V^{IJ}, \tag{23.182}$$

as the generalization of (23.52) is

$$(Q_a^{\mathrm{M}I})^\dagger = Q_a^{\mathrm{M}I} \tag{23.183}$$

(for $I = 1, 2, \ldots, N$ and $a = 1, \ldots, 4$), it follows from (23.14), (23.180) and (23.181) that

$$(U^{IJ})^\dagger = U^{IJ} \tag{23.184}$$

and

$$(V^{IJ})^\dagger = V^{IJ} \tag{23.185}$$

(for $I, J = 1, 2, \ldots, N$). As $i\mathbf{C}^{\mathrm{M}}$ and $\gamma_5^{\mathrm{M}}\mathbf{C}^{\mathrm{M}}$ are both purely real, with the Majorana representation the structure constants on the right-hand side of (23.182) are purely *real*.

From these considerations it is clear that the internal Lie algebra $\mathscr{L}_{\mathrm{int}}$ is in general the direct sum of three subalgebras: one subalgebra spanned by not more than $N(N-1)$ central charges U^{IJ} and V^{IJ}, a second consisting of the semi-simple part of $\mathscr{L}_{\mathrm{int}}$, and a third containing the remaining Abelian part of $\mathscr{L}_{\mathrm{int}}$. (Each of these subalgebras may be trivial in special cases.) Assuming that the second and third subalgebras together have dimension n_0, if $n_0 > 0$, the basis elements of the second and third subalgebras of $\mathscr{L}_{\mathrm{int}}$ may be denoted by B_j (for $j = 1, 2, \ldots, n_0$) and may be assumed to possess the commutation relations

$$[B_j, B_k]_- = \sum_{l=1}^{n_0} c_{jk}^l B_l, \tag{23.186}$$

where the structure constants c_{jk}^l may be assumed to be all *real*. It may also be assumed that if B_j is a member of the third subalgebra of $\mathscr{L}_{\mathrm{int}}$ then $\boldsymbol{\Gamma}^{\mathrm{int}}(B_j) \neq \mathbf{0}$ (for otherwise B_j would not only commute with every element of the Lie superalgebra $\mathscr{L}_s$ but also B_j would never appear in any commutator or anticommutator).

For future convenience the complete set of commutation and anticommutation relations of the Lie superalgebra $\mathscr{L}_s$ in their final form (and for a general representation of the Dirac matrices) may be summarized as follows:

$$[M_{\alpha\beta}, M_{\lambda\mu}]_- = \frac{\hbar}{i}(\eta_{\alpha\lambda}M_{\beta\mu} - \eta_{\alpha\mu}M_{\beta\lambda} - \eta_{\beta\lambda}M_{\alpha\mu} + \eta_{\beta\mu}M_{\alpha\lambda}), \tag{23.39}$$

$$[M_{\alpha\beta}, P_\mu]_- = \frac{\hbar}{i}(\eta_{\alpha\mu}P_\beta - \eta_{\beta\mu}P_\alpha), \tag{23.40}$$

$$[P_\lambda, P_\mu]_- = 0, \tag{23.41}$$

$$[B_j, B_k]_- = \sum_{l=1}^{n_0} c_{jk}^l B_l, \tag{23.186}$$

$$[B_j, M_{\alpha\beta}]_- = [B_j, P_\mu]_- = 0, \tag{23.187}$$

$$[U^{IJ}, B_j]_- = [V^{IJ}, B_j] = 0, \tag{23.179}$$

$$[U^{IJ}, U^{KL}]_- = [U^{IJ}, V^{KL}]_- = [V^{IJ}, V^{KL}]_- = 0, \qquad (23.177), (23.178)$$

$$[U^{IJ}, M_{\alpha\beta}]_- = [V^{IJ}, M_{\alpha\beta}]_- = 0, \tag{23.175}$$

$$[U^{IJ}, P_\mu]_- = [V^{IJ}, P_\mu]_- = 0, \tag{23.176}$$

$$[M_{\alpha\beta}, Q_a^I]_- = \frac{1}{2}\frac{\hbar}{i}\sum_{b=1}^{4}(\gamma_\alpha\gamma_\beta)_{ab}Q_b^I \tag{23.156}$$

(for $\alpha \neq \beta$),

$$[P_\mu, Q_a^I]_- = 0, \tag{23.157}$$

$$[B_j, Q_a^I]_- = \sum_{J=1}^{N} \sum_{b=1}^{4} \{\delta_{ab} \operatorname{Re} \boldsymbol{\Gamma}^{\text{int}}(B_j)_{JI} + (\gamma_5)_{ab} \operatorname{Im} \boldsymbol{\Gamma}^{\text{int}}(B_j)_{JI}\} Q_b^J, \tag{23.170}$$

$$[U^{IJ}, Q_a^K]_- = [V^{IJ}, Q_a^K]_- = 0, \tag{23.174}$$

$$[Q_a^I, Q_b^J]_+ = -2\delta^{IJ} \sum_{\mu=1}^{4} (\gamma^\mu \mathbf{C})_{ab} P_\mu + i(\mathbf{C})_{ab} U^{IJ} + (\gamma_5 \mathbf{C})_{ab} V^{IJ} \tag{23.173}$$

(here $a, b, \alpha, \beta, \mu = 1, \ldots, 4; j, k = 1, \ldots, n_0$ and $I, J, K, L = 1, 2, \ldots, N$). In the Majorana representation of the Dirac matrices and with $Q_a^I = Q_a^{MI}$ this is a *real* Lie superalgebra.

As noted in the derivation of these results, some of these equations simplify in the modified chiral representation of (23.25). Indeed in this representation (23.156) reduce to (23.163) and (23.164), which can be written more explicitly using (23.58) and (23.59) as

$$[M_{\alpha\beta}, Q_A^I]_- = -\frac{1}{2}\frac{\hbar}{i} \sum_{B=1}^{2} (\boldsymbol{\sigma}_\alpha \boldsymbol{\sigma}_\beta)_{AB} Q_B^I \tag{23.188}$$

and

$$[M_{\alpha\beta}, \bar{Q}_A^I]_- = -\frac{1}{2}\frac{\hbar}{i} \sum_{B=1}^{2} (\boldsymbol{\sigma}_\alpha \boldsymbol{\sigma}_\beta)^*_{AB} \bar{Q}_B^I \tag{23.189}$$

(for $\alpha, \beta = 1, \ldots, 4$ and $A = 1, 2$). Similarly (23.170) reduces to

$$[B_j, Q_A^I]_- = \sum_{J=1}^{N} \boldsymbol{\Gamma}^{\text{int}}(B_j)_{JI} Q_A^J \tag{23.190}$$

and

$$[B_j, \bar{Q}_A^I]_- = \sum_{J=1}^{N} \boldsymbol{\Gamma}^{\text{int}}(B_j)^*_{JI} \bar{Q}_A^J \tag{23.191}$$

(which are essentially just (23.161) and (23.162) with (23.168) and (23.169) being applied). Likewise, by (23.25) and (23.26), (23.173) reduces to

$$[Q_A^I, Q_B^J]_+ = \varepsilon_{AB}(U^{IJ} - iV^{IJ}), \tag{23.192}$$

$$[\bar{Q}_A^I, \bar{Q}_B^J]_+ = -\varepsilon_{AB}(U^{IJ} + iV^{IJ}), \tag{23.193}$$

$$[Q_A^I, \bar{Q}_B^J]_+ = 2\delta^{IJ} \sum_{\mu=1}^{4} (\boldsymbol{\sigma}_\mu)_{AB} P^\mu \tag{23.194}$$

(for $A, B = 1, 2$ and $I, J = 1, 2, \ldots, N$), where $\boldsymbol{\varepsilon}$ is the 2×2 matrix defined in

(23.72) (cf. (23.66)). Finally

$$[P_\mu, Q^I_A]_- = 0 \tag{23.195}$$

and

$$[P_\mu, \bar{Q}^I_A]_- = 0 \tag{23.196}$$

(for $\mu = 1, \ldots, 4; I, J = 1, 2, \ldots, N$ and $A = 1, 2$).

The only sets of Jacobi identity conditions (21.14) that have not been fully investigated are those for which $a = B_j$, $b = Q^I_A$, $c = Q^J_B$ and $a = B_j$, $b = \bar{Q}^I_A$, $c = \bar{Q}^J_B$. In the case in which $\mathscr{L}_s$ contains *no* central charges both of these sets of Jacobi identity conditions are identically satisfied with the commutation and anticommutation relations that have been listed. Consequently if there are *no* central charges present then there are no further conditions on the representation $\boldsymbol{\Gamma}^{\text{int}}$ of $\mathscr{L}_{\text{int}}$, and, as the matrices of $\boldsymbol{\Gamma}^{\text{int}}$ may be assumed to be anti-Hermitian, $\mathscr{L}_{\text{int}}$ is isomorphic to u(N) or some subalgebra of u(N). However, if $\mathscr{L}_s$ contains *at least one* central charge then there are further conditions that restrict the internal symmetry algebra. To see this, suppose that there are N_U linearly independent central charges of the type U^{IJ}, which may be denoted by U^S for $S = 0, 1, 2, \ldots, N_U$, and that there are N_V linearly independent central charges of the type V^{IJ}, which may be denoted by V^S for $S = 0, 1, 2, \ldots, N_V$. Necessarily $N_U \leqslant \frac{1}{2}N(N-1)$ and $N_V \leqslant \frac{1}{2}N(N-1)$, and the assumption will be made that $1 \leqslant N_U + N_V$. For $S = 1, 2, \ldots, N_U$ let $\mathbf{A}^S$ be the real antisymmetric $N \times N$ matrix defined by

$$U^{IJ} = \sum_{S=1}^{N_U} (\mathbf{A}^S)_{IJ} U^S, \tag{23.197}$$

and similarly for $S = 1, 2, \ldots, N_V$ let $\mathbf{B}^S$ be the real antisymmetric $N \times N$ matrix defined by

$$V^{IJ} = \sum_{S=1}^{N_V} (\mathbf{B}^S)_{IJ} V^S. \tag{23.198}$$

Then, by (23.179), (23.188) and (23.192), the Jacobi identity conditions (21.14) with $a = B_j, b = Q^I_A$ and $c = Q^J_B$ give

$$\mathbf{A}^S \boldsymbol{\Gamma}^{\text{int}}(B_j) = \tilde{\boldsymbol{\Gamma}}^{\text{int}}(B_j)\mathbf{A}^S \tag{23.199}$$

for $S = 1, 2, \ldots, N_U$ and $j = 1, 2, \ldots, n_0$ (provided that $N_U \geqslant 1$) and

$$\dot{\mathbf{B}}^S \boldsymbol{\Gamma}^{\text{int}}(B_j) = \tilde{\boldsymbol{\Gamma}}^{\text{int}}(B_j)\mathbf{B}^S \tag{23.200}$$

for $S = 1, 2, \ldots, N_V$ and $j = 1, 2, \ldots, n_0$ (provided that $N_V \geqslant 1$). The Jacobi identity conditions (21.14) with $a = B_j, b = \bar{Q}^J_A$ and $c = \bar{Q}^J_B$ give the complex conjugates of (23.199) and (23.200), and so have essentially the same content. Equations (23.199) and (23.200) together give $N_U + N_V$ conditions, there being one for each central charge that is present. Then, for example, if $\mathscr{L}_s$ possesses only one central charge U^1 and N is even, and if the U^{IJ} are chosen so that

$$U^{12} = U^{34} = \cdots = U^{N-1,N} = U^1, \tag{23.201}$$

which implies by (23.197) that

$$\mathbf{A}^1 = \begin{bmatrix} \mathbf{0} & \mathbf{1}_{N/2} \\ -\mathbf{1}_{N/2} & \mathbf{0} \end{bmatrix}, \tag{23.202}$$

then reference to Table 10.1 shows that (23.199) is the condition for the matrices $\mathbf{\Gamma}^{\text{int}}$ to form an sp($\frac{1}{2}N$) simple real Lie algebra, and hence, with this one central charge, $\mathscr{L}_{\text{int}}$ must be isomorphic to sp($\frac{1}{2}N$) or to a proper subalgebra of sp($\frac{1}{2}N$). As another example suppose that $\mathscr{L}_{\text{int}} = \text{su}(2)$ (which is isomorphic to sp(1), cf. Chapter 14, Section 2) and that $\mathbf{\Gamma}^{\text{int}}$ is the 2-dimensional irreducible representation of su(2) in which $\mathbf{\Gamma}^{\text{int}}(\mathbf{a}) = \mathbf{a}$ for all $\mathbf{a} \in \text{su}(2)$. In this case (23.199) each have only one linearly independent solution in which $\mathbf{A}^1 = \mathbf{B}^1 = \boldsymbol{\varepsilon}$, where $\boldsymbol{\varepsilon}$ is the 2×2 matrix defined in (23.72), so $\mathscr{L}_{\text{int}} = \text{su}(2)$ admits *two* linearly independent central charges $U^{12}(= -U^{21} = U^1)$ and $V^{12}(= -V^{21} = V^1)$.

In the case in which $N = 1$ these considerations show that there are no central charges and that the maximal internal symmetry algebra is isomorphic to the 1-dimensional Abelian Lie algebra u(1). Denoting the one basis element B_1 of u(1) by R, and choosing the 1-dimensional anti-Hermitian representation $\mathbf{\Gamma}^{\text{int}}$ to be such that $\mathbf{\Gamma}^{\text{int}}(R) = [i]$, (23.170) reduces to

$$[R, Q_a]_- = \sum_{b=1}^{4} i(\gamma_5)_{ab} Q_b \tag{23.203}$$

(for $a = 1, \ldots, 4$), where the redundant superscript I on the Q_a^I has been omitted. Because of the form of (23.203), R is called a *chiral rotation generator*.

It is sometimes convenient to note that the forms of (23.192) and (23.193) can be simplified by making the definition

$$Z^{IJ} = \tfrac{1}{2}(U^{IJ} - iV^{IJ}), \tag{23.204}$$

which, by (23.184) and (23.185), implies that

$$Z^{IJ\dagger} = \tfrac{1}{2}(U^{IJ} + iV^{IJ}). \tag{23.205}$$

Then (23.192) and (23.193) become

$$[Q_A^I, Q_B^J]_+ = 2\varepsilon_{AB} Z^{IJ} \tag{23.206}$$

and

$$[\bar{Q}_A^I, \bar{Q}_B^J]_+ = -2\varepsilon_{AB} Z^{IJ\dagger} \tag{23.207}$$

(for $A, B = 1, 2$ and $I, J = 1, 2, \ldots, N$), where $\boldsymbol{\varepsilon}$ is the 2×2 matrix defined in (23.72). Of course (23.180) and (23.181) imply that

$$Z^{IJ} = -Z^{JI}, \tag{23.208}$$

so that

$$Z^{II} = 0 \tag{23.209}$$

for $I = 1, 2, \ldots, N$, and

$$Z^{IJ\dagger} = -Z^{JI\dagger} \tag{23.210}$$

for $I, J = 1, 2, \ldots, N$, so that at most $\frac{1}{2}N(N-1)$ of the Z^{IJ} and $\frac{1}{2}N(N-1)$ of the $Z^{IJ\dagger}$ are linearly independent.

There is no difficulty in generalizing the Lie *supergroup* construction given for the $N = 1$ case in Sections 2(*c*) and 2(*d*) to the N-extended situation. A graded representation $\boldsymbol{\Gamma}^N$ of the N-extended $D = 4$ Poincaré Lie superalgebra $\mathscr{L}_s$ with $d_0 = 5$ and $d_1 = 4N$ can be defined by

$$\boldsymbol{\Gamma}^N_{00}(M_{\alpha\beta}) = \mathbf{1}_1 \otimes \boldsymbol{\Gamma}_{00}(M_{\alpha\beta}), \tag{23.211}$$

$$\boldsymbol{\Gamma}^N_{00}(P_\mu) = \mathbf{1}_1 \otimes \boldsymbol{\Gamma}_{00}(P_\mu), \tag{23.212}$$

$$\boldsymbol{\Gamma}^N_{00}(B_j) = \mathbf{0}, \tag{23.213}$$

$$\boldsymbol{\Gamma}^N_{11}(M_{\alpha\beta}) = \mathbf{1}_N \otimes \boldsymbol{\Gamma}_{11}(M_{\alpha\beta}), \tag{23.214}$$

$$\boldsymbol{\Gamma}^N_{11}(P_\mu) = \mathbf{1}_N \otimes \boldsymbol{\Gamma}_{11}(P_\mu), \tag{23.215}$$

$$\boldsymbol{\Gamma}^N_{11}(A_j) = \{\boldsymbol{\Gamma}^{\text{int}}(A_j) \otimes \mathbf{1}_2\} \oplus \{\boldsymbol{\Gamma}^{\text{int}}(A_j)^* \otimes \mathbf{1}_2\}, \tag{23.216}$$

$$\boldsymbol{\Gamma}^N_{01}(Q^I_a) = \boldsymbol{\delta}_I \otimes \boldsymbol{\Gamma}_{01}(Q^I_a), \tag{23.217}$$

$$\boldsymbol{\Gamma}^N_{10}(Q^I_a) = \boldsymbol{\delta}_I \otimes \boldsymbol{\Gamma}_{10}(Q^I_a), \tag{23.218}$$

where $\boldsymbol{\Gamma}_{00}(M_{\alpha\beta}), \boldsymbol{\Gamma}_{00}(P_\mu), \boldsymbol{\Gamma}_{11}(M_{\alpha\beta}), \boldsymbol{\Gamma}_{11}(P_\mu), \boldsymbol{\Gamma}_{01}(Q^I_a)$ and $\boldsymbol{\Gamma}_{10}(Q^I_a)$ are as defined in (23.83), (23.85), (23.87), (23.88), (23.89) and (23.90) respectively, and $\boldsymbol{\delta}_I$ is the $N \times 1$ matrix defined by

$$(\boldsymbol{\delta}_I)_{J1} = \delta_{IJ} \tag{23.219}$$

(for $I, J = 1, 2, \ldots, N$). It is easily checked that these satisfy the commutation and anticommutation relations of $\mathscr{L}_s$.

The corresponding "super" Lie algebra is obtained by generalizing (23.97) and consists of the set of supermatrices of the form

$$\begin{aligned}\mathbf{M} = {} & \frac{i}{\hbar} \sum_{\alpha,\beta=1;\alpha<\beta}^{4} \Omega^{\alpha\beta} \boldsymbol{\Gamma}^N(M_{\alpha\beta}) + \frac{i}{\hbar} \sum_{\mu=1}^{4} T^\mu \boldsymbol{\Gamma}^N(P_\mu) \\ & + \sum_{j=1} Y^j \boldsymbol{\Gamma}^N(A_j) \\ & + \frac{i}{\hbar^{1/2}} \sum_{I=1}^{N} \sum_{a,b=1}^{4} (\zeta^{Ia})^\# (\gamma_4)_{ab} \boldsymbol{\Gamma}^N(Q^I_b),\end{aligned} \tag{23.220}$$

where $\mathbf{\Omega}^{\alpha\beta}$ and T^{μ} are exactly as in the $N=1$ case, Y^{j} (for $j=1,2,\ldots$) are dimensionless copies of the even part $\mathbb{R}\mathrm{B}_{L0}$ of $\mathbb{R}\mathrm{B}_{L}$, and the ζ^{Ia} (for $a=1,\ldots,4$ and $I=1,2,\ldots,N$) are $4N$ copies of the odd part $\mathbb{C}\mathrm{B}_{L1}$ of $\mathbb{C}\mathrm{B}_{L}$, which have the dimensions of $(\text{length})^{1/2}$ and satisfy the constraints

$$\boldsymbol{\gamma}_4\mathbf{C}(\boldsymbol{\zeta}^I)^{\#}=-\boldsymbol{\zeta}^I, \tag{23.221}$$

(cf. (23.100)). Here the $4N$ odd parameters ζ^{Ia} have been formed into N sets of 4-component quantities by defining $\boldsymbol{\zeta}^I=(\zeta^{I1},\zeta^{I2},\zeta^{I3},\zeta^{I4})$ for $I=1,2,\ldots,N$.

As the rest of the analysis given earlier for the $N=1$ case generalizes in the obvious way, the details will be omitted. The main point to note is that the superspace on which the N-extended $D=4$ Poincaré supergroup acts is essentially $\mathrm{B}_L^{4,4N}$, where the $4N$ odd Grassmann coordinates Θ^{Ia} can be formed into N sets of 4-component quantities by defining $\mathbf{\Theta}^I=(\Theta^{I1},\Theta^{I2},\Theta^{I3},\Theta^{I4})$ for $I=1,2,\ldots,N$. (For an explicit treatment of the example in which $\mathscr{L}_{\text{int}}=\text{su}(2)$ see Dondi and Sohnius (1974).)

4 The Poincaré superalgebras and supergroups for Minkowski space–times of general dimension D

(a) *The unextended Poincaré superalgebras and supergroups for general dimension D*

The theory given in the previous two sections can be generalized to a Minkowski space–time of general dimension D, and for most values of D this generalization is very straightforward. The covariant metric tensor $\eta_{\alpha\beta}$ of D-dimensional Minkowski space–time will be assumed to be such that

$$\eta_{\alpha\beta}=\begin{cases}-1 & \text{for } \alpha=\beta=1,2,\ldots,D-1,\\ +1 & \text{for } \alpha=\beta=D,\\ 0 & \text{for } \alpha\neq\beta.\end{cases} \tag{23.222}$$

The analysis depends heavily on the properties of the matrices $\boldsymbol{\gamma}_\alpha$ of the D-dimensional Minkowski Clifford algebra, which are assumed to satisfy the anticommutation relations

$$\boldsymbol{\gamma}_\alpha\boldsymbol{\gamma}_\beta+\boldsymbol{\gamma}_\beta\boldsymbol{\gamma}_\alpha=2\eta_{\alpha\beta}\mathbf{1}_d \tag{23.223}$$

(for $\alpha,\beta=1,2,\ldots,D$), the relevant properties of which are summarized in Appendix L.

The D-dimensional Minkowski Clifford algebra for *even* D has one irreducible representation of dimension

$$d=2^{D/2} \tag{23.224}$$

(cf. (L.15)), and for *odd D* it has two irreducible representations of dimension

$$d = 2^{(D-1)/2} \tag{23.225}$$

(cf. (L.73)) which merely differ in overall sign, so that it does not matter which is selected. For $D \bmod 8 = 0, 1, 2, 3$ and 4 there exists a $d \times d$ generalized charge conjugation matrix $\mathbf{C}$ with the properties that

$$\tilde{\gamma}_\beta \tilde{\gamma}_\alpha = \mathbf{C}^{-1} \gamma_\beta \gamma_\alpha \mathbf{C} \tag{23.226}$$

(for $\alpha, \beta = 1, 2, \ldots, D$) (cf. (L.72) and (L.109)), that $\gamma^\alpha \mathbf{C}$ is *symmetric* (for $\alpha = 1, 2, \ldots, D$) and that in the Majorana representation the corresponding matrices $\gamma_{\mathrm{M}}^\alpha \mathbf{C}^{\mathrm{M}}$ are all *real*. Moreover for these values of D the d-dimensional spinor representation Γ^{spin} of so$(D-1, 1)$ defined by (L.10) has a Majorana representation

$$\Gamma_{\mathrm{M}}^{\mathrm{spin}}(L_{\alpha\beta}) = -\tfrac{1}{2}\gamma_\alpha^{\mathrm{M}} \gamma_\beta^{\mathrm{M}} \tag{23.227}$$

(for $p, q = 1, 2, \ldots, D$ with $p \neq q$) (cf. (L.14)), which is also *real*. By contrast if $D \bmod 8 = 5, 6$ or 7 there does *not* exist a matrix $\mathbf{C}$ with the property (23.226) that is also symmetric, *nor* does there exist a Majorana representation. This case will be briefly revisited later in this section, but for the present it will be assumed that $D \bmod 8 = 0, 1, 2, 3$ or 4.

In these cases the generalization of the 4-component formulation of the "unextended" ($N = 1$) $D = 4$ Poincaré superalgebra theory is simply given by allowing in Section 2(*a*) all the space–time indices $\alpha, \beta, \mu, \ldots$ to assume the values $1, 2, \ldots, D$, and all the spinor indices $a, b, c, \ldots$ to take the values $1, 2, \ldots, d$. (It should be noted that (23.224) shows that $d = D$ *only* for $D = 2$ and $D = 4$). In particular there exist d supertranslation generators Q_a, and (23.39), (23.40), (23.41), (23.43), (23.44) and (23.45) become

$$[M_{\alpha\beta}, M_{\lambda\mu}]_- = \frac{\hbar}{i}(\eta_{\alpha\lambda} M_{\beta\mu} - \eta_{\alpha\mu} M_{\beta\lambda} - \eta_{\beta\lambda} M_{\alpha\mu} + \eta_{\beta\mu} M_{\alpha\lambda}) \tag{23.228}$$

(for $\alpha, \beta = 1, 2, \ldots, D$),

$$[M_{\alpha\beta}, P_\mu]_- = \frac{\hbar}{i}(\eta_{\alpha\mu} P_\beta - \eta_{\beta\mu} P_\alpha) \tag{23.229}$$

(for $\alpha, \beta, \mu = 1, 2, \ldots, D$),

$$[P_\lambda, P_\mu]_- = 0 \tag{23.230}$$

(for $\lambda, \mu = 1, 2, \ldots, D$),

$$[M_{\alpha\beta}, Q_a]_- = -\frac{\hbar}{i} \sum_{b=1}^{d} \Gamma^{\mathrm{spin}}(L_{\alpha\beta})_{ab} Q_b \tag{23.231}$$

$$= \frac{1}{2}\frac{\hbar}{i} \sum_{b=1}^{d} (\gamma_\alpha \gamma_\beta)_{ab} Q_b \tag{23.232}$$

(for $\alpha, \beta = 1, 2, \ldots, D$ with $\alpha \neq \beta$, and for $a = 1, 2, \ldots, d$),

$$[P_\mu, Q_a]_- = 0 \tag{23.233}$$

(for $\mu = 1, 2, \ldots, D$ and $a = 1, 2, \ldots, d$), and

$$[Q_a, Q_b]_+ = -2 \sum_{\mu=1}^{D} (\gamma^\mu \mathbf{C})_{ab} P_\mu \tag{23.234}$$

(for $a, b = 1, 2, \ldots, d$). (The proof that this forms a Lie superalgebra proceeds exactly along the lines presented in Section 2(*a*), only the ranges of the space–time and spinor indices needing modification in the manner indicated above.)

In the *Majorana* representation this Poincaré superalgebra is again a *real* Lie superalgebra (provided that the basis elements are taken to be $iM_{\alpha\beta}$ (for $\alpha, \beta = 1, 2, \ldots, D$), iP_μ (for $\mu = 1, 2, \ldots, D$), and $e^{-i\pi/4} Q_a^{\mathrm{M}}$ (for $a = 1, 2, \ldots, d$), and it may be assumed that

$$(Q_a^{\mathrm{M}})^\dagger = Q_a^{\mathrm{M}} \tag{23.235}$$

for $a = 1, 2, \ldots, d$ (cf. (23.52)). The sign of the right-hand side of (23.234) has to be negative (as in (23.45)) in order to be compatible with the assumption that the operator P_D is positive definite and the assumption that $\mathbf{C}^{\mathrm{M}}$ is chosen so that

$$\mathbf{C}^{\mathrm{M}} = -\gamma_D^{\mathrm{M}} \tag{23.236}$$

(cf. (L.61), (L.71), (L.98) and (L.108)). (It follows from (23.235) that the adjointness condition for a general representation of the Minkowski Clifford algebra is

$$\sum_{b=1}^{d} (\gamma_4 \mathbf{C})_{ab} Q_b^\dagger = -Q_a \tag{23.237}$$

(cf. (23.53)).) Then, for any ψ of the physical Hilbert space, (23.235) implies that

$$(\psi, [Q_a^{\mathrm{M}}, Q_a^{\mathrm{M}}]_+ \psi) = (Q_a^{\mathrm{M}}\psi, Q_a^{\mathrm{M}}\psi) \geqslant 0,$$

and, by (23.234), (L.62) and (L.99),

$$\sum_{a=1}^{d} (\psi, [Q_a, Q_a]_+ \psi) = -2 \sum_{\mu=1}^{D} \operatorname{tr}(\gamma_{\mathrm{M}}^\mu \mathbf{C}^{\mathrm{M}})(\psi, P_\mu \psi) = 2D(\psi, P_D \psi).$$

The factor of 2 in the right-hand side of (23.234) is merely a matter of convention, and has been selected to give agreement with the choice made in the $D = 4$ theory.

For *even* values of D the spinor representation $\mathbf{\Gamma}^{\mathrm{spin}}$ of so$(D-1, 1)$ defined in (L.10) is reducible and in the chiral representation can be written as the direct sum of two irreducible representations $\mathbf{\Gamma}_{\mathrm{L}}$ and $\mathbf{\Gamma}_{\mathrm{R}}$, both of which have dimension $\frac{1}{2}d$ (cf. (L.25)). Consequently for *even* values of D there is a "$\frac{1}{2}d$-

component formulation", which generalizes the two-component formulation of the $D = 4$ case that was given in Section 2(*b*), but there is *no* corresponding formulation for *odd* values of D.

It should be noted that (23.224) and (23.225) imply that both the $N = 1$, $D = 10$ and $N = 1$, $D = 11$ Poincaré superalgebras contain 32 spinor generators; that is, exactly the *same* number as in the $N = 8$, $D = 4$ Poincaré superalgebra.

For $D \bmod 8 = 2$, that is, for $D = 2, 10, \ldots$, Theorem IV of Appendix L, Section 2 shows that the Minkowski Clifford algebra possesses a *Majorana–Weyl* representation in which *all* the matrices $\boldsymbol{\gamma}^p\mathbf{C}$ of (23.234) *and* both the irreducible representations $\boldsymbol{\Gamma}_L$ and $\boldsymbol{\Gamma}_R$ are *real*. The simplest case is that of $D = 2$, for which all the details are very easily derived, as the following Example shows.

Example I *The unextended $D = 2$ Poincaré superalgebra*

The unextended $D = 2$ Poincaré superalgebra has only the three even basis elements M_{12}, P_1 and P_2, and as $d = 2$ it has only two supertranslation generators. The $D = 2$ Minkowski Clifford algebra matrices $\boldsymbol{\gamma}_p^B$ of (L.16) and (L.17) for $p = 1, 2$ are themselves *imaginary and chiral*, no similarity transformation being required to produce this effect, so

$$\boldsymbol{\gamma}_1^B = \boldsymbol{\gamma}_1^C = \boldsymbol{\gamma}_1^M = i\boldsymbol{\sigma}_1 \tag{23.238}$$

and

$$\boldsymbol{\gamma}_2^B = \boldsymbol{\gamma}_2^C = \boldsymbol{\gamma}_2^M = \boldsymbol{\sigma}_2 \tag{23.239}$$

(cf. (L.37)). Then

$$\boldsymbol{\Gamma}_B^{spin}(M_{12}) = \boldsymbol{\Gamma}_C^{spin}(M_{12}) = \boldsymbol{\Gamma}_M^{spin}(M_{12}) = -\frac{1}{2}\frac{\hbar}{i}\boldsymbol{\sigma}_3, \tag{23.240}$$

(cf. (L.48)) and so the irreducible representations $\boldsymbol{\Gamma}_L$ and $\boldsymbol{\Gamma}_R$ are 1-dimensional and are given by

$$\boldsymbol{\Gamma}_L(M_{12}) = \frac{1}{2}\frac{\hbar}{i}[-1], \quad \boldsymbol{\Gamma}_R(M_{12}) = \frac{1}{2}\frac{\hbar}{i}[1], \tag{23.241}$$

(cf. (L.49)). With the choice (23.236), $\mathbf{C}^M = -\boldsymbol{\gamma}_2^M = -\boldsymbol{\sigma}_2$, and so

$$\boldsymbol{\gamma}_1^M\mathbf{C}^M = \boldsymbol{\sigma}_3, \quad \boldsymbol{\gamma}_2^M\mathbf{C}^M = -\mathbf{1}_2. \tag{23.242}$$

Denoting the two supertranslation generators by Q_{L1} and Q_{R1} (so that $Q_{L1} = Q_1^C = Q_1^M$ and $Q_{R1} = Q_2^C = Q_2^M$), (23.232) becomes

$$[M_{12}, Q_{L1}]_- = \frac{1}{2}\frac{\hbar}{i}Q_{L1}, \quad [M_{12}, Q_{R1}]_- = -\frac{1}{2}\frac{\hbar}{i}Q_{R1}, \tag{23.243}$$

and (23.234) reduce to

$$[Q_{L1}, Q_{L1}]_+ = 2(P_1 + P_2), \quad [Q_{R1}, Q_{R1}]_+ = 2(-P_1 + P_2) \qquad (23.244)$$

and

$$[Q_{L1}, Q_{R1}]_+ = 0. \qquad (23.245)$$

(In this superalgebra the odd basis elements Q_{L1} and Q_{R1} are not coupled to each other.)

The construction of the $N = 1$ Poincaré *supergroup* for a general value of D proceeds in exactly the same way as the construction of the $N = 1$, $D = 4$ Poincaré supergroup that was described in Sections 2(*c*) and 2(*d*). All that has to be modified are the dimensions of the matrices involved and the numbers of even and odd Grassmann parameters and coordinates. In particular the dimensions of the matrices $\mathbf{\Gamma}$, $\mathbf{\Gamma}_{00}$ and $\mathbf{\Gamma}_{11}$ of (23.80) become $(D + d + 1) \times (D + d + 1)$, $(D + 1) \times (D + 1)$ and $d \times d$ respectively, those of $\mathbf{\Gamma}_{01}$ and $\mathbf{\Gamma}_{10}$ of (23.81) become $(D + 1) \times d$ and $d \times (D + 1)$ respectively, while $\mathbf{\Gamma}^{\text{spin}}$ of (23.87) becomes a $d \times d$ matrix, the $\mathbf{U}_a$ of (23.90) become a set of d $D \times d$ matrices, the $\boldsymbol{\delta}_r$ of (23.86) become a set of D $D \times 1$ matrices, and the $\boldsymbol{\delta}_a$ of (23.91) become a set of d $d \times 1$ matrices. The corresponding "super" Lie algebra consists of the set of supermatrices of the form

$$\mathbf{M} = \frac{i}{\hbar} \sum_{\alpha,\beta=1;\alpha<\beta}^{D} \Omega^{\alpha\beta} \mathbf{\Gamma}(M_{\alpha\beta}) + \frac{i}{\hbar} \sum_{\mu=1}^{D} T^{\mu} \mathbf{\Gamma}(P_\mu) + \frac{i}{\hbar^{1/2}} \sum_{a,b=1}^{d} \zeta^{a\#} (\gamma_4)_{ab} \mathbf{\Gamma}(Q_b) \qquad (23.246)$$

(cf. (23.97)), where $\Omega^{\alpha\beta}$ (for $\alpha, \beta = 1, 2, \dots, D$, with $\alpha < \beta$) are $\frac{1}{2}D(D-1)$ dimensionless copies of the even part $\mathbb{R}\mathbf{B}_{L0}$ of the real Grassmann algebra $\mathbb{R}\mathbf{B}_L$, T^μ (for $\mu = 1, \dots, 4$) are another D copies of $\mathbb{R}\mathbf{B}_{L0}$ (but with the dimensions of length), and the ζ^a (for $a = 1, 2, \dots, d$) are d copies of the odd part $\mathbb{C}\mathbf{B}_{L1}$ of $\mathbb{C}\mathbf{B}_L$ (with the dimensions of $(\text{length})^{1/2}$), which satisfy the condition

$$\gamma_4 \mathbf{C} \boldsymbol{\zeta}^\# = -\boldsymbol{\zeta},$$

where $\boldsymbol{\zeta}$ is the $d \times 1$ column matrix with entries ζ^a $(a = 1, 2, \dots, d)$ (cf. (23.100)). The elements of the $N = 1$ Poincaré supergroup near the identity are then supermatrices of the form $\exp \mathbf{M}$, and they have the properties given in (23.105)–(23.120) (provided that the ranges of the space–time and spinor indices are modified in the obvious way). The superspace on which this supergroup acts is essentially $\mathbb{R}\mathbf{B}_L^{D,d}$, which has D real even Grassmann coordinates X^p (with $p = 1, 2, \dots, D$) and d odd Grassmann coordinates Θ^a (with $a = 1, 2, \dots, d$), which satisfy the condition

$$\gamma_4 \mathbf{C} \boldsymbol{\Theta}^\# = -\boldsymbol{\Theta},$$

where $\mathbf{\Theta}$ is the $d \times 1$ column matrix with entries Θ^a $(a = 1, 2, \ldots, d)$. The supergroup action on superspace is again given by (23.127) and (23.128) (provided of course that the ranges of the space–time and spinor indices are modified in the obvious way).

(b) *The extended Poincaré superalgebras and supergroups for general dimension D*

The construction of N-extended Poincaré superalgebras for general values of D is slightly complicated because the situation is different for different values of $D \bmod 8$. Attention will be largely concentrated on the class $D \bmod 8 = 4$, which is the simplest because it is a straightforward generalization of the case $D = 4$ that was considered in Section 3, and on the class $D \bmod 8 = 2$, which includes the interesting cases of $D = 2$, 10 and 26. The other classes be discussed in outline without giving all the details for all the possible cases. (For further information see Strathdee (1987).)

Example I *Extended Poincaré superalgebras for* $D \bmod 8 = 4$

This is the simplest class because the N-extended Poincaré superalgebras can be simply obtained from those for the $D = 4$ case considered in Section 3 by merely allowing all the space–time indices $\alpha, \beta, \mu, \ldots$ to assume the values $1, 2, \ldots, D$, and all the spinor indices $a, b, c, \ldots$ to take the values $1, 2, \ldots, d$. With $I, J, \ldots$ taking values $1, 2, \ldots, N$ and j, k taking values $1, 2, \ldots, n_0$ (n_0 being the dimension of the internal symmetry algebra $\mathscr{L}_{\text{int}}$), the commutation and anticommutation relations consist of (23.157), (23.174)–(23.179), (23.186), (23.187), (23.228), (23.229) and (23.230), together with

$$[M_{\alpha\beta}, Q_a^I]_- = \frac{1}{2}\frac{\hbar}{i} \sum_{b=1}^{d} (\gamma_\alpha \gamma_\beta)_{ab} Q_b^I \tag{23.248}$$

(for $\alpha \neq \beta$),

$$[B_j, Q_a^I]_- = \sum_{J=1}^{N} \sum_{b=1}^{d} \{\delta_{ab} \operatorname{Re} \mathbf{\Gamma}^{\text{int}}(B_j)_{JI} + (\gamma_{D+1})_{ab} \operatorname{Im} \mathbf{\Gamma}^{\text{int}}(B_j)_{JI}\} Q_b^J, \tag{23.249}$$

$$[Q_a^I, Q_b^J]_+ = -2\delta^{IJ} \sum_{\mu=1}^{D} (\gamma^\mu \mathbf{C})_{ab} P_\mu + i(\mathbf{C})_{ab} U^{IJ} + (\gamma_{D+1} \mathbf{C})_{ab} V^{IJ}. \tag{23.250}$$

In (23.249) and (23.250) γ_{D+1} is defined for even D by

$$\gamma_{D+1} = i^{(D-2)/2} \gamma_1 \gamma_2 \gamma_3 \cdots \gamma_{D-1} \gamma_D \tag{23.251}$$

(cf. (L.20)). In (23.249) $\mathbf{\Gamma}^{\text{int}}$ is an N-dimensional anti-Hermitian representation of the internal symmetry algebra $\mathscr{L}_{\text{int}}$. The maximum number of central charges U^{IJ} and V^{IJ} together is $N(N-1)$, and all are self-adjoint. If there are no central charges present then $\mathscr{L}_{\text{int}}$ is isomorphic to $\text{u}(N)$ or a subalgebra of $\text{u}(N)$. These properties depend on the fact that for $D \bmod 8 = 4$ both $\mathbf{C}$ and $\gamma_{D+1}\mathbf{C}$ are antisymmetric, and in the Majorana representation $\mathbf{C}^{\text{M}}$ is imaginary.

Example II *Extended Poincaré superalgebras for* $D \bmod 8 = 2$

As shown in Appendix L, Section 2, for $D \bmod 8 = 2$ a *Majorana–Weyl* representation can be constructed for the D-dimensional Minkowski Clifford algebra. For $D = 2$ explicit expressions for the matrices are given in (23.238) and (23.239), and Theorem III of Appendix L, Section 4, exhibits expressions for these matrices for $D \geqslant 10$. All the discussion in this example will be in terms of these representations.

Suppose that there exist N_{L} sets of left-handed supertranslation operators $Q^I_{\text{L}A}$, where $I = 1, 2, \dots, N_{\text{L}}$ and $A = 1, 2, \dots, \frac{1}{2}d$, and that there also exist N_{R} sets of right-handed supertranslation operators $Q^I_{\text{R}A}$, where $I = 1, 2, \dots, N_{\text{R}}$ and $A = 1, 2, \dots, \frac{1}{2}d$. By contrast to the situation in Example I, there is no need here to assume that N_{L} and N_{R} are equal. The generalizations of (23.163) and (23.164) are

$$[M_{\alpha\beta}, Q^I_{\text{L}A}]_- = -\sum_{B=1}^{d/2} \Gamma_{\text{L}}(M_{\alpha\beta})_{AB} Q^I_{\text{L}B} \tag{23.252}$$

(for $\alpha, \beta = 1, 2, \dots, D$ $(\alpha \neq \beta)$; $I = 1, 2, \dots, N_{\text{L}}$ and $A = 1, 2, \dots, \frac{1}{2}d$), and

$$[M_{\alpha\beta}, Q^I_{\text{R}A}]_- = -\sum_{B=1}^{d/2} \Gamma_{\text{R}}(M_{\alpha\beta})_{AB} Q^I_{\text{R}B} \tag{23.253}$$

(for $\alpha, \beta = 1, 2, \dots, D$ $(\alpha \neq \beta)$; $I = 1, 2, \dots, N_{\text{R}}$ and $A = 1, 2, \dots, \frac{1}{2}d$), where $\mathbf{\Gamma}_{\text{L}}$ and $\mathbf{\Gamma}_{\text{R}}$ are the $\frac{1}{2}d$-dimensional irreducible representations of $\text{so}(D-1, 1)$ that are defined by (L.25). Similarly the generalizations of (23.161) and (23.162) for this case are

$$[B, Q^I_{\text{L}A}]_- = \sum_{J=1}^{N_{\text{L}}} \text{L}(B)_{JI} Q^J_{\text{L}A} \tag{23.254}$$

(for all $B \in \mathscr{L}_{\text{int}}$, $I = 1, 2, \dots, N_{\text{L}}$ and $A = 1, 2, \dots, \frac{1}{2}d$) and

$$[B, Q^I_{\text{R}A}]_- = \sum_{J=1}^{N_{\text{R}}} \text{R}(B)_{JI} Q^J_{\text{R}A} \tag{23.255}$$

(for all $B \in \mathscr{L}_{\text{int}}$, $I = 1, 2, \dots, N_{\text{R}}$ and $A = 1, 2, \dots, \frac{1}{2}d$), where $\mathbf{L}$ and $\mathbf{R}$ provide two anti-Hermitian representations of the internal symmetry algebra $\mathscr{L}_{\text{int}}$ of dimensions N_{L} and N_{R} respectively. However, as the $Q^I_{\text{L}A}$ and $Q^I_{\text{R}A}$ belong not only to the chiral representation but *also* to the *Majorana* representation, the adjoint conditions are

$$(Q^I_{\mathrm{L}A})^\dagger = Q^I_{\mathrm{L}A} \tag{23.256}$$

(for $I = 1, 2, \ldots, N_\mathrm{L}$ and $A = 1, 2, \ldots, \frac{1}{2}d$) and

$$(Q^I_{\mathrm{R}A})^\dagger = Q^I_{\mathrm{R}A} \tag{23.257}$$

(for $I = 1, 2, \ldots, N_\mathrm{R}$ and $A = 1, 2, \ldots, \frac{1}{2}d$). By contrast with (23.55), these do *not* couple $Q^I_{\mathrm{L}A}$ and $Q^I_{\mathrm{R}A}$. Assuming again that the internal symmetry operators B are such that $B^\dagger = -B$ (for all $B \in \mathscr{L}_{\mathrm{int}}$), it follows that all the matrices $\mathbf{L}(B)$ and $\mathbf{R}(B)$ must be *real*, but the representations $\mathbf{L}$ and $\mathbf{R}$ of $\mathscr{L}_{\mathrm{int}}$ are *not* related to each other in any way.

It remains to consider the anticommutators of the supertranslation operators. The natural generalization of (23.234) and (23.250) is

$$[Q^I_{\mathrm{L}A}, Q^J_{\mathrm{L}B}]_+ = -2\delta^{IJ} \sum_{\mu=1}^{D} (\gamma^\mu_{\mathrm{M}} \mathbf{C}^{\mathrm{M}})_{AB} P_\mu \tag{23.258}$$

(for $I, J = 1, 2, \ldots, N_\mathrm{L}$ and $A, B = 1, 2, \ldots, \frac{1}{2}d$),

$$[Q^I_{\mathrm{R}A}, Q^J_{\mathrm{R}B}]_+ = -2\delta^{IJ} \sum_{\mu=1}^{D} (\gamma^\mu_{\mathrm{M}} \mathbf{C}^{\mathrm{M}})_{A+d/2, B+d/2} P_\mu \tag{23.259}$$

(for $I, J = 1, 2, \ldots, N_\mathrm{R}$ and $A, B = 1, 2, \ldots, \frac{1}{2}d$), and

$$[Q^I_{\mathrm{L}A}, Q^J_{\mathrm{R}B}]_+ = 2i\delta_{AB} Z^{IJ} \tag{23.260}$$

(for $I = 1, 2, \ldots, N_\mathrm{L}$; $J = 1, 2, \ldots, N_\mathrm{R}$ and $A, B = 1, 2, \ldots, \frac{1}{2}d$), where the central charges Z^{IJ} are again members of $\mathscr{L}_{\mathrm{int}}$ and again commute with every basis element of the Poincaré superalgebra. As the ranges of I and J in (23.260) are different, there can be *no* symmetry (or antisymmetry) relations for the Z^{IJ} that are analogous to those of (23.208). (In deducing these from (23.250), the particular forms of the matrices of the Majorana–Weyl representation (L.118) and (L.119) have been used, together with the relation $\mathbf{C}^{\mathrm{M}} = -\gamma^{\mathrm{M}}_D$ (as, for example, in (L.116)).) The self-adjointness of the $Q^I_{\mathrm{L}A}$ and $Q^I_{\mathrm{R}A}$ implies that

$$(Z^{IJ})^\dagger = -Z^{IJ}. \tag{23.261}$$

In the absence of central charges the internal symmetry algebra $\mathscr{L}_{\mathrm{int}}$ must be isomorphic to $\mathrm{so}(N_\mathrm{L}) \oplus \mathrm{so}(N_\mathrm{R})$ or to a proper subalgebra of $\mathrm{so}(N_\mathrm{L}) \oplus \mathrm{so}(N_\mathrm{R})$.

For $D \bmod 8 = 0$ the same general line of argument that is given in Example I applies, but some of the properties mentioned at the end of Example I are not true for this case, forcing some modifications to the details of the results quoted. For example, the analysis of Appendix L, Section 2 shows that for $D \bmod 8 = 0$ $\mathbf{C}$ and $\gamma_{D+1}\mathbf{C}$ are symmetric and in the Majorana representation $\mathbf{C}^{\mathrm{M}}$ is real. Much of this analysis also applies to the N-extended theory for general *odd* values of D, but, in addition to the points just mentioned, two further effects follow from the fact that there is *no* chiral representation for odd D. First the matrix γ_{D+1} is not defined, and secondly the representation of the even part $\mathscr{L}_0$ of the superalgebra on the odd part $\mathscr{L}_1$ must have the form of a *single* direct product $\mathbf{\Gamma}^{\text{spin}} \otimes \mathbf{\Gamma}^{\text{int}}$. The adjointness condition $B^\dagger = -B$ for the operators B of $\mathscr{L}_{\text{int}}$ then implies that $\mathbf{\Gamma}^{\text{int}}$ is a *real* N-dimensional representation of $\mathscr{L}_{\text{int}}$. Consequently the last terms on the right-hand sides of (23.249) and (23.250) do not appear if D is odd. The result is that the Lie superalgebra can contain fewer central charges, and, in the special case in which there are no central charges, $\mathscr{L}_{\text{int}}$ has to be isomorphic to so(N) or to a subalgebra of so(N).

Lie *supergroups* corresponding to $N > 1$ and $D \bmod 8 = 0, 1, 2, 3$ or 4 can be constructed by combining the ideas for the $N = 1$ case discussed above with those for the $N > 1$, $D = 4$ case of the previous section in the obvious way.

For $D \bmod 8 = 5, 6$ or 7 a "non-extended" Poincaré superalgebra with just d supertranslation operators Q_a does *not* exist, but Poincaré superalgebras possessing Nd supertranslation operators with $N > 2$ can be constructed. Unlike the ones considered above, in these the anticommutation relations analogous to (23.250) do not involve the generalized charge conjugation matrices $\mathbf{C}$ directly. Strathdee (1987) has given a comprehensive account for all values of D. For further details of the $D = 5$ case see Breitenlohner and Kabelschacht (1979) and Cremmer (1981), and for the $D = 6$ case see Breitenlohner and Kabelschacht (1979) and Howe *et al.* (1983b).

5 Irreducible representations of the unextended $D = 4$ Poincaré superalgebra

As the $N = 1$, $D = 4$ Poincaré superalgebra contains the $D = 4$ Poincaré algebra in its even part, every irreducible representation of the Poincaré superalgebra must provide a representation of the Poincaré algebra, which may be reducible (and indeed always is reducible, as closer examination shows). The irreducible representations of the $D = 4$ Poincaré algebra are deducible from those of the $D = 4$ proper orthochronous Poincaré group, which were first investigated by Wigner (1937), an account of whose results can be found in Chapter 17, Section 8. The analysis of the irreducible representations of the $N = 1$, $D = 4$ Poincaré superalgebra was first given by Salam and Strathdee (1974b) (see also Grosser 1975a, Salam and Strathdee 1978). As *all* the representations of the superalgebras that appear in the next three sections are *graded*, to avoid unnecessary repetition the word "graded" will be omitted for the rest of this chapter.

Attention will be confined to those representations that are physically significant. These are of two types, each of which can be associated with the rest mass M of a free particle. One corresponds to M being real and positive, and the other to M being zero. They will be referred to as the "massive" and "massless" cases respectively. Both are infinite-dimensional.

Each irreducible representation of the Poincaré group of these two types is also specified by a further parameter j, which can be identified with the spin of the associated particle. As noted in (17.195), (17.196), (17.225), and (17.226), the Poincaré algebra has two Casimir operators P and W, and the eigenvalue of P is simply M^2c^2 (where c is the speed of light), whereas that of W depends on both M and j. As

$$P = \sum_{\alpha,\beta=1}^{4} \eta^{\alpha\beta} P_\alpha P_\beta, \tag{23.262}$$

it follows from (23.44) that this is a Casimir operator of the Poincaré superalgebra as well, but the corresponding result is not true for W (because of its more complicated structure). Consequently each irreducible representation of the $D = 4$ Poincaré superalgebra will be associated with a *single* mass M, but it is to be expected that several *different* values of spin will occur within the same irreducible representation. These considerations suggest that the cases $M > 0$ and $M = 0$ should be considered separately.

The Casimir operator W of the Poincaré algebra defined by (17.196) has a simple generalization that is obtained by letting

$$C^\alpha = \frac{1}{2\hbar} \sum_{\beta,\lambda,\mu=1}^{4} \varepsilon^{\alpha\beta\lambda\mu} M_{\beta\lambda} P_\mu - \frac{1}{8} \sum_{a,b=1}^{4} Q_a(\gamma^\alpha \gamma_5 C^{-1})_{ab} Q_b$$

and

$$C^{\lambda\mu} = P^\lambda C^\mu - P^\mu C^\lambda,$$

and defining the operator C by

$$C = \sum_{\alpha,\beta,\lambda,\mu=1}^{4} \eta_{\alpha\lambda}\eta_{\beta\mu} C^{\alpha\beta} C^{\lambda\mu}.$$

Then C is a Casimir operator of the Poincaré superalgebra (cf. Salam and Strathdee 1974b, Sokatchev 1975).

(a) *Irreducible representations of the unextended $D = 4$ Poincaré superalgebra corresponding to $M > 0$*

In Chapter 17, Section 8 the basis vectors of the carrier space of an irreducible representation of the proper orthochronous Poincaré group corresponding to rest mass M (> 0) and spin j ($= 0, \frac{1}{2}, 1, \ldots$) were denoted by $\phi_{p,m}$, where $p = (p^1, p^2, p^3, p^4)$ will henceforth be identified with the particle's *contra-*

variant 4-momentum and so necessarily satisfies the constraint

$$\sum_{\alpha,\beta=1}^{4} \eta_{\alpha\beta} p^\alpha p^\beta = M^2 c^2, \tag{23.263}$$

and where m takes the values $j, j-1, j-2, \ldots, -j$. In particular, if $p = \hat{p}$, where

$$\hat{p} = (0, 0, 0, Mc), \tag{23.264}$$

which is the contravariant 4-momentum in the rest frame of the particle, then the values of $\hbar m$ are the components of spin in the Oz direction. It is convenient now to modify this notation and to write

$$|p, k, j, m\rangle = \phi_{p,m}, \tag{23.265}$$

it being understood implicitly that this corresponds to rest mass M. The extra label k in (23.265) is included to allow for the possibility that different vectors can occur with the same values of p, j and m. It is natural to allow k to take values $1, 2, \ldots$. Then (17.223) implies that

$$P^\alpha |p, k, j, m\rangle = p^\alpha |p, k, j, m\rangle \tag{23.266}$$

(for $\alpha = 1, \ldots, 4$). (The quantities $|p, k, j, m\rangle$ have to be used with a certain amount of caution because they cannot form a discrete orthonormal basis of the Hilbert space of physical states, since the 4-momenta satisfying (23.263) constitute a continuous set, but this will not pose any difficulties in the treatment that follows.)

On removing the labels $\hat{p}$ and j from the operators corresponding to each proper orthochronous Poincaré transformation $\{\mathbf{\Lambda} | \mathbf{t}\}$, (17.220) becomes

$$\Phi([\mathbf{\Lambda}|\mathbf{t}])|p, k, j', m'\rangle = \exp\left\{\frac{i}{\hbar} \sum_{\alpha,\beta=1}^{4} \eta_{\alpha\beta} (\mathbf{\Lambda p})^\alpha t^\beta \right\}$$

$$\times \sum_{m''=-j'}^{j'} D^{j'}([\mathbf{B}(\Lambda p, \hat{p})^{-1} \mathbf{\Lambda B}(p, \hat{p}) | \mathbf{0}])_{m''m'} |p, k, j', m''\rangle \tag{23.267}$$

for all $m' = j', j'-1, \ldots, -j'$. Here $\mathbf{B}(p, \hat{p})$ is the "Lorentz boost" that transforms $\hat{p}$ to p, so that (cf. (17.200))

$$\mathbf{B}(p, \hat{p})\hat{\mathbf{p}} = \mathbf{p}. \tag{23.268}$$

In matrix expressions such as those in (23.267) and (23.268) $\mathbf{p}$ denotes the 4×1 column matrix whose matrix elements are the components p^1, p^2, p^3 and p^4 of the *contravariant* 4-momentum p. In (23.267) $\mathbf{D}^{j'}$ is the $(2j'+1)$-dimensional irreducible representation of the universal covering group SU(2) of the SO(3) group of proper rotation in $\mathbb{R}^3$.

In particular, for $p = \hat{p}$ and

$$\boldsymbol{\Lambda} = \begin{bmatrix} \mathbf{R} & \mathbf{0} \\ \mathbf{0} & \mathbf{1}_1 \end{bmatrix}, \tag{23.269}$$

where $\mathbf{R}$ is a 3×3 real orthogonal matrix of SO(3), so that

$$\boldsymbol{\Lambda}\hat{\mathbf{p}} = \hat{\mathbf{p}}, \tag{23.270}$$

(23.267) reduces to

$$\Phi([\boldsymbol{\Lambda}|\mathbf{0}])|\hat{p},k,j',m'\rangle = \sum_{m''=-j'}^{j'} D^{j'}([\boldsymbol{\Lambda}|\mathbf{0}])_{m''m'}|\hat{p},k,j',m''\rangle, \tag{23.271}$$

(cf. (17.222)). Also, since (23.269) implies that $\mathbf{B}(\Lambda\hat{p},\hat{p}) = \boldsymbol{\Lambda}$, (23.267) shows that

$$\Phi([\boldsymbol{\Lambda}|\mathbf{0}])|\hat{p},k,j',m'\rangle = |\Lambda\hat{p},k,j',m'\rangle. \tag{23.272}$$

Of course the simple form of (23.271) is not coincidental but can be traced back to the fact that the set of Lorentz transformations $\boldsymbol{\Lambda}$ that satisfy (23.270) form the *little group* of the 4-momentum $\hat{p}$ from which the representation (23.267) has been constructed by the induced-representation method. The corresponding *little algebra* of the homogeneous Lorentz algebra then consists of those elements $\mathbf{M}$ of the Lorentz algebra such that

$$\mathbf{M}\hat{\mathbf{p}} = \mathbf{0}. \tag{23.273}$$

This little algebra has dimension 3, basis elements M_{12}, M_{23} and M_{31}, and is isomorphic to su(2).

The first stage in applying these results to the $N = 1$, $D = 4$ Poincaré superalgebra is to note that if Q_a (for $a = 1, \ldots, 4$) are the supertranslation operators then (23.44) implies that $Q_a|p,k,j',m'\rangle$ is an eigenvector of P_α with eigenvalue p_α (for $\alpha = 1, \ldots, 4$ and all 4-momenta p_α satisfying (23.263)); that is, the Q_a do not change the 4-momenta. (To simplify the notation, $\boldsymbol{\Phi}(Q_a)$, $\boldsymbol{\Phi}(P_\alpha)$, and $\boldsymbol{\Phi}(M_{\alpha\beta})$ are written here as Q_a, P_α and $M_{\alpha\beta}$ respectively, which is a convention that has already been adopted in (23.267).) Consequently it is natural to consider as the superalgebra generalization of the little algebra the subspace of elements of the Poincaré superalgebra generated by M_{12}, M_{23}, M_{31} and the four supertranslation operators Q_1, Q_2, Q_3 and Q_4. The commutation and anticommutation relations of the basis elements of this subspace take their simplest form if the supertranslation operators are taken to be the set Q_A and $\bar{Q}_A$ (for $A = 1, 2$) of Section 2(*b*) and if M_+, M_-, and M_3 are defined by

$$M_+ = M_{23} + iM_{31}, \tag{23.274}$$

$$M_- = M_{23} - iM_{31}, \tag{23.275}$$

$$M_3 = M_{12} \tag{23.276}$$

(cf. (17.8), (17.10), (17.17) and (17.23)). Apart from an extra factor of $\hbar$, these satisfy the same commutation relations as the operators A_+, A_- and A_3 of Chapter 12, Section 3, so that all the results and conventions of Chapter 12 may be applied immediately. In particular, the second-order Casimir operator M^2 of su(2) may be defined by

$$M^2 = (M_{12})^2 + (M_{23})^2 + (M_{31})^2, \tag{23.277}$$

(cf. (12.18)) and may be rewritten as

$$M^2 = (M_3)^2 - M_3 + M_+ M_- \tag{23.278}$$

(cf. (12.22)). Moreover, the basis vectors $|\hat{p}, k, j', m'\rangle$ may be chosen to be an orthonormal set so that

$$M^2 |\hat{p}, k, j', m'\rangle = \hbar^2 j'(j'+1) |\hat{p}, k, j', m'\rangle, \tag{23.279}$$

$$M_3 |\hat{p}, k, j', m'\rangle = \hbar m' |\hat{p}, k, j', m'\rangle, \tag{23.280}$$

$$M_+ |\hat{p}, k, j', m'\rangle = \hbar \{(j'-m')(j'+m'+1\}]^{1/2} |\hat{p}, k, j', m'\rangle, \tag{23.281}$$

$$M_- |\hat{p}, k, j', m'\rangle = \hbar \{(j'+m')(j'-m'+1)\}^{1/2} |\hat{p}, k, j', m'\rangle, \tag{23.282}$$

(cf. (12.25)–(12.28)).

The strategy is to investigate these rest-frame state vectors first, and then "boost" to any other frame using (23.272). As the supertranslation operators form the basis of the carrier space of the representation $-\tilde{\mathbf{\Gamma}}^{\text{spin}}$ of the homogeneous Lorentz algebra, they also form a carrier space when this is exponentiated to give the representation $(\tilde{\mathbf{\Gamma}}^{\text{spin}})^{-1}$ of the covering group of the proper orthochronous Lorentz group. Thus for a general representation of the Dirac matrices,

$$[\Phi([\mathbf{\Lambda}|\mathbf{0}]), Q_a]_- = \sum_{b=1}^{4} (\mathbf{\Gamma}^{\text{spin}}([\mathbf{\Lambda}|\mathbf{0}])^{-1})_{ab} Q_b \tag{23.283}$$

(for $a = 1, \ldots, 4$), which, by reference to (23.56) and (23.57), reduces in the two-component formulation to

$$[\Phi([\mathbf{\Lambda}|\mathbf{0}]), Q_A]_- = \sum_{B=1}^{2} (\mathbf{\Gamma}_{\text{L}}([\mathbf{\Lambda}|\mathbf{0}])^{-1})_{AB} Q_B \tag{23.284}$$

and

$$[\Phi([\mathbf{\Lambda}|\mathbf{0}]), \bar{Q}_A]_- = \sum_{B=1}^{2} (\mathbf{\Gamma}_{\text{R}}([\mathbf{\Lambda}|\mathbf{0}])^{-1})_{AB} \bar{Q}_B \tag{23.285}$$

(for $A = 1, 2$). Then

$$\begin{aligned} Q_A |\Lambda\hat{p}, k, j', m'\rangle = {} & \Phi([\mathbf{\Lambda}|\mathbf{0}])\{Q_A |\hat{p}, k, j', m'\rangle\} \\ & - \sum_{B=1}^{2} (\mathbf{\Gamma}_{\text{L}}([\mathbf{\Lambda}|\mathbf{0}])^{-1})_{AB} Q_B |\hat{p}, k, j', m'\rangle \end{aligned} \tag{23.286}$$

and

$$\bar{Q}_A|\Lambda\hat{p},k,j',m'\rangle = \Phi([\mathbf{\Lambda}|\mathbf{0}])\{\bar{Q}_A|\hat{p},k,j',m'\rangle\} - \sum_{B=1}^{2}(\mathbf{\Gamma}_{\mathrm{R}}([\mathbf{\Lambda}|\mathbf{0}])^{-1})_{AB}\bar{Q}_B|\hat{p},k,j',m'\rangle \quad (23.287)$$

(for $A = 1, 2$). Thus a knowledge of how the Q_A and $\bar{Q}_A$ act on the rest-frame state vectors $|\hat{p},k,j',m'\rangle$ enables their effect on any other state vector to be determined. The rest of the argument will therefore concentrate on the rest-frame state vectors

The first point to note is that, on using (23.264), (23.66) gives the simple action

$$[Q_A, \bar{Q}_B) + |\hat{p},k,j',m'\rangle,$$

so that, provided attention is restricted to the subspace of state vectors with contravariant 4-momentum $\hat{p}$ given by (23.264),

$$[Q_A, \bar{Q}_B]_+ = 2Mc\delta_{AB}1 \quad (23.288)$$

(for $A, B = 1, 2$), where 1 is the identity operator, which, of course, commutes with all the other operators. Consequently the *little superalgebra* can be taken to have basis elements M_{12}, M_{23}, M_{31}, Q_1, Q_2, $\bar{Q}_3$, $\bar{Q}_4$ *and* the identity 1. Clearly this is a Lie superalgebra (although it is not a subalgebra of the Poincaré superalgebra). The complete set of commutation and anti-commutation relations of the little superalgebra then consist of (23.288) together with

$$[Q_A, Q_B]_+ = 0, \quad (23.289)$$

$$[\bar{Q}_A, \bar{Q}_B]_+ = 0 \quad (23.290)$$

(cf. (23.64) and (23.65)).

Moreover, it follows from (23.56)–(23.59), (23.274)–(23.276) and (23.280)–(23.282) that

$$M_3Q_1|\hat{p},k,j',m'\rangle = \hbar(m' - \tfrac{1}{2})Q_1|\hat{p},k,j',m'\rangle, \quad (23.291)$$

$$M_3Q_2|\hat{p},k,j',m'\rangle = \hbar(m' + \tfrac{1}{2})Q_2|\hat{p},k,j',m'\rangle, \quad (23.292)$$

$$M_+Q_1|\hat{p},k,j',m'\rangle = -\hbar Q_2|\hat{p},k,j',m'\rangle + \hbar\{(j'-m')(j'+m'+1)\}^{1/2}Q_1|\hat{p},k,j',m'+1\rangle, \quad (23.293)$$

$$M_+Q_2|\hat{p},k,j',m'\rangle = \hbar\{(j'-m')(j'+m'+1)\}^{1/2}Q_2|\hat{p},k,j',m'+1\rangle, \quad (23.294)$$

$$M_-Q_1|\hat{p},k,j',m'\rangle = \hbar\{(j'+m')(j'-m'+1)\}^{1/2}Q_1|\hat{p},k,j',m'-1\rangle, \quad (23.295)$$

$$M_-Q_2|\hat{p},k,j',m'\rangle = -\hbar Q_1|\hat{p},k,j',m'\rangle + \hbar\{(j'+m')(j'-m'+1)\}^{1/2}Q_2|\hat{p},k,j',m'-1\rangle, \quad (23.296)$$

and

$$M_3\bar{Q}_1|\hat{p},k,j',m'\rangle = \hbar(m'+\tfrac{1}{2})\bar{Q}_1|\hat{p},k,j',m'\rangle \tag{23.297}$$

$$M_3\bar{Q}_2|\hat{p},k,j',m'\rangle = \hbar(m'-\tfrac{1}{2})\bar{Q}_2|\hat{p},k,j',m'\rangle, \tag{23.298}$$

$$M_+\bar{Q}_1|\hat{p},k,j',m'\rangle = \hbar\{(j'-m')(j'+m'+1)\}^{1/2}\bar{Q}_1|\hat{p},k,j',m'+1\rangle, \tag{23.299}$$

$$M_+\bar{Q}_2|\hat{p},k,j',m'\rangle = \hbar\bar{Q}_1|\hat{p},k,j',m'\rangle + \hbar\{(j'-m')(j'+m'+1)\}^{1/2}\bar{Q}_2|\hat{p},k,j',m'+1\rangle, \tag{23.300}$$

$$M_-\bar{Q}_1|\hat{p},k,j',m'\rangle = \hbar\bar{Q}_2|\hat{p},k,j',m'\rangle + \hbar\{(j'+m')(j'-m'+1)\}^{1/2}\bar{Q}_1|\hat{p},k,j',m'-1\rangle, \tag{23.301}$$

$$M_-\bar{Q}_2|\hat{p},k,j',m'\rangle = \hbar\{(j'+m')(j'-m'+1)\}^{1/2}\bar{Q}_2|\hat{p},k,j',m'-1\rangle. \tag{23.302}$$

To construct an irreducible representation, first choose any allowed value of j $(=0,\frac{1}{2},1,\ldots)$ and suppose that

$$Q_A|\hat{p},1,j,m\rangle = 0 \tag{23.303}$$

for $A=1,2$ and $m=j,j-1,\ldots,-j$. Then the set of $4(2j+1)$ vectors $|\hat{p},1,j,m\rangle$, $\bar{Q}_1|\hat{p},1,j,m\rangle$, $\bar{Q}_2|\hat{p},1,j,m\rangle$ and $\bar{Q}_1\bar{Q}_2|\hat{p},1,j,m\rangle$ (for $m=j,j-1,\ldots,-j$) span a subspace of the space of physical states that is left invariant by the action of Q_1, Q_2, $\bar{Q}_1$, $\bar{Q}_2$, M_+, M_- and M_3 (in the usual sense that every one of these operators acting on a vector of this set produces a linear combination of vectors of this set). Closer examination will show that this subspace is irreducible; that is, there is no non-trivial proper subset of the vectors with the same property.

It will now be shown that appropriate linear combinations of the $4(2j+1)$ members of this set can be identified with state vectors $|\hat{p},k,j',m'\rangle$ satisfying (23.279)–(23.282) and these fall into four subsets:

(i) one subset corresponding to $j'=j+\frac{1}{2}$ (for which k may be assigned the value 1), whose $2j+2$ normalized basis elements may be denoted by $|\hat{p},1,j+\frac{1}{2},m'\rangle$ (for $m'=j+\frac{1}{2},j-\frac{1}{2},\ldots,-j-\frac{1}{2}$);

(ii) one subset corresponding to $j'=j-\frac{1}{2}$ (for whick k may be assigned the value 1), whose $2j$ normalized basis elements may be denoted by $|\hat{p},1,j-\frac{1}{2},m'\rangle$ (for $m'=j-\frac{1}{2},\ldots,-j+\frac{1}{2}$);

(iii) two subsets corresponding to $j'=j$ (for which k may be assigned the values 1 and 2), whose $2(2j+1)$ normalized basis elements may be denoted by $|\hat{p},1,j,m'\rangle$ and $|\hat{p},2,j,m'\rangle$ (for $m'=j,j-1,\ldots,-j$).

If $j=0$, the subset (ii) is empty. Put more concisely, a typical massive particle multiplet for the $N=1$, $D=4$ Poincaré superalgebra corresponds to spins $j+\frac{1}{2}$, j and $j-\frac{1}{2}$, with j occurring twice (the spin $j-\frac{1}{2}$ appearing only if $j\geqslant\frac{1}{2}$).

It should be noted that such a multiplet always contains states with *integer* spin *and* states with *half-integer* spin and that the number of integer-spin states and the number of half-integer-spin states are always *equal*.

Before proceeding with the proof of these results, it is worth noting that the anticommutation relations (23.288)–(23.290) can be viewed as being those of two fermion annihilation operators $(2Mc)^{-1/2}Q_A$ and their corresponding two fermion creation operators $(2Mc)^{-1/2}\bar{Q}_A$ (for $A = 1, 2$), an interpretation that is reinforced by (23.55), which shows that $((2Mc)^{-1/2}Q_A)^{\dagger} = (2Mc)^{-1/2}\bar{Q}_A$. Moreover, (23.303) indicates that for each chosen value of j and each m $(= j, j-1, \ldots, -j)$ the state vector $|\hat{p}, 1, j, m\rangle$ can be regarded as being a "vaccum state" of a fermion Fock space whose basis consists of the four state vectors $|\hat{p}, 1, j, m\rangle$, $\bar{Q}_1|\hat{p}, 1, j, m\rangle$, $\bar{Q}_2|\hat{p}, 1, j, m\rangle$ and $\bar{Q}_1\bar{Q}_2|\hat{p}, 1, j, m\rangle$.

To derive the results on the spin content of the irreducible representation, consider first $\bar{Q}_1|\hat{p}, 1, j, j\rangle$. By (23.276), (23.297), (23.299) and (23.301),

$$M_3\bar{Q}_1|\hat{p}, 1, j, j\rangle = \hbar(j + \tfrac{1}{2})\bar{Q}_1|\hat{p}, 1, j, j\rangle \tag{23.304}$$

and

$$M^2\bar{Q}_1|\hat{p}, 1, j, j\rangle = \hbar^2(j + \tfrac{1}{2})\{(j + \tfrac{1}{2}) + 1\}\bar{Q}_1|\hat{p}, 1, j, j\rangle, \tag{23.305}$$

and since $\bar{Q}_1|\hat{p}, 1, j, j\rangle$ cannot be zero (if it were then (23.303) would imply that (23.288) is violated), it follows that $\bar{Q}_1|\hat{p}, 1, j, j\rangle$ is a simultaneous eigenvector of M_3 with eigenvalue $\hbar(j + \frac{1}{2})$ and of M^2 with eigenvalue $\hbar^2(j + \frac{1}{2})\{(j + \frac{1}{2}) + 1\}$. Let

$$|\hat{p}, 1, j + \tfrac{1}{2}, j + \tfrac{1}{2}\rangle = \eta\bar{Q}_1|\hat{p}, 1, j, j\rangle, \tag{23.306}$$

where η is some complex number. By (23.55), (23.288) and (23.303), the normalization assumption for state vectors implies that $2Mc|\eta|^2 = 1$, and since the most convenient choice is to take η to be real and positive, so that $\eta = (2Mc)^{-1/2}$, (23.306) becomes

$$|\hat{p}, 1, j + \tfrac{1}{2}, j + \tfrac{1}{2}\rangle = (2Mc)^{-1/2}\bar{Q}_1|\hat{p}, 1, j, j\rangle. \tag{23.307}$$

Applying the operators $M_-, (M_-)^2, (M_-)^3, \ldots$ to $|\hat{p}, 1, j + \frac{1}{2}, j + \frac{1}{2}\rangle$ and using (23.282) gives its partners $|\hat{p}, 1, j + \frac{1}{2}, m'\rangle$ for $m' = j - \frac{1}{2}, \ldots, -j - \frac{1}{2}$. In particular,

$$\begin{aligned}|\hat{p}, 1, j + \tfrac{1}{2}, j - \tfrac{1}{2}\rangle &= \hbar^{-1}\{2Mc(2j + 1)\}^{-1/2}M_-\bar{Q}_1|\hat{p}, 1, j, j\rangle \\ &= \{2Mc(2j + 1)\}^{-1/2}\{\bar{Q}_2|\hat{p}, 1, j, j\rangle \\ &\quad + (2j)^{1/2}\bar{Q}_1|\hat{p}, 1, j, j - 1\rangle\}.\end{aligned} \tag{23.308}$$

Indeed, (23,297) and (23.298) show that both $\bar{Q}_1|\hat{p}, 1, j, j - 1\rangle$ and $\bar{Q}_2|\hat{p}, 1, j, j\rangle$ are eigenvectors of M_3 with eigenvalue $\hbar(j - \frac{1}{2})$. By construction, the linear combination on the right-hand side of (23.308) is also an eigenvector of M^2 with eigenvalue $\hbar^2(j + \frac{1}{2})\{(j + \frac{1}{2}) + 1\}$, so if $j \geqslant \frac{1}{2}$ then there must also be a linear combination of $\bar{Q}_2|\hat{p},1,j,j\rangle$ and $\bar{Q}_1|\hat{p},1,j,j-1\rangle$ that is an eigenvector of M^2 with eigenvalue $\hbar^2(j-\frac{1}{2})\{(j-\frac{1}{2})+1\}$. Denoting this by $|\hat{p},1,j-\frac{1}{2},j-\frac{1}{2}\rangle$,

$$|\hat{p},1,j-\tfrac{1}{2},j-\tfrac{1}{2}\rangle=\alpha\bar{Q}_2|\hat{p},1,j,j\rangle+\beta\bar{Q}_1|\hat{p},1,j,j-1\rangle, \qquad (23.309)$$

where α and β are complex numbers that may be determined (up to a common multiplicative factor) by the condition that

$$M_+|\hat{p},1,j-\tfrac{1}{2},j-\tfrac{1}{2}\rangle=0. \qquad (23.310)$$

Applying the operators M_-, $(M_-)^2$, $(M_-)^3,\ldots$ to $|\hat{p},1,j-\frac{1}{2},j-\frac{1}{2}\rangle$ and using (23.282) then gives its partners $|\hat{p},1,j-\frac{1}{2},m'\rangle$ for $m'=(j-\frac{1}{2})-1,\ldots,-(j-\frac{1}{2})$.

As (23.301) implies that $\bar{Q}_1\bar{Q}_1=0$, (23.307) shows that acting on $|\hat{p},1,j+\frac{1}{2},j+\frac{1}{2}\rangle$ with $\bar{Q}_1$ will give zero.

Finally, by (23.276) and (23.297)–(23.302),

$$[M_3,\bar{Q}_1\bar{Q}_2]_-=0 \qquad (23.311)$$

and

$$[M^2,\bar{Q}_1\bar{Q}_2]_-=0, \qquad (23.312)$$

so $\bar{Q}_1\bar{Q}_2|\hat{p},1,j,m\rangle$ is a simultaneous eigenvector of M_3 with eigenvalue $\hbar m$ and of M^2 with eigenvalue $\hbar^2 j(j+1)$. Consequently one may define

$$|\hat{p},2,j,j\rangle=(2Mc)^{-1}\bar{Q}_1\bar{Q}_2|\hat{p},1,j,j\rangle,$$

the factor $(2Mc)^{-1}$ being inserted to normalize $|\hat{p},2,j,j\rangle$, from which it follows that

$$|\hat{p},2,j,m\rangle=(2Mc)^{-1}\bar{Q}_1\bar{Q}_2|\hat{p},1,j,m\rangle \qquad (23.313)$$

for all $m=j,j-1,\ldots,-j$. This completes the establishment of the spin content of the multiplet.

It is easily shown that the Casimir operator C has eigenvalue $-2M^4c^4j(j+1)$ on these rest-frame state vectors, and hence on all the state vectors of the irreducible representation. Consequently j is known as the *superspin*.

It is not difficult to derive explicit expressions for the actions of all the operators Q_1, Q_2, $\bar{Q}_1$ and $\bar{Q}_2$ on all the rest-frame states. As the arguments involved are similar to those given in Chapter 12, it should come as no surprise that the resulting expressions involve the Clebsch–Gordan coefficients $(j_1,m_1,j_2,m_2|j_1,j_2,j,m)$ of su(2). for example

$$\begin{aligned}\bar{Q}_1|\hat{p},1,j,m\rangle=&(2Mc)^{1/2}(\tfrac{1}{2},\tfrac{1}{2},j,m|\tfrac{1}{2},j,j+\tfrac{1}{2},m+\tfrac{1}{2})|\hat{p},1,j+\tfrac{1}{2},m+\tfrac{1}{2}\rangle\\&+(2Mc)^{1/2}(\tfrac{1}{2},\tfrac{1}{2},j,m|\tfrac{1}{2},j,j-\tfrac{1}{2},m+\tfrac{1}{2})|\hat{p},1,j-\tfrac{1}{2},m+\tfrac{1}{2}\rangle\end{aligned} \qquad (23.314)$$

and

$$\bar{Q}_2|\hat{p},1,j,m\rangle = (2Mc)^{1/2}(\tfrac{1}{2},-\tfrac{1}{2},j,m|\tfrac{1}{2},j,j+\tfrac{1}{2},m-\tfrac{1}{2})|\hat{p},1,j+\tfrac{1}{2},m-\tfrac{1}{2}\rangle + (2Mc)^{1/2}(\tfrac{1}{2},-\tfrac{1}{2},j,m|\tfrac{1}{2},j,j-\tfrac{1}{2},m-\tfrac{1}{2})|\hat{p},1,j-\tfrac{1}{2},m-\tfrac{1}{2}\rangle. \quad (23.315)$$

Before leaving this case, it is interesting to note that if the Majorana supertranslation operators Q_a^{M} (for $a = 1,\ldots,4$) are used instead of the supertranslation operators Q_A and $\bar{Q}_A$ (for $A = 1,2$) of the two-component formulation then the analogues of (23.288)–(23.290) are the single set of equations

$$Q_a^{\mathrm{M}}Q_b^{\mathrm{M}} + Q_b^{\mathrm{M}}Q_a^{\mathrm{M}} = 2Mc\delta_{ab}1, \quad (23.316)$$

(for $a,b = 1,\ldots,4$). (These follow from applying (23.265) and (23.266) to (23.264), using (17.167) (which is a special case of (L.71).) Comparison with (L.2) and (L.4) shows that the operators $(Mc)^{-1/2}Q_a^M$ form a 4-dimensional Euclidean Clifford algebra. For each j and m specifying a vacuum state vector the 4-dimensional fermion Fock space mentioned above is then the basis for the 4-dimensional irreducible representation of this Clifford algebra.

(b) *Irreducible representations of the unextended $D = 4$ Poincaré superalgebra corresponding to $M = 0$*

The general strategy for this case is the same as that for the massive case in that the effect of the supertranslation operators is first determined on the states corresponding to a special 4-momentum $\hat{p}$, from which the corresponding results for any other allowed 4-momentum p can be deduced using (23.286) and (23.287). In the $M = 0$ case it is convenient to assume that the special contravariant 4-momentum $\hat{p}$ is given by

$$\hat{p} = (0,0,\kappa,\kappa), \quad (23.317)$$

where κ is a positive real constant with the dimensions of linear momentum. The corresponding little group, defined as in (23.270), is isomorphic to the covering group of the 2-dimensional Euclidean group (cf. Chapter 17, Section 8(D)). Its rotational part has generator M_{12} (which will continue to be denoted by M_3, as in (23.276), and its translational parts have generators $M_{23} + M_{24}$ and $M_{31} + M_{41}$. The massless irreducible representations of the $D=4$ Poincaré group are induced from irreducible representations of this little group in which the translational parts are represented trivially. As the rotational part of the little group is Abelian, all these irreducible representations of the little group are one-dimensional. They may be specified by a parameter j', which takes both positive and negative integer and half-integer values (i.e. $j' = 0, \pm\frac{1}{2}, \pm 1,\ldots$), $\hbar j'$ being identified with the helicity of the particle. The corresponding state vector may be denoted by $|\hat{p}, j'\rangle$, so that

$$(M_{23} + M_{24})|\hat{p}, j'\rangle = 0, \tag{23.318}$$

$$(M_{31} + M_{41})|\hat{p}, j'\rangle = 0, \tag{23.319}$$

$$M_3|\hat{p}, j'\rangle = M_{12}|\hat{p}, j'\rangle = \hbar j'|\hat{p}, j'\rangle. \tag{23.320}$$

(In terms of the notation of Chapter 17, Section 8(D), this implies that

$$\mathbf{\Gamma}^{j}_{L(\hat{p})}(\mathbf{a}(\xi, \eta; \phi)) = [\mathrm{e}^{-j'\phi}], \tag{23.321}$$

where a superfluous factor of $\frac{1}{2}$ has been removed from the exponent and the sign of j' has been chosen so that (23.320) has the same sign as in the corresponding equation (23.280) of the massive case. Moreover, it should also be noted that the matrices

$$\mathbf{a}(\xi, \eta; \phi) = \begin{bmatrix} \mathrm{e}^{-(1/2)i\phi} & (\xi + i\eta)\mathrm{e}^{-(1/2)i\phi} \\ 0 & \mathrm{e}^{-(1/2)i\phi} \end{bmatrix}, \tag{23.322}$$

of the two-dimensional Euclidean group correspond to proper orthochronous Lorentz transformations $\mathbf{\Lambda}(\mathbf{a})$ that leave invariant the *contravariant* 4-momentum $\hat{p}$ of (23.317), for which $\hat{p}^1 = 0, \hat{p}^2 = 0, \hat{p}^3 = \kappa$ and $\hat{p}^4 = \kappa$ (and *not* the *covariant* 4-momentum for which $\hat{p}_1 = 0$, $\hat{p}_2 = 0$, $\hat{p}_3 = \kappa$ and $\hat{p}_4 = \kappa$).

On using (23.317), (23.66) gives

$$[Q_1, \bar{Q}_2]_+|\hat{p}, j'\rangle = 0,$$
$$[Q_2, \bar{Q}_1]_+|\hat{p}, j'\rangle = 0,$$
$$[Q_2, \bar{Q}_2]_+|\hat{p}, j'\rangle = 0,$$
$$[Q_1, \bar{Q}_1]_+|\hat{p}, j'\rangle = 4\kappa.$$

Thus, when restricted to the set of state vectors $|\hat{p}, j'\rangle$ with $\hat{p}$ given by (23.317), the supertranslation operators satisfy the anticommutation relations

$$[Q_1, \bar{Q}_2]_+ = 0, \tag{23.323}$$

$$[Q_2, \bar{Q}_1]_+ = 0, \tag{23.324}$$

$$[Q_2, \bar{Q}_2]_+ = 0, \tag{23.325}$$

$$[Q_1, \bar{Q}_1]_+ = 4\kappa 1, \tag{23.326}$$

where 1 is the identity operator. Thus in this massless case the little superalgebra consists of $M_{23} + M_{24}, M_{31} + M_{41}, M_3, Q_1, Q_2, \bar{Q}_1, \bar{Q}_2$ and the identity 1. The anticommutation relations (23.323)–(23.326) may be supplemented by (23.64) and (23.65); that is,

$$[Q_A, Q_B]_+|\hat{p}, j'\rangle = 0 \tag{23.327}$$

and

$$[\bar{Q}_A, \bar{Q}_B]_+|\hat{p}, j'\rangle = 0, \tag{23.328}$$

According to (23.56)–(23.59), the remaining non trivial commutation relations of the little superalgebra are

$$[M_{23}+M_{24},Q_1]_- = -\hbar Q_2, \tag{23.329}$$

$$[M_{23}+M_{24},Q_2]_- = 0, \tag{23.330}$$

$$[M_{23}+M_{24},\bar{Q}_1]_- = \hbar\bar{Q}_2, \tag{23.331}$$

$$[M_{23}+M_{24},\bar{Q}_2]_- = 0, \tag{23.332}$$

$$[M_{31}+M_{41},Q_1]_- = -i\hbar Q_2, \tag{23.333}$$

$$[M_{31}+M_{41},Q_2]_- = 0, \tag{23.334}$$

$$[M_{31}+M_{41},\bar{Q}_1]_- = -i\hbar\bar{Q}_2, \tag{23.335}$$

$$[M_{31}+M_{41},\bar{Q}_2]_- = 0, \tag{23.336}$$

and

$$[M_3,Q_1]_- = -\tfrac{1}{2}\hbar Q_1, \tag{23.337}$$

$$[M_3,Q_2]_- = \tfrac{1}{2}\hbar Q_2, \tag{23.338}$$

$$[M_3,\bar{Q}_1]_- = \tfrac{1}{2}\hbar\bar{Q}_1, \tag{23.339}$$

$$[M_3,\bar{Q}_2]_- = -\tfrac{1}{2}\hbar\bar{Q}_2. \tag{23.340}$$

The anticommutation relation (23.325) taken with (23.55) implies that $Q_2|\hat{p},j'\rangle$ and $\bar{Q}_2|\hat{p},j'\rangle$ have norms of opposite sign. The assumption that no state vector norms are negative then requires that

$$Q_2|\hat{p},j'\rangle = 0 \tag{23.341}$$

and

$$\bar{Q}_2|\hat{p},j'\rangle = 0 \tag{23.342}$$

(for all values of j').

For any $j = 0, \pm\frac{1}{2}, \pm 1, \ldots$, suppose that $|\hat{p},j\rangle$ is a normalized eigenvector of M_3 with eigenvalue $\hbar j$ that is such that

$$Q_1|\hat{p},j\rangle = 0. \tag{23.343}$$

Then (23.326) implies that $\bar{Q}_1|\hat{p},j\rangle \neq 0$, and since (23.339) demonstrates that

$$M_3\bar{Q}_1|\hat{p},j\rangle = \hbar(j+\tfrac{1}{2})\bar{Q}_1|\hat{p},j\rangle, \tag{23.344}$$

it follows that $\bar{Q}_1|\hat{p},j\rangle$ is an eigenvector of M_3 with eigenvalue $\hbar(j+\frac{1}{2})$. Denoting the corresponding normalized eigenvector by $|\hat{p},j+\frac{1}{2}\rangle$, with an appropriate choice of phase,

$$\bar{Q}_1|\hat{p},j\rangle = 2\kappa^{1/2}|\hat{p},j+\tfrac{1}{2}\rangle \tag{23.345}$$

and

$$Q_1|\hat{p},j+\tfrac{1}{2}\rangle = 2\kappa^{1/2}|\hat{p},j\rangle. \tag{23.346}$$

Finally, since (23.328) shows that $\bar{Q}_1\bar{Q}_1|\hat{p},j\rangle = 0$, it follows that

$$\bar{Q}_1|\hat{p},j+\tfrac{1}{2}\rangle = 0. \tag{23.347}$$

It is easily checked that (23.341)–(23.346) are consistent with the commutation relations (23.329)–(23.340).

The conclusion of this argument for the massless case is that for each $j = 0, \pm\frac{1}{2}, \pm 1, \ldots$, the corresponding multiplet consists of only *two* non-degenerate spin states j and $j + \frac{1}{2}$. Nevertheless, it is again true that such a multiplet always contains states with *integer* spin *and* states with *half-integer* spin and that the number of integer-spin states and the number of half-integer-spin states are always *equal*.

It should be noted that $(4\kappa)^{-1/2}\bar{Q}_1$ and $(4\kappa)^{-1/2}Q_1$ act as a pair of fermion creation and annihilation operators, with $|\hat{p}, j\rangle$ being the corresponding Fock space vacuum state vector.

6 Irreducible representations of the extended $D = 4$ Poincaré superalgebras

The general approach for the N-extended $D = 4$ Poincaré superalgebra is the same as for the $N = 1$ case since the operator P of (17.195) remains a Casimir operator. Consequently the massive and massless representations have to be considered separately and the irreducible representations are again completely determined by the action of the little superalgebras on the state vectors with the special contravariant 4-momenta (23.264) and (23.317). However, in the general case there are three separate complicating features: the number of supertranslation operators is increased to $4N$, there is an internal symmetry algebra with non-trivial commutation relations with the supertranslation operators, and there are also central charges. The notation for the N-extended $D = 4$ Poincaré superalgebra is as defined in Section 3. The Casimir operator C may be defined as in Section 5, the only change being that the expression for C^α becomes

$$C^\alpha = \frac{1}{2\hbar} \sum_{\beta,\lambda,\mu=1}^{4} \varepsilon^{\alpha\beta\lambda\mu} M_{\beta\lambda} P_\mu - \frac{1}{8} \sum_{a,b=1}^{4} \sum_{I=1}^{N} Q_a^I (\gamma^\alpha \gamma_5 \mathbf{C}^{-1})_{ab} Q_b^I$$

(cf. Taylor 1980). For an investigation of the Casimir operators of the extended $D = 4$ Poincaré superalgebras containing central charges see Galperin *et al.* (1983).

(a) *Irreducible representations of the N-extended Poincaré superalgebra corresponding to $M > 0$*

There are two subcases, depending on whether or not the Poincaré superalgebra contains central charges.

(i) No central charges in the Poincaré superalgebra

If there are no central charges in the Poincaré superalgebra then (23.192) and (23.193) reduce to

$$[Q^I_A, Q^J_B]_+ = 0 \tag{23.348}$$

and

$$[\bar{Q}^I_A, \bar{Q}^J_B]_+ = 0 \tag{23.349}$$

(for $A, B = 1, 2$ and $I, J = 1, \ldots, N$). Moreover, when restricted to acting on state vectors with the special contravariant 4-momentum $\hat{p}$ given by (23.264), that is with $\hat{p} = (0, 0, 0, Mc)$, (23.194) reduces to

$$[Q^I_A, \bar{Q}^J_B]_+ = 2Mc\delta^{IJ}\delta_{AB}1 \tag{23.350}$$

(for $A, B = 1, 2$ and $I, J = 1, 2, \ldots, N$). Consequently the little superalgebra of $\hat{p}$ consists of M_+, M_- and M_3 (which form an so(3) Lie algebra), together with all $4N$ supertranslation operators Q^I_A and $\bar{Q}^I_A$ (for $A = 1, 2$ and $I = 1, 2, \ldots, N$), the identity 1 and the internal symmetry algebra operators B_k (for $k = 1, 2, \ldots, n_0$) of $\mathscr{L}_{\text{int}}$. In addition to (23.238)–(23.240), its non-trivial commutation and anticommutation relations consist of (23.163), (23.164), (23.186), (23.188) and (23.189).

At first sight these appear to give a complicated structure, but this can be very considerably simplified by the following definitions. First (by analogy with (17.10)) define the three operators M_1, M_2 and M_3 by

$$M_1 = M_{23}, \quad M_2 = M_{31}, \quad M_3 = M_{12} \tag{23.351}$$

(which includes (23.276)), so that

$$[M_p, M_q]_- = i\hbar \sum_{r=1}^{3} \varepsilon_{pqr} M_r \tag{23.352}$$

(cf. (17.7)), where ε_{pqr} is the permutation symbol of (10.20). Then define the three operators $\bar{M}_1, \bar{M}_2$ and $\bar{M}_3$ by

$$\bar{M}_p = M_p - \frac{\hbar}{4Mc} \sum_{I=1}^{N} \sum_{A,B=1}^{2} \bar{Q}^I_A Q^I_B (\boldsymbol{\sigma}_p)_{AB} \tag{23.353}$$

for $p = 1, 2, 3$. Finally define the operators

$$\bar{B}_k = B_k - \frac{\hbar}{2Mc} \sum_{I,J=1}^{N} \sum_{A=1}^{2} \bar{Q}^I_A Q^J_A \boldsymbol{\Gamma}^{\text{int}}(B_k)^*_{IJ} \tag{23.354}$$

for $k = 1, 2, \ldots, n_0$, where $\boldsymbol{\Gamma}^{\text{int}}$ is the representation of $\mathscr{L}_{\text{int}}$ appearing in (23.188) and (23.189). Then it is easily shown that

$$[\bar{M}_p, \bar{M}_q]_- = i\hbar \sum_{r=1}^{3} \varepsilon_{pqr} \bar{M}_r \tag{23.355}$$

(for $p, q = 1, 2, 3$) and

$$[\bar{B}_j, \bar{B}_k]_- = \sum_{l=1}^{n_0} c^l_{jk} \bar{B}_l \tag{23.356}$$

(for $j, k = 1, 2, \ldots, n_0$). Moreover,

$$[\bar{M}_p, Q_A]_- = 0, \quad [\bar{M}_p, \bar{Q}_A]_- = 0, \tag{23.357}$$

$$[\bar{B}_k, Q_A]_- = 0, \quad [\bar{B}_k, \bar{Q}_A]_- = 0, \tag{23.358}$$

$$[\bar{M}_p, \bar{B}_k]_- = 0 \tag{23.359}$$

(for $A = 1, 2$; $p = 1, 2, 3$ and $k = 1, 2, \ldots, n_0$). Thus the little superalgebra of $\hat{p}$ is the *direct sum* of three algebras. One is the Lie algebra with basis elements $\bar{M}_1, \bar{M}_2$ and $\bar{M}_3$, which is isomorphic to the so(3) angular momentum Lie algebra, the second with basis elements $\bar{B}_k$ (for $k = 1, 2, \ldots, n_0$) is isomorphic to $\mathscr{L}_{\text{int}}$, and the third is the Lie superalgebra consisting of the $4N$ supertranslation operators Q^I_A and $\bar{Q}^I_A$ (for $A = 1, 2$ and $I = 1, 2, \ldots, N$) and the identity 1. Consequently the irreducible representations of the little superalgebra are just direct products of irreducible representations of its three constituent parts.

To construct such an irreducible representation of the little superalgebra, choose *any* irreducible representation $\mathbf{D}^j$ of so(3) (where $j = 0, \frac{1}{2}, 1, \ldots$) (cf. Chapter 12, Section 3), and *any* irreducible representation $\mathbf{\Gamma}^{\text{int}}_{\text{V}}$ of $\mathscr{L}_{\text{int}}$ of dimension d_{V}, and let $|\hat{p}, j, m, r\rangle$ be a set of basis vectors for the direct product of these representations, chosen so that

$$\bar{M}_p|\hat{p}, j, m, r\rangle = \sum_{m'=-j}^{j} D^j(\bar{M}_p)_{m'm}|\hat{p}, j, m', r\rangle \tag{23.360}$$

and

$$\bar{B}_k|\hat{p}, j, m, r\rangle = \sum_{r'=1}^{d_{\text{V}}} \Gamma^{\text{int}}_{\text{V}}(\bar{B}_k)_{r'r}|\hat{p}, j, m, r'\rangle \tag{23.361}$$

(for $p = 1, 2, 3$; $k = 1, 2, \ldots, n_0$; $m = j, j-1, \ldots, -j$ and $r = 1, 2, \ldots, d_{\text{V}}$). Assume also that

$$Q^I_A|\hat{p}, j, m, r\rangle = 0 \tag{23.362}$$

(for $I = 1, 2, \ldots, N$; $A = 1, 2$; $m = j, j-1, \ldots, -j$ and $r = 1, 2, \ldots, d_{\text{V}}$). Then each vector $|\hat{p}, j, m, r\rangle$ is a "vacuum state vector" for the set of fermion creation and annihilation operators $\bar{Q}^I_A$ and Q^I_A, and the complete set of basis vectors of the corresponding irreducible representation of the little superalgebra is given by the $|\hat{p}, j, m, r\rangle$ and all vectors obtained from the $|\hat{p}, j, m, r\rangle$ by acting with every possible product of the creation operators $\bar{Q}^I_A$. Consequently the resulting irreducible representation of the little superalgebra has dimension $(2j+1)d_{\text{V}}2^{2N}$.

It is of interest to determine which direct products of irreducible representations of the "true" so(3) angular momentum algebra (with basis elements M_p) and the "true" internal symmetry algebra $\mathscr{L}_{\text{int}}$ (with basis elements B_j) appear in this irreducible representation of the little superalgebra. For the "vacuum" basis vectors $|\hat{p}, j, m, r\rangle$ themselves (23.353), (23.354) and (23.360)–(23.362) imply that

$$M_p|\hat{p}, j, m, r\rangle = \sum_{m'=-j}^{j} D^j(M_p)_{m'm}|\hat{p}, j, m', r\rangle \tag{23.363}$$

and

$$B_k|\hat{p}, j, m, r\rangle = \sum_{r'=1}^{d_V} \Gamma_V^{\text{int}}(B_k)_{r'r}|\hat{p}, j, m, r'\rangle \tag{23.364}$$

(for $p = 1, 2, 3$; $k = 1, 2, \ldots, n_0$; $m = j, j-1, \ldots, -j$ and $r = 1, 2, \ldots, d_V$), so that the vacuum state vectors $|\hat{p}, j, m, r\rangle$ actually transform as the direct product of the irreducible representation $\mathbf{D}^j$ of the "true" so(3) angular momentum Lie algebra with the irreducible representation $\boldsymbol{\Gamma}_V^{\text{int}}$ of the "true" internal symmetry algebra $\mathscr{L}_{\text{int}}$.

For the other basis vectors the situation is more complicated, and other irreducible representations of the true angular momentum and internal symmetry algebras appear. For example, for $\bar{Q}_A^I|\hat{p}, j, m\, r\rangle$ (23.353) shows that

$$\begin{aligned} M_p\bar{Q}_A^I|\hat{p}, j, m, r\rangle &= \bar{M}_p\bar{Q}_A^I|\hat{p}, j, m, r\rangle \\ &\quad + \frac{\hbar}{4Mc}\sum_{I'=1}^{N}\sum_{A',B'=1}^{2} \bar{Q}_{A'}^{I'} Q_{B'}^{I'} \bar{Q}_A^I(\boldsymbol{\sigma}_p)_{A'B'}|\hat{p}, j, m, r\rangle, \end{aligned}$$

and so, by (23.350), (23.357), (23.360) and (23.362),

$$\begin{aligned} M_p\bar{Q}_A^I|\hat{p}, j, m, r\rangle = &\sum_{m'=-j}^{j} D^j(M_p)_{m'm}\bar{Q}_A^I|\hat{p}, j, m', r\rangle \\ &+ \tfrac{1}{2}\hbar\sum_{A'=1}^{2}(\boldsymbol{\sigma}_p)_{A'A}\bar{Q}_{A'}^I|\hat{p}, j, m, r\rangle. \end{aligned} \tag{23.365}$$

As $\mathbf{D}^{1/2}(M_p) = \frac{1}{2}\hbar\boldsymbol{\sigma}_p$ (for $p = 1, 2, 3$), (11.26) implies that the $\bar{Q}_A^I|\hat{p}, j, m, r\rangle$ are basis vectors of the direct product representation $\mathbf{D}^j \otimes \mathbf{D}^{1/2}$ of so(3). With the assumption that $j \geqslant \frac{1}{2}$, because $\mathbf{D}^j \otimes \mathbf{D}^{1/2} \approx \mathbf{D}^{j+1/2} \oplus \mathbf{D}^{j-1/2}$ (cf. (12.41), appropriate linear combinations of the $\bar{Q}_A^I|\hat{p}, j, m, r\rangle$ transform as $\mathbf{D}^{j+1/2}$ and $\mathbf{D}^{j-1/2}$, and so correspond to spins $j+\frac{1}{2}$ and $j-\frac{1}{2}$. (If $j = 0$, the situation is simpler, since $\mathbf{D}^0 \otimes \mathbf{D}^{1/2} \approx \mathbf{D}^{1/2}$, implying that only the irreducible representation $\mathbf{D}^{1/2}$ appears.) Indeed, since the generalizations of (23.294) and (23.295) show that each $\bar{Q}_1^I$ increases the eigenvalue of M_3 by $\frac{1}{2}\hbar$, and $\bar{Q}_2^I$ decreases the eigenvalue of M_3 by $\frac{1}{2}\hbar$, it is clear that $\bar{Q}_1^I|\hat{p}, j, m, r\rangle$ belongs to $\mathbf{D}^{j+1/2}$ and has eigenvalue $\hbar(j+\frac{1}{2})$ with M_3. A similar argument shows that $\bar{Q}_1^I\bar{Q}_1^J|\hat{p}, j, j, r\rangle$ (for $I \neq J$) belongs to $\mathbf{D}^{j+1}$ and has eigenvalue $\hbar(j+1)$ with M_3. Clearly the *highest* spin that arises in this way is $j+\frac{1}{2}N$, for $(\bar{Q}_1^1\bar{Q}_1^2\cdots\bar{Q}_1^{N-1}\bar{Q}_1^N)|\hat{p}, j, j, r\rangle$

belongs to $\mathbf{D}^{j+N/2}$ and has the eigenvalue $\hbar(j+\frac{1}{2}N)$ with M_3. It is also obvious that exactly one-half of these $(2j+1)d_V 2^{2N}$ state vectors will have spins that differ from j by an integer, and for the other half the spins differ from j by a half-integer. Thus again the numbers of bosonic and fermionic states in the corresponding irreducible representation of the Poincaré superalgebra are *equal.*

The reduction with respect to the "true" $\mathscr{L}_{\text{int}}$ can be discussed in a similar way. For example, for $\bar{Q}^I_A|\hat{p}, j, m, r\rangle$ (23.354) shows that

$$B_k\bar{Q}^I_A|\hat{p}, j, m, r\rangle = \bar{B}_k\bar{Q}^I_A|\hat{p}, j, m, r\rangle + \frac{1}{2Mc}\sum_{I',J'=1}^{N}\sum_{A'=1}^{2} \bar{Q}^{I'}_{A'}\bar{Q}^{J'}_{A'}\bar{Q}^I_A\Gamma^{\text{int}}(B_k)^*_{I'J'}|\hat{p}, j, m, r\rangle,$$

and so, by (23.350), (23.358), (23.361) and (23.362)

$$B_k\bar{Q}^I_A|\hat{p}, j, m, r\rangle = \sum_{r'=1}^{d_V} \Gamma^{\text{int}}_V(B_k)_{r'r}\bar{Q}^I_A|\hat{p}, j, m, r'\rangle + \sum_{I'=1}^{N} \Gamma^{\text{int}}(B_k)_{I'I}\bar{Q}^{I'}_A|\hat{p}, j, m, r\rangle \tag{23.366}$$

(for $A = 1, 2$; $I = 1, 2, \ldots, N$ and $k = 1, 2, \ldots, n_0$). Then (11.26) implies that the $\bar{Q}^I_A|\hat{p}, j, m, r\rangle$ are basis vectors of the direct product representation $\mathbf{\Gamma}^{\text{int}}_V \otimes \mathbf{\Gamma}^{\text{int}}$ of the "true" $\mathscr{L}_{\text{int}}$. A similar argument shows that for each fixed pair of values of A and B the $\bar{Q}^I_A\bar{Q}^J_B|\hat{p}, j, m, r\rangle$ (for $I \neq J$) are basis vectors of the direct product representation $\mathbf{\Gamma}^{\text{int}}_V \otimes \mathbf{\Gamma}^{\text{int}} \otimes \mathbf{\Gamma}^{\text{int}}$ of the "true" $\mathscr{L}_{\text{int}}$, and so on.

(ii) *Central charges in the Poincaré superalgebra*

For convenience the discussion will be largely limited to the simplest case of the $N = 2$ Poincaré superalgebra, in which there are two central charges U and V, where $U^{12} = -U^{21} = U$ and $V^{12} = -V^{21} = V$. Then (23.173) can be written as

$$[Q^I_a, Q^J_b]_+ = -2\delta^{IJ}\sum_{\mu=1}^{4}(\gamma^\mu\mathbf{C})_{ab}P_\mu + i(\mathbf{C})_{ab}\varepsilon^{IJ}U + (\gamma_5\mathbf{C})_{ab}\varepsilon^{IJ}V, \tag{23.367}$$

where $\boldsymbol{\varepsilon}$ is the 2×2 antisymmetric matrix defined in (23.72). As the central charges U and V commute with every element of the Poincaré superalgebra, it follows that in every irreducible representation of the Poincaré superalgebra

$$U = u1, \quad V = v1, \tag{23.368}$$

where u and v are constants, which, as U and V are self-adjoint, must both be real. Thus, when acting on the state vectors corresponding to the special

4-momentum $\hat{p}$ of (23.264), (23.367) reduces to

$$[Q_a^I, Q_b^J]_+ = -2Mc\delta^{IJ}(\gamma_4\mathbf{C})_{ab}1 + \varepsilon^{IJ}iu(\mathbf{C})_{ab}1 + \varepsilon^{IJ}v(\gamma_5\mathbf{C})_{ab}1. \quad (23.369)$$

This can be simplified by making a "chiral transformation" and defining a set of operators $\hat{Q}_a^I$ by

$$\hat{Q}_a^I = \sum_{b=1}^{4} (\exp(-i\gamma_5\xi))_{ab} Q_b^I \quad (23.370)$$

(for $I = 1, 2$ and $a = 1, \ldots, 4$), where ξ is a real number chosen so that

$$u\mathbf{1}_4 - iv\gamma_5 = r\exp(-i\gamma_5\xi). \quad (23.371)$$

Using the definition (10.1) and the fact that $(\gamma_5)^2 = \mathbf{1}_4$, it follows that the real number r introduced in (23.371) is such that $u = r\sin\xi$ and $v = r\sin\xi$. It may be assumed that r is non-negative, so that

$$r = (u^2 + v^2)^{1/2}. \quad (23.372)$$

As (17.152) implies that $\mathbf{C}\tilde{\gamma}_5 = \gamma_5\mathbf{C}$, and as $\gamma_4\gamma_5 = -\gamma_5\gamma_4$, (23.369) reduces to

$$[\hat{Q}_a^I, \hat{Q}_b^J]_+ = -2Mc\delta^{IJ}(\gamma_4\mathbf{C})_{ab}1 + \varepsilon^{IJ}ir(\mathbf{C})_{ab}1. \quad (23.373)$$

The definition (23.370) implies that the operators $\hat{Q}_a^I$ transform in the same way as the Q_a^I under any similarity transformation of the Dirac matrices, so the analogue of (23.46) is

$$\hat{Q}'_a = \sum_{b=1}^{4} S_{ab}\hat{Q}_b. \quad (23.374)$$

As $\exp(-i\gamma_5^{\mathrm{M}}\xi)$ is a real orthogonal matrix, (23.183) implies that

$$(\hat{Q}_a^{\mathrm{M}I})^\dagger = \hat{Q}_a^{\mathrm{M}I}, \quad (23.375)$$

and hence in *any* representation of the Dirac matrices

$$\sum_{b=1}^{4} (\gamma_4\mathbf{C})_{ab}\hat{Q}_b^\dagger = -\hat{Q}_a \quad (23.376)$$

(for $a = 1, \ldots, 4$) (cf. (23.53)).

The general expressions of the analysis thus far are valid in *any* representation of the 4-dimensional Minkowski Clifford algebra. To analyse the situation further, it is convenient to use the "Dirac representation" of (17.117). In this case (23.376) becomes

$$\sum_{b=1}^{4} (\gamma_4^{\mathrm{D}}\mathbf{C}^{\mathrm{D}})_{ab}(\hat{Q}_b^{\mathrm{D}})^\dagger = -\hat{Q}_a^{\mathrm{D}}, \quad (23.377)$$

where, by (17.117) and (17.172),

$$\gamma_4^{\mathrm{D}}\mathbf{C}^{\mathrm{D}} = \begin{bmatrix} \mathbf{0} & \boldsymbol{\sigma}_2 \\ -\boldsymbol{\sigma}_2 & \mathbf{0} \end{bmatrix}. \quad (23.378)$$

The final stage of the analysis is to define the four operators $\hat{Q}_a^{D+}$ by

$$\hat{Q}_a^{D+} = 2^{-1/2}(\hat{Q}_a^{D1} + i\hat{Q}_a^{D2}) \tag{23.379}$$

(for $a = 1, \ldots, 4$). Then, by (23.373) and (23.377), it follows from (17.154) and (17.116) that

$$[\hat{Q}_a^{D+}, (\hat{Q}_b^{D+})^\dagger]_+ = \{2Mc(\mathbf{1}_4)_{ab} + r(\gamma_4^D)_{ab}\}1, \tag{23.380}$$

$$[\hat{Q}_a^{D+}, \hat{Q}_b^{D+}]_+ = 0, \tag{23.381}$$

$$[(\hat{Q}_a^{D+})^\dagger, (\hat{Q}_b^{D+})^\dagger]_+ = 0 \tag{23.382}$$

(for $a, b = 1, \ldots, 4$). The 4×4 matrix on the right-hand side of (23.380) is diagonal, with its first two diagonal elements having value $2Mc + r$ and its last two diagonal elements having value $2Mc - r$.

Acting on any state vector ψ with contravariant 4-momentum $\hat{p}$, for $a = b$ the left-hand side of (23.380) is $2\|\hat{Q}_a^{D+}\psi\|^2$, while the right-hand side is $(2Mc + r)\|\psi\|^2$ if $a = b = 1$ or 2, and is $(2Mc - r)\|\psi\|^2$ if $a = b = 3$ or 4. As $\|\hat{Q}_a^{D+}\psi\|^2$ and $\|\psi\|^2$ are both non-negative, this implies that

$$r \leqslant 2Mc, \tag{23.383}$$

and so, by (23.372), the eigenvalues of u and v of the central charges U and V are restricted by

$$(u^2 + v^2)^{1/2} \leqslant 2Mc. \tag{23.384}$$

If $r < 2Mc$ then the interpretation of (23.380)–(23.382) is that for $a = 1$ and 2 the operators $(2Mc + r)^{-1/2}(\hat{Q}_a^{D+})^\dagger$ and $(2Mc + r)^{-1/2}\hat{Q}_a^{D+}$ are a pair of fermion creation and annihilation operators, and for $a = 3$ and 4 the operators $(2Mc - r)^{-1/2}(\hat{Q}_a^{D+})^\dagger$ and $(2Mc - r)^{-1/2}\hat{Q}_a^{D+}$ are another pair of fermion creation and annihilation operators. A 16-dimensional representation of these operators can be constructed by taking a vacuum state $|0\rangle$ for which $\hat{Q}_a^{D+}|0\rangle = 0$ for $a = 1, \ldots, 4$ and setting up the Fock space by acting on this with all linearly independent products of creation operators. If $|0\rangle$ corresponds to spin $j\hbar$ then the resulting states have spin between $(j + 1)\hbar$ and $(j - 1)\hbar$; that is, if $r < 2Mc$ then the spin spectrum is the *same* as for the case of no central charges.

By contrast, if $r = 2Mc$ then the interpretation of (23.380)–(23.382) is that for $a = 1$ and 2 the operators $(4Mc)^{-1/2}(\hat{Q}_a^{D+})^\dagger$ and $(4Mc)^{-1/2}\hat{Q}_a^{D+}$ are again a pair of fermion creation and annihilation operators, but since the right-hand side of (23.380) is zero for $a = 3$ and 4, the operators $(\hat{Q}_a^{D+})^\dagger$ and $\hat{Q}_a^{D+}$ for $a = 3$ and 4 acting on any state vector with 4-momentum $\hat{p}$ give *zero*. Consequently the representation of these operators is 4-dimensional in this case, and if the vacuum state $|0\rangle$ (defined in this case by $\hat{Q}_a^{D+}|0\rangle = 0$ for $a = 1, 2$) has spin $j\hbar$ then the states in the corresponding Fock space have spins that range only from $(j+\frac{1}{2})\hbar$ to $(j-\frac{1}{2})\hbar$; that is, the range is *reduced* if the upper bound on r in (23.383) is attained.

Further information on the "massive" irreducible representations of extended $D=4$ Poincaré superalgebra may be found in the papers and reviews of Dondi and Sohnius (1974), Firth and Jenkins (1975), Grosser (1975b), Salam and Strathdee (1975a), Nahm (1978), Sohnius (1978, 1985), Witten and Olive (1978), Fayet (1979), Freedman (1979), Ferrara *et al.* (1981), Łopuszanski and Wolf (1981), Rittenberg and Sokatchev (1981), Ferrara and Savoy (1982), Taylor (1982) and Galperin *et al.* (1983).

(b) *Irreducible representations of the N-extended Poincaré superalgebra corresponding to $M=0$*

Again there are two sub-cases, depending on whether the Poincaré superalgebra contains central charges or not.

(i) *No central charges in the Poincaré superalgebra*

If there are no central charges in the Poincaré superalgebra, (23.192) and (23.193) again reduce to

$$[Q^I_A, Q^J_B]_+ = 0 \tag{23.385}$$

and

$$[\bar{Q}^I_A, \bar{Q}^J_B]_+ = 0 \tag{23.386}$$

(for $A, B = 1, 2$ and $I, J = 1, 2, \ldots, N$). Moreover, when restricted to acting on state vectors with the special contravariant 4-momentum $\hat{p}$ given by (23.317), that is with $\hat{p} = (0, 0, \kappa, \kappa)$, (23.194) reduces to

$$[Q^I_1, \bar{Q}^J_2]_+ = 0, \tag{23.387}$$

$$[Q^I_2, \bar{Q}^J_1]_+ = 0, \tag{23.388}$$

$$[Q^I_2, \bar{Q}^J_2]_+ = 0, \tag{23.389}$$

$$[Q^I_1, \bar{Q}^J_1]_+ = 4\kappa\delta^{IJ}1 \tag{23.390}$$

(for $I, J = 1, 2, \ldots, N$) (cf. (23.323)–(23.326)). Consequently in the massless case the little superalgebra of $\hat{p}$ consists of $M_{23} + M_{24}$, $M_{31} + M_{41}, M_3$ (which again form the basis for a 2-dimensional Euclidean Lie algebra), together with all $4N$ supertranslation operators Q^I_A and $\bar{Q}^I_A$ (for $A = 1, 2$ and $I = 1, 2, \ldots, N$), the identity 1, and the internal symmetry algebra operators B_k (for $k = 1, 2, \ldots, n_0$) of $\mathscr{L}_{\text{int}}$. For the reasons given in the $N = 1$ case the operators $M_{23} + M_{24}, M_{31} + M_{41}, Q^I_2$ and Q^I_2 (for $I = 1, 2, \ldots, N$) may all be assumed to produce the zero vector when acting on any state vector with 4-momentum $\hat{p}$. Consequently the *effective little superalgebra* of $\hat{p}$ consists only of M_3, together with the $2N$ supertranslation operators Q_1^I and $\bar{Q}_1^I$ (for $I = 1, 2, \ldots, N$), the identity 1, and the internal symmetry algebra operators B_k (for $k = 1, 2, \ldots, n_0$) of $\mathscr{L}_{\text{int}}$. Its irreducible representations are most easily discussed using the simplifications introduced in the massive case, since $\bar{M}_3$ and the $\bar{B}_k$ can again be defined by (23.353) and (23.354) (with $2Mc$ replaced by 4κ), and the commutation relations (23.356)–(23.359) remain valid. Thus the effective little superalgebra of $\hat{p}$ is also the *direct sum* of three algebras. One is the one-dimensional Abelian Lie algebra with basis element $\bar{M}_3$, the second with basis elements $\bar{B}_k$ (for $k = 1, 2, \ldots, n_0$) is isomorphic to $\mathscr{L}_{\text{int}}$, and the third is the Lie superalgebra consisting of the $2N$ supertranslation operators Q_1^I and $\bar{Q}_1^I$ (for $I = 1, 2, \ldots, N$) and the identity 1. Of course the irreducible representations of this effective little superalgebra are just direct products of irreducible representations of its three constituent parts.

To construct such an irreducible representation of the effective little superalgebra, choose *any* one-dimensional irreducible representation $\Gamma^j(\bar{M}_3) = [\hbar j]$ of the Abelian Lie algebra, where $j = 0, \pm\frac{1}{2}, \pm 1, \ldots,$ and *any* irreducible representation $\Gamma_{\text{V}}^{\text{int}}$ of $\mathscr{L}_{\text{int}}$ of dimension d_{V}, and let $|\hat{p}, j, r\rangle$ be a set of basis vectors for the direct product of these representations chosen so that

$$\bar{M}_3 |\hat{p}, j, r\rangle = \hbar j |\hat{p}, j, r\rangle \tag{23.391}$$

and

$$\bar{B}_k |\hat{p}, j, r\rangle = \sum_{r'=1}^{d_{\text{V}}} \Gamma_{\text{V}}^{\text{int}}(\bar{B}_k)_{r'r} |\hat{p}, j, r'\rangle \tag{23.392}$$

(for $k = 1, 2, \ldots, n_0$ and $r = 1, 2, \ldots, d_{\text{V}}$). Assume also that

$$Q_1^I |\hat{p}, j, r\rangle = 0 \tag{23.393}$$

(for $I = 1, 2, \ldots, N$ and $r = 1, 2, \ldots, d_{\text{V}}$). Then each vector $|\hat{p}, j, r\rangle$ is a "vacuum state vector" for the set of fermion creation and annihilation operators $\bar{Q}_1^I$ and Q_1^I, and the complete set of basis vectors of the corresponding irreducible representation of the effective little superalgebra is given by the $|\hat{p}, j, r\rangle$ and all vectors obtained from the $|\hat{p}, j, r\rangle$ by acting with every possible product of the creation operators $\bar{Q}_1^I$. Consequently the resulting irreducible representation of the effective little superalgebra has dimension $d_{\text{V}} 2^N$.

It is again of interest to determine the eigenvalues of the "true" helicity operator M_3 and the irreducible representations of the "true" internal symmetry algebra $\mathscr{L}_{\text{int}}$ (with basis elements B_j) that appear in this irreducible representation of the effective little superalgebra. Equations (23.353), (23.354) and (23.391)–(23.393) imply that the "vacuum" basis vectors $|\hat{p}, j, r\rangle$ themselves actually have eigenvalue $\hbar j$ with the "true" helicity operator M_3 and are basis vectors of the irreducible representation $\mathbf{\Gamma}_{\mathrm{V}}^{\mathrm{int}}$ of the "true" internal symmetry algebra $\mathscr{L}_{\mathrm{int}}$.

For the other basis vectors the situation is only slightly more complicated. Equations (23.353), (23.357), (23.390), (23.391) and (23.393) show that

$$\bar{M}_3 \bar{Q}_1^I |\hat{p}, j, r\rangle = \hbar(j+\tfrac{1}{2})\bar{Q}_1^I |\hat{p}, j, r\rangle \tag{23.394}$$

(for $I = 1, 2, \ldots, N$) and

$$\bar{M}_3 \bar{Q}_1^I \bar{Q}_1^J |\hat{p}, j, r\rangle = \hbar(j+1)\bar{Q}_1^I \bar{Q}_1^J |\hat{p}, j, r\rangle \tag{23.395}$$

(for $I, J = 1, 2, \ldots, N$, with $I \neq J$). Clearly if a product of n different creation operators is applied to $|\hat{p}, j, r\rangle$ then the resulting vector is an eigenvector of $\bar{M}_3$ with eigenvalue $\hbar(j+\frac{1}{2}n)$, the number of such state vectors formed in this way from each single vector $|\hat{p}, j, r\rangle$ being $N!/n!(N-n)!$ Obviously the maximum value of the helicity is $\hbar(j+\frac{1}{2}N)$, which occurs for the vectors $\bar{Q}_1^1\bar{Q}_1^2\cdots\bar{Q}_1^{N-1}\bar{Q}_1^N|\hat{p}, j, r\rangle$. Exactly one-half of these $d_{\mathrm{V}}2^N$ state vectors will have helicities that differ from j by an integer, and for the other half the helicities differ from j by a half-integer. Thus again the numbers of bosonic and fermionic states in the corresponding irreducible representation of the Poincaré superalgebra are *equal.*

The reduction with respect to the "true" $\mathscr{L}_{\mathrm{int}}$ can be discussed in the same way as for the massive case. For example, a trivial modification of (23.366) (in which the label m is removed from the state vectors) shows that the $\bar{Q}_1^I|\hat{p}, j, r\rangle$ are basis vectors of the direct product representation $\mathbf{\Gamma}_{\mathrm{V}}^{\mathrm{int}} \otimes \mathbf{\Gamma}^{\mathrm{int}}$ of the "true" $\mathscr{L}_{\mathrm{int}}$. A similar argument shows that the $\bar{Q}_1^I\bar{Q}_1^J|\hat{p}, j, r\rangle$ (for $I \neq J$) are basis vectors of the direct product representation $\mathbf{\Gamma}_{\mathrm{V}}^{\mathrm{int}} \otimes \{\mathbf{\Gamma}^{\mathrm{int}} \otimes \mathbf{\Gamma}^{\mathrm{int}}\}_{\mathrm{antisymm}}$ of the "true" $\mathscr{L}_{\mathrm{int}}$, where $\{\mathbf{\Gamma}^{\mathrm{int}} \otimes \mathbf{\Gamma}^{\mathrm{int}}\}_{\mathrm{antisymm}}$ is the antisymmetric part of $\mathbf{\Gamma}^{\mathrm{int}} \otimes \mathbf{\Gamma}^{\mathrm{int}}$ (cf. Chapter 12, Section 4) (only the antisymmetric part appears because $\bar{Q}_1^I\bar{Q}_1^J|\hat{p}, j, r\rangle = -\bar{Q}_1^J\bar{Q}_1^I|\hat{p}, j, r\rangle$ by (23.386)). Clearly the action of n different creation operators applied to $|\hat{p}, j, r\rangle$ produces basis vectors of the direct product of $\mathbf{\Gamma}_{\mathrm{V}}^{\mathrm{int}}$ with the totally antisymmetric direct product of n copies of $\mathbf{\Gamma}^{\mathrm{int}}$.

Example I *Massless irreducible representations of the $N = 4$, $D = 4$ Poincaré superalgebra*

With $N = 4$ and $j = -1$ the helicities take the values $-\hbar$, $-\frac{1}{2}\hbar$, 0, $\frac{1}{2}\hbar$ and $\hbar$; that is, within this *single* irreducible representation of the $N = 4$, $D = 4$ Poincaré superalgebra all the values required to describe a massless spin-1 theory appear. Consequently these representations play an important role in supersymmetric Yang–Mills theories.

The simplest situation is that in which there is a *single* vacuum state vector $|\hat{p}, -1, 1\rangle$ belonging to the trivial one-dimensional irreducible representation

Γ_V^{int} of $\mathscr{L}_{\text{int}}$ in which $\Gamma_V^{\text{int}}(B_k) = [0]$ for $k = 1, 2, \ldots, n_0$. Then the four vectors $\bar{Q}_1^I|\hat{p}, -1, 1\rangle$ (with helicity $-\frac{1}{2}\hbar$) are basis vectors of the irreducible representation Γ^{int} of the "true" $\mathscr{L}_{\text{int}}$, the six vectors $\bar{Q}_1^I\bar{Q}_1^J|\hat{p}, -1, 1\rangle$ (for $I \neq J$) (with helicity 0) are basis vectors of the representation $\{\Gamma^{\text{int}} \otimes \Gamma^{\text{int}}\}_{\text{antisymm}}$ of $\mathscr{L}_{\text{int}}$, the four vectors $\bar{Q}_1^I\bar{Q}_1^J\bar{Q}_1^K|\hat{p}, -1, 1\rangle$ (for I, J, K all different) are basis vectors (with helicity $\frac{1}{2}\hbar$) of the representation $\{\Gamma^{\text{int}} \otimes \Gamma^{\text{int}} \otimes \Gamma^{\text{int}}\}_{\text{antisymm}}$ of $\mathscr{L}_{\text{int}}$, and the single vector $\bar{Q}_1^1\bar{Q}_1^2\bar{Q}_1^3\bar{Q}_1^4\ |\hat{p}, -1, 1\rangle$ has helicity $\hbar$ and belongs to the representation $\{\Gamma^{\text{int}} \otimes \Gamma^{\text{int}} \otimes \Gamma^{\text{int}} \otimes \Gamma^{\text{int}}\}_{\text{antisymm}}$ of $\mathscr{L}_{\text{int}}$, which is necessarily one-dimensional.

For $N = 4$ the maximal internal symmetry algebra $\mathscr{L}_{\text{int}}$ in u(4). As u(4) = u(1) $\oplus$ su(4), the irreducible representations of u(4) are direct products of one-dimensional irreducible representations of u(1) and irreducible representations of su(4). In particular, if Γ^{int} is taken to coincide on su(4) with the 4-dimensional defining representation of su(4) that corresponds to the irreducible representation $\Gamma\{1, 0, 0\}$ of its complexification A_3 then the 6-dimensional irreducible representation corresponds to $\Gamma\{0, 1, 0\}$.

Example II *Massless irreducible representations of the $N = 8$, $D = 4$ Poincaré superalgebra*

With $N = 8$ and $j = -2$ the helicities take the values $-2\hbar$, $-\frac{3}{2}\hbar$, $-\hbar$, $-\frac{1}{2}\hbar$, 0, $\frac{1}{2}\hbar$, $\hbar$, $\frac{3}{2}\hbar$ and $2\hbar$; that is, within this *single* irreducible representation of the $N = 8$, $D = 4$ Poincare′ superalgebra all the values required to describe a massless spin-2 theory appear. These representations therefore play an important role in supergravity theories.

As in the previous example, the simplest situation is that in which there is a *single* vacuum state vector $|\hat{p}, -2, 1\rangle$ belonging to the trivial one-dimensional irreducible representation Γ_V^{int} of $\mathscr{L}_{\text{int}}$ in which $\Gamma_V^{\text{int}}(B_k) = [0]$ for $k = 1, 2, \ldots, n_0$. The multiplicity of the states with helicity $\pm 2\hbar$ is 1, that of the states with helicity $\pm\frac{3}{2}\hbar$ is 8, that of the states with helicity $\pm\hbar$ is 28, that of the states with helicity $\pm\frac{1}{2}\hbar$ is 56, and that of the states with helicity 0 is 70. The discussion of the representations of $\mathscr{L}_{\text{int}}$ proceeds along the same lines as in Example I.

(*ii*) *The Poincaré superalgebra has central charges*

For a "massless" irreducible representation of an extended 4-dimensional Poincaré superalgebra all the central charges have to be represented by *zero* operators. Consequently the "massless" theory just reduces to that of the case where the central charges are absent.

This is most easily seen by considering the example of the $N = 2$ Poincaré superalgebra with two central charges U and V that was examined in the "massive" case in the previous subsection. Most of the analysis given there still applies, but when acting on the state vectors corresponding to the special 4-momentum $\hat{p}$ of (23.317) (instead of (23.264)) the analogue of (23.369) becomes

$$[Q_a^I, Q_b^J]_+ = -2\kappa\delta^{IJ}((\gamma_3+\gamma_4)\mathbf{C})_{ab}1 + \varepsilon^{IJ}iu(\mathbf{C})_{ab}1 + \varepsilon^{IJ}v(\gamma_5\mathbf{C})_{ab}1.$$

This implies that the first term of (23.373) has to be modified to give

$$[\hat{Q}_a^I, \hat{Q}_b^J]_+ = -2\kappa\delta^{IJ}((\gamma_3+\gamma_4)\mathbf{C})_{ab}1 + \varepsilon^{IJ}ir(\mathbf{C})_{ab}1. \qquad (23.396)$$

In particular, using the Majorana representation of Theorem II(b) of Appendix L, Section 4, for which (L.113) and (L.114) give

$$\gamma_3^{(M,4)} = \begin{bmatrix} \mathbf{0} & i\mathbf{1} \\ i\mathbf{1} & \mathbf{0} \end{bmatrix}$$

and

$$\gamma_4^{(M,4)} = \begin{bmatrix} \mathbf{0} & i\mathbf{1} \\ -i\mathbf{1} & \mathbf{0} \end{bmatrix},$$

and with $\mathbf{C} = -\gamma_4^{(M,4)}$, the matrix on the right-hand side of (23.396) becomes

$$\begin{bmatrix} 4\kappa\delta^{IJ}\mathbf{1}_2 & -r\varepsilon^{IJ}\mathbf{1}_2 \\ r\varepsilon^{IJ}\mathbf{1}_2 & \mathbf{0} \end{bmatrix}.$$

Thus, with $a=b=3$ (or 4) with $I=J$, on forming the expectation value of both sides of (23.396) with any state vector ψ with contravariant 4-momentum $\hat{p}$ given by (23.317), it follows from (23.376) that $\| Q_a^I\psi \|^2 = 0$, implying that $\hat{Q}_a^I\psi = 0$. Consequently, on again forming the expectation value of both sides of (23.396) with ψ, but this time taking $a=3$, $b=1$, $I=1$ and $J=2$, it follows in a similar fashion that $r=0$. Thus, by (23.372), the eigenvalues u and v of the central charges U and V are both zero.

For further details of the "massless" irreducible representations of the extended $D=4$ Poincaré superalgebra see Nahm (1978), Salam and Strathdee (1975a), Freedman (1979), Ferrara and Savoy (1982), Taylor (1982) and Sohnius (1985).

7 Irreducible representations of the Poincaré superalgebras for general space–time dimensions

(a) *Irreducible representations of the unextended D-dimensional Poincaré superalgebras*

Attention will now be focused on the irreducible representations of Poincaré superalgebras for *general* values of the space–time dimension D. The unextended superalgebras will be considered in this subsection, and the representations of their extensions will be discussed in the next subsection. The notation is that of Section 4, where it was shown that $N = 1$ Poincaré superalgebras exist only if $D \bmod 8 = 0$, 1, 2, 3 or 4. The general line of argument of the previous two sections can be reapplied. In particular, there are again two classes of physically relevant irreducible representations, which will continue to be described as "massive" and "massless", and the whole analysis again reduces to a study of the irreducible representations of little superalgebras. The discussion that follows will be concentrated on those features that are different from those of the $D = 4$ case that was described in detail in Section 5(a) and 5(b).

(A) *Irreducible representations of the unextended D-dimensional Poincaré superalgebra corresponding to $M > 0$*

It is fairly easy to extend the representation theory of the $D = 4$ Poincaré group to general space–time dimensions. In particular, the natural generalization of (23.264) is to work in the rest frame, where the contravariant D-momentum $\hat{p}$ is given by

$$\hat{p} = (0, 0, \ldots, 0, Mc). \tag{23.397}$$

The corresponding little group is then Spin$(D-1)$, which is the universal covering group of SO$(D-1)$. Its corresponding little Lie algebra is so$(D-1)$, whose basis elements may be taken to be $M_{\alpha\beta}$ (for $\alpha, \beta = 1, 2, \ldots, D-1$ and $\alpha < \beta$).

Let Q_a^{M} be the supertranslation operators in the Majorana representation, where $a = 1, 2, \ldots, d$, d being the dimension of the irreducible representation of the D-dimensional Minkowski Clifford algebra, so that $d = 2^{D/2}$ if D is even and $d = 2^{(D-1)/2}$ if D is odd (cf. (23.224) and (23.225)), implying that d is always even (except in the trivial case where $D = 1$). Acting on states with D-momentum $\hat{p}$ given by (23.397) and invoking (23.236), the relation (23.234) gives

$$Q_a^{\mathrm{M}} Q_b^{\mathrm{M}} + Q_b^{\mathrm{M}} Q_a^{\mathrm{M}} = 2Mc\delta_{ab} 1 \tag{23.398}$$

(for $a, b = 1, 2, \ldots, d$), which is the D-dimensional generalization of (23.316). Comparison with (L.2) and (L.4) shows that the operators $(Mc)^{-1/2} Q_a^{\mathrm{M}}$ are generators of a d-dimensional Euclidean Clifford algebra. The little superalgebra for $\hat{p}$ is then spanned by the $\frac{1}{2}(D-1)(D-2)$ basis elements $M_{\alpha\beta}$ of so$(D-1)$, the d supertranslation operators Q_a^{M} (for $a = 1, 2, \ldots, d$) and the identity 1. Its commutation and anticommutation relations are given by (23.228), (23.232) and (23.398). Its essential structure can be exposed on defining a set of operators $\bar{M}_{\alpha\beta}$ by

$$\bar{M}_{\alpha\beta} = M_{\alpha\beta} - \frac{1}{4Mc} \sum_{a,b=1}^{d} Q_a^{\mathrm{M}} Q_b^{\mathrm{M}} \Gamma_{\mathrm{M}}^{\mathrm{spin}}(M_{\alpha\beta})_{ab} \tag{23.399}$$

(for $\alpha, \beta = 1, 2, \ldots, D-1$, with $\alpha > \beta$), where $\Gamma_{\mathrm{M}}^{\mathrm{spin}}$ is the d-dimensional spinor representation in its Majorana form (cf. (23.227)), thereby generalizing (23.353). It is easily shown that

$$[\bar{M}_{\alpha\beta}, \bar{M}_{\lambda\mu}]_- = \frac{\hbar}{i}(\eta_{\alpha\lambda}\bar{M}_{\beta\mu} - \eta_{\alpha\mu}\bar{M}_{\beta\lambda} - \eta_{\beta\lambda}\bar{M}_{\alpha\mu} + \eta_{\beta\mu}\bar{M}_{\alpha\lambda}). \tag{23.400}$$

(for $\alpha, \beta, \lambda, \mu = 1, 2, \ldots, D-1$). Comparison with (23.228) then shows that the set of operators $\bar{M}_{\alpha\beta}$ form a Lie algebra that is isomorphic to the Lie algebra $\mathrm{so}(D-1)$ of the $M_{\alpha\beta}$. Henceforth the $\mathrm{so}(D-1)$ Lie algebra formed by the $\bar{M}_{\alpha\beta}$ will be denoted by $\mathrm{so}(D-1)_{\bar{M}}$, and the $\mathrm{so}(D-1)$ Lie algebra formed by the $M_{\alpha\beta}$ will be denoted by $\mathrm{so}(D-1)_M$. Moreover,

$$[\bar{M}_{\alpha\beta}, Q_a^{\mathrm{M}}]_- = 0 \tag{23.401}$$

(for $\alpha, \beta = 1, 2, \ldots, D-1$ and $a = 1, 2, \ldots, d$). Thus the little superalgebra is the *direct sum* of the $\mathrm{so}(D-1)_{\bar{M}}$ Lie algebra and the d-dimensional Euclidean Clifford algebra, and so its irreducible representations are the direct products of those of these two constituent parts. (If $D \leqslant 2$ then the $\mathrm{so}(D-1)$ algebras are zero-dimensional and can be omitted from the argument.)

Although this provides a complete specification of all of the irreducible representations of the little superalgebra, it is desirable to be able to find which irreducible representations of the $\mathrm{so}(D-1)_M$ subalgebra appear. This can be done by reformulating the question in purely Lie algebraic terms. First it may be noted that the terms with $a = b$ in the summation on the right-hand side of (23.399) give a total contribution of zero (since, by (23.398), their sum is $Mc\{\mathrm{tr}\,\Gamma_{\mathrm{M}}^{\mathrm{spin}}(M_{\alpha\beta})\}$, which is zero by Theorem VI of Chapter 15, Section 2). Consequently these terms may be excluded from the summation on the right-hand side of (23.399). As d is even, with the definition

$$Q_{d+1}^{\mathrm{M}} = Q_1^{\mathrm{M}} Q_2^{\mathrm{M}} Q_3^{\mathrm{M}} \cdots Q_{d-1}^{\mathrm{M}} Q_d^{\mathrm{M}}, \tag{23.402}$$

the set of operators Q_a^{M} for $a = 1, 2, \ldots, d+1$ are generators for the $(d+1)$-dimensional Euclidean Clifford algebra, and the $2^{d/2}$-dimensional irreducible representation of the d-dimensional Euclidean Clifford algebra provides a $2^{d/2}$-dimensional irreducible representation of the $(d+1)$-dimensional Euclidean Clifford algebra. With the further definitions

$$S_{ab} = \begin{cases} -\dfrac{1}{2Mc} Q_a^{\mathrm{M}} Q_b^{\mathrm{M}} & \text{for } a \neq b, \\ 0 & \text{for } a = b, \end{cases} \tag{23.403}$$

(cf. (17.134)), the operators S_{ab} (for $a, b = 1, 2, \ldots, d+1$ with $a < b$) form a basis for the simple Lie algebra $\text{so}(d+1)$, which has a spinor irreducible representation whose dimension is also $2^{d/2}$. Thus each irreducible representation of the little superalgebra corresponds to an irreducible representation of the same dimension of the Lie algebra $\text{so}(D-1)_{\bar{M}} \oplus \text{so}(d+1)$. By virtue of (23.399) and (23.403), the elements $M_{\alpha\beta}$ are now members of the Lie algebra $\text{so}(D-1)_{\bar{M}} \oplus \text{so}(d+1)$, so the problem reduces to finding which irreducible representations of the $\text{so}(D-1)_M$ subalgebra appear in the irreducible representations of $\text{so}(D-1)_{\bar{M}} \oplus \text{so}(d+1)$. For further details see Yang and Wybourne (1986) and Strathdee (1987), whose analyses extend to the N-extended case as well.

(*B*) *Irreducible representations of the unextended D-dimensional Poincaré superalgebra corresponding to $M = 0$*

Many of the features of the $M > 0$ case reappear in the $M = 0$ situation. For $M = 0$ the natural generalization of the special 4-momentum of (23.317) is the contravariant D-momentum $\hat{p}$ given by

$$\hat{p} = (0, 0, \ldots, 0, \kappa, \kappa). \tag{23.404}$$

The corresponding little group is then the covering group of the Euclidean group in $D-2$ dimensions, whose translation generators can be assumed to act trivially. Consequently the non-trivial part of the corresponding little Lie algebra is $\text{so}(D-2)$, whose basis elements may be taken to be $M_{\alpha\beta}$ (for $\alpha, \beta = 1, 2, \ldots, D-2$ and $\alpha < \beta$). (For $D \leqslant 3$ this $\text{so}(D-2)$ algebra is zero-dimensional and can be left out of the discussion.)

It is convenient to adopt the Majorana representation $\gamma_\alpha^{(\mathrm{M},D)}$ of the D-dimensional Minkowski Clifford algebra that appears in Theorem II of Appendix L, Section 4. Let Q_a^{M} be the supertranslation operators in this representation, where $a = 1, 2, \ldots, d$, where $d = 2^{D/2}$ if D is even and $d = 2^{(D-1)/2}$ if D is odd (cf. (23.224) and (23.225)). Equation (23.234) can be rewritten as

$$[Q_a^{\mathrm{M}}, Q_b^{\mathrm{M}}]_+ = -2 \sum_{\mu=1}^{D} (\gamma_\mu^{(\mathrm{M},D)} \mathbf{C}^{(\mathrm{M},D)})_{ab} P^\mu, \tag{23.405}$$

which, when restricted to acting on states with D-momentum $\hat{p}$ given by (23.404), reduces to

$$Q_a^{\mathrm{M}} Q_b^{\mathrm{M}} + Q_b^{\mathrm{M}} Q_a^{\mathrm{M}} = -2\kappa \{(\gamma_{D-1}^{(\mathrm{M},D)} + \gamma_D^{(\mathrm{M},D)}) \mathbf{C}^{(\mathrm{M},D)}\}_{ab} 1 \tag{23.406}$$

(for $a, b = 1, 2, \ldots, d$). However, by (L.117),

$$\{\gamma_{D-1}^{(\mathrm{M},D)} + \gamma_D^{(\mathrm{M},D)}\} \mathbf{C}^{(\mathrm{M},D)} = \begin{bmatrix} -2\mathbf{1}_{d/2} & \mathbf{0} \\ \mathbf{0} & \mathbf{0} \end{bmatrix}. \tag{23.407}$$

Thus if the set of d supertranslation operators Q_a^{M} is split into two subsets each containing $\frac{1}{2}d$ members by the definitions

$$Q_a^+ = Q_a^{\mathrm{M}} \qquad \text{for } a = 1, 2, \ldots, \tfrac{1}{2}d \tag{23.408}$$

and

$$Q_a^- = Q_{a+d/2}^{\mathrm{M}} \quad \text{for } a = 1, 2, \ldots, \tfrac{1}{2}d \tag{23.409}$$

then (23.406) and (23.407) reduce to

$$Q_a^+ Q_b^+ + Q_b^+ Q_a^+ = 4\kappa\delta_{ab} 1, \tag{23.410}$$

$$Q_a^- Q_b^- + Q_b^- Q_a^- = 0, \tag{23.411}$$

$$Q_a^+ Q_b^- + Q_b^- Q_a^+ = 0 \tag{23.412}$$

(for $a, b = 1, 2, \ldots, \frac{1}{2}d$). Moreover, because the matrices $\gamma_\alpha^{(\mathrm{M},D)}$ for $\alpha = 1, 2, \ldots, D-2$ of (L.110) and (L.113) are direct sums of $\frac{1}{2}d \times \frac{1}{2}d$ matrices, the d-dimensional representation $\boldsymbol{\Gamma}_{\mathrm{M}}^{\mathrm{spin}}$ of so$(D-1, 1)$ reduces to the direct sum of two $\frac{1}{2}d$-dimensional irreducible representations of its so$(D-2)$ subalgebra, and the Q_a^+ and Q_a^- (for $a = 1, 2, \ldots, \frac{1}{2}d$) form carrier spaces for them (or—rather more precisely—for their negative transposes). Denoting these two irreducible representations of so$(D-2)$ by $\boldsymbol{\Gamma}^+$ and $\boldsymbol{\Gamma}^-$, (23.232) reduces to

$$[M_{\alpha\beta}, Q_a^+]_- = -\sum_{b=1}^{d/2} \Gamma^+(M_{\alpha\beta})_{ab} Q_b^+ \tag{23.413}$$

and

$$[M_{\alpha\beta}, Q_a^-]_- = -\sum_{b=1}^{d/2} \Gamma^-(M_{\alpha\beta})_{ab} Q_b^- \tag{23.414}$$

(for $\alpha, \beta = 1, 2, \ldots, D-2$ and $a = 1, 2, \ldots, \frac{1}{2}d$). Thus the two subsets of supertranslation operators are decoupled from each other. As (23.411) has the same form as the equation (23.325) for the $D = 4$ case, it implies that Q_a^- (for $a = 1, 2, \ldots, \frac{1}{2}d$) gives zero when acting on any state vector with D-momentum $\hat{p}$ (otherwise negative-norm states would appear). Comparison of (23.410) with (L.2) and (L.4) shows that the operators $(2\kappa)^{-1/2} Q_a^+$ are generators of a $\frac{1}{2}d$-dimensional Euclidean Clifford algebra. The effective little superalgebra for $\hat{p}$ is then spanned by the $\frac{1}{2}(D-2)(D-3)$ basis elements $M_{\alpha\beta}$ of so$(D-2)$, the $\frac{1}{2}d$ supertranslation operators Q_a^+ (for $a = 1, 2, \ldots, \frac{1}{2}d$) and the identity 1. Its commutation and anticommutation relations are given by (23.228), (23.410) and (23.413). Its structure can be analysed in the same way as in the massive case. By analogy with (23.399), a set of operators $\bar{M}_{\alpha\beta}$ may be defined by

$$\bar{M}_{\alpha\beta} = M_{\alpha\beta} - \frac{1}{8\kappa} \sum_{a,b=1}^{d/2} Q_a^+ Q_b^+ \Gamma^+(M_{\alpha\beta})_{ab} \tag{23.415}$$

(for $\alpha, \beta = 1, 2, \ldots, D-2$, with $\alpha < \beta$), where $\boldsymbol{\Gamma}^+$ is the $\frac{1}{2}d$-dimensional spinor representation (23.413) of so($D-2$). Again it is easily shown that

$$[\bar{M}_{\alpha\beta}, \bar{M}_{\lambda\mu}]_- = \frac{\hbar}{i}(\eta_{\alpha\lambda}\bar{M}_{\beta\mu} - \eta_{\alpha\mu}\bar{M}_{\beta\lambda} - \eta_{\beta\lambda}\bar{M}_{\alpha\mu} + \eta_{\beta\mu}\bar{M}_{\alpha\lambda}) \tag{23.416}$$

(for $\alpha, \beta, \lambda, \mu = 1, 2, \dots, D-2$), and so the set of operators $\bar{M}_{\alpha\beta}$ form a Lie algebra that is isomorphic to the Lie algebra so($D-2$) of the $M_{\alpha\beta}$. Henceforth the so($D-2$) Lie algebra formed by the $\bar{M}_{\alpha\beta}$ will be denoted by so$(D-2)_{\bar{M}}$, and the so($D-2$) Lie algebra formed by the $M_{\alpha\beta}$ will be denoted by so$(D-2)_M$. Moreover,

$$[\bar{M}_{\alpha\beta}, Q_a^+]_- = 0 \tag{23.417}$$

(for $\alpha, \beta = 1, 2, \dots, D-2$, and for $a = 1, 2, \dots, \frac{1}{2}d$). Thus the effective little superalgebra is the *direct sum* of the so$(D-2)_{\bar{M}}$ Lie algebra and the $\frac{1}{2}d$-dimensional Euclidean Clifford algebra, and so again its irreducible representations are the direct products of these two constituent parts.

For $D \geqslant 4$ the determination of which "massless" irreducible representations of the so$(D-2)_M$ subalgebra appear in this representation can be carried out along the same lines as in the massive case. As $\frac{1}{2}d$ is even for $D \geqslant 4$, with the analogue

$$Q_{d/2+1}^+ = Q_1^+ Q_2^+ Q_3^+ \cdots Q_{d/2-1}^+ Q_{d/2}^+ \tag{23.418}$$

of (23.402), the set of operators Q_a^{M} for $a = 1, 2, \dots, \frac{1}{2}d + 1$ are generators for the $(\frac{1}{2}d + 1)$-dimensional Euclidean Clifford algebra, and the $2^{d/4}$-dimensional irreducible representation of the $\frac{1}{2}d$-dimensional Euclidean Clifford algebra provides a $2^{d/4}$-dimensional irreducible representation of the $(\frac{1}{2}d + 1)$-dimensional Euclidean Clifford algebra. Similarly the analogues of (23.402) are the definitions

$$S_{ab} = \begin{cases} -\dfrac{1}{4\kappa} Q_a^+ Q_b^+ & \text{for } a \neq b, \\ 0 & \text{for } a = b, \end{cases} \tag{23.419}$$

which give a set of operators S_{ab} (for $a, b = 1, 2, \dots, \frac{1}{2}d + 1$), with $a < b$) that form a basis for the simple Lie algebra so$(\frac{1}{2}d + 1)$, which itself has a spinor irreducible representation whose dimension is also $2^{d/4}$. Thus in the $M = 0$ case each irreducible representation of the effective little superalgebra corresponds to an irreducible representation of the same dimension as the Lie algebra so$(D-2)_{\bar{M}} \oplus$ so$(\frac{1}{2}d + 1)$. Because of (23.415) and (23.419), the elements $M_{\alpha\beta}$ are now members of the Lie algebra so$(D-2)_{\bar{M}} \oplus$ so$(\frac{1}{2}d + 1)$, so the problem reduces to finding which irreducible representations of the so$(D-2)_M$ subalgebra appear in the irreducible representations of so$(D-2)_{\bar{M}} \oplus$ so$(\frac{1}{2}d + 1)$, which is again a purely Lie algebraic question. For further details see Yang and Wybourne (1986) and Strathdee (1987), whose analyses again extend to the N-extended case as well.

Example I *"Massless" irreducible representations of the unextended Poincaré superalgebras for $D = 10$ and $D = 11$*

For both $D = 10$ and $D = 11$, (23.224) and (23.225) show that $d = 32$, and hence $\frac{1}{2}d = 16$ and $\frac{1}{4}d = 8$. The irreducible representations of the effective little superalgebra of smallest dimension are those in which the subalgebras $so(D-2)_{\bar{M}}$ are represented by the 1-dimensional trivial representations. In both cases the direct product representation of the effective little superalgebra has dimension $2^{d/4} = 2^8 = 256$, *exactly* as for the smallest massless irreducible representation of the $N = 8$, $D = 4$ Poincaré superalgebra. Indeed the effective little superalgebra of the $N = 8$, $D = 4$ Poincaré superalgebra is a subalgebra of those of the $N = 1$ Poincaré superalgebras for $D = 10$ and 11, since all three contain the same Clifford algebra. Consequently, when the reduction to $D = 4$ is considered, the resulting helicities of the simplest massless irreducible representations of the unextended $D = 10$ and $D = 11$ Poincaré superalgebras are exactly the same as those of the simplest massless irreducible representation of the $N = 8$, $D = 4$ Poincaré superalgebra.

For $D \bmod 8 = 2$ the above arguments can be further refined, but the presentation of these will be deferred until the next subsection, where they will be discussed as special cases of the corresponding extended Poincaré superalgebras.

(b) Irreducible representations of the extended D-dimensional Poincaré superalgebras

The basic idea here is to combine the techniques used in discussing the extended $D = 4$ Poincaré superalgebra with those outlined above for the unextended superalgebras of general dimension D. As the description of all the possible cases would be too lengthy, the discussion will be restricted to just three cases, which will be presented in the form of three examples.

Example II *The "massless" irreducible representations of the extended Poincaré superalgebras for D* $\bmod 8 = 4$

The class $D \bmod 8 = 4$ is the simplest to consider because it is just a straightforward generalization of the $D = 4$ case. The commutation and anti-commutation relations are those of Example I of Section 4(*b*), which can be simply obtained from those for the $D = 4$ case considered in Section 3 by merely allowing all the space–time indices $\alpha, \beta, \mu, \ldots$ to assume the values $1, 2, \ldots, D$, and all the spinor indices $a, b, c, \ldots$ to take the values $1, 2, \ldots, d$. Concentrating again on the special contravariant D-momentum $\hat{p}$ given by (23.404), the space–time part of the corresponding little group is still the covering group of the Euclidean group in $D-2$ dimensions, whose translation generators can be assumed to act trivially. As any central charges that are present will have zero eigenvalues, they may be omitted from the discussion. Consequently the non-trivial part of the little Lie algebra is the direct sum of so$(D-2)$ (whose basis elements may be taken to be $M_{\alpha\beta}$ (for $\alpha, \beta = 1, 2, \ldots, D-2$ and $\alpha < \beta$)) and the internal symmetry algebra $\mathscr{L}_{\text{int}}$. Again adopting the Majorana representation $\gamma_\alpha^{(\mathrm{M},D)}$ of the D-dimensional Minkowski Clifford algebra that appears in Theorem II of Appendix L, Section 4, the generalizations of (23.408) and (23.409) are

$$Q_a^{I+} = Q_a^{\mathrm{M}I} \qquad \text{for } a = 1, 2, \ldots, \tfrac{1}{2}d \tag{23.420}$$

and

$$Q_a^{I-} = Q_{a+d/2}^{\mathrm{M}I} \qquad \text{for } a = 1, 2, \ldots, \tfrac{1}{2}d \tag{23.421}$$

(where now $I = 1, 2, \ldots, N$), so that (23.410)–(23.412) become

$$Q_a^{I+}Q_b^{J+} + Q_b^{J+}Q_a^{I+} = 4\kappa\delta^{IJ}\delta_{ab}1, \tag{23.422}$$

$$Q_a^{I-}Q_b^{J-} + Q_b^{J-}Q_a^{I-} = 0, \tag{23.423}$$

$$Q_a^{I+}Q_b^{J-} + Q_b^{J-}Q_a^{I+} = 0 \tag{23.424}$$

(for $a, b = 1, 2, \ldots, \frac{1}{2}d$ and $I, J = 1, 2, \ldots, N$). Moreover, the generalizations of (23.413) and (23.414) are

$$[M_{\alpha\beta}, Q_a^{I+}]_- = -\sum_{b=1}^{d/2} \Gamma^+(M_{\alpha\beta})_{ab} Q_b^{I+} \tag{23.425}$$

and

$$[M_{\alpha\beta}, Q_a^{I-}]_- = -\sum_{b=1}^{d/2} \Gamma^-(M_{\alpha\beta})_{ab} Q_b^{I-} \tag{23.426}$$

(for $\alpha, \beta = 1, 2, \ldots, D-2$; $I = 1, 2, \ldots, N$ and $a = 1, 2, \ldots, \frac{1}{2}d$). Consequently the two subsets of supertranslation operators are again decoupled from each other, and (23.423) implies that each Q_a^{I-} (for $a = 1, 2, \ldots, \frac{1}{2}d$ and $I = 1, 2, \ldots, N$) gives zero when acting on any state vector with D-momentum $\hat{p}$ (as otherwise negative-norm states would appear). The operators $(2\kappa)^{-1/2}Q_a^{I+}$ are generators of a $\frac{1}{2}Nd$-dimensional Euclidean Clifford algebra.

The effective little superalgebra for $\hat{p}$ is then spanned by the $\frac{1}{2}(D-2)(D-3)$ basis elements $M_{\alpha\beta}$ of so$(D-2)$, the n_0 basis elements B_k of the internal symmetry algebra $\mathscr{L}_{\text{int}}$, the $\frac{1}{2}Nd$ supertranslation operators Q_a^{I+} (for $a = 1, 2, \ldots, \frac{1}{2}d$ and $I = 1, 2, \ldots, N$) and the identity 1. Defining the operators

$\bar{M}_{\alpha\beta}$ and $\bar{B}_k$ by

$$\bar{M}_{\alpha\beta} = M_{\alpha\beta} - \frac{1}{2\kappa} \sum_{I=1}^{N} \sum_{a,b=1}^{d/2} Q_a^{MI} Q_b^{MI} \Gamma^+(M_{\alpha\beta})_{ab} \tag{23.427}$$

(for $\alpha, \beta = 1, 2, \ldots, D-2$, with $\alpha \neq \beta$) and

$$\bar{B}_k = B_k - \frac{1}{2\kappa} \sum_{I,J=1}^{N} \sum_{a=1}^{d/2} Q_a^{I+} Q_a^{J+} \Gamma^{\text{int}}(B_k)^*_{IJ} \tag{23.428}$$

(for $k = 1, 2, \ldots, n_0$), it follows that the effective little superalgebra is the *direct sum* of the Lie algebra with basis elements $\bar{M}_{\alpha\beta}$ (which is isomorphic to so$(D-2)$), the Lie algebra with basis elements $\bar{B}_k$ (which is isomorphic to $\mathscr{L}_{\text{int}}$), and the $\frac{1}{2}Nd$-dimensional Euclidean Clifford algebra, and so its irreducible representations are the direct products of those of these three constituent parts.

In considering the irreducible representations of the class of extended Poincaré superalgebras with D mod $8 = 2$ it is necessary to separate the special case $D = 2$ from those with $D > 2$.

Example III *"Massless" irreducible representations of the extended Poincaré superalgebras for $D = 2$*

The commutation and anticommutation relations of the unextended and extended Poincaré superalgebras for $D = 2$ are given in Example I of Section 4(*a*) and Example II of Section 4(*b*) respectively. It will be assumed that the internal symmetry algebra $\mathscr{L}_{\text{int}}$ is so$(N_{\text{L}}) \oplus$ so(N_{R}). As any central charges that are present will have zero eigenvalues, they may be omitted from the discussion. (The unextended $D = 2$ Poincaré superalgebra can be regarded as the special case in which $N_{\text{L}} = N_{\text{R}} = 1$, for which $\mathscr{L}_{\text{int}}$ consists only of the zero element.)

$D = 2$ Minkowski space–time is exceptional in that the contravariant 2-momenta $\hat{p} = (\kappa, \kappa)$ and $\hat{p} = (-\kappa, \kappa)$ lie in *different* orbits of the Poincaré group, and so they give rise to different little superalgebras, which have to be considered separately.

(*i*) *Little superalgebra corresponding to the contravariant 2-momenta $\hat{p} = (\kappa, \kappa)$*

For the contravariant 2-momenta $\hat{p} = (\kappa, \kappa)$, (23.258)–(23.260) reduce (on using (23.238) and (23.239)) to

$$[Q^I_{\text{L}1}, Q^J_{\text{L}1}]_+ = 4\kappa\delta^{IJ} 1 \tag{23.429}$$

(for $I, J = 1, 2, \ldots, N_L$),

$$[Q^I_{R1}, Q^J_{R1}]_+ = 0 \tag{23.430}$$

(for $I, J = 1, 2, \ldots, N_R$) and

$$[Q^I_{L1}, Q^J_{R1}]_+ = 0 \tag{23.431}$$

(for $I = 1, 2, \ldots, N_L$ and $J = 1, \ldots, N_R$). It follows from (23.430) that when acting on state vectors corresponding to the 2-momentum $\hat{p}$ the operators Q^I_{R1} all give zero. Thus the effective little superalgebra consists of the N_L operators $(2\kappa)^{-1/2}Q^I_{L1}$ (for $I = 1, 2, \ldots, N_L$), which generate an N_L-dimensional Euclidean Clifford algebra, together with the $\frac{1}{2}N_L(N_L - 1)$ basis elements of the so(N_L) subalgebra of $\mathscr{L}_{\text{int}}$ (which consists of those elements B of $\mathscr{L}_{\text{int}}$ for which $\mathbf{R}(B) = \mathbf{0}$). With the definition

$$\bar{B}_k = B_k - \frac{1}{2\kappa} \sum_{I,J=1}^{N_L} Q^I_{L1} Q^J_{L1} L(B_k)_{IJ}, \tag{23.432}$$

the set of operators $\bar{B}_k$ form a Lie algebra that is isomorphic to so(N_L) and each of them commutes with all the Q^I_{L1}. Thus the effective little superalgbra is the direct sum of the N_L-dimensional Euclidean Clifford algebra and an so(N_L) Lie algebra, and its irreducible representations are the direct products of those of its constituent parts. As the effective little superalgebra contains *no* part that corresponds to space–time rotations, any identification of these irreducible representations with helicity is lost, and so any breakdown into fermion and boson states is at most purely conventional. (See Strathdee (1987) for further discussion of this point.)

(ii) Little superalgebra corresponding to the contravariant 2-momenta $\hat{p} = (-\kappa, \kappa)$

For the contravariant 2-momenta $\hat{p} = (-\kappa, \kappa)$ the analogues of (23.429)–(23.431) are

$$[Q^I_{R1}, Q^J_{R1}]_+ = 4\kappa\delta^{IJ}1 \tag{23.433}$$

(for $I, J = 1, 2, \ldots, N_R$),

$$[Q^I_{L1}, Q^J_{L1}]_+ = 0 \tag{23.434}$$

(for $I, J = 1, 2, \ldots, N_L$) and

$$[Q^I_{L1}, Q^J_{R1}]_+ = 0 \tag{23.435}$$

(for $I = 1, 2, \ldots, N_L$ and $J = 1, \ldots, N_R$). Consequently the theory of case (i) can be repeated with all left-handed objects replaced by right-handed objects. The effective little superalgebra that emerges is thus the direct sum of the

N_R-dimensional Euclidean Clifford algebra and an so(N_R) Lie algebra, and its irreducible representations are the direct products of those of its constituent parts.

Example IV *"Massless" irreducible representations of the extended Poincaré superalgebras for $D \bmod 8 = 2$ and $D \geqslant 10$*

The commutation and anticommutation relations of the extended Poincaré superalgebra for $D \bmod 8 = 2$ are given in Example II of Section 4(*b*). It will be assumed that the internal symmetry algebra $\mathscr{L}_{\text{int}}$ is so(N_L)$\oplus$so(N_R). As any central charges that are present will have zero eigenvalues, they may be omitted from the discussion. (As the unextended Poincaré superalgebra for $D \bmod 8 = 2$ can be regarded as the special case in which $N_L = N_R = 1$ and for which $\mathscr{L}_{\text{int}}$ consists only of the zero element, the results that follow provide a refinement of those already given for the unextended case in Section 7(*a*)(*B*).)

For state vectors corresponding to the contravariant D-momenta $\hat{p} = (0, 0, \ldots, 0, \kappa, \kappa)$, on using the *Majorana–Weyl* representation of the D-dimensional Minkowski Clifford algebra that appears in part (b) of Theorem III of Appendix L, Section 4, and in particular by invoking (L.127), equations (23.258)–(23.260) reduce to

$$[Q^I_{LA}, Q^J_{LB}]_+ = 4\kappa\delta^{IJ}\delta_{AB}1 \tag{23.436}$$

(for $I, J = 1, 2, \ldots, N_L$ and $A, B = 1, 2, \ldots, \frac{1}{4}d$).

$$[Q^I_{RA}, Q^J_{RB}]_+ = 4\kappa\delta^{IJ}\delta_{AB}1 \tag{23.437}$$

(for $I, J = 1, 2, \ldots, N_R$ and $A, B = \frac{1}{4}d + 1, \frac{1}{4}d + 2, \ldots, \frac{1}{2}d$),

$$[Q^I_{LA}, Q^J_{LB}]_+ = 0 \tag{23.438}$$

(for $I, J = 1, 2, \ldots, N_L$ and $A, B = \frac{1}{4}d + 1, \frac{1}{4}d + 2, \ldots, \frac{1}{2}d$), and

$$[Q^I_{RA}, Q^J_{RB}]_+ = 0 \tag{23.439}$$

(for $I, J = 1, 2, \ldots, N_R$ and $A, B = 1, 2, \ldots, \frac{1}{4}d$), with all other anticommutators of the supertranslation operators being zero. It follows from (23.438) and (23.439) that the action of Q^I_{LA} (for $I = 1, 2, \ldots, N_L$ and $A = \frac{1}{4}d + 1, \frac{1}{4}d + 2, \ldots, \frac{1}{2}d$) and the action of Q^I_{RA} (for $I = 1, 2, \ldots, N_R$ and $A = 1, \ldots, \frac{1}{4}d$) on any state vector with the special D-momentum $\hat{p}$ gives zero. Moreover, (23.436) and (23.437) show that the set of operators $(2\kappa)^{-1/2}Q^I_{LA}$ (for $I = 1, 2, \ldots, N_L$ and $A = 1, 2, \ldots, \frac{1}{4}d$) and the set of operators $(2\kappa)^{-1/2}Q^I_{RA}$ (for $I = 1, 2, \ldots, N_R$ and $A = \frac{1}{4}d + 1$, $\frac{1}{4}d + 2, \ldots, \frac{1}{2}d$) generate two Euclidean Clifford algebras of dimensions $\frac{1}{4}N_L d$ and $\frac{1}{4}N_R d$ respectively. Also, (L.128) and (L.129) indicate that these two latter sets are basis vectors for carrier spaces of the

$\frac{1}{4}d$- dimensional irreducible representations $\Gamma_{\mathrm{L}}^{\mathrm{so}(D-2)}$ and $\Gamma_{\mathrm{R}}^{\mathrm{so}(D-2)}$ of so$(D-2)$. With the definitions

$$\bar{M}_{\alpha\beta} = M_{\alpha\beta} - \frac{1}{2\kappa} \sum_{I=1}^{N_{\mathrm{L}}} \sum_{A,B=1}^{d/4} Q_{\mathrm{L}A}^{I} Q_{\mathrm{L}B}^{I} \Gamma_{\mathrm{L}}^{\mathrm{so}(D-2)}(M_{\alpha\beta})_{AB}$$
$$- \frac{1}{2\kappa} \sum_{I=1}^{N_{\mathrm{R}}} \sum_{A,B=d/4+1}^{d/2} Q_{\mathrm{R}A}^{I} Q_{\mathrm{R}B}^{I} \Gamma_{\mathrm{R}}^{\mathrm{so}(D-2)}(M_{\alpha\beta})_{AB} \tag{23.440}$$

(for $\alpha, \beta = 1, 2, \ldots, D-2$),

$$\bar{B}_k = B_k - \frac{1}{2\kappa} \sum_{I,J=1}^{N_{\mathrm{L}}} \sum_{A=1}^{d/4} Q_{\mathrm{L}A}^{I} Q_{\mathrm{L}A}^{J} L(B_k)_{IJ} \tag{23.441}$$

(for each basis element B_k of so(N_{L})) and

$$\bar{B}_k = B_k - \frac{1}{2\kappa} \sum_{I,J=1}^{N_{\mathrm{R}}} \sum_{A=d/4+1}^{d/2} Q_{\mathrm{R}A}^{I} Q_{\mathrm{R}A}^{J} R(B_k)_{IJ} \tag{23.442}$$

(for each basis element B_k of so(N_{R})), it follows that the effective little superalgebra is the direct sum of five parts: the two Euclidean Clifford algebras of dimensions $\frac{1}{4}N_{\mathrm{L}}d$ and $\frac{1}{4}N_{\mathrm{R}}d$, the Lie algebras so$(N_{\mathrm{L}})$ and so(N_{L}) with basis elements defined by (23.441) and (23.442), and the so$(D-2)$ Lie algebra with basis elements defined by (23.440). Its irreducible representations are then given by the direct products of these constituent subalgebras.

(By contrast to the situation in the previous example, the contravariant D-momentum $\hat{p} = (0, 0, \ldots, 0, \kappa, \kappa)$ is here in the *same* orbit of the D-dimensional Poincaré group as $\hat{p} = (0, 0, \ldots, 0, -\kappa, \kappa)$.)

24

Poincaré Supersymmetric Fields

1 Supersymmetric field theory

The object of this chapter is to discuss field theoretical models for elementary particles whose actions are invariant under the transformations of Poincaré superalgebras. For brevity the normal practice of simply calling these "supersymmetric field theories" will be followed. Two different (but related) formalisms have been widely employed. They are known as the "component formalism" and the "superfield formalism", and each has its advantages and disadvantages, as will become apparent from the developments that follow.

In Section 2 an introduction is given to the component formalism. The first stage is the setting up in Section 2(*a*) of the simplest "supermultiplet", which is known as the "chiral multiplet" (or "Wess–Zumino multiplet"). This is then used in Section 2(*b*) to construct the "Wess–Zumino model", which is the simplest supersymmetric field theory. Not only does this model exhibit general features that keep on occurring in supersymmetric models, such as the existence of "auxiliary fields", but also individual parts reappear in more realistic models, such as the supersymmetric Yang–Mills gauge theories of Section 4. Section 2(*c*) describes the properties of the more elaborate "general multiplet" and an action associated with it.

In Section 3 essentially the same ground is covered in the superfield formalism, although for convenience the ordering is different. The "scalar superfield" is investigated and related to the general multiplet in Section 3(*a*). The "chiral superfields", which are associated with the chiral multiplet, are first introduced in Section 3(*b*), and are then used in Section 3(*c*) to cast the action of the Wess–Zumino model in the form of superspace integrals.

Section 4 is devoted to the description of supersymmetric gauge theories. The first stage is the construction of a supersymmetric Abelian theory, first (in Section 4(*a*)) in a component formalism, and then (in Section 4(*b*)) in its superfield version. This latter form is then generalized in Section 4(*c*) to produce the non-Abelian "supersymmetric Yang–Mills theories".

Finally, in Section 5, the effects of the spontaneous breaking of symmetries in both global and gauge theories are described, the account covering both the breaking of internal symmetries and of the supersymmetry itself.

Because the literature on this subject is so vast, it is quite impossible to give a comprehensive set of references, and so the references that do appear are restricted almost entirely to the original papers of the mid-1970s in which the foundations of the subject were laid and to some more specialized reviews, which themselves contain many other references.

2 Supersymmetric multiplets

(*a*) *The chiral multiplet in component form*

The first supersymmetric multiplet of fields was set up by Wess and Zumino (1974a, b). Although they gave it the name *scalar multiplet*, subsequent developments led Salam and Strathdee (1975b) to call it the *chiral multiplet*, an appellation that has since been widely adopted and that will be used here. (The term *Wess–Zumino multiplet* is an alternative designation that is also commonly used in the literature.)

The chiral multiplet consists of four fields $A(\mathbf{x})$, $B(\mathbf{x})$, $F(\mathbf{x})$ and $G(\mathbf{x})$, which are assumed to transform as scalars with respect to proper Lorentz transformations, together with a 4-component spinor field $\boldsymbol{\psi}(\mathbf{x})$ (with components $\psi_a(\mathbf{x})$, where a takes values $1,\ldots,4$), which is assumed to transform as the 4-dimensional spinor representation $\boldsymbol{\Gamma}^{\text{spin}}$ of (23.28). Here $\mathbf{x} = (x^1, x^2, x^3, x^4)$ is a set of four real variables specifying a point in Minkowski space–time, and $A(\mathbf{x})$, $B(\mathbf{x})$, $F(\mathbf{x})$, $G(\mathbf{x})$ and the components $\psi_a(\mathbf{x})$ are *operators* acting on the Hilbert space of physical states.

Consider the commutation and anticommutation relations

$$[Q_a, A(\mathbf{x})]_- = i\hbar^{1/2}\psi_a(\mathbf{x}), \tag{24.1}$$

$$[Q_a, B(\mathbf{x})]_- = -\hbar^{1/2} \sum_{b=1}^{4} (\gamma_5)_{ab}\psi_b(\mathbf{x}), \tag{24.2}$$

$$[Q_a, F(\mathbf{x})]_- = \hbar^{1/2} \sum_{b=1}^{4} \sum_{\alpha=1}^{4} (\gamma^\alpha)_{ab}\partial_\alpha\psi_b(\mathbf{x}), \tag{24.3}$$

$$[Q_a, G(\mathbf{x})]_- = i\hbar^{1/2} \sum_{b=1}^{4} \sum_{\alpha=1}^{4} (\gamma_5\gamma^\alpha)_{ab}\partial_\alpha\psi_b(\mathbf{x}) \tag{24.4}$$

(for $a = 1, \ldots, 4$), and

$$[Q_a, \psi_b(\mathbf{x})]_+ = \hbar^{1/2} \sum_{\alpha=1}^{4} \{(\gamma^\alpha \mathbf{C})_{ba} \partial_\alpha A(\mathbf{x}) + (i\gamma_5\gamma^\alpha \mathbf{C})_{ba} \partial_\alpha B(\mathbf{x})\} + \hbar^{1/2}\{i(\mathbf{C})_{ba} F(\mathbf{x}) - (\gamma_5 \mathbf{C})_{ba} G(\mathbf{x})\} \quad (24.5)$$

(for $a, b = 1, \ldots, 4$). Then Q_a (for $a = 1, \ldots, 4$) are the supertranslation operators of the unextended $D = 4$ Poincaré superalgebra that was introduced in Chapter 23, Section 2(a), the notations and conventions of which will be employed again here. Also, ∂_α will henceforth be used as an abbreviation for the partial derivative $\partial/\partial x^\alpha$.

It should first be noted that these equations are *form-invariant* under any similarity transformation of the Dirac matrices γ^α of the form (23.13), *provided* that the spinor field components transform as

$$\psi'_a(\mathbf{x}) = \sum_{b=1}^{4} S_{ab} \psi_b(\mathbf{x}), \quad (24.6)$$

that the supertranslation operators transform as (23.46), and that the scalar fields $A(\mathbf{x})$, $B(\mathbf{x})$, $F(\mathbf{x})$ and $G(\mathbf{x})$ are left unchanged. In particular, in the *Majorana* representation of (23.14) the corresponding supertranslation operators, which may be denoted by Q_a^{M}, are *self-adjoint*; that is,

$$[Q_a^{\mathrm{M}})^\dagger = Q_a^{\mathrm{M}} \quad (24.7)$$

(for $a = 1, \ldots, 4$) (cf. (23.52)). As the corresponding Dirac matrices $\gamma_\alpha^{\mathrm{M}}$ are all purely imaginary, and as the charge conjugation matrix $\mathbf{C}^{\mathrm{M}}$ $(= -\gamma_4^{\mathrm{M}})$ is also purely imaginary, it follows that the field operators $A(\mathbf{x})$, $B(\mathbf{x})$, $F(\mathbf{x})$ and $G(\mathbf{x})$ and spinor field components $\psi_a^{\mathrm{M}}(\mathbf{x})$ in the Majorana representation can also be chosen to be *self-adjoint*; that is,

$$(A(\mathbf{x}))^\dagger = A(\mathbf{x}), \quad (B(\mathbf{x}))^\dagger = B(\mathbf{x}), \quad (F(\mathbf{x}))^\dagger = F(\mathbf{x}), \quad (G(\mathbf{x}))^\dagger = G(\mathbf{x}) \quad (24.8)$$

and (for $a = 1, \ldots, 4$)

$$(\psi_a^{\mathrm{M}}(\mathbf{x}))^\dagger = \psi_a^{\mathrm{M}}(\mathbf{x}). \quad (24.9)$$

As the scalar fields $A(\mathbf{x})$, $B(\mathbf{x})$, $F(\mathbf{x})$ and $G(\mathbf{x})$ are left invariant by similarity transformations of the Dirac matrices, (24.8) are valid for *every* choice of the representation of the Dirac matrices, while the generalization of (24.9) to any representation is

$$\sum_{b=1}^{4} (\gamma_4 \mathbf{C})_{ab} \psi_b(\mathbf{x})^\dagger = -\psi_a(\mathbf{x}), \quad (24.10)$$

(cf. (23.53)), or, more succinctly, in a notation in which $\boldsymbol{\psi}(\mathbf{x})$ denotes the 4×1 column vector with entries $\psi_1(\mathbf{x})$, $\psi_2(\mathbf{x})$, $\psi_3(\mathbf{x})$ and $\psi_4(\mathbf{x})$,

$$\gamma_4 \mathbf{C} \boldsymbol{\psi}(\mathbf{x})^\dagger = -\boldsymbol{\psi}(\mathbf{x}). \quad (24.11)$$

(A 4-component spinor $\psi(\mathbf{x})$ that satisfies (24.11) is said to be a "Majorana" spinor.)

It is convenient to denote the collection of components of such a chiral multiplet by Φ_C, so that

$$\Phi_C = \{A(\mathbf{x}), B(\mathbf{x}), F(\mathbf{x}), G(\mathbf{x}); \psi(\mathbf{x})\}. \tag{24.12}$$

If the components satisfy the adjoint conditions (24.8)–(24.11), the chiral multiplet Φ_C is said to be *real*. A *complex* chiral multiplet Φ_C can be constructed from two real chiral multiplets $\Phi_{C1} = \{A_1(\mathbf{x}), B_1(\mathbf{x}), F_1(\mathbf{x}), G_1(\mathbf{x}); \psi_1(\mathbf{x})\}$ and $\Phi_{C2} = \{A_2(\mathbf{x}), B_2(\mathbf{x}), F_2(\mathbf{x}), G_2(\mathbf{x}); \psi_2(\mathbf{x})\}$ by defining

$$A(\mathbf{x}) = A_1(\mathbf{x}) + iA_2(\mathbf{x}), \tag{24.13}$$

$$\psi(\mathbf{x}) = \psi_1(\mathbf{x}) + i\psi_2(\mathbf{x}), \tag{24.14}$$

and so on, and it is convenient to express this concisely by writing

$$\Phi_C = \Phi_{C1} + i\Phi_{C2}. \tag{24.15}$$

The second point to note is that there are two classes of field. One consists of $A(\mathbf{x})$, $B(\mathbf{x})$, $F(\mathbf{x})$ and $G(\mathbf{x})$, and may be called "even", and the other consists of the $\psi_a(\mathbf{x})$, and may be called "odd". The generalized Lie product of any Q_a with an *even* field must be taken to be a *commutator*, whereas that with an *odd* field must be assumed to be an anticommutator. It is easily checked that only with this assignment can the generalized Jacobi identities (21.3) be satisfied.

The commutation and anticommutation relations (24.1)–(24.5) have been chosen to give the result

$$[[Q_a, Q_b]_+, \Phi(\mathbf{x})]_- = -2 \sum_{\alpha=1}^{4} (\gamma^\alpha \mathbf{C})_{ab} \frac{\hbar}{i} \partial_\alpha \Phi(\mathbf{x}) \tag{24.16}$$

for $a, b = 1, \ldots, 4$, where $\Phi(\mathbf{x}) = A(\mathbf{x})$, $B(\mathbf{x})$, $F(\mathbf{x})$, $G(\mathbf{x})$ and $\psi_{a'}(\mathbf{x})$ (for $a' = 1, \ldots, 4$). With the identification

$$\frac{\hbar}{i} \partial_\alpha \Phi(\mathbf{x}) = [P_\alpha, \Phi(\mathbf{x})]_- \tag{24.17}$$

for any operator-valued field or field component $\Phi(x)$, it follows that

$$[[Q_a, Q_b]_+, \Phi(\mathbf{x})]_- = -2 \sum_{\alpha=1}^{4} (\gamma^\alpha \mathbf{C})_{ab} [P_\alpha, \Phi(\mathbf{x})]_-, \tag{24.18}$$

which is consistent with the anticommutation relations (23.45); that is, with

$$[Q_a, Q_b]_+ = -2 \sum_{\alpha=1}^{4} (\gamma^\alpha \mathbf{C})_{ab} P_\alpha.$$

Thus the fields $\Phi(\mathbf{x}) = A(\mathbf{x}), B(\mathbf{x}), F(\mathbf{x}), G(\mathbf{x})$ and $\psi_a(\mathbf{x})$ (for $a = 1, \ldots, 4$) provide a carrier space on which the supertranslation operators act in accordance with the appropriate anticommutation relations.

(The equality (24.17) itself follows from the assumption that for each of these fields and for every space–time translation $[\mathbf{1}|\mathbf{t}]$ through a contravariant 4-vector $\mathbf{t} = (t^1, t^2, t^3, t^4)$ there is an associated operator $U([\mathbf{1}|\mathbf{t}])$ such that

$$U([\mathbf{1}|\mathbf{t}])\Phi(\mathbf{x})U([\mathbf{1}|\mathbf{t}])^{-1} = \Phi(\mathbf{x} + \mathbf{t}) \tag{24.19}$$

(cf. Weinberg 1964a, b, 1969) and that is related to the 4-momentum operators P_α by

$$U([\mathbf{1}|\mathbf{t}]) = \exp\left\{\sum_{\alpha=1}^{4} \frac{t^\alpha}{\hbar} iP_\alpha\right\}.$$

That is, for pure space–time *translations*, the transformation law for *all* operator-valued fields is the same as for the operator-valued *scalar* fields considered in Chapter 23, Section 2(*d*), so the explicit form (24.17) follows immediately from (23.155).)

There is no difficulty in checking the anticommutation relations (24.16) for $\Phi(\mathbf{x}) = A(\mathbf{x}), B(\mathbf{x}), F(\mathbf{x})$ and $G(\mathbf{x})$, but for $\Phi(\mathbf{x}) = \psi_a(\mathbf{x})$ (for $a = 1, \ldots, 4$) the algebraic manipulations are more complicated. Although the original method of checking used by Wess and Zumino (1974a) involved invoking the matrix identity (L.136) that is presented in Appendix L, Section 5, probably the simplest way is to employ the two-component formalism by writing (24.1)–(24.5) in the modified chiral representation in which the Dirac matrices have the form (23.25). Using the definition (23.54) of Q_A and $\bar{Q}_A$ (for $A = 1, 2$), similarly defining $\psi_{\mathrm{L}A}(\mathbf{x})$ and $\lambda_{\mathrm{R}A}(\mathbf{x})$ (for $A = 1, 2$) by

$$\psi_{\mathrm{L}A}(\mathbf{x}) = \psi_A^{\mathrm{C}'}(\mathbf{x}), \quad \psi_{\mathrm{R}A}(\mathbf{x}) = \psi_{A+2}^{\mathrm{C}'}(\mathbf{x}) \tag{24.20}$$

(for $A = 1, 2$), and letting

$$A_{\mathrm{L}}(\mathbf{x}) = A(\mathbf{x}) - iB(\mathbf{x}), \quad A_{\mathrm{R}}(\mathbf{x}) = A(\mathbf{x}) + iB(\mathbf{x}) \tag{24.21}$$

and

$$F_{\mathrm{L}}(\mathbf{x}) = F(\mathbf{x}) + iG(\mathbf{x}), \quad F_{\mathrm{R}}(\mathbf{x}) = F(\mathbf{x}) - iG(\mathbf{x}), \tag{24.22}$$

(24.1)–(24.5) reduce to

$$[Q_A, A_{\mathrm{L}}(\mathbf{x})]_- = 2i\hbar^{1/2}\psi_{\mathrm{L}A}(\mathbf{x}), \tag{24.23}$$

$$[Q_A, F_{\mathrm{L}}(\mathbf{x})]_- = 0, \tag{24.24}$$

$$[Q_A, \psi_{\mathrm{L}B}(\mathbf{x})]_+ = -i\hbar^{1/2}(\boldsymbol{\sigma}_2)_{AB}F_{\mathrm{L}}(\mathbf{x}), \tag{24.25}$$

$$[\bar{Q}_A, A_{\mathrm{L}}(\mathbf{x})]_- = 0, \tag{24.26}$$

$$[\bar{Q}_A, F_{\mathrm{L}}(\mathbf{x})]_- = -2\hbar^{1/2}\sum_{\alpha=1}^{4}\sum_{B=1}^{2}(\boldsymbol{\sigma}_2\boldsymbol{\sigma}_\alpha)_{AB}\,\partial_\alpha\psi_{\mathrm{L}B}(\mathbf{x}), \tag{24.27}$$

$$[\bar{Q}_A, \psi_{\mathrm{L}B}(\mathbf{x})]_+ = -\hbar^{1/2}\sum_{\alpha,\beta=1}^{4}\eta^{\alpha\beta}(\boldsymbol{\sigma}_\alpha)_{BA}\,\partial_\beta A_{\mathrm{L}}(\mathbf{x}), \tag{24.28}$$

$$[Q_A, A_{\mathrm{R}}(\mathbf{x})]_- = 0, \tag{24.29}$$

$$[Q_A, F_{\mathrm{R}}(\mathbf{x})]_- = -2\hbar^{1/2}\sum_{\alpha,\beta=1}^{4}\sum_{B=1}^{2}\eta^{\alpha\beta}(\boldsymbol{\sigma}_\alpha\boldsymbol{\sigma}_2)_{AB}\,\partial_\beta\psi_{\mathrm{R}B}(\mathbf{x}), \tag{24.30}$$

$$[Q_A, \psi_{\mathrm{R}B}(\mathbf{x})]_+ = -\hbar^{1/2}\sum_{\alpha,\beta=1}^{4}\eta^{\alpha\beta}(\boldsymbol{\sigma}_\alpha)_{AB}\,\partial_\beta A_{\mathrm{R}}(\mathbf{x}), \tag{24.31}$$

$$[\bar{Q}_A, A_{\mathrm{R}}(\mathbf{x})]_- = 2i\hbar^{1/2}\psi_{\mathrm{R}A}(\mathbf{x}), \tag{24.32}$$

$$[\bar{Q}_A, F_{\mathrm{R}}(\mathbf{x})]_- = 0, \tag{24.33}$$

$$[\bar{Q}_A, \psi_{\mathrm{R}B}(\mathbf{x})]_+ = -i\hbar^{1/2}(\boldsymbol{\sigma}_2)_{AB}F_{\mathrm{R}}(\mathbf{x}) \tag{24.34}$$

(for $A, B = 1, 2$). The checking of (24.16) now involves only manipulating 2×2 matrices, and so is quite straightforward.

It is worth noting that (24.11) and (23.27) imply that

$$(\psi_{\mathrm{L}A}(\mathbf{x}))^\dagger = \psi_{\mathrm{R}A}(\mathbf{x}), \quad (\psi_{\mathrm{R}A}(\mathbf{x}))^\dagger = \psi_{\mathrm{L}A}(\mathbf{x}) \tag{24.35}$$

(for $A = 1, 2$), and (24.8) gives

$$(A_{\mathrm{L}}(\mathbf{x}))^\dagger = A_{\mathrm{R}}(\mathbf{x}), \quad (F_{\mathrm{L}}(\mathbf{x}))^\dagger = F_{\mathrm{R}}(\mathbf{x}). \tag{24.36}$$

On recalling that

$$(Q_A)^\dagger = \bar{Q}_A, \quad (\bar{Q}_A)^\dagger = Q_A$$

(for $A = 1, 2$) (cf. (23.55)), it becomes clear that (24.29)–(24.34) are the adjoints of (24.23)–(24.28).

Equations (24.23)–(24.34) show that the chiral multiplet splits into two *disjoint invariant* parts, one consisting of $A_{\mathrm{L}}(\mathbf{x})$, $F_{\mathrm{L}}(\mathbf{x})$ and $\psi_{\mathrm{L}A}(\mathbf{x})$ (for $A = 1, 2$), and the other consisting of $A_{\mathrm{R}}(\mathbf{x})$, $F_{\mathrm{R}}(\mathbf{x})$ and $\psi_{\mathrm{R}A}(\mathbf{x})$ (for $A = 1, 2$). These will be referred to as the "left-handed" and "right-handed" multiplets respectively. It is apparent that the spin content of both of the the left-handed multiplet and the right-handed multiplet is the same as that of the "massive" irreducible representation of the unextended $D = 4$ Poincaré superalgebra with $j = 0$ that was discussed in Chapter 23, Section 5(*a*). (This correspondence will be discussed further in Section 2(*c*).)

In the literature a commonly encountered way of expressing the above commutation and anticommutation relations is to introduce four odd Grassmann variables ζ^1, ζ^2, ζ^3 and ζ^4, which are assumed to commute with the even fields and to anticommute with the odd fields, and to define the operator δ_ζ by

$$\delta_\zeta \Phi(\mathbf{x}) = \hbar^{-1/2} \sum_{a,b=1}^{4} [(\zeta^b)^\#(i\gamma_4)_{ba}Q_a, \Phi(\mathbf{x})]_- \tag{24.37}$$

for any field or field component $\Phi(\mathbf{x})$. Here # denotes the Grassmann adjoint operation of (20.30) and (20.31). Then if $\Phi(\mathbf{x})$ is *even*

$$\delta_\zeta \Phi(\mathbf{x}) = \hbar^{-1/2} \sum_{a,b=1}^{4} [(\zeta^b)^\#(i\gamma_4)_{ba}Q_a, \Phi(\mathbf{x})]_-, \tag{24.38}$$

whereas if $\Phi(\mathbf{x})$ is *odd*

$$\delta_\zeta \Phi(\mathbf{x}) = \hbar^{-1/2} \sum_{a,b=1}^{4} (\zeta^b)^\#(i\gamma_4)_{ba}[Q_a, \Phi(\mathbf{x})]_+. \tag{24.39}$$

The combination

$$\hbar^{-1/2} \sum_{a,b=1}^{4} (\zeta^b)^\#(i\gamma_4)_{ba}Q_a \tag{24.40}$$

that appears in these expressions is precisely that which was introduced in (23.97). As noted in Chapter 23, Section 2(*c*), with the assumption that the ζ^a transform in the same way as spinors under similarity transformations and with the assumption that they satisfy (23.99) in the Majorana representation, the combination (24.40) is always a member of the *real* super Lie algebra corresponding to the unextended $D = 4$ Poincaré superalgebra. (Indeed, the operator δ_ζ is just the "component" version of the superspace operator defined in (23.145).)

In this notation (24.1) can be rewritten as

$$\delta_\zeta A(\mathbf{x}) = - \sum_{a,b=1}^{4} (\zeta^b)^\#(\gamma_4)_{ba}\psi_a(\mathbf{x}),$$

or, more succinctly, as

$$\delta_\zeta A(\mathbf{x}) = - \tilde{\boldsymbol{\zeta}}^\#\gamma_4\boldsymbol{\psi}(\mathbf{x}). \tag{24.41}$$

where $\tilde{\boldsymbol{\zeta}}$ is the transpose of the 4×1 column matrix $\boldsymbol{\zeta}$ with entries ζ_1, ζ_2, ζ_3 and ζ_4. Similarly (24.2)–(24.5) give

$$\delta_\zeta B(\mathbf{x}) = - i\tilde{\boldsymbol{\zeta}}^\#\gamma_4\gamma_5\boldsymbol{\psi}(\mathbf{x}), \tag{24.42}$$

$$\delta_\zeta F(\mathbf{x}) = i \sum_{\alpha=1}^{4} \tilde{\boldsymbol{\zeta}}^\#\gamma_4\gamma^\alpha\partial_\alpha\boldsymbol{\psi}(\mathbf{x}), \tag{24.43}$$

$$\delta_\zeta G(\mathbf{x}) = - \sum_{\alpha=1}^{4} \tilde{\boldsymbol{\zeta}}^\#\gamma_4\gamma_5\gamma^\alpha\partial_\alpha\boldsymbol{\psi}(\mathbf{x}), \tag{24.44}$$

$$\delta_\zeta \psi_a(\mathbf{x}) = \sum_{b=1}^{4} (\zeta^b)^\# \sum_{\alpha=1}^{4} \{(i\gamma_4 \tilde{\mathbf{C}} \tilde{\gamma}^\alpha)_{ba}\, \partial_\alpha A(\mathbf{x}) + (\gamma_4 \tilde{\mathbf{C}} \tilde{\gamma}^\alpha \tilde{\gamma}_5)_{ba}\, \partial_\alpha B(\mathbf{x})\}$$

$$- \sum_{b=1}^{4} (\zeta^b)^\# \{(\gamma_4 \tilde{\mathbf{C}})_{ba} F(\mathbf{x}) + (i\gamma_4 \tilde{\mathbf{C}} \tilde{\gamma}_5)_{ba} G(\mathbf{x})\}. \tag{24.45}$$

Using (23.100), this last equation can be rewritten as

$$\delta_\zeta \boldsymbol{\psi}(\mathbf{x}) = \sum_{\alpha=1}^{4} \{i\gamma^\alpha \boldsymbol{\zeta}\, \partial_\alpha A(\mathbf{x}) + \gamma_5 \gamma^\alpha \boldsymbol{\zeta}\, \partial_\alpha B(\mathbf{x})\} - \boldsymbol{\zeta} F(\mathbf{x}) - i\gamma_5 \boldsymbol{\zeta} G(\mathbf{x}). \tag{24.46}$$

(It was essentially in the form of (24.41)–(24.44) and (24.46) that the chiral multiplet originally appeared (Wess and Zumino 1974a), although there are several differences in conventions between the original version and the present account.)

Similarly, since (24.40) reduces in the two-component notation to

$$-\hbar^{-1/2} \sum_{A,B=1}^{2} (\zeta_A \varepsilon_{AB} Q_B + \bar{\zeta}_A \varepsilon_{AB} \bar{Q}_B) \tag{24.47}$$

(cf. (23.103)), (24.23)–(24.34) give

$$\delta_\zeta A_{\mathrm{L}}(\mathbf{x}) = -2i \sum_{A,B=1}^{2} \zeta_A \varepsilon_{AB} \psi_{\mathrm{L}B}(\mathbf{x}), \tag{24.48}$$

$$\delta_\zeta F_{\mathrm{L}}(\mathbf{x}) = 2i \sum_{\alpha=1}^{4} \sum_{A,B=1}^{2} \bar{\zeta}_A (\boldsymbol{\sigma}_\alpha)_{AB}\, \partial_\alpha \psi_{\mathrm{L}B}(\mathbf{x}), \tag{24.49}$$

$$\delta_\zeta \psi_{\mathrm{L}A}(\mathbf{x}) = -i \sum_{\alpha,\beta=1}^{4} \sum_{B=1}^{2} \eta^{\alpha\beta} (\boldsymbol{\sigma}_\alpha \boldsymbol{\sigma}_2)_{AB} \bar{\zeta}_B\, \partial_\beta A_{\mathrm{L}}(\mathbf{x}) - \zeta_A F_{\mathrm{L}}(\mathbf{x}), \tag{24.50}$$

$$\delta_\zeta A_{\mathrm{R}}(\mathbf{x}) = -2i \sum_{A,B=1}^{2} \bar{\zeta}_A \varepsilon_{AB} \psi_{\mathrm{R}B}(\mathbf{x}), \tag{24.51}$$

$$\delta_\zeta F_{\mathrm{R}}(\mathbf{x}) = 2i \sum_{\alpha=1}^{4} \sum_{A,B=1}^{2} \zeta_A (\boldsymbol{\sigma}_\alpha)^*_{AB} \partial_\alpha \psi_{\mathrm{R}B}(\mathbf{x}), \tag{24.52}$$

$$\delta_\zeta \psi_{\mathrm{R}A}(\mathbf{x}) = i \sum_{\alpha,\beta=1}^{4} \sum_{B=1}^{2} \eta^{\alpha\beta} (\boldsymbol{\sigma}_\alpha \boldsymbol{\sigma}_2)^*_{AB} \zeta_B \partial_\beta A_{\mathrm{R}}(\mathbf{x}) - \bar{\zeta}_A F_{\mathrm{R}}(\mathbf{x}), \tag{24.53}$$

where $A = 1, 2$ in (24.50) and (24.53). (Of course these also follows from substituting the modified chiral representation matrices of (23.25) into (24.41)–(24.44), and (24.46). Here $\zeta_A = \zeta^{C'}_A$, $\bar{\zeta}_A = \zeta^{C'}_{A+2}$ $(A = 1, 2)$ (cf. (23.54)).

One advantage of this presentation is that, as the right-hand side of (24.43) is a 4-divergence, (24.43) indicates that the 4-dimensional integral of $F(\mathbf{x})$ over Minkowski space–time is *invariant* under the operations induced by supertranslation generators (as well as under the ordinary Poincaré transformations). When combined with the results on products of chiral multiplets, this provides a method for constructing invariant actions, as will be demonstrated in detail in Section 2(*b*). Incidentally it is clear that there exists *no* non-constant field $\Phi(\mathbf{x})$ that is invariant in the sense that $[Q_a, \Phi(\mathbf{x})] = 0$ for $a = 1, \ldots, 4$, because this requirement taken with (24.16) implies that $\partial_\alpha \Phi(\mathbf{x}) = 0$ for $\alpha = 1, \ldots, 4$. Consequently the invariance of a space–time integral (such as that just mentioned) is the most that can be achieved.

From (24.1)–(24.5) it is clear that the fields $A(\mathbf{x})$ and $B(\mathbf{x})$ have the same dimensions, and the fields $F(\mathbf{x})$ and $G(\mathbf{x})$ have the same dimensions, but the ratios $\psi_a(\mathbf{x})/A(\mathbf{x})$ and $F(\mathbf{x})/\psi_a(\mathbf{x})$ are both of dimension $(\text{length})^{-1/2}$.

If $\Phi_{C1} = \{A_1(\mathbf{x}), B_1(\mathbf{x}), F_1(\mathbf{x}), G_1(\mathbf{x}); \boldsymbol{\psi}_1(\mathbf{x})\}$ and $\Phi_{C2} = \{A_2(\mathbf{x}), B_2(\mathbf{x}), F_2(\mathbf{x}), G_2(\mathbf{x}); \boldsymbol{\psi}_2(\mathbf{x})\}$ are any two chiral multiplets then a third chiral multiplet $\Phi_{C3} = \{A_3(\mathbf{x}), B_3(\mathbf{x}), F_3(\mathbf{x}), G_3(\mathbf{x}); \boldsymbol{\psi}_3(\mathbf{x})\}$ may be constructed from them by defining

$$A_3(\mathbf{x}) = A_1(\mathbf{x})A_2(\mathbf{x}) - B_1(\mathbf{x})B_2(\mathbf{x}), \tag{24.54}$$

$$B_3(\mathbf{x}) = A_1(\mathbf{x})B_2(\mathbf{x}) + B_1(\mathbf{x})A_2(\mathbf{x}), \tag{24.55}$$

$$\begin{aligned} F_3(\mathbf{x}) = {} & F_1(\mathbf{x})A_2(\mathbf{x}) + A_1(\mathbf{x})F_2(\mathbf{x}) + B_1(\mathbf{x})G_2(\mathbf{x}) + G_1(\mathbf{x})B_2(\mathbf{x}) \\ & - \tilde{\boldsymbol{\psi}}_1(\mathbf{x})^\dagger \gamma_4 \boldsymbol{\psi}_2(\mathbf{x}), \end{aligned} \tag{24.56}$$

$$\begin{aligned} G_3(\mathbf{x}) = {} & G_1(\mathbf{x})A_2(\mathbf{x}) + A_1(\mathbf{x})G_2(\mathbf{x}) - B_1(\mathbf{x})F_2(\mathbf{x}) - F_1(\mathbf{x})B_2(\mathbf{x}) \\ & - i\tilde{\boldsymbol{\psi}}_1(\mathbf{x})^\dagger \gamma_4 \gamma_5 \boldsymbol{\psi}_2(\mathbf{x}), \end{aligned} \tag{24.57}$$

$$\boldsymbol{\psi}_3(\mathbf{x}) = \{A_1(\mathbf{x})\mathbf{1} - iB_1(\mathbf{x})\gamma_5\}\boldsymbol{\psi}_2(\mathbf{x}) + \{A_2(\mathbf{x})\mathbf{1} - iB_2(\mathbf{x})\gamma_5\}\boldsymbol{\psi}_1(\mathbf{x}). \tag{24.58}$$

These are most easily checked in the modified chiral representation (23.25), where they simply reduce to

$$A_{\mathrm{L}3}(\mathbf{x}) = A_{\mathrm{L}1}(\mathbf{x})A_{\mathrm{L}2}(\mathbf{x}), \tag{24.59}$$

$$\begin{aligned} F_{\mathrm{L}3}(\mathbf{x}) = {} & A_{\mathrm{L}1}(\mathbf{x})F_{\mathrm{L}2}(\mathbf{x}) + A_{\mathrm{L}2}(\mathbf{x})F_{\mathrm{L}1}(\mathbf{x}) \\ & - 2i \sum_{A,B=1}^{2} \psi_{\mathrm{L}1A}(\mathbf{x})\varepsilon_{AB}\psi_{\mathrm{L}2B}(\mathbf{x}), \end{aligned} \tag{24.60}$$

$$\psi_{\mathrm{L}3A}(\mathbf{x}) = A_{\mathrm{L}1}(\mathbf{x})\psi_{\mathrm{L}2A}(\mathbf{x}) + A_{\mathrm{L}2}(\mathbf{x})\psi_{\mathrm{L}1A}(\mathbf{x}) \tag{24.61}$$

(for $A = 1, 2$), the corresponding right-handed expressions just being obtained by replacing every subscript "L" by a subscript "R". Clearly the resulting expressions for $A_3(\mathbf{x})$, $B_3(\mathbf{x})$, $F_3(\mathbf{x})$ and $G_3(\mathbf{x})$ are invariant under similarity transformations of the Dirac matrices, while $\boldsymbol{\psi}_3(\mathbf{x})$ transforms as a spinor. This construction of this product may be indicated by writing

$$\Phi_{C3} = \Phi_{C1}\Phi_{C2}, \tag{24.62}$$

and it is obvious that

$$\Phi_{C1}\Phi_{C2} = \Phi_{C2}\Phi_{C1}. \tag{24.63}$$

It is also easily checked that if $\Phi_C = \{A(\mathbf{x}), B(\mathbf{x}), F(\mathbf{x}), G(\mathbf{x}); \psi(\mathbf{x})\}$ is a chiral multiplet then so too is $\Phi'_C = \{A'(\mathbf{x}), B'(\mathbf{x}), F'(\mathbf{x}), G'(\mathbf{x}); \psi'(\mathbf{x})\}$, where

$$A'(\mathbf{x}) = -F(\mathbf{x}), \tag{24.64}$$

$$B'(\mathbf{x}) = -G(\mathbf{x}), \tag{24.65}$$

$$F'(\mathbf{x}) = \Box A(\mathbf{x}) \equiv \sum_{\alpha,\beta=1}^{4} \eta^{\alpha\beta}\frac{\partial^2 A(\mathbf{x})}{\partial x^\alpha \partial x^\beta}, \tag{24.66}$$

$$G'(\mathbf{x}) = \Box B(\mathbf{x}) \equiv \sum_{\alpha,\beta=1}^{4} \eta^{\alpha\beta}\frac{\partial^2 B(\mathbf{x})}{\partial x^\alpha \partial x^\beta}, \tag{24.67}$$

$$\psi'(\mathbf{x}) = i\sum_{\alpha=1}^{4} \gamma^\alpha \partial_\alpha \psi(\mathbf{x}). \tag{24.68}$$

It follows that applying this process *twice* to the chiral multiplet $\Phi_C = \{A(\mathbf{x}), B(\mathbf{x}), F(\mathbf{x}), G(\mathbf{x}); \psi(\mathbf{x})\}$ will give the chiral multiplet $\Phi''_C = \{-\Box A(\mathbf{x}), -\Box B(\mathbf{x}), -\Box F(\mathbf{x}), -\Box G(\mathbf{x}); -\Box\psi(\mathbf{x})\}$. Because of the role that it plays in the construction of the kinetic-energy Lagrangian density $\mathscr{L}_{KE}(\mathbf{x})$ in Wess–Zumino model (cf. Section 2(*b*)), the multiplet Φ'_C is sometimes known as the *kinetic chiral multiplet* associated with the chiral multiplet $\Phi_C = \{A(\mathbf{x}), B(\mathbf{x}), F(\mathbf{x}), G(\mathbf{x}); \psi(\mathbf{x})\}$.

(*b*) *The Wess–Zumino model in component form*

This model, which was first studied by Wess and Zumino (1974a, b), involves the fields $A(\mathbf{x})$, $B(\mathbf{x})$, $F(\mathbf{x})$, $G(\mathbf{x})$ and $\psi(\mathbf{x})$ of the real chiral multiplet, as set up in Section 2(*a*). Its invariant action is the space–time integral of a Lagrangian density that consists of three parts, which may be constructed by considering "*F*-terms" of appropriate chiral multiplet products as follows:

(*i*) *"Kinetic-energy Lagrangian density"* $\mathscr{L}_{KE}(\mathbf{x})$:

Forming the product of the chiral multiplet $\{A(\mathbf{x}), B(\mathbf{x}), F(\mathbf{x}), G(\mathbf{x}); \psi(\mathbf{x})\}$ and its associated kinetic multiplet $\{A'(\mathbf{x}), B'(\mathbf{x}), F'(\mathbf{x}), G'(\mathbf{x}); \psi'(\mathbf{x})\}$, where the latter is defined in (24.64)–(24.68), and taking $-\hbar c$ times the "*F*-term" of (24.56) (where c is the speed of light) gives

$$\hbar c\{F(\mathbf{x})^2 + G(\mathbf{x})^2\} - \hbar c\{A(\mathbf{x})\Box A(\mathbf{x}) + B(\mathbf{x})\Box B(\mathbf{x})\} + i\hbar c\sum_{\alpha=1}^{4} \tilde{\psi}(\mathbf{x})^\dagger \gamma_4 \gamma^\alpha \partial_\alpha \psi(\mathbf{x}). \tag{24.69}$$

On rewriting the second pair of terms, this gives

$$\mathscr{L}_{\mathrm{KE}}(\mathbf{x}) = \hbar c\{F(\mathbf{x})^2 + G(\mathbf{x})^2\} + \hbar c \sum_{\alpha,\beta=1}^{4} \eta^{\alpha\beta}\{\partial_\alpha A(\mathbf{x})\partial_\beta A(\mathbf{x}) + \partial_\alpha B(\mathbf{x})\partial_\beta B(\mathbf{x})\}$$

$$+ i\hbar c \sum_{\alpha=1}^{4} \tilde{\psi}(\mathbf{x})^\dagger \gamma_4 \gamma^\alpha \partial_\alpha \psi(\mathbf{x}), \qquad (24.70)$$

together with a term that may be neglected since it is a 4-divergence (and so gives a constant on integration). Then, in the notation of (24.35), $\delta_\zeta \mathscr{L}_{\mathrm{KE}}(\mathbf{x})$ is a 4-divergence, and hence its space–time integral is invariant under transformations generated by the supertranslation operators. The description "kinetic energy" is given to $\mathscr{L}_{\mathrm{KE}}(\mathbf{x})$ because the second pair of terms and the last term are the derivative terms of the Lagrangian densities of a pair of free uncharged particles with spin 0 and of a spin-$\frac{1}{2}$ particle respectively (cf. (18.1) and (18.3)).

For an equivalent expression for $\mathscr{L}_{\mathrm{KE}}(\mathbf{x})$ in terms of the "D-term" of the general multiplet formed from a symmetric product of the chiral multiplet with itself see (24.136).

(ii) "Mass Lagrangian density" $\mathscr{L}_{\mathrm{mass}}(\mathbf{x})$

Forming the product of the chiral multiplet $\{A(\mathbf{x}), B(\mathbf{x}), F(\mathbf{x}), G(\mathbf{x}); \psi(\mathbf{x})\}$ with itself and taking Mc^2 times the "F-term" of (24.56), where M has the dimensions of mass, gives

$$\mathscr{L}_{\mathrm{mass}}(\mathbf{x}) = 2Mc^2\{A(\mathbf{x})F(\mathbf{x}) + B(\mathbf{x})G(\mathbf{x}) - \tfrac{1}{2}\tilde{\psi}(\mathbf{x})^\dagger \gamma_4 \psi(\mathbf{x})\}. \qquad (24.71)$$

Then $\delta_\zeta \mathscr{L}_{\mathrm{mass}}(\mathbf{x})$ is also a 4-divergence, and hence its space–time integral too is invariant under transformations generated by the supertranslation operators. $\mathscr{L}_{\mathrm{mass}}(\mathbf{x})$ gets its name from the fact that the last term is the mass term of the Lagrangian density of a spin-$\frac{1}{2}$ particle (cf. (18.3)).

(iii) "interaction Lagrangian density" $\mathscr{L}_{\mathrm{int}}(\mathbf{x})$

Forming the product of the chiral multiplet $\{A(\mathbf{x}), B(\mathbf{x}), F(\mathbf{x}), G(\mathbf{x}); \psi(\mathbf{x})\}$ with itself *twice* (i.e. taking the "triple" product) and evaluating $-\frac{2}{3}g$ times the resulting "F-term" of (24.56), where g is a coupling constant, gives

$$\mathscr{L}_{\mathrm{int}}(\mathbf{x}) = 2g\{F(\mathbf{x})A(\mathbf{x})^2 - F(\mathbf{x})B(\mathbf{x})^2 + 2G(\mathbf{x})A(\mathbf{x})B(\mathbf{x})\}$$
$$- 2g\tilde{\psi}(\mathbf{x})^\dagger\{A(\mathbf{x})\gamma_4 - iB(\mathbf{x})\gamma_4\gamma_5\}\psi(\mathbf{x}). \qquad (24.72)$$

Then $\delta_\zeta \mathscr{L}_{\mathrm{int}}(\mathbf{x})$ is also a 4-divergence, and hence its space–time integral too is invariant under transformations generated by the supertranslation operators.

(*iv*) "*The F-term*" $\mathscr{L}_F(\mathbf{x})$

The only remaining expressions that can be formed from the chiral multiplet $\{A(\mathbf{x}), B(\mathbf{x}), F(\mathbf{x}), G(\mathbf{x}); \psi(\mathbf{x})\}$ and whose space–time integrals are invariant under transformations generated by the supertranslation operators are just proportional to the $F(\mathbf{x})$ component itself; that is, they are of the form

$$\mathscr{L}_F(\mathbf{x}) = \lambda F(\mathbf{x}), \tag{24.73}$$

where λ is a constant. (As will become apparent in Section 5, this term can eventually be removed by an appropriate field redefinition, but it is desirable to retain it at this stage to ensure that no essential features are overlooked.)

The total Lagrangian density $\mathscr{L}_{\text{tot}}(\mathbf{x})$ is then assumed to be given by

$$\mathscr{L}_{\text{tot}}(\mathbf{x}) = \mathscr{L}_{\text{KE}}(\mathbf{x}) + \mathscr{L}_{\text{mass}}(\mathbf{x}) + \mathscr{L}_{\text{int}}(\mathbf{x}) + \mathscr{L}_{\text{F}}(\mathbf{x}), \tag{24.74}$$

and has the supersymmetry property possessed by each of its parts. The Euler–Lagrange equations corresponding to the variation of $\mathscr{L}_{\text{tot}}(\mathbf{x})$ with respect to $F(\mathbf{x})$ and $G(\mathbf{x})$ are

$$F(\mathbf{x}) = -\frac{\lambda}{2\hbar c} - \frac{Mc}{\hbar}A(\mathbf{x}) - \frac{g}{\hbar c}\{A(\mathbf{x})^2 - B(\mathbf{x})^2\} \tag{24.75}$$

and

$$G(\mathbf{x}) = -\frac{Mc}{\hbar}B(\mathbf{x}) - \frac{2g}{\hbar c}A(\mathbf{x})B(\mathbf{x}), \tag{24.76}$$

which are unusual in that they are both *algebraic* equations, whereas those corresponding to the variation of $\mathscr{L}_{\text{tot}}(\mathbf{x})$ with respect to the other fields and field components are the more usual *differential* equations. Substituting (24.75) and (24.76) back into (24.74) gives a Lagrangian density that depends *only* on the fields $A(\mathbf{x})$, $B(\mathbf{x})$ and $\psi(\mathbf{x})$, its explicit expression being

$$\begin{aligned}
\mathscr{L}_{\text{tot}}(\mathbf{x}) = {} & \frac{c^3}{\hbar}\left\{\frac{\hbar^2}{c^2}\sum_{\alpha,\beta=1}^{4} \eta^{\alpha\beta}\,\partial_\alpha A(\mathbf{x})\,\partial_\beta A(\mathbf{x}) - M^2\left(A(\mathbf{x}) + \frac{\lambda}{2Mc^2}\right)^2\right\} \\
& + \frac{c^3}{\hbar}\left\{\frac{\hbar^2}{c^2}\sum_{\alpha,\beta=1}^{4} \eta^{\alpha\beta}\partial_\alpha B(\mathbf{x})\partial_\beta B(\mathbf{x}) - M^2 B(\mathbf{x})^2\right\} \\
& + \tilde{\psi}(\mathbf{x})^\dagger\left\{\sum_{\alpha=1}^{4} i\hbar c\gamma_4\gamma^\alpha\partial_\alpha\psi(\mathbf{x}) - Mc^2\gamma_4\psi(\mathbf{x})\right\} \\
& - 2g\tilde{\psi}(\mathbf{x})^\dagger\{A(\mathbf{x})\gamma_4 - iB(\mathbf{x})\gamma_4\gamma_5\}\psi(\mathbf{x}) \\
& - \frac{2\lambda g}{\hbar c}\{A(\mathbf{x})^2 - B(\mathbf{x})^2\} \\
& - \frac{2Mcg}{\hbar}A(\mathbf{x})\{A(\mathbf{x})^2 + B(\mathbf{x})^2\} - \frac{g^2}{\hbar c}\{A(\mathbf{x})^2 + B(\mathbf{x})^2\}^2.
\end{aligned} \tag{24.77}$$

The first two pairs of terms in (24.77) will be recognized as including the Lagrangian density of a pair of free uncharged particles of spin 0 and mass M, and moreover reduce precisely to this Lagrangian density when $\lambda = 0$. The third pair corresponds to a free spin-$\frac{1}{2}$ particle whose mass is also M, and the remaining terms describe interactions between these particles. The Euler–Lagrange equations that follow from the Lagrangian density (24.77) are (for the case $\lambda = 0$)

$$\left(\Box + \frac{M^2c^2}{\hbar^2}\right)A(\mathbf{x}) = -\frac{Mg}{\hbar^2}\{3A(\mathbf{x})^2 + B(\mathbf{x})^2\} - \frac{2g^2}{c^2\hbar^2}A(\mathbf{x})\{A(\mathbf{x})^2 + B(\mathbf{x})^2\} - \frac{g}{\hbar c}\tilde{\psi}(\mathbf{x})^\dagger\gamma_4\psi(\mathbf{x}), \tag{24.78}$$

$$\left(\Box + \frac{M^2c^2}{\hbar^2}\right)B(\mathbf{x}) = -\frac{2Mg}{\hbar^2}A(\mathbf{x})B(\mathbf{x}) - \frac{2g^2}{c^2\hbar^2}B(\mathbf{x})\{A(\mathbf{x})^2 + B(\mathbf{x})^2)\} - \frac{ig}{\hbar c}\tilde{\psi}(\mathbf{x})^\dagger\gamma_4\gamma_5\psi(\mathbf{x}), \tag{24.79}$$

$$i\sum_{\alpha=1}^{4}\gamma^\alpha\partial_\alpha\psi(\mathbf{x}) - \frac{Mc}{\hbar}\psi(\mathbf{x}) = \frac{2g}{\hbar c}\{A(\mathbf{x})\mathbf{1} - iB(\mathbf{x})\gamma_5\}\psi(\mathbf{x}). \tag{24.80}$$

It is easily checked that if (24.75) and (24.76) are substituted into the right-hand side of (24.1), (24.2) and (24.5), giving

$$[Q_a, A(\mathbf{x})]_- = i\hbar^{1/2}\psi_a(\mathbf{x}), \tag{24.81}$$

$$[Q_a, B(\mathbf{x})]_- = -\hbar^{1/2}\sum_{b=1}^{4}(\gamma_5)_{ab}\psi_b(\mathbf{x}), \tag{24.82}$$

$$[Q_a, \psi_b(\mathbf{x})]_+ = \hbar^{1/2}\left(\left\{\sum_{\alpha=1}^{4}\gamma^\alpha\partial_\alpha - \frac{iMc}{\hbar}\mathbf{1} - \frac{ig}{\hbar c}\{\mathbf{1}A(\mathbf{x}) + i\gamma_5B(\mathbf{x})\}\right\}\{\mathbf{1}A(\mathbf{x}) + i\gamma_5B(\mathbf{x})\}\right)_{ba}, \tag{24.83}$$

then the relation (24.16) is only valid for $\Phi(\mathbf{x}) = A(\mathbf{x}), B(\mathbf{x})$, and $\psi_{a'}(\mathbf{x})$ (for $a' = 1, \ldots, 4$) *provided* that these fields satisfy their equations of motion (24.78)–(24.80). Then the action associated with the Lagrangian density (24.77) is invariant with respect to the supersymmetry transformations (24.81)–(24.83). By contrast with the situation for (24.74), the Lagrangian density (24.77) *cannot* be written as the sum of parts that are themselves invariant up to a 4-divergence.

Because the algebraic nature of the equations of motion (24.75) and (24.76) of the fields $F(\mathbf{x})$ and $G(\mathbf{x})$ allows them to be eliminated in the manner that has just been described, $F(\mathbf{x})$ and $G(\mathbf{x})$ are known as *auxiliary fields*. As the supersymmetry transformations (24.1)–(24.5) are valid without any restriction on the fields, and in particular produce (24.16) without requiring that any equations of motion be satisfied, they are described as being valid *off-mass-shell*. By contrast, the supersymmetry transformations (24.81)–(24.83) are said to be valid only *on-mass-shell*, because their closure *does* require that the equations of motion be satisfied, implying they must be associated with particles of a particular mass M.

Although the interaction part of (24.77) involves all the possible parity-invariant renormalizable interaction terms that can be formed from $A(\mathbf{x})$, $B(\mathbf{x})$ and $\psi(\mathbf{x})$, it is a set of very specific couplings that depend on M, g and λ. It has been shown by Wess and Zumino (1974b) and Iliopoulos and Zumino (1974) that this model can be renormalized in a way that preserves the supersymmetry. Moreover, close examination (cf. Nandi 1975) shows that the theory is less divergent than might have been anticipated.

The spin content of the "on-mass-shell" model that contains only the "physical" fields $A(\mathbf{x})$, $B(\mathbf{x})$ and $\psi(\mathbf{x})$ is the same as that of the "massive" irreducible representation of the unextended $D = 4$ Poincaré superalgebra with $j = 0$ that was discussed in Chapter 23, Section 5(*a*). This correspondence can be made precise for the non-interacting situation by expanding the mass-M free fields in terms of creation and annihilation operators, and noting that for any particle the set of creation operators acting on the vacuum state forms the basis for an irreducible representation of the 4-dimensional Poincaré group (cf. Weinberg 1964a, b).

(*c*) *The general multiplet in component form*

Wess and Zumino (1974a, b) also introduced another supersymmetric multiplet, which they called the "vector multiplet" because it contains a vector field, but which, in conformity with more recent practice, will here be referred to as the *general multiplet*. This consists of four fields $C(\mathbf{x})$, $D(\mathbf{x})$, $M(\mathbf{x})$ and $N(\mathbf{x})$ that are assumed to transform as scalars with respect to proper Lorentz transformations, together with two 4-component spinor fields $\chi_a(\mathbf{x})$ and $\lambda_a(\mathbf{x})$ (where $a = 1, \ldots, 4$) and a 4-component vector field $V_\rho(\mathbf{x})$ (where $\rho = 1, \ldots, 4$), so that the multiplet contains 16 component fields in all. The analogues of (24.1)–(24.5) are

$$[Q_a, C(\mathbf{x})]_- = -\hbar^{1/2} \sum_{b=1}^{4} (\gamma_5)_{ab} \chi_b(\mathbf{x}), \tag{24.84}$$

$$[Q_a, D(\mathbf{x})]_- = i\hbar^{1/2} \sum_{b=1}^{4} \sum_{\alpha=1}^{4} (\gamma_5 \gamma^\alpha)_{ab} \partial_\alpha \lambda_b(\mathbf{x}), \tag{24.85}$$

$$[Q_a, M(\mathbf{x})]_- = i\hbar^{1/2} \lambda_a(\mathbf{x}) + \hbar^{1/2} \sum_{b=1}^{4} \sum_{\alpha=1}^{4} (\gamma^\alpha)_{ab} \partial_\alpha \chi_b(\mathbf{x}), \tag{24.86}$$

$$\begin{aligned}[Q_a, N(\mathbf{x})]_- = & - \hbar^{1/2} \sum_{b=1}^{4} (\gamma_5)_{ab} \lambda_b(\mathbf{x}) \\ & + i\hbar^{1/2} \sum_{b=1}^{4} \sum_{\alpha=1}^{4} (\gamma_5 \gamma^\alpha)_{ab} \partial_\alpha \chi_b(\mathbf{x})\end{aligned} \tag{24.87}$$

(for $a = 1, \ldots, 4$) and

$$\begin{aligned}[Q_a, \chi_b(\mathbf{x})]_+ = & \hbar^{1/2} \sum_{\alpha=1}^{4} \{(\gamma^\alpha \mathbf{C})_{ba} V_\alpha(\mathbf{x}) - (i\gamma_5 \gamma^\alpha \mathbf{C})_{ba} \partial_\alpha C(\mathbf{x})\} \\ & + \hbar^{1/2} \{i(\mathbf{C})_{ba} M(\mathbf{x}) - (\gamma_5 \mathbf{C})_{ba} N(\mathbf{x})\},\end{aligned} \tag{24.88}$$

$$\begin{aligned}[Q_a, \lambda_b(\mathbf{x})]_+ = & - \tfrac{1}{2} i\hbar^{1/2} \sum_{\alpha,\beta=1}^{4} (\gamma^\alpha \gamma^\beta \mathbf{C})_{ba} \{\partial_\beta V_\alpha(\mathbf{x}) - \partial_\alpha V_\beta(\mathbf{x})\} \\ & - \hbar^{1/2} (\gamma_5 \mathbf{C})_{ba} D(\mathbf{x})\end{aligned} \tag{24.89}$$

(for $a, b = 1, \ldots, 4$), together with

$$[Q_a, V_\rho(\mathbf{x})]_- = - \hbar^{1/2} \sum_{b=1}^{4} (\gamma_\rho)_{ab} \lambda_b(\mathbf{x}) + i\hbar^{1/2} \partial_\rho \chi_a(\mathbf{x}) \tag{24.90}$$

(for $a = 1, \ldots, 4$ and $\rho = 1, \ldots, 4$). Again the Q_a (for $a = 1, \ldots, 4$) are the supertranslation operators of the unextended $D = 4$ Poincaré superalgebra.

The general multiplet has the same sort of properties as the chiral multiplet, so it will suffice to merely summarize them. One of the most important is that these equations are again *form-invariant* under any similarity transformation (23.11) *provided* that both sets of spinor field components $\boldsymbol{\chi}(\mathbf{x})$ and $\boldsymbol{\lambda}(\mathbf{x})$ transform as (24.6), that the supertranslation operators transform as (23.46), and that the scalar fields $C(\mathbf{x})$, $D(\mathbf{x})$, $M(\mathbf{x})$ and $N(\mathbf{x})$ and the components of the vector field $\mathbf{V}(\mathbf{x})$ are left unchanged. In particular, in the *Majorana* representation of (23.14), for which the corresponding supertranslation operators Q_a^{M} are self-adjoint (cf. (24.7)) and the corresponding Dirac matrices $\gamma_\alpha^{\mathrm{M}}$ and charge conjugation matrix $\mathbf{C}^{\mathrm{M}}$ $(= -\gamma_4^{\mathrm{M}})$ are all purely imaginary, it follows from (24.84)–(24.90) that the field operators $C(\mathbf{x})$, $D(\mathbf{x})$, $M(\mathbf{x})$, $N(\mathbf{x})$ and $V_\rho(\mathbf{x})$ (for $\rho = 1, \ldots, 4$) and prior field components $\chi_a^{\mathrm{M}}(\mathbf{x})$ and $\lambda_a^{\mathrm{M}}(\mathbf{x})$ (for $a = 1, \ldots, 4$) of the Majorana representation can also be chosen to be *self-adjoint*. That is,

$$(C(\mathbf{x}))^\dagger = C(\mathbf{x}), \quad (D(\mathbf{x}))^\dagger = D(\mathbf{x}), \quad (M(\mathbf{x}))^\dagger = M(\mathbf{x}), \quad (N(\mathbf{x}))^\dagger = N(\mathbf{x}), \tag{24.91}$$

$$(V_\rho(\mathbf{x}))^\dagger = V_\rho(\mathbf{x}) \tag{24.92}$$

(for $\rho = 1, \ldots, 4$) and

$$(\chi_a^{\mathrm{M}}(\mathbf{x}))^\dagger = \chi_a^{\mathrm{M}}(\mathbf{x}), \quad (\lambda_a^{\mathrm{M}}(\mathbf{x}))^\dagger = \lambda_a^{\mathrm{M}}(\mathbf{x}) \tag{24.93}$$

(for $a = 1, \ldots, 4$), (24.91) and (24.92) being valid for *every* choice of representation of the Dirac matrices. The condition (24.93) generalizes in any representation to give

$$\gamma_4 \mathbf{C} \boldsymbol{\chi}(\mathbf{x})^\dagger = -\boldsymbol{\chi}(\mathbf{x}), \quad \gamma_4 \mathbf{C} \boldsymbol{\lambda}(\mathbf{x})^\dagger = -\boldsymbol{\lambda}(\mathbf{x}) \tag{24.94}$$

(cf. (24.11)), so that $\boldsymbol{\chi}(\mathbf{x})$ and $\boldsymbol{\lambda}(\mathbf{x})$ are Majorana spinors.

The collection of components of such a general multiplet may be denoted by Φ_{G}, so that

$$\Phi_{\mathrm{G}} = \{C(\mathbf{x}), D(\mathbf{x}), M(\mathbf{x}), N(\mathbf{x}); \boldsymbol{\chi}(\mathbf{x}), \boldsymbol{\lambda}(\mathbf{x}); \mathbf{V}(\mathbf{x})\}. \tag{24.95}$$

If the components satisfy the adjoint conditions (24.91)–(24.94), the general multiplet Φ_{G} is said to be *real*. A *complex* general multiplet Φ_{G} can be constructed from two real general multiplets

$$\Phi_{\mathrm{G}1} = \{C_1(\mathbf{x}), D_1(\mathbf{x}), M_1(\mathbf{x}), N_1(\mathbf{x}); \boldsymbol{\chi}_1(\mathbf{x}), \boldsymbol{\lambda}_1(\mathbf{x}); \mathbf{V}_1(\mathbf{x})\}$$

and

$$\Phi_{\mathrm{G}2} = \{C_2(\mathbf{x}), D_2(\mathbf{x}), M_2(\mathbf{x}), N_2(\mathbf{x}); \boldsymbol{\chi}_2(\mathbf{x}), \boldsymbol{\lambda}_2(\mathbf{x}); \mathbf{V}_2(\mathbf{x})\}$$

by defining

$$C(\mathbf{x}) = C_1(\mathbf{x}) + iC_2(\mathbf{x}), \tag{24.96}$$

$$\boldsymbol{\chi}(\mathbf{x}) = \boldsymbol{\chi}_1(\mathbf{x}) + i\boldsymbol{\chi}_2(\mathbf{x}), \tag{24.97}$$

and so on, and this can be expressed concisely by writing

$$\Phi_{\mathrm{G}} = \Phi_{\mathrm{G}1} + i\Phi_{\mathrm{G}2}. \tag{24.98}$$

For the general multiplet the spinor fields $\boldsymbol{\chi}(\mathbf{x})$ and $\boldsymbol{\lambda}(\mathbf{x})$ are "odd", and all the others are "even". Moreover, (24.16) is valid for every field and field component of the general multiplet.

From the relations (24.84)–(24.90) it is clear that the fields $V_\rho(\mathbf{x})$, $M(\mathbf{x})$ and $N(\mathbf{x})$ have the same dimensions, but the ratios $C(\mathbf{x})/V_\rho(\mathbf{x})$, $\chi_a(\mathbf{x})/V_\rho(\mathbf{x})$, $\lambda_a(\mathbf{x})/V_\rho(\mathbf{x})$ and $D(\mathbf{x})/V_\rho(\mathbf{x})$ are of dimensions $(\text{length})^{+1}$, $(\text{length})^{+1/2}$, $(\text{length})^{-1/2}$ and $(\text{length})^{-1}$ respectively.

In terms of the operator δ_ζ of (24.37), equations (24.84)–(24.90) imply that

$$\delta_\zeta C(\mathbf{x}) = -i\tilde{\boldsymbol{\zeta}}^\# \gamma_4 \gamma_5 \boldsymbol{\chi}(\mathbf{x}), \tag{24.99}$$

$$\delta_\zeta D(\mathbf{x}) = -\sum_{\alpha=1}^{4} \tilde{\boldsymbol{\zeta}}^\# \gamma_4 \gamma_5 \gamma^\alpha \partial_\alpha \boldsymbol{\lambda}(\mathbf{x}), \tag{24.100}$$

$$\delta_\zeta M(\mathbf{x}) = -\tilde{\boldsymbol{\zeta}}^\# \gamma_4 \boldsymbol{\lambda}(\mathbf{x}) + i\sum_{\alpha=1}^{4} \tilde{\boldsymbol{\zeta}}^\# \gamma_4 \gamma^\alpha \partial_\alpha \boldsymbol{\chi}(\mathbf{x}), \tag{24.101}$$

$$\delta_\zeta N(\mathbf{x}) = -i\tilde{\boldsymbol{\zeta}}^\# \gamma_4 \gamma_5 \boldsymbol{\lambda}(\mathbf{x}) - \sum_{\alpha=1}^{4} \tilde{\boldsymbol{\zeta}}^\# \gamma_4 \gamma_5 \gamma^\alpha \partial_\alpha \boldsymbol{\chi}(\mathbf{x}) \tag{24.102}$$

and

$$\delta_\zeta \chi_a(\mathbf{x}) = \sum_{b=1}^{4} (\zeta^b)^\# \sum_{\alpha=1}^{4} \{(i\gamma_4 \tilde{\mathbf{C}} \tilde{\gamma}^\alpha)_{ba} V_\alpha(\mathbf{x}) + (\gamma_4 \tilde{\mathbf{C}} \tilde{\gamma}^\alpha \tilde{\gamma}_5)_{ba} \partial_\alpha C(\mathbf{x})\}$$

$$- \sum_{b=1}^{4} (\zeta^b)^\# \{(\gamma_4 \tilde{\mathbf{C}})_{ba} M(\mathbf{x}) + (i\gamma_4 \tilde{\mathbf{C}} \tilde{\gamma}_5)_{ba} N(\mathbf{x})\}, \qquad (24.103)$$

$$\delta_\zeta \lambda_a(\mathbf{x}) = \frac{1}{2} \sum_{b=1}^{4} (\zeta^b)^\# \sum_{\alpha,\beta=1}^{4} (\gamma_4 \tilde{\mathbf{C}} \tilde{\gamma}^\alpha \tilde{\gamma}^\beta)_{ba} \{\partial_\beta V_\alpha(\mathbf{x}) - \partial_\alpha V_\beta(\mathbf{x})\}$$

$$- \sum_{b=1}^{4} (\zeta^b)^\# (i\gamma_4 \tilde{\mathbf{C}} \tilde{\gamma}_5)_{ba} D(\mathbf{x}) \qquad (24.104)$$

(for $a = 1, \ldots, 4$), together with

$$\delta_\zeta V_\rho(\mathbf{x}) = -i\tilde{\boldsymbol{\zeta}}^\# \gamma_4 \gamma_\rho \boldsymbol{\lambda}(\mathbf{x}) - \tilde{\boldsymbol{\zeta}}^\# \gamma_4 \partial_\rho \boldsymbol{\chi}(\mathbf{x}) \qquad (24.105)$$

(for $\rho = 1, \ldots, 4$). Equations (24.103) and (24.104) can be rewritten using (23.100) to give

$$\delta_\zeta \boldsymbol{\chi}(\mathbf{x}) = \sum_{\alpha=1}^{4} \{i\gamma^\alpha \boldsymbol{\zeta} V_\alpha(\mathbf{x}) + \gamma_5 \gamma^\alpha \boldsymbol{\zeta} \partial_\alpha C(\mathbf{x})\} - \boldsymbol{\zeta} M(\mathbf{x}) - i\gamma_5 \boldsymbol{\zeta} N(\mathbf{x}) \qquad (24.106)$$

and

$$\delta_\zeta \boldsymbol{\lambda}(\mathbf{x}) = \frac{1}{2} \sum_{\alpha,\beta=1}^{4} \gamma^\beta \gamma^\alpha \boldsymbol{\zeta} \{\partial_\beta V_\alpha(\mathbf{x}) - \partial_\alpha V_\beta(\mathbf{x})\} - i\gamma_5 \boldsymbol{\zeta} D(\mathbf{x}). \qquad (24.107)$$

As the right-hand side of (24.100) is a 4-divergence, the 4-dimensional integral of $D(\mathbf{x})$ over Minkowski space–time is *invariant* under the operations induced by supertranslation generators (as well as under the ordinary Poincaré transformations).

It is clear that the spin content of the general multiplet is much more elaborate than that of the irreducible representations of the $N = 1$, $D = 4$ Poincaré superalgebra. Indeed, the subset of fields

$$\left\{ M(\mathbf{x}), N(\mathbf{x}), -\sum_{\alpha,\beta=1}^{4} \eta^{\alpha\beta} \partial_\alpha V_\beta(\mathbf{x}), D(\mathbf{x}) + \Box C(\mathbf{x}); \boldsymbol{\lambda}(\mathbf{x}) - i \sum_{\alpha=1}^{4} \gamma^\alpha \partial_\alpha \boldsymbol{\chi}(\mathbf{x}) \right\} \qquad (24.108)$$

themselves form a chiral multiplet. Moreover, the subset of fields

$$\{\boldsymbol{\lambda}(\mathbf{x}); F_{\alpha\beta}(\mathbf{x}); D(\mathbf{x})\} \qquad (24.109)$$

(for $\alpha, \beta = 1, \ldots, 4$, with $\alpha < \beta$), where

$$F_{\alpha\beta}(\mathbf{x}) = \partial_\alpha V_\beta(\mathbf{x}) - \partial_\beta V_\alpha(\mathbf{x}), \qquad (24.110)$$

also form an invariant set—which, because of the form of (24.110), is often called the *curl multiplet*. Its transformation rules are

$$\delta_\zeta \boldsymbol{\lambda}(\mathbf{x}) = \frac{1}{2} \sum_{\alpha,\beta=1}^{4} \gamma^\beta \gamma^\alpha \zeta F_{\alpha\beta}(\mathbf{x}) - i\gamma_5 \zeta D(\mathbf{x}), \tag{24.111}$$

$$\delta_\zeta D(\mathbf{x}) = \sum_{\alpha=1}^{4} \tilde{\zeta}^{\#} \gamma_4 \gamma_5 \gamma^\alpha \partial_\alpha \boldsymbol{\lambda}(\mathbf{x}), \tag{24.112}$$

$$\delta_\zeta F_{\alpha\beta}(\mathbf{x}) = i\tilde{\zeta}^{\#} \gamma_4 \{\gamma_\alpha \partial_\beta \boldsymbol{\lambda}(\mathbf{x}) - \gamma_\beta \partial_\alpha \boldsymbol{\lambda}(\mathbf{x})\} \tag{24.113}$$

(for $\alpha, \beta = 1, \ldots, 4$).

If all the components of the curl multiplet (24.109) are set equal to zero, the remaining fields constitute another chiral multiplet of the form

$$\{V(\mathbf{x}), C(\mathbf{x}), M(\mathbf{x}), N(\mathbf{x}); \boldsymbol{\chi}(\mathbf{x})\}, \tag{24.114}$$

where $V(\mathbf{x})$ is defined to be such that $V_\rho(\mathbf{x}) = \partial_\rho V(\mathbf{x})$ for $\rho = 1, \ldots, 4$, a set of equations that have a consistent solution if $F_{\alpha\beta}(\mathbf{x}) = 0$, as is being assumed. This can be expressed alternatively but equivalently by the statement that if $\Phi_C = \{A(\mathbf{x}), B(\mathbf{x}), F(\mathbf{x}), G(\mathbf{x}); \boldsymbol{\psi}(\mathbf{x})\}$ is a chiral multiplet then the components of the set

$$\partial\Phi_C = \{B(\mathbf{x}), 0, F(\mathbf{x}), G(\mathbf{x}); \boldsymbol{\psi}(\mathbf{x}), \mathbf{0}; \mathbf{V}(\mathbf{x})\}, \tag{24.115}$$

with $V_\rho(\mathbf{x}) = \partial_\rho A(\mathbf{x})$ (for $\rho = 1, \ldots, 4$) transform as components of a general multiplet.

Finally, if all the components of the chiral multiplet (24.103) are set equal to zero, the remaining fields form another invariant multiplet consisting of

$$\{C(\mathbf{x}); \boldsymbol{\chi}(\mathbf{x}); \mathbf{V}(\mathbf{x})\}, \tag{24.116}$$

which is known as the *linear multiplet*, and for which the transformation rules analogous to (24.99)–(24.107) are

$$\delta_\zeta C(\mathbf{x}) = -i\tilde{\zeta}^{\#} \gamma_4 \gamma_5 \boldsymbol{\chi}(\mathbf{x}), \tag{24.117}$$

$$\delta_\zeta \boldsymbol{\chi}(\mathbf{x}) = \sum_{\alpha=1}^{4} \{i\gamma^\alpha \zeta V_\alpha(\mathbf{x}) + \gamma_5 \gamma^\alpha \zeta \partial_\alpha C(\mathbf{x})\}, \tag{24.118}$$

$$\delta_\zeta V_\rho(\mathbf{x}) = \frac{1}{2} \sum_{\alpha,\beta=1}^{4} \eta_{\rho\beta} \tilde{\zeta}^{\#} \gamma_4 [\gamma^\beta, \gamma^\alpha]_- \partial_\alpha \boldsymbol{\chi}(\mathbf{x}) \tag{24.119}$$

(for $\rho = 1, \ldots, 4$), provided that

$$\sum_{\alpha,\beta=1}^{4} \eta^{\alpha\beta} \partial_\alpha V_\beta(\mathbf{x}) = 0. \tag{24.120}$$

With the Lagrangian density $\mathscr{L}_0^{\text{v}}(\mathbf{x})$ defined by

$$\mathscr{L}_0^{\text{v}}(\mathbf{x}) = -\frac{1}{4}\sum_{\alpha,\beta,\rho,\kappa=1}^{4} \eta^{\alpha\rho}\eta^{\beta\kappa}F_{\alpha\rho}(\mathbf{x})F_{\beta\kappa}(\mathbf{x}) + \tfrac{1}{2}i\sum_{\alpha=1}^{4}\tilde{\boldsymbol{\lambda}}(\mathbf{x})^{\dagger}\gamma_4\gamma^{\alpha}\partial_{\alpha}\boldsymbol{\lambda}(\mathbf{x}) + \tfrac{1}{2}D(\mathbf{x})^2, \tag{24.121}$$

where $F_{\alpha\rho}(\mathbf{x})$ and $F_{\beta\kappa}(\mathbf{x})$ are as given in (24.110), it is easily shown that $\delta_{\zeta}\mathscr{L}_0^{\text{v}}(\mathbf{x})$ is equal to the 4-divergence of a certain quantity, so that the action given by its space–time integral is supersymmetric. Clearly the Lagrangian density $\mathscr{L}_0^{\text{v}}(\mathbf{x})$ describes free massless spin-1 particles whose field is $\mathbf{V}(\mathbf{x})$ and free massless spin-$\frac{1}{2}$ particles whose field is $\boldsymbol{\lambda}(\mathbf{x})$. The remaining field $D(\mathbf{x})$ is an auxiliary field since its Euler–Lagrange equation is simply the algebraic equation $D(\mathbf{x}) = 0$. Only the fields of the "curl multiplet" of (24.109) appear in $\mathscr{L}_0^{\text{v}}(\mathbf{x})$.

If $\Phi_{C1} = \{A_1(\mathbf{x}),\ B_1(\mathbf{x}),\ F_1(\mathbf{x}),\ G_1(\mathbf{x}); \boldsymbol{\psi}_1(\mathbf{x})\}$ and $\Phi_{C2} = \{A_2(\mathbf{x}),\ B_2(\mathbf{x}), F_2(\mathbf{x}), G_2(\mathbf{x}); \boldsymbol{\psi}_2(\mathbf{x})\}$ are any two chiral multiplets then two different general multiplets may be formed from them. The first, which may be called the *symmetric product* and which may be denoted by $\Phi_{C1} \times \Phi_{C2}$, has components defined by

$$C(\mathbf{x}) = A_1(\mathbf{x})A_2(\mathbf{x}) + B_1(\mathbf{x})B_2(\mathbf{x}), \tag{24.122}$$

$$\begin{aligned} D(\mathbf{x}) = {} & 2F_1(\mathbf{x})F_2(\mathbf{x}) + 2G_1(\mathbf{x})G_2(\mathbf{x}) \\ & + 2\sum_{\alpha,\beta=1}^{4} \eta^{\alpha\beta}\{\partial_{\alpha}A_1(\mathbf{x})\partial_{\beta}A_2(\mathbf{x}) + \partial_{\alpha}B_1(\mathbf{x})\partial_{\beta}(\mathbf{x})\} \\ & + i\sum_{\alpha=1}^{4}\{\tilde{\boldsymbol{\psi}}_1(\mathbf{x})^{\dagger}\gamma_4\gamma^{\alpha}\partial_{\alpha}\boldsymbol{\psi}_2(\mathbf{x}) + \tilde{\boldsymbol{\psi}}_2(\mathbf{x})^{\dagger}\gamma_4\gamma^{\alpha}\partial_{\alpha}\boldsymbol{\psi}_1(\mathbf{x})\}, \end{aligned} \tag{24.123}$$

$$M(\mathbf{x}) = F_1(\mathbf{x})B_2(\mathbf{x}) + F_2(\mathbf{x})B_1(\mathbf{x}) + G_1(\mathbf{x})A_2(\mathbf{x}) + G_2(\mathbf{x})A_1(\mathbf{x}), \tag{24.124}$$

$$N(\mathbf{x}) = G_1(\mathbf{x})B_2(\mathbf{x}) + G_2(\mathbf{x})B_1(\mathbf{x}) - F_1(\mathbf{x})A_2(\mathbf{x}) - F_2(\mathbf{x})A_1(\mathbf{x}), \tag{24.125}$$

$$\boldsymbol{\chi}(\mathbf{x}) = \{B_1(\mathbf{x})\mathbf{1} - iA_1(\mathbf{x})\gamma_5\}\boldsymbol{\psi}_2(\mathbf{x}) + \{B_2(\mathbf{x})\mathbf{1} - iA_2(\mathbf{x})\gamma_5\}\boldsymbol{\psi}_1(\mathbf{x}), \tag{24.126}$$

$$\begin{aligned} \boldsymbol{\lambda}(\mathbf{x}) = {} & \{G_1(\mathbf{x})\mathbf{1} + iF_1(\mathbf{x})\gamma_5\}\boldsymbol{\psi}_2(\mathbf{x}) + \{G_2(\mathbf{x})\mathbf{1} + iF_2(\mathbf{x})\gamma_5\}\boldsymbol{\psi}_1(\mathbf{x}) \\ & + i\sum_{\alpha=1}^{4}\partial_{\alpha}\{B_2(\mathbf{x}) + i\gamma_5A_2(\mathbf{x})\}\gamma^{\alpha}\boldsymbol{\psi}_1(\mathbf{x}) + \partial_{\alpha}\{B_1(\mathbf{x}) + i\gamma_5A_1(\mathbf{x})\}\gamma^{\alpha}\boldsymbol{\psi}_2(\mathbf{x}), \end{aligned} \tag{24.127}$$

$$\begin{aligned} V_{\rho}(\mathbf{x}) = {} & B_1(\mathbf{x})\partial_{\rho}A_2(\mathbf{x}) + B_2(\mathbf{x})\partial_{\rho}A_1(\mathbf{x}) + A_1(\mathbf{x})\partial_{\rho}B_2(\mathbf{x}) \\ & + A_2(\mathbf{x})\partial_{\rho}B_1(\mathbf{x}) - \tilde{\boldsymbol{\psi}}_1(\mathbf{x})^{\dagger}\gamma_4\gamma_5\gamma_{\rho}\boldsymbol{\psi}_2(\mathbf{x}) \end{aligned} \tag{24.128}$$

(for $\rho = 1, \ldots, 4$). Clearly

$$\Phi_{C1} \times \Phi_{C2} = \Phi_{C2} \times \Phi_{C1}.$$

The other combination, which may be called the *antisymmetric product* and denoted by $\Phi_{C1} \wedge \Phi_{C2}$, has components defined by

$$C(\mathbf{x}) = A_1(\mathbf{x})B_2(\mathbf{x}) - A_2(\mathbf{x})B_1(\mathbf{x}), \tag{24.129}$$

$$\begin{aligned} D(\mathbf{x}) = {} & 2F_2(\mathbf{x})G_1(\mathbf{x}) - 2F_1(\mathbf{x})G_2(\mathbf{x}) \\ & + 2\sum_{\alpha,\beta=1}^{4} \eta^{\alpha\beta}\{\partial_\alpha A_2(\mathbf{x})\,\partial_\beta B_1(\mathbf{x}) - \partial_\alpha A_1(\mathbf{x})\,\partial_\beta B_2(\mathbf{x})\} \\ & - \sum_{\alpha=1}^{4}\{\tilde{\boldsymbol{\psi}}_1(\mathbf{x})^\dagger \gamma_4\gamma_5\gamma^\alpha\,\partial_\alpha \boldsymbol{\psi}_2(\mathbf{x}) - \tilde{\boldsymbol{\psi}}_2(\mathbf{x})^\dagger \gamma_4\gamma_5\gamma^\alpha\,\partial_\alpha \boldsymbol{\psi}_1(\mathbf{x})\}, \end{aligned} \tag{24.130}$$

$$M(\mathbf{x}) = A_1(\mathbf{x})F_2(\mathbf{x}) - A_2(\mathbf{x})F_1(\mathbf{x}) - B_1(\mathbf{x})G_2(\mathbf{x}) + B_2(\mathbf{x})G_1(\mathbf{x}), \tag{24.131}$$

$$N(\mathbf{x}) = A_1(\mathbf{x})G_2(\mathbf{x}) - A_2(\mathbf{x})G_1(\mathbf{x}) + B_1(\mathbf{x})F_2(\mathbf{x}) - B_2(\mathbf{x})F_1(\mathbf{x}), \tag{24.132}$$

$$\boldsymbol{\chi}(\mathbf{x}) = \{A_1(\mathbf{x})\mathbf{1} + iB_1(\mathbf{x})\gamma_5\}\boldsymbol{\psi}_2(\mathbf{x}) - \{A_2(\mathbf{x})\mathbf{1} + iB_2(\mathbf{x})\gamma_5\}\boldsymbol{\psi}_1(\mathbf{x}), \tag{24.133}$$

$$\begin{aligned} \boldsymbol{\lambda}(\mathbf{x}) = {} & \{F_2(\mathbf{x})\mathbf{1} - iG_2(\mathbf{x})\gamma_5\}\boldsymbol{\psi}_1(\mathbf{x}) - \{F_1(\mathbf{x})\mathbf{1} - iG_1(\mathbf{x})\gamma_5\}\boldsymbol{\psi}_2(\mathbf{x}) \\ & - i\sum_{\alpha=1}^{4} \partial_\alpha\{A_2(\mathbf{x}) - i\gamma_5 B_2(\mathbf{x})\}\,\gamma^\alpha\boldsymbol{\psi}_1(\mathbf{x}) - \partial_\alpha\{A_1(\mathbf{x}) - i\gamma_5 B_1(\mathbf{x})\}\,\gamma^\alpha\boldsymbol{\psi}_2(\mathbf{x}), \end{aligned} \tag{24.134}$$

$$\begin{aligned} V_\rho(\mathbf{x}) = {} & A_1(\mathbf{x})\,\partial_\rho A_2(\mathbf{x}) - A_2(\mathbf{x})\,\partial_\rho A_1(\mathbf{x}) + B_1(\mathbf{x})\,\partial_\rho B_2(\mathbf{x}) \\ & - B_2(\mathbf{x})\,\partial_\rho B_1(\mathbf{x}) + i\tilde{\boldsymbol{\psi}}_1(\mathbf{x})^\dagger\gamma_4\gamma_\rho\boldsymbol{\psi}_2(\mathbf{x}) \end{aligned} \tag{24.135}$$

(for $\rho = 1, \ldots, 4$). It is obvious that

$$\Phi_{C1} \wedge \Phi_{C2} = -\Phi_{C2} \wedge \Phi_{C1}.$$

It will prove useful to note that an alternative expression for the kinetic-energy term $\mathscr{L}_{KE}$ of the Lagrangian density of the Wess–Zumino model (24.70) is given by

$$\mathscr{L}_{KE} = \tfrac{1}{2}\hbar c\{\text{“}D\text{” term of } \Phi_C \wedge \Phi_C\}, \tag{24.136}$$

where $\Phi_C = \{A(\mathbf{x}), B(\mathbf{x}), F(\mathbf{x}), G(\mathbf{x}); \boldsymbol{\psi}(\mathbf{x})\}$.

Starting from two general multiplets $\Phi_{G1} = \{C_1(\mathbf{x}), D_1(\mathbf{x}), M_1(\mathbf{x}), N_1(\mathbf{x}); \boldsymbol{\chi}_1(\mathbf{x}), \boldsymbol{\lambda}_1(\mathbf{x}); \mathbf{V}_1(\mathbf{x})\}$ and $\Phi_{G2} = \{C_2(\mathbf{x}), D_2(\mathbf{x}), M_2(\mathbf{x}), N_2(\mathbf{x}); \boldsymbol{\chi}_2(\mathbf{x}), \boldsymbol{\lambda}_2(\mathbf{x}); \mathbf{V}_2(\mathbf{x})\}$, it is also possible to construct a third general multiplet $\Phi_{G3} = \{C_3(\mathbf{x}), D_3(\mathbf{x}), M_3(\mathbf{x}), N_3(\mathbf{x}); \boldsymbol{\chi}_3(\mathbf{x}), \boldsymbol{\lambda}_3(\mathbf{x}); \mathbf{V}_3(\mathbf{x})\}$ by defining

$$C_3(\mathbf{x}) = C_1(\mathbf{x})C_2(\mathbf{x}), \tag{24.137}$$

$$\begin{aligned} D_3(\mathbf{x}) = {} & C_1(\mathbf{x})D_2(\mathbf{x}) + C_2(\mathbf{x})D_1(\mathbf{x}) + M_1(\mathbf{x})M_2(\mathbf{x}) + N_1(\mathbf{x})N_2(\mathbf{x}) \\ & - \sum_{\alpha,\beta=1}^{4} \eta^{\alpha\beta}\{\partial_\alpha C_1(\mathbf{x})\,\partial_\beta C_2(\mathbf{x}) + V_\alpha(\mathbf{x})V_\beta(\mathbf{x})\} \\ & - \tfrac{1}{2}i\sum_{\alpha=1}^{4}\{\partial_\alpha\tilde{\boldsymbol{\chi}}_1(\mathbf{x})^\dagger\,\gamma_4\gamma^\alpha\boldsymbol{\chi}_2(\mathbf{x}) + \partial_\alpha\tilde{\boldsymbol{\chi}}_2(\mathbf{x})^\dagger\,\gamma_4\gamma^\alpha\boldsymbol{\chi}_1(\mathbf{x})\} \\ & + \tilde{\boldsymbol{\chi}}_1(\mathbf{x})^\dagger\,\gamma_4\boldsymbol{\lambda}_2(\mathbf{x}) + \tilde{\boldsymbol{\chi}}_2(\mathbf{x})^\dagger\,\gamma_4\boldsymbol{\lambda}_1(\mathbf{x}), \end{aligned} \tag{24.138}$$

$$M_3(\mathbf{x}) = C_1(\mathbf{x})M_2(\mathbf{x}) + C_2(\mathbf{x})M_1(\mathbf{x}) + \tfrac{1}{2}i\tilde{\chi}_1(\mathbf{x})^\dagger \gamma_4\gamma_5\chi_2(\mathbf{x}), \tag{24.139}$$

$$N_3(\mathbf{x}) = C_1(\mathbf{x})N_2(\mathbf{x}) + C_2(\mathbf{x})N_1(\mathbf{x}) - \tfrac{1}{2}\tilde{\chi}_1(\mathbf{x})^\dagger \gamma_4\chi_2(\mathbf{x}), \tag{24.140}$$

$$\chi_3(\mathbf{x}) = C_1(\mathbf{x})\chi_2(\mathbf{x}) + C_2(\mathbf{x})\chi_1(\mathbf{x}), \tag{24.141}$$

$$\begin{aligned}\lambda_3(\mathbf{x}) = {} & C_1(\mathbf{x})\lambda_2(\mathbf{x}) + C_2(\mathbf{x})\lambda_1(\mathbf{x}) + \tfrac{1}{2}N_1(\mathbf{x})\chi_2(\mathbf{x}) + \tfrac{1}{2}N_2(\mathbf{x})\chi_1(\mathbf{x}) \\ & + \frac{1}{2}\sum_{\alpha=1}^{4} \gamma^\alpha\{\partial_\alpha C_1(\mathbf{x})\,\chi_2(\mathbf{x}) + \partial_\alpha C_2(\mathbf{x})\,\chi_1(\mathbf{x})\} \\ & + \frac{1}{2}\sum_{\alpha=1}^{4} \{V_{\alpha 1}\gamma_5\gamma^\alpha\chi_2(\mathbf{x}) + V_{\alpha 2}\gamma_5\gamma^\alpha\chi_1(\mathbf{x})\} \\ & + \tfrac{1}{2}iM_1(\mathbf{x})\gamma_5\chi_2(\mathbf{x}) + \tfrac{1}{2}iM_2(\mathbf{x})\gamma_5\chi_1(\mathbf{x}),\end{aligned} \tag{24.142}$$

$$V_{\rho 3}(\mathbf{x}) = C_1(\mathbf{x})V_{\rho 2}(\mathbf{x}) + C_2(\mathbf{x})V_{\rho 1}(\mathbf{x}) + \tfrac{1}{2}\tilde{\chi}_1(\mathbf{x})^\dagger \gamma_4\gamma_5\gamma_\rho\chi_2(\mathbf{x}) \tag{24.143}$$

(for $\rho = 1, \ldots, 4$). This may be indicated by writing

$$\Phi_{G3} = \Phi_{G1}\cdot\Phi_{G2}. \tag{24.144}$$

It is clear that

$$\Phi_{G1}\cdot\Phi_{G2} = \Phi_{G2}\cdot\Phi_{G1}, \tag{24.145}$$

and, if the process is repeated for any three general multiplets Φ_{G1}, Φ_{G2} and Φ_{G3},

$$(\Phi_{G1}\cdot\Phi_{G2})\cdot\Phi_{G3} = \Phi_{G1}\cdot(\Phi_{G2}\cdot\Phi_{G3}). \tag{24.146}$$

Also, if Φ_{C1}, Φ_{C2} and Φ_{C3} are any three chiral multiplets then

$$(\Phi_{C1}\Phi_{C2}) \times \Phi_{C3} - (\Phi_{C1}\Phi_{C3}) \times \Phi_{C2} = 2(\Phi_{C1} \wedge \Phi_{C2})\cdot\partial\Phi_{C3} \tag{24.147}$$

and

$$(\Phi_{C1}\Phi_{C2}) \wedge \Phi_{C2} = -(\Phi_{C2} \times \Phi_{C2})\cdot\partial\Phi_{C1}, \tag{24.148}$$

$\partial\Phi_{C1}$ and $\partial\Phi_{C3}$ being defined as in (24.115). All of these properties of products of chiral and general multiplets were established by Wess and Zumino (1974c).

3 Superfields

In their 4-component form the scalar and chiral superfields were first discussed by Salam and Strathdee (1974a,b, d, 1975b), the slightly later review (Salam and Strathdee 1978) giving a comprehensive account. The first published work on the 2-component version was by Ferrara *et al.* (1974). The discussion that follows is restricted to the superfields associated with the $N = 1$ Poincaré superalgebra of 4-dimensional space–time. For an introduction to its generalization to the N-extended case see Taylor (1982).

(a) The scalar superfield

As noted in (23.144), the transformation law for a scalar superfield $\Phi_s(\mathbf{X}, \mathbf{\Theta})$ is

$$U([\mathbf{\Lambda}|\mathbf{T}|\zeta])\Phi_s(\mathbf{X};\mathbf{\Theta})\,U([\mathbf{\Lambda}|\mathbf{T}|\zeta])^{-1} = \Phi_s([\mathbf{\Lambda}|\mathbf{T}|\zeta](\mathbf{X};\mathbf{\Theta})) \quad (24.149)$$

for every Poincaré supergroup transformation $[\mathbf{\Lambda}|\mathbf{T}|\zeta]$. Assuming that $\Phi_s(\mathbf{X};\mathbf{\Theta})$ is an operator-valued *superanalytic* function on the superspace $\mathbb{R}B_L^{4,4}$, Theorem II of Chapter 22, Section 2(*c*) shows that $\Phi_s(\mathbf{X};\mathbf{\Theta})$ has an expansion of the form

$$\Phi_s(\mathbf{X};\mathbf{\Theta}) = \sum_{\Lambda} \Phi_s^{\Lambda}(\mathbf{X})\Theta^{\Lambda}, \quad (24.150)$$

(cf. (22.33)), where Θ^{Λ} is a product of Θ^k factors, as defined in (22.18), the sum being over all 16 ($= 2^4$) sets Λ of the type (22.16) that satisfy (22.17), and where each $\Phi_s^{\Lambda}(\mathbf{X})$ is an operator-valued function that is independent of the odd variables $\Theta^1, \Theta^2, \Theta^3$ and Θ^4. Moreover, each $\Phi_s^{\Lambda}(\mathbf{X})$ is the "continuation" of a unique function $\Phi_s^{\Lambda}(X^1_{\varnothing}, X^2_{\varnothing}, X^3_{\varnothing}, X^4_{\varnothing})$, where $(X^1_{\varnothing}, X^2_{\varnothing}, X^3_{\varnothing}, X^4_{\varnothing})$ is a set of four real variables, which may here be thought of as specifying the coordinates of points in Minkowski space–time. It is convenient to simplify the notation by indicating such a point by $\mathbf{x}$, so that $\mathbf{x} = (x^1, x^2, x^3, x^4) = (X^1_{\varnothing}, X^2_{\varnothing}, X^3_{\varnothing}, X^4_{\varnothing})$, and by denoting a function on Minkowski space–time *and* its "continuation" into $\mathbb{R}\mathrm{B}_L^{4,4}$ by the *same* symbols, so that $\Phi_s^{\Lambda}(\mathbf{x}) = \Phi_s^{\Lambda}(\mathbf{x})$ and $\Phi_s^{\Lambda}(\mathbf{x}) = Z(\Phi_s^{\Lambda}(\mathbf{x}))$. The expansion (24.150) contains one term that is independent of the Θ^a, four terms that are linear in the Θ^a, six terms that are of second order, four that are of third order, and one that is of fourth order in these odd Grassmann variables. It is easily checked that there are only six linearly independent matrix products involving Dirac matrices or products of Dirac matrices of the form $(\tilde{\mathbf{\Theta}}^{\#}\gamma_{\alpha}\cdots\gamma_{\beta}\mathbf{\Theta})$ and that these may be taken to be $(\tilde{\mathbf{\Theta}}^{\#}\gamma_4\mathbf{\Theta})$, $(\tilde{\mathbf{\Theta}}^{\#}\gamma_4\gamma_5\mathbf{\Theta})$ and $(\tilde{\mathbf{\Theta}}^{\#}\gamma_4\gamma_5\gamma^{\alpha}\mathbf{\Theta})$ (for $\alpha = 1, \ldots, 4$). (All of these matrix products are invariant under similarity transformations of the Dirac matrices, so it is sufficient to consider them in the Majorana representation, where (23.134) and the symmetry and antisymmetry properties of the $\gamma_{\alpha}^{\mathrm{M}}$ that are apparent from (23.14) may be used in conjunction with the assumed anticommuting property of the Θ^a.) These six matrix products then provide the six second-order terms.

Taking account of these observations, it is very convenient to choose the coefficient functions $\Phi_s^{\Lambda}(\mathbf{X})$ of (24.150) so that $\Phi_s(\mathbf{X};\mathbf{\Theta})$ has the expansion

$$\begin{aligned}\Phi_s(\mathbf{X};\mathbf{\Theta}) = {} & C(\mathbf{X}) - i(\tilde{\mathbf{\Theta}}^{\#}\gamma_4\gamma_5\boldsymbol{\chi}(\mathbf{X})) - \tfrac{1}{2}i(\tilde{\mathbf{\Theta}}^{\#}\gamma_4\gamma_5\mathbf{\Theta})M(\mathbf{X}) + \tfrac{1}{2}(\tilde{\mathbf{\Theta}}^{\#}\gamma_4\mathbf{\Theta})N(\mathbf{X}) \\ & - \frac{1}{2}\sum_{\alpha=1}^{4}(\tilde{\mathbf{\Theta}}^{\#}\gamma_4\gamma_5\gamma^{\alpha}\mathbf{\Theta})\,V_{\alpha}(\mathbf{X}) - (\tilde{\mathbf{\Theta}}^{\#}\gamma_4\mathbf{\Theta})\sum_{\alpha=1}^{4}(\tilde{\mathbf{\Theta}}^{\#}\gamma_4\gamma_5\gamma^{\alpha}\,\partial_{\alpha}\boldsymbol{\chi}(\mathbf{X})) \\ & - i(\tilde{\mathbf{\Theta}}^{\#}\gamma_4\mathbf{\Theta})(\tilde{\mathbf{\Theta}}^{\#}\gamma_4\gamma_5\boldsymbol{\lambda}(\mathbf{X})) \\ & + \tfrac{1}{4}(\tilde{\mathbf{\Theta}}^{\#}\gamma_4\mathbf{\Theta})(\tilde{\mathbf{\Theta}}^{\#}\gamma_4\mathbf{\Theta})\{D(\mathbf{X}) - C(\mathbf{X})\}. \end{aligned} \quad (24.151)$$

With this choice,

$$\delta_\zeta \Phi_s(\mathbf{X};\mathbf{\Theta}) = \sum_{\Lambda} \{\delta_\zeta \Phi_s^\Lambda(\mathbf{X})\}\Theta^\Lambda, \tag{24.152}$$

where the operator δ_ζ on the left-hand side of (24.152) is the scalar superfield operator of (23.145) and (23.146) whereas the operator δ_ζ on the right-hand side is the general multiplet operator of (24.99)–(24.107). *Thus the coefficient functions of the scalar superfield expansion* (24.151) *are just the "continuations" of the "component" functions* $C(\mathbf{x})$, $D(\mathbf{x})$, $M(\mathbf{x})$, $N(\mathbf{x})$, $\boldsymbol{\chi}(\mathbf{x})$, $\boldsymbol{\lambda}(\mathbf{x})$ and $V_\rho(\mathbf{x})$ *of the general multiplet of Section* 2(*c*). Here it has been assumed that the odd Grassmann variables Θ^a commute with the operators $C(\mathbf{x})$, $D(\mathbf{x})$, $M(\mathbf{x})$, $N(\mathbf{x})$ and $V_\rho(\mathbf{x})$ (for $\rho = 1,\dots,4$), but the Θ^a *anticommute* with the $\chi_b(\mathbf{x})$ and $\lambda_b(\mathbf{x})$ (for $b = 1,\dots,4$).

The demonstration of this identification can either be done in a general 4-component formulation using the matrix identity (L.136) of Appendix L, Section 5 or by using the two-component formulation and thereby involving odd variables Θ_A and $\bar{\Theta}_A$ (for $A = 1, 2$). In this latter approach application of (23.25), (23.135) and (23.136) reduces (24.152) to the form

$$\begin{aligned}
\Phi_s(\mathbf{X};\mathbf{\Theta}) = {} & C(\mathbf{X}) - \bar{\Theta}\bar{\chi}(\mathbf{X}) + \Theta\chi(\mathbf{X}) \\
& - \tfrac{1}{2}\bar{\Theta}\bar{\Theta}\{M(\mathbf{X}) - iN(\mathbf{X})\} + \tfrac{1}{2}\Theta\Theta\{M(\mathbf{X}) + iN(\mathbf{X})\} \\
& + \sum_{\alpha,\beta=1}^{4} \sum_{A,B=1}^{2} \eta^{\alpha\beta}\Theta_A(\boldsymbol{\varepsilon}\boldsymbol{\sigma}_\alpha\boldsymbol{\varepsilon})_{AB}\bar{\Theta}_B V_\beta(\mathbf{X}) \\
& - i(\bar{\Theta}\bar{\Theta} + \Theta\Theta)\{\bar{\Theta}\bar{\lambda}(\mathbf{X}) - \Theta\lambda(\mathbf{X})\} \\
& + i(\bar{\Theta}\bar{\Theta} + \Theta\Theta)\sum_{\alpha,\beta=1}^{4} \sum_{A,B=1}^{2} \eta^{\alpha\beta}\{\partial_\beta\chi_A(\mathbf{X})(\boldsymbol{\varepsilon}\boldsymbol{\sigma}_\alpha\boldsymbol{\varepsilon})_{AB}\bar{\Theta}_B \\
& + \Theta_A(\boldsymbol{\varepsilon}\boldsymbol{\sigma}_\alpha\boldsymbol{\varepsilon})_{AB}\partial_\beta\bar{\chi}_B(\mathbf{X})\} + \tfrac{1}{4}(\bar{\Theta}\bar{\Theta} + \Theta\Theta)^2\{-D(\mathbf{X}) + \Box C(\mathbf{X})\}.
\end{aligned} \tag{24.153}$$

Here an abbreviated notation has been used in which

$$\Theta\chi = \sum_{A,B=1}^{2} \Theta_A\varepsilon_{AB}\chi_B, \tag{24.154}$$

$\boldsymbol{\varepsilon}$ being the 2×2 antisymmetric matrix of (23.72), and in which

$$\chi_A(\mathbf{x}) = \chi_A^{C'}(\mathbf{x}), \quad \bar{\chi}_A(\mathbf{x}) = \chi_{A+2}^{C'}(\mathbf{x}) \tag{24.155}$$

and

$$\lambda_A(\mathbf{x}) = \lambda_A^{C'}(\mathbf{x}), \quad \bar{\lambda}_A(\mathbf{x}) = \lambda_{A+2}^{C'}(\mathbf{x}) \tag{24.156}$$

(for $A = 1, 2$) are the spinors of the modified chiral representation of (23.25).

It is worth noting that if $\Phi_{1s}(\mathbf{X};\mathbf{\Theta})$ and $\Phi_{2s}(\mathbf{X};\mathbf{\Theta})$ are two scalar superfields then $\Phi_{1s}(\mathbf{X};\mathbf{\Theta}) = \Phi_{2s}(\mathbf{X};\mathbf{\Theta})$ if and only if $\Phi_{1s}(\mathbf{X};\mathbf{0}) = \Phi_{2s}(\mathbf{X};\mathbf{0})$; that is, if and only if the $\mathbf{\Theta}$-independent terms $C_1(\mathbf{X})$ and $C_2(\mathbf{X})$ in their respective expansion are equal (in the usual sense of operator equality). (The necessity of this condition is obvious, and its sufficiency follows from the fact that (24.84)–(24.90) relate all the components in an expansion of the form (24.151) to the $\mathbf{\Theta}$-independent term.)

As the component function of the highest allowed product $\Theta^1\Theta^2\Theta^3\Theta^4$ of the odd Grassmann variables differs from a term proportional to $D(\mathbf{x})$ by a 4-divergence, and as the 4-dimensional integral of $D(\mathbf{x})$ over Minkowski space–time is invariant under the operations induced by supertranslation generators (as well as under the ordinary Poincarés transformations), these properties are also possessed by the component function $\Theta^1\Theta^2\Theta^3\Theta^4$.

It is useful to define an adjoint operator ‡ for operator-valued scalar superfields by requiring that if B_1 and B_2 are any Grassmann-valued parameters and Ψ_1 and Ψ_2 are any "component" operators (in the above sense) then

$$(B_1\Psi_1 + B_2\Psi_2)^{\ddagger} = \Psi_1^{\dagger}B_1^{\#} + \Psi_1^{\dagger}B_2^{\#}, \tag{24.157}$$

where † denotes the ordinary adjoint of an operator (cf. Appendix B, Section 4) and # indicates the Grassmann adjoint operation of (20.30) and (20.31), where, it should be noted, the order of factors in each product is reversed. With this definition, it follows from (24.151) that if the component fields satisfy (24.91)–(24.93) then

$$(\Phi_s(\mathbf{X};\mathbf{\Theta}))^{\ddagger} = \Phi_s(\mathbf{X};\Theta). \tag{24.158}$$

Such a scalar superfield is then said to be "real".

Clearly if $\Phi_{1s}(\mathbf{X};\mathbf{\Theta})$ and $\Phi_{2s}(\mathbf{X};\mathbf{\Theta})$ are any two scalar superfields then their product $\Phi_{1s}(\mathbf{X};\mathbf{\Theta})\Phi_{2s}(\mathbf{X};\mathbf{\Theta})$ is also a scalar superfield. Denoting this product scalar superfield by $\Phi_{3s}(\mathbf{X};\mathbf{\Theta})$, it is obvious that $\Phi_{3s}(\mathbf{X};\mathbf{\Theta})$ also has an expansion of the form (24.151), which necessarily contains only 16 terms, so the component functions of $\Phi_{3s}(\mathbf{X};\mathbf{\Theta})$ can be deduced from those of $\Phi_{1s}(\mathbf{X};\mathbf{\Theta})$ and $\Phi_{2s}(\mathbf{X};\mathbf{\Theta})$ by simply equating coefficients of powers of the odd Grassmann variables Θ^a. Thus, for example,

$$C_3(\mathbf{X}) = C_1(\mathbf{X})C_2(\mathbf{X}) \tag{24.159}$$

and

$$\chi_3(\mathbf{X}) = C_1(\mathbf{X})\chi_2(\mathbf{X}) + C_1(\mathbf{X})\chi_2(\mathbf{X}).$$

Comparison of (24.159) with (24.137) shows that if the superfields $\Phi_{1s}(\mathbf{X};\mathbf{\Theta}), \Phi_{2s}(\mathbf{X};\mathbf{\Theta})$ and $\Phi_{3s}(\mathbf{X};\mathbf{\Theta})$ correspond to multiplets Φ_{G1}, Φ_{G2} and Φ_{G3} respectively then the $\mathbf{\Theta}$-independent term of the product superfield $\Phi_{3s}(\mathbf{X};\mathbf{\Theta})$ is equal to the $\mathbf{\Theta}$-independent term of the superfield corresponding to the product $\Phi_{G3} = \Phi_{G1}\cdot\Phi_{G2}$ of (24.144). Thus

$$\Phi_{3s}(\mathbf{X};\mathbf{\Theta}) = \Phi_{1s}(\mathbf{X};\mathbf{\Theta})\Phi_{2s}(\mathbf{X};\mathbf{\Theta})$$

if and only if

$$\Phi_{G3} = \Phi_{G1} \cdot \Phi_{G2}.$$

The transformation law (24.149) for a scalar superfield can be generalized to that for spinor or tensor superfields by requiring that

$$U([\mathbf{\Lambda}|\mathbf{T}|\zeta])\Phi_m(\mathbf{X};\mathbf{\Theta})U([\mathbf{\Lambda}|\mathbf{T}|\zeta])^{-1} = \sum_{n=1}^{d} \mathbf{\Gamma}([\mathbf{\Lambda}|\mathbf{0}|\mathbf{0}])_{mn}\Phi_n(\mathbf{X};\mathbf{\Theta}) \quad (24.160)$$

for all $[\mathbf{\Lambda}|\mathbf{T}|\zeta]$, where $\mathbf{\Gamma}$ is the appropriate representation (of some dimension d) of the covering group of the homogeneous Lorentz group. It should be noted that even though the scalar superfield is the simplest superfield that could be constructed, it does *not* correspond to an *irreducible* representation of the $N = 1, D = 4$ Poincaré superalgebra. As will be seen in the next subsection, further conditions have to be applied to it to get a connection with the chiral multiplet. For a general treatment of the reduction of general superfields into the sum of irreducible superfields see Sokatchev (1975) and Ogievetsky and Sokatchev (1977).

For the N-extended $D = 4$ Poincaré supergroup the corresponding superspace has $4N$ odd Grassmann coordinates, so the expansion for the corresponding scalar superfield contains 2^{4N} terms. Although this means that there are only 16 terms for $N = 1$, for $N = 2, 4$ and 8 it implies the existence of 256 terms, 65 536 terms, and 4 294 967 296 terms respectively! For general introductions to the techniques for handling such large numbers of terms and constructing extended superfields see Sohnius *et al.* (1981), Taylor (1982) and Wess and Bagger (1983). Sierra and Townsend (1983) have given an introduction to the $N = 2$ case, where the supermultiplets are often referred to as "hypermultiplets" (following the suggestion of Fayet (1976)). The decomposition of extended superfields into their irreducible parts has been studied by Taylor (1980), Pickup and Taylor (1981), Rittenberg and Sokatchev (1981), Kim (1984) and Ketov (1988). Constructions of off-shell superfield formalisms in the $N = 2$ case have been given by Galperin *et al.* (1984a, b, 1985a, b) (developing the harmonic superspace ideas of Salam and Strathdee 1982) and Yamron and Siegel (1986). Saidi (1988a, b) has considered the SU(2) representations of the $N = 2$ on-shell supersymmetric multiplets, and put in a more general context the hypermultiplets of Salam and Strathdee (1975c), Fayet (1976), Sohnius (1978) and Howe *et al.* (1983a).

(*b*) *The chiral superfields*

Let D_a (for $a = 1, \ldots, 4$) be the set of operators defined by

$$[\mathrm{D}_a, \Phi_s(\mathbf{X};\mathbf{\Theta})]_- = 2^{-1/2}\left\{ i \sum_{\alpha=1}^{4} (\gamma^\alpha \mathbf{\Theta})_a \frac{\partial}{\partial X^\alpha} + \sum_{b=1}^{4} C_{ab} \frac{\partial}{\partial \Theta^b} \right\} \Phi_s(\mathbf{X};\mathbf{\Theta}) \quad (24.161)$$

that act on scalar superfields. For brevity the left-hand side of (24.161) is often written as $D_a\Phi_s(\mathbf{X};\mathbf{\Theta})$. Then it is quite easily shown that

$$[U(M_{\alpha\beta}), D_a]_- = \sum_{b=1}^{4} \Gamma^{\text{spin}}(M_{\alpha\beta})_{ab} D_b, \tag{24.162}$$

(for $\alpha, \beta = 1,\dots,4$ and $a = 1,\dots,4$),

$$[U(P_\mu), D_a]_- = 0 \tag{24.163}$$

(for $\mu = 1,\dots,4$ and $a = 1,\dots,4$) and

$$[U(Q_a), D_b]_+ = 0 \tag{24.164}$$

(for $a, b = 1,\dots,4$), where the $U(M_{\alpha\beta}), U(P_\mu)$, and $U(Q_a)$ are as given in (23.150), (23.152), and (23.148) respectively. Moreover,

$$[D_a, D_b]_+ = -i \sum_{\alpha=1}^{4} (\gamma^\alpha \mathbf{C})_{ab} \frac{\partial}{\partial X^\alpha} \tag{24.165}$$

(for $a, b = 1,\dots,4$). The definition (24.161) implies that the quantities transform in the same way as a spinor under a similarity transformation of the Dirac matrices of the form (23.11), that is

$$D'_a = \sum_{b=1}^{4} S_{ab} D_b \tag{24.166}$$

(cf. (23.46)), and clearly (24.162)–(24.165) are form-invariant under such transformations.

Using the two-component formulation, for which the corresponding operators may be defined (by analogy with (23.54)) by

$$D_1 = D_1^{C'}, \quad D_2 = D_2^{C'}, \quad \bar{D}_1 = D_3^{C'}, \quad \bar{D}_2 = D_4^{C'}; \tag{24.167}$$

that is, on using (23.25) and (23.26), (24.161) gives

$$D_A = -2^{-1/2} \sum_{B=1}^{2} \left\{ i \sum_{\alpha=1}^{4} (\boldsymbol{\sigma}_2 \boldsymbol{\sigma}_\alpha^*)_{AB} \bar{\Theta}_B \frac{\partial}{\partial X^\alpha} - (\boldsymbol{\sigma}_2)_{AB} \frac{\partial}{\partial \Theta_B} \right\} \tag{24.168}$$

and

$$\bar{D}_A = -2^{-1/2} \sum_{B=1}^{2} \left\{ i \sum_{\alpha=1}^{4} (\boldsymbol{\sigma}_2 \boldsymbol{\sigma}_\alpha)_{AB} \Theta_B \frac{\partial}{\partial X^\alpha} - (\boldsymbol{\sigma}_2)_{AB} \frac{\partial}{\partial \bar{\Theta}_B} \right\} \tag{24.169}$$

(for $A = 1, 2$). Moreover, because of the block form (23.31) of $\mathbf{\Gamma}^{\text{spin}}_{C'}$, (24.162)–(24.164) imply that if

$$[D_A, \Phi_s(\mathbf{X};\mathbf{\Theta})]_- = 0 \tag{24.170}$$

(for $A = 1, 2$) then the commutators of $[D_A, \Phi_s(\mathbf{X};\mathbf{\Theta})]_-$ with the $U(M_{\alpha\beta})$ and

$U(P_\mu)$ and the anticommutator with the $U(Q_a)$ are all zero. Thus (24.170) is a pair of *invariant* conditions; that is, they are invariant under all transformations of the Poincaré supergroup. The same is true of the pair of conditions

$$[\bar{\mathrm{D}}_A, \Phi_s(\mathbf{X};\mathbf{\Theta})]_- = 0 \tag{24.171}$$

(for $A = 1, 2$). For this reason Salam and Strathdee (1974a, b, d, 1975b) called these operators D_A and $\bar{\mathrm{D}}_A$ *covariant derivatives.*

The effect of imposing the condition (24.171) will be investigated first. Clearly this condition requires that

$$i \sum_{\alpha=1}^{4} (\boldsymbol{\sigma}_\alpha)_{AB}\Theta_B \frac{\partial \Phi_s(\mathbf{X};\mathbf{\Theta})}{\partial X^\alpha} - \frac{\partial \Phi_s(\mathbf{X};\mathbf{\Theta})}{\partial \bar{\Theta}_B} = 0 \tag{24.172}$$

(for $B = 1, 2$). It is easily checked that the most general solution of this pair of differential equations is of the form

$$\Phi_L(\mathbf{X};\mathbf{\Theta}) = \Psi(\mathbf{Z}_L; \Theta_1, \Theta_2), \tag{24.173}$$

where $\mathbf{Z}_L$ is a 4-component vector, each of whose entries are the members of the *even* Grassmann subspace that are defined by

$$Z_L^\alpha = X^\alpha + i \sum_{A,B=1}^{2} \bar{\Theta}_A(\boldsymbol{\sigma}_\alpha)_{AB}\Theta_B \tag{24.174}$$

(for $\alpha = 1, \ldots, 4$). The expansion of the most general scalar superfield $\Phi_L(\mathbf{X};\mathbf{\Theta})$ that satisfies the condition (24.171) can then be written as

$$\Phi_L(\mathbf{X};\mathbf{\Theta}) = A_L(\mathbf{Z}_L) - 2i \sum_{A,B=1}^{2} \Theta_A \varepsilon_{AB} \psi_{LB}(\mathbf{Z}_L) - iF_L(\mathbf{Z}_L) \sum_{A,B=1}^{2} \Theta_A \varepsilon_{AB} \Theta_B, \tag{24.175}$$

where $A_L(\mathbf{Z}_L), F_L(\mathbf{Z}_L)$ and $\psi_{LA}(\mathbf{Z}_L)$ (for $A = 1, 2$) are functions of $\mathbf{Z}_L$ alone, and the expansion terminates after quadratic terms because there are only *two* odd Grassmann variables Θ_1 and Θ_2 involved. Application of (24.152) and (23.149) and comparison with (24.48)–(24.50) then shows that with the coefficients chosen as in (24.175) the components $A_L(\mathbf{Z}_L)$, $F_L(\mathbf{Z}_L)$ and $\psi_{LA}(\mathbf{Z}_L)$ (with $A = 1, 2$) are precisely the *continuations* of the operators $A_L(\mathbf{x}), F_L(\mathbf{x})$ and $\psi_{LA}(\mathbf{x})$ (with $A = 1, 2$) of the left-handed chiral multiplet. For this reason $\Phi_L(\mathbf{X};\mathbf{\Theta})$ may be called the *left-handed chiral superfield* (although it is sometimes known just as the "chiral superfield").

Similarly the most general scalar superfield $\Phi_R(\mathbf{X};\mathbf{\Theta})$ that satisfies the condition (24.170) has the form

$$\Phi_R(\mathbf{X};\mathbf{\Theta}) = \Psi(\mathbf{Z}_R; \bar{\Theta}_1, \bar{\Theta}_2), \tag{24.176}$$

where $\mathbf{Z}_{\mathrm{R}}$ is a 4-component vector, each of whose entries are again members of the *even* Grassmann subspace but are now defined by

$$Z_{\mathrm{R}}^{\alpha} = X^{\alpha} - i \sum_{A,B=1}^{2} \bar{\Theta}_A (\boldsymbol{\sigma}_{\alpha})_{AB} \Theta_B \tag{24.177}$$

(for $\alpha = 1, \ldots, 4$). In this case the expansion of the most general scalar superfield $\Phi_{\mathrm{R}}(\mathbf{X}; \boldsymbol{\Theta})$ that satisfies the condition (24.170) can be written as

$$\Phi_{\mathrm{R}}(\mathbf{X}; \boldsymbol{\Theta}) = A_{\mathrm{R}}(\mathbf{Z}_{\mathrm{R}}) - 2i \sum_{A,B=1}^{2} \bar{\Theta}_A \varepsilon_{AB} \psi_{\mathrm{R}B}(\mathbf{Z}_{\mathrm{R}}) - iF_{\mathrm{R}}(\mathbf{Z}_{\mathrm{R}}) \sum_{A,B=1}^{2} \bar{\Theta}_A \varepsilon_{AB} \bar{\Theta}_B, \tag{24.178}$$

where $A_{\mathrm{R}}(\mathbf{Z}_{\mathrm{R}}), F_{\mathrm{R}}(\mathbf{Z}_{\mathrm{R}})$ and $\psi_{\mathrm{R}A}(\mathbf{Z}_{\mathrm{R}})$ (for $A = 1, 2$) are now functions of $\mathbf{Z}_{\mathrm{R}}$ alone, and the expansion terminates again after quadratic terms because only *two* odd Grassmann variables are involved, these now being $\bar{\Theta}_1$ and $\bar{\Theta}_2$. Application of (24.152) and (23.149) and comparison with (24.51)–(24.53) then shows that with the coefficients chosen as in (24.178) the components $A_{\mathrm{R}}(\mathbf{Z}_{\mathrm{R}}), F_{\mathrm{R}}(\mathbf{Z}_{\mathrm{R}})$ and $\psi_{\mathrm{R}A}(\mathbf{Z}_{\mathrm{R}})$ (with $A = 1, 2$) are precisely the *continuations* of the operators $A_{\mathrm{R}}(\mathbf{x}), F_{\mathrm{R}}(\mathbf{x})$ and $\psi_{\mathrm{R}A}(\mathbf{x})$ (with $A = 1, 2$) of the right-handed chiral multiplet. For this reason $\Phi_{\mathrm{R}}(\mathbf{X}, \boldsymbol{\Theta})$ may be called the *right-handed chiral superfield*, although the designation "antichiral superfield" sometimes appears in the literature.

It should be noted that (24.174), (24.177) and (23.136) imply that

$$(\mathbf{Z}_{\mathrm{L}})^{\#} = \mathbf{Z}_{\mathrm{R}}, \quad (\mathbf{Z}_{\mathrm{R}})^{\#} = \mathbf{Z}_{\mathrm{L}}. \tag{24.179}$$

Thus

$$(\Phi_{\mathrm{L}}(\mathbf{X}; \boldsymbol{\Theta}))^{\ddagger} = \Phi_{\mathrm{R}}(\mathbf{X}; \boldsymbol{\Theta}), \quad (\Phi_{\mathrm{R}}(\mathbf{X}; \boldsymbol{\Theta}))^{\ddagger} = \Phi_{\mathrm{L}}(\mathbf{X}; \boldsymbol{\Theta}), \tag{24.180}$$

where $\ddagger$ is defined in (24.157).

Let D^2 and $\bar{\mathrm{D}}^2$ be two operators defined in terms of the covariant derivatives by

$$\mathrm{D}^2 = \sum_{A,B=1}^{2} \varepsilon_{AB} \mathrm{D}_A \mathrm{D}_B \tag{24.181}$$

and

$$\bar{\mathrm{D}}^2 = \sum_{A,B=1}^{2} \varepsilon_{AB} \bar{\mathrm{D}}_A \bar{\mathrm{D}}_B, \tag{24.182}$$

where $\boldsymbol{\varepsilon}$ is the 2×2 antisymmetric matrix of (23.72). Then, for any scalar superfield $\Phi_{\mathrm{s}}(\mathbf{X}; \boldsymbol{\Theta})$,

$$\mathrm{D}_A\{\mathrm{D}^2 \Phi_{\mathrm{s}}(\mathbf{X}; \boldsymbol{\Theta})\} = 0 \tag{24.183}$$

and

$$\bar{\mathrm{D}}_A\{\bar{\mathrm{D}}^2\Phi_s(\mathbf{X};\Theta)\} = 0 \tag{24.184}$$

(for $A = 1, 2$). (These results follow immediately from the fact that substitution of matrices (23.25) and (23.26) of the modified chiral representation into (24.165) gives

$$[\mathrm{D}_A, \mathrm{D}_B]_+ = 0 \tag{24.185}$$

and

$$[\bar{\mathrm{D}}_A, \bar{\mathrm{D}}_B]_+ = 0 \tag{24.186}$$

(for $A, B = 1, 2$)). The results (24.183) and (24.184) when taken with the defining conditions (24.170) and (24.171) imply that if $\Phi_s(\mathbf{X};\mathbf{\Theta})$ is any scalar superfield then $\mathrm{D}^2\Phi_s(\mathbf{X};\mathbf{\Theta})$ and $\bar{\mathrm{D}}^2\Phi_s(\mathbf{X};\mathbf{\Theta})$ are right-handed and left-handed chiral superfields respectively.

In particular, if $\Phi_L(\mathbf{X};\mathbf{\Theta})$ is the left-handed chiral superfield of (24.175) and if $\Phi_R(\mathbf{X};\mathbf{\Theta})$ is its associated right-handed chiral superfield (24.178) then $\mathrm{D}^2\Phi_L(\mathbf{X};\mathbf{\Theta})$ is a right-handed chiral superfield and $\bar{\mathrm{D}}^2\Phi_R(\mathbf{X};\mathbf{\Theta})$ is a left-handed chiral superfield. Moreover, it is easily shown that the components of $-\frac{1}{2}i\mathrm{D}^2\Phi_L(\mathbf{X};\mathbf{\Theta})$ and $-\frac{1}{2}i\bar{\mathrm{D}}^2\Phi_R(\mathbf{X};\mathbf{\Theta})$ are respectively the right- and left-handed parts of the components of the kinetic multiplet (which are given in (24.64)–(24.68)) of the associated original chiral multiplet. That is, for example, the right-handed chiral superfield $\Phi'_R(\mathbf{X};\mathbf{\Theta})$ defined by

$$\Phi'_R(\mathbf{X};\mathbf{\Theta}) = -\tfrac{1}{2}i\mathrm{D}^2\Phi_L(\mathbf{X};\mathbf{\Theta}) \tag{24.187}$$

has an expansion of the form

$$\Phi'_R(\mathbf{X};\mathbf{\Theta}) = A'_R(\mathbf{Z}_R) - 2i\sum_{A,B=1}^{2}\bar{\Theta}_A\varepsilon_{AB}\psi'_{RB}(\mathbf{Z}_R) - iF'_R(\mathbf{Z}_R)\sum_{A,B=1}^{2}\bar{\Theta}_A\varepsilon_{AB}\bar{\Theta}_B, \tag{24.188}$$

where

$$A'_R(\mathbf{x}) = -F_L(\mathbf{x}), \tag{24.189}$$

$$F'_R(\mathbf{x}) = \Box A_L(\mathbf{x}), \tag{24.190}$$

and so on.

Clearly if $\Phi_{L1}(\mathbf{X};\mathbf{\Theta})$ and $\Phi_{L2}(\mathbf{X};\mathbf{\Theta})$ are any two left-handed chiral superfields then their product

$$\Phi_{L3}(\mathbf{X};\mathbf{\Theta}) = \Phi_{L1}(\mathbf{X};\mathbf{\Theta})\Phi_{L2}(\mathbf{X};\mathbf{\Theta}) \tag{24.191}$$

is also a left-handed chiral superfield. Comparison with (24.59)–(24.61) shows that the components of $\Phi_{L3}(\mathbf{X};\mathbf{\Theta})$ are just those of the product of the corresponding left-handed multiplets. Similarly if $\Phi_{R1}(\mathbf{X};\mathbf{\Theta})$ and $\Phi_{R2}(\mathbf{X};\mathbf{\Theta})$ are any two right-handed chiral superfields then their product

$$\Phi_{R3}(\mathbf{X};\mathbf{\Theta}) = \Phi_{R1}(\mathbf{X};\mathbf{\Theta})\Phi_{R2}(\mathbf{X};\mathbf{\Theta}) \qquad (24.192)$$

is the right-handed chiral superfield corresponding to the product of the two corresponding right-handed multiplets.

It is also easily shown that if $\Phi_{L1}(\mathbf{X};\mathbf{\Theta})$ and $\Phi_{L2}(\mathbf{X};\mathbf{\Theta})$ are any two left-handed chiral superfields and if $\Phi_{R1}(\mathbf{X};\mathbf{\Theta})$ and $\Phi_{R2}(\mathbf{X};\mathbf{\Theta})$ are their associated right-handed chiral superfields, so that

$$(\Phi_{L1}(\mathbf{X};\mathbf{\Theta}))^{\ddagger} = \Phi_{R1}(\mathbf{X};\mathbf{\Theta}), \quad (\Phi_{L2}(\mathbf{X};\mathbf{\Theta}))^{\ddagger} = \Phi_{R2}(\mathbf{X};\mathbf{\Theta})$$

(cf. (24.180)), then the combinations

$$\tfrac{1}{2}\{\Phi_{L1}(\mathbf{X};\mathbf{\Theta})\Phi_{R2}(\mathbf{X};\mathbf{\Theta}) + \Phi_{R1}(\mathbf{X};\mathbf{\Theta})\Phi_{L2}(\mathbf{X};\mathbf{\Theta})\} \qquad (24.193)$$

and

$$\tfrac{1}{2}i\{\Phi_{L1}(\mathbf{X};\mathbf{\Theta})\Phi_{R2}(\mathbf{X};\mathbf{\Theta}) - \Phi_{R1}(\mathbf{X};\mathbf{\Theta})\Phi_{L2}(\mathbf{X};\mathbf{\Theta})\} \qquad (24.194)$$

are both scalar superfields that satisfy (24.158). Moreover, if the components of $\Phi_{L1}(\mathbf{X};\mathbf{\Theta})$ and $\Phi_{R1}(\mathbf{X};\mathbf{\Theta})$ belong to the chiral multiplet Φ_{C1}, and if the components of $\Phi_{L2}(\mathbf{X};\mathbf{\Theta})$ and $\Phi_{R2}(\mathbf{X};\mathbf{\Theta})$ belong to the chiral multiplet Φ_{C2} then the components of (24.193) and (24.194) are those of the symmetric and antisymmetric products $\Phi_{C1} \times \Phi_{C2}$ and $\Phi_{C1} \wedge \Phi_{C2}$ of (24.122)–(24.128) and (24.129)–(24.135) respectively.

Covariant quantization of chiral superfields has been investigated by de Azcárraga *et al.* (1988).

(c) *Superfield formulation of the action of the Wess–Zumino model*

This subsection will be devoted to demonstrating that all the terms in the action corresponding to the Wess–Zumino model that was introduced in Section 2(*b*) can be written as superspace integrals involving various combinations of the chiral superfields.

(*i*) *The "kinetic" term S_{KE} of the action*:

Consider the superspace integral

$$S_{KE} = \frac{1}{i\hbar c}\int d^4X \int d^2\Theta\, \Phi_L(\mathbf{X};\mathbf{\Theta})\, D^2\Phi_R(\mathbf{X};\mathbf{\Theta})$$

$$+ \frac{1}{i\hbar c}\int d^4X \int d^2\bar{\Theta}\, \Phi_R(\mathbf{X};\mathbf{\Theta}) D^2\Phi_L(\mathbf{X};\mathbf{\Theta}), \qquad (24.195)$$

where the first term contains a Berezin integral over the two odd Grassmann variables Θ_1 and Θ_2, the corresponding Berezin integral in the second term being over $\bar{\Theta}_1$ and $\bar{\Theta}_2$. As implied by the definition of the Berezin integral that was given in (22.61), the result of "carrying out the integration" in the first term is simply to project out the component function of the $\Theta_1\Theta_2$ term in the expansion of the integrand, i.e. the component function of the highest power in the odd Grassmann variables. A similar result is true for the second term of the integral. The integration over the even Grassmann variables in the first term can then be defined as the four-fold integration over the set of four real variables (x^1, x^2, x^3, x^4) $(= \mathbf{x})$, where

$$x^\alpha = X^\alpha_\varnothing = Z^\alpha_{L\varnothing} \tag{24.196}$$

(for $\alpha = 1, \ldots, 4$). Here $X^\alpha_\varnothing$ and $Z^\alpha_{L\varnothing}$ indicate the real-valued coefficients of X^α and Z^α_L respectively that correspond to the Grassmann identity element $\mathcal{E}_\varnothing$, and the (operator-valued) component functions are now being treated as functions of these real variables. The even-variable integration of the second term is defined in a similar way, with

$$x^\alpha = X^\alpha_\varnothing = Z^\alpha_{R\varnothing} \tag{24.197}$$

(for $\alpha = 1, \ldots, 4$). Then

$$S_{KE} = \int d^4x\, \mathscr{L}_{KE}(\mathbf{x}), \tag{24.198}$$

where $\mathscr{L}_{KE}(\mathbf{x})$ is the kinetic-energy part of the Lagrangian density of the Wess–Zumino model, as displayed in (24.69).

An alternative superspace integral for S_{KE} is given by

$$S_{KE} = -\frac{1}{4\hbar c}\int d^4X \int d^2\Theta \int d^2\bar{\Theta}\, \Phi_R(\mathbf{X};\boldsymbol{\Theta})\Phi_L(\mathbf{X};\boldsymbol{\Theta}), \tag{24.199}$$

which now involves a four-fold Berezin integral.

(*ii*) *The "mass" term S_{mass} of the action*:

With a similar interpretation of the superspace integrals, the action S_{mass} defined by

$$S_{mass} = \tfrac{1}{4}iMc^2 \int d^4X \int d^2\Theta\, \Phi_L(\mathbf{X};\boldsymbol{\Theta})\Phi_L(\mathbf{X};\boldsymbol{\Theta}) + \tfrac{1}{4}iMc^2 \int d^4X \int d^2\bar{\Theta}\, \Phi_R(\mathbf{X};\boldsymbol{\Theta})\Phi_R(\mathbf{X};\boldsymbol{\Theta}) \tag{24.200}$$

is such that

$$S_{mass} = \int d^4x \mathscr{L}_{mass}(\mathbf{x}). \tag{24.201}$$

Here $\mathscr{L}_{\text{mass}}(\mathbf{x})$ is the "mass" part of the Lagrangian density of the Wess–Zumino model, as given in (24.71).

(*iii*) *The "interaction" term S_{int} of the action*:

Finally, by a similar argument, the action S_{int} defined by

$$S_{\text{int}} = ig\int d^4X \int d^2\Theta\, \Phi_L(\mathbf{X};\mathbf{\Theta})\Phi_L(\mathbf{X};\mathbf{\Theta})\Phi_L(\mathbf{X};\mathbf{\Theta}) + ig\int d^4X \int d^2\bar{\Theta}\, \Phi_R(\mathbf{X};\mathbf{\Theta})\Phi_R(\mathbf{X};\mathbf{\Theta})\Phi_R(\mathbf{X};\mathbf{\Theta}) \tag{24.202}$$

is such that

$$S_{\text{int}} = \int d^4x \mathscr{L}_{\text{int}}(\mathbf{x}), \tag{24.203}$$

where $\mathscr{L}_{\text{int}}(\mathbf{x})$ is the "interaction" part of the Lagrangian density of the Wess–Zumino model, as given in (24.72).

4 Supersymmetric gauge theories

It is probably simplest to start the study of supersymmetric gauge theories by examining their formulation in terms of components before looking at the superfield version. The first stage is the setting up in Section 4(*a*) of a supersymmetric extension of the Abelian gauge theory that was presented in Chapter 19, Section 2. As quantum electrodynamics (QED) can be regarded as an example of such an Abelian gauge theory, its supersymmetric extension is often referred to as *super-QED*. Section 4(*b*) will be devoted to the reformulation of this in superfield language, which will then be generalized in Section 4(*c*) to give the supersymmetric extension of *non*-Abelian gauge theories; that is, the so-called *supersymmetric Yang–Mills theories*. The component formulation of super-QED was first discussed by Wess and Zumino (1974c), the superfield extensions being developed by Salam and Strathdee (1974d) and Ferrara and Zumino (1974).

(*a*) *The component formulation of super-QED*

First the essential features of Abelian gauge theories will be recalled, and some of the notation used in the earlier discussion in Chapter 19, Section 2 will be modified and extended.

Ordinary QED is concerned with the interaction of two fields. One is a

4-component spinor field, which describes spin-$\frac{1}{2}$ fermions, such as electrons. This was denoted by $\psi(\mathbf{x})$, and will be called later the *matter field*. The other is a 4-component vector field, which describes the photons. The components of this were denoted previously by $A_\mu(\mathbf{x})$, but henceforth they will be written as $V_\mu(\mathbf{x})$. The total Lagrangian density $\mathscr{L}^{\mathrm{QED}}(\mathbf{x})$ for QED is the sum

$$\mathscr{L}^{\mathrm{QED}}(\mathbf{x}) = \mathscr{L}^{\mathrm{f}}_0(\mathbf{x}) + \mathscr{L}^{\mathrm{v}}_0(\mathbf{x}) + \mathscr{L}_{\mathrm{int}}(\mathbf{x}). \tag{24.204}$$

Here $\mathscr{L}^{\mathrm{f}}_0(\mathbf{x})$ is the Lagrangian density of free electrons with mass M, which is given by

$$\mathscr{L}^{\mathrm{f}}_0(\mathbf{x}) = \tilde{\psi}(\mathbf{x})^\dagger \left\{ i\hbar c \sum_{\rho=1}^{4} \gamma^\rho \partial_\rho \psi(\mathbf{x}) - Mc^2\psi(\mathbf{x}) \right\}, \tag{24.205}$$

(cf. (18.3) and (19.1)), where (as in the previous two sections) $\tilde{\psi}(\mathbf{x})^\dagger$ denotes the transpose of the 4×1 column matrix whose entries are the operator adjoints of the components of $\psi(\mathbf{x})$. Similarly $\mathscr{L}^{\mathrm{v}}_0(\mathbf{x})$ is the Lagrangian density of free photons, which is given by

$$\mathscr{L}^{\mathrm{v}}_0(x) = -\tfrac{1}{4} \sum_{\rho,\kappa,\tau,\mu=1}^{4} \eta^{\rho\tau}\eta^{\kappa\mu} F_{\rho\tau}(\mathbf{x}) F_{\kappa\mu}(\mathbf{x}), \tag{24.206}$$

where

$$F_{\lambda\rho}(\mathbf{x}) = \partial_\lambda V_\rho(\mathbf{x}) - \partial_\rho V_\lambda(\mathbf{x}) \tag{24.207}$$

(cf. (18.4) and (18.5)). Finally $\mathscr{L}_{\mathrm{int}}(\mathbf{x})$, which describes the electron–photon interaction, has the form

$$\mathscr{L}_{\mathrm{int}}(\mathbf{x}) = -g \sum_{\rho=1}^{4} \tilde{\psi}(\mathbf{x})^\dagger \gamma^4 \gamma^\rho V_\rho(\mathbf{x}) \psi(\mathbf{x}), \tag{24.208}$$

where the coupling constant g is given by the electron's charge.

The gauge invariance of this theory is embodied in the statement that the Lagrangian density $\mathscr{L}^{\mathrm{QED}}(\mathbf{x})$ defined by (24.204) is unchanged by the pair of substitutions

$$\psi(\mathbf{x}) \to \psi'(\mathbf{x}) = \psi(\mathbf{x}) \exp\{i\alpha(\mathbf{x})\} \tag{24.209}$$

(cf. (19.2)) and

$$V_\rho(\mathbf{x}) \to V'_\rho(\mathbf{x}) = V_\rho(\mathbf{x}) - \frac{\hbar c}{g} \partial_\rho \alpha(\mathbf{x}) \tag{24.210}$$

(for $\rho = 1, \ldots, 4$), where $\alpha(\mathbf{x})$ is an arbitrary real dimensionless differentiable scalar function of space–time. As the quantities $[\exp\{i\alpha(\mathbf{x})\}]$ form (at each point of space–time) a group that is isomorphic to the one-dimensional Abelian Lie group U(1), QED is said to exhibit "U(1) gauge invariance", or, more precisely, "local U(1) gauge invariance". As noted in Chapter 19, Section 2, the procedure for constructing such gauge-invariant Lagrangian densities is simply to replace every space–time derivative ∂_ρ in the total free Lagrangian density by a "covariant derivative", which will now be denoted by ∇_ρ to avoid confusion with the superfield covariant derivatives of (24.161), and which is defined by

$$\nabla_\rho = \partial_\rho + i\frac{g}{\hbar c} V_\rho(\mathbf{x}) \tag{24.211}$$

for $\rho = 1, \ldots, 4$ (cf. (19.4)).

It is obvious that the electron's field *cannot* be described by a Majorana spinor, for $\psi(\mathbf{x})$ and $\psi(\mathbf{x})\exp\{i\alpha(\mathbf{x})\}$ cannot both satisfy the Majorana condition (24.11) (except in the trivial case $\alpha(\mathbf{x}) = 0$). However, the electron's field $\psi(\mathbf{x})$ may be written in the form

$$\psi(\mathbf{x}) = \psi_1(\mathbf{x}) + i\psi_2(\mathbf{x}), \tag{24.212}$$

where $\psi_1(\mathbf{x})$ and $\psi_2(\mathbf{x})$ both satisfy the Majorana condition (24.11).

To express the Lie algebraic version of (24.209) and (24.210), it is convenient to define $\delta_{\text{gauge}}\psi(\mathbf{x})$ and $\delta_{\text{gauge}}V_\rho(\mathbf{x})$ by

$$\delta_{\text{gauge}}\psi(\mathbf{x}) = \psi'(\mathbf{x}) - \psi(\mathbf{x}) \tag{24.213}$$

and

$$\delta_{\text{gauge}}V_\rho(\mathbf{x}) = V'_\rho(\mathbf{x}) - V_\rho(\mathbf{x}) \tag{24.214}$$

(for $\rho = 1, \ldots, 4$), it being understood that in both of the right-hand sides only the *first-order* terms, i.e. only the terms *linear* in $\alpha(\mathbf{x})$, are retained. Thus, from (24.209) and (24.212),

$$\delta_{\text{gauge}}\psi_1(\mathbf{x}) = -\alpha(\mathbf{x})\psi_2(\mathbf{x}) \tag{24.215}$$

and

$$\delta_{\text{gauge}}\psi_2(\mathbf{x}) = \alpha(\mathbf{x})\psi_1(\mathbf{x}), \tag{24.216}$$

while (24.210) gives

$$\delta_{\text{gauge}}V_\rho(\mathbf{x}) = -\frac{\hbar c}{g}\partial_\rho\alpha(\mathbf{x}) \tag{24.217}$$

(for $\rho = 1, \ldots, 4$). (The subscript "gauge" on this variation δ_{gauge} has been added to avoid any confusion with the supersymmetry operators δ_ζ introduced in Section 2.)

In the component form of super-QED investigated by Wess and Zumino (1974c)

(i) the two Majorana spinor fields $\psi_1(\mathbf{x})$ and $\psi_2(\mathbf{x})$ are both replaced by real chiral multiplets, which will be denoted by Φ_{C1} and Φ_{C2}, their components being taken to be

$$\Phi_{C1} = \{A_1(\mathbf{x}), B_1(\mathbf{x}), F_1(\mathbf{x}), G_1(\mathbf{x}); \boldsymbol{\psi}_1(\mathbf{x})\} \tag{24.218}$$

and

$$\Phi_{C2} = \{A_2(\mathbf{x}), B_2(\mathbf{x}), F_2(\mathbf{x}), G_2(\mathbf{x}); \boldsymbol{\psi}_2(\mathbf{x})\}; \tag{24.219}$$

(ii) the real scalar field $\alpha(\mathbf{x})$ is replaced by a third real chiral multiplet, which will be denoted by Λ, its components being taken to be

$$\Lambda = \{\alpha(\mathbf{x}), \beta(\mathbf{x}), \omega(\mathbf{x}), \xi(\mathbf{x}); \boldsymbol{\phi}(\mathbf{x})\}; \tag{24,220}$$

and

(iii) the vector field $\mathbf{V}(\mathbf{x})$ is replaced by a real general multiplet, which will be denoted by V, its components being taken to be

$$V = \{C(\mathbf{x}), D(\mathbf{x}), M(\mathbf{x}), N(\mathbf{x}); \boldsymbol{\chi}(\mathbf{x}), \boldsymbol{\lambda}(\mathbf{x}); \mathbf{V}(\mathbf{x})\}. \tag{24.221}$$

Then (24.215), (24.216) and (24.217) are replaced by

$$\delta_{\text{gauge}}\Phi_{C1} = -\Lambda\Phi_{C2}, \tag{24.222}$$

$$\delta_{\text{gauge}}\Phi_{C2} = \Lambda\Phi_{C1}, \tag{24.223}$$

$$\delta_{\text{gauge}}V = -\frac{\hbar c}{g}\partial\Lambda \tag{24.224}$$

respectively. In (24.222) and (24.223) the products on the right-hand sides are those of the symmetric product of two chiral multiplets as defined in (24.150)–(24.155), and the $\partial\Lambda$ on the right-hand side of (24.224) is the general multiplet defined in terms of the chiral multiplet Λ by (24.115). (The coupling constant g of (24.224) is not related in any way to the coupling constant of the interaction term (24.72) of the Wess–Zumino model, but instead will play a role similar to that of (24.210).) The interpretation of (24.222)–(24.224) is that the corresponding components on the left- and right-hand sides have to be equated. Thus, for example, when written out in detail, (24.222) becomes

$$\delta_{\text{gauge}}A_1(\mathbf{x}) = -\{\alpha(\mathbf{x})A_2(\mathbf{x}) - \beta(\mathbf{x})B_2(\mathbf{x})\}, \tag{24.225}$$

$$\delta_{\text{gauge}}B_1(\mathbf{x}) = -\{\beta(\mathbf{x})A_2(\mathbf{x}) + \alpha(\mathbf{x})B_2(\mathbf{x})\}, \tag{24.226}$$

$$\begin{aligned}\delta_{\text{gauge}}F_1(\mathbf{x}) = -\{&\omega(\mathbf{x})A_2(\mathbf{x}) + \alpha(\mathbf{x})F_2(\mathbf{x}) + \beta(\mathbf{x})G_2(\mathbf{x}) + \xi(\mathbf{x})B_2(\mathbf{x})\\ &- \tilde{\boldsymbol{\phi}}(\mathbf{x})^\dagger\gamma_4\boldsymbol{\psi}_2(\mathbf{x})\},\end{aligned} \tag{24.227}$$

$$\begin{aligned}\delta_{\text{gauge}}G_1(\mathbf{x}) = -\{&\xi(\mathbf{x})A_2(\mathbf{x}) + \alpha(\mathbf{x})G_2(\mathbf{x}) - \beta(\mathbf{x})F_2(\mathbf{x}) - \omega(\mathbf{x})B_2(\mathbf{x})\\ &- i\tilde{\boldsymbol{\phi}}(\mathbf{x})^\dagger\gamma_4\gamma_5\boldsymbol{\psi}_2(\mathbf{x})\},\end{aligned} \tag{24.228}$$

$$\delta_{\text{gauge}}\boldsymbol{\psi}_1(\mathbf{x}) = -\{\{\alpha(\mathbf{x})\mathbf{1} - i\beta(\mathbf{x})\gamma_5\}\boldsymbol{\psi}_2(\mathbf{x}) + \{A_2(\mathbf{x})\mathbf{1} - iB_2(\mathbf{x})\gamma_5\}\boldsymbol{\phi}(\mathbf{x})\}. \tag{24.229}$$

Similarly, in component form, (24.224) becomes

$$\delta_{\text{gauge}} C(\mathbf{x}) = -\frac{\hbar c}{g}\beta(\mathbf{x}), \tag{24.230}$$

$$\delta_{\text{gauge}} D(\mathbf{x}) = 0, \tag{24.231}$$

$$\delta_{\text{gauge}} M(\mathbf{x}) = -\frac{\hbar c}{g}\omega(\mathbf{x}), \tag{24.232}$$

$$\delta_{\text{gauge}} N(\mathbf{x}) = -\frac{\hbar c}{g}\xi(\mathbf{x}), \tag{24.233}$$

$$\delta_{\text{gauge}} \boldsymbol{\chi}(\mathbf{x}) = -\frac{\hbar c}{g}\boldsymbol{\phi}(\mathbf{x}), \tag{24.234}$$

$$\delta_{\text{gauge}} \boldsymbol{\lambda}(\mathbf{x}) = \mathbf{0}, \tag{24.235}$$

$$\delta_{\text{gauge}} V_\rho(\mathbf{x}) = -\frac{\hbar c}{g}\partial_\rho \alpha(\mathbf{x}) \tag{24.236}$$

(for $\rho = 1,\ldots,4$).

Clearly (24.229) and (24.236) reduce to (24.215) and (24.217) if all the components of the chiral multiplet other than $\alpha(\mathbf{x})$ are taken to be zero. Consequently in the above generalization it is natural to choose the dimensions of the chiral multiplets Φ_{C1} and Φ_{C2} so that their spinor components $\boldsymbol{\psi}_1(\mathbf{x})$ and $\boldsymbol{\psi}_2(\mathbf{x})$ have the same dimensions as the electron's field of Chapter 19, Section 2, to choose the dimensions of the general multiplet V so that its vector component $\mathbf{V}(\mathbf{x})$ has the same dimensions as the photons' field of Chapter 19, Section 2, and to choose the dimensions of the chiral multiplet Λ so that its scalar component $\alpha(\mathbf{x})$ is dimensionless.

A Lagrangian density that is invariant under these gauge transformations, that provides a supersymmetric action, and that reduces in the limit $g \to 0$ to the kinetic-energy Lagrangian density of the Wess–Zumino model for the two chiral multiplets will now be constructed using the very ingenious procedure of Wess and Zumino (1974c). First two general multiplets V^1 and V^2 are defined in terms of the chiral multiplets Φ_{C1} and Φ_{C2} by the prescriptions

$$V^1 = \tfrac{1}{2}(\Phi_{C1} \times \Phi_{C1} + \Phi_{C2} \times \Phi_{C2}) \tag{24.237}$$

and

$$V^2 = \Phi_{C1} \wedge \Phi_{C2}, \tag{24.238}$$

these products being defined in (24.122)–(24.128) and (24.129)–(24.135). Then, by (24.222)–(24.224),

$$\delta_{\text{gauge}} V^1 = -(\Lambda\Phi_{C2}) \times \Phi_{C1} + (\Lambda\Phi_{C1}) \times \Phi_{C2}$$
$$= (\Phi_{C1} \times \Phi_{C1} + \Phi_{C2} \times \Phi_{C2}) \cdot \partial\Lambda$$

(using (24.147))

$$= 2V^1 \cdot \partial\Lambda. \tag{24.239}$$

Similarly

$$\delta_{\text{gauge}} V^2 = -(\Lambda\Phi_{C2}) \wedge \Phi_{C2} + \Phi_{C1} \wedge (\Lambda\Phi_{C1})$$
$$= 2(\Phi_{C1} \wedge \Phi_{C2}) \cdot \partial\Lambda$$

(using (24.138))

$$= 2V^2 \cdot \partial\Lambda. \tag{24.240}$$

Thus, with the general multiplets V^+ and V^- defined by

$$V^+ = V^1 + V^2 \tag{24.241}$$

and

$$V^- = V^1 - V^2, \tag{24.242}$$

it follows immediately that V^+ and V^- obey the simple transformation rules

$$\delta_{\text{gauge}} V^+ = 2V^+ \cdot \partial\Lambda \tag{24.243}$$

and

$$\delta_{\text{gauge}} V^- = -2V^- \cdot \partial\Lambda. \tag{24.244}$$

(Again these are to be interpreted as equalities between corresponding components.)

As the first component field $(2g/\hbar c)C(\mathbf{x})$ of the multiplet $(2g/\hbar c)V$ is dimensionless, in every product of $(2g/\hbar c)V$ with itself the first ("C") components are *all* dimensionless, the second ("D") components all have the *same* dimension (length)$^{-2}$, and so on, each particular component of *any* power always having the *same* dimension. Moreover, since the multiplication law (24.144) for general multiplets is associative (cf. (24.146)), it is possible to define the exponential of $(2g/\hbar c)V$ by

$$\exp\left(\frac{2gV}{\hbar c}\right) = 1 + \frac{2g}{\hbar c} V + \frac{1}{2!}\left(\frac{2g}{\hbar c}\right)^2 V \cdot V + \frac{1}{3!}\left(\frac{2g}{\hbar c}\right)^3 V \cdot V \cdot V + \cdots.$$

Then, by (24.224),

$$\delta_{\text{gauge}} \exp\left(\frac{2gV}{\hbar c}\right) = -2\exp\left(\frac{2gV}{\hbar c}\right) \cdot \partial\Lambda, \tag{24.245}$$

and similarly

$$\delta_{\text{gauge}} \exp\left(-\frac{2gV}{\hbar c}\right) = 2\exp\left(-\frac{2gV}{\hbar c}\right)\cdot\partial\Lambda. \tag{24.246}$$

Thus

$$\delta_{\text{gauge}}\left\{V^{+}\cdot\exp\left(\frac{2gV}{\hbar c}\right)\right\} = 0 \tag{24.247}$$

and

$$\delta_{\text{gauge}}\left\{V^{-}\cdot\exp\left(-\frac{2gV}{\hbar c}\right)\right\} = 0 \tag{24.248}$$

(because, for example, by (24.243) and (24.245),

$$\delta_{\text{gauge}}\left\{V^{+}\cdot\exp\left(\frac{2gV}{\hbar c}\right)\right\} = (2V^{+}\cdot\partial\Lambda)\cdot\exp\left(\frac{2gV}{\hbar c}\right) + V^{+}\cdot\left\{-2\exp\left(\frac{2gV}{\hbar c}\right)\cdot\partial\Lambda\right\},$$

which vanishes by virtue of the commutative and associative properties (24.145) and (24.146)). The most appropriate combination of these two gauge-invariant quantities is

$$\tfrac{1}{2}\hbar c\left\{V^{+}\cdot\exp\left(\frac{2gV}{\hbar c}\right) + V^{-}\cdot\exp\left(-\frac{2gV}{\hbar c}\right)\right\}, \tag{24.249}$$

because for $g = 0$ it reduces, by (24.241), to

$$\tfrac{1}{2}\hbar c(V^{+} + V^{-}) = \hbar c V^{1}, \tag{24.250}$$

whose "D" term is (by (24.136)) the sum of the kinetic-energy terms of the Wess–Zumino model Lagrangian densities associated with the two chiral multiplets Φ_{C1} and Φ_{C2}.

To this may be added the sum of the Wess–Zumino mass terms corresponding to the two chiral multiplets Φ_{C1} and Φ_{C2}, which by (24.71) is given by

$$Mc^{2}\{\text{“}F\text{” term of } (\Phi_{C1}\Phi_{C1} + \Phi_{C2}\Phi_{C2})\}. \tag{24.251}$$

Application of (24.222) and (24.223) shows immediately that this is gauge-invariant, and the same result is true of the Lagrangian density $\mathscr{L}_{0}^{\text{v}}(\mathbf{x})$ of (24.121). However, it is easily seen that the sum of the Wess–Zumino interaction terms corresponding to the two chiral multiplets Φ_{C1} and Φ_{C2}, which by (24.72) is proportional to the "F" term of $\Phi_{C1}\Phi_{C1}\Phi_{C1} + \Phi_{C2}\Phi_{C2}\Phi_{C2}$, is *not* gauge-invariant. Thus the Lagrangian density $\mathscr{L}^{\text{SQED}}(\mathbf{x})$ of super-QED may be taken to be the sum of (24.251) and (24.121) with the "D" term of (24.249).

Although (24.249) contains an infinite number of terms, and so at first sight would appear to be intractable, Wess and Zumino (1974c) observed that, because of the form of the gauge transformations (24.230)–(24.236), the fields $C(\mathbf{x})$, $M(\mathbf{x})$, $N(\mathbf{x})$ and $\chi(\mathbf{x})$ can all be made to be identically *zero* by an appropriate finite gauge transformation. In the resulting gauge, which has subsequently been called the *Wess–Zumino gauge*, all that remain of the multiplet V are the fields $D(\mathbf{x})$, $\boldsymbol{\lambda}(\mathbf{x})$ and $\mathbf{V}(\mathbf{x})$. Moreover, in this gauge only the "D" term of the product $V \cdot V$ is non-zero, its expression being

$$-\sum_{\rho,\kappa=1}^{4} \eta^{\rho\kappa} V_\rho(\mathbf{x}) V_\kappa(\mathbf{x}), \tag{24.252}$$

and all the components of $V \cdot V \cdot V$ and all higher powers of V are zero. Thus in the Wess–Zumino gauge the series in (24.249) contains only a *finite* number of terms, its explicit expansion being

$$\tfrac{1}{2}\hbar c(V^+ + V^-) + g(V^+ \cdot V - V^- \cdot V) + \frac{g^2}{\hbar c}(V^+ \cdot V \cdot V + V^- \cdot V \cdot V),$$

which reduces by (24.241) and (24.242) to

$$\hbar c V^1 + 2g V^2 \cdot V + \frac{2g^2}{\hbar c} V^1 \cdot V \cdot V. \tag{24.253}$$

Thus in the Wess–Zumino gauge the Lagrangian density $\mathscr{L}^{\mathrm{SQED}}(\mathbf{x})$ may be taken to be the sum of (24.251) and (24.121) with the "D" term of (24.253). That is,

$$\mathscr{L}^{\mathrm{SQED}}(\mathbf{x}) = \mathscr{L}^{\mathrm{SQED}}_{\mathrm{free}}(\mathbf{x}) + \mathscr{L}^{\mathrm{SQED}}_{\mathrm{int}}(\mathbf{x}), \tag{24.254}$$

where the "free" part $\mathscr{L}^{\mathrm{SQED}}_{\mathrm{free}}(\mathbf{x})$ is given by

$$\mathscr{L}^{\mathrm{SQED}}_{\mathrm{free}}(\mathbf{x}) = \mathscr{L}^{\mathrm{WZ}}_{\mathrm{KE1}}(\mathbf{x}) + \mathscr{L}^{\mathrm{WZ}}_{\mathrm{KE2}}(\mathbf{x}) + \mathscr{L}^{\mathrm{WZ}}_{\mathrm{mass\,1}}(\mathbf{x}) + \mathscr{L}^{\mathrm{WZ}}_{\mathrm{mass\,2}}(\mathbf{x}) + \mathscr{L}^{\mathrm{v}}_{0}(\mathbf{x}), \tag{24.255}$$

$\mathscr{L}^{\mathrm{WZ}}_{\mathrm{KE1}}(\mathbf{x})$ and $\mathscr{L}^{\mathrm{WZ}}_{\mathrm{KE2}}(\mathbf{x})$ being the kinetic-energy Lagrangian densities of the Wess–Zumino model associated with the chiral multiplets Φ_{C1} and Φ_{C2}, so that

$$\mathscr{L}^{\mathrm{WZ}}_{\mathrm{KE}j}(\mathbf{x}) = \hbar c\{F_j(\mathbf{x})^2 + G_j(\mathbf{x})^2\} + \hbar c \sum_{\rho,\kappa=1}^{4} \eta^{\rho\kappa}\{\partial_\rho A_j(\mathbf{x})\,\partial_\kappa A_j(\mathbf{x}) + \partial_\rho B_j(\mathbf{x})\,\partial_\kappa B_j(\mathbf{x})\}$$

$$+ i\hbar c \sum_{\rho=1}^{4} \tilde{\boldsymbol{\psi}}_j(\mathbf{x})^\dagger \boldsymbol{\gamma}_4 \boldsymbol{\gamma}^\rho\, \partial_\rho \boldsymbol{\psi}_j(\mathbf{x}) \tag{24.256}$$

(cf. (24.70)), $\mathscr{L}^{\mathrm{WZ}}_{\mathrm{mass\,1}}(\mathbf{x})$ and $\mathscr{L}^{\mathrm{WZ}}_{\mathrm{mass\,2}}(\mathbf{x})$ being the corresponding mass terms, so that

$$\mathscr{L}^{\mathrm{WZ}}_{\mathrm{mass}j}(\mathbf{x}) = 2Mc^2\{A_j(\mathbf{x})F_j(\mathbf{x}) + B_j(\mathbf{x})G_j(\mathbf{x}) - \tfrac{1}{2}\tilde{\boldsymbol{\psi}}_j(\mathbf{x})^\dagger \boldsymbol{\gamma}_4 \boldsymbol{\psi}_j(\mathbf{x})\} \tag{24.257}$$

(cf. (24.71)), and $\mathscr{L}^{\mathrm{v}}_{0}(\mathbf{x})$ being the free Lagrangian density of the non-zero fields $D(\mathbf{x})$, $\boldsymbol{\lambda}(\mathbf{x})$ and $\mathbf{V}(\mathbf{x})$ of the general multiplet V (as defined in (24.121)). The part $\mathscr{L}_{\text{int}}^{\text{SQED}}(\mathbf{x})$ describes the interactions and is given by

$$\begin{aligned}\mathscr{L}_{\text{int}}^{\text{SQED}}(\mathbf{x}) = 2g\Bigg\{ & \{A_1(\mathbf{x})B_2(\mathbf{x}) - A_2(\mathbf{x})B_1(\mathbf{x})\}D(\mathbf{x}) \\ & + \tilde{\boldsymbol{\psi}}_2(\mathbf{x})^\dagger\{A_1(\mathbf{x})\mathbf{1} + iB_1(\mathbf{x})\gamma_5\}\gamma_4\boldsymbol{\lambda}(\mathbf{x}) \\ & - \tilde{\boldsymbol{\psi}}_1(\mathbf{x})^\dagger\{A_2(\mathbf{x})\mathbf{1} + iB_2(\mathbf{x})\gamma_5\}\gamma_4\boldsymbol{\lambda}(\mathbf{x}) \\ & - \sum_{\rho,\kappa=1}^{4} \eta^{\rho\kappa}\{A_1(\mathbf{x})\,\partial_\rho A_2(\mathbf{x}) - A_2(\mathbf{x})\,\partial_\rho A_1(\mathbf{x}) \\ & + B_1(\mathbf{x})\,\partial_\rho B_2(\mathbf{x}) - B_2(\mathbf{x})\,\partial_\rho B_1(\mathbf{x})\}\,V_\kappa(\mathbf{x}) \\ & - i\sum_{\rho=1}^{4} \tilde{\boldsymbol{\psi}}_1(\mathbf{x})^\dagger\gamma_4\gamma^\rho\boldsymbol{\psi}_2(\mathbf{x})V_\rho(\mathbf{x})\Bigg\} \\ & - \frac{g^2}{\hbar c}\{A_1(\mathbf{x})^2 + B_1(\mathbf{x})^2 + A_2(\mathbf{x})^2 + B_2(\mathbf{x})^2\} \\ & \times \sum_{\rho,\kappa=1}^{4} \eta^{\rho\kappa}V_\rho(\mathbf{x})V_\kappa(\mathbf{x}). \end{aligned} \tag{24.258}$$

(Equations (24.129), (24.133), (24.135) and (24.138) have been used in deriving (24.258) from the "D" component of the last two terms of (24.253).)

After eliminating the auxiliary fields $F_1(\mathbf{x})$, $G_1(\mathbf{x})$, $F_2(\mathbf{x})$, $G_2(\mathbf{x})$ and $D(\mathbf{x})$ by using their Euler–Lagrange equations (which are again purely algebraic in form), the "dynamical" fields that remain are as follows:

(i) the field $\boldsymbol{\psi}(\mathbf{x})$ of the electron, which is assumed to have mass M,

(ii) the two complex fields $A_1(\mathbf{x}) + iB_1(\mathbf{x})$ and $A_2(\mathbf{x}) + iB_2(\mathbf{x})$ of the "electron's supersymmetric partners", which are often called *sleptons*, and which also have mass M, but are bosonic;

(iii) the field $\mathbf{V}(\mathbf{x})$ of the photon, which is massless; and

(iv) the field $\boldsymbol{\lambda}(\mathbf{x})$ of the "photon's supersymmetric partner", which is often referred to as a *photino*, and which is also massless, but is fermionic.

It should be noted that the penultimate term of (24.258) reduces on using (24.212) to the usual term that describes electron–photon interactions (cf. (24.208)).

The Lagrangian density of (24.254) is not itself invariant under the gauge transformations (24.222)–(24.224) because it has been derived in a particular choice of gauge. However, there remain some residual combined gauge and supersymmetry transformations under which it is invariant. For details see Wess and Zumino (1974c).

Although $D(\mathbf{x})$ is only an auxiliary field, it was shown by Fayet and Iliopoulos (1974) that it plays as important role when spontaneous breaking of supersymmetry is considered (cf. Section 5). Indeed, it is possible to add to the Lagrangian density an additional *Fayet–Iliopoulos term* $\mathscr{L}_{\mathrm{FI}}(\mathbf{x})$ of the form

$$\mathscr{L}_{\mathrm{FI}}(\mathbf{x}) = \xi D(\mathbf{x}), \tag{24.259}$$

where ξ is a constant. This is gauge-invariant by virtue of (24.231).

(b) Super-QED in a superfield formulation

As the first stage in setting up the superfield formulation of general local gauge theories, the super-QED theory of the previous subsection will be rewritten in superfield language.

The first step consists of forming appropriate superfields from each of the multiplets in the following way.

(i) From the real chiral multiplet $\Lambda = \{\alpha(\mathbf{x}), \beta(\mathbf{x}), \omega(\mathbf{x}), \xi(\mathbf{x}); \boldsymbol{\phi}(\mathbf{x})\}$ form the left-hand and right-handed chiral superfields

$$\Lambda_{\mathrm{L}}(\mathbf{X};\boldsymbol{\Theta}) = \alpha_{\mathrm{L}}(\mathbf{Z}_{\mathrm{L}}) - 2i \sum_{A,B=1}^{2} \Theta_A \varepsilon_{AB} \phi_{LB}(\mathbf{Z}_{\mathrm{L}}) - i\omega_{\mathrm{L}}(\mathbf{Z}_{\mathrm{L}}) \sum_{A,B=1}^{2} \Theta_A \varepsilon_{AB} \Theta_B \tag{24.260}$$

and

$$\Lambda_{\mathrm{R}}(\mathbf{X};\boldsymbol{\Theta}) = \alpha_{\mathrm{R}}(\mathbf{Z}_{\mathrm{R}}) - 2i \sum_{A,B=1}^{2} \bar{\Theta}_A \varepsilon_{AB} \phi_{RB}(\mathbf{Z}_{\mathrm{R}}) - i\omega_{\mathrm{R}}(\mathbf{Z}_{\mathrm{R}}) \sum_{A,B=1}^{2} \bar{\Theta}_A \varepsilon_{AB} \bar{\Theta}_B \tag{24.261}$$

(cf. (24.175) and (24.178)), where $\mathbf{Z}_{\mathrm{L}}$ and $\mathbf{Z}_{\mathrm{R}}$ are defined in (24.174) and (24.177) respectively, and the component functions are defined in terms of those of the real chiral multiplet Λ by

$$\alpha_{\mathrm{L}}(\mathbf{x}) = \alpha(\mathbf{x}) - i\beta(\mathbf{x}), \quad \alpha_{\mathrm{R}}(\mathbf{x}) = \alpha(\mathbf{x}) + i\beta(\mathbf{x}), \tag{24.262}$$

$$\omega_{\mathrm{L}}(\mathbf{x}) = \omega(\mathbf{x}) + i\xi(\mathbf{x}), \quad \omega_{\mathrm{R}}(\mathbf{x}) = \omega(\mathbf{x}) - i\xi(\mathbf{x}), \tag{24.263}$$

$$\phi_{\mathrm{L}A}(\mathbf{x}) = \phi_A^{\mathrm{C}'}(\mathbf{x}), \quad \phi_{\mathrm{R}A}(\mathbf{x}) = \phi_{A+2}^{\mathrm{C}'}(\mathbf{x}) \tag{24.264}$$

(for $A = 1, 2$), where C' indicates that the components of $\boldsymbol{\phi}(\mathbf{x})$ are taken to be those of the modified chiral representation of (23.25). Moreover,

$$\Lambda_{\mathrm{L}}(\mathbf{X};\boldsymbol{\Theta})^{\ddagger} = \Lambda_{\mathrm{R}}(\mathbf{X};\boldsymbol{\Theta}), \quad \Lambda_{\mathrm{R}}(\mathbf{X};\boldsymbol{\Theta})^{\ddagger} = \Lambda_{\mathrm{L}}(\mathbf{X};\boldsymbol{\Theta}) \tag{24.265}$$

(cf. (24.180)).

(ii) From the two real chiral multiplets $\Phi_{\mathrm{C}1} = \{A_1(\mathbf{x}), B_1(\mathbf{x}), F_1(\mathbf{x}), G_1(\mathbf{x}); \boldsymbol{\psi}_1(\mathbf{x})\}$ and $\Phi_{\mathrm{C}2} = \{A_2(\mathbf{x}), B_2(\mathbf{x}), F_2(\mathbf{x}), G_2(\mathbf{x}); \boldsymbol{\psi}_2(\mathbf{x})\}$ form the pair of left-hand

and right-handed chiral superfields

$$\Phi_{\mathrm{L}j}(\mathbf{X};\boldsymbol{\Theta}) = A_{\mathrm{L}j}(\mathbf{Z}_{\mathrm{L}}) - 2i \sum_{A,B=1}^{2} \Theta_A \varepsilon_{AB} \psi_{\mathrm{L}jB}(\mathbf{Z}_{\mathrm{L}}) - iF_{\mathrm{L}j}(\mathbf{Z}_{\mathrm{L}}) \sum_{A,B=1}^{2} \Theta_A \varepsilon_{AB} \Theta_B \tag{24.266}$$

and

$$\Phi_{\mathrm{R}j}(\mathbf{X};\boldsymbol{\Theta}) = A_{\mathrm{R}j}(\mathbf{Z}_{\mathrm{R}}) - 2i \sum_{A,B=1}^{2} \bar{\Theta}_A \varepsilon_{AB} \psi_{\mathrm{R}jB}(\mathbf{Z}_{\mathrm{R}}) - iF_{\mathrm{R}j}(\mathbf{Z}_{\mathrm{R}}) \sum_{A,B=1}^{2} \bar{\Theta}_A \varepsilon_{AB} \bar{\Theta}_B \tag{24.267}$$

for $j = 1, 2$ (cf. (24.175) and (24.178)), where the component functions are defined for $j = 1, 2$ in terms of those of the real chiral multiplets $\Phi_{\mathrm{C}j}$ by

$$A_{\mathrm{L}j}(\mathbf{x}) = A_j(\mathbf{x}) - iB_j(\mathbf{x}), \quad A_{\mathrm{R}j}(\mathbf{x}) = A_j(\mathbf{x}) + iB_j(\mathbf{x}), \tag{24.268}$$

$$F_{\mathrm{L}j}(\mathbf{x}) = F_j(\mathbf{x}) + iG_j(\mathbf{x}), \quad F_{\mathrm{R}j}(\mathbf{x}) = F_j(\mathbf{x}) - iG_j(\mathbf{x}), \tag{24.269}$$

$$\psi_{\mathrm{L}jA}(\mathbf{x}) = \psi_{jA}^{\mathrm{C}'}(\mathbf{x}), \qquad \psi_{\mathrm{R}jA}(\mathbf{x}) = \psi_{jA+2}^{\mathrm{C}'}(\mathbf{x}) \tag{24.270}$$

(for $A = 1, 2$), where again C′ indicates that the components of $\boldsymbol{\psi}_j(\mathbf{x})$ are taken to be those of the modified chiral representation of (23.25). Then

$$\Phi_{\mathrm{L}j}(\mathbf{X};\boldsymbol{\Theta})^{\ddagger} = \Phi_{\mathrm{R}j}(\mathbf{X};\boldsymbol{\Theta}), \quad \Phi_{\mathrm{R}j}(\mathbf{X};\boldsymbol{\Theta})^{\ddagger} = \Phi_{\mathrm{L}j}(\mathbf{X};\boldsymbol{\Theta}) \tag{24.271}$$

(for $j = 1, 2$) (cf. (24.180)). It is convenient to form from these the "complex" chiral superfields $\Phi_{\mathrm{L}}(\mathbf{X};\boldsymbol{\Theta})$ and $\Phi_{\mathrm{R}}(\mathbf{X};\boldsymbol{\Theta})$ by the definitions

$$\Phi_{\mathrm{L}}(\mathbf{X};\boldsymbol{\Theta}) = \Phi_{\mathrm{L}1}(\mathbf{X};\boldsymbol{\Theta}) + i\Phi_{\mathrm{L}2}(\mathbf{X};\boldsymbol{\Theta}) \tag{24.272}$$

and

$$\Phi_{\mathrm{R}}(\mathbf{X};\boldsymbol{\Theta}) = \Phi_{\mathrm{R}1}(\mathbf{X};\boldsymbol{\Theta}) + i\Phi_{\mathrm{R}2}(\mathbf{X};\boldsymbol{\Theta}). \tag{24.273}$$

Then, by (24.271),

$$\Phi_{\mathrm{L}}(\mathbf{X};\boldsymbol{\Theta})^{\ddagger} = \Phi_{\mathrm{R}1}(\mathbf{X};\boldsymbol{\Theta}) - i\Phi_{\mathrm{R}2}(\mathbf{X};\boldsymbol{\Theta}) \neq \Phi_{\mathrm{R}}(\mathbf{X};\boldsymbol{\Theta}), \tag{24.274}$$

and similarly

$$\Phi_{\mathrm{R}}(\mathbf{X};\boldsymbol{\Theta})^{\ddagger} = \Phi_{\mathrm{L}1}(\mathbf{X};\boldsymbol{\Theta}) - i\Phi_{\mathrm{L}2}(\mathbf{X};\boldsymbol{\Theta}) \neq \Phi_{\mathrm{L}}(\mathbf{X};\boldsymbol{\Theta}). \tag{24.275}$$

(iii) From the real general multiplet $V = \{C(\mathbf{x}), D(\mathbf{x}), M(\mathbf{x}), N(\mathbf{x}); \boldsymbol{\chi}(\mathbf{x}), \boldsymbol{\lambda}(\mathbf{x}); \mathbf{V}(\mathbf{x})\}$ form the real scalar superfield $V(\mathbf{X};\boldsymbol{\Theta})$ by the definition

$$\begin{aligned} V(\mathbf{X};\boldsymbol{\Theta}) = {} & C(\mathbf{X}) - i(\tilde{\boldsymbol{\Theta}}^{\#}\gamma_4\gamma_5\boldsymbol{\chi}(\mathbf{X})) - \tfrac{1}{2}i(\tilde{\boldsymbol{\Theta}}^{\#}\gamma_4\gamma_5\boldsymbol{\Theta})M(\mathbf{X}) + \tfrac{1}{2}(\tilde{\boldsymbol{\Theta}}^{\#}\gamma_4\boldsymbol{\Theta})N(\mathbf{X}) \\ & - \frac{1}{2}\sum_{\rho=1}^{4}(\tilde{\boldsymbol{\Theta}}^{\#}\gamma_4\gamma_5\gamma^{\rho}\boldsymbol{\Theta})V_{\rho}(\mathbf{X}) - (\tilde{\boldsymbol{\Theta}}^{\#}\gamma_4\boldsymbol{\Theta})\sum_{\rho=1}^{4}(\tilde{\boldsymbol{\Theta}}^{\#}\gamma_4\gamma_5\gamma^{\rho}\,\partial_{\rho}\boldsymbol{\chi}(\mathbf{X})) \\ & - i(\tilde{\boldsymbol{\Theta}}^{\#}\gamma_4\boldsymbol{\Theta})(\tilde{\boldsymbol{\Theta}}^{\#}\gamma_4\gamma_5\boldsymbol{\lambda}(\mathbf{X})) \\ & + \tfrac{1}{4}(\tilde{\boldsymbol{\Theta}}^{\#}\gamma_4\boldsymbol{\Theta})(\tilde{\boldsymbol{\Theta}}^{\#}\gamma_4\boldsymbol{\Theta})\{D(\mathbf{X}) - \Box C(\mathbf{X})\} \end{aligned} \tag{24.276}$$

(cf. (24.151)).

The second step is to rewrite the gauge transformations (24.222) and (24.223) in superfield form, where they become

$$\delta_{\text{gauge}}\Phi_{L1}(\mathbf{X};\mathbf{\Theta}) = -\Lambda_L(\mathbf{X};\mathbf{\Theta})\Phi_{L2}(\mathbf{X};\mathbf{\Theta}), \tag{24.277}$$

$$\delta_{\text{gauge}}\Phi_{R1}(\mathbf{X};\mathbf{\Theta}) = -\Lambda_R(\mathbf{X};\mathbf{\Theta})\Phi_{R2}(\mathbf{X};\mathbf{\Theta}), \tag{24.278}$$

$$\delta_{\text{gauge}}\Phi_{L2}(\mathbf{X};\mathbf{\Theta}) = \Lambda_L(\mathbf{X};\mathbf{\Theta})\Phi_{L1}(\mathbf{X};\mathbf{\Theta}), \tag{24.279}$$

$$\delta_{\text{gauge}}\Phi_{R2}(\mathbf{X};\mathbf{\Theta}) = \Lambda_R(\mathbf{X};\mathbf{\Theta})\Phi_{R1}(\mathbf{X};\mathbf{\Theta}). \tag{24.280}$$

These can be written more simply in terms of the "complex" chiral superfields $\Phi_L(\mathbf{X};\mathbf{\Theta})$ and $\Phi_R(\mathbf{X};\mathbf{\Theta})$ of (24.272) and (24.273), since then they just become

$$\delta_{\text{gauge}}\Phi_L(\mathbf{X};\mathbf{\Theta}) = i\Lambda_L(\mathbf{X};\mathbf{\Theta})\Phi_L(\mathbf{X};\mathbf{\Theta}) \tag{24.281}$$

and

$$\delta_{\text{gauge}}\Phi_R(\mathbf{X};\mathbf{\Theta}) = i\Lambda_R(\mathbf{X};\mathbf{\Theta})\Phi_R(\mathbf{X};\mathbf{\Theta}), \tag{24.282}$$

which clearly correspond to the "finite" gauge transformations

$$\Phi_L(\mathbf{X};\mathbf{\Theta}) \to \Phi'_L(\mathbf{X};\mathbf{\Theta}) = \Phi_L(\mathbf{X};\mathbf{\Theta})\exp\{i\Lambda_L(\mathbf{X};\mathbf{\Theta})\} \tag{24.283}$$

and

$$\Phi_R(\mathbf{X};\mathbf{\Theta}) \to \Phi'_R(\mathbf{X};\mathbf{\Theta}) = \Phi_R(\mathbf{X};\mathbf{\Theta})\exp\{i\Lambda_R(\mathbf{X};\mathbf{\Theta})\} \tag{24.284}$$

(which are the supersymmetric generalizations of (24.209)). However, by (24.274), (24.277) and (24.278),

$$\delta_{\text{gauge}}\Phi_L(\mathbf{X};\mathbf{\Theta})^{\ddagger} = -i\Lambda_R(\mathbf{X};\mathbf{\Theta})\Phi_L(\mathbf{X};\mathbf{\Theta})^{\ddagger}, \tag{24.285}$$

and similarly

$$\delta_{\text{gauge}}\Phi_R(\mathbf{X};\mathbf{\Theta})^{\ddagger} = -i\Lambda_L(\mathbf{X};\mathbf{\Theta})\Phi_R(\mathbf{X};\mathbf{\Theta})^{\ddagger}, \tag{24.286}$$

for which the corresponding "finite" gauge transformations are

$$\Phi_L(\mathbf{X};\mathbf{\Theta})^{\ddagger} \to \Phi'_L(\mathbf{X};\mathbf{\Theta})^{\ddagger} = \Phi_L(\mathbf{X};\mathbf{\Theta})^{\ddagger}\exp\{-i\Lambda_R(\mathbf{X};\mathbf{\Theta})\} \tag{24.287}$$

and

$$\Phi_R(\mathbf{X};\mathbf{\Theta})^{\ddagger} \to \Phi'_R(\mathbf{X};\mathbf{\Theta})^{\ddagger} = \Phi_R(\mathbf{X};\mathbf{\Theta})^{\ddagger}\exp\{-i\Lambda_L(\mathbf{X};\mathbf{\Theta})\}. \tag{24.288}$$

Turning first to the sum of the chiral multiplet kinetic-energy terms in the action, (24.199) shows that this can be written in superfield form as

$$\begin{aligned} S^{WZ}_{KE1} + S^{WZ}_{KE2} = &-\frac{1}{4\hbar c}\int d^4X \int d^2\Theta \int d^2\bar{\Theta}\, \Phi_{R1}(\mathbf{X};\mathbf{\Theta})\Phi_{L1}(\mathbf{X};\mathbf{\Theta}) \\ &-\frac{1}{4\hbar c}\int d^4X \int d^2\Theta \int d^2\bar{\Theta}\, \Phi_{R2}(\mathbf{X};\mathbf{\Theta})\Phi_{L2}(\mathbf{X};\mathbf{\Theta}). \end{aligned} \tag{24.289}$$

This can be reexpressed using (24.274) and (24.275) as

$$S_{\mathrm{KE1}}^{\mathrm{WZ}} + S_{\mathrm{KE2}}^{\mathrm{WZ}} = -\frac{1}{8\hbar c}\int \mathrm{d}^4X \int \mathrm{d}^2\Theta \int \mathrm{d}^2\bar{\Theta}\, \Phi_{\mathrm{L}}(\mathbf{X};\boldsymbol{\Theta})^{\ddagger}\Phi_{\mathrm{L}}(\mathbf{X};\boldsymbol{\Theta}) - \frac{1}{8\hbar c}\int \mathrm{d}^4X \int \mathrm{d}^2\Theta \int \mathrm{d}^2\bar{\Theta}\, \Phi_{\mathrm{R}}(\mathbf{X};\boldsymbol{\Theta})^{\ddagger}\Phi_{\mathrm{R}}(\mathbf{X};\boldsymbol{\Theta}). \quad (24.290)$$

Under the gauge transformations (24.283), (24.284), (24.287) and (24.288), the "integrands" of (24.290) transform as

$$\Phi_{\mathrm{L}}(\mathbf{X};\boldsymbol{\Theta})^{\ddagger}\Phi_{\mathrm{L}}(\mathbf{X};\boldsymbol{\Theta}) \to \Phi_{\mathrm{L}}(\mathbf{X};\boldsymbol{\Theta})^{\ddagger} \exp\{-i\Lambda_{\mathrm{R}}(\mathbf{X};\boldsymbol{\Theta})\} \exp\{i\Lambda_{\mathrm{L}}(\mathbf{X};\boldsymbol{\Theta})\}\, \Phi_{\mathrm{L}}(\mathbf{X};\boldsymbol{\Theta})$$

and

$$\Phi_{\mathrm{R}}(\mathbf{X};\boldsymbol{\Theta})^{\ddagger}\Phi_{\mathrm{R}}(\mathbf{X};\boldsymbol{\Theta}) \to \Phi_{\mathrm{R}}(\mathbf{X};\boldsymbol{\Theta})^{\ddagger} \exp\{-i\Lambda_{\mathrm{L}}(\mathbf{X};\boldsymbol{\Theta})\} \exp\{i\Lambda_{\mathrm{R}}(\mathbf{X};\boldsymbol{\Theta})\}\, \Phi_{\mathrm{R}}(\mathbf{X};\boldsymbol{\Theta}),$$

and so they are not invariant. However, the combinations

$$\Phi_{\mathrm{L}}(\mathbf{X};\boldsymbol{\Theta})^{\ddagger} \exp\left\{\frac{2g}{\hbar c} V(\mathbf{X};\boldsymbol{\Theta})\right\} \Phi_{\mathrm{L}}(\mathbf{X};\boldsymbol{\Theta}) \quad (24.291)$$

and

$$\Phi_{\mathrm{R}}(\mathbf{X};\boldsymbol{\Theta})^{\ddagger} \exp\left\{-\frac{2g}{\hbar c} V(\mathbf{X};\boldsymbol{\Theta})\right\} \Phi_{\mathrm{R}}(\mathbf{X};\boldsymbol{\Theta}) \quad (24.292)$$

are invariant under the gauge transformations (24.283), (24.284), (24.287) and (24.288) provided that $V(\mathbf{X};\boldsymbol{\Theta})$ transforms as

$$\exp\left\{\frac{2g}{\hbar c} V(\mathbf{X};\boldsymbol{\Theta})\right\} \to \exp\{i\Lambda_{\mathrm{R}}(\mathbf{X};\boldsymbol{\Theta})\} \exp\left\{\frac{2g}{\hbar c} V(\mathbf{X};\boldsymbol{\Theta})\right\} \exp\{-i\Lambda_{\mathrm{L}}(\mathbf{X};\boldsymbol{\Theta})\}. \quad (24.293)$$

To first order in $V(\mathbf{X};\boldsymbol{\Theta})$, $\Lambda_{\mathrm{L}}(\mathbf{X};\boldsymbol{\Theta})$ and $\Lambda_{\mathrm{R}}(\mathbf{X};\boldsymbol{\Theta})$, this reduces to

$$V(\mathbf{X};\boldsymbol{\Theta}) \to V(\mathbf{X};\boldsymbol{\Theta}) + \frac{i\hbar c}{2g}\{\Lambda_{\mathrm{R}}(\mathbf{X};\boldsymbol{\Theta}\} - \Lambda_{\mathrm{L}}(\mathbf{X};\Theta)\},$$

that is, to

$$\delta_{\mathrm{gauge}} V(\mathbf{X};\boldsymbol{\Theta}) = \frac{i\hbar c}{2g}\{\Lambda_{\mathrm{R}}(\mathbf{X};\boldsymbol{\Theta}) - \Lambda_{\mathrm{L}}(\mathbf{X};\boldsymbol{\Theta})\}. \quad (24.294)$$

On equating the component functions of each of the products of the odd Grassmann variables, the gauge transformations (24.230)–(24.236) are recovered. Thus (24.294) is the superfield analogue of (24.224), and consequently in the Wess–Zumino gauge all the component functions of $V(\mathbf{X};\boldsymbol{\Theta})$ can be taken to be identically zero except for $\mathbf{V}(\mathbf{X})$, $\boldsymbol{\lambda}(\mathbf{X})$ and $D(\mathbf{X})$, and

$$V(\mathbf{X};\boldsymbol{\Theta})^n = 0 \quad (24.295)$$

for $n \geqslant 3$. Moreover, the superfield version of the gauge-invariant and supersymmetric term in the action that incorporates both the kinetic-energy terms and the interactions between the matter fields and the gauge fields is

$$S_{\mathrm{KE+int}} = -\frac{1}{8\hbar c}\int \mathrm{d}^4X \int \mathrm{d}^2\Theta \int \mathrm{d}^2\bar{\Theta}\, \Phi_{\mathrm{L}}(\mathbf{X};\mathbf{\Theta})^{\ddagger} \exp\left\{\frac{2g}{\hbar c}V(\mathbf{X};\mathbf{\Theta})\right\}\Phi_{\mathrm{L}}(\mathbf{X};\mathbf{\Theta})$$
$$-\frac{1}{8\hbar c}\int \mathrm{d}^4X \int \mathrm{d}^2\Theta \int \mathrm{d}^2\bar{\Theta}\, \Phi_{\mathrm{R}}(\mathbf{X};\mathbf{\Theta})^{\ddagger} \exp\left\{-\frac{2g}{\hbar c}V(\mathbf{X};\mathbf{\Theta})\right\}\Phi_{\mathrm{R}}(\mathbf{X};\mathbf{\Theta}). \tag{24.296}$$

A similar but simpler argument applies to the mass terms. By (24.200) and using (24.274) and (24.275), the sum of the chiral multiplet mass terms in the action can be written in superfield form as

$$S^{\mathrm{WZ}}_{\mathrm{mass\,1}} + S^{\mathrm{WZ}}_{\mathrm{mass\,2}} = \tfrac{1}{4}iMc^2 \int \mathrm{d}^4X \int \mathrm{d}^2\Theta\, \Phi_{\mathrm{R}}(\mathbf{X};\mathbf{\Theta})^{\ddagger}\Phi_{\mathrm{L}}(\mathbf{X};\mathbf{\Theta})$$
$$+\tfrac{1}{4}iMc^2 \int \mathrm{d}^4X \int \mathrm{d}^2\bar{\Theta}\, \Phi_{\mathrm{L}}(\mathbf{X};\mathbf{\Theta})^{\ddagger}\Phi_{\mathrm{R}}(\mathbf{X};\mathbf{\Theta}), \tag{24.297}$$

which is obviously invariant under the gauge transformations (24.283), (24.284), (24.287) and (24.288).

The final stage is to write down the action for the fields in $V(\mathbf{X};\mathbf{\Theta})$ as a superspace integral. This can be done by introducing a pair of left-handed chiral superfields $W_{\mathrm{L}A}(\mathbf{X};\mathbf{\Theta})$ by the definition

$$W_{\mathrm{L}A}(\mathbf{X};\mathbf{\Theta}) = \bar{\mathrm{D}}^2\mathrm{D}_A V(\mathbf{X};\mathbf{\Theta}) \tag{24.298}$$

(for $A = 1, 2$), where D_A and $\bar{\mathrm{D}}^2$ are defined in (24.168) and (24.182) respectively. (That these are left-handed chiral superfields follows immediately from (24.184).) Then

$$\delta_{\mathrm{gauge}} W_{\mathrm{L}A}(\mathbf{X};\mathbf{\Theta}) = 0; \tag{24.299}$$

that is, these superfields themselves are invariant under the gauge transformation (24.294). (This follows because (24.294) and (24.298) imply that

$$\delta_{\mathrm{gauge}} W_{\mathrm{L}A}(\mathbf{X};\mathbf{\Theta}) = \frac{i\hbar c}{2g}\bar{\mathrm{D}}^2\mathrm{D}_A\{\Lambda_{\mathrm{R}}(\mathbf{X};\mathbf{\Theta}) - \Lambda_{\mathrm{L}}(\mathbf{X};\mathbf{\Theta})\},$$

but $\mathrm{D}_A\Lambda_{\mathrm{R}}(\mathbf{X};\mathbf{\Theta}) = 0$ by (24.170), and by using (24.165) in conjunction with (23.25) the remaining expression can be reduced to an operator that acts on a term of the form $\bar{\mathrm{D}}_A\Lambda_{\mathrm{L}}(\mathbf{X};\mathbf{\Theta})$, this latter term being zero by virtue of (24.171).) The desired gauge-invariant action is then provided by

$$S^{\mathrm{v}}_0 = \tfrac{1}{32}\,\mathrm{Re} \int \mathrm{d}^4X \int \mathrm{d}^2\Theta \sum_{A,B=1}^{2} W_{\mathrm{L}A}(\mathbf{X};\mathbf{\Theta})\varepsilon_{AB}W_{\mathrm{L}B}(\mathbf{X};\mathbf{\Theta}), \tag{24.300}$$

which corresponds to the free Lagrangian density $\mathscr{L}_0^{\mathrm{v}}(\mathbf{x})$ of the non-zero fields $D(\mathbf{x})$, $\boldsymbol{\lambda}(\mathbf{x})$ and $\mathbf{V}(\mathbf{x})$ of the general multiplet V that was defined in (24.121).

Finally it may be noted that the superfield action corresponding to the Fayet–Iliopoulos Lagrangian density term $\mathscr{L}_{\mathrm{FI}}(\mathbf{x})$ of (24.259) is simply

$$S_{\mathrm{FI}} = \xi \int \mathrm{d}^4 X \int \mathrm{d}^2\Theta \int \mathrm{d}^2\bar{\Theta}\; V(\mathbf{X};\boldsymbol{\Theta}). \tag{24.301}$$

(c) *Supersymmetric Yang–Mills theories in a superfield formulation*

It is quite easy to generalize the superfield formulation of the supersymmetric Abelian gauge theory that was given in the previous subsection to the non-Abelian situation. This was first done by Ferrara and Zumino (1974).

Let $\mathscr{G}$ be any simply connected compact Lie group and suppose that $n_{\mathscr{G}}$ is its dimension. Then, as noted in Chapter 15, Section 6, $\mathscr{G}$ is linear and its real Lie algebra $\mathscr{L}$ is Abelian or semi-simple, or is isomorphic to a direct sum of such algebras. For convenience the theory will be developed in detail on the assumption that $\mathscr{L}$ contains no invariant subalgebras, which implies that $\mathscr{L}$ is either simple or is a one-dimensional Abelian Lie algebra. The complications involved in removing this restriction will be mentioned briefly at the end of this subsection.

There is no loss in generality in working with a faithful finite-dimensional representation of $\mathscr{G}$ and thence of $\mathscr{L}$. Suppose that in this representation $\mathbf{a}_I$ (for $I = 1, 2, \ldots, n_{\mathscr{G}}$) form a basis for $\mathscr{L}$, the choice of which will be deferred until later in the discussion. A typical element of $\mathscr{G}$ will be denoted by $\mathbf{u}$, so that there exists a set of $n_{\mathscr{G}}$ real numbers α_I (for $I = 1, 2, \ldots, n_{\mathscr{G}}$) such that

$$\mathbf{u} \equiv \mathbf{u}(\alpha_1, \alpha_2, \ldots) = \exp\left\{ \sum_{I=1}^{n_{\mathscr{G}}} \alpha_I \mathbf{a}_I \right\} \tag{24.302}$$

(cf. (19.12)). In "ordinary" local gauge theory these transformations are replaced by transformations of the form

$$\mathbf{u}(\mathbf{x}) \equiv \mathbf{u}(\alpha_1(\mathbf{x}), \alpha_2(\mathbf{x}), \ldots) = \exp\left\{ \sum_{I=1}^{n_{\mathscr{G}}} \alpha_I(\mathbf{x}) \mathbf{a}_I \right\}, \tag{24.303}$$

where $\alpha_I(\mathbf{x})$ (for $I = 1, 2, \ldots, n_{\mathscr{G}}$) are $n_{\mathscr{G}}$ arbitrary real dimensionless, differentiable functions of the space–time coordinates $\mathbf{x} = (x^1, x^2, x^3, x^4)$ (cf. (19.16)).

In the supersymmetric Abelian theory of the previous subsection the first step consisted of setting up a real chiral multiplet $\Lambda = \{\alpha(\mathbf{x}), \beta(\mathbf{x}), \omega(\mathbf{x}), \xi(\mathbf{x}); \boldsymbol{\phi}(\mathbf{x})\}$ and forming from it a real left-handed chiral superfield $\Lambda_{\mathrm{L}}(\mathbf{X};\boldsymbol{\Theta})$ and a real right-handed chiral superfield $\Lambda_R(\mathbf{X};\boldsymbol{\Theta})$ by means of the definitions (24.260)–(24.264). The natural generalization of this for a Lie group $\mathscr{G}$ of dimension $n_{\mathscr{G}}$ is to set up $n_{\mathscr{G}}$ real chiral multiplets $\Lambda_I = \{\alpha_I(\mathbf{x}), \beta_I(\mathbf{x}), \omega_I(\mathbf{x}), \xi_I(\mathbf{x}); \boldsymbol{\phi}_I(\mathbf{x})\}$ and form from them $n_{\mathscr{G}}$ real left-handed chiral superfields $\Lambda_{LI}(\mathbf{X};\boldsymbol{\Theta})$ and $n_{\mathscr{G}}$ real right-handed chiral superfields $\Lambda_{RI}(\mathbf{X};\boldsymbol{\Theta})$ by merely inserting subscript I everywhere in the obvious way. From these can be formed the combinations

$$\Lambda_L(\mathbf{X};\boldsymbol{\Theta}) = \sum_{I=1}^{n_{\mathscr{G}}} \Lambda_{LI}(\mathbf{X};\boldsymbol{\Theta})\mathbf{a}_I \tag{24.304}$$

and

$$\Lambda_R(\mathbf{X};\boldsymbol{\Theta}) = \sum_{I=1}^{n_{\mathscr{G}}} \Lambda_{RI}(\mathbf{X};\boldsymbol{\Theta})\mathbf{a}_I. \tag{24.305}$$

Although these are essentially matrices in which each component is a chiral superfield, for convenience they will be referred to as being "Lie-algebra-valued", for they will play the role taken previously by the Lie-algebra-valued field

$$\sum_{I=1}^{n_{\mathscr{G}}} \alpha_I(\mathbf{x})\mathbf{a}_I$$

that appears in the exponent of (24.303). The generalizations of (24.265) are

$$\Lambda_{LI}(\mathbf{X};\boldsymbol{\Theta})^{\ddagger} = \Lambda_{RI}(\mathbf{X};\boldsymbol{\Theta}), \quad \Lambda_{RI}(\mathbf{X};\boldsymbol{\Theta})^{\ddagger} = \Lambda_{LI}(\mathbf{X};\boldsymbol{\Theta}) \tag{24.306}$$

(for $I = 1, 2, \ldots, n_{\mathscr{G}}$).

The next stage is to deal with the matter fields. In an "ordinary" local gauge theory of scalar bosons these consist of d scalar fields $\phi_1(\mathbf{x})$, $\phi_2(\mathbf{x}), \ldots, \phi_d(\mathbf{x})$ (which are either all real or are all complex) that are assumed to transforms as

$$\phi_j(\mathbf{x}) \to \phi'_j(\mathbf{x}) = \sum_{k=1}^{d} \Gamma(\mathbf{u}(\mathbf{x}))_{kj}\phi_k(\mathbf{x}) \tag{24.307}$$

(for $j = 1, 2, \ldots, d$), where Γ is a d-dimensional unitary representation of $\mathscr{G}$ (cf. (19.14) and (19.21)), the transformations for spinors being similar (cf. (19.41)). The first step in formulating the supersymmetric generalization is to replace each of these scalar fields by a chiral multiplet Φ_{Cj}, where $\Phi_{Cj} = \{A_j(\mathbf{x}), B_j(\mathbf{x}), F_j(\mathbf{x}), G_j(\mathbf{x}); \boldsymbol{\psi}_j(\mathbf{x})\}$ (for $j = 1, \ldots, d$). The next step is to form d left-handed chiral superfields $\Phi_{Lj}(\mathbf{X};\boldsymbol{\Theta})$ and d right-handed chiral superfields $\Phi_{Lj}(\mathbf{X};\boldsymbol{\Theta})$ from these using (24.266)–(24.270). Their gauge transformations may be assumed to be

$$\Phi_{Lj}(\mathbf{X};\boldsymbol{\Theta}) \to \Phi'_{Lj}(\mathbf{X};\boldsymbol{\Theta}) = \sum_{k=1}^{d} \Gamma(\exp \Lambda_L(\mathbf{X};\boldsymbol{\Theta}))_{kj}\Phi_{Lk}(\mathbf{X};\boldsymbol{\Theta}) \tag{24.308}$$

and

$$\Phi_{Rj}(\mathbf{X};\boldsymbol{\Theta}) \to \Phi'_{Rj}(\mathbf{X};\boldsymbol{\Theta}) = \sum_{k=1}^{d} \Gamma(\exp \Lambda_R(\mathbf{X};\boldsymbol{\Theta}))_{kj}\Phi_{Rk}(\mathbf{X};\boldsymbol{\Theta}) \quad (24.309)$$

(for $j = 1, 2, \ldots, d$), which generalize both the gauge transformation (24.307) of the non-Abelian non-supersymmetric case and the gauge transformations (24.283) and (24.284) of the Abelian supersymmetric case.

If $\boldsymbol{\Gamma}$ is a *real* representation of $\mathcal{G}$ then all the chiral multiplets Φ_{Cj} may be taken to be real. If $\boldsymbol{\Gamma}$ is not a *real* representation then there are two ways of proceeding. One is just to take all the chiral multiplets Φ_{Cj} to be complex. The other is to form a real representation from $\boldsymbol{\Gamma}$ of twice the dimension of $\boldsymbol{\Gamma}$ and proceed with real chiral multiplets. Thus, for example, the Abelian case of the previous subsection can be recovered by assuming that $d = 1$, that

$$\boldsymbol{\Gamma}(\mathbf{a}_1) = [i], \quad (24.310)$$

and by taking the chiral superfields of (24.308) and (24.309) to be complex, *or* by working with the corresponding 2-dimensional real representation $\boldsymbol{\Gamma}'$ defined by

$$\boldsymbol{\Gamma}'(\mathbf{a}_1) = \begin{bmatrix} 0 & 1 \\ -1 & 0 \end{bmatrix},$$

and using real chiral superfields in (24.308) and (24.309). (Essentially these two choices just correspond to the two choices discussed previously for the simplest non-supersymmetric case in Example I of Chapter 18, Section 5.) There is an advantage in working entirely with real superfields when the effects of spontaneous symmetry breaking are investigated.

It is convenient to define $\boldsymbol{\Phi}_L(\mathbf{X};\boldsymbol{\Theta})$ and $\boldsymbol{\Phi}_R(\mathbf{X};\boldsymbol{\Theta})$ as the $d \times 1$ column matrices whose $(j, 1)$ entries are $\Phi_{Lj}(\mathbf{X};\boldsymbol{\Theta})$ and $\Phi_{Rj}(\mathbf{X};\boldsymbol{\Theta})$ respectively (for $j = 1, 2, \ldots, d$). Then (24.308) and (24.309) can be written more concisely as

$$\boldsymbol{\Phi}_L(\mathbf{X};\boldsymbol{\Theta}) \to \boldsymbol{\Phi}'_L(\mathbf{X};\boldsymbol{\Theta}) = \tilde{\boldsymbol{\Gamma}}(\exp \Lambda_L(\mathbf{X};\boldsymbol{\Theta}))\boldsymbol{\Phi}_L(\mathbf{X};\boldsymbol{\Theta}) \quad (24.311)$$

and

$$\boldsymbol{\Phi}_R(\mathbf{X};\boldsymbol{\Theta}) \to \boldsymbol{\Phi}'_R(\mathbf{X};\boldsymbol{\Theta}) = \tilde{\boldsymbol{\Gamma}}(\exp \Lambda_R(\mathbf{X};\boldsymbol{\Theta}))\boldsymbol{\Phi}_R(\mathbf{X};\boldsymbol{\Theta}). \quad (24.312)$$

As the unitary representation $\boldsymbol{\Gamma}$ of $\mathcal{G}$ provides a representation of $\mathcal{L}$ that is composed of anti-Hermitian matrices, (24.304)–(24.306) imply that

$$\boldsymbol{\Gamma}(\Lambda_L(\mathbf{X};\boldsymbol{\Theta}))^{\ddagger} = -\boldsymbol{\Gamma}(\Lambda_R(\mathbf{X};\boldsymbol{\Theta})), \quad \boldsymbol{\Gamma}(\Lambda_R(\mathbf{X};\boldsymbol{\Theta}))^{\ddagger} = -\boldsymbol{\Gamma}(\Lambda_L(\mathbf{X};\boldsymbol{\Theta})). \quad (24.313)$$

Thus (24.311) and (24.312) require that

$$\boldsymbol{\Phi}_L(\mathbf{X};\boldsymbol{\Theta})^{\ddagger} \to \boldsymbol{\Phi}'_L(\mathbf{X};\boldsymbol{\Theta})^{\ddagger} = \boldsymbol{\Phi}_L(\mathbf{X};\boldsymbol{\Theta})^{\ddagger}\boldsymbol{\Gamma}(\exp(-\tilde{\Lambda}_R(\mathbf{X};\boldsymbol{\Theta}))) \quad (24.314)$$

and

$$\boldsymbol{\Phi}_R(\mathbf{X};\boldsymbol{\Theta})^{\ddagger} \to \boldsymbol{\Phi}'_R(\mathbf{X};\boldsymbol{\Theta})^{\ddagger} = \boldsymbol{\Phi}_R(\mathbf{X};\boldsymbol{\Theta})^{\ddagger}\boldsymbol{\Gamma}(\exp(-\tilde{\Lambda}_L(\mathbf{X};\boldsymbol{\Theta}))), \quad (24.315)$$

which of course reduce to (24.287) and (24.288) in the Abelian case.

The final constituents that must be considered are the gauge fields themselves. In the supersymmetric Abelian theory of the previous subsection these were put in a single real general multiplet $V = \{C(\mathbf{x}), D(\mathbf{x}), M(\mathbf{x}), N(\mathbf{x}); \boldsymbol{\chi}(\mathbf{x}), \boldsymbol{\lambda}(\mathbf{x}); \mathbf{V}(\mathbf{x})\}$, from which was formed a single real scalar superfield $V(\mathbf{X};\boldsymbol{\Theta})$ by the definition (24.276). The natural generalization of this for a Lie group $\mathscr{G}$ of dimension $n_{\mathscr{G}}$ is to set up $n_{\mathscr{G}}$ real general multiplets $V_I = \{C_I(\mathbf{x}), D_I(\mathbf{x}), M_I(\mathbf{x}), N_I(\mathbf{x}); \boldsymbol{\chi}_I(\mathbf{x}), \boldsymbol{\lambda}_I(\mathbf{x}); \mathbf{V}_I(\mathbf{x})\}$ and form from them $n_{\mathscr{G}}$ real scalar superfields $V_I(\mathbf{X};\boldsymbol{\Theta})$ by merely inserting subscript I everywhere in the obvious way. From these can be formed the combination

$$\mathbf{V}(\mathbf{X};\boldsymbol{\Theta}) = \sum_{I=1}^{n_{\mathscr{G}}} V_I(\mathbf{X};\boldsymbol{\Theta})\mathbf{a}_I. \tag{24.316}$$

Again, although $\mathbf{V}(\mathbf{X};\boldsymbol{\Theta})$ is essentially a matrix in which each component is a real general superfield, for convenience it will be described as being *Lie-algebra-valued*, since it will play the role taken previously by the Lie-algebra-valued Yang–Mills potential $\mathbf{A}_\mu(\mathbf{x})$ of (19.18).

In order to generalize the gauge transformation (24.293) in a natural way, it is necessary that the dimensions of the matrix $\mathbf{V}(\mathbf{X};\boldsymbol{\Theta})$ and those of the matrices $\boldsymbol{\Gamma}(\boldsymbol{\Lambda}_{\mathrm{L}}(\mathbf{X};\boldsymbol{\Theta}))$ and $\boldsymbol{\Gamma}(\boldsymbol{\Lambda}_{\mathrm{R}}(\mathbf{X};\boldsymbol{\Theta}))$ should be the same. This can be achieved most conveniently by assuming that the $\mathbf{a}_I$ of the original basis of $\mathscr{L}$ are chosen so that

$$-\tilde{\boldsymbol{\Gamma}}(\mathbf{a}_I) = \mathbf{a}_I \tag{24.317}$$

(for $I = 1, \ldots, n_{\mathscr{G}}$). With this choice,

$$\tilde{\boldsymbol{\Gamma}}(\boldsymbol{\Lambda}_{\mathrm{L}}(\mathbf{X};\boldsymbol{\Theta})) = -\boldsymbol{\Lambda}_{\mathrm{L}}(\mathbf{X};\boldsymbol{\Theta}), \quad \tilde{\boldsymbol{\Gamma}}(\boldsymbol{\Lambda}_{\mathrm{R}}(\mathbf{X};\boldsymbol{\Theta})) = -\boldsymbol{\Lambda}_{\mathrm{R}}(\mathbf{X};\boldsymbol{\Theta}), \tag{24.318}$$

and the matter superfield gauge transformations (24.311), (24.312), (24.314) and (24.315) can be rewritten more simply as

$$\boldsymbol{\Phi}_{\mathrm{L}}(\mathbf{X};\boldsymbol{\Theta}) \to \boldsymbol{\Phi}'_{\mathrm{L}}(\mathbf{X};\boldsymbol{\Theta}) = \exp(-\boldsymbol{\Lambda}_{\mathrm{L}}(\mathbf{X};\boldsymbol{\Theta}))\,\boldsymbol{\Phi}_{\mathrm{L}}(\mathbf{X};\boldsymbol{\Theta}), \tag{24.319}$$

$$\boldsymbol{\Phi}_{\mathrm{R}}(\mathbf{X};\boldsymbol{\Theta}) \to \boldsymbol{\Phi}'_{\mathrm{R}}(\mathbf{X};\boldsymbol{\Theta}) = \exp(-\boldsymbol{\Lambda}_{\mathrm{R}}(\mathbf{X};\boldsymbol{\Theta}))\,\boldsymbol{\Phi}_{\mathrm{R}}(\mathbf{X};\boldsymbol{\Theta}), \tag{24.320}$$

$$\boldsymbol{\Phi}_{\mathrm{L}}(\mathbf{X};\boldsymbol{\Theta})^\ddagger \to \boldsymbol{\Phi}'_{\mathrm{L}}(\mathbf{X};\boldsymbol{\Theta})^\ddagger = \boldsymbol{\Phi}_{\mathrm{L}}(\mathbf{X};\boldsymbol{\Theta})^\ddagger \exp(\boldsymbol{\Lambda}_{\mathrm{R}}(\mathbf{X};\boldsymbol{\Theta})), \tag{24.321}$$

$$\boldsymbol{\Phi}_{\mathrm{R}}(\mathbf{X};\boldsymbol{\Theta})^\ddagger \to \boldsymbol{\Phi}'_{\mathrm{R}}(\mathbf{X};\boldsymbol{\Theta})^\ddagger = \boldsymbol{\Phi}_{\mathrm{R}}(\mathbf{X};\boldsymbol{\Theta})^\ddagger \exp(\boldsymbol{\Lambda}_{\mathrm{L}}(\mathbf{X};\boldsymbol{\Theta})). \tag{24.322}$$

The natural generalization of (24.293) is then

$$\exp\left\{\frac{2gi}{\hbar c}\mathbf{V}(\mathbf{X};\boldsymbol{\Theta})\right\} \to \exp\{-\boldsymbol{\Lambda}_{\mathrm{R}}(\mathbf{X};\boldsymbol{\Theta})\}\exp\left\{\frac{2gi}{\hbar c}\mathbf{V}(\mathbf{X};\boldsymbol{\Theta})\right\}\exp\{\boldsymbol{\Lambda}_{\mathrm{L}}(\mathbf{X};\boldsymbol{\Theta})\}, \tag{24.323}$$

where g is some coupling constant, since in the one-dimensional Abelian case the assumptions (24.310) and (24.317) imply that $\mathbf{a}_1 = [-i]$. Application of the Campbell–Baker–Hausdorff formula (10.5) shows that to first order in $\mathbf{V}(\mathbf{X};\mathbf{\Theta})$, $\mathbf{\Lambda}_{\mathrm{L}}(\mathbf{X};\mathbf{\Theta})$ and $\mathbf{\Lambda}_{\mathrm{R}}(\mathbf{X};\mathbf{\Theta})$ this reduces to

$$\mathbf{V}(\mathbf{X};\mathbf{\Theta}) \to \mathbf{V}(\mathbf{X};\mathbf{\Theta}) + \frac{i\hbar c}{2g}\{\mathbf{\Lambda}_{\mathrm{R}}(\mathbf{X};\mathbf{\Theta}) - \mathbf{\Lambda}_{\mathrm{L}}(\mathbf{X};\mathbf{\Theta})\} + \tfrac{1}{2}[\mathbf{V}(\mathbf{X};\mathbf{\Theta}), \mathbf{\Lambda}_{\mathrm{R}}(\mathbf{X};\mathbf{\Theta}) + \mathbf{\Lambda}_{\mathrm{L}}(\mathbf{X};\mathbf{\Theta})]_-;$$

that is, to

$$\delta_{\text{gauge}}\mathbf{V}(\mathbf{X};\mathbf{\Theta}) = \frac{i\hbar c}{2g}\{\mathbf{\Lambda}_{\mathrm{R}}(\mathbf{X};\mathbf{\Theta}) - \mathbf{\Lambda}_{\mathrm{L}}(\mathbf{X};\mathbf{\Theta})\} + \tfrac{1}{2}[\mathbf{V}(\mathbf{X};\mathbf{\Theta}), \mathbf{\Lambda}_{\mathrm{R}}(\mathbf{X};\mathbf{\Theta}) + \mathbf{\Lambda}_{\mathrm{L}}(\mathbf{X};\mathbf{\Theta})]_-. \quad (24.324)$$

In particular, examination of the component in the expansion of $\mathbf{V}(\mathbf{X};\mathbf{\Theta})$ that involves the Yang–Mills potential $\mathbf{V}(\mathbf{X})$ shows that (24.326) contains as a special case the Lie algebraic limit of the usual Yang–Mills transformation law (19.22). Just as in the Abelian case, the gauge transformation (24.323) can be used to set to zero all the component functions of the $V_I(\mathbf{X};\mathbf{\Theta})$ except for $\mathbf{V}_I(\mathbf{X})$, $\boldsymbol{\lambda}_I(\mathbf{X})$ and $D_I(\mathbf{X})$ (for all $I = 1, 2, \ldots, n_{\mathscr{G}}$). This choice is again known as the *Wess–Zumino gauge*. In this gauge

$$\mathbf{V}(\mathbf{X};\mathbf{\Theta})^n = \mathbf{0} \quad (24.325)$$

for $n \geqslant 3$ (cf. (24.295)), so all exponential series involving $\mathbf{V}(\mathbf{X};\mathbf{\Theta})$ terminate after three terms.

The sum of the chiral multiplet kinetic-energy terms in the action can be written in matrix superfield form as

$$S_{\mathrm{KE}}^{\mathrm{WZ}} = -\frac{1}{8\hbar c}\int \mathrm{d}^4X \int \mathrm{d}^2\Theta \int \mathrm{d}^2\bar{\Theta}\, \mathbf{\Phi}_{\mathrm{L}}(\mathbf{X};\mathbf{\Theta})^{\ddagger}\mathbf{\Phi}_{\mathrm{L}}(\mathbf{X};\mathbf{\Theta}) - \frac{1}{8\hbar c}\int \mathrm{d}^4X \int \mathrm{d}^2\Theta \int \mathrm{d}^2\bar{\Theta}\, \mathbf{\Phi}_{\mathrm{R}}(\mathbf{X};\mathbf{\Theta})^{\ddagger}\mathbf{\Phi}_{\mathrm{R}}(\mathbf{X};\mathbf{\Theta}), \quad (24.326)$$

which generalizes (24.290). Under the gauge transformations (24.319)–(24.322), the "integrands" of (24.326) transform as

$$\mathbf{\Phi}_{\mathrm{L}}(\mathbf{X};\mathbf{\Theta})^{\ddagger}\mathbf{\Phi}_{\mathrm{L}}(\mathbf{X};\mathbf{\Theta}) \to \mathbf{\Phi}_{\mathrm{L}}(\mathbf{X};\mathbf{\Theta})^{\ddagger}\exp\{\mathbf{\Lambda}_{\mathrm{R}}(\mathbf{X};\mathbf{\Theta})\}\exp\{-\mathbf{\Lambda}_{\mathrm{L}}(\mathbf{X};\mathbf{\Theta})\}\,\mathbf{\Phi}_{\mathrm{L}}(\mathbf{X};\mathbf{\Theta})$$

and

$$\mathbf{\Phi}_{\mathrm{R}}(\mathbf{X};\mathbf{\Theta})^{\ddagger}\mathbf{\Phi}(\mathbf{X};\mathbf{\Theta}) \to \mathbf{\Phi}_{\mathrm{R}}(\mathbf{X};\mathbf{\Theta})^{\ddagger}\exp\{\mathbf{\Lambda}_{\mathrm{L}}(\mathbf{X};\mathbf{\Theta})\}\exp\{-\mathbf{\Lambda}_{\mathrm{R}}(\mathbf{X};\mathbf{\Theta})\}\,\mathbf{\Phi}_{\mathrm{R}}(\mathbf{X};\mathbf{\Theta}),$$

and so they are not invariant. However, the combinations

$$\mathbf{\Phi}_{\mathrm{L}}(\mathbf{X};\mathbf{\Theta})^{\ddagger}\exp\left\{\frac{2gi}{\hbar c}\mathbf{V}(\mathbf{X};\mathbf{\Theta})\right\}\mathbf{\Phi}_{\mathrm{L}}(\mathbf{X};\mathbf{\Theta}) \quad (24.327)$$

and

$$\mathbf{\Phi}_{\mathrm{R}}(\mathbf{X};\mathbf{\Theta})^{\ddagger}\exp\left\{-\frac{2gi}{\hbar c}\mathbf{V}(\mathbf{X};\mathbf{\Theta})\right\}\mathbf{\Phi}_{\mathrm{R}}(\mathbf{X};\mathbf{\Theta}) \tag{24.328}$$

are invariant under the gauge transformations (24.319)–(24.322) provided that $\mathbf{V}(\mathbf{X};\mathbf{\Theta})$ transforms as in (24.323). Thus the gauge-invariant and supersymmetric term in the action that incorporates both the kinetic-energy terms and the interactions between the matter fields and the gauge fields is

$$\begin{aligned} S_{\mathrm{KE+int}} = &-\frac{1}{8\hbar c}\int \mathrm{d}^4X \int \mathrm{d}^2\Theta \int \mathrm{d}^2\bar{\Theta}\, \mathbf{\Phi}_{\mathrm{L}}(\mathbf{X};\mathbf{\Theta})^{\ddagger}\exp\left\{\frac{2gi}{\hbar c}\mathbf{V}(\mathbf{X};\mathbf{\Theta})\right\}\mathbf{\Phi}_{\mathrm{L}}(\mathbf{X};\mathbf{\Theta}) \\ &-\frac{1}{8\hbar c}\int \mathrm{d}^4X \int \mathrm{d}^2\Theta \int \mathrm{d}^2\bar{\Theta}\, \mathbf{\Phi}_{\mathrm{R}}(\mathbf{X};\mathbf{\Theta})^{\ddagger}\exp\left\{-\frac{2gi}{\hbar c}\mathbf{V}(\mathbf{X};\mathbf{\Theta})\right\}\mathbf{\Phi}_{\mathrm{R}}(\mathbf{X};\mathbf{\Theta}). \end{aligned} \tag{24.329}$$

Again a similar but simpler argument applies to the mass terms. The sum of the chiral multiplet mass terms in the action can be written in superfield form as

$$\begin{aligned} S_{\mathrm{mass}} = &\tfrac{1}{4}iMc^2\int \mathrm{d}^4X \int \mathrm{d}^2\Theta\, \mathbf{\Phi}_{\mathrm{R}}(\mathbf{X};\mathbf{\Theta})^{\ddagger}\mathbf{\Phi}_{\mathrm{L}}(\mathbf{X};\mathbf{\Theta}) \\ &+\tfrac{1}{4}iMc^2\int \mathrm{d}^4X \int \mathrm{d}^2\bar{\Theta}\, \mathbf{\Phi}_{\mathrm{L}}(\mathbf{X};\mathbf{\Theta})^{\ddagger}\mathbf{\Phi}_{\mathrm{R}}(\mathbf{X};\mathbf{\Theta}) \end{aligned} \tag{24.330}$$

(cf. (24.297)), which is obviously invariant under the gauge transformations (24.319)–(24.322).

The last stage is to write down the action for the fields in $\mathbf{V}(\mathbf{X};\mathbf{\Theta})$ as a superspace integral and thereby generalize (24.300). This can be achieved by introducing a pair of left-handed chiral Lie-algebra-valued superfields $\mathbf{W}_{\mathrm{L}A}(\mathbf{X};\mathbf{\Theta})$, for which the definition (24.298) has to be replaced by

$$\mathbf{W}_{\mathrm{L}A}(\mathbf{X};\mathbf{\Theta}) = -\frac{\hbar c}{2g}\bar{\mathrm{D}}^2\left\{\exp\left\{-\frac{2gi}{\hbar c}\mathbf{V}(\mathbf{X};\mathbf{\Theta})\right\}\mathrm{D}_A\exp\left\{\frac{2gi}{\hbar c}\mathbf{V}(\mathbf{X};\mathbf{\Theta})\right\}\right\} \tag{24.331}$$

(for $A = 1, 2$), where D_A and $\bar{\mathrm{D}}^2$ are defined in (24.168) and (24.182) respectively, which reduces to (24.298) in the Abelian case. By analogy with (24.316), $\mathbf{W}_{\mathrm{L}A}(\mathbf{X};\mathbf{\Theta})$ has an expansion of the form

$$\mathbf{W}_{\mathrm{L}A}(\mathbf{X};\mathbf{\Theta}) = \sum_{I=1}^{n_{\mathcal{G}}} W_{I\mathrm{L}A}(\mathbf{X};\mathbf{\Theta})\mathbf{a}_I. \tag{24.332}$$

Then (24.323) taken with the fact that $\mathrm{D}_A\mathbf{\Lambda}_{\mathrm{R}}(\mathbf{X};\mathbf{\Theta}) = 0$ (cf. (24.170)) implies that in the non-Abelian case the $\mathbf{W}_{\mathrm{L}A}(\mathbf{X};\mathbf{\Theta})$ are not gauge-invariant but obey the transformation law

$$\mathbf{W}_{\mathrm{L}A}(\mathbf{X};\mathbf{\Theta}) \rightarrow \exp\{-\mathbf{\Lambda}_{\mathrm{L}}(\mathbf{X};\mathbf{\Theta})\}\,\mathbf{W}_{\mathrm{L}A}(\mathbf{X};\mathbf{\Theta})\exp\{\mathbf{\Lambda}_{\mathrm{L}}(\mathbf{X};\mathbf{\Theta})\} \quad (24.333)$$

(for $A = 1, 2$). A gauge-invariant and supersymmetric action for the gauge fields is then given by

$$S_0^{\mathrm{v}} = -\tfrac{1}{32}\,\mathrm{Re}\int \mathrm{d}^4X \int \mathrm{d}^2\Theta\,\mathrm{tr}\left\{\sum_{A,B=1}^{2} \mathbf{W}_{\mathrm{L}A}(\mathbf{X};\mathbf{\Theta})\varepsilon_{AB}\mathbf{W}_{\mathrm{L}B}(\mathbf{X};\mathbf{\Theta})\right\}. \quad (24.334)$$

(The gauge invariance follows immediately from (24.333) because of the invariance of a trace under a similarity transformation.)

If $\mathscr{L}$ is Abelian then (24.334) reduces to (24.300). If $\mathscr{L}$ is simple then (16.12) and (24.332) imply that

$$\mathrm{tr}\{\mathbf{W}_{\mathrm{L}A}(\mathbf{X};\mathbf{\Theta})\mathbf{W}_{\mathrm{L}B}(\mathbf{X};\mathbf{\Theta})\} = \gamma B(\mathbf{W}_{\mathrm{L}A}(\mathbf{X};\mathbf{\Theta}), \mathbf{W}_{\mathrm{L}B}(\mathbf{X};\mathbf{\Theta})),$$

where $B(\ ,\ )$ is the Killing form of $\mathscr{L}$ and γ is the Dynkin index of the representation of $\mathscr{L}$ formed by the matrices $\mathbf{a}_I$. This can be further simplified by assuming that the basis of $\mathscr{L}$ is chosen so that

$$B(\mathbf{a}_I, \mathbf{a}_J) = -\delta_{IJ}$$

(for $I, J = 1, 2, \ldots, n_{\mathscr{g}}$) (see Chapter 14, Section 2). After rescaling S_0^{v} by dividing by γ, and applying the Wess–Zumino gauge, it can be shown that S_0^{v} corresponds to the Lagrangian density

$$\begin{aligned}\mathscr{L}_0^{\mathrm{v}}(\mathbf{x}) = &-\frac{1}{4}\sum_{\rho,\kappa,\tau,\mu=1}^{4}\sum_{I=1}^{n_{\mathscr{g}}} \eta^{\rho\kappa}\eta^{\tau\mu}F_{I\rho\kappa}(\mathbf{x})F_{I\tau\mu}(\mathbf{x})\\ &+\tfrac{1}{2}i\sum_{I=1}^{n_{\mathscr{g}}}\sum_{\rho=1}^{4}\tilde{\boldsymbol{\lambda}}_I(\mathbf{x})^{\dagger}\boldsymbol{\gamma}_4\boldsymbol{\gamma}^{\rho}\left\{\partial_\rho\boldsymbol{\lambda}_I(\mathbf{x}) - \frac{g}{\hbar c}\sum_{J,K=1}^{n_{\mathscr{g}}} c_{JK}^{I}V_{J\rho}(\mathbf{x})\boldsymbol{\lambda}_K(\mathbf{x})\right\}\\ &+\frac{1}{2}\sum_{I=1}^{n_{\mathscr{g}}} D_I(x)^2, \end{aligned} \quad (24.335)$$

(cf. (24.121)), where

$$F_{I\rho\kappa}(\mathbf{x}) = \partial_\rho V_{I\kappa}(\mathbf{x}) - \partial_\kappa V_{I\rho}(\mathbf{x}) - \frac{g}{\hbar c}\sum_{J,K=1}^{n_{\mathscr{g}}} c_{JK}^{I}V_{J\rho}(\mathbf{x})V_{K\kappa}(\mathbf{x}), \quad (24.336)$$

c_{JK}^{I} are the structure constants of $\mathscr{L}$ as defined by

$$[\mathbf{a}_J, \mathbf{a}_K]_- = \sum_{I=1}^{n_{\mathscr{g}}} c_{JK}^{I}\mathbf{a}_I,$$

and $\mathbf{V}_I(\mathbf{x})$, $\boldsymbol{\lambda}_I(\mathbf{x})$ and $D_I(\mathbf{x})$ are the remaining component functions of the scalar superfields $V_I(\mathbf{X};\mathbf{\Theta})$ (for $I = 1, \ldots, n_{\mathscr{g}}$).

It is very straightforward to extend this formulation to the case in which the Lie algebra $\mathscr{L}$ is the direct sum of S subalgebras:

$$\mathscr{L} = \mathscr{L}^1 \oplus \mathscr{L}^2 \oplus \cdots \oplus \mathscr{L}^S.$$

For $J = 1, 2, \ldots, S$ suppose that $\mathscr{L}^J$ has dimension $n^J_{\mathscr{G}}$ and has basis $\mathbf{a}^J_1, \mathbf{a}^J_2, \ldots$, and that $V^J_I = \{C^J_I(\mathbf{x}), D^J_I(\mathbf{x}), M^J_I(\mathbf{x}), N^J_I(\mathbf{x}); \chi^J_I(\mathbf{x}), \lambda^J_I(\mathbf{x}); \mathbf{V}^J_I(\mathbf{x})\}$ are a set of $n^J_{\mathscr{G}}$ real general multiplets and from which may be formed $n^J_{\mathscr{G}}$ real scalar superfields $V^J_I(\mathbf{X}; \mathbf{\Theta})$. From these can be formed the combinations

$$\mathbf{V}^J(\mathbf{X}; \mathbf{\Theta}) = \sum_{I=1}^{n^J_{\mathscr{G}}} V^J_I(\mathbf{X}; \mathbf{\Theta})\mathbf{a}^J_I \tag{24.337}$$

(for $J = 1, 2, \ldots, S$). Then all that has to be done is to replace the quantity $(2gi/\hbar c)\mathbf{V}(\mathbf{X}; \mathbf{\Theta})$ in (24.323) and (24.327)–(24.329) by

$$\frac{2i}{\hbar c} \sum_{J=1}^{S} g^J \mathbf{V}^J(\mathbf{X}; \mathbf{\Theta}),$$

where $g^1, g^2, \ldots, g^S$ are S independent coupling constants, and to modify (24.324) in the obvious way. Moreover, with this substitution the term

$$\bar{\mathrm{D}}^2 \left\{ \exp\left\{ -\frac{2gi}{\hbar c} \mathbf{V}(\mathbf{X}; \mathbf{\Theta}) \right\} \mathrm{D}_A \exp\left\{ \frac{2gi}{\hbar c} \mathbf{V}(\mathbf{X}; \mathbf{\Theta}) \right\} \right\}$$

that appears on the right-hand side of (24.331) becomes the *sum* of S terms of the form

$$\bar{\mathrm{D}}^2 \left\{ \exp\left\{ -\frac{2g^J i}{\hbar c} \mathbf{V}^J(\mathbf{X}; \mathbf{\Theta}) \right\} \mathrm{D}_A \exp\left\{ \frac{2g^J i}{\hbar c} \mathbf{V}^J(\mathbf{X}; \mathbf{\Theta}) \right\} \right\},$$

each of which can be multiplied by a corresponding factor $-(\hbar c/2g^J)$. That is, as in the non-supersymmetric Yang–Mills case, if $\mathscr{L}$ has S invariant subalgebras then the theory contains S *independent* coupling constants.

For each one-dimensional Abelian invariant subalgebra of $\mathscr{L}$ it is possible to add to the action a Fayet–Iliopoulos term S_{FI} of the form (24.301), but there are no corresponding gauge-invariant terms for non-Abelian subalgebras. On the other hand, although it was noted in Section 4(*a*) that the Wess–Zumino self-interaction term of the one-dimensional Abelian theory is not gauge-invariant, in the non-Abelian situation it is possible to construct gauge-invariant self-interaction terms involving the matter multiplets $\mathbf{\Phi}_{\mathrm{L}}(\mathbf{X}; \mathbf{\Theta})$ and $\mathbf{\Phi}_{\mathrm{R}}(\mathbf{X}; \mathbf{\Theta})$.

The construction of extended supersymmetric Yang–Mills theories has been greatly complicated by the difficulties involved in setting the appropriate supermultiplets and superfields that were referred to at the end of Section 3(*a*). There is a very extensive literature on the subject, which has been reviewed by Scherk (1979), Sohnius *et al.* (1980, 1981), Wess and Bagger (1983), Sohnius

(1985), Stelle (1985), Ferrara (1987), Mohapatra (1986), West (1986c), Dragon *et al.* (1987) and Volkov (1987). Although the results are most nearly complete for the $N = 2$ case, the $N = 4$ situation has been closely studied because in this case all the particles must lie in the gauge multiplet and because of its exceptional divergence-free properties.

5 Spontaneous symmetry breaking

In supersymmetric models two different types of spontaneous symmetry breaking can be distinguished. First there is the spontaneous breaking of supersymmetry itself, which should manifest itself in the breaking of the equality of fermion and bosonic masses. Secondly there is the breaking of internal symmetries of the type considered in Chapters 18 and 19, which for supersymmetric models should make itself apparent by breaking equalities between fermion masses in each fermionic internal symmetry multiplet and between bosonic masses in each bosonic internal symmetry multiplet, but with each bosonic state being associated with a fermionic state of the same mass, and *vice versa*. It is to be expected that both types of breaking would be essential in any model that describes the fundamental particles of the real world.

The simplest model to examine is the Wess–Zumino model that was discussed in Section 2(*b*). The study of spontaneous supersymmetry breaking in this model was first made by Salam and Strathdee (1974c) and Iliopoulos and Zumino (1974). All that has to be done is to apply the standard spontaneous symmetry breaking arguments (as described in Chapter 18, Section 5) to the total Lagrangian density $\mathscr{L}_{\text{tot}}(\mathbf{x})$ of (24.77) that is obtained by eliminating the auxiliary fields $F(\mathbf{x})$ and $G(\mathbf{x})$. The argument is facilitated by writing

$$\mathscr{L}_{\text{tot}}(\mathbf{x}) = \mathscr{L}^{AB}_{\text{KE}}(\mathbf{x}) + \mathscr{L}_{\text{fermion}}(\mathbf{x}) - V(A(\mathbf{x}), B(\mathbf{x})), \tag{24.338}$$

where

$$\mathscr{L}^{AB}_{\text{KE}}(\mathbf{x}) = \frac{c^3}{\hbar}\left(\frac{\hbar^2}{c^2}\right) \sum_{\alpha,\beta=1}^{4} \eta^{\alpha\beta}\{\partial_\alpha A(\mathbf{x})\,\partial_\beta A(\mathbf{x}) - \partial_\alpha B(\mathbf{x})\,\partial_\beta B(\mathbf{x})\} \tag{24.339}$$

is the sum of the kinetic-energy terms of the $A(\mathbf{x})$ and $B(\mathbf{x})$ fields alone,

$$\begin{aligned}\mathscr{L}_{\text{fermion}}(\mathbf{x}) = \tilde{\boldsymbol{\psi}}(\mathbf{x})^\dagger \left\{ \sum_{\alpha=1}^{4} i\hbar c\gamma_4\gamma^\alpha\,\partial_\alpha\psi_b(\mathbf{x}) - Mc^2\gamma_4\boldsymbol{\psi}(\mathbf{x}) \right\} \\ - 2g\tilde{\boldsymbol{\psi}}(\mathbf{x})^\dagger\{A(\mathbf{x})\gamma_4 - iB(\mathbf{x})\gamma_4\gamma_5\}\boldsymbol{\psi}(\mathbf{x})\end{aligned} \tag{24.340}$$

is the sum of all the parts that contain fermion terms, and

$$V(A(\mathbf{x}), B(\mathbf{x})) = \hbar c M^2 \left\{ \left\{ A(\mathbf{x}) + \frac{\lambda}{2Mc^2} \right\}^2 + B(\mathbf{x})^2 \right\}$$
$$+ \frac{2Mcg}{\hbar} A(\mathbf{x}) \left\{ A(\mathbf{x})^2 + B(\mathbf{x})^2 \right\} + \frac{2\lambda g}{\hbar c} \{ A(\mathbf{x})^2 - B(\mathbf{x})^2 \}$$
$$+ \frac{g^2}{\hbar c} \{ A(\mathbf{x})^2 + B(\mathbf{x})^2 \}^2 \qquad (24.341)$$

is a potential term that depends only on the fields $A(\mathbf{x})$ and $B(\mathbf{x})$. Denoting the values of the fields $A(\mathbf{x})$ and $B(\mathbf{x})$ at which $V(A(\mathbf{x}), B(\mathbf{x}))$ takes local extrema by $\langle A \rangle$ and $\langle B \rangle$, it was shown by Salam and Strathdee (1974c) and Iliopoulos and Zumino (1974) that there are essentially only two sets of values for $\langle A \rangle$ and $\langle B \rangle$ that correspond to physically acceptable solutions, and that for both of these the bosonic and fermionic masses are equal. Consequently neither is associated with a spontaneous breakdown of supersymmetry. Detailed calculation shows that for both of these extrema $\langle F \rangle = 0$ and $\langle G \rangle = 0$.

The general conclusion that can be drawn from the study of this model is that it is difficult to obtain spontaneous breakdown of supersymmetry because there is very little freedom in the form of the potential. Indeed, the potential can be written as

$$V(A(\mathbf{x}), B(\mathbf{x})) = \frac{\hbar^3}{c} \{ F(\mathbf{x}) + B(\mathbf{x})^2 \},$$

indicating that $V(A(\mathbf{x}), B(\mathbf{x})) \geqslant 0$. Moreover, since $\langle F \rangle = 0$ and $\langle G \rangle = 0$ at the extrema, it follows that the extrema are not merely local minima but are actually *global* minima at $V = 0$.

It is appropriate to mention at this point that it is easily shown that making a pair of "field redefinitions" in the Wess–Zumino model in which $A(\mathbf{x})$ is replaced by $A(\mathbf{x}) - \langle A \rangle$ and $B(\mathbf{x})$ is replaced by $B(\mathbf{x}) - \langle B \rangle$ is equivalent to choosing a Lagrangian density in which $\lambda = 0$, i.e. to a Lagrangian density in which the $\lambda F(\mathbf{x})$ term is eliminated.

O'Raifeartaigh (1975a, b, c) has extended the analysis of spontaneous symmetry breaking in the simple Wess–Zumino model by considering N coupled chiral superfields, and showed that spontaneous breakdown of supersymmetry can *never* occur when the potential minimum is at $V = 0$, but it can occur at potential minima for which $V > 0$. Moreover, if $F_j(\mathbf{x})$ and $G_j(\mathbf{x})$ (for $j = 1, 2, \ldots, N$) are the N auxiliary fields then the potential V is proportional to

$$\sum_{j=1}^{N} \{ F_j(\mathbf{x})^2 + G_j(\mathbf{x})^2 \},$$

so spontaneous breakdown of supersymmetry *never* occurs when $\langle F_j \rangle = 0$ and $\langle G_j \rangle = 0$ for $j = 1, 2, \ldots, N$; that is, the *non-vanishing* of at least one $\langle F_j \rangle$ or $\langle G_j \rangle$ is the signal for spontaneous breakdown of supersymmetry. Although this is difficult to achieve, it is not impossible, as O'Raifeartaigh (1975b, c) has shown by constructing an explicit example. A different model exhibiting spontaneous breakdown of supersymmetry has been discussed by Fayet (1975a).

By contrast, the spontaneous breaking of *internal* symmetries in supersymmetric theories is easily obtained, as a number of studies of generalizations of the Wess–Zumino model have shown (for details see Iliopoulos and Zumino 1974, Fayet 1975a, O'Raifeartaigh 1975a, b, c, Salam and Strathdee 1975b, Fayet and Ferrara 1977). If an internal global symmetry is spontaneously broken but the supersymmetry is preserved then *equal* numbers of zero-mass bosons and fermions appear. However, by contrast with the non-supersymmetric situation discussed in Theorem I of Chapter 18, Section 5, in the supersymmetric case the number of zero-mass bosons is no longer equal to $n_{\mathscr{G}} - n_{\mathscr{G}'}$ (where $n_{\mathscr{G}}$ is the order of the global internal symmetry group $\mathscr{G}$ and $n_{\mathscr{G}'}$ is the order of the residual symmetry group $\mathscr{G}'$). A detailed investigation by Buchmüller *et al.* (1983) showed that the number of zero-mass bosons, n_{B}, lies between $n_{\min}$ and $n_{\max}$, where

$$n_{\min} = \begin{cases} n_{\mathscr{G}} - n_{\mathscr{G}'} & \text{if } n_{\mathscr{G}} - n_{\mathscr{G}'} \text{ is even,} \\ n_{\mathscr{G}} - n_{\mathscr{G}'} + 1 & \text{if } n_{\mathscr{G}} - n_{\mathscr{G}'} \text{ is odd,} \end{cases}$$

and $n_{\max} = 2(n_{\mathscr{G}} - n_{\mathscr{G}'})$, depending on the details of the potential, and that in general the zero-mass bosons can be considered to fall into two types, namely the "Goldstone bosons" (of which there are $n_{\mathscr{G}} - n_{\mathscr{G}'}$) and the "quasi-Goldstone bosons" (which do not appear if $n_{\mathrm{B}} = n_{\min} = n_{\mathscr{G}} - n_{\mathscr{G}'}$). The zero-mass fermions have been given various names in the literature, such as "Goldstone fermions", "Goldstone spinors" and "quasi-Goldstone fermions". This analysis has been further refined by Lerche (1983), Bando *et al.* (1984), Kugo *et al.* (1984) and Lerche and Lust (1984). If $n_{\mathrm{B}} = n_{\max} = 2(n_{\mathscr{G}} - n_{\mathscr{G}'})$ then "doubling" or "total doubling" is said to have occurred, whereas the other situations are said to involve some "non-doubling".

For supersymmetric gauge theories Fayet and Iliopoulos (1974) have shown that an alternative source of spontaneous breaking of supersymmetry can be obtained by adding a Fayet–Iliopoulos term to the Lagrangian density for each Abelian invariant subalgebra. In the simplest situation, which was that orginally considered by Fayet and Iliopoulos (1974), the gauge group consists only of a one-dimensional Abelian group. Applying the theory of Section 4(*a*), on adding the Fayet–Iliopoulos term (24.259) to the Lagrangian density and eliminating the auxiliary fields using their equations of motions, it is found that the spinor and scalar components of the chiral multiplets correspond to different masses, the mass differences being proportional to ξg, g being the gauge coupling constant. The signal of the spontaneous supersymmetry breaking is now $\langle D \rangle \neq 0$.

In the general case the situation is quite complicated. The simplest class of models are those in which "total doubling" occurs at the global symmetry level. In these models the zero-mass particles in chiral multiplets form *complete* real chiral supermultiplets, and gauging the symmetry results in the total disappearance of these zero-mass states into the states of the gauge fields, thereby providing a supersymmetric generalization of the "Higgs–Kibble" mechanism described in Chapter 19, Section 4. More precisely, recalling that after the elimination of the auxiliary fields each chiral gauge multiplet Λ_I consists only of a vector $\mathbf{V}_I(\mathbf{x})$ and a Majorana spinor $\boldsymbol{\lambda}_I(\mathbf{x})$ and that each real chiral multiplet Φ_{Ck} consists only of two spin-0 states $A_k(\mathbf{x})$ and $B_k(\mathbf{x})$ and a Majorana spinor $\boldsymbol{\psi}_k(\mathbf{x})$, for each chiral gauge multiplet Λ_I that acquires a mass the following occur:

(i) the massless vector $\mathbf{V}_I(\mathbf{x})$ of Λ_I combines with a zero-mass chiral multiplet bosonic state to provide a set of massive spin-1 states;

(ii) the massless Majorana spinor $\boldsymbol{\lambda}_I(\mathbf{x})$ of Λ_I combines with the zero-mass Majorana spinor of the chiral multiplet to provide a set of massive spin-$\frac{1}{2}$ states; and

(iii) the remaining spin-0 massless state of the chiral multiplet acquires a mass, and so gives a massive spin-0 state.

All the massive states that are formed in this way possess the same mass, so supersymmetry is still preserved. Moreover, the resulting structure has precisely the spin content predicted in Chapter 23, Section 5(*a*) for a massive $N = 1$, $D = 4$ supermultiplet with spin $j = \frac{1}{2}$; namely, in the rest-mass frame there is just one spin-0 state, four spin-$\frac{1}{2}$ states, and three spin-1 states. (For a detailed presentation of the formation of such a supermultiplet in the context of spontaneous symmetry breaking of an Abelian U(1) group see Fayet (1975b) and Mainland and Tanaka (1975).)

For models in which "non-doubling" occurs it has been shown by Shore (1984) that the situation is very different, for the spectrum that results from the generalized Higgs–Kibble mechanism breaks supersymmetry not only in the masses of the particles but also in their spins.

For some reviews of the very considerable literature on the application of spontaneous symmetry breaking in supersymmetric models that might have some relevance to the fundamental particles of the physical world see Ellis *et al.* (1980), Fayet (1980, 1984a, b), Nath and Arnowitt (1980), Raby (1983), Ferrara (1984, 1985), Llewellyn Smith (1984), Nanopoulos (1984), Nilles (1984), Ross (1984), Savoy-Navarro (1984), Haber and Kane (1985), Kane (1985), Mohapatra (1986) and Volkas and Joshi (1988).

25

Simple Lie Superalgebras

1 An outline of the presentation

Although they have not been used so extensively in physical applications as the Poincaré superalgebras, the simple Lie superalgebras have been employed in various contexts. Their theory involves many of the elegant features that were discussed in Chapters 13–16 for semi-simple Lie algebras, such as the concepts of the Cartan subalgebra, roots, weights, Dynkin diagrams and Cartan matrices. On the other hand, several new features appear, particularly the possibility of vanishing Killing forms and the occurrence of "atypical" representations.

The development that is given here follows on directly from the general description of Lie superalgebras that was given in Chapter 21, the material of the intervening chapters not being required (except for some aspects of Section 6).

The first stage, which is described in Section 2, is to define the concept of a simple Lie superalgebra and investigate the consequences that flow immediately from the definition. To proceed further, the additional assumption is made in Section 3 that the superalgebra is "classical". In Section 3(*b*) this is refined further with the introduction of the idea of a "basic" classical simple Lie superalgebra, the properties of which are developed in detail in Section 3(*c*). Section 4 is devoted to the study of graded representations. It is shown in Section 4(*a*) that the weights have properties similar to those of the semi-simple Lie algebras, but nevertheless there are differences, the most important being the distinction in the superalgebra case between "typical" and "atypical" representations, which is investigated in some detail in Section 4(*b*). This account of representations concludes in Section 4(*c*) with a description of Casimir operators and indices.

All of these considerations apply on the assumption that the Lie superalgebras involved are *complex*. The classification of the corresponding *real* Lie superalgebras is presented in Section 5. Finally, in Section 6, the conformal, de Sitter and anti-de Sitter superalgebras are discussed in the context of being mainly special cases of simple real Lie superalgebras.

It is worth pointing out at this stage that the relationship between semi-simple Lie superalgebras and simple Lie superalgebras is quite different from that of the Lie algebra case. For details see the end of Section 2.

2 The definition of a simple Lie superalgebra and some immediate consequences

Although the definition of a simple Lie superalgebra is a straightforward generalization of the corresponding definition for a simple Lie algebra that was given in Chapter 13, Section 2, many of the consequences are quite different. All of the considerations of this section apply equally to real and complex Lie superalgebras.

Definition *Simple Lie superalgebra*

A Lie superalgebra $\mathscr{L}_s$ is said to be *simple* if it is not Abelian and it does not possess a proper invariant graded subalgebra.

The terms "Abelian" and "proper invariant graded subalgebra" in the context of Lie superalgebras were introduced in Chapter 21, Sections 2 and 3. Nahm and Scheunert (1976) have shown that if $\mathscr{L}_s$ is a simple Lie superalgebra then, as well as not possessing any proper invariant *graded* subalgebras, $\mathscr{L}_s$ does not possess any proper invariant *non-graded* subalgebras either, so the word "graded" is not really needed in the definition.

If $\mathscr{L}_s$ is a simple Lie superalgebra with a zero odd part (i.e. if $n = 0$) then $\mathscr{L}_s$ is also a simple Lie algebra. As the properties of these have already been discussed in earlier chapters, the main interest now lies in determining the properties of those simple Lie superalgebras that have non-trivial odd parts.

In the theory of semi-simple Lie algebras Cartan's "second criterion" (Theorem V of Chapter 13, Section 2) plays a pivotal role. Unfortunately in the Lie superalgebra case the corresponding results are appreciably weaker, as the following two theorems show.

Theorem I If $\mathscr{L}_s$ is a simple Lie superalgebra then its Killing form $B(\ ,\ )$ is either non-degenerate or else it is identically zero.

Proof The definition of non-degeneracy for bilinear forms was given in Appendix B, Section 5. $\mathscr{L}_s$ is an invariant graded subalgebra of itself, so applying Theorem III of Chapter 21, Section 5, with $\mathscr{L}'_s = \mathscr{L}_s$ shows that $\mathscr{L}_s^\perp$ ($= \mathscr{L}_s'^\perp$) is an invariant graded subalgebra of $\mathscr{L}_s$. Here $\mathscr{L}_s^\perp$ is the set of elements $a \in \mathscr{L}_s$ such that $B(a,b) = 0$ for all $b \in \mathscr{L}_s$. With $\mathscr{L}_s$ assumed simple, $\mathscr{L}_s^\perp$ can only be the trivial set $\{0\}$ consisting only of the zero element 0, in which case the Killing form is non-degenerate, or else $\mathscr{L}_s^\perp = \mathscr{L}_s$, in which case $B(a,b) = 0$ for all $a, b \in \mathscr{L}_s$.

It will be demonstrated later that there *do* exist simple Lie superalgebras with identically zero Killing forms. Consequently the situation is very different from the Lie algebra case, where Cartan's "first criterion" (Theorem IV of Chapter 13, Section 2) shows that if a Lie algebra has zero Killing form then it must be solvable, and hence it cannot be semi-simple. Of course the theory of simple Lie superalgebras is most like that of simple Lie algebras for those superalgebras that do have non-degenerate bilinear supersymmetric consistent invariant forms.

Theorem II If $\mathscr{L}_s$ is a Lie superalgebra whose Killing form is non-degenerate then $\mathscr{L}_s$ is either a *simple* Lie superalgebra with non-degenerate Killing form or else it is a direct sum of such Lie superalgebras.

Proof See Appendix K, Section 4.

As the proof of Theorem I involves only the use of the Killing form properties (21.81)–(21.84) and (21.86), it can be generalized to show that *if* $B'(\ ,\ )$ is *any* bilinear supersymmetric consistent invariant form on a simple Lie superalgebra $\mathscr{L}_s$ *then* $B'(\ ,\ )$ is either non-degenerate or is identically zero. Such forms occurs naturally, for if $\boldsymbol{\Gamma}$ is *any* graded representation of $\mathscr{L}_s$ and $B'(\ ,\ )$ is defined for all $a, b \in \mathscr{L}_s$ by

$$B'(a,b) = \operatorname{str}\{\boldsymbol{\Gamma}(a)\boldsymbol{\Gamma}(b)\} \tag{25.1}$$

then a minor modification of the proof of Theorem II of Chapter 21, Section 5 shows that $B'(\ ,\ )$ is indeed a bilinear sypersymmetric consistent invariant form on $\mathscr{L}_s$. The following theorem then provides a generalization to Lie superalgebras of Theorem I of Chapter 16, Section 3.

Theorem III If $B'(\ ,\ )$ and $B''(\ ,\ )$ are any two non-degenerate bilinear supersymmetric consistent invariant forms on a simple Lie superalgebra $\mathscr{L}_s$ then

$$B''(a,b) = \lambda B'(a,b) \tag{25.2}$$

for all $a, b \in \mathscr{L}_s$, where λ is a constant independent of a and b.

Proof See Appendix K, Section 4.

The next theorem gives some useful information on the structure of simple Lie superalgebras.

Theorem IV If $\mathscr{L}_s$ is a simple Lie superalgebra with even subspace $\mathscr{L}_0$ and non-trivial odd subspace $\mathscr{L}_1$ then

(a) the representation of $\mathscr{L}_0$ on $\mathscr{L}_1$ is faithful,

(b) $[\mathscr{L}_1, \mathscr{L}_1] = \mathscr{L}_0$, and

(c) $[\mathscr{L}_0, \mathscr{L}_1] = \mathscr{L}_1$.

Proof See Appendix K, Section 4.

The statement in (b) that $[\mathscr{L}_1, \mathscr{L}_1] = \mathscr{L}_0$ means that not only is $[a, b] \in \mathscr{L}_0$ for all $a, b \in \mathscr{L}_1$ but also *every* element $c \in \mathscr{L}_0$ can be written as $c = [a, b]$ for some $a, b \in \mathscr{L}_1$. The meaning of $[\mathscr{L}_0, \mathscr{L}_1] = \mathscr{L}_1$ is similar.

A partial converse of these results is provided by the next theorem.

Theorem V If $\mathscr{L}_s$ is a Lie superalgebra with non-trivial even subspace $\mathscr{L}_0$ and non-trivial odd subspace $\mathscr{L}_1$ and if

(a) the representation of $\mathscr{L}_0$ on $\mathscr{L}_1$ is faithful and irreducible, and

(b) $[\mathscr{L}_1, \mathscr{L}_1] = \mathscr{L}_0$,

then $\mathscr{L}_s$ is simple.

Proof See Appendix K, Section 4.

Definition *Semi-simple Lie superalgebra*

A Lie superalgebra $\mathscr{L}_s$ is said to be *semi-simple* if it does not possess a solvable invariant graded subalgebra.

This definition is suggested by the result quoted in Theorem III of Chapter 13, Section 2 to the effect that a Lie algebra $\mathscr{L}$ is semi-simple if and only if $\mathscr{L}$ does not possess a solvable invariant subalgebra. (The idea of a solvable Lie superalgebra was defined in Chapter 21, Section 3.) Unfortunately the analogue of Theorem VI of Chapter 13, Section 2 is not true for semi-simple Lie superalgebras, for there do exist semi-simple Lie superalgebras that are *not* expressible as the direct sum of simple Lie superalgebras. Such superalgebras have been investigated by Kac (1977a) and Hurni (1987).

A Lie superalgebra with a non-degenerate Killing form is sometimes referred to as a "strictly semi-simple Lie superalgebra" (cf. Nahm *et al.* (1976)). Theorem II above shows that such superalgebras are either simple Lie superalgebras with non-degenerate Killing forms or are the direct sums of such simple Lie superalgebras.

3 Classical simple Lie superalgebras

(*a*) *Definition and basic theorems*

Definition *Classical simple Lie superalgebra*

A simple Lie superalgebra is said to be *classical* if the representation of its even part $\mathscr{L}_0$ on its odd part $\mathscr{L}_1$ is either *irreducible*, or, if it is reducible, it is *completely* reducible.

Only this *classical* type of simple Lie superalgebra will be discussed in detail in this book, for with the non-classical simple Lie superalgebras, which are known as "Cartan Lie superalgebras", not only is the analysis rather more involved but these algebras have not found significant applications in mathematical physics. (For details of the Cartan Lie superalgebras see Kac (1977a, b), Rittenberg (1978) and Scheunert (1979).)

The even parts of classical simple Lie superalgebras involve a type of Lie algebra known as a "reductive" Lie algebra that was not considered explicitly in Volume II, but which is closely related to the semi-simple Lie algebras that were considered in detail there. The Theorems I–IV that follow can all be found in Bourbaki (1971), which also contains a more extended discussion of this type of Lie algebra than the presentation that will be given here, which will be confined to the essential properties only.

Definition *Reductive Lie algebra*

A Lie algebra is said to be *reductive* if its adjoint representation is either irreducible, or, if it is reducible, it is completely reducible.

Theorem I A Lie algebra $\mathscr{L}$ is reductive if and only if it is Abelian, or it is semi-simple, or it is the direct sum of an Abelian and a semi-simple Lie algebra.

Theorem II A Lie algebra is reductive if and only if it possesses a faithful representation that is either irreducible, or, if it is reducible, it is completely reducible.

Definition *Kernel of a representation* $\mathbf{\Gamma}$ of a Lie algebra $\mathscr{L}$

The *kernel* $\ker_\Gamma$ of a representation $\mathbf{\Gamma}$ of a Lie algebra $\mathscr{L}$ is the set of elements of $\mathscr{L}$ that are represented in this representation by the zero matrix.

Definition *Nilpotent radical of a Lie algebra* $\mathscr{L}$

The *nilpotent radical* of a Lie algebra $\mathscr{L}$ is the intersection of the kernels $\ker_\Gamma$ of all the finite-dimensional irreducible representations $\mathbf{\Gamma}$ of $\mathscr{L}$.

Theorem III A Lie algebra $\mathscr{L}$ is reductive if and only if its nilpotent radical consists solely of the zero element 0 of $\mathscr{L}$.

For every representation $\mathbf{\Gamma}$ of a Lie algebra $\mathscr{L}$ an invariant bilinear form $B_\Gamma(\ ,\ )$ of $\mathscr{L}$ may be defined by

$$B_\Gamma(a, b) = \mathrm{tr}\,\{\mathbf{\Gamma}(a)\mathbf{\Gamma}(b)\} \tag{25.3}$$

(cf. Chapter 16, Section 3). The "orthogonal complement of $\mathscr{L}$ with respect to $B_\Gamma(\ ,\ )$", $\mathscr{L}_\Gamma^\perp$, can be defined as the subset of elements $a \in \mathscr{L}$ with the property that

$$B_\Gamma(a, b) = 0 \tag{25.4}$$

for all $b \in \mathscr{L}$.

Theorem IV The nilpotent radical of a Lie algebra $\mathscr{L}$ is the intersection of the orthogonal complements $\mathscr{L}_\Gamma^\perp$ of $\mathscr{L}$ with respect to the bilinear forms $B_\Gamma(\ ,\ )$ associated with *all* the finite-dimensional representations $\mathbf{\Gamma}$ of $\mathscr{L}$.

Armed with this collection of results on reductive Lie algebras, it is now possible to proceed with the development of the theory of classical simple Lie superalgebras.

Theorem V A simple Lie superalgebra $\mathscr{L}_s$ is *classical* if and only if its even subspace $\mathscr{L}_0$ is a *reductive* Lie algebra.

Proof Suppose that $\mathscr{L}_s$ is a classical simple Lie superalgebra. By Theorem IV of Section 2, since $\mathscr{L}_s$ is simple, the representation of $\mathscr{L}_0$ on the odd part $\mathscr{L}_1$ is faithful, and since $\mathscr{L}_s$ is classical, the definition requires that this representation be either irreducible or completely reducible. Thus, by Theorem II, $\mathscr{L}_0$ must be reductive.

For a proof of the converse proposition (which is appreciably longer) see Scheunert (1979a).

Theorems I and IV taken together imply that if $\mathscr{L}_s$ is a classical simple Lie algebra then its even part $\mathscr{L}_0$ has the direct sum form

$$\mathscr{L}_0 = \mathscr{L}_0^{A} \oplus \mathscr{L}_0^{ss}, \tag{25.5}$$

where $\mathscr{L}_0^{A}$ is Abelian and $\mathscr{L}_0^{ss}$ is semi-simple. The elements of $\mathscr{L}_0^{A}$ commute with all the elements of $\mathscr{L}_0$ and form the "centre" of $\mathscr{L}_0$. Examples of classical simple Lie superalgebras occur with both $\mathscr{L}_0^{A}$ and $\mathscr{L}_0^{ss}$ non-trivial, and in other examples $\mathscr{L}_0^{A} = \{0\}$ (in which case $\mathscr{L}_0 = \mathscr{L}_0^{ss}$, implying that $\mathscr{L}_0$ is semi-simple), but *none* occur with $\mathscr{L}_0^{ss} = \{0\}$ (as will be demonstrated later in this section).

Theorem IV If $\mathscr{L}_s$ is a classical simple Lie superalgebra and if the representation of $\mathscr{L}_0$ on $\mathscr{L}_1$ is *irreducible* then its even part $\mathscr{L}_0$ is *semi-simple.*

Proof See Appendix K, Section 5.

It should be noted that the converse of Theorem VI is *not* true, for there do exist examples of classical simple Lie superalgebras whose even parts are semi-simple but whose representations of even parts on odd parts are not irreducible. (One class of examples is provided by $\mathscr{L}_s = A(r/r)$ for $r \geqslant 1$. See Appendix M, Section 2.)

Theorem VII If $\mathscr{L}_s$ is a classical simple Lie superalgebra then the representation of $\mathscr{L}_0$ on $\mathscr{L}_1$ is *either*

(i) irreducible

or

(ii) equivalent to the direct sum of *two* irreducible representations of $\mathscr{L}_0$.

Moreover, in case (ii) if $\mathscr{L}_1^1$ and $\mathscr{L}_1^2$ are the invariant subspaces of $\mathscr{L}_1$ corresponding to these two irreducible representations then

$$[\mathscr{L}_1^1, \mathscr{L}_1^1] = [\mathscr{L}_1^2, \mathscr{L}_1^2] = 0 \tag{25.6}$$

and

$$[\mathscr{L}_1^1, \mathscr{L}_1^2] = \mathscr{L}_0. \tag{25.7}$$

Proof See Appendix K, Section 5.

In developing the representation theory, it is convenient to use this theorem to divide the classical simple Lie superalgebras with non-trivial odd parts into two types.

Definition *Type I and type II classical simple Lie superalgebras*

A classical simple Lie superalgebra with non-trivial odd part $\mathscr{L}_1$ is said to be of *type I* if the representation of $\mathscr{L}_0$ on $\mathscr{L}_1$ is equivalent to the direct sum of *two* irreducible representations of $\mathscr{L}_0$, and is said to be of *type II* if the representation of $\mathscr{L}_0$ on $\mathscr{L}_1$ is *irreducible.*

Theorem VIII If $\mathscr{L}_s$ is a complex classical simple Lie superalgebra and the centre of its even part $\mathscr{L}_0$ is *non-trivial* then

(i) $\mathscr{L}_0^A$ is *one*-dimensional, and

(ii) the representation of $\mathscr{L}_0$ on $\mathscr{L}_1$ is equivalent to the direct sum of *two* irreducible representations of $\mathscr{L}_0$.

Moreover, if $\mathscr{L}_1^1$ and $\mathscr{L}_1^2$ are the two invariant subspaces of $\mathscr{L}_1$ corresponding to the two irreducible representations then there exists an element c of $\mathscr{L}_0^A$ such that

$$[c, b^1] = b^1 \tag{25.8}$$

for all $b^1 \in \mathscr{L}_1^1$, and

$$[c, b^2] = -b^2 \tag{25.9}$$

for all $b^2 \in \mathscr{L}_1^2$.

Proof See Appendix K, Section 5.

Using this theorem, it is easily shown that there are *no* classical simple Lie superalgebras with $\mathscr{L}_0^{ss} = \{0\}$, that is, with $\mathscr{L}_0$ Abelian. (For, if such a superalgebra were to exist, Theorem VIII demonstrates that its $\mathscr{L}_0^A$ would be one-dimensional, and as all the irreducible representations of an Abelian Lie algebra are one-dimensional, both $\mathscr{L}_1^1$ and $\mathscr{L}_1^2$ would be one-dimensional. It is then easily checked that (25.6)–(25.9) are incompatible in this case.)

Theorem IX If $\mathscr{L}_s$ is a simple Lie superalgebra and the Killing form of $\mathscr{L}_s$ is *non-degenerate* then $\mathscr{L}_s$ is *classical.*

Proof See Appendix K, Section 5.

It will be shown by producing examples that there also exist classical simple Lie superalgebras with *zero* Killing form, which is the other possibility allowed by Theorem I of Section 2.

For the rest of this section only *complex* Lie superalgebras will be considered. Such a superalgebra will be denoted by $\tilde{\mathscr{L}}_s$. If $\tilde{\mathscr{L}}_s$ is a classical simple complex Lie superalgebra then its even part $\tilde{\mathscr{L}}_0$ is a reductive complex

Lie algebra whose form is

$$\tilde{\mathscr{L}}_0 = \tilde{\mathscr{L}}_0^{\mathrm{A}} \oplus \tilde{\mathscr{L}}_0^{\mathrm{ss}}, \tag{25.10}$$

according to (25.5), where $\tilde{\mathscr{L}}_0^{\mathrm{A}}$ is an Abelian complex Lie algebra and $\tilde{\mathscr{L}}_0^{\mathrm{ss}}$ is a semi-simple complex Lie algebra. Of course Theorem VIII implies that if $\tilde{\mathscr{L}}_0^{\mathrm{A}}$ is non-trivial then $\tilde{\mathscr{L}}_0^{\mathrm{A}}$ is one-dimensional.

Definition *Cartan subalgebra $\mathscr{H}_{\mathrm{s}}$ of a classical simple complex Lie superalgebra $\tilde{\mathscr{L}}_{\mathrm{s}}$*

Let $\mathscr{H}_0^{\mathrm{ss}}$ be a Cartan subalgebra of the semi-simple complex Lie algebra $\tilde{\mathscr{L}}_0^{\mathrm{ss}}$ of (25.10). Then

$$\mathscr{H}_{\mathrm{s}} = \tilde{\mathscr{L}}_0^{\mathrm{A}} \oplus \mathscr{H}_0^{\mathrm{ss}} \tag{25.11}$$

is said to be a *Cartan subalgebra of* $\tilde{\mathscr{L}}_{\mathrm{s}}$.

Of course, if $\tilde{\mathscr{L}}_0^{\mathrm{A}} = \{0\}$ then (25.11) implies that

$$\mathscr{H}_{\mathrm{s}} = \mathscr{H}_0^{\mathrm{ss}}. \tag{25.12}$$

It can be shown (cf. Kac 1977a) that every inner automorphism of $\tilde{\mathscr{L}}_0$ can be extended to become an automorphism of $\tilde{\mathscr{L}}_{\mathrm{s}}$, and since all Cartan subalgebras of $\tilde{\mathscr{L}}_0$ are related by inner automorphisms of $\tilde{\mathscr{L}}_0$, it follows that any two Cartan subalgebras of $\tilde{\mathscr{L}}_{\mathrm{s}}$ are related by an automorphism of $\tilde{\mathscr{L}}_{\mathrm{s}}$. Consequently all Cartan subalgebras of $\tilde{\mathscr{L}}_{\mathrm{s}}$ have the same dimension.

Definition *The rank of a classical simple complex Lie superalgebra $\tilde{\mathscr{L}}_{\mathrm{s}}$*

The *rank* l of a classical simple complex Lie superalgebra $\tilde{\mathscr{L}}_{\mathrm{s}}$ is defined to be the dimension of its Cartan subalgebras.

As all Cartan subalgebras of $\tilde{\mathscr{L}}_{\mathrm{s}}$ are equivalent, it is sufficient to make a specific choice of one Cartan subalgebra $\mathscr{H}_{\mathrm{s}}$, and to always work with this. Clearly

$$[h, h'] = 0 \tag{25.13}$$

for all $h, h' \in \mathscr{H}_{\mathrm{s}}$.

If $\tilde{\mathscr{L}}_0^{\mathrm{A}} = \{0\}$, so that $\tilde{\mathscr{L}}_0 = \tilde{\mathscr{L}}_0^{\mathrm{ss}}$, then the considerations of Chapter 15, Section 2 imply that, since the odd part $\tilde{\mathscr{L}}_1$ of $\tilde{\mathscr{L}}_{\mathrm{s}}$ is the carrier space of a representation of $\tilde{\mathscr{L}}_0$, $\tilde{\mathscr{L}}_1$ can be written as the direct sum of a set of weight subspaces spanned by the eigenvectors of (15.3). Taken with the root property (13.8) of $\tilde{\mathscr{L}}_0^{\mathrm{ss}}$ itself, this shows that the *whole* of $\tilde{\mathscr{L}}_{\mathrm{s}}$ can be regarded as being the vector space direct sum of root subspaces $\tilde{\mathscr{L}}_{\mathrm{s}\alpha}$. That is, if $\alpha(h)$ is any linear functional defined on $\mathscr{H}_{\mathrm{s}}$ for which there exists at least one element

$a_\alpha \in \tilde{\mathscr{L}}_s$ such that

$$[h, a_\alpha] = \alpha(h) a_\alpha \tag{25.14}$$

for all $h \in \mathscr{H}_s$ then α is said to be a "root" of $\tilde{\mathscr{L}}_s$, and the set of all $a_\alpha \in \tilde{\mathscr{L}}_s$ that satisfy (25.14) forms the "root subspace" $\tilde{\mathscr{L}}_{s\alpha}$.

The same results are true for the other case in which $\tilde{\mathscr{L}}_0^A$ is non-trivial, for then the (one-dimensional) basis of $\tilde{\mathscr{L}}_0^A$ may be taken to be the element c of Theorem VIII. Since c commutes with every element of $\tilde{\mathscr{L}}_0^{ss}$, every root α of $\tilde{\mathscr{L}}_0^{ss}$ can be extended to become a root of $\tilde{\mathscr{L}}_0$ by the definition $\alpha(c) = 0$, and then (13.8), and hence (25.14), becomes valid for all $h \in \mathscr{H}_s$ $(= \tilde{\mathscr{L}}_0^A \oplus \mathscr{H}_0^{ss})$ and for every corresponding element a_α of $\tilde{\mathscr{L}}_0^{ss}$. Moreover, since (25.8) and (25.9) have the same form as (25.14) (with $\alpha(c) = \pm 1$), it is clear that $\tilde{\mathscr{L}}_s$ can again be written as the vector space direct sum of root subspaces $\tilde{\mathscr{L}}_{s\alpha}$, the elements of which satisfy (25.14).

If an element a_α of (25.14) is a member of $\tilde{\mathscr{L}}_0$ then α is said to be an *even root* of $\tilde{\mathscr{L}}_s$, whereas if $a_\alpha \in \tilde{\mathscr{L}}_1$ then α is said to be an *odd root*. (It is possible for a root to be both even and odd, but this situation only occurs for $\tilde{\mathscr{L}}_s = Q(r)$ with $r \geqslant 2$ (cf. Appendix M, Section 10)). The set of all distinct non-zero even roots will be denoted by Δ_0, and the set of distinct odd roots will be denoted by Δ_1. (This latter definition does not exclude the possibility that Δ_1 contains an odd root that is identically zero, but again this situation only occurs for $\tilde{\mathscr{L}}_s = Q(r)$ with $r \geqslant 2$ (cf. Appendix M, Section 10).) The set of distinct roots of $\tilde{\mathscr{L}}_s$ contained in Δ_0 or Δ_1 (or both) will be indicated by Δ_s, so $\Delta_s = \Delta_0 \cap \Delta_1$.

It is sometimes convenient to regard $\mathscr{H}_s$ as being the subspace of $\tilde{\mathscr{L}}_s$ corresponding to *zero even root*, in which case $\mathscr{H}_s$ becomes a subset of $\tilde{\mathscr{L}}_{s0}$.

Theorem X If $\alpha \in \Delta_0$ then $\dim \tilde{\mathscr{L}}_{s\alpha} = 1$, and $k\alpha \in \Delta_0$ only if $k = 1$ or -1.

Proof Clearly if $\tilde{\mathscr{L}}_0 = \tilde{\mathscr{L}}_0^{ss}$ then the even roots of $\tilde{\mathscr{L}}_s$ are the roots of $\tilde{\mathscr{L}}_0^{ss}$. If $\tilde{\mathscr{L}}_0^A$ is non-trivial then every root α of $\tilde{\mathscr{L}}_0^{ss}$, being a linear functional on $\mathscr{H}_0^{ss}$, can (as above) be trivially extended to the whole of $\mathscr{H}_s$ by assuming that $\alpha = 0$ for $h \in \tilde{\mathscr{L}}_0^A$. With this extension, there is again a one-to-one correspondence between the roots of $\tilde{\mathscr{L}}_0^{ss}$ and the even roots of $\tilde{\mathscr{L}}_s$. The stated result then follows immediately from Theorem VIII of Chapter 13, Section 5.

Examination of the results given in Appendix M shows that $\dim \tilde{\mathscr{L}}_{s\alpha} = 1$ for every $\alpha \in \Delta_1$ and for every classical simple complex Lie superalgebra $\tilde{\mathscr{L}}_s$ *except* for all the odd roots of $\tilde{\mathscr{L}}_s = A(1/1)$ and the odd root $\alpha = 0$ of $\tilde{\mathscr{L}}_s = Q(r)$ with $r \geqslant 2$.

Let $\mathscr{H}_{0R}^{ss}$ be the real vector space that was introduced in Chapter 13, Section 5, and called there $\mathscr{H}_R$. The basis elements of $\mathscr{H}_{0R}^{ss}$ provide a basis for $\mathscr{H}_0^{ss}$, and, as shown by the Theorem VI of Chapter 13, Section 5, for every root α of $\tilde{\mathscr{L}}_0^{ss}$ the quantity $\alpha(h)$ is *real* for all $h \in \mathscr{H}_{0\mathbf{R}}^{ss}$. Now define the corresponding *real* space $\mathscr{H}_{s\mathbf{R}}$ for the Cartan subalgebra $\mathscr{H}_s$ of $\tilde{\mathscr{L}}_s$ by putting $\mathscr{H}_{s\mathbf{R}} = \mathscr{H}_{0\mathbf{R}}^{ss}$ if $\tilde{\mathscr{L}}_0^{A} = \{0\}$, and, in the case in which $\tilde{\mathscr{L}}_0^{A}$ is non-trivial, by taking $\mathscr{H}_{s\mathbf{R}}$ to be the direct sum of $\mathscr{H}_{0\mathbf{R}}^{ss}$ with the one-dimensional real Lie algebra with the basis element c of $\tilde{\mathscr{L}}_0^{A}$ that was introduced in Theorem VIII above.

Theorem XI For any $\alpha \in \Delta_s$, $\alpha(h)$ is *real* for all $h \in \mathscr{H}_{s\mathbf{R}}$.

Proof If α is an *even* root of $\tilde{\mathscr{L}}_s$ (i.e. if $\alpha \in \Delta_0$) then α is a root of $\tilde{\mathscr{L}}_0^{ss}$ and so (as noted above) $\alpha(h)$ is real on $\mathscr{H}_{0\mathbf{R}}^{ss}$. If $\tilde{\mathscr{L}}_0^{A} \neq \{0\}$ then $\alpha(c) = 0$, so $\alpha(h)$ is real for all $h \in \mathscr{H}_{s\mathbf{R}}$.

If α is an *odd* root of $\tilde{\mathscr{L}}_s$ (i.e. if $\alpha \in \Delta_1$) then α is a weight of $\tilde{\mathscr{L}}_0^{ss}$, and so, by Theorem III of Chapter 15, Section 2, $\alpha(h)$ is real for all $h \in \mathscr{H}_{0\mathbf{R}}^{ss}$. If $\tilde{\mathscr{L}}_0^{A} \neq \{0\}$ then Theorem VIII above implies that $\alpha(c) = \pm 1$, and so $\alpha(h)$ is again real for all $h \in \mathscr{H}_{s\mathbf{R}}$.

Theorem XII If $a_\alpha \in \tilde{\mathscr{L}}_{s\alpha}$ and $a_\beta \in \tilde{\mathscr{L}}_{s\beta}$ then $[a_\alpha, a_\beta] \in \tilde{\mathscr{L}}_{s(\alpha+\beta)}$ if $\alpha + \beta$ is a root and $[a_\alpha, a_\beta] = 0$ if $\alpha + \beta$ is not a root.

Proof This is a straightforward generalization of the corresponding result for semi-simple Lie algebras (Theorem I of Chapter 13, Section 4). By the generalized Jacobi identity (21.3), if $a_\alpha \in \tilde{\mathscr{L}}_{s\alpha}$, $a_\beta \in \tilde{\mathscr{L}}_{s\beta}$ and $h \in \mathscr{H}_s$ (so that $\deg h = 0$) then

$$[h, [a_\alpha, a_\beta]] + [a_\alpha, [a_\beta, h]](-1)^{(\deg a_\alpha)(\deg a_\beta)} + [a_\beta, [h, a_\alpha]] = 0,$$

so, by (25.14) and (21.2),

$$[h, [a_\alpha, a_\beta]] = \{\alpha(h) + \beta(h)\}[a_\alpha, a_\beta], \tag{25.15}$$

from which the stated result follows immediately.

As every semi-simple complex Lie algebra is either simple or is the direct sum of simple complex Lie algebras, $\tilde{\mathscr{L}}_0^{ss}$ can be written either as

$$\tilde{\mathscr{L}}_0^{ss} = \tilde{\mathscr{L}}_0^{s1} \tag{25.16}$$

or as

$$\tilde{\mathscr{L}}_0^{ss} = \tilde{\mathscr{L}}_0^{s1} \oplus \tilde{\mathscr{L}}_0^{s2} \oplus \cdots, \tag{25.17}$$

where the $\tilde{\mathscr{L}}_0^{sj}$ $(j = 1, \ldots)$ are all *simple*. For any $a \in \tilde{\mathscr{L}}_0$ its adjoint representation with respect to $\tilde{\mathscr{L}}_s$ (as defined in (21.75)) has the form

$$\mathbf{ad}(a) = \begin{bmatrix} \mathbf{ad}_0(a) & \mathbf{0} \\ \mathbf{0} & \mathbf{D}(a) \end{bmatrix}, \tag{25.18}$$

where $\mathbf{ad}_0$ is the adjoint representation of $\tilde{\mathscr{L}}_0$ and $\mathbf{D}$ (as defined in (21.16)) is the representation of $\tilde{\mathscr{L}}_0$ on $\tilde{\mathscr{L}}_1$. Thus, by (21.78) and (21.80), the Killing form $B(\ ,\ )$ of $\tilde{\mathscr{L}}_s$ is related to the Killing form $B_0(\ ,\ )$ of $\tilde{\mathscr{L}}_0$ by

$$B(a, b) = B_0(a, b) - \mathrm{tr}\,\{\mathbf{D}(a)\mathbf{D}(b)\} \tag{25.19}$$

for all $a, b \in \tilde{\mathscr{L}}_0$. In particular, for $a, b \in \tilde{\mathscr{L}}_0^{sj}$ $(j = 1, \ldots)$, if $\gamma_{\mathbf{D}}^j$ is the Dynkin index of $\mathbf{D}$ regarded as a representation of $\tilde{\mathscr{L}}_0^{sj}$, so that by (16.2)

$$\mathrm{tr}\,\{\mathbf{D}(a)\mathbf{D}(b)\} = \gamma_{\mathbf{D}}^j B_0^j(a, b) \tag{25.20}$$

for all $a, b \in \tilde{\mathscr{L}}_0^{sj}$, then it follows that

$$B(a, b) = (1 - \gamma_{\mathbf{D}}^j) B_0^j(a, b) \tag{25.21}$$

for all $a, b \in \tilde{\mathscr{L}}_0^{sj}$, where $B_0^j(\ ,\ )$ is the Killing form of $\tilde{\mathscr{L}}_0^{sj}$. (It should be noted that the factor $1 - \gamma_{\mathbf{D}}^j$ in (25.21) is always *real*.)

Theorem XIII If the Killing form $B(\ ,\ )$ of a classical simple complex Lie superalgebra $\tilde{\mathscr{L}}_s$ is identically *zero* then

$$\gamma_{\mathbf{D}}^j = 1$$

for every simple complex Lie algebra $\tilde{\mathscr{L}}_0^{sj}$ in $\tilde{\mathscr{L}}_0$.

Proof This is an immediate consequence of (25.21).

To produce a theory of roots similar to that presented in Chapter 13 for semi-simple complex Lie algebras, it is necessary to make a *further* assumption and restrict attention to the so-called *basic* classical simple complex Lie superalgebras. These form the subject of the next subsection.

(*b*) *Basic classical simple complex Lie superalgebras*

The set of classical simple Lie superalgebras can be divided into two mutually exclusive subsets.

Definition *Basic classical simple Lie superalgebra*

A classical simple Lie superalgebra is said to be *basic* if it possesses a *non-degenerate* bilinear supersymmetric consistent invariant form $B'(\ ,\ )$.

Definition *Strange classical simple Lie superalgebra*

A classical simple Lie superalgebra is said to be *strange* if it does *not* possess any *non-degenerate* bilinear supersymmetric consistent invariant form.

The implication of this definition is that if a classical simple Lie superalgebra is *strange* then *every* bilinear supersymmetric consistent invariant form that it possesses is identically *zero*.

The subset of *basic* classical simple complex Lie superalgebras can itself be divided into two categories:

(a) those for which the *Killing form* is *non-degenerate*—for these the form $B'(\ ,\ )$ of the above definition can be taken to be the Killing form $B(\ ,\ )$;

(b) those for which the *Killing form* is *identically zero*—for these the form $B'(\ ,\ )$ is necessarily completely independent of the Killing form $B(\ ,\ )$.

Examples exist of both of these categories, and of the strange classical simple Lie superalgebras. A complete list of the classical simple complex Lie superalgebras is given in Table 25.1. Definitions and details of all of these may be found in Appendix M. The division into types I and II is summarized in Table 25.2.

Many of the results on the roots of semi-simple complex Lie algebras that were given in Sections 5–7 of Chapter 13 generalize in a straightforward way when the basic classical simple complex Lie superalgebras are considered,

Table 25.1 Classical simple complex Lie superalgebras.

(1) Basic classical simple complex Lie superalgebras

(a) With non-degenerate Killing form:

(i) Simple complex Lie algebras;

(ii) $A(r/s)$, $r > s \geqslant 0$;
$B(r/s)$, $r \geqslant 0$, $s \geqslant 1$;
$C(s)$, $s \geqslant 2$;
$D(r/s)$, $r \geqslant 2$, $s \geqslant 1$, $r \neq s+1$;
$F(4)$;
$G(3)$.

(b) With zero Killing form:
$A(r/r)$, $r \geqslant 1$;
$D(s+1/s)$, $s \geqslant 1$;
$D(2/1;\alpha)$, α being any complex number except 0, -1 or ∞.

(2) Strange classical simple complex Lie superalgebras:
$P(r)$, $r \geqslant 2$;
$Q(r)$, $r \geqslant 2$.

Table 25.2 Division of the classical simple complex Lie superalgebras with non-trivial odd parts into types I and II.

(1) Type I:

$A(r/s)$, $r > s \geqslant 0$;
$A(r/r)$, $r \geqslant 1$:
$C(s)$, $s \geqslant 2$;
$P(r)$, $r \geqslant 2$.

(2) Type II:

$B(r/s)$, $r \geqslant 0$, $s \geqslant 1$;
$D(r/s)$, $r \geqslant 2$, $s \geqslant 1$;
$D(2/1; \alpha)$, α being any complex number except 0, -1 or ∞;
$F(4)$;
$G(3)$;
$Q(r)$, $r \geqslant 2$.

so attention will be confined to these superalgebras for the rest of this subsection and the whole of the next subsection.

Theorem XIV If $\tilde{\mathscr{L}}_s$ is a basic classical simple complex Lie superalgebra with non-degenerate bilinear supersymmetric consistent invariant form $B'(\ ,\ )$, and if $a_\alpha \in \tilde{\mathscr{L}}_{s\alpha}$, $a_\beta \in \tilde{\mathscr{L}}_{s\beta}$ and $\alpha + \beta \neq 0$, then

$$B'(a_\alpha, a_\beta) = 0. \tag{25.22}$$

Proof As $B'(\ ,\ )$ is invariant, for all $h \in \mathscr{H}_s$

$$B'([a_\alpha, h], a_\beta) = B'(a_\alpha, [h, a_\beta])$$

(cf. (21.86)), and so, from (25.14),

$$\{\alpha(h) + \beta(h)\} B'(a_\alpha, a_\beta) = 0.$$

Thus $B'(a_\alpha, a_\beta) = 0$ if $\alpha(h) + \beta(h) \neq 0$ for some $h \in \mathscr{H}_s$.

Of course, if $a_\alpha \in \tilde{\mathscr{L}}_0$ and $a_\beta \in \tilde{\mathscr{L}}_1$ then the consistency property (cf. (21.84)) automatically implies that

$$B'(a_\alpha, a_\beta) = 0. \tag{25.23}$$

As $\mathscr{H}_s$ is a subset of $\tilde{\mathscr{L}}_{s0}$, it follows that for any $\alpha \in \Delta_s$

$$B'(h, a_\alpha) = 0, \tag{25.24}$$

the result for even non-zero roots being implied by (25.22) and for odd roots by (25.23). Also, if $a_\alpha \in \tilde{\mathscr{L}}_{s\alpha}$ and $\alpha \neq 0$ then (25.22) shows that

$$B'(a_\alpha, a_\alpha) = 0.$$

Theorem XV If $\tilde{\mathscr{L}}_s$ is a basic classical simple complex Lie superalgebra with non-degenerate bilinear supersymmetric consistent invariant form $B'(\ ,\ )$ then $B'(\ ,\ )$ provides a non-degenerate symmetric bilinear form on $\mathscr{H}_s$.

Proof All that has to be shown is that $B'(\ ,\ )$ is *non-degenerate* on $\mathscr{H}_s$; that is, if $h' \in \mathscr{H}_s$ and $B'(h', h) = 0$ for all $h \in \mathscr{H}_s$ then $h' = 0$ (for it is obvious that $B'(\ ,\ )$ is symmetric and bilinear on $\mathscr{H}_s$). Suppose therefore that $h' \in \mathscr{H}_s$ and $B'(h', h) = 0$ for all $h \in \mathscr{H}_s$. Then, by (25.24), $B'(h', a) = 0$ for all $a \in \tilde{\mathscr{L}}_s$, and, since $B'(\ ,\ )$ is assumed to be non-degenerate on $\tilde{\mathscr{L}}_s$, it follow that $h' = 0$.

Theorem XVI The Killing form of a basic classical simple complex Lie superalgebra $\tilde{\mathscr{L}}_s$ is *non-degenerate* if and only if

$$\gamma_D^j \neq 1$$

for every simple Lie algebra $\tilde{\mathscr{L}}_0^{sj}$ in $\tilde{\mathscr{L}}_0$ (cf. (25.16) and (25.17)), γ_D^j being the Dynkin index of (25.20) and (25.21).

Proof See Appendix K, Section 5.

It follows from Theorem XV that to every linear functional $\alpha(h)$ on $\mathscr{H}_s$, and in particular to every root $\alpha \in \Delta_s$, a unique element h_α of $\mathscr{H}_s$ can be defined by the prescription

$$B'(h, h_\alpha) = \alpha(h) \tag{25.25}$$

for all $h \in \mathscr{H}_s$ (cf. Theorem I of Appendix B, Section 6). If $\alpha(h)$ and $\beta(h)$ are any two linear functionals on $\mathscr{H}_s$ then

$$h_{\alpha+\beta} = h_\alpha + h_\beta, \tag{25.26}$$

and, since $B'(\ ,\ )$ is symmetric on $\mathscr{H}_s$,

$$\alpha(h_\beta) = \beta(h_\alpha) = B'(h_\alpha, h_\beta).$$

With $\langle \alpha, \beta \rangle$ defined by

$$\langle \alpha, \beta \rangle = B'(h_\alpha, h_\beta), \tag{25.27}$$

it follows that

$$\langle \alpha, \beta \rangle = \langle \beta, \alpha \rangle \tag{25.28}$$

and

$$\langle \alpha, \beta \rangle = \alpha(h_\beta) = \beta(h_\alpha). \tag{25.29}$$

Theorem XVII Let $\tilde{\mathscr{L}}_s$ be a basic classical simple complex Lie superalgebra with *non-degenerate* Killing form $B(\ ,\ )$. If the form $B'(\ ,\ )$ is taken to be identical with $B(\ ,\ )$ then the quantity $\langle\alpha,\beta\rangle$ is *real* for every pair of roots α and β of $\tilde{\mathscr{L}}_s$.

Proof Theorem XI of the previous subsection and the superalgebra extension of (13.16) show that $B'(h,h')$ $(=B(h,h'))$ is real for all $h,h'\in\mathscr{H}_{sR}$. A further application of Theorem XI taken with the definition (25.25) demonstrates that $h_\alpha\in\mathscr{H}_{sR}$ for every $\alpha\in\Delta_s$, and so, by (25.27), $\langle\alpha,\beta\rangle$ is real for all $\alpha,\beta\in\Delta_s$.

Inspection of the results given in Appendix M, Sections 2, 5 and 6 shows that for the basic classical simple complex Lie superalgebras with *identically zero* Killing form the situation is more complicated. For $\tilde{\mathscr{L}}_s=A(r/r)$ (for $r\geqslant 1$) and for $\tilde{\mathscr{L}}_s=D(s+1/s)$ (for $s\geqslant 1$) it remains true that $\langle\alpha,\beta\rangle$ is real for all $\alpha,\beta\in\Delta_s$, *but* for $\tilde{\mathscr{L}}_s=D(2/1;\alpha)$ with the parameter α non-real there is *no* choice of $B'(\ ,\ )$ for which all the $\langle\alpha,\beta\rangle$ corresponding to roots are real.

Theorem XVIII If $\tilde{\mathscr{L}}_s$ is a basic classical simple complex Lie superalgebra and if $\alpha\in\Delta_s$ then $-\alpha\in\Delta_s$.

Proof Suppose (to the contrary) that $\alpha\in\Delta_s, a_\alpha\in\tilde{\mathscr{L}}_{s\alpha}$, but $-\alpha\notin\Delta_s$. Then Theorem XIV above implies that $B'(a_\alpha,a)=0$ for all $a\in\tilde{\mathscr{L}}_s$, which is not possible since $B'(\ ,\ )$ is assumed to be non-degenerate. Thus $-\alpha\notin\Delta_s$.

Inspection of Appendix M, Sections 9 and 10 shows that this result is also true for the *strange* superalgebras $Q(r)$ for $r\geqslant 2$, but *not* for the *strange* superalgebras $P(r)$ for $r\geqslant 2$.

Theorem XIX If $\tilde{\mathscr{L}}_s$ is a basic classical simple complex Lie superalgebra with non-degenerate bilinear supersymmetric consistent invariant form $B'(\ ,\ )$, and if $a_\alpha\in\tilde{\mathscr{L}}_{s\alpha}$ and $a_{-\alpha}\in\tilde{\mathscr{L}}_{s(-\alpha)}$, then

$$[a_\alpha,a_{-\alpha}]=B'(a_\alpha,a_{-\alpha})h_\alpha. \tag{25.30}$$

Moreover, for each $\alpha\in\Delta_s$ and any $a_\alpha\in\tilde{\mathscr{L}}_{s\alpha}$ there exists an element $a_{-\alpha}\in\tilde{\mathscr{L}}_{s(-\alpha)}$ such that

$$B'(a_\alpha,a_{-\alpha})\neq 0. \tag{25.31}$$

Proof This is just an obvious generalization of the proof of the corresponding results for semi-simple complex Lie algebras (Theorems I and II of Chapter 13, Section 5).

For semi-simple complex Lie algebras Theorem III of Chapter 13,

Section 5 shows that $\langle \alpha, \alpha \rangle > 0$ for *every* non-zero root α. However, this result does *not* generalize to the basic classical simple complex Lie superalgebras, since for these there are two new features.

(i) If α is an *even* non-zero root of $\tilde{\mathscr{L}}_s$ (i.e. if $\alpha \in \Delta_0$) then

$$\langle \alpha, \alpha \rangle \neq 0, \tag{25.32}$$

but $\langle \alpha, \alpha \rangle$ need not be real and positive. If $\tilde{\mathscr{L}}_s$ has *non-degenerate* Killing form, Theorem XVII shows that $\langle \alpha, \alpha \rangle$ is real, but $\langle \alpha, \alpha \rangle$ may be positive *or* negative. The reason for this is that if $\alpha \in \Delta_0$ then there exists a simple Lie algebra $\tilde{\mathscr{L}}_0^{sj}$ in $\tilde{\mathscr{L}}_0^{ss}$ (with Cartan subalgebra $\mathscr{H}_0^j$) and a root α^j of $\tilde{\mathscr{L}}_0^{sj}$ such that

$$\alpha(h) = \begin{cases} \alpha^j(h) & \text{for } h \in \mathscr{H}_0^j, \\ 0 & \text{for } h \in \mathscr{H}_s \text{ but } h \notin \mathscr{H}_0^j. \end{cases} \tag{25.33}$$

(The root α may be called the *extension* of the root α^j, and no confusion will arise if the same symbol is used for both (as is done in Appendix M).) It then follows from (25.21) that

$$\langle \alpha, \alpha \rangle = \frac{\langle \alpha^j, \alpha^j \rangle_0^j}{1 - \gamma_D^j}, \tag{25.34}$$

where $\langle \alpha^j, \alpha^j \rangle_0^j$ is evaluated for $\tilde{\mathscr{L}}_0^{sj}$ from (13.14) using the Killing form $B_0^j(\ ,\ )$ of $\tilde{\mathscr{L}}_0^{sj}$. Although $\langle \alpha^j, \alpha^j \rangle_0^j > 0$, the quantity $1 - \gamma_D^j$ can be positive or negative. For example, for $\tilde{\mathscr{L}}_s = A(r/s)$ with $r > s \geqslant 0$, taking $\tilde{\mathscr{L}}_0^{s1} = A_r$ and $\tilde{\mathscr{L}}_0^{s2} = A_s$, (M.4) shows that $1 - \gamma_D^1 = (r-s)/(r+1) > 0$ but $1 - \gamma_D^2 = -(r-s)/(r+1) < 0$.

(ii) If α is an odd root of $\tilde{\mathscr{L}}_s$ (i.e. if $\alpha \in \Delta_1$) it is possible to have

$$\langle \alpha, \alpha \rangle = 0 \tag{25.35}$$

even when α is not identically zero. This is illustrated by the following example.

Example I *Roots of the basic classical simple complex Lie superalgebra* $\tilde{\mathscr{L}}_s = A(1/0)$

$A(1/0)$ is the complex Lie superalgebra sl$(2/1;\mathbb{C})$ (cf. Appendix M, Section 1, and Example II of Chapter 21, Section 2). Its even part $\tilde{\mathscr{L}}_0$ is given by

$$\tilde{\mathscr{L}}_0 = \tilde{\mathscr{L}}_0^A \oplus \tilde{\mathscr{L}}_0^{ss}, \tag{25.36}$$

where $\tilde{\mathscr{L}}_0^A$ is a one-dimensional Abelian Lie algebra and

$$\tilde{\mathscr{L}}_0^{ss} = \tilde{\mathscr{L}}_0^{s1} = A_1. \tag{25.37}$$

The Cartan subalgebra $\mathscr{H}_s$ is 2-dimensional and may be taken to have basis elements $\mathbf{h}_1^1$ and $\mathbf{c}$, where

$$\mathbf{h}_1^1 = \left[\begin{array}{cc|c} 1 & 0 & 0 \\ 0 & -1 & 0 \\ \hline 0 & 0 & 0 \end{array}\right] = \mathbf{e}_{11} - \mathbf{e}_{22}$$

and

$$\mathbf{c} = \left[\begin{array}{cc|c} -1 & 0 & 0 \\ 0 & -1 & 0 \\ \hline 0 & 0 & -2 \end{array}\right] = -\mathbf{e}_{11} - \mathbf{e}_{22} - 2\mathbf{e}_{33}.$$

Here $\mathbf{e}_{pq}$ denotes the 3×3 matrix whose (p, q) element has value 1 and whose other elements are all zero, and the grading partitioning has been displayed for extra clarity. Clearly $\mathbf{c}$ provides the basis element of $\tilde{\mathscr{L}}_0^A$ and $\mathbf{h}_1^1$ is the basis element of the Cartan subalgebra $\mathscr{H}_0^{ss}$ of $\tilde{\mathscr{L}}_0^{ss}$ $(= \tilde{\mathscr{L}}_0^{s1} = A_1)$. The remaining basis elements of $\tilde{\mathscr{L}}_0^{s1}$ may be taken to be $\mathbf{e}_{12}$ and $\mathbf{e}_{21}$, and $\mathbf{e}_{13}, \mathbf{e}_{23}, \mathbf{e}_{31}$ and $\mathbf{e}_{32}$ may be taken as the basis elements of the odd part $\tilde{\mathscr{L}}_1$. It is easily shown that $\tilde{\mathscr{L}}_s = A(1/0)$ is simple, and (25.36) taken with Theorem V of the previous subsection imply that it is classical.

As

$$[\mathbf{h}_1^1, \mathbf{e}_{13}] = \mathbf{e}_{13}, \quad [\mathbf{c}, \mathbf{e}_{13}] = \mathbf{e}_{13},$$

it follows that $\mathbf{e}_{13}$ is a basis element of a root subspace with odd root α' such that

$$\alpha'(\mathbf{h}_1^1) = 1, \quad \alpha'(\mathbf{c}) = 1. \tag{25.38}$$

Similarly $\mathbf{e}_{23}$ is an element of a root subspace with odd root α'' such that

$$\alpha''(\mathbf{h}_1^1) = -1, \quad \alpha''(\mathbf{c}) = 1, \tag{25.39}$$

and $\mathbf{e}_{12}$ belongs to the root subspace with even root α''' such that

$$\alpha'''(\mathbf{h}_1^1) = 2, \quad \alpha'''(\mathbf{c}) = 0. \tag{25.40}$$

Likewise $\mathbf{e}_{31}$, $\mathbf{e}_{32}$ and $\mathbf{e}_{31}$ are elements of root subspaces with odd root $-\alpha'$, odd root $-\alpha''$ and even root $-\alpha'''$ respectively. With the basis elements of $\tilde{\mathscr{L}}_s$ listed in the order $\mathbf{c}, \mathbf{h}_1^1, \mathbf{e}_{12}, \mathbf{e}_{21}, \mathbf{e}_{13}, \mathbf{e}_{31}, \mathbf{e}_{23}$ and $\mathbf{e}_{32}$, (21.75) shows that $\mathbf{ad}(\mathbf{c})$ and $\mathbf{ad}(\mathbf{h}_1^1)$ are diagonal with diagonal entries $\{0, 0, 0, 0, -1, 1, -1, 1\}$ and $\{0, 0, 2, -2, 1, -1, -1, 1\}$ respectively. Thus, by (21.80),

$$B(\mathbf{c}, \mathbf{c}) = -4, \quad B(\mathbf{h}_1^1, \mathbf{h}_1^1) = 4, \quad B(\mathbf{c}, \mathbf{h}_1^1) = 0. \tag{25.41}$$

Consequently (25.25) implies that the elements $\mathbf{h}_\alpha$ corresponding to the roots

$\alpha = \alpha', \alpha''$ and α''' are given by

$$\mathbf{h}_{\alpha'} = -\tfrac{1}{4}(\mathbf{c} - \mathbf{h}_1^1), \quad \mathbf{h}_{\alpha''} = -\tfrac{1}{4}(\mathbf{c} + \mathbf{h}_1^1), \quad \mathbf{h}_{\alpha'''} = \tfrac{1}{2}\mathbf{h}_1^1. \tag{25.42}$$

Thus, by (25.27) (with $B'(a,b) = B(a,b)$ for all $a, b \in \tilde{\mathscr{L}}_s$)

$$\langle \alpha', \alpha' \rangle = B(-\tfrac{1}{4}(\mathbf{c} - \mathbf{h}_1^1), -\tfrac{1}{4}(\mathbf{c} - \mathbf{h}_1^1)),$$

and so, by (25.41), for the odd root α'

$$\langle \alpha', \alpha' \rangle = 0. \tag{25.43}$$

Similarly for the other odd root α''

$$\langle \alpha'', \alpha'' \rangle = 0, \tag{25.44}$$

but for the even root α'''

$$\langle \alpha''', \alpha''' \rangle = 1. \tag{25.45}$$

Likewise

$$\langle \alpha', \alpha'' \rangle = -\tfrac{1}{2}, \quad \langle \alpha', \alpha''' \rangle = \tfrac{1}{2}, \quad \langle \alpha'', \alpha''' \rangle = -\tfrac{1}{2}. \tag{25.46}$$

It should be noted that (25.38)–(25.40) demonstrate that

$$\alpha' = \alpha'' + \alpha'''. \tag{25.47}$$

Clearly $\mathbf{c}$ can be identified with the basis element c of the Abelian subalgebra $\tilde{\mathscr{L}}_0^A$ that appears in Theorem VIII of the previous subsection. Finally, (25.41) shows that the Killing form $B(\ ,\ )$ is not identically zero on $\tilde{\mathscr{L}}_s$, so $\tilde{\mathscr{L}}_s$ is basic. Moreover, since $\operatorname{str}(\mathbf{MN})$ provides a bilinear supersymmetric consistent invariant form on $\tilde{\mathscr{L}}_s$, Theorem III of Section 2 implies that

$$B(\mathbf{M}, \mathbf{N}) = \mu \operatorname{str}(\mathbf{MN})$$

for all $\mathbf{M}$ and $\mathbf{N}$ of $\tilde{\mathscr{L}}_s$, μ being a constant independent of $\mathbf{M}$ and $\mathbf{N}$. As $\operatorname{str}(\mathbf{cc}) = -2$, (25.41) indicates that $\mu = 2$, so

$$B(\mathbf{M}, \mathbf{N}) = 2 \operatorname{str}(\mathbf{MN}) \tag{25.48}$$

for all $\mathbf{M}$ and $\mathbf{N}$ of $\tilde{\mathscr{L}}_s$.

The Lie superalgebras $A(r/s)$ for all $r > s \geqslant 0$ are analysed along the same lines in Appendix M, Section 1.

Theorem XX If $\tilde{\mathscr{L}}_s$ is a basic classical simple complex Lie superalgebra and α is an *odd* root of $\tilde{\mathscr{L}}_s$ such that $\langle \alpha, \alpha \rangle \neq 0$ then 2α is an *even* root of $\tilde{\mathscr{L}}_s$ and

$$\dim \tilde{\mathscr{L}}_{s\alpha} = \dim \tilde{\mathscr{L}}_{s(-\alpha)} = 1. \tag{25.49}$$

However, 3α is never a root of $\tilde{\mathscr{L}}_s$.

Proof See Appendix K, Section 5.

Examination of the results given in Appendix M shows that $\langle \alpha, \alpha \rangle = 0$ for every odd root α of every basic classical simple complex Lie superalgebra $\tilde{\mathscr{L}}_s$, except for some odd roots of $B(r/s)$ (for $r \geqslant 0$ and $s > 0$) and of $G(3)$.

(c) *Simple roots, Cartan matrices, generalized Dynkin diagrams and the Weyl group*

The next stage is to introduce for the *basic* classical simple complex Lie superalgebras the concepts of positive, negative and simple roots in such a way that as many as possible of the corresponding results on semi-simple complex Lie algebras are generalized. Further details and proofs may be found in the work of Kac (1978b).

Let $\mathscr{H}_0^{ss}$ be the Cartan subalgebra of the semi-simple Lie algebra $\tilde{\mathscr{L}}_0^{ss}$ that appears in the even part $\tilde{\mathscr{L}}_0$ of the Lie superalgebra $\tilde{\mathscr{L}}_s$ (cf. (25.10)), and suppose that a choice of positive roots of $\tilde{\mathscr{L}}_0^{ss}$ has been made (cf. Chapter 13, Section 7). Then the subalgebra $\mathscr{B}_0^{ss}$ of $\tilde{\mathscr{L}}_0^{ss}$ that consists of all the elements of $\mathscr{H}_0^{ss}$ together with all the elements of the root subspaces $\tilde{\mathscr{L}}_{0\alpha}^{ss}$ corresponding to positive roots α of $\tilde{\mathscr{L}}_0^{ss}$ forms a maximal solvable subalgebra of $\tilde{\mathscr{L}}_0^{ss}$. If $\tilde{\mathscr{L}}_0$ contains a one-dimensional Abelian Lie algebra $\tilde{\mathscr{L}}_0^A$ then define the subalgebra $\mathscr{B}_0$ to be the union of $\mathscr{B}_0^{ss}$ and $\tilde{\mathscr{L}}_0^A$, but if $\tilde{\mathscr{L}}_0^A = \{0\}$ then define $\mathscr{B}_0$ to be $\mathscr{B}_0^{ss}$. In both cases $\mathscr{B}_0$ is a maximal solvable subalgebra of $\tilde{\mathscr{L}}_0$, and is known as a "Borel subalgebra" of $\tilde{\mathscr{L}}_0$. (Other choices of the positive roots of $\tilde{\mathscr{L}}_0^{ss}$ would lead to different Borel subalgebras, but it can be shown that any two can be mapped into each by an automorphism of $\tilde{\mathscr{L}}_0$ and so are equivalent.)

Let $\mathscr{B}$ be a maximal solvable subalgebra of the Lie superalgebra $\tilde{\mathscr{L}}_s$ that contains $\mathscr{B}_0$. (Even with $\mathscr{B}_0$ kept fixed, there are several different choices of $\mathscr{B}$ that are inequivalent.) As the Cartan subalgebra $\mathscr{H}_s$ of $\tilde{\mathscr{L}}_s$ is contained in $\mathscr{B}_0$, $\mathscr{H}_s$ must be contained in $\mathscr{B}$, and so a subspace $\mathscr{N}^+$ of $\tilde{\mathscr{L}}_s$ can be defined such that

$$\mathscr{B} = \mathscr{H}_s \oplus \mathscr{N}^+ \tag{25.50}$$

(this being a vector space direct sum). Finally a further subspace $\mathscr{N}^-$ of $\tilde{\mathscr{L}}_s$ can be introduced by the prescription

$$\tilde{\mathscr{L}}_s = \mathscr{N}^- \oplus \mathscr{H}_s \oplus \mathscr{N}^+ \tag{25.51}$$

(where again the vector space direct sums are implied).

Definition *Positive and negative roots of a basic classical simple complex Lie superalgebra* $\tilde{\mathscr{L}}_s$

A root α of $\tilde{\mathscr{L}}_s$ is said to be *positive* if the intersection $\tilde{\mathscr{L}}_{s\alpha} \cap \mathscr{N}^+$ of the corresponding root subspace $\tilde{\mathscr{L}}_{s\alpha}$ with $\mathscr{N}^+$ is non-trivial. Similarly α is said to be *negative* if $\tilde{\mathscr{L}}_{s\alpha} \cap \mathscr{N}^-$ is non-trivial.

This definition implies that if α is positive then $-\alpha$ is negative and vice versa. Moreover, every positive even root is an extension of a positive root of $\tilde{\mathscr{L}}_0^{ss}$ (cf. (25.33)), and every positive root of $\tilde{\mathscr{L}}_0^{ss}$ extends to a positive even root of $\tilde{\mathscr{L}}_s$.

If the root subspace $\mathscr{L}_{s\alpha}$ is *one*-dimensional and α is positive then α cannot be negative as well. Detailed examination (cf. Appendix M) shows that this is the situation for every root α of every basic classical simple complex Lie superalgebra $\tilde{\mathscr{L}}_s$ *except* for $\tilde{\mathscr{L}}_s = A(1/1)$. In the exceptional case $\tilde{\mathscr{L}}_s = A(1/1)$, $\dim \tilde{\mathscr{L}}_{s\alpha} = 2$ for every odd root α, and, moreover, $\tilde{\mathscr{L}}_{s\alpha} \cap \mathscr{N}^+$ and $\tilde{\mathscr{L}}_{s\alpha} \cap \mathscr{N}^-$ are *both* non-trivial, implying that for $\tilde{\mathscr{L}}_s = A(1/1)$ every odd root is *both* positive and negative. (See Appendix M, Section 2, for more details of $A(1/1)$.)

Definition *Simple roots of a basic classical simple complex Lie superalgebra* $\tilde{\mathscr{L}}_s$

A non-zero root α of $\tilde{\mathscr{L}}_s$ is said to be *simple* if α is positive but α cannot be expressed in the form $\alpha = \beta + \gamma$, where β and γ are both positive roots of $\tilde{\mathscr{L}}_s$.

Suppose that $\tilde{\mathscr{L}}_s$ has L simple roots, which will be denoted by $\alpha_1, \alpha_2, \ldots, \alpha_L$.

Theorem XXI Every positive root α of a basic classical simple complex Lie superalgebra $\tilde{\mathscr{L}}_s$ can be written in terms of the simple roots of $\tilde{\mathscr{L}}_s$ as

$$\alpha = \sum_{j=1}^{L} k_j \alpha_j, \tag{25.52}$$

where the coefficients k_j are a set of *non-negative integers.*

Proof This is essentially the same proof as for the corresponding part of Theorem II of Chapter 13, Section 7.

That different choices of $\mathscr{B}$ lead to different sets of simple roots is clearly demonstrated by the following example.

Example II *The positive, negative and simple roots of* $\tilde{\mathscr{L}}_s = A(1/0)$

The roots of $A(1/0)$ as determined in Example I of the previous subsection are $\alpha = \alpha', \alpha'', \alpha''', -\alpha', -\alpha''$ and $-\alpha'''$. Their corresponding root subspaces $\tilde{\mathscr{L}}_{s\alpha}$ are all one-dimensional and have basis elements $\mathbf{e}_{13}$, $\mathbf{e}_{23}$, $\mathbf{e}_{12}$, $\mathbf{e}_{31}$, $\mathbf{e}_{32}$, and $\mathbf{e}_{21}$ respectively.

(i) One choice of $\mathscr{B}$ has basis $\{\mathbf{c}, \mathbf{h}_1^1, \mathbf{e}_{12}, \mathbf{e}_{13}, \mathbf{e}_{23}\}$, so that $\mathscr{N}^+$ has basis $\{\mathbf{e}_{12}, \mathbf{e}_{13}, \mathbf{e}_{23}\}$ and hence $\mathscr{N}^-$ has basis $\{\mathbf{e}_{21}, \mathbf{e}_{31}, \mathbf{e}_{32}\}$. With this choice, α', α'' and α''' are all positive roots, and $-\alpha'$, $-\alpha''$ and $-\alpha'''$ are all negative roots. As $\alpha' = \alpha'' + \alpha'''$ (cf. (25.47)), the corresponding simple roots are α'' and α''', and so $L=2$. With

$$\alpha_1 = \alpha''', \quad \alpha_2 = \alpha'', \tag{25.53}$$

(25.47) shows that

$$\alpha' = \alpha_1 + \alpha_2. \tag{25.54}$$

It should be noted that (25.45) gives

$$\langle \alpha_1, \alpha_1 \rangle = 1, \tag{25.55}$$

but (25.44) implies that

$$\langle \alpha_2, \alpha_2 \rangle = 0. \tag{25.56}$$

Also (25.46) shows that

$$\langle \alpha_1, \alpha_2 \rangle = -\tfrac{1}{2}. \tag{25.57}$$

(ii) Another inequivalent choice of $\mathscr{B}$ has basis $\{\mathbf{c}, \mathbf{h}_1^1, \mathbf{e}_{12}, \mathbf{e}_{13}, \mathbf{e}_{32}\}$, so that $\mathscr{N}^+$ has basis $\{\mathbf{e}_{12}, \mathbf{e}_{13}, \mathbf{e}_{32}\}$ and hence $\mathscr{N}^-$ has basis $\{\mathbf{e}_{21}, \mathbf{e}_{31}, \mathbf{e}_{23}\}$. In this case the positive roots are α', $-\alpha''$ and α''', and the negative roots are $-\alpha', \alpha''$ and $-\alpha'''$. Rewriting (25.47) as $\alpha''' = \alpha' - \alpha''$, it is clear that again $L=2$, but the simple roots are now α' and $-\alpha''$. With

$$\alpha_1 = \alpha', \quad \alpha_2 = -\alpha'', \tag{25.58}$$

(25.47) gives

$$\alpha''' = \alpha_1 + \alpha_2. \tag{25.59}$$

Also, by (25.43) and (25.44),

$$\langle \alpha_1, \alpha_1 \rangle = \langle \alpha_2, \alpha_2 \rangle = 0 \tag{25.60}$$

and, by (25.46),

$$\langle \alpha_1, \alpha_2 \rangle = \tfrac{1}{2}. \tag{25.61}$$

Detailed examination (cf. Appendix M) shows that L (the number of simple roots) and l (the rank of $\tilde{\mathscr{L}}_s$) are related by

$$L = \begin{cases} l+1 & \text{if } \tilde{\mathscr{L}}_s = A(r/r) \text{ for } r>0, \\ l & \text{for every other basic } \tilde{\mathscr{L}}_s. \end{cases} \tag{25.62}$$

Moreover, the set of simple roots $\alpha_1, \alpha_2, \ldots, \alpha_L$ is always a linearly independent set *except* for $\tilde{\mathscr{L}}_s = A(r/r)$ for $r>0$.

If $\alpha = \alpha_j$ is a simple root such that $\langle \alpha_j, \alpha_j \rangle \neq 0$, an element H_α of $\mathscr{H}_s$ may be defined by

$$H_{\alpha_j} = \frac{2}{\langle \alpha_j, \alpha_j \rangle} h_{\alpha_j}, \tag{25.63}$$

where h_α (for $\alpha = \alpha_j$) is given by (25.25) (cf. (13.25)). If α_j is a simple root such that $\langle \alpha_j, \alpha_j \rangle = 0$ then choose another simple root $\alpha_{j'}$ such that $\langle \alpha_j, \alpha_{j'} \rangle \neq 0$, and define

$$H_{\alpha_j} = \frac{1}{\langle \alpha_j, \alpha_{j'} \rangle} h_{\alpha_j}, \tag{25.64}$$

Definition *Cartan matrix* **A** *of a basic classical simple complex Lie superalgebra* $\tilde{\mathscr{L}}_s$

The *Cartan matrix* **A** of $\tilde{\mathscr{L}}_s$ is a $L \times L$ matrix whose elements A_{jk} are defined in terms of the simple roots $\alpha_1, \alpha_2, \ldots, \alpha_L$ by

$$A_{jk} = \alpha_k(H_{\alpha_j}), \quad j, k = 1, 2, \ldots, L, \tag{25.65}$$

where H_α (for $\alpha = \alpha_j$) is defined by (25.63) or (25.64), as appropriate.

Clearly if α_j is such that $\langle \alpha_j, \alpha_j \rangle \neq 0$ then

$$A_{jk} = \frac{2\langle \alpha_j, \alpha_k \rangle}{\langle \alpha_j, \alpha_j \rangle}, \tag{25.66}$$

so that the expression for A_{jk} is the same as that given in (13.61) for a semi-simple complex Lie algebra. In this case it is clear that $A_{jj} = 2$, and it can be shown that the only possible values of A_{jk} for $k \neq j$ are 0, -1, -2 and -3. *However*, if $\langle \alpha_j, \alpha_j \rangle = 0$ then (25.64) and (25.65) imply that

$$A_{jk} = \frac{\langle \alpha_j, \alpha_k \rangle}{\langle \alpha_j, \alpha_{j'} \rangle}, \tag{25.67}$$

so that, in particular,

$$A_{jj} = 0 \tag{25.68}$$

and

$$A_{jj'} = 1. \tag{25.69}$$

Equation (25.62) shows that **A** is an $l \times l$ matrix except for $\tilde{\mathscr{L}}_s = A(r/r)$ with $r > 0$. It should be noted that for $\tilde{\mathscr{L}}_s = D(2/1; \alpha)$, with the parameter α non-real at least one off-diagonal element of **A** is *not real* (cf. Appendix M, Section 6).

The following example demonstrates that the Cartan matrix of a given Lie superalgebra $\tilde{\mathscr{L}}_s$ depends on the choice of the simple roots (and hence on the choice of $\mathscr{B}$).

Example III *Cartan matrices of* $\tilde{\mathscr{L}}_s = A(1/0)$

(i) With the choice (25.53) for the simple roots, since $\langle \alpha_2, \alpha_2 \rangle = 0$ and $\langle \alpha_2, \alpha_1 \rangle \neq 0$, one can take $\alpha_{2'} = \alpha_1$, implying, by (25.55)–(25.57), (25.68) and (25.69), that

$$\mathbf{A} = \begin{bmatrix} 2 & -1 \\ 1 & 0 \end{bmatrix}. \tag{25.70}$$

(ii) With the choice (25.58) for the simple roots, and taking $\alpha_{2'} = \alpha_1$ and $\alpha_{1'} = \alpha_2$, it follows from (25.60) and (25.69) that

$$\mathbf{A} = \begin{bmatrix} 0 & 1 \\ 1 & 0 \end{bmatrix}. \tag{25.71}$$

It is convenient to associate with each Cartan matrix a *generalized Dynkin diagram*. The rules for constructing such a diagram are as follows:

(i) Assign to each simple root α_j $(j = 1, 2, \ldots, L)$ a *vertex*, which is drawn as a small circle ○ if α_j is even, as ⊗ if α_j is odd and $\langle \alpha_j, \alpha_j \rangle = 0$, or as ● if α_j is odd and $\langle \alpha_j, \alpha_j \rangle \neq 0$—these types of vertex will be called "white", "grey" and "black" respectively;

(ii) consider all pairs of vertices labelled by α_j and α_k $(j, k = 1, 2, \ldots, L)$, in turn;

(iii) draw L_{jk} lines from the α_j vertex to the α_k vertex, where

$$L_{jk} = \max\{|A_{jk}|, |A_{kj}|\},$$

except in the case $\tilde{\mathscr{L}}_s = D(2/1; \alpha)$ (for $\alpha \neq 0, -1, \infty$), for which L_{jk} is defined by

$$L_{jk} = \begin{cases} 1 & \text{if } A_{jk} \neq 0, \\ 0 & \text{if } A_{jk} = 0; \end{cases}$$

(iv) add an arrow pointing from the α_j vertex to the α_k vertex if $|A_{kj}| > 1$, *except* in the case $\tilde{\mathscr{L}}_s = D(2/1; \alpha)$ (for $\alpha \neq 0, -1, \infty$).

Example IV *Generalized Dynkin diagrams of* $\tilde{\mathscr{L}}_s = A(1/0)$

The generalized Dynkin diagram corresponding to the Cartan matrix (25.70) is shown in Figure 25.1, and that corresponding to the Cartan matrix (25.71) is displayed in Figure 25.2.

Figure 25.1 Generalized Dynkin diagram for $A(1/0)$ corresponding to the Cartan matrix (25.70).

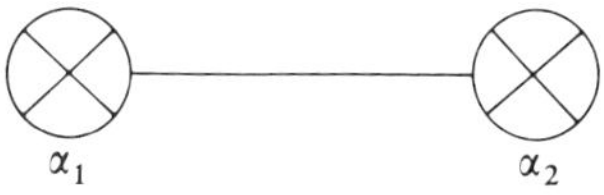

Figure 25.2 Generalized Dynkin diagram for $A(1/0)$ corresponding to the Cartan matrix (25.71).

For the simple complex Lie algebras the generalized Dynkin diagrams constructed according to these rules are displayed later in Figure 26.2. Although each simple complex Lie algebra has a *unique* generalized Dynkin diagram (up to labelling of the vertices), Example IV demonstrated that in the superalgebra case this uniqueness disappears. Clearly the superalgebras $D(2/1;\alpha)$ are thoroughly anomalous, and, moreover, both for these superalgebras and for $D(2/s)$ (with $s > 0$) the above prescription cannot be reversed to allow the Cartan matrix to be uniquely determined by its generalized Dynkin diagram. For these reasons the generalized Dynkin diagrams play a much less significant role in the classification of the basic classical simple complex Lie superalgebras than their counterparts do in the classification of the simple complex Lie algebras.

For each basic classical simple complex Lie superalgebra $\tilde{\mathscr{L}}_s$ with non-trivial odd part $\tilde{\mathscr{L}}_1$ there is a *distinguished* choice of simple roots, which consists of *one odd* simple root, with all the other $L-1$ simple roots being *even*. The corresponding generalized Dynkin diagram has $L-1$ white vertices, the remaining vertex being either grey or black. (For $\tilde{\mathscr{L}}_s = A(1/0)$ this "distinguished" set of simple roots is the set (25.53), the generalized Dynkin diagram for which is given in Figure 25.1.) The simple roots listed in Appendix M for all the basic classical simple complex Lie superalgebras are all chosen to be of this "distinguished" type. Their generalized Dynkin diagrams are displayed in Figure 25.3.

In the case of a basic classical simple complex Lie superalgebra $\tilde{\mathscr{L}}_s$ that is of type I *every* extension of a *simple* root of $\tilde{\mathscr{L}}_0^{ss}$ is also a *simple* root of $\tilde{\mathscr{L}}_s$, but if $\tilde{\mathscr{L}}_s$ is of type II then this is true for every simple root of $\tilde{\mathscr{L}}_0^{ss}$ *except* for one, which will subsequently be denoted by β. (See Appendix M for details.)

The concept of the "*Weyl group*" $\mathscr{W}$ for a basic classical simple complex Lie superalgebra $\tilde{\mathscr{L}}_s$ is very similar to that given for semi-simple complex Lie algebras in Chapter 13, Section 9. For any linear functional λ defined on $\mathscr{H}_s$, and for any non-zero *even* root α of $\tilde{\mathscr{L}}_s$, the linear functional $S_\alpha\lambda$ may be defined by

$$S_\alpha\lambda = \lambda(h) - \frac{2\langle \lambda, \alpha\rangle}{\langle \alpha, \alpha\rangle}\alpha(h) \tag{25.72}$$

for all $h \in \mathscr{H}_s$. This defines an operator S_α, which has the following properties

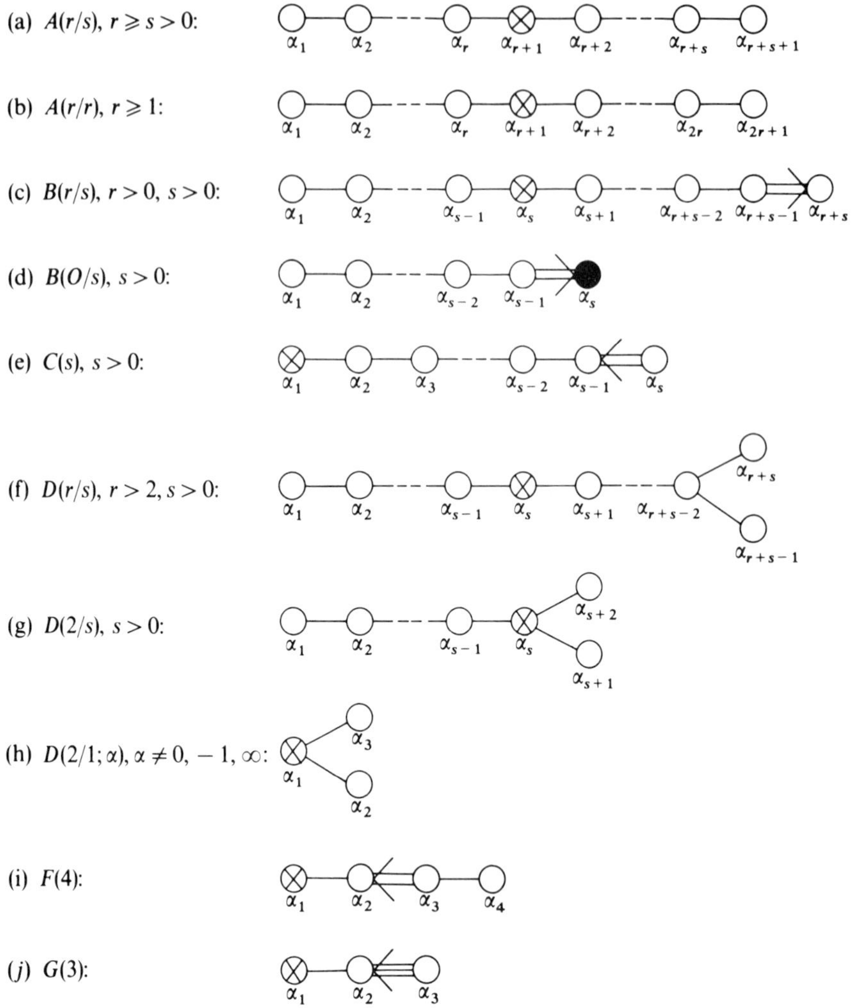

Figure 25.3 Generalized Dynkin diagrams for the basic classical simple complex Lie superalgebras corresponding to the "distinguished" choice of simple roots.

(which are valid for any $\alpha \in \Delta_0$):

(a) $S_\alpha \alpha = -\alpha$;

(b) $S_\alpha(S_\alpha \lambda) = \lambda$ for any linear functional λ on $\mathscr{H}_s$;

(c) for any linear functionals λ and λ' on $\mathscr{H}_s$

$$\langle S_\alpha \lambda, S_\alpha \lambda' \rangle = \langle \lambda, \lambda' \rangle;$$

(d) for any two linear functionals λ and λ' on $\mathscr{H}_s$ and any two complex numbers c and c'

$$S_\alpha(c\lambda + c'\lambda') = cS_\alpha\lambda + c'S_\alpha\lambda'.$$

S_α is again called a *Weyl reflection*, and the set consisting of the identity, the Weyl reflections and all their distinct products forms the *Weyl group* $\mathscr{W}$ of $\tilde{\mathscr{L}}_s$.

Theorem XXII The Weyl group of $\tilde{\mathscr{L}}_s$ is isomorphic to the Weyl group of its semi-simple Lie subalgebra $\tilde{\mathscr{L}}_0^{ss}$.

Proof See Appendix K, Section 5.

Theorem XXIII For any element S of the Weyl group $\mathscr{W}$ of $\tilde{\mathscr{L}}_s$, if $\gamma\in\Delta_0$ then $S\gamma\in\Delta_0$, and if $\gamma\in\Delta_1$ then $S\gamma\in\Delta_1$.

Proof See Appendix K, Section 5.

Further information on the classification and properties of basic classical simple complex Lie superalgebras may be found in the papers of Kac (1975, 1977a, b, 1978a), Pais and Rittenberg (1975), Freund and Kaplansky (1976), Nahm *et al.* (1976), Nahm and Scheunert (1976), Scheunert *et al.* (1976a, b), Backhouse (1977) and Hurni and Morel (1982).

4 Graded representations of basic classical simple complex Lie superalgebras

(a) *Weights and highest weights*

Attention will be confined to representations that are both graded and finite-dimensional.

Consider first a representation of a simple Lie superalgebra of "total" dimension 1; that is, one for which $d_0 + d_1 = 1$, where d_0 and d_1 are the dimensions of the even and odd parts respectively. It is easily shown that there is only one such representation, and in it every element of the superalgebra is represented by the 1×1 matrix $\mathbf{0}$. (For the odd elements of this superalgebra this is implied by (21.53) since $d_0 d_1 = 0$, and the corresponding result for the even elements follows by part (b) of Theorem IV of Section 2.) Although this representation may be regarded as being rather trivial, it is nevertheless an example of a graded irreducible representation.

The algebras $A(r/r)$ (for $r \geqslant 1$) are to some extent exceptional, as some of the results of this section do not apply to them. However, these results do apply to the complex Lie superalgebras $\mathrm{sl}(r+1/r+1;\mathbb{C})$ (for $r \geqslant 1$), to which $A(r/r)$ is intimately related. (As shown in Appendix M, Section 2, $\mathrm{sl}(r+1/r+1;\mathbb{C})$ is the direct sum of $A(r/r)$ and a one-dimensional Abelian algebra.) Even when it is not stated explicitly, it will be implicit in the following account that $A(r/r)$ has to be replaced everywhere by $\mathrm{sl}(r+1/r+1;\mathbb{C})$.

Theorem I For any graded representation $\mathbf{\Gamma}$ of any simple complex Lie superalgebra $\tilde{\mathscr{L}}_s$

$$\mathrm{str}\,\mathbf{\Gamma}(a) = 0$$

for all $a \in \tilde{\mathscr{L}}_s$.

Proof By Theorem IV of Section 2, for any $a \in \tilde{\mathscr{L}}_s$ there exist elements a' and $a'' \in \tilde{\mathscr{L}}_s$ such that $a = [a', a'']$, and so $\mathbf{\Gamma}(a) = [\mathbf{\Gamma}(a'), \mathbf{\Gamma}(a'')]$. But $\mathrm{str}\,[\mathbf{\Gamma}(a'), \mathbf{\Gamma}(a'')] = 0$ by (20.69) and (21.20), so $\mathrm{str}\,\mathbf{\Gamma}(a) = 0$.

As in the semi-simple Lie algebra situation (cf. Chapter 15, Section 2), a similarity transformation can be applied to any graded representation $\mathbf{\Gamma}$ of a basic classical simple complex Lie superalgebra $\tilde{\mathscr{L}}_s$ in such a way that $\mathbf{\Gamma}(h)$ becomes a diagonal matrix for every h of the Cartan subalgebra $\mathscr{H}_s$ of $\tilde{\mathscr{L}}_s$. Henceforth it will always be assumed that any such necessary similarity transformation has already been applied. The carrier space V of the graded representation $\mathbf{\Gamma}$ then has as its basis elements a set of *weight vectors* $\psi(\lambda)$ that are such that

$$\Phi(h)\psi(\lambda) = \lambda(h)\psi(\lambda) \tag{25.73}$$

for all $h \in \mathscr{H}_s$ (cf. (15.3)), the linear functional λ on $\mathscr{H}_s$ being described as a *weight* of the graded representation $\mathbf{\Gamma}$. If the subspace of V corresponding to the weight λ has dimension $m(\lambda)$ then $m(\lambda)$ is said to be the *multiplicity* of λ, and if $m(\lambda) = 1$ then λ is said to be a *simple weight*. The weight λ may be described as being "even" if $\psi(\lambda)$ is a member of the even subspace V_0 of V, and "odd" if $\psi(\lambda)$ is a member of the odd subspace V_1.

Theorem II If λ is a weight of a graded representation of a basic classical simple complex Lie superalgebra $\tilde{\mathscr{L}}_s$ then $\lambda + \alpha$ is also a weight of the same representation for each $\alpha \in \Delta_s$ such that $\Phi(e_\alpha)\psi(\lambda) \neq 0$.

Proof This is the same as that of Theorem I of Chapter 15, Section 2, only commutators being involved since every element h of $\mathscr{H}_s$ is necessarily even.

For every basic classical simple complex Lie superalgebra $\tilde{\mathscr{L}}_s$ except $\tilde{\mathscr{L}}_s = A(r/r)$ (for $r \geqslant 1$) $L = l$ and the simple roots $\alpha_1, \alpha_2, \ldots, \alpha_l$ form a basis for the dual space $\mathscr{H}_s^*$ of $\mathscr{H}_s$. Consequently every weight λ of such a superalgebra can be expressed in the form

$$\lambda = \sum_{j=1}^{l} \mu_j \alpha_j,$$

where $\mu_1, \mu_2, \ldots, \mu_l$ are a set of coefficients. However, by contrast with the Lie algebra result of Theorem III of Chapter 15, Section 2, these coefficients μ_j need *not* be real, as will be demonstrated explicitly in Example I below. Nevertheless, Theorem II implies that in the corresponding expansion of the difference $\lambda - \lambda'$ of any two weights λ and λ' of any representation the coefficients are all *integers*.

Definition *Highest weight Λ of a graded representation*

If Λ is a weight of graded representation of a basic classical simple complex Lie superalgebra $\tilde{\mathscr{L}}_s$ such that

$$\Phi(e_\alpha)\psi(\lambda) = 0 \tag{25.74}$$

for every *positive* root α of $\tilde{\mathscr{L}}_s$ then Λ is said to be the *highest weight* of the representation.

The implication is that if Λ is the highest weight of a graded representation of $\tilde{\mathscr{L}}_s$ then $\Lambda + \alpha$ is *not* a weight of this representation for any *positive* root α of $\tilde{\mathscr{L}}_s$. (The corresponding definition for semi-simple Lie algebras that was given in Chapter 15, Section 3 could be formulated equivalently in these terms.) As Example I will show explicitly, the highest weight can be either even or odd.

Definition *Numerical marks of a graded representation*

Suppose that a basic classical simple complex Lie superalgebra $\tilde{\mathscr{L}}_s$ has L distinguished simple roots $\alpha = \alpha_1, \alpha_2, \ldots, \alpha_L$, and the corresponding elements H_α of $\mathscr{H}_s$ are as defined in (25.63) and (25.64). Then the set of L numbers $n_1, n_2, \ldots, n_L$ defined in terms of the highest weight Λ of a graded representation of $\tilde{\mathscr{L}}_s$ by

$$n_j = \Lambda(H_{\alpha_j}), \quad j = 1, 2, \ldots, L, \tag{25.75}$$

are said to be the *numerical marks* of the representation. (These numerical marks are also sometimes referred to as "Dynkin" or "Kac–Dynkin" weights, parameters, coefficients or labels.)

From (25.63) and (25.64)

$$n_j = \begin{cases} \dfrac{2\langle \Lambda, \alpha_j \rangle}{\langle \alpha_j, \alpha_j \rangle} & \text{if } \langle \alpha_j, \alpha_j \rangle \neq 0, \\[2ex] \dfrac{\langle \Lambda, \alpha_j \rangle}{\langle \alpha_j, \alpha_{j'} \rangle} & \text{if } \langle \alpha_j, \alpha_j \rangle = 0. \end{cases} \tag{25.76}$$

Moreover, from (25.73) and (25.75)

$$\Phi(H_{\alpha_j})\psi(\Lambda) = n_j\psi(\Lambda) \tag{25.77}$$

for $j = 1, 2, \ldots, L$.

Except in the case $\tilde{\mathscr{L}}_s = A(r/r)$ (for $r \geqslant 1$), it is easy to reexpress these ideas in terms of "fundamental weights". Let $\Lambda_1, \Lambda_2, \ldots, \Lambda_l$ be the l fundamental weights defined (cf. (15.10)) by

$$\Lambda_j(h) = \sum_{k=1}^{l} (\mathbf{A}^{-1})_{kj}\alpha_k(h).$$

Then, by (25.65), for every simple root α_j

$$\Lambda_j(H_{\alpha_k}) = \delta_{jk},$$

and hence

$$\Lambda = \sum_{j=1}^{l} n_j\Lambda_j,$$

where $n_1, n_2, \ldots, n_l$ are the numerical marks.

Theorem III If $\tilde{\mathscr{L}}_s$ is a *type I* basic classical simple complex Lie superalgebra $A(r/s)$ (for $r > s \geqslant 0$) or $C(s)$ (for $s \geqslant 2$), or if $\tilde{\mathscr{L}}_s = \mathrm{sl}(r+1/r+1;\mathbb{C})$ (for $r \geqslant 1$), then the graded representation with highest weight Λ is of *finite* dimension if and only if the numerical mark n_j is a *non-negative integer* for every distinguished sample root α_j of $\tilde{\mathscr{L}}_s$ such that $\langle \alpha_j, \alpha_j \rangle \neq 0$.

Proof See Appendix K, Section 6.

It should be noted that Theorem III does not impose any condition on the numerical mark n_j corresponding to the distinguished simple root α_j for which $\langle \alpha_j, \alpha_j \rangle = 0$, and indeed *this* n_j is allowed to take *any complex value*, as the following example shows. This example also demonstrates that the question of complete reducibility is non-trivial, and that complications occur for the so-called "atypical" representations, a matter that will be taken up in more detail in the next subsection.

Example I *Graded representations of* $A(1/0)$ *for which* $d_0 = d_1 = 2$

The structure of the type I basic classical simple complex Lie superalgebra $A(1/0)$ was considered in detail in Examples I and II of Sections 3(*b*) and 3(*c*). With the distinguished choice of simple roots α_1 and α_2 of $A(1/0)$ (as given in set (i) of the Example II just mentioned), α_1 is the only positive even root, and α_2 and $\alpha_1 + \alpha_2$ are its two positive odd roots. The corresponding basis elements may be taken to be

$$e_{\alpha_1} = \mathbf{e}_{12}, \quad e_{-\alpha_1} = \mathbf{e}_{21}, \quad e_{\alpha_2} = \mathbf{e}_{23}, \quad e_{-\alpha_2} = \mathbf{e}_{32},$$
$$e_{\alpha_1+\alpha_2} = \mathbf{e}_{13}, \quad e_{-(\alpha_1+\alpha_2)} = \mathbf{e}_{31},$$

and

$$h_{\alpha_1} = \tfrac{1}{2}(\mathbf{e}_{11} - \mathbf{e}_{22}), \quad h_{\alpha_2} = \tfrac{1}{2}(\mathbf{e}_{22} + \mathbf{e}_{33}).$$

It may then be checked that all the commutation and anticommutation relations are satisfied with the following representation (in which $d_0 = d_1 = 2$):

$$\mathbf{\Gamma}(e_{\alpha_1}) = \begin{bmatrix} 0 & 0 & 0 & 0 \\ 0 & 0 & 0 & 0 \\ 0 & 0 & 0 & 1 \\ 0 & 0 & 0 & 0 \end{bmatrix}, \quad \mathbf{\Gamma}(e_{-\alpha_1}) = \begin{bmatrix} 0 & 0 & 0 & 0 \\ 0 & 0 & 0 & 0 \\ 0 & 0 & 0 & 0 \\ 0 & 0 & 1 & 0 \end{bmatrix},$$

$$\mathbf{\Gamma}(e_{\alpha_2}) = \begin{bmatrix} 0 & 0 & \kappa_+ & 0 \\ 0 & 0 & 0 & 0 \\ 0 & 0 & 0 & 0 \\ 0 & \mu_+ & 0 & 0 \end{bmatrix}, \quad \mathbf{\Gamma}(e_{-\alpha_2}) = \begin{bmatrix} 0 & 0 & 0 & 0 \\ 0 & 0 & 0 & \mu_- \\ \kappa_- & 0 & 0 & 0 \\ 0 & 0 & 0 & 0 \end{bmatrix},$$

$$\mathbf{\Gamma}(e_{\alpha_1+\alpha_2}) = \begin{bmatrix} 0 & 0 & 0 & \kappa_+ \\ 0 & 0 & 0 & 0 \\ 0 & \mu_+ & 0 & 0 \\ 0 & 0 & 0 & 0 \end{bmatrix}, \quad \mathbf{\Gamma}(e_{-(\alpha_1+\alpha_2)}) = \begin{bmatrix} 0 & 0 & 0 & 0 \\ 0 & 0 & \mu_- & 0 \\ 0 & 0 & 0 & 0 \\ \kappa_- & 0 & 0 & 0 \end{bmatrix},$$

$$\mathbf{\Gamma}(h_{\alpha_1}) = \frac{1}{2}\begin{bmatrix} 0 & 0 & 0 & 0 \\ 0 & 0 & 0 & 0 \\ 0 & 0 & 1 & 0 \\ 0 & 0 & 0 & -1 \end{bmatrix}, \quad \mathbf{\Gamma}(h_{\alpha_2}) = \frac{1}{2}\begin{bmatrix} \kappa_+\kappa_- & 0 & 0 & 0 \\ 0 & \mu_+\mu_- & 0 & 0 \\ 0 & 0 & \kappa_+\kappa_- & 0 \\ 0 & 0 & 0 & \mu_+\mu_- \end{bmatrix},$$

where κ_+, κ_-, μ_+ and μ_- are any four complex numbers that are subject only to the single condition

$$\mu_+\mu_- = \kappa_+\kappa_- + 1. \tag{25.78}$$

Each weight λ of this representation can be expressed in the form

$$\lambda(h) = \mu_1\alpha_1(h) + \mu_2\alpha_2(h),$$

where α_1, α_2 are the simple roots and μ_1, μ_2 are complex numbers. As $\langle\alpha_1, \alpha_1\rangle = 1$, $\langle\alpha_1, \alpha_2\rangle = -\tfrac{1}{2}$ and $\langle\alpha_2, \alpha_2\rangle = 0$, these coefficients μ_1 and μ_2 are given by

$$\mu_1 = 2\lambda(h_{\alpha_2})$$

and

$$\mu_2 = -2\lambda(h_{\alpha_1}) - 4\lambda(h_{\alpha_2}).$$

Thus if $\lambda_j(h) = \Gamma(h)_{jj}$ (for $j = 1, \ldots, 4$) then

$$\lambda_1 = (-\kappa_+\kappa_-)\alpha_1 + (-2\kappa_+\kappa_-)\alpha_2,$$

$$\lambda_2 = (-\kappa_+\kappa_- - 1)\alpha_1 + (-2\kappa_+\kappa_- - 2)\alpha_2,$$

$$\lambda_3 = (-\kappa_+\kappa_-)\alpha_1 + (-2\kappa_+\kappa_- - 1)\alpha_2,$$

$$\lambda_4 = (-\kappa_+\kappa_- - 1)\alpha_1 + (-2\kappa_+\kappa_- - 1)\alpha_2.$$

(It should be noted that in general these are *not* real.) Clearly the *highest weight* Λ is given by

$$\Lambda = \lambda_1,$$

so that

$$\Lambda = (-\kappa_+\kappa_-)\alpha_1 + (-2\kappa_+\kappa_-)\alpha_2$$

and

$$\lambda_2 = \Lambda - \alpha_1 - 2\alpha_2,$$

$$\lambda_3 = \Lambda - \alpha_2,$$

$$\lambda_4 = \Lambda - \alpha_1 - \alpha_2.$$

Here λ_1 and λ_2 are even weights, whereas λ_3 and λ_4 are odd weights. Then, by (25.76), the numerical marks n_1 and n_2 have the values.

$$n_1 = 0, \quad n_2 = -\kappa_+\kappa_-. \tag{25.79}$$

That is, n_2 can take *any* complex value.

If *none* of the parameters $\kappa_+, \kappa_-, \mu_+$ or μ_- is *zero* then this representation is *irreducible*, but if one or two of them are *zero* then the representation becomes *reducible*. For example, if $\kappa_- = 0$ then all the entries in the first three rows of the first column are zero for every element of the Lie superalgebra, and so the 4-dimensional representation reduces to two representations. One of these is given by the elements of the first row and first column, and can be identified with the irreducible representation for which $d_0 + d_1 = 1$ that was considered right at the beginning of this section. Its numerical marks are $n_1 = 0$ and $n_2 = 0$. The other is given by the last three rows and last three columns, and so its weights (with the previous labelling and $\kappa_+\kappa_- = 0$) are

$$\lambda_2 = -\alpha_1 - 2\alpha_2,$$

$$\lambda_3 = -\alpha_2,$$

$$\lambda_4 = -\alpha_1 - \alpha_2.$$

Clearly λ_3 is the highest weight of this set and corresponds to the numerical marks $n_1 = 1$ and $n_2 = 0$. Closer inspection shows that it too is irreducible. (Clearly the highest weight λ_3 of this 3-dimensional representation is odd, whereas the highest weight λ_1 of the original 4-dimensional representation is even.)

It should be noted that if $\kappa_+ \neq 0$ (and $\kappa_- = 0$), the 4-dimensional representation is reducible but it is *not completely reducible*! The situation is similar if $\kappa_+ = 0$, the irreducible representations appearing in the reduction being the same as for the case $\kappa_- = 0$. It will be seen in the next subsection that these two irreducible representations that appear in these reductions are of the type that is known as "atypical". A very similar situation arises when either μ_- or μ_+ is zero.

Theorem IV If $\tilde{\mathscr{L}}_s$ is a *type II* basic classical simple complex Lie superalgebra then the graded representation with highest weight Λ is of *finite* dimension if and only if *all* the following conditions are satisfied.

(1) For *every* distinguished simple root α_j such that $\langle \alpha_j, \alpha_j \rangle \neq 0$ the numerical mark n_j must be a *non-negative integer*.

(2) If β is the extension of a simple root of $\tilde{\mathscr{L}}_0^{ss}$ that is not a simple root of $\tilde{\mathscr{L}}_s$ and if N_β is defined by

$$N_\beta = \frac{2\langle \Lambda, \beta \rangle}{\langle \beta, \beta \rangle} \tag{25.80}$$

then N_β must be a *non-negative integer*.

(3) (a) If $\tilde{\mathscr{L}}_s = B(r/s)$ (for $r \geqslant 1$ and $s \geqslant 1$) and if $N_\beta < r$ then

$$n_{s+N_\beta+1} = n_{s+N_\beta+2} = \cdots = n_{s+r} = 0.$$

(b) If $\tilde{\mathscr{L}}_s = D(r/s)$ (for $r \geqslant 2$ and $s \geqslant 1$) and

(i) if $N_\beta \leqslant r - 2$ then

$$n_{s+N_\beta+1} = n_{s+N_\beta+2} = \cdots = n_{s+r} = 0,$$

whereas

(ii) if $N_\beta = r - 1$ then

$$n_{r+s-1} = n_{r+s}.$$

(c) If $\tilde{\mathscr{L}}_s = D(2/1; \alpha)$ and

(i) if $N_\beta = 0$ then

$$n_2 = n_3 = 0,$$

whereas

(ii) if $N_\beta = 1$ then

$$(n_3 + 1)\alpha = n_2 + 1.$$

(d) If $\tilde{\mathscr{L}}_s = F(4)$ then it is not possible for N_β to take the value 1, and

(i) if $N_\beta = 0$ then

$$n_2 = n_3 = n_4 = 0,$$

(ii) if $N_\beta = 2$ then

$$n_2 = n_4 = 0,$$

(iii) if $N_\beta = 3$ then

$$n_2 = 2n_4 + 1.$$

(e) If $\tilde{\mathscr{L}}_s = G(3)$ then it is not possible for N_β to take the value 1, and

(i) if $N_\beta = 0$ then

$$n_2 = n_3 = 0,$$

(ii) if $N_\beta = 2$ then

$$n_2 = 0.$$

Proof See Appendix K, Section 6.

In the conditions (3), which are known as *supplementary* or *consistency* conditions, the distinguished roots are labelled in the manner specified in Appendix M. (For $\tilde{\mathscr{L}}_s = B(0/s)$ (for $s \geqslant 1$) there are *no* conditions of this type.)

The quantities N_β are easily calculated in terms of the numerical marks using (25.76), (25.80), the expressions for β in terms of the simple roots of $\tilde{\mathscr{L}}_s$, and the quantities $\langle \alpha_j, \alpha_k \rangle$ that are listed in Appendix M. The results are as follows:

(a) for $\tilde{\mathscr{L}}_s = B(0/s)$ (for $s \geqslant 1$)

$$N_\beta = \tfrac{1}{2} n_s;$$

(b) for $\tilde{\mathscr{L}}_s = B(r/s)$ (for $r \geqslant 1$ and $s \geqslant 1$)

$$N_\beta = n_s - n_{s+1} - \cdots - n_{r+s-1} - \tfrac{1}{2} n_{r+s};$$

(c) for $\tilde{\mathscr{L}}_s = D(r/s)$ (for $r \geqslant 2$ and $s \geqslant 1$)

$$N_\beta = n_s - n_{s+1} - \cdots - n_{r+s-2} - \tfrac{1}{2} n_{r+s-1} - \tfrac{1}{2} n_{r+s};$$

(d) for $\tilde{\mathscr{L}}_s = D(2/1;\alpha)$

$$N_\beta = \frac{2n_1 - n_2 - \alpha n_3}{1+\alpha};$$

(e) for $\tilde{\mathscr{L}}_s = F(4)$

$$N_\beta = \tfrac{1}{3}(2n_1 - 3n_2 - 4n_3 - 2n_4);$$

(f) for $\tilde{\mathscr{L}}_s = G(3)$

$$N_\beta = \tfrac{1}{2}(n_1 - 2n_2 - 3n_3).$$

Thus, by condition (2) of Theorem IV, the numerical mark n_j corresponding to the distinguished simple root α_j such that $\langle \alpha_j, \alpha_j \rangle = 0$ is determined by the other numerical marks and N_β. Moreover, for all the type II superalgebras except $B(0/s)$ ($s \geqslant 1$), for certain small values of N_β the conditions (3) impose restrictions on some of the numerical marks.

Although Theorems II–IV apply to *both* reducible and irreducible graded representations, the following theorem makes more precise the specification of the *irreducible* graded representations.

Theorem V If Λ is the highest weight of an *irreducible* graded representation of a basic classical simple complex Lie superalgebra $\tilde{\mathscr{L}}_s$ then

(a) Λ is a *simple* weight (i.e. $m(\Lambda) = 1$);

(b) every other weight λ of the representation has the form

$$\lambda = \Lambda - \sum_{j=1}^{L} q_j \alpha_j,$$

where $q_1, q_2, \dots, q_L$ are a set of *non-negative integers* and $\alpha_1, \alpha_2, \dots, \alpha_L$ are the simple roots of $\tilde{\mathscr{L}}_s$; and

(c) two irreducible graded representations of $\tilde{\mathscr{L}}_s$ are *equivalent* if and only if they have *equal* highest weights.

Proof This is essentially the same as for the corresponding theorem for complex semi-simple Lie algebras (Theorem I of Chapter 15, Section 3), part (c) being implied by the construction used there.

(b) "Typical" and "atypical" irreducible representations

Whereas in the case of a semi-simple complex Lie algebra every reducible representation is *completely* reducible (cf. Theorem I of Chapter 15, Section 1), the corresponding result is *not* true for the basic classical simple complex

Lie superalgebras. For these it is necessary to divide the irreducible representations into two classes, as follows.

Definition *"Typical" and "atypical" irreducible representations*

An irreducible graded representation with highest weight Λ of a basic classical simple complex Lie superalgebra $\tilde{\mathscr{L}}_s$ is said to be *typical* if any reducible graded representation of $\tilde{\mathscr{L}}_s$ with highest weight Λ can be put in the form of a direct sum of this irreducible graded representation with some other graded representation of $\tilde{\mathscr{L}}_s$. An *atypical* irreducible graded representation is defined to be an irreducible graded representation that is not "typical".

This definition implies that if $\tilde{\mathscr{L}}_s$ is an "atypical" irreducible graded representation with highest weight Λ then there exists at least one reducible graded representation of $\tilde{\mathscr{L}}_s$ with highest weight Λ that is not completely reducible.

Theorem VI An irreducible graded representation with highest weight Λ of a basic classical simple complex Lie superalgebra $\tilde{\mathscr{L}}_s$ is *atypical* if and only if

$$\langle \Lambda + \rho, \alpha \rangle = 0 \tag{25.81}$$

for some *positive odd root* α of $\tilde{\mathscr{L}}_s$ that is such that 2α is not a root of $\tilde{\mathscr{L}}_s$. Here

$$\rho = \frac{1}{2}\sum_{\alpha\in\Delta_0^+} \alpha - \frac{1}{2}\sum_{\alpha\in\Delta_1^+} \alpha, \tag{25.82}$$

Δ_0^+ and Δ_1^+ being the sets of positive even and positive odd roots of $\tilde{\mathscr{L}}_s$ respectively.

Proof See Kac (1978b).

This theorem implies that in general the numerical marks n_j must satisfy certain constraints if the corresponding irreducible graded representation is to be "typical", and so too must the quantities N_β for the type II superalgebras. It should be noted that for $\tilde{\mathscr{L}}_s = B(0/s)$ (for $s \geqslant 1$) *every* irreducible graded representation is "typical", since the set of roots involved in (25.81) is empty in this case, but this result is *not* true for any other basic classical simple complex Lie superalgebra.

Example II *Conditions for the irreducible graded representations of* $A(1/0)$ *to be "atypical"*

As $A(1/0)$ has one positive even root α_1 and two positive odd roots α_2 and

$\alpha_1 + \alpha_2$, by (25.82)

$$\rho = \tfrac{1}{2}\alpha_1 - \tfrac{1}{2}\{\alpha_2 + (\alpha_1 + \alpha_2)\} = -\alpha_2.$$

For neither of these positive odd roots α is 2α a root. Thus (25.81) shows that the graded irreducible representation with highest weight Λ is "atypical" if

$$\langle \Lambda - \alpha_2, \alpha_2 \rangle = 0 \tag{25.83}$$

or

$$\langle \Lambda - \alpha_2, \alpha_1 + \alpha_2 \rangle = 0. \tag{25.84}$$

But since $\langle \alpha_2, \alpha_2 \rangle = 0$ and $n_2 = \langle \Lambda, \alpha_2 \rangle / \langle \alpha_1, \alpha_2 \rangle$, the condition (25.83) reduces to

$$n_2 = 0.$$

Similarly, since $n_1 = 2\langle \Lambda, \alpha_1 \rangle / \langle \alpha_1, \alpha_1 \rangle$, on using (25.55) and (25.57), the condition (25.84) becomes

$$-n_2 + n_1 + 1 = 0.$$

Thus the graded irreducible representation of $A(1/0)$ with numerical marks n_1 and n_2 is *typical* only if *neither* of the two conditions $n_2 = 0$ or $n_2 = n_1 + 1$ are satisfied.

It is interesting to reexamine the results contained in Example I in the light of these conditions. For all values of the parameters κ_+, κ_-, μ_+ and μ_- (25.79) shows that $n_1 = 0$ for the 4-dimensional graded representation, so this representation is "typical" provided that $n_2 \neq 0$ or 1. Thus by (25.78) this representation is "typical" provided that $\kappa_+\kappa_- \neq 0$ and $\mu_+\mu_- \neq 0$, which are precisely the conditions found in Example I for this representation to be irreducible. It should also be noted that in the case when $\kappa_+\kappa_- = 0$ the two irreducible constituents were found in Example I to correspond to $n_1 = 0$, $n_2 = 0$ and to $n_1 = 1$, $n_2 = 0$, and the conditions just derived indicate that both of these irreducible representations are "atypical". The situation for the case $\mu_+\mu_- = 0$ is similar.

For "typical" graded irreducible representations formulae for the dimensions d_0 and d_1 of the even and odd parts (as defined in Chapter 21, Section 4) are provided by the following two theorems.

Theorem VII The total dimension $d_0 + d_1$ of a *typical* irreducible graded representation with highest weight Λ of a basic classical simple complex Lie superalgebra $\tilde{\mathscr{L}}_s$ is given by

$$d_0 + d_1 = 2^{N_1^+} \prod_{\alpha \in \Delta_0^+} \frac{\langle \Lambda + \rho, \alpha \rangle}{\langle \rho_0, \alpha \rangle}, \tag{25.85}$$

where Δ_0^+ is the set of positive even roots of $\tilde{\mathscr{L}}_s$, ρ is defined by (25.82),

$$\rho_0 = \frac{1}{2} \sum_{\alpha \in \Delta_0^+} \alpha, \tag{25.86}$$

and N_1^+ is the number of positive odd roots.

Proof See Kac (1978b).

Theorem VIII (a) The even and odd dimensions d_0 and d_1 of a *typical* irreducible graded representation of any basic classical simple complex Lie superalgebra $\tilde{\mathscr{L}}_s$ other than $B(0/s)$ $(s \geqslant 1)$ are such that

$$d_0 = d_1.$$

That is, the "superdimension" $d_0 - d_1$ is zero for these Lie superalgebras.

(b) If $\tilde{\mathscr{L}}_s = B(0/s)$ $(s \geqslant 1)$ then for an irreducible graded representation with highest weight Λ (which is necessarily "typical") and with the convention that Λ is an *even* weight

$$d_0 - d_1 = 2^{1-s} \prod_{\alpha \in \bar{\Delta}_0^+} \frac{\langle \Lambda + \rho, \alpha \rangle}{\langle \bar{\rho}_0, \alpha \rangle} \tag{25.87}$$

where $\bar{\Delta}_0^+$ is the set of positive even roots α of $\tilde{\mathscr{L}}_s$ such that $\frac{1}{2}\alpha$ is not an odd root of $\tilde{\mathscr{L}}_s$, and

$$\bar{\rho}_0 = \frac{1}{2} \sum_{\alpha \in \bar{\Delta}_0^+} \alpha. \tag{25.88}$$

Proof See Kac (1977c, 1978b) and Tsohantjis and Cornwell (1989).

Example III *Dimensions of the "typical" irreducible graded representations of $A(1/0)$*

As noted in Example II above, $A(1/0)$ has only one positive even root α_1, and only two positive odd roots α_2 and $\alpha_1 + \alpha_2$. Thus $\rho = -\alpha_2$, $\rho_0 = \frac{1}{2}\alpha_1$ and $N_1^+ = 2$, and hence, by (25.85),

$$d_0 + d_1 = \frac{8\langle \Lambda - \alpha_2, \alpha_1 \rangle}{\langle \alpha_1, \alpha_1 \rangle}.$$

Consequently, since $2\langle \Lambda, \alpha_1 \rangle / \langle \alpha_1, \alpha_1 \rangle = n_1$, $\langle \alpha_1, \alpha_1 \rangle = 1$ and $\langle \alpha_1, \alpha_2 \rangle = -\frac{1}{2}$, it follows that $d_0 + d_1 = 4(n_1 + 1)$, and so, by part (a) of Theorem VIII,

$$d_0 = d_1 = 2(n_1 + 1).$$

As is to be expected, these results do *not* depend on the numerical mark n_2

(which can take any complex value). For the case $n_1 = 0$ they reduce to $d_0 = d_1 = 2$, and so correspond to the representation that was discussed in Examples I and II (which is "typical" provided that $n_2 \neq 0$ or 1).

For a basic classical simple complex Lie superalgebra $\tilde{\mathscr{L}}_s$ of type II the even subalgebra $\tilde{\mathscr{L}}_0$ is a complex *semi-simple* Lie algebra $\tilde{\mathscr{L}}_0^{ss}$, the representation theory of which was described in Chapters 14 and 15. Every irreducible representation of $\tilde{\mathscr{L}}_0^{ss}$ is determined by its highest weight, from which all the other weights of the representation can be found by the methods described in Chapter 14 (particularly Section 3). If $\tilde{\mathscr{L}}_s$ is of type I then either $\tilde{\mathscr{L}}_0 = \tilde{\mathscr{L}}_0^{ss}$ or $\tilde{\mathscr{L}}_0 = \tilde{\mathscr{L}}_0^A \oplus \tilde{\mathscr{L}}_0^{ss}$ (where $\tilde{\mathscr{L}}_0^A$ is a one-dimensional Abelian Lie algebra). In the first case the situation is exactly as just described, and in the second it is only slightly different, since the representation theory for $\tilde{\mathscr{L}}_0^{ss}$ is easily extended to such an $\tilde{\mathscr{L}}_0$, and clearly the concepts of weights and highest weights still apply. Moreover, in each irreducible representation of such an $\tilde{\mathscr{L}}_0$ the elements of $\tilde{\mathscr{L}}_0^A$ are represented by diagonal matrices with all diagonal entries being equal. Consequently the highest weight of an irreducible representation of $\tilde{\mathscr{L}}_0$ again determines the complete set of weights of this representation of $\tilde{\mathscr{L}}_0$, those of $\tilde{\mathscr{L}}_0^A$ all being the same, and those of $\tilde{\mathscr{L}}_0^{ss}$ being found by the methods of Chapter 14. Application of the following theorem then allows *all* the weights of a "typical" irreducible graded representation of $\tilde{\mathscr{L}}_s$ to be deduced for any basic classical simple complex Lie superalgebra $\tilde{\mathscr{L}}_s$.

Theorem IX Let Λ be the highest weight of a *typical* irreducible graded representation of a basic classical simple complex Lie superalgebra $\tilde{\mathscr{L}}_s$. The number of times $m_\Lambda(\Lambda')$ that the irreducible representation with highest weight Λ' of the even subalgebra $\tilde{\mathscr{L}}_0$ of $\tilde{\mathscr{L}}_s$ appears in the reduction of this representation when considered as a representation of $\tilde{\mathscr{L}}_0$ alone is given by

$$m_\Lambda(\Lambda') = \sum_{S \in \mathscr{W}} (\det S) K\{S(\Lambda + \rho) - (\Lambda' + \rho)\}, \tag{25.89}$$

where the sum on the right-hand side is over all the elements S of the Weyl group $\mathscr{W}$ of $\tilde{\mathscr{L}}_s$, and $K(\lambda)$ is the number of different ways in which a linear functional λ defined on the Cartan subalgebra $\mathscr{H}_s$ of $\tilde{\mathscr{L}}_s$ can be written in terms of the odd positive roots in the form

$$\lambda = \sum_{\alpha \in \Delta_1^+} k_\alpha \alpha, \tag{25.90}$$

with each k_α being restricted to the values 0 and 1. Also in (25.90), det S is defined to be $+1$ if the decomposition of S into a product of Weyl reflections contains an even number of Weyl reflections and is defined to be -1 if it contains an odd number, and ρ is given by (25.82).

Proof See Kac (1978b).

Example IV *Reduction of the "typical" irreducible graded representations of $A(1/0)$*

As noted in the previous examples, each irreducible graded representation of $A(1/0)$ is specified by two numerical marks n_1 and n_2, where n_1 is a non-negative integer and n_2 can take any complex value. The Weyl group $\mathscr{W}$ of $A(1/0)$ contains only the identity E and the Weyl reflection S_α with $\alpha = \alpha_1$, for which $\det S = 1$ and -1 respectively. Equation (25.89) then indicates that $m_\Lambda(\Lambda')$ can be positive only if $K\{(\Lambda + \rho) - (\Lambda' + \rho)\}$ is non-zero. But, since Δ_1^+ contains only α_2 and $\alpha_1 + \alpha_2$, $K(\lambda) = 1$ only for $\lambda = 0$, α_2, $\alpha_1 + \alpha_2$ and $\alpha_1 + 2\alpha_2$ $(= \alpha_2 + (\alpha_1 + \alpha_2))$, with $K(\lambda) = 0$ for all other λ. Thus the only possible highest weights Λ' are those for which $\Lambda - \Lambda' = 0$, α_2, $\alpha_1 + \alpha_2$ or $\alpha_1 + 2\alpha_2$; that is, those for which $\Lambda' = \Lambda$, $\Lambda - \alpha_2$, $\Lambda - (\alpha_1 + \alpha_2)$ or $\Lambda - (\alpha_1 + 2\alpha_2)$. Of course, consider for example the possibility $\Lambda' = \Lambda - (\alpha_1 + \alpha_2)$. As

$$S_{\alpha_1}(\Lambda + \rho) - (\Lambda - \alpha_1 - \alpha_2 + \rho) = -n_1\alpha_1 + \alpha_2,$$

which is only in the set of λ for which $K(\lambda) = 1$ if $n_1 = 0$, it follows from (25.89) that $m_\Lambda(\Lambda - (\alpha_1 + \alpha_2)) = 1$ if $n_1 > 0$, but $m_\Lambda(\Lambda - (\alpha_1 + \alpha_2)) = 0$ if $n_1 = 0$. However, a similar analysis of the other cases shows that $m_\Lambda(\Lambda) = 1$, $m_\Lambda(\Lambda - \alpha_2) = 1$ and $m_\Lambda(\Lambda - (\alpha_1 + 2\alpha_2)) = 1$ for *all* values of n_1. Thus for $n_1 = 0$ the highest weights of $\tilde{\mathscr{L}}_0$ are $\Lambda, \Lambda - \alpha_2$ and $\Lambda - (\alpha_1 + 2\alpha_2)$, whereas if $n_1 \geqslant 1$ they are Λ, $\Lambda - \alpha_2$, $\Lambda - (\alpha_1 + \alpha_2)$ and $\Lambda - (\alpha_1 + 2\alpha_2)$.

For $A(1/0)$ the even subalgebra $\tilde{\mathscr{L}}_0$ is given by $\tilde{\mathscr{L}}_0 = \tilde{\mathscr{L}}_0^A \oplus \tilde{\mathscr{L}}_0^{ss}$, where $\tilde{\mathscr{L}}_0^{ss} = \tilde{\mathscr{L}}_0^{s1} = A_1$ (cf. (25.36) and (25.37)), α_1 being the positive root of A_1. For $\alpha = \alpha_1$ the eigenvalues $\Lambda'(H_\alpha)$ for $\Lambda' = \Lambda$, $\Lambda - \alpha_2$, $\Lambda - (\alpha_1 + \alpha_2)$ and $\Lambda - (\alpha_1 + 2\alpha_2)$ are $n_1, n_1 + 1$, $n_1 - 1$ and $n_1 + 1$ respectively, so these irreducible representations of $\tilde{\mathscr{L}}_0$ have dimensions $n_1 + 1$, $n_1 + 2$, n_1 (for $n_1 \geqslant 1$) and $n_1 + 1$ respectively. These dimensions are in agreement with the results $d_0 = d_1 = 2(n_1 + 1)$ obtained in Example III, since $\Lambda' = \Lambda$ and $\Lambda - (\alpha_1 + 2\alpha_2)$ belong to the even subspace of the graded representation of $\tilde{\mathscr{L}}_s$, and $\Lambda' = \Lambda - \alpha_2$ and $\Lambda - (\alpha_1 + \alpha_2)$ belong to the odd subspace of this representation.

Most of the proofs by Kac (1978b) that are referred to above make use of his supersymmetry generalization of Weyl's character formula for the representations of semi-simple Lie algebras, the basic ideas of which are discussed briefly in Chapter 27, Section 4. Other discussions of character formulae for "typical" irreducible representations of simple Lie superalgebras have been given by Kac (1977c), Balantekin (1984), and Wybourne (1984).

The "typical" irreducible graded representations have received more attention here than the "atypical" ones only because comparatively little is known in the way of general formulae for the "atypical" representations. There is no reason to regard the "typical" representations as being more important either physically or mathematically, and indeed they do not even form a closed set, since the direct product of two "typical" irreducible representations can be an "atypical" representation. Thierry-Mieg (1983, 1984b) has made some progress with the general theory, obtaining an expression for the superdimension $d_0 - d_1$ of "atypical" representations and a tabulation of *all* the irreducible graded representations of $A(r/s)$ for $(r+1)(s+1) \leqslant 8$, $A(r/r)$ for $r \leqslant 2$, $B(0/s)$ for $s \leqslant 3$, $C(s)$ for $s \leqslant 4$, $D(2/1;\alpha)$ for all allowed α, $F(4)$ and $G(3)$. Hughes and King (1987) have conjectured a character formula for the "atypical" irreducible representations of $A(r/s)$, and have discussed the relation of this formula to those obtained previously by Bernstein and Leites (1980) and Van der Jeugt (1987a). Leites (1980a, b) and Sharp *et al.* (1985) have proposed corresponding formulae for Lie superalgebras of the types $B(r/s)$, $C(s)$ and $D(r/s)$.

These comments on the importance of "atypical" representations may be illustrated by considering the adjoint representation, for which the following theorem applies.

Theorem X *If $\tilde{\mathscr{L}}_s$ is a simple Lie superalgebra then its adjoint representation is irreducible.*

Proof This goes through in exactly the same way as the proof of the corresponding result for simple Lie algebras (Theorem VIII of Chapter 13, Section 2).

However, in most cases the *adjoint* representation is actually *atypical*! This follows because for the adjoint representation $d_0 = m$ and $d_1 = n$, where m is the dimension of $\tilde{\mathscr{L}}_0$ and n is the dimension of $\tilde{\mathscr{L}}_1$. For most simple Lie superalgebras $m \neq n$, implying that $d_0 \neq d_1$, and in such a situation (and provided that $\tilde{\mathscr{L}}_s \neq B(0/s)$ $(s \geqslant 1)$) Theorem VIII shows that the adjoint representation *must* be "atypical".

The generalization of the Young tableaux method (cf. Chapter 16, Section 7) to the representation theory of simple Lie superalgebras has been the subject of papers by Dondi and Jarvis (1981), Balantekin and Bars (1982), Bars *et al.* (1983), Berele and Regev (1983), Jarvis and Green (1983), King (1983, 1986), Ruegg (1983), Delduc and Gourdin (1984), Farmer and Jarvis (1984), Hurni *et al.* (1984), Thierry-Mieg (1984a), Morel *et al.* (1985), Gourdin (1986a, b, 1987), Sciarrino and Sorba (1986) and Cummins and King (1987a, b).

As $\tilde{\mathscr{L}}_s = B(0/s)$ (for $s \geqslant 1$) are the only basic classical simple complex Lie superalgebras for which *all* reducible representations are completely reducible, it is only for these that the reduction of direct product representations is straightforward and for which the discussions of Clebsch–Gordan coefficients, the Wigner–Eckart theorem and related matters that were given earlier for semi-simple Lie algebras can be extended without difficulty. Nevertheless, investigations of these ideas for other simple Lie superalgebras have been made by Mezincescu (1977), Agrawala (1979), Marcu (1980b), Chen *et al.* (1983a, 1984a, b), Li and Sun (1986) and Zeng (1987).

Papers dealing with other aspects of the representation theory of basic classical simple complex Lie superalgebras or presenting more detailed investigations of specific cases include those of Scheunert *et al.* (1977a, b), Palev (1978, 1980, 1981, 1985, 1986, 1987a, b), Marcu (1980a), Song and Han (1980), Song *et al.* (1980), Sun *et al.* (1980), Han (1981), Han *et al.* (1981), Sun and Han (1981), Hughes (1981), Backhouse (1982), Hurni and Morel (1982, 1983), Lemire and Patera (1982), Palev and Stoytchev (1982), Rittenberg and Scheunert (1982), Chen and Chen (1983), Farmer and Jarvis (1983), Han and Sun (1983), Ruegg (1983), Gruber *et al.* (1984), Nicolai and Sezgin (1984), Wybourne (1984), Scheunert (1985), Farmer (1986a, b), Mitra (1987) and Van der Jeugt (1987b). Of course, every representation of a simple complex Lie superalgebra provides a corresponding representation of each of its real forms in the usual way (cf. Chapter 13, Section 3), and so a number of these papers are couched in terms of representations of real forms.

(c) *Casimir operators and indices of representations*

The following theorem provides a refinement of the generalization of Schur's Lemma that was given in Theorem I of Chapter 21, Section 4.

Theorem XI Suppose that the matrices $\mathbf{\Gamma}(a)$ form a (d_0/d_1)-dimensional *irreducible* graded representation of a basic classical simple complex Lie superalgebra $\tilde{\mathscr{L}}_s$ and that $\mathbf{M}$ is a matrix is $\mathrm{M}(d_0/d_1;\,\mathbb{C})$ such that

$$[\mathbf{M}, \mathbf{\Gamma}(a)] = \mathbf{0} \tag{25.91}$$

for all $a \in \tilde{\mathscr{L}}_s$. Then $\mathbf{M}$ must have the form $\mathbf{M} = \kappa\mathbf{1}$, where κ is any constant and $\mathbf{1}$ is the $(d_0 + d_1)$-dimensional unit matrix.

Proof Theorem I of Chapter 21, Section 4 shows that for a general Lie superalgebra there are two possible forms for a matrix $\mathbf{M}$ that satisfies the requirements of this theorem, but the possibility (b) is excluded here because even if the representations $\mathbf{\Gamma}_{00}$ and $\mathbf{\Gamma}_{11}$ of the even part $\tilde{\mathscr{L}}_0$ have the same dimension they cannot be equivalent in this case.

Before discussing the Casimir operators, it is necessary to look more closely at the non-degenerate bilinear supersymmetric consistent invariant form $B'(\ ,\ )$ that was introduced at the beginning of Section 3(*b*). Let $a_1, a_2, \ldots, a_{m+n}$ be a basis of $\tilde{\mathscr{L}}_s$ chosen so that $a_1, a_2, \ldots, a_m$ are all even and $a_{m+1}, a_{m+2}, \ldots, a_{m+n}$ are all odd, and let $\mathbf{B}'$ be the $(m+n) \times (m+n)$ matrix defined by

$$(\mathbf{B}')_{pq} = B'(a_p, a_q) \tag{25.92}$$

for $p, q = 1, 2, \ldots, m+n$. The consistency property (cf. (21.84)) then implies that $\mathbf{B}'$ must have the form

$$\mathbf{B}' = \begin{bmatrix} \mathbf{B}'_0 & \mathbf{0} \\ \mathbf{0} & \mathbf{B}'_1 \end{bmatrix}, \tag{25.93}$$

where $\mathbf{B}'_0$ is an $m \times m$ matrix and $\mathbf{B}'_1$ is an $n \times n$ matrix. Moreover, the supersymmetry property (cf. (21.81)) shows that $\mathbf{B}'_0$ is symmetric and $\mathbf{B}'_1$ is antisymmetric. As $B'(\ ,\ )$ is assumed to be non-degenerate, $(\mathbf{B}')^{-1}$ exists, and it follows immediately that

$$(\mathbf{B}')^{-1} = \begin{bmatrix} (\mathbf{B}'_0)^{-1} & \mathbf{0} \\ \mathbf{0} & (\mathbf{B}'_1)^{-1} \end{bmatrix}, \tag{25.94}$$

where again $(\mathbf{B}'_0)^{-1}$ is symmetric and $(\mathbf{B}'_1)^{-1}$ is antisymmetric. This implies the useful identity

$$\begin{aligned} \sum_{p,q=1}^{m+n} (\mathbf{B}'^{-1})_{pq}(\mathbf{B}')_{pq} &= \sum_{p,q=1}^{m} (\mathbf{B}_0'^{-1})_{pq}(\mathbf{B}'_0)_{qp} - \sum_{p,q=1}^{n} (\mathbf{B}_1'^{-1})_{pq}(\mathbf{B}'_1)_{qp} \\ &= m - n. \end{aligned} \tag{25.95}$$

It is always possible to choose a basis $a_1, a_2, \ldots, a_m$ for the even part $\tilde{\mathscr{L}}_0$ such that

$$B'(a_p, a_q) = B'_0(a_p, a_q) = \delta_{pq} \tag{25.96}$$

for $p, q = 1, 2, \ldots, m$ (that is, it is orthonormal with respect to $B'(\ ,\ )$). A convenient choice would consist of l orthonormal elements $H_1, H_2, \ldots, H_l$ from the Cartan subalgebra $\mathscr{H}_s$, together with $\{2B'(a_\alpha, a_{-\alpha})\}^{-1/2}(a_\alpha + a_{-\alpha})$ and $i\{2B'(a_\alpha, a_{-\alpha})\}^{-1/2}(a_\alpha - a_{-\alpha})$ for each positive even root α (the orthonormality properties of which depend on (25.22)–(25.24)). With this choice of basis of $\tilde{\mathscr{L}}_0$, it follows that

$$(\mathbf{B}'_0)^{-1} = \mathbf{B}'_0 = \mathbf{1}_m. \tag{25.97}$$

For the odd part $\tilde{\mathscr{L}}_1$ it is not possible to choose an orthonormal basis because $B'(a_p, a_p) = 0$ for every odd basis element a_p, but it is convenient to define for every positive odd root α the pair of elements

$$b_{\alpha+} = \frac{a_\alpha + a_{-\alpha}}{\{2B'(a_{-\alpha}, a_\alpha)\}^{1/2}} \tag{25.98}$$

and

$$b_{\alpha-} = \frac{a_\alpha - a_{-\alpha}}{\{2B'(a_{-\alpha}, a_\alpha)\}^{1/2}}. \tag{25.99}$$

These have the properties that

$$B'(b_{\alpha+}, b_{\alpha-}) = B'_1(b_{\alpha+}, b_{\alpha-}) = -B'(b_{\alpha-}, b_{\alpha+}) = -B'_1(b_{\alpha-}, b_{\alpha+}) = 1 \tag{25.100}$$

for every positive odd root α, that

$$B'(b_{\alpha+}, b_{\beta-}) = B'_1(b_{\alpha+}, b_{\beta-}) = B'(b_{\alpha-}, b_{\beta+}) = B'_1(b_{\alpha-}, b_{\beta+}) = 0 \tag{25.101}$$

for every pair of non-equal positive odd roots α and β, and that

$$B'(b_{\alpha+}, b_{\beta+}) = B'_1(b_{\alpha+}, b_{\beta+}) = B'(b_{\alpha-}, b_{\beta-}) = B'_1(b_{\alpha-}, b_{\beta-}) = 0 \tag{25.102}$$

for every pair of positive odd roots α and β. With the basis of $\tilde{\mathscr{L}}_1$ being taken to consist of $b_{\alpha+}$ and $b_{\alpha-}$ for each positive odd root α, it follows that

$$(\mathbf{B}'_1)^{-1} = -\mathbf{B}'_1. \tag{25.103a}$$

It then follows from the supersymmetric property $B'([a_{m+p}, a_{m+q}], a_r) = B'(a_{m+p}, [a_{m+q}, a_r])$ and (21.16) that for $p, q = 1, 2, \ldots, n$

$$[a_{m+p}, a_{m+q}] = -\sum_{r=1}^{m} (\mathbf{B}'_1\mathbf{D}(a_r))_{pq} a_r. \tag{25.103b}$$

Definition *The second-order Casimir operator C_2*

Let $a_1, a_2, \ldots, a_{m+n}$ be the basis of a basic classical simple complex Lie superalgebra $\tilde{\mathscr{L}}_s$ with non-degenerate bilinear supersymmetric consistent invariant form $B'(\ ,\ )$ and let V be the carrier space of some graded representation Γ of $\tilde{\mathscr{L}}_s$ whose linear operators are $\Phi(a)$ $(a \in \tilde{\mathscr{L}}_s)$. The second-order Casimir operator C_2 for this representation of $\tilde{\mathscr{L}}_s$ is defined by

$$C_2 = \sum_{p,q=1}^{m+n} (\mathbf{B}'^{-1})_{pq} \Phi(a_p)\Phi(a_q). \tag{25.104}$$

Because of the form of (25.94), the elements a_p and a_q that appear explicitly in the expression (25.104) for C_2 are either *both even* or *both odd*. Consequently C_2 is an *even* operator, in that it maps the even subspace V_0 of the carrier space V into itself and maps the odd subspace V_1 of V into itself, and so it is natural to consider the *commutator* of C_2 with the operators $\Phi(a)$. It will be noted that (25.104) has the same form as the corresponding semi-simple Lie algebra expression (16.1).

With the basis of $\tilde{\mathscr{L}}_s$ that has just been described, the formula (25.104) for C_2 reduces to

$$C_2 = \sum_{j=1}^{l} \Phi(H_j)^2 + \sum_{\alpha \in \Delta^+} \frac{\Phi(a_\alpha)\Phi(a_{-\alpha}) + \Phi(a_{-\alpha})\Phi(a_\alpha)}{B'(a_{-\alpha}, a_\alpha)}. \tag{25.105}$$

But

$$a_\alpha a_{-\alpha} = [a_\alpha, a_{-\alpha}] + (-1)^{\deg a_\alpha} a_{-\alpha} a_\alpha,$$

and so, by (25.30),

$$a_\alpha a_{-\alpha} = B'(a_\alpha, a_{-\alpha}) h_\alpha + (-1)^{\deg a_\alpha} a_{-\alpha} a_\alpha;$$

hence, again using the fact that

$$B'(a_\alpha, a_{-\alpha}) = (-1)^{\deg a_\alpha} B'(a_{-\alpha}, a_\alpha)$$

(cf. (21.81) with $\deg a_\alpha = \deg a_{-\alpha}$), (25.105) becomes

$$C_2 = \sum_{j=1}^{l} \Phi(H_j)^2 + \sum_{\alpha\in\Delta_0^+} \Phi(h_\alpha) - \sum_{\alpha\in\Delta_1^+} \Phi(h_\alpha) + \sum_{\alpha\in\Delta^+} \frac{2\Phi(a_{-\alpha})\Phi(a_\alpha)}{B'(a_\alpha, a_{-\alpha})}. \quad (25.106)$$

The Dynkin index of (16.12) also has a supersymmetric generalization.

Definition *The Dynkin superindex of a representation*

Let $\mathbf{\Gamma}$ be a graded representation of a basic classical simple complex Lie superalgebra $\tilde{\mathscr{L}}_s$ with non-degenerate bilinear supersymmetric consistent invariant form $B'(\ ,\ )$. Then the *Dynkin superindex* γ of $\mathbf{\Gamma}$ is defined by

$$\gamma = \frac{\operatorname{str}\{\mathbf{\Gamma}(a)\mathbf{\Gamma}(b)\}}{B'(a,b)} \quad (25.107)$$

for all a and b of $\tilde{\mathscr{L}}_s$.

As noted previously, $\operatorname{str}\{\mathbf{\Gamma}(a)\mathbf{\Gamma}(b)\}$ is a bilinear supersymmetric consistent invariant form, so the fact that the right-handed side of (25.107) is independent of a and b is an immediate consequence of Theorem III of Section 2.

The generalization of Theorem I of Chapter 16, Section 2 and Theorem II of Chapter 16, Section 3 is as follows.

Theorem XII Let $\tilde{\mathscr{L}}_s$ be a basic classical simple complex Lie superalgebra, let C_2 be the second-order Casimir operator defined in (25.106) for the graded representation $\mathbf{\Gamma}$ of $\tilde{\mathscr{L}}_s$ whose carrier space is V and whose linear operators are $\Phi(a)$ ($a\in\tilde{\mathscr{L}}_s$), and let γ the corresponding Dynkin superindex (as defined in (25.107)). Then the following hold.

(a) C_2 is independent of the choice of basis elements $a_1, a_2, \ldots, a_{m+n}$ of $\tilde{\mathscr{L}}_s$.

(b) For all $a\in\tilde{\mathscr{L}}_s$

$$[C_2, \Phi(a)]_- = 0. \quad (25.108)$$

(c) If Γ is a graded *irreducible* representation of $\tilde{\mathscr{L}}_s$ then C_2 is a constant times the identity operator. If Γ has highest weight Λ then this constant will again be written as $C_2(\Lambda)$. Then, for any $\psi \in V$,

$$C_2\psi = C_2(\Lambda)\psi, \tag{25.109}$$

so that again $C_2(\Lambda)$ may be described as the "eigenvalue of C_2 in the irreducible representation with highest weight Λ".

(d) This eigenvalue is given by the expression

$$C_2(\Lambda) = \langle \Lambda, \Lambda + 2\rho \rangle, \tag{25.110}$$

where, as in (25.82),

$$\rho = \frac{1}{2}\sum_{\alpha \in \Delta_0^+} \alpha - \frac{1}{2}\sum_{\alpha \in \Delta_1^+} \alpha,$$

Δ_0^+ and Δ_1^+ being the sets of positive even and positive odd roots of $\tilde{\mathscr{L}}_s$ respectively.

(e) If Γ is a graded *irreducible* representation of $\tilde{\mathscr{L}}_s$, and m and n denote the dimensions of $\tilde{\mathscr{L}}_0$ and $\tilde{\mathscr{L}}_1$ respectively, and d_0 and d_1 denote the dimensions of the even and odd parts of V respectively, then

$$(m - n)\gamma = (d_0 - d_1)C_2(\Lambda). \tag{25.111}$$

(f) If Γ is the adjoint representation **ad** of $\tilde{\mathscr{L}}_s$ and (i) if $\tilde{\mathscr{L}}_s$ possesses a non-zero Killing form $B(\ ,\ )$, and $B'(\ ,\ )$ is chosen so that $B'(a, b) = B(a, b)$ for all $a, b \in \tilde{\mathscr{L}}_s$, then

$$C_2(\Lambda) = 1, \tag{25.112}$$

whereas
(ii) if $\tilde{\mathscr{L}}_s$ has Killing form that is identically zero then

$$C_2(\Lambda) = 0.$$

(g) If Γ is a graded *irreducible* representation of $\tilde{\mathscr{L}}_s$ and if $d_0 = d_1$ and $m \neq n$ then $\gamma = 0$.

(h) For *any* graded representation Γ of $\tilde{\mathscr{L}}_s$ the Dynkin superindex γ can be written in the form

$$\gamma = \frac{1}{l}\left\{ \sum_{\lambda \,\text{even}} \langle \lambda, \lambda \rangle - \sum_{\lambda \,\text{odd}} \langle \lambda, \lambda \rangle \right\}, \tag{25.113}$$

where l is the rank of $\tilde{\mathscr{L}}_s$ and the summations are over all the even and odd weights λ of the representation, a weight λ of multiplicity $m(\lambda)$ being included $m(\lambda)$ times.

Proof (*a*) This is exactly the same as the proof of the corresponding part of Theorem I of Chapter 16, Section 2.

(b) This is most easily checked using the special basis of (25.96)–(25.102) using (25.103) and (14.3).

(c) This follows immediately from the refinement of Schur's Lemma given in Theorem XI.

(d) Applying the expression for C_2 given in (25.106) to $\psi(\Lambda)$, the basis element of V corresponding to the highest weight Λ, since $\Phi(H_j)\psi(\Lambda) = \Lambda(H_j)\psi(\Lambda)$, $\Phi(h_\alpha)\psi(\Lambda) = \Lambda(h_\alpha)\psi(\Lambda)$ and $\Phi(a_\alpha)\psi(\Lambda) = 0$ for positive roots,

$$C_2(\Lambda) = \sum_{j=1}^{l} \Lambda(H_j)^2 + \sum_{\alpha\in\Delta_0^+} \Lambda(h_\alpha) - \sum_{\alpha\in\Delta_1^+} \Lambda(h_\alpha).$$

Thus, by (13.58) (which is still valid here), (25.29) and (25.82), $C_2(\Lambda) = \langle \Lambda, \Lambda + 2\rho \rangle$.

(e) Equations (25.104) and (25.109) give

$$\sum_{p,q=1}^{m+n} (\mathbf{B}'^{-1})_{pq} \mathbf{\Gamma}(a_p)\mathbf{\Gamma}(a_q) = C_2(\Lambda)\mathbf{1},$$

which, on taking the supertrace of both sides, gives

$$\sum_{p,q=1}^{m+n} (\mathbf{B}'^{-1})_{pq} \operatorname{str}\{\mathbf{\Gamma}(a_p)\mathbf{\Gamma}(a_q)\} = C_2(\Lambda)(d_0 - d_1),$$

from which (25.111) follows on applying (25.107) and (25.95).

(f) If $B'(\ ,\ ) = B(\ ,\ )$ and $\mathbf{\Gamma} = \mathbf{ad}$ then (25.107) shows immediately that $\gamma = 1$. Moreover, when $\mathbf{\Gamma} = \mathbf{ad}$ the dimensions are related by $m = d_0$ and $n = d_1$. Equation (25.111) then becomes $m - n = (m - n)C_2(\Lambda)$, which immediately gives $C_2(\Lambda) = 1$ if $m - n \neq 0$. The only cases in which $\tilde{\mathscr{L}}_s$ has a non-zero Killing form and $m - n = 0$ are $\tilde{\mathscr{L}}_s = A(r + 1/r)$ (for $r \geqslant 0$), $c(2)$, and $D(r/r)$ (for $r \geqslant 2$), for which direct calculation, using (25.110), shows again that $C_2(\Lambda) = 1$. If $\tilde{\mathscr{L}}_s$ has zero Killing form then (25.107) shows that $\gamma = 0$. As $m - n \neq 0$ in all such cases, (25.111) implies that $C_2(\Lambda) = 0$ for such a superalgebra.

(g) This follows immediately from (25.111).

(h) With $H_1, H_2, \ldots, H_l$ being the l orthonormal elements of the Cartan subalgebra $\mathscr{H}_s$, (25.107) gives $\operatorname{str}\{\mathbf{\Gamma}(H_k)\mathbf{\Gamma}(H_k)\} = \gamma B'(H_k, H_k) = \gamma$ for each $k = 1, 2, \ldots, l$. However,

$$\operatorname{str}\{\mathbf{\Gamma}(H_k)\mathbf{\Gamma}(H_k)\} = \sum_{\lambda\ \text{even}} \lambda(H_k)^2 - \sum_{\lambda\ \text{odd}} \lambda(H_k)^2,$$

from which (25.113) follows by a further application of (13.58).

One consequence of part (e) (when taken with part (a) of Theorem VIII) is that for *any typical* graded irreducible representation of *any* basic classical simple complex Lie superalgebra $\tilde{\mathscr{L}}_s$ for which $m \neq n$, *other than* $B(0/s)$ (for $s \geqslant 0$), the Dynkin superindex γ must be zero! This makes the Dynkin superindex a rather less useful object than the Dynkin index of a semi-simple Lie algebra. However, Morel *et al.* (1987) have shown that it is possible to define (and evaluate) useful indices

$$I_\Lambda^{(2n)} = \sum_{\lambda \,\text{even}} \langle \lambda, \lambda \rangle^n + \sum_{\lambda \,\text{odd}} \langle \lambda, \lambda \rangle^n, \tag{25.114}$$

and superindices

$$S_\Lambda^{(2n)} = \sum_{\lambda \,\text{even}} \langle \lambda, \lambda \rangle^n - \sum_{\lambda \,\text{odd}} \langle \lambda, \lambda \rangle^n,$$

for all $n = 0, 1, 2, \ldots,$ but with $\langle \lambda, \lambda \rangle$ being defined differently from (25.27).

The *higher*-order Casimir operators for a basic classical simple complex Lie superalgebra $\tilde{\mathscr{L}}_s$ can be found in much the same way as for the semi-simple Lie algebras. With the basis $a_1, a_2, \ldots, a_{m+n}$ of $\tilde{\mathscr{L}}_s$ in which $a_1, a_2, \ldots, a_m$ are all even and $a_{m+1}, a_{m+2}, \ldots, a_{m+n}$ are all odd that was introduced above, define $b_1, b_2, \ldots, b_{m+n}$ of $\tilde{\mathscr{L}}_s$ by

$$b_p = \sum_{r=1}^{m+n} (\mathbf{B}'^{-1})_{rp} a_r. \tag{25.115}$$

It follows from (25.94) that $b_1, b_2, \ldots, b_m \in \tilde{\mathscr{L}}_0$ and $b_{m+1}, b_{m+2}, \ldots, b_{m+n} \in \tilde{\mathscr{L}}_1$. The second-order Casimir operator C_2 of (25.104) can then be reexpressed as

$$C_2 = \gamma^{-1} \sum_{p_1, p_2 = 1}^{m+n} \operatorname{str}\{\boldsymbol{\Gamma}(a_{p_1})\boldsymbol{\Gamma}(a_{p_2})\}\, \Phi(b_{p_1})\Phi(b_{p_2}), \tag{25.116}$$

where $\boldsymbol{\Gamma}$ is a representation of $\tilde{\mathscr{L}}_s$ with non-zero Dynkin super index γ, so that $\operatorname{str}\{\boldsymbol{\Gamma}(a)\boldsymbol{\Gamma}(b)\} = \gamma B'(a, b) \in \tilde{\mathscr{L}}_s$. (Of course, if the Killing form is non-degenerate and $B'(\,,)$ is chosen so that $B'(a, b) = B(a, b)$ then $\boldsymbol{\Gamma}$ can be taken to be the adjoint representation and γ will have value 1.) The jth-order Casimir operator C_j may then be defined by

$$C_j = \gamma^{-j/2} \sum_{p_1, p_2, \ldots, p_j = 1}^{m+n} \operatorname{str}\{\boldsymbol{\Gamma}(a_{p_1})\boldsymbol{\Gamma}(a_{p_2}) \cdots \boldsymbol{\Gamma}(a_{p_j})\}\, \Phi(b_{p_1})\Phi(b_{p_2}) \cdots \Phi(b_{p_j}) \tag{25.117}$$

(which generalizes (16.10)). Backhouse (1977) has discussed these Casimir operators in detail. They have also been investigated by Bednar and Sachl

(1978), Backhouse (1979, 1980), Jarvis and Green (1979), Hlavaty and Niederle (1980), McAvaty and Niederle (1980), Agrawala (1981), Balantekin and Bars (1981a, b), Balantekin (1982), Bincer (1983), Chen *et al.* (1983a, b), Scheunert (1983b, c, 1984), Green and Jarvis (1983), Bednar *et al.* (1984) and Nwachuku and Rashid (1985). Jarvis and Murray (1983) have studied the Casimir operators for the *strange* classical simple complex Lie superalgebras.

5 The classical simple real Lie superalgebras

The relationship between the real and complex classical Lie superalgebras is very similar to that between the real and complex semi-simple Lie algebras. In the following $\mathscr{L}_s$ will denote a real Lie superalgebra, with $\mathscr{L}_0$ and $\mathscr{L}_1$ being its even and odd parts, while $\tilde{\mathscr{L}}_s$ will denote the complexification of $\mathscr{L}_s$, with $\tilde{\mathscr{L}}_0$ and $\tilde{\mathscr{L}}_1$ being the even and odd parts of $\tilde{\mathscr{L}}_s$. Clearly $\mathscr{L}_0$ is a real form of $\tilde{\mathscr{L}}_0$, but not every real form of $\tilde{\mathscr{L}}_0$ need occur in this way.

The definitions of several of the more important simple real Lie superalgebras will be given in the three examples that follow. The other relevant Lie superalgebras $\mathrm{sl}(p/q;\mathbb{R})$, $\mathrm{sl}(p/q;\mathbb{C})$ and $\mathrm{sl}(p/q;\mathbb{H})$ were defined in Example II of Chapter 21, Section 2.

Example I *The orthosymplectic Lie superalgebra* $\mathrm{osp}(p-p', p'/q;\mathbb{R})$

Let p and p' be two integers such that $p \geqslant 2p', p \geqslant 2$ and $p' \geqslant 0$, and let q be a positive even integer. With the generalized Lie product being specified by (21.20), the real Lie superalgebra $\mathrm{osp}(p-p', p'/q;\mathbb{R})$ is defined as the set of matrices $\mathbf{M}$ of $\mathrm{M}(p/q;\mathbb{R})$ that satisfy the condition

$$\mathbf{M}^{\mathrm{st}}\mathbf{K} + (-1)^{\deg \mathbf{M}}\mathbf{K}\mathbf{M} = \mathbf{0}, \tag{25.118}$$

where

$$\mathbf{K} = \begin{bmatrix} \mathbf{G}_{p,p'} & \mathbf{0} \\ \mathbf{0} & \mathbf{J} \end{bmatrix},$$

$\mathbf{J}$ being given by

$$\mathbf{J} = \begin{bmatrix} \mathbf{0} & \mathbf{1}_{q/2} \\ -\mathbf{1}_{q/2} & \mathbf{0} \end{bmatrix}$$

(i.e. by (21.33)) and where

$$\mathbf{G}_{p,p'} = \begin{bmatrix} \mathbf{1}_{p-p'} & \mathbf{0} \\ \mathbf{0} & -\mathbf{1}_{p'} \end{bmatrix}. \tag{25.119}$$

(That is, (25.118) is of the form (21.34) but with $\mathbf{G} = \mathbf{1}_p$ being replaced by

$\mathbf{G} = \mathbf{G}_{p,p'}$.) In generalizing these superalgebras, as in the next two examples, it is useful to note that (25.118) can be rewritten equivalently in terms of ordinary transposes instead of supertransposes as

$$\tilde{\mathbf{M}}\mathbf{K} + i^{\deg \mathbf{M}}\mathbf{K}\mathbf{M} = \mathbf{0}. \tag{25.120}$$

Following the same line of argument as that given in Example III of Chapter 21, Section 2, the even part $\mathscr{L}_0$ of $\mathscr{L}_s = \operatorname{osp}(p - p', p'/q; \mathbb{R})$ consists of all those matrices that have the form

$$\mathbf{M} = \begin{bmatrix} \mathbf{A} & \mathbf{0} \\ \mathbf{0} & \mathbf{D} \end{bmatrix}, \tag{25.121}$$

where $\mathbf{A}$ is a $p \times p$ real matrix that satisfies the constraint

$$\tilde{\mathbf{A}}\mathbf{G}_{p,p'} + \mathbf{G}_{p,p'}\mathbf{A} = \mathbf{0} \tag{25.122}$$

and $\mathbf{D}$ is a $q \times q$ real matrix that satisfies the constraint

$$\tilde{\mathbf{D}}\mathbf{J} + \mathbf{J}\mathbf{D} = \mathbf{0}. \tag{25.123}$$

Thus

$$\mathscr{L}_0 = \operatorname{so}(p - p', p') \oplus \operatorname{sp}(\tfrac{1}{2}q, \mathbb{R}),$$

where $\operatorname{so}(p - p', p')$ and $\operatorname{sp}(\frac{1}{2}q, \mathbb{R})$ are simple real Lie algebras defined in Table 10.1.

The elements of the odd part $\mathscr{L}_1$ of $\mathscr{L}_s = \operatorname{osp}(p - p', p'/q; \mathbb{R})$ have the form

$$\mathbf{M} = \begin{bmatrix} \mathbf{0} & \mathbf{B} \\ \mathbf{C} & \mathbf{0} \end{bmatrix}, \tag{25.124}$$

where $\mathbf{B}$ and $\mathbf{C}$ are $p \times q$ and $q \times p$ real matrices that satisfy the constraint

$$\tilde{\mathbf{B}}\mathbf{G}_{p,p'} = \mathbf{J}\mathbf{C} \tag{25.125}$$

(cf. (21.39)).

In the special case in which $p' = q' = 0$ the matrix $\mathbf{G}_{p,p'}$ reduces to $\mathbf{G}_{p,p'} = \mathbf{1}_p$, and consequently $\operatorname{osp}(p, 0/q; \mathbb{R})$ is just the real Lie superalgebra $\operatorname{osp}(p/q; \mathbb{R})$ that was defined in Example III of Chapter 21, Section 2.

Example II *The quaternionic ortho symplectic Lie superalgebra* $\operatorname{osp}(p/q - q', q'; \mathbb{H})$

Let p, q and q' be three non-negative integers such that $p \geqslant 2, q \geqslant 2q', q \geqslant 2$ and $q' \geqslant 0$. Let $\mathbf{M}(p/q; \mathbb{H})$ denote the set of $(p/q) \times (p/q)$ supermatrices $\mathbf{M}$ with *quaternionic* entries. (See Appendix B, Section 8 for a brief introduction to quaternions.) With the generalized Lie product being specified by (21.20), the real Lie superalgebra $\text{osp}(p/q-q',q';\mathbb{H})$ is defined as the subset of supermatrices $\mathbf{M}$ of $\text{M}(p/q;\mathbb{H})$ that satisfy the condition

$$\tilde{\mathbf{M}}\mathbf{K}' + \varepsilon_1^{\deg \mathbf{M}}\mathbf{K}'\mathbf{M}^* = \mathbf{0} \tag{25.126}$$

(whose form is a quaternionic generalization of (25.120)). In (25.126) $\mathbf{M}^*$ denotes the matrix obtained from $\mathbf{M}$ by replacing every element by its conjugate in the quaternionic sense of equation (B.43), ε_1 is a non-identity quaternionic basis element, and

$$\varepsilon_1^{\deg \mathbf{M}} = \begin{cases} 1 & \text{if } \deg \mathbf{M} = 0, \\ \varepsilon_1 & \text{if } \deg \mathbf{M} = 1. \end{cases} \tag{25.127}$$

ıso $\mathbf{K}'$ is defined by

$$\mathbf{K}' = \begin{bmatrix} \mathbf{1}_p\varepsilon_0 & \mathbf{0} \\ \mathbf{0} & \mathbf{G}_{q,q'}\varepsilon_1 \end{bmatrix},$$

where

$$\mathbf{G}_{q,q'} = \begin{bmatrix} \mathbf{1}_{q-q'} & \mathbf{0} \\ \mathbf{0} & -\mathbf{1}_{q'} \end{bmatrix}. \tag{25.128}$$

Following again the same line of argument as that given in Example III of Chapter 21, Section 2, the even part $\mathscr{L}_0$ of $\mathscr{L}_s = \text{osp}(p/q-q',q';\mathbb{H})$ consists of all those matrices that have the form

$$\mathbf{M} = \begin{bmatrix} \mathbf{A} & \mathbf{0} \\ \mathbf{0} & \mathbf{D} \end{bmatrix},$$

where $\mathbf{A}$ is a $p \times p$ matrix with *quaternionic* entries that satisfies the constraint

$$\tilde{\mathbf{A}}\varepsilon_1 + \varepsilon_1\mathbf{A}^* = \mathbf{0}, \tag{25.129}$$

and $\mathbf{D}$ is a $q \times q$ matrix with *quaternionic* entries that satisfies the constraint

$$\tilde{\mathbf{D}}\mathbf{G}_{q,q'} + \mathbf{G}_{q,q'}\mathbf{D}^* = \mathbf{0}. \tag{25.130}$$

The argument given in Appendix G, Section 3 for $\text{sp}(\frac{1}{2}N)$ can easily be modified to show that the matrices $\mathbf{A}$ form a real Lie algebra that is isomorphic to $\text{so}^*(2p)$ and the matrices $\mathbf{D}$ form a real Lie algebra that is isomorphic to $\text{sp}(q-q',q')$, both of these algebras being as defined in Table 10.1, so

$$\mathscr{L}_0 = \text{so}^*(2p) \oplus \text{sp}(q-q',q').$$

In the notation of Kac (1977a, b), $\text{osp}(p/q-q',q';\mathbb{H}) = \text{hosp}(p,q;q')$.

Example III *The special unitary Lie superalgebras* $\mathrm{su}(p-p', p'/q-q', q')$ *and* $\mathrm{su}(p/q)$

Let p, p', q and q' be four integers such that $p \geqslant 2p'$, $p' \geqslant 0$, $p \geqslant 2$ and $q \geqslant 2q'$, $q' \geqslant 0$, $q \geqslant 2$. With the generalized Lie product being specified by (21.20), the real Lie superalgebra $\mathrm{su}(p-p', p'/q-q', q')$ is defined as the subset of matrices $\mathbf{M}$ of $\mathrm{sl}(p/q;\mathbb{C})$ that satisfy the condition

$$\tilde{\mathbf{M}}\mathbf{K}'' + i^{\deg \mathbf{M}}\mathbf{K}''\mathbf{M}^* = \mathbf{0}, \tag{25.131}$$

where $\mathbf{M}^*$ denotes the complex conjugate of $\mathbf{M}$ and $\mathbf{K}''$ is defined by

$$\mathbf{K}'' = \begin{bmatrix} \mathbf{G}_{p,p'} & \mathbf{0} \\ \mathbf{0} & \mathbf{G}_{q,q'} \end{bmatrix},$$

$\mathbf{G}_{p,p'}$ and $\mathbf{G}_{q,q'}$ being defined by (25.119) and (25.128) respectively.

Following yet again the same line of argument as that given in Example III of Chapter 21, Section 2, the even part $\mathscr{L}_0$ of $\mathscr{L}_s = \mathrm{su}(p-p', p'/q-q', q')$ consists of all those matrices that have the form

$$\mathbf{M} = \begin{bmatrix} \mathbf{A} & \mathbf{0} \\ \mathbf{0} & \mathbf{D} \end{bmatrix}, \tag{25.132}$$

where $\mathbf{A}$ is a $p \times p$ complex matrix that satisfies the constraint

$$\tilde{\mathbf{A}}\mathbf{G}_{p,p'} + \mathbf{G}_{p,p'}\mathbf{A}^* = \mathbf{0} \tag{25.133}$$

and $\mathbf{D}$ is a $q \times q$ complex matrix that satisfies the constraint

$$\tilde{\mathbf{D}}\mathbf{G}_{q,q'} + \mathbf{G}_{q,q'}\mathbf{D}^* = \mathbf{0}, \tag{25.134}$$

and where the zero supertrace condition of $\mathrm{sl}(p/q;\mathbb{C})$ requires that

$$\mathrm{tr}\,\mathbf{A} = \mathrm{tr}\,\mathbf{D}.$$

Thus

$$\mathscr{L}_0 = \mathrm{su}(p-p', p') \oplus \mathrm{su}(q-q', q') \oplus \mathscr{L}_0^{\mathrm{A}}, \tag{25.135}$$

where $\mathscr{L}_0^{\mathrm{A}}$ is a one-dimensional Abelian real Lie algebra that consists of matrices of the form (25.132) with $\mathbf{A} = (i\mu/p)\mathbf{1}_p$ and $\mathbf{D} = (i\mu/q)\mathbf{1}_q$ and with μ being any real number. Also in (25.135), the real Lie algebra $\mathrm{su}(p-p', p')$ consists of the matrices (25.132) whose submatrices $\mathbf{A}$ satisfy (25.133) and $\mathrm{tr}\,\mathbf{A} = 0$ and whose submatrices $\mathbf{D}$ are such that $\mathbf{D} = \mathbf{0}$. Likewise $\mathrm{su}(q-q', q')$ consists of the matrices (25.132) whose submatrices $\mathbf{D}$ satisfy (25.134) and $\mathrm{tr}\,\mathbf{D} = 0$ and whose submatrices $\mathbf{A}$ are such that $\mathbf{A} = \mathbf{0}$.

The elements of the odd part $\mathscr{L}_1$ of $\mathscr{L}_s = \mathrm{su}(p-p', p'/q-q', q')$ have the form

$$\mathbf{M} = \begin{bmatrix} \mathbf{0} & \mathbf{B} \\ \mathbf{C} & \mathbf{0} \end{bmatrix}, \tag{25.136}$$

where **B** and **C** are $p \times q$ and $q \times p$ complex matrices that satisfy the constraint

$$\tilde{\mathbf{B}}\mathbf{G}_{p,p'} = -i\mathbf{G}_{q,q'}\mathbf{C}^*. \tag{25.137}$$

In the special case in which $p' = q' = 0$ the matrices $\mathbf{G}_{p,p'}$ and $\mathbf{G}_{q,q'}$ reduce to $\mathbf{G}_{p,p'} = \mathbf{1}_p$ and $\mathbf{G}_{q,q'} = \mathbf{1}_q$. Consequently (25.133) and (25.134) become

$$\tilde{\mathbf{A}} + \mathbf{A}^* = \mathbf{0}$$

and

$$\tilde{\mathbf{D}} + \mathbf{D}^* = \mathbf{0},$$

so that (25.135) reduces to

$$\mathscr{L}_0 = \mathrm{su}(p) \oplus \mathrm{su}(q) \oplus \mathscr{L}_0^{\mathrm{A}},$$

and (25.137) simplifies to

$$\tilde{\mathbf{B}} = -i\mathbf{C}^*.$$

The resulting real Lie superalgebra $\mathrm{su}(p, 0/q, 0)$ is often denoted by $\mathrm{su}(p/q)$.

In Chapter 13 an argument was given that allowed all the semi-simple real Lie algebras to be found from the semi-simple complex Lie algebras. This argument made use of the involutive automorphisms of the compact real forms, but could be recast in terms of the involutive semimorphisms of the complex Lie algebras, a *semimorphism* of a complex Lie algebra $\tilde{\mathscr{L}}_0$ being a mapping of $\tilde{\mathscr{L}}_s$ onto itself such that

$$\psi(\alpha a + \beta b) = \alpha^*\psi(a) + \beta^*\psi(b) \tag{25.138}$$

for all a and b of $\tilde{\mathscr{L}}_0$ and all complex numbers α and β, and such that

$$\psi([a, b]) = [\psi(a), \psi(b)] \tag{25.139}$$

for all a and b of $\tilde{\mathscr{L}}_0$. (It will be noted that whereas (25.139) is identical with (11.2), (25.138) differs from (11.1) in that in (25.138) the complex conjugates α^* and β^* appear.) By extending such a semimorphism analysis to the Lie superalgebra situation, Parker (1980, 1981) obtained a *complete* classification of all the basic classical simple real Lie superalgebras starting from the basic classical simple complex Lie superalgebras. The following theorem is fundamental to the analysis.

Theorem I If $\tilde{\mathscr{L}}_s$ is a classical simple real Lie superalgebra then its complexification $\tilde{\mathscr{L}}_s$ is *either* a classical simple complex Lie superalgebra *or* is a direct sum of *two* classical simple complex Lie superalgebras that are *isomorphic*.

Proof See Parker (1980).

The present account will be restricted to listing the real forms of the families $A(r/s)$, $B(r/s)$, $C(s)$ and $D(r/s)$, and identifying their even parts. For information on the real forms of the other simple complex Lie superalgebras see Kac (1977a, b), Parker (1980, 1981) and Serganova (1984). (Serganova (1985) has also studied the automorphisms of the simple Lie superalgebras.)

(1) $\tilde{\mathscr{L}}_s = A(r/s)$ *for* $r > s \geqslant 0$

This Lie superalgebra has three families of real forms.

(a) $\mathscr{L}_s = \mathrm{sl}(r+1/s+1; \mathbb{R})$. As noted in Example II of Chapter 21, Section 2, in this case

$$\mathscr{L}_0 = \mathrm{sl}(r+1, \mathbb{R}) \oplus \mathrm{sl}(s+1, \mathbb{R}) \oplus \mathscr{L}_0^{\mathrm{A}},$$

where $\mathscr{L}_0^{\mathrm{A}}$ is a one-dimensional Abelian Lie algebra and $\mathrm{sl}(r+1, \mathbb{R})$ is defined in Table 10.1.

(b) $\mathscr{L}_s = \mathrm{sl}(r+1/s+1; \mathbb{H})$, provided that r and s are both odd integers. A similar argument to that for $\mathscr{L}_s = \mathrm{sl}(r+1/s+1; \mathbb{R})$ gives

$$\mathscr{L}_0 = q_{(r+1)/2} \oplus q_{(s+1)/2} \oplus \mathscr{L}_0^{\mathrm{A}},$$

where $\mathscr{L}_0^{\mathrm{A}}$ is essentially the same one-dimensional Abelian Lie algebra and where $q_{(r+1)/2}$ (for odd r) is the quaternionic Lie algebra defined on p. 556 of Chapter 14, Section 4, and which is shown there to be isomorphic to $\mathrm{su}^*(r+1)$. Thus $\mathscr{L}_0$ can be written alternatively as

$$\mathscr{L}_0 = \mathrm{su}^*(r+1) \oplus \mathrm{su}^*(s+1) \oplus \mathscr{L}_0^{\mathrm{A}}.$$

(c) $\mathscr{L}_s = \mathrm{su}(r+1-r', r'/s+1-s', s')$, where r' and s' are two non-negative integers such that $r+1 \geqslant 2r' \geqslant 0$ and $s+1 \geqslant 2s' \geqslant 0$ (and $r > s \geqslant 0$). A similar argument to that for $\mathscr{L}_s = \mathrm{sl}(r+1/s+1; \mathbb{R})$ gives

$$\mathscr{L}_0 = \mathrm{su}(r+1-r', r') \oplus \mathrm{su}(s+1-s', s') \oplus \mathscr{L}_0^{\mathrm{A}},$$

where $\mathscr{L}_0^{\mathrm{A}}$ is essentially the same one-dimensional Abelian Lie algebra, and where $\mathrm{su}(r+1-r', r')$ is defined in Table 10.1 (and where it is now to be understood for convenience that $\mathrm{su}(r+1, 0) = \mathrm{su}(r+1)$).

(2) $\tilde{\mathscr{L}}_s = A(r/r)$ *for* $r \geqslant 1$

This Lie superalgebra has four families of real forms.

(a) $\mathscr{L}_s = \mathrm{sl}(r+1/r+1; \mathbb{R})$ factored out by the subspace with basis element $\mathbf{1}_{2r+2}$ (cf. Appendix M, Section 2), for which

$$\mathscr{L}_0 = \mathrm{sl}(r+1, \mathbb{R}) \oplus \mathrm{sl}(r+1, \mathbb{R}),$$

where $\mathrm{sl}(r+1, \mathbb{R})$ is defined in Table 10.1.

(b) $\mathscr{L}_s = \mathrm{sl}(r+1/r+1; \mathbb{H})$ factored out by the subspace with basis element $\mathbf{1}_{2r+2}$ (cf. Appendix M, Section 2), provided that r is an odd integer. Then

$$\mathscr{L}_0 = q_{(r+1)/2} \oplus q_{(r+1)/2},$$

where $q_{(r+1)/2}$ (for odd r) is the quaternionic Lie algebra defined on p. 556 of Chapter 14, Section 4, and which is shown there to be isomorphic to $\mathrm{su}^*(r+1)$. Thus $\mathscr{L}_0$ can be written alternatively as

$$\mathscr{L}_0 = \mathrm{su}^*(r+1) \oplus \mathrm{su}^*(r+1).$$

(c) $\mathscr{L}_s = \mathrm{su}(r+1-r', r'/r+1-r', r')$ factored out by the subspace with basis element $\mathbf{1}_{2r+2}$ (cf. Appendix M, Section 2), where r' is such that $r+1 \geqslant 2r' \geqslant 0$ (and $r \geqslant 1$). Then

$$\mathscr{L}_0 = \mathrm{su}(r+1-r', r') \oplus \mathrm{su}(r+1-r', r'),$$

where $\mathrm{su}(r+1-r', r')$ is defined in Table 10.1 (and where it is again to be understood that $\mathrm{su}(r+1, 0) = \mathrm{su}(r+1)$).

(d) $\mathscr{L}_s = H(4; r; \mathbb{R})$, which is a subalgebra of a real form of one of the Cartan superalgebras ($W(n)$) (cf. Kac 1977a), for which

$$\mathscr{L}_0 = \mathrm{sl}(r+1, \mathbb{C}),$$

considered as a real Lie algebra.

(3) $\tilde{\mathscr{L}}_s = B(r/s)$ *for* $r \geqslant 0$, $s \geqslant 1$

For $r \geqslant 1$ and $s \geqslant 1$, $B(r/s)$ has one family of real forms:

$$\mathscr{L}_s = \mathrm{osp}(2r+1-r', r'/2s; \mathbb{R}), \quad \text{where } 2r+1 \geqslant 2r' \geqslant 0.$$

Then

$$\mathscr{L}_0 = \mathrm{so}(2r+1-r', r') \oplus \mathrm{sp}(s, \mathbb{R}),$$

where $so(2r+1-r', r')$ and $sp(s, \mathbb{R})$ are defined in Table 10.1 (and where it is to be understood that $so(2r+1, 0) = so(2r+1)$).

However, $B(0/s)$ (for $s \geqslant 1$) only has one real form:

$$\mathscr{L}_s = osp(1/2s; \mathbb{R}),$$

for which

$$\mathscr{L}_0 = sp(s, \mathbb{R}).$$

(4) $\tilde{\mathscr{L}}_s = C(s)$ *for* $s \geqslant 2$

This Lie superalgebra has two families of real forms.

(a) $\mathscr{L}_s = osp(2, 0/2s-2); \mathbb{R})$, for which

$$\mathscr{L}_0 = so(2) \oplus sp(s-1, \mathbb{R}),$$

where $sp(s-1, \mathbb{R})$ is defined in Table 10.1.

(b) $\mathscr{L}_s = osp(2/s-1-s', s'; \mathbb{H})$, where $s' = 0, 1, \ldots, [\frac{1}{2}(s-1)]$, where $[\frac{1}{2}s]$ denotes the largest integer not greater than $\frac{1}{2}(s-1)$. It follows from Example II above that

$$\mathscr{L}_0 = so^*(2) \oplus sp(s-1-s', s'),$$

where $so^*(2)$ is a one-dimensional real Abelian Lie algebra (which is isomorphic to $so(2)$), and where $sp(s-1-s', s')$ is defined in Table 10.1 (here it is to be understood that $sp(s-1, 0) = sp(s-1)$).

(5) $\tilde{\mathscr{L}}_s = D(r/s)$ *for* $r \geqslant 2$, $s \geqslant 1$

This Lie superalgebra has two families of real forms.

(a) $\mathscr{L}_s = osp(2r-r', r'/2s; \mathbb{R})$, where $r \geqslant r' \geqslant 0$, for which

$$\mathscr{L}_0 = so(2r-r', r') \oplus sp(s, \mathbb{R}),$$

where $so(2r-r', r')$ and $sp(N, \mathbb{R})$ are defined in Table 10.1 (and where it is to be understood that $so(2r, 0) = so(2r)$).

(b) $\mathscr{L}_s = osp(2r/s-s', s'; \mathbb{H})$, where $s \geqslant 2s' \geqslant 0$. It follows from Example II above that

$$\mathscr{L}_0 = so^*(2r) \oplus sp(s-s', s')$$

(where it is to be understood that $sp(s, 0) = sp(s)$).

It is easy to identify the real Lie superalgebras whose complexifications are the direct sum of *two* isomorphic simple complex Lie superalgebras. Consider, for example, the case of $\mathrm{sl}(r+1/s+1;\mathbb{C})$ (for $r > s \geqslant 0$). Considered as a *complex* Lie superalgebra, this is isomorphic to $A(r/s)$, but it can also be considered as a *real* Lie superalgebra, which has twice the dimension and has $A(r/s) \oplus A(r/s)$ as its complexification. Similarly $\mathrm{osp}(2r+1/2s;\mathbb{C})$, $\mathrm{osp}(2/2s-2;\mathbb{C})$ and $\mathrm{osp}(2r/2s;\mathbb{C})$ considered as real Lie superalgebras are the complexifications of $B(r/s) \oplus B(r/s)$, $C(s) \oplus C(s)$ and $D(r/s) \oplus D(r/s)$ respectively.

A simple real Lie superalgebra $\mathscr{L}_s$ is said to be the *compact form* of its complexification $\tilde{\mathscr{L}}_s$ if the semi-simple part $\mathscr{L}_0^{ss}$ of the even part $\mathscr{L}_0$ of $\mathscr{L}_s$ is a compact Lie algebra in the sense of Chapter 14, Section 2. As Example III of Chapter 22, Section 3(*b*) demonstrates, this does *not* imply that any Lie supergroup $\mathscr{G}_s(\mathbb{C}B_L)$ corresponding to $\mathscr{L}_s$ is a compact Lie group, since the real Lie superalgebra $\mathrm{su}(p/q)$ is the compact form of $A(p-1/q-1)$ but the Lie supergroup $\mathrm{SU}(p/q)$ is not compact.

6 The conformal, de Sitter and anti-de Sitter superalgebras

In addition to the Poincaré superalgebras, three other classes of superalgebras associated with space–time symmetries have also been quite widely used in the literature. These are the conformal superalgebras, the de Sitter superalgebras and the anti-de Sitter superalgebras.

A brief general account of conformal groups and their corresponding algebras appears later in Chapter 28, Section 1. As mentioned there, for 4-dimensional space–time the conformal group is a 15-parameter Lie group, 10 of these parameters being associated with Poincaré transformations, 4 with "special conformal transformations" and 1 with "dilatations". The corresponding real Lie algebra is isomorphic to $\mathrm{su}(2,2)$ (and to $\mathrm{so}(4,2)$) (cf. Mack and Salam 1969)), its complexification being isomorphic to the simple complex Lie algebras A_3 $(=D_3)$. The smallest real Lie superalgebra that contains $\mathrm{su}(2,2)$ in its even subalgebra is $\mathrm{su}(2,2/1,0)$. As shown in Example III of the previous section, $\mathrm{su}(2,2/1,0)$ also contains one other even basis element, which is often called the "chiral charge" and which commutes with all the elements of the conformal Lie algebra. The odd part of this $\mathrm{su}(2,2/1,0)$ conformal superalgebra is 8-dimensional. Its complexification is the simple Lie superalgebra $A(3/0)$. The "N-extended" version is $\mathrm{su}(2,2/N,0)$ (which is often denoted in the literature by $\mathrm{su}(2,2/N)$), whose even subalgebra is isomorphic to su(2, 2)$\oplus$u(N) and whose complexification is $A(3/N-1)$ ($=\mathrm{sl}(4/N;\mathbb{C})$). The structures of these superalgebras have been discussed in detail by Ferrara (1974), Haag *et al.* (1975) and Bedding (1984, 1985), the corresponding analysis for other space–time dimensions being given by Van Holten and Van Proeyen (1982) and Van Proeyen (1987). Their superfields and infinite-dimensional "unitary" representations have been studied by Freund (1976), Grosser (1976), Molotkov *et al.* (1976), Aneva *et al.* (1976, 1977, 1978), Ivanov and Sorin (1980), Mansouri and Schaer (1981), Schwarz (1981), Sohnius (1983), Flato and Fronsdal (1984), Dobrev and Petkova (1985, 1986a, b), Günaydin and Marcus (1985), Binegar (1986) and Morel *et al.* (1986). Their gauge theories have been examined by Crispim-Romão *et al.* (1977), Kaku *et al.* (1977a, b) and Sohnius and West (1981).

The only 4-dimensional space–times that have constant curvature other than that of Minkowski are those that were introduced by de Sitter (1917). One of these has positive curvature and the other negative curvature. Both can be embedded in flat 5-dimensional pseudo-Euclidean spaces with metric tensors η_{AB} that are diagonal and such that $\eta_{AB}=\mathrm{diag}(1,1,1,1,-1)$ and $\eta_{AB}=\mathrm{diag}(1,1,1,-1,-1)$ (cf. Synge 1960), the former corresponding to positive curvature and the latter to negative curvature. Consequently the groups of transformations that leave these metrics invariant are isomorphic to the real Lie algebras so(4, 1) and so(3, 2). These are often called the "de Sitter algebra" and the "anti-de Sitter algebra" respectively. Both have complexifications that are isomorphic to the simple complex Lie algebras B_2 and C_2. Moreover, both are 10-dimensional and contain the curvature in their structure constants as a parameter, and in the limit in which the curvature tends to zero both of these Lie algebras "contract" onto the Poincaré Lie algebra.

The supersymmetric extension is most straightforward for the anti-de Sitter algebra so(3, 2), since the simple complex Lie superalgebra $B(0/2)$ whose even subalgebra is B_2 ($=C_2$) has the real form osp(1/4; $\mathbb{R}$), whose even subalgebra is sp(2, $\mathbb{R}$), which, as noted on p. 548, is isomorphic to so(3, 2). Closer examination shows that osp(1/4; $\mathbb{R}$) is the smallest Lie superalgebra that contains so(3, 2) (cf. Keck (1975, Pilch *et al.* 1985). The even part of osp(1/4; $\mathbb{R}$) consists entirely of so(3, 2), and the odd part is 4-dimensional. The "N-extended" version is osp(N/4; $\mathbb{R}$), whose even subalgebra is isomorphic to so(3, 2)$\oplus$so(N), and whose complexification is osp(N/4; $\mathbb{C}$) (which is isomorphic to $B(\frac{1}{2}(N-1)/2)$ if N is odd and to $D(\frac{1}{2}N/2)$ if N is even). The structure and infinite-dimensional "unitary" irreducible representations of these Lie superalgebras have been comprehensively reviewed by Nicolai (1984). For further details of these representations see Breitenlohner and Freedman (1982), Heidereich (1982), Freedman and Nicolai (1984) and Günaydin *et al.* (1985, 1986).

The construction of Lie superalgebras that contain so(4, 1) in their even parts is more complicated. One manifestation of this is that although $B(0/2)$ contains B_2 (the complexification of so(4, 1)) as its even subalgebra, the only real form osp(1/4; $\mathbb{R}$) of $B(0/2)$ does not contain so(4, 1) in its even part. For a through discussion of the Lie superalgebras containing so(4, 1) see Lukierski and Nowicki (1982, 1985) and Pilch *et al.* (1985).

In addition to the studies in the papers mentioned above, the construction of superfields and supersymmetric gauge theories and the occurrence of spontaneous symmetry breaking for de Sitter and anti-de Sitter space–times have been examined by Zumino (1977), Sakai and Tanii (1984), Burgess (1985), Burges *et al.* (1986), Bellucci (1987a, b, c) and Vasiliev (1987, 1988).

Part E

Infinite-Dimensional Lie Algebras and Superalgebras and their Applications

26

The Structure of Kac–Moody Algebras

1 Introduction to infinite-dimensional Lie algebras

Up to this point the generalizations of Lie algebras and Lie groups that have been described have all involved gradings to produce Lie superalgebras and Lie supergroups. Henceforth the generalization will proceed in a different direction, with the previous *finite*-dimension restriction on the Lie algebras being relaxed to allow the consideration of *infinite*-dimensional Lie algebras. Two particular kinds of such algebras will be discussed in detail. In this chapter and the one that follows an account will be given of "Kac–Moody algebras", and in Chapter 28 the "Virasoro algebra" will be described. It is actually possible to go further and incorporate both forms of generalization, thereby giving "Kac–Moody superalgebras" and "Virasoro superalgebras", brief descriptions of which appear in Section 6 of this chapter and Section 4 of Chapter 28.

The account here of Kac–Moody algebras is arranged in two parts. Their structure is described in this chapter, and their representations in the next. It will be seen that many of the features of semi-simple Lie algebras again reappear, and indeed such algebras can be regarded as being just particular cases of Kac–Moody algebras with the special property of being finite-dimensional! After laying the general foundations in Sections 2, 3 and 4, the construction and properties of the most important type of infinite-dimensional Kac–Moody algebra, the "affine Kac–Moody algebras", are described in detail in Section 5.

Although the only application of Kac–Moody algebras that will be treated specifically is that to heterotic string theory (in Chapter 29, Section 5), a variety of other direct applications exist. In particular, Bars (1984), Bhattacharya and Chau (1984), Dolan (1984a, b, 1985), Julia (1985) and Affleck (1987) have reviewed the use of Kac–Moody algebras in integrable quantum systems, hadron physics, Yang–Mills theories, gravity and supergravity theories, and quantum spin systems. In addition, there are also the indirect applications of Kac–Moody algebras to conformally invariant systems using the Sugawara construction that will be described in Chapter 28, Section 3.

For reviews of properties of the associated groups see Kac (1985b), Mickelsson (1986, 1987) and Pressley and Segal (1986).

2 Construction of Kac–Moody algebras

As the concept of a Kac–Moody algebra is based on the generalization of a certain set of properties of semi-simple complex Lie algebras, it is useful to start by summarizing the relevant properties of these algebras. (More details may be found in Chapter 13.)

Each semi-simple complex Lie algebra $\tilde{\mathscr{L}}$ of dimension n and rank l possesses a Cartan subalgebra $\mathscr{H}$, which is of dimension l and is such that

$$[h, h']_- = 0 \quad \text{for all } h, h' \in \mathscr{H}. \tag{26.1}$$

The remaining $n - l$ basis elements of $\tilde{\mathscr{L}}$ may each be associated with a different non-zero root α, this being a linear functional defined on $\mathscr{H}$. The set of non-zero roots is denoted by Δ. For each $\alpha \in \Delta$ the set of elements $a_\alpha \in \tilde{\mathscr{L}}$ such that

$$[h, a_\alpha]_- = \alpha(h) a_\alpha \tag{26.2}$$

(for all $h \in \mathscr{H}$) form a subspace of $\tilde{\mathscr{L}}$, which is denoted by $\tilde{\mathscr{L}}_\alpha$ (cf. (13.8)), and each of these subspaces is one-dimensional. Δ contains a subset of l "simple roots" $\alpha_1, \alpha_2, \dots, \alpha_l$ that has the property that every $\alpha \in \Delta$ can be written as

$$\alpha = \sum_{j=1}^{l} \kappa_j^a \alpha_j, \tag{26.3}$$

where the coefficients $\kappa_1^\alpha, \kappa_2^\alpha, \dots, \kappa_l^\alpha$ are integers. Moreover, either every member of the set $\{\kappa_1^\alpha, \kappa_2^\alpha, \dots, \kappa_l^\alpha\}$ is non-negative, in which case α is said to be "positive", or each member is non-positive, when α is described as being "negative". The sets of positive and negative roots are denoted by Δ_+ and Δ_- respectively, so that $\Delta = \Delta_+ \cup \Delta_-$. A unique element $h_\alpha \in \mathscr{H}$ can be assigned to each $\alpha \in \Delta$ by the requirement that

$$B(h_\alpha, h) = \alpha(h) \tag{26.4}$$

for all $h \in \mathscr{H}$ (cf. (13.11)), $B(\ ,\ )$ being the Killing form of $\tilde{\mathscr{L}}$ (cf. (13.1)).

Using these elements, one defines the real numbers

$$\langle \alpha, \beta \rangle = B(h_\alpha, h_\beta) \tag{26.5}$$

(for each $\alpha, \beta \in \Delta$), and these in turn give the Cartan matrix $\mathbf{A}$, which is an $l \times l$ matrix whose elements are defined (cf. (13.61)) by

$$A_{jk} = \frac{2\langle \alpha_j, \alpha_k \rangle}{\langle \alpha_j, \alpha_j \rangle}, \quad j, k = 1, 2, \ldots, l. \tag{26.6}$$

$\mathbf{A}$ has four important properties:

(a) $A_{jj} = 2$ for $j = 1, 2, \ldots, l$;

(b) $A_{jk} = 0, -1, -2$ or -3 if $j \neq k$;

(c) for $j \neq k$, $A_{jk} = 0$ if and only if $A_{kj} = 0$;

(d) $\det \mathbf{A}$ and all proper principal minors of $\mathbf{A}$ are positive.

(A *principal minor of* $\mathbf{A}$ is the determinant of a *principal submatrix of* $\mathbf{A}$, which is itself defined as a submatrix consisting of elements A_{jk} in which j and k both run over the *same* subset of the index set of $\mathbf{A}$. These quantities are called *proper* if this subset is a *proper* subset of the index set.)

As discussed in detail in Chapter 13, several different choices for the basis of $\tilde{\mathscr{L}}$ appear in the literature. One choice is the "Weyl basis", for which the basis elements of $\mathscr{H}$ are taken to be h_α (for $\alpha = \alpha_j$, where $j = 1, 2, \ldots, l$), and the basis elements e_α and $e_{-\alpha}$ of $\tilde{\mathscr{L}}_\alpha$ and $\tilde{\mathscr{L}}_{-\alpha}$ are chosen so that

$$B(e_\alpha, e_{-\alpha}) = -1 \tag{26.7}$$

(cf. (13.37)). This implies (cf. (13.39)) that

$$[e_\alpha, e_{-\alpha}]_- = -h_\alpha \tag{26.8}$$

for each $\alpha \in \Delta$. Although most of the discussion in Chapter 13 and the subsequent chapters of Volume II was formulated in terms of this basis, it is more convenient for the introduction of Kac–Moody algebras to use the alternative "Chevalley basis", which consists of basis elements H_α (for $\alpha = \alpha_j$ where $j = 1, 2, \ldots, l$) and E_α ($\alpha \in \Delta$), where

$$H_\alpha = \frac{2}{\langle \alpha, \alpha \rangle} h_\alpha, \qquad \alpha \in \Delta, \tag{26.9}$$

$$E_\alpha = \left(\frac{2}{\langle \alpha, \alpha \rangle} \right)^{1/2} e_\alpha, \qquad \alpha \in \Delta_+, \tag{26.10}$$

$$E_{-\alpha} = -\left(\frac{2}{\langle \alpha, \alpha \rangle} \right)^{1/2} e_{-\alpha}, \quad -\alpha \in \Delta_- \tag{26.11}$$

(cf. (13.25) and (13.51)). With this choice,

$$\alpha_k(H_{\alpha_j}) = A_{jk} \tag{26.12}$$

(for $j, k = 1, 2, \ldots, l$), so that for $j, k = 1, 2, \ldots, l$

$$[H_{\alpha_j}, E_{\alpha_k}]_- = A_{jk} E_{\alpha_k}, \tag{26.13}$$

$$[H_{\alpha_j}, E_{-\alpha_k}]_- = -A_{jk} E_{-\alpha_k}, \tag{26.14}$$

and, since $\alpha_j - \alpha_k \notin \Delta$,

$$[E_{\alpha_j}, E_{-\alpha_k}]_- = \delta_{jk} H_{\alpha_j}. \tag{26.15}$$

From the set of elements E_α with $\alpha = \alpha_j$ and $j = 1, 2, \ldots, l$ can be constructed commutators of the form

$$[E_{\alpha_j}, E_{\alpha_{j'}}]_-, \quad [E_{\alpha_j}, [E_{\alpha_{j'}}, E_{\alpha_{j''}}]_-]_-, \ldots$$

and so on. Each of these is either identically zero or is a member of $\tilde{\mathscr{L}}_\alpha$ for some $\alpha \in \Delta_+$, this α being the sum of the α_j that appear in the commutator, and *every* $\tilde{\mathscr{L}}_\alpha$ for $\alpha \in \Delta_+$ and α non-simple appears in this process. The basis elements

$$E_{\alpha_1}, E_{\alpha_2}, \ldots, E_{\alpha_l}$$

are said to be *generators* of $\tilde{\mathscr{L}}$. (It should be noted that the word "generator" has a more specific meaning here (and in the rest of this chapter) than that of its usual usage in the mathematical physics literature, where it is often used synonymously with the words "basis element"). As a semi-simple Lie algebra has finite dimension, only a finite number of these commutators are non-zero. Indeed, it can be shown that

$$(\text{ad } E_{\alpha_j})^{(1-A_{jk})} E_{\alpha_k} = 0 \tag{26.16}$$

for all $j \neq k$ ($j, k = 1, 2, \ldots, l$), where "ad" is the adjoint operator defined in Chapter 11, Section 5. That is, for $j \neq k$,

$$[E_{\alpha_j}, [E_{\alpha_j}, [\ldots, [E_{\alpha_j}, E_{\alpha_k}]_-]_- \cdots]_- = 0,$$

where the commutator has $(1 - A_{jk}) E_\alpha$ factors with $\alpha = \alpha_j$ and one E_α factor with $\alpha = \alpha_k$. (It should be noted that $A_{jk} \leqslant 0$ for $j \neq k$, so that here $1 - A_{jk}$ is a positive integer.) The situation for commutators involving the E_α with $\alpha = -\alpha_j$ is similar. In this case they generate all the subspaces $\tilde{\mathscr{L}}_\alpha$ for $\alpha \in \Delta_-$, and (26.16) has as its analogue

$$(\text{ad } E_{-\alpha_j})^{(1-A_{jk})} E_{-\alpha_k} = 0 \tag{26.17}$$

for all $j \neq k$ ($j, k = 1, 2, \ldots, l$).

In the "classical" approach to the structure of semi-simple complex Lie algebras that has just been outlined the starting point is the definition of semi-simplicity, from which, after proceeding through various intermediate theorems, the Cartan matrix is reached as an end point. However, it was noted by Serre (1966) that the line of argument can be *reversed*, and that every semi-simple complex Lie algebra can actually be *constructed* from its Cartan matrix **A** using the properties (26.1) and (26.13)–(26.17) *alone*. This led Kac (1968) and Moody (1967, 1968, 1969) to enquire whether similar constructions are possible if the conditions on the Cartan matrix **A** are slightly relaxed. The resulting algebras (or minor extensions of them) are appropriately called *Kac–Moody algebras*. Not only do they contain the semi-simple Lie algebras as special cases, but many of the features of the theory of semi-simple Lie algebras are retained. (The notation that will be used in the remainder of this chapter has been chosen to make this relationship as clear as possible.) However, there is one vital difference, which is that every Kac–Moody algebra that is not semi-simple is of *infinite* dimension.

Close analysis shows that the richest theory is based on the following definition.

Definition *Generalized Cartan matrix*

A *generalized Cartan matrix* **A** is defined to be a $d_{\mathrm{A}} \times d_{\mathrm{A}}$ matrix such that

(a) $A_{jj} = 2$ for $j \in I$;

(b) for $j \neq k$ $(j, k \in I)$, A_{jk} is either zero or a negative integer;

(c) for $j \neq k$ $(j, k \in I)$, $A_{jk} = 0$ if and only if $A_{kj} = 0$.

(Here the index set I that labels the rows and columns of **A** will be taken to be $\{0, 1, \ldots, d_{\mathrm{A}} - 1\}$ in the general discussions of Kac–Moody algebras, but, to ensure compatibility with the conventions of Chapter 13, I will be taken to be $\{1, 2, \ldots, d_{\mathrm{A}}\}$ for semi-simple Lie algebras).

It should be noted that there is *no* requirement that $\det \mathbf{A}$ be non-zero, although this possibility is not excluded. Suppose that **A** has rank l, i.e. the largest square submatrix of **A** with non-zero determinant has dimensions $l \times l$. Clearly $l \leqslant d_{\mathrm{A}}$, equality occurring only when $\det \mathbf{A} \neq 0$.

Although some progress can be made with these conditions alone, it is convenient to impose the further condition that **A** is "symmetrizable".

Definition *Symmetrizable generalized Cartan matrix*

A generalized Cartan matrix **A** (of dimensions $d_{\mathrm{A}} \times d_{\mathrm{A}}$) is said to be *symmetrizable* if there exists a non-singular diagonal $d_{\mathrm{A}} \times d_{\mathrm{A}}$ matrix **D** such that the $d_{\mathrm{A}} \times d_{\mathrm{A}}$ matrix **B** defined by

$$\mathbf{A} = \mathbf{D}\mathbf{B} \qquad (26.18)$$

is symmetric. With the diagonal elements of **D** being denoted by ε_j ($j \in I$), so that $D_{jk} = \varepsilon_j \delta_{jk}$ for $j, k \in I$, (26.18) implies that

$$A_{jk} = \varepsilon_j B_{jk} \tag{26.19}$$

for $j, k \in I$.

The condition $\det \mathbf{D} \neq 0$ implies that $\varepsilon_j \neq 0$ for all $j \in I$. It can be shown that the ε_j may all be chosen to be real and positive.

Example I *Some generalized Cartan matrices*

The matrices **A** given by

$$\text{(i)} \begin{bmatrix} 2 & -2 \\ -2 & 2 \end{bmatrix}, \quad \text{(ii)} \begin{bmatrix} 2 & -4 \\ -1 & 2 \end{bmatrix}, \quad \text{(iii)} \begin{bmatrix} 2 & -3 \\ -3 & 2 \end{bmatrix}$$

are all generalized Cartan matrices with $d_{\mathbf{A}} = 2$. For both (i) and (ii) $\det \mathbf{A} = 0$ and $l = 1$, whereas for (iii) $\det \mathbf{A} = -5$ and $l = 2$. In cases (i) and (iii) **A** is already symmetric, so $\mathbf{D} = \varepsilon \mathbf{1}$ and $\mathbf{B} = \varepsilon^{-1}\mathbf{A}$, whereas for (ii)

$$\mathbf{D} = \varepsilon \begin{bmatrix} 2 & 0 \\ 0 & \frac{1}{2} \end{bmatrix}, \quad \mathbf{B} = \frac{1}{\varepsilon} \begin{bmatrix} 1 & -2 \\ -2 & 4 \end{bmatrix},$$

where ε can be taken to be real and positive in all cases. (In the classification of Kac–Moody algebras that follows later (i) corresponds to the "untwisted affine" algebra $A_1^{(1)}$, (ii) to the "twisted affine" algebra $A_2^{(2)}$ and (iii) to an "indefinite" algebra.)

(In the case of a semi-simple complex Lie algebra the Cartan matrix **A** is always symmetrizable, since the expression (26.6) for **A** can be cast in the form (26.19) with **B** defined by $B_{jk} = \langle \alpha_j, \alpha_k \rangle$ and with $\varepsilon_j = 2/\langle \alpha_j, \alpha_j \rangle$. This **B** is symmetric by (13.25).)

The next stage in the construction of a Kac–Moody algebra is to introduce a complex vector space $\mathscr{H}$ whose dimension $n_{\mathscr{H}}$ is defined by

$$n_{\mathscr{H}} = 2d_{\mathbf{A}} - l. \tag{26.20}$$

As $d_{\mathbf{A}} \geqslant l$, it follows that $n_{\mathscr{H}} \geqslant d_{\mathbf{A}}$, with equality occurring only when $\det \mathbf{A} \neq 0$, in which case $n_{\mathscr{H}} = d_{\mathbf{A}} = l$. Let h_j ($j \in I$) be *defined* to be *any* $d_{\mathbf{A}}$ linearly independent elements of $\mathscr{H}$. As always, the dual space $\mathscr{H}^*$ of $\mathscr{H}$ is the set of linear functionals defined on $\mathscr{H}$ and has the same dimension $n_{\mathscr{H}}$. Let α_j ($j \in I$) be $d_{\mathbf{A}}$ linear functionals on $\mathscr{H}$ that are both linearly independent of each other and are such that

$$\alpha_k(h_j) = A_{jk}$$

for $j, k \in I$. When $n_{\mathscr{H}} = d_{\mathbf{A}}$ the elements h_j ($j \in I$) form a basis for $\mathscr{H}$ and the functionals $\alpha_j\ (j \in I)$ form a basis for $\mathscr{H}^*$, but when $n_{\mathscr{H}} > d_A$ further elements are required to complete both bases. The notation for these basis elements can be refined by writing

$$H_{\alpha_j} = h_j$$

for $j \in I$, so that

$$\alpha_k(H_{\alpha_j}) = A_{jk} \tag{26.21}$$

for $j, k \in I$, thereby indicating the link that these basis elements have acquired with the linear functionals α_j.

Next construct the complex Lie algebra $\tilde{\mathscr{L}}'$ whose set of generators (in the sense described above) consists of

(i) the basis elements of $\mathscr{H}$,

together with

(ii) $2d_A$ elements E_α defined for $\alpha = \alpha_j$ and $-\alpha_j$ and for each $j \in I$,

with the whole set being assumed to satisfy the following relations:

$$[h, h']_- = 0 \quad \text{for all } h, h' \in \mathscr{H}, \tag{26.22}$$

$$[E_{\alpha_j}, E_{-\alpha_k}]_- = \delta_{jk} H_{\alpha_j} \quad \text{for all } j, k \in I, \tag{26.23}$$

$$[h, E_{\alpha_k}]_- = \alpha_k(h) E_{\alpha_k} \quad \text{for all } h \in \mathscr{H} \text{ and } k \in I, \tag{26.24}$$

$$[h, E_{-\alpha_k}]_- = -\alpha_k(h) E_{-\alpha_k} \quad \text{for all } h \in \mathscr{H} \text{ and } k \in I. \tag{26.25}$$

This algebra contains *all* the commutators of the form

$$[E_{\alpha_j}, E_{\alpha_{j'}}]_-, \quad [E_{\alpha_j}, [E_{\alpha_{j'}}, E_{\alpha_{j''}}]_-]_-, \ldots$$

and so on, together with those of the form

$$[E_{-\alpha_j}, E_{-\alpha_{j'}}]_-, \quad [E_{-\alpha_j}, [E_{-\alpha_{j'}}, E_{-\alpha_{j''}}]_-]_-, \ldots$$

etc. It may be noted that with $h = H_{\alpha_j}$ (26.24) and (26.25) reduce to

$$[H_{\alpha_j}, E_{\alpha_k}]_- = A_{jk} E_{\alpha_k} \tag{26.26}$$

and

$$[H_{\alpha_j}, E_{-\alpha_k}]_- = -A_{jk} E_{-\alpha_k} \tag{26.27}$$

respectively by virtue of (26.21). The similarity of (26.21), (26.22), (26.23), (26.26) and (26.27) to (26.12), (26.1), (26.15), (26.13) and (26.14) will be obvious.

At this stage the higher-order commutators

$$[E_{\alpha_j}, [E_{\alpha_{j'}}, E_{\alpha_{j''}}]_-]_-,$$

etc., have not been restricted in any way. This will now be done in a slightly indirect fashion. First let $\mathscr{R}$ be the maximal invariant subalgebra of $\tilde{\mathscr{L}}'$ whose intersection with $\mathscr{H}$ is trivial; that is, the intersection contains only the zero element of $\tilde{\mathscr{L}}'$. Then define $\tilde{\mathscr{L}}$ to be the complex Lie algebra that is the factor algebra $\tilde{\mathscr{L}}'/\mathscr{R}$. Such an object can be defined from any invariant subalgebra $\mathscr{R}$ of $\tilde{\mathscr{L}}'$ by a procedure analogous to that given in Chapter 2, Section 5 for groups. The details of the construction of $\tilde{\mathscr{L}}'/\mathscr{R}$ are as follows. Each element of $\tilde{\mathscr{L}}'/\mathscr{R}$ is defined to be a *set* of elements of $\tilde{\mathscr{L}}'$ of the form $\{a+r\}$, where a is fixed and r runs over all members of $\mathscr{R}$. Such an element of $\tilde{\mathscr{L}}/\mathscr{R}$ may be denoted by $a+\mathscr{R}$. Clearly two elements $a+\mathscr{R}$ and $b+\mathscr{R}$ of $\tilde{\mathscr{L}}'/\mathscr{R}$ are distinct if and only if $a-b$ is not a member of $\mathscr{R}$. The set of elements of $\tilde{\mathscr{L}}'/\mathscr{R}$ can be made into a complex Lie algebra if addition, scalar multiplication and the commutator for $\tilde{\mathscr{L}}'/\mathscr{R}$ are defined by

$$(a+\mathscr{R})+(b+\mathscr{R})=(a+b)+\mathscr{R} \quad \text{for all } a, b\in\tilde{\mathscr{L}}',$$

$$\lambda(a+\mathscr{R})=(\lambda a)+\mathscr{R} \quad \text{for all } \lambda\in\mathbb{C} \text{ and } a\in\tilde{\mathscr{L}}',$$

$$[a+\mathscr{R}, b+\mathscr{R}]_-=[a,b]_-+\mathscr{R} \quad \text{for all } a, b\in\tilde{\mathscr{L}}',$$

where the operations of addition and scalar multiplication and the commutator of the right-hand sides are those of $\tilde{\mathscr{L}}'$. It is easily shown that there exists a homomorphic mapping ϕ of $\tilde{\mathscr{L}}'$ onto $\tilde{\mathscr{L}}'/\mathscr{R}$ defined by $\phi(a)=a+\mathscr{R}$.

In the present context, since $\mathscr{R}$ is assumed to have only trivial intersection with $\mathscr{H}$, the elements $\phi(h_j)$ for $j\in I$ are all distinct. It can be shown that the same is true for $\phi(E_\alpha)$ for $\alpha=\pm\alpha_j\,(j\in I)$. The notation may then be simplified by denoting the elements $\phi(H_{\alpha_j})\ (=\phi(h_j))$, $\phi(E_{\alpha_j})$ and $\phi(E_{-\alpha_j})$ of $\tilde{\mathscr{L}}=\tilde{\mathscr{L}}'/\mathscr{R}$ by H_{α_j}, E_{α_j} and $E_{-\alpha_j}$ respectively. This will not cause any confusion, since the algebra $\tilde{\mathscr{L}}'$ will not appear on its own again. As ϕ is a homomorphic mapping, the commutation relations (26.22)–(26.25) (and (26.26) and (26.27)) remain valid when all the elements that are involved are regarded as being members of $\tilde{\mathscr{L}}$ (rather than of $\tilde{\mathscr{L}}'$ as previously). The effect of this construction is to ensure that $\tilde{\mathscr{L}}$ has *no* non-trivial invariant subalgebra with non-trivial intersection with $\mathscr{H}$. The resulting complex Lie algebra $\tilde{\mathscr{L}}$ will be called *the Kac–Moody algebra based on the generalized Cartan matrix* **A**.

It is not difficult to show that for the elements E_α with $\alpha=\pm\alpha_j\,(j\in I)$ of the Kac–Moody algebra $\tilde{\mathscr{L}}$ it is indeed true that

$$(\operatorname{ad} E_{\alpha_j})^{(1-A_{jk})}E_{\alpha_k}=0 \tag{26.28}$$

and

$$(\operatorname{ad} E_{-\alpha_j})^{(1-A_{jk})}E_{-\alpha_k}=0 \tag{26.29}$$

for $j\neq k$ $(j,k\in I)$. (The proof (cf. Kac 1985a) involves showing that if the left-hand sides of (26.28) and (26.29) are assumed to be non-zero then they generate an invariant subalgebra of $\tilde{\mathscr{L}}$ with non-trivial intersection on $\mathscr{H}$, a situation that has just been excluded by the construction of $\tilde{\mathscr{L}}$ as a factor algebra.)

The precise relationship between the above definition of a Kac–Moody algebra and the definitions in the original papers of Kac (1968) and Moody (1967, 1968, 1969) will be discussed in Section 3.

It can be shown quite easily that if the index set I is *reordered*—that is, if the rows of $\mathbf{A}$ are permuted, and, simultaneously, the columns of $\mathbf{A}$ are permuted in exactly the same way—then the resulting Kac–Moody algebra is isomorphic to that based on the original $\mathbf{A}$. Conversely, it can also be shown that two Kac–Moody algebras are isomorphic only if their generalized Cartan matrices are related in this way. If $\mathbf{A}$ has the block form

$$\mathbf{A} = \begin{bmatrix} \mathbf{A}^{11} & \mathbf{0} \\ \mathbf{0} & \mathbf{A}^{22} \end{bmatrix}, \tag{26.30}$$

where the submatrices $\mathbf{A}^{11}$ and $\mathbf{A}^{22}$ are both square, or if $\mathbf{A}$ can be put in this form by a reordering of its index set I, then $\mathbf{A}$ is said to be *decomposable*. If this is not true then $\mathbf{A}$ is described as being *indecomposable*. It can be shown that the Kac–Moody algebra corresponding to the $\mathbf{A}$ of (26.30) is the direct sum of Kac–Moody algebras corresponding to $\mathbf{A}^{11}$ and $\mathbf{A}^{22}$. Consequently there is no loss in generality in assuming henceforth that $\mathbf{A}$ is indecomposable. With this assumption, the set $\{\varepsilon_j | j \in I\}$ of (26.19) is unique up to a common multiplicative factor.

(The notation introduced above has been chosen to conform with that appearing in the account of semi-simple Lie algebra given in Chapters 13–17. The generators $\alpha_j^\vee$, e_j and f_j used by Kac (1985a) are related to the generators defined above by

$$\alpha_j^\vee = H_{\alpha_j}, \quad e_j = E_{\alpha_j}, \quad f_j = E_{-\alpha_j},$$

and the quantities $\langle \alpha, h \rangle$ and $\langle h, \alpha \rangle$ employed by Kac (1985a) are such that

$$\langle \alpha, h \rangle = \langle h, \alpha \rangle = \alpha(h)$$

(for any $h \in \mathscr{H}$ and any linear functional α defined on $\mathscr{H}$).)

3 Properties of general Kac–Moody algebras

The basic ideas and terminology for roots and root subspaces for a complex Kac–Moody algebra $\tilde{\mathscr{L}}$ are very similar to those for a semi-simple complex Lie algebra. In particular, the commutative subalgebra $\mathscr{H}$ of $\tilde{\mathscr{L}}$ is still called the *Cartan subalgebra of* $\tilde{\mathscr{L}}$, and the set of elements a_α of $\tilde{\mathscr{L}}$ possessing the property that

$$[h, a_\alpha]_- = \alpha(h) a_\alpha \tag{26.31}$$

for all $h \in \mathscr{H}$ is again said to form the root subspace $\tilde{\mathscr{L}}_\alpha$ corresponding to the root α. Because of (26.24) and (26.25) for $\alpha = \alpha_k$ and $k \in I$, the generators E_α and $E_{-\alpha}$ are members of $\tilde{\mathscr{L}}_\alpha$ and $\tilde{\mathscr{L}}_{-\alpha}$ respectively. The set of linear functionals $\alpha_j(h)$ (for $j \in I$) are called the *simple roots* of $\tilde{\mathscr{L}}$. Also, non-zero commutators of the form

$$[E_{\alpha_j}, E_{\alpha_{j'}}]_-, \quad [E_{\alpha_j}, [E_{\alpha_{j'}}, E_{\alpha_{j''}}]_-]_-, \ldots$$

and so on, all belong to root subspaces for which the corresponding root α has the form

$$\alpha = \sum_{j \in I} \kappa_j^\alpha \alpha_j, \tag{26.32}$$

where each κ_j^α $(j \in I)$ is a non-negative integer. Here κ_j^α is the number of times the generator E_α with $\alpha = \alpha_j$ appears in the corresponding commutator. Such a root is called a *positive root*. In particular, the simple roots are positive roots. Similarly commutators of the form

$$[E_{-\alpha_j}, E_{-\alpha_{j'}}]_-, \quad [E_{-\alpha_j}, [E_{-\alpha_{j'}}, E_{-\alpha_{j''}}]_-]_-, \ldots$$

all belong to root subspaces for which the root α is given by (26.32) but with each κ_j^α a non-positive integer, such a root being called a *negative root*. The set of all non-zero roots of $\tilde{\mathscr{L}}$ will be denoted by Δ, the set of positive roots by Δ_+, and the set of negative roots by Δ_-. If $\alpha \in \Delta_+$ then $-\alpha \in \Delta_-$ and vice versa. The direct sums of the positive and negative root subspaces will be denoted by $\tilde{\mathscr{L}}_+$ and $\tilde{\mathscr{L}}_-$ respectively, so that

$$\tilde{\mathscr{L}} = \tilde{\mathscr{L}}_+ \oplus \mathscr{H} \oplus \tilde{\mathscr{L}}_-, \tag{26.33}$$

where the $\oplus$ symbol indicates only a vector space direct sum, and does not imply that separate parts mutually commute.

Three points where Kac–Moody and simple Lie algebra properties do not coincide are worth noting. Firstly, for a Kac–Moody algebra the root subspaces $\tilde{\mathscr{L}}_\alpha$ can be of more than one dimension. Secondly, for a Kac–Moody algebra the Cartan subalgebra $\mathscr{H}$ can be divided into two parts $\mathscr{H}'$ and $\mathscr{H}''$, $\mathscr{H}'$ being defined to be the complex vector space that has H_{α_j} $(j \in I)$ as its basis and $\mathscr{H}''$ being defined to be the complementary subspace of $\mathscr{H}'$ in $\mathscr{H}$. Clearly $\mathscr{H}'$ has dimension $d_{\mathbf{A}}$, and, since $\mathscr{H}$ has dimension $n_{\mathscr{H}} = 2d_{\mathbf{A}} - l$, $\mathscr{H}''$ must have dimension $d_{\mathbf{A}} - l$. Thus $\mathscr{H}''$ is non-trivial only when $\det \mathbf{A} = 0$. The third difference is embodied in the following theorem.

Theorem I If $\mathscr{C}$ is defined to be the subset of $\mathscr{H}$ such that

$$\alpha_k(h) = 0 \quad \text{for all } k \in I \tag{26.34}$$

then $\mathscr{C}$ is a subspace of $\mathscr{H}'$ of dimension $d_{\mathbf{A}} - l$.

Proof Consider first the set of elements $h \in \mathscr{H}'$ satisfying (26.34). Let

$$h = \sum_{j \in I} \mu_j H_{\alpha_j}.$$

Then (26.34) implies that

$$\sum_{j \in I} \mu_j \alpha_k(H_{\alpha_j}) = 0;$$

that is, by (26.21),

$$\tilde{\mathbf{A}}\boldsymbol{\mu} = \mathbf{0}, \qquad (26.35)$$

where $\boldsymbol{\mu}$ is the column matrix with entries μ_j $(j \in I)$. Elementary matrix theory shows that (26.35) has $d_{\mathbf{A}} - l$ linearly independent solutions, so there exist $d_{\mathbf{A}} - l$ linearly independent members of $\mathscr{H}'$ that satisfy (26.34). It remains to show that *no* non-trivial element of $\mathscr{H}''$ satisfies (26.34). To this end, define the $d_{\mathbf{A}}$-dimensional subspace $\mathscr{H}_+$ of $\mathscr{H}$ whose basis elements h_j $(j \in I)$ have the property that $\alpha_k(h_j) = \delta_{jk}$ $(j, k \in I)$. Then each $h_+ \in \mathscr{H}_+$ can be written in the form $h_+ = \sum_{j \in I} \lambda_j h_j$, so $\lambda_k = \alpha_k(h_+)$ for $k \in I$ and hence *no* non-trivial element of $\mathscr{H}_+$ satisfies (26.34). Thus the $d_{\mathbf{A}}$ elements h_j $(j \in I)$ of $\mathscr{H}_+$ and the $d_{\mathbf{A}} - l$ elements of $\mathscr{H}'$ that satisfy (26.34) are linearly independent. Together they form a basis for $\mathscr{H}$, implying that $\mathscr{H}''$ is contained in $\mathscr{H}_+$, and so no non-trivial member of $\mathscr{H}''$ satisfies (26.34).

In the case in which $\tilde{\mathscr{L}}$ is simple, $d_{\mathbf{A}} = l$, so that $\mathscr{C}$ has zero dimension. For a general Kac–Moody algebra it is clear that the elements of $\mathscr{C}$ commute with *all* the members of $\tilde{\mathscr{L}}$. It is not difficult to show that they are the only elements of $\tilde{\mathscr{L}}$ with this property, so $\mathscr{C}$ is the *centre* of $\tilde{\mathscr{L}}$.

The construction of a bilinear form $B(\ , \ )$ that possesses both the *symmetry* property

$$B(a, b) = B(b, a) \qquad (26.36)$$

(for all $a, b \in \tilde{\mathscr{L}}$) and the *invariance* property

$$B([a, b]_-, c) = B(a, [b, c]_-) \qquad (26.37)$$

(for all $a, b, c \in \tilde{\mathscr{L}}$) will now be discussed. For a *finite*-dimensional Lie algebra these properties are possessed by the Killing form (as was shown in Theorem I of Chapter 13, Section 2), but since the Killing form involves taking a "trace" it is not well-defined for an infinite-dimensional algebra, so the approach for a Kac–Moody algebra has to be different.

The first stage is to define $B(\ , \ )$ on the Cartan subalgebra $\mathscr{H}$ by the requirements that B be bilinear on $\mathscr{H}'$, that

$$B(h, H_{\alpha_k}) = \alpha_k(h)\varepsilon_k \quad \text{for all } h \in \mathscr{H} \text{ and } k \in I, \qquad (26.38)$$

that

$$B(h_1'', h_2'') = 0 \qquad \text{for all } h_1'', h_2'' \in \mathscr{H}'', \qquad (26.39)$$

and that

$$B(H_{\alpha_k}, h'') = B(h'', H_{\alpha_k}) \quad \text{for all } h'' \in \mathscr{H}'' \text{ and } k \in I. \tag{26.40}$$

In (26.38) the ε_j $(j \in I)$ are the diagonal elements of the "diagonalizing" matrix **D** of (26.18), and are all assumed to be positive. Then, with

$$h = H_{\alpha_j},$$

(26.38), (26.21) and (26.19) show that

$$B(H_{\alpha_j}, H_{\alpha_k}) = B_{jk}\varepsilon_j\varepsilon_k \tag{26.41}$$

(for $j, k \in I$), which, together with (26.39) and (26.40), implies that $B(\ ,\ )$ is symmetric on $\mathscr{H}$. The invariance property (26.37) is trivially satisfied on $\mathscr{H}$ because $\mathscr{H}$ is commutative.

As noted previously, the set $\{\varepsilon_j | j \in I\}$ is only determined up to a common multiplicative factor, so that $B(\ ,\ )$ is only determined by (26.38) up to a multiplicative constant. A natural choice of this constant for "affine" Kac–Moody algebras will be given in Section 5.

Theorem II The bilinear from $B(\ ,\ )$ is *non-degenerate* on $\mathscr{H}$.

Proof Suppose that there exists an element $h_0 \in \mathscr{H}$ such that $B(h_0, h) = 0$ for all $h \in \mathscr{H}$. To establish that $B(\ ,\ )$ is non-degenerate on $\mathscr{H}$, it has merely to be shown that $h_0 = 0$ (cf. Appendix B, Section 5). With $h = H_{\alpha_k}$, this condition, together with (26.38), implies that $\alpha_k(h_0)\varepsilon_k = 0$, and, since $\varepsilon_k \neq 0$, this means that $\alpha_k(h_0) = 0$ for all $k \in I$. Thus $h_0 \in \mathscr{C}$, and so, by Theorem I, $h_0 \in \mathscr{H}'$. Let

$$h_0 = \sum_{j \in I} \mu_j H_{\alpha_j}.$$

Then it is required that

$$\sum_{j \in I} \mu_j B(H_{\alpha_j}, h) = 0$$

for all $h \in \mathscr{H}$; that is, by (26.38), $\sum_{j \in I} \mu_j \varepsilon_j \alpha_j(h) = 0$ for all $h \in \mathscr{H}$. However, the linear functionals α_j $(j \in I)$ are assumed to be linearly independent on $\mathscr{H}$, so $\mu_j \varepsilon_j = 0$ and hence $\mu_j = 0$ for each $j \in I$; that is, $h_0 = 0$.

It follows from this theorem that corresponding to each linear functional α on $\mathscr{H}$ there exists an element h_α of $\mathscr{H}$ that is defined by

$$B(h_\alpha, h) = \alpha(h) \tag{26.42}$$

for all $h \in \mathscr{H}$ and which, for a given bilinear form $B(\ ,\ )$, is unique (cf. Appendix B,

Section 5). Comparison of (26.38) and (26.42) shows that for the simple roots,

$$h_{\alpha_k} = \varepsilon_k^{-1} H_{\alpha_k} \tag{26.43}$$

for $k \in I$. As $B(\ ,\ )$ is bilinear, (26.42) implies that

$$h_{\alpha+\beta} = h_\alpha + h_\beta \tag{26.44}$$

for any two linear functionals α and β on $\mathscr{H}$.

The next stage is to introduce a symmetric bilinear form on the dual space $\mathscr{H}^*$ of $\mathscr{H}$ by the definition

$$\langle \alpha, \beta \rangle = B(h_\alpha, h_\beta) \tag{26.45}$$

for any two linear functionals α and β on $\mathscr{H}$. Then, by (26.42),

$$\langle \alpha, \beta \rangle = \alpha(h_\beta) = \beta(h_\alpha) = \langle \beta, \alpha \rangle. \tag{26.46}$$

(Clearly (26.42), (26.44), (26.45) and (26.46) have the same forms as the corresponding equations (13.11), (13.12), (13.14) and (13.15) for semi-simple complex Lie algebras.)

From (26.45), (26.43) and (26.41) it follows that

$$\langle \alpha_j, \alpha_k \rangle = B_{jk} \tag{26.47}$$

for all $j, k \in I$. By virtue of (26.19), this can be rewritten as

$$\langle \alpha_j, \alpha_k \rangle = \varepsilon_j^{-1} A_{jk}, \tag{26.48}$$

and, since $A_{jj} = 2$, this implies that

$$\langle \alpha_j, \alpha_j \rangle = 2/\varepsilon_j \tag{26.49}$$

for $j \in I$. Substituting (26.49) back in (26.48) then gives

$$A_{jk} = \frac{2\langle \alpha_j, \alpha_k \rangle}{\langle \alpha_j, \alpha_j \rangle}, \tag{26.50}$$

for $j, k \in I$ (which has the same form as (13.61) and (26.6)). Moreover, by (26.43) and (26.49),

$$H_{\alpha_k} = \frac{2}{\langle \alpha_k, \alpha_k \rangle} h_{\alpha_k} \tag{26.51}$$

for $k \in I$ (which is the same form as (13.25) and (26.9) for $\alpha = \alpha_k$).

As the ε_j have been chosen to be positive, (26.49) shows that $\langle \alpha_j, \alpha_j \rangle$ is real and positive for all $j \in I$. However, in *contrast* with the situation described for semi-simple complex Lie algebras in Chapter 13, Section 5, for complex Kac–Moody algebras it is possible to find roots α such that $\langle \alpha, \alpha \rangle \leqslant 0$. Examples of this are provided by the root $\alpha = \alpha_0 + \alpha_1$ of the algebras corresponding to the generalized Cartan matrices of cases (i) and (iii) of

Example I of Section 2. As $\mathbf{B} = \varepsilon^{-1}\mathbf{A}$ in both of these cases, (26.47) shows that $\langle \alpha, \alpha \rangle = 0$ for (i) and $\langle \alpha, \alpha \rangle = -2$ for (iii). The existence of this type of root implies that in general neither $\mathscr{H}$ nor its subspace $\mathscr{H}'$ are inner product spaces. This type of root will be examined in some detail after the extension of $B(\ ,\ )$ from $\mathscr{H}$ to the whole of $\tilde{\mathscr{L}}$ has been discussed.

Theorem III The bilinear form $B(\ ,\ )$ defined on $\mathscr{H}$ by (26.38)–(26.40) may be extended to provide a symmetric invariant bilinear form on the whole of $\tilde{\mathscr{L}}$. With this form,

$$B(a_\alpha, a_\beta) = 0 \tag{26.52a}$$

for all $a_\alpha \in \tilde{\mathscr{L}}_\alpha$ and $a_\beta \in \tilde{\mathscr{L}}_\beta$ unless $\alpha + \beta = 0$, and in this case

$$[a_\alpha, a_{-\alpha}]_- = B(a_\alpha, a_{-\alpha})h_\alpha \tag{26.52b}$$

for all $a_\alpha \in \tilde{\mathscr{L}}_\alpha$ and all $a_{-\alpha} \in \tilde{\mathscr{L}}_{-\alpha}$. Moreover, up to an arbitrary multiplicative constant, this is the *unique* symmetric invariant bilinear form on $\tilde{\mathscr{L}}$.

Clearly (26.52a) and (26.52b) have the same form as Theorem II of Chapter 13, Section 4 and (13.22) respectively. (For a proof see Kac (1985a).)

The result (26.52b) permits clarification of the relationship between the Kac–Moody algebra $\tilde{\mathscr{L}}$ defined in Section 2 and the Lie algebras that appeared in the original papers of Kac (1968) and Moody (1967, 1968, 1969). Recalling that $\mathscr{H} = \mathscr{H}' \oplus \mathscr{H}''$, where $\mathscr{H}''$ is the set of all linear combinations of H_{α_j} $(j \in I)$ and $\mathscr{H}''$ is the complementary subspace, (26.52b) shows that $[a_\alpha, a_{-\alpha}]_- \in \mathscr{H}'$ for all $\alpha \in \Delta_+$, *no* element of $\mathscr{H}''$ appearing in this way. Thus the subspace of $\tilde{\mathscr{L}}$ that is spanned by the basis elements

$$H_{\alpha_j}, \quad E_{\alpha_j}, \quad E_{-\alpha_j} \quad (j \in I) \tag{26.53}$$

and all their commutators of all orders is

$$\tilde{\mathscr{L}}_+ \oplus \mathscr{H}' \oplus \tilde{\mathscr{L}}_-,$$

in which $\mathscr{H}''$ does *not* appear. Thus $\tilde{\mathscr{L}}_+ \oplus \mathscr{H}' \oplus \tilde{\mathscr{L}}_-$ is isomorphic to an algebra $\tilde{\mathscr{L}}''$ that has the set (26.53) as its generators and satisfies (26.23) and (26.26)–(26.29). This means that $\tilde{\mathscr{L}}''$ is the "derived" subalgebra $[\tilde{\mathscr{L}}, \tilde{\mathscr{L}}]_-$ of $\mathscr{L}$. $\mathscr{L}''$ is the type of algebra considered in the original papers. The converse proposition that every algebra of the type $\tilde{\mathscr{L}}''$ is isomorphic to $\tilde{\mathscr{L}}_+ \oplus \mathscr{H}' \oplus \tilde{\mathscr{L}}_-$ was established by Gabber and Kac (1981). Thus if $\det \mathbf{A} \neq 0$ (so that $\mathscr{H}''$ is trivial) then both definitions are equivalent, but if $\det \mathbf{A} = 0$ then $\tilde{\mathscr{L}}$ is isomorphic to $\tilde{\mathscr{L}}'' \oplus \mathscr{H}''$ (a vector space direct sum).

Weyl reflections can be introduced in the same way as in Chapter 13, Section 9. For any $\alpha \in \Delta$ such that $\langle \alpha, \alpha \rangle \neq 0$ and for any linear functional β

on $\mathscr{H}$ the linear functional $S_\alpha\beta$ on $\mathscr{H}$ is defined by

$$(S_\alpha\beta)(h) = \beta(h) - \frac{2\langle\beta,\alpha\rangle}{\langle\alpha,\alpha\rangle}\alpha(h) \qquad (26.54)$$

for all $h\in\mathscr{H}$ (cf. (13.73)), thereby defining the Weyl reflection S_α as an operator that acts on $\mathscr{H}^*$. Then, as in the "simple" case,

(a) $S_\alpha\alpha = -\alpha$;

(b) $S_\alpha(S_\alpha\beta) = \beta$ for every linear functional β on $\mathscr{H}$;

(c) for any linear functionals β and γ on $\mathscr{H}$

$$\langle S_\alpha\beta, S_\alpha\gamma\rangle = \langle\beta,\gamma\rangle; \qquad (26.55)$$

(d) for any two linear functionals β and γ on $\mathscr{H}$ and any two complex numbers μ and λ

$$S_\alpha(\lambda\beta + \mu\gamma) = \lambda(S_\alpha\beta) + \mu(S_\alpha\gamma). \qquad (26.56)$$

The Weyl reflections S_α for $\alpha = \alpha_j$ ($j\in I$) corresponding to the simple roots are known as the "fundamental reflections".

Definition *Weyl group $\mathscr{W}$ of $\tilde{\mathscr{L}}$*

With the identity operator E defined by $E\alpha = \alpha$ (for any linear functional α on $\mathscr{H}$) and with the products such as $S_\alpha S_\beta$ defined by $S_\alpha S_\beta\gamma = S_\alpha(S_\beta\gamma)$ (for any linear functional γ on $\mathscr{H}$), the set consisting of the identity operator, the fundamental reflections and all products of the fundamental reflections forms a group called the *Weyl group of* $\tilde{\mathscr{L}}$, which will be denoted by $\mathscr{W}$. A typical element of $\mathscr{W}$ will be denoted by S.

In the case in which $\tilde{\mathscr{L}}$ is a simple Lie algebra this definition is equivalent to that given in Chapter 13, Section 9, since in this case every Weyl reflection is a product of fundamental reflections.

Definition *Real and imaginary roots of $\tilde{\mathscr{L}}$*

A root $\alpha\in\Delta$ is said to be a *real* root of a Kac–Moody algebra $\tilde{\mathscr{L}}$ if there exists an element $S\in\mathscr{W}$ and a simple root α_j of $\tilde{\mathscr{L}}$ such that

$$\alpha = S\alpha_j. \qquad (26.57)$$

If $\alpha\in\Delta$ is not real then α is described as being *imaginary*.

The sets of real roots, positive real roots, negative real roots, imaginary roots, positive imaginary roots and negative imaginary roots will be denoted by $\Delta^r, \Delta^r_+, \Delta^r_-, \Delta^i, \Delta^i_+$ and Δ^i_- respectively. Because of their origin, real roots are sometimes alternatively known as "Weyl roots". Clearly every simple root is a real root. It can be shown that for a *simple* complex Lie algebra *every* root is a real root. Moreover, the real roots of a general Kac–Moody algebra always have the same properties as the roots of a simple Lie algebra, as the following theorems show.

Theorem IV If $\alpha \in \Delta^r$ then

$$\langle \alpha, \alpha \rangle > 0. \tag{26.58}$$

Proof This is an immediate consequence of (26.55), the fact that $\langle \alpha_j, \alpha_j \rangle > 0$ for all $j \in I$ and the definition (26.55) of a real root.

Theorem V Let $\alpha \in \Delta^r$ and $\beta \in \Delta$. Then there exist two non-negative integers p and q (which depend on α and β) such that $\beta + k\alpha$ is in the α-string containing β for *every* integer k that satisfies the relation $-p \leqslant k \leqslant q$. Moreover, p and q are such that

$$p - q = \frac{2\langle \beta, \alpha \rangle}{\langle \alpha, \alpha \rangle}. \tag{26.59}$$

Also,

$$S_\alpha \beta = \beta - \frac{2\langle \beta, \alpha \rangle}{\langle \alpha, \alpha \rangle}\alpha \tag{26.60}$$

is a non-zero root and

$$\dim \tilde{\mathscr{L}}_{S_\alpha \beta} = \dim \tilde{\mathscr{L}}_\beta. \tag{26.61}$$

Proof The proof is almost exactly the same as that of the corresponding theorem for semi-simple Lie algebras (Theorem IX of Chapter 13, Section 5) that was given in Appendix E, Section 7. The only difference is that now $\mathrm{ad}_{\tilde{\mathscr{L}}'}$ need not be an *irreducible* representation of the A_1 subalgebra that appears in that proof, but j can be defined now to be such that the largest irreducible component of this representation has dimension $2j + 1$. The result (26.61) follows from the fact that the multiplicities of the eigenvalues $\langle \beta, \alpha \rangle / \langle \alpha, \alpha \rangle$ and $-\langle \beta, \alpha \rangle / \langle \alpha, \alpha \rangle$ of $\mathrm{ad}_{\tilde{\mathscr{L}}'}(\frac{1}{2}H_\alpha)$ must always be equal, and it remains true that the former eigenvalue corresponds to the root α and the latter to the root $\beta - (2\langle \beta, \alpha \rangle / \langle \alpha, \alpha \rangle)\alpha$.

Theorem VI If $\beta \in \Delta$ and $S \in \mathscr{W}$, then $S\beta \in \Delta$ and

$$\dim \tilde{\mathscr{L}}_{S\beta} = \dim \tilde{\mathscr{L}}_\beta. \tag{26.62}$$

Proof The previous theorem shows that this is true for every fundamental reflection. As every element of $\mathscr{W}$ is a product of fundamental reflections, the stated result follows immediately.

Theorem VII If $\alpha \in \Delta_+$ and $\alpha \neq \alpha_j$ (for all $j \in I$) then

$$S_{\alpha_j}\alpha \in \Delta_+ .$$

Proof The proof is exactly as for Theorem I of Chapter 13, Section 9.

Theorem VIII If $\alpha \in \Delta^r$ then

(i) $\dim \tilde{\mathscr{L}}_\alpha = 1$, and (26.63)

(ii) $k\alpha \in \Delta$ only if $k = +1$ or -1. (26.64)

Proof (i) This is obviously true for the simple roots, and Theorem VI then implies that it must be true for all real roots.

(ii) This is certainly true for every simple root α_j because $k\alpha_j$ for $k > 1$ and $k < -1$ correspond to commutators consisting solely of k copies of E_α with $\alpha = \alpha_j$ alone and k copies of E_α with $\alpha = -\alpha_j$ alone respectively, and such commutators are identically zero. For the case $\alpha \in \Delta^r$ and α not simple, there exists a $S \in \mathscr{W}$ such that $\alpha = S\alpha_j$, and so $k\alpha = S(k\alpha_j)$. Theorem VI then shows that

$$\dim \tilde{\mathscr{L}}_{k\alpha} = \dim \tilde{\mathscr{L}}_{k\alpha_j},$$

where the quantity on the right-hand side has just been shown to be zero for $k \neq \pm 1$.

For *imaginary* roots the situation is quite *different*.

Theorem IX (a) If $\alpha \in \Delta^i_+$ and $S \in \mathscr{W}$ then $S\alpha \in \Delta^i_+$.

(b) If $\alpha \in \Delta^i_+$ then there exists an element $S \in \mathscr{W}$ such that $S\alpha \in \Delta^i_{+-}$, where Δ^i_{+-} is the subset of Δ^i_+ consisting of elements β with the property that

$$\langle \beta, \alpha_j \rangle \leqslant 0 \quad \text{for all } j \in I. \tag{26.65}$$

(c) If $\alpha \in \Delta$ then $\alpha \in \Delta^i$ if and only if

$$\langle \alpha, \alpha \rangle \leqslant 0. \tag{26.66}$$

Proof (a) Suppose that $\alpha \in \Delta^i_+$. As $\alpha \in \Delta_+$ and $\alpha \in \Delta^i$, Theorem VII implies that for *all* $j \in I$

$$S_{\alpha_j}\alpha \in \Delta_+ .$$

Moreover, this latter root is not a simple root (since $\alpha \in \Delta^i$), so successive applications of any fundamental reflections will always produce members of Δ_+. As every $S \in \mathscr{W}$ is a product of fundamental reflections, it follows that $S\alpha \in \Delta_+$ for all $S \in \mathscr{W}$. It remains to show that $S\alpha \in \Delta^{\mathrm{i}}$ for all $S \in \mathscr{W}$. If $S\alpha \in \Delta^{\mathrm{r}}$ then $\alpha = S^{-1}\beta$, where $\beta \in \Delta^{\mathrm{r}}$, and since there exists an $S' \in \mathscr{W}$ such that $\beta = S'\alpha_j$ for some $j \in I$, one gets $\alpha = (S^{-1}S')\alpha_j$, implying that $\alpha \in \Delta^{\mathrm{r}}$. As this is contrary to the original assumption, the only possibility is $S\alpha \in \Delta^{\mathrm{i}}$.

(b) The "height" ht(α) of any root α may be defined by

$$\mathrm{ht}(\alpha) = \sum_{j \in I} \kappa_j^{\alpha},$$

where κ_j^{α} ($j \in I$) are the integers of the expansion (26.32). Now suppose that $\alpha \in \Delta^{\mathrm{i}}_+$, and let β be an element in the set $\{S\alpha; S \in \mathscr{W}\}$ with minimum height. (It will be noted that $\mathrm{ht}(S\alpha) > 0$ by (a).) Then for each $j \in I$

$$\mathrm{ht}(S_{\alpha_j}\beta - \beta) \leqslant 0.$$

But

$$S_{\alpha_j}\beta - \beta = \left\{\frac{2\langle \beta, \alpha_j \rangle}{\langle \alpha_j, \alpha_j \rangle}\right\}\alpha_j,$$

and, since $\langle \alpha_j, \alpha_j \rangle > 0$, this implies that $\langle \beta, \alpha_j \rangle \leqslant 0$ for all $j \in I$. As $\beta \in \Delta^{\mathrm{i}}_+$ by part (a), $\beta \in \Delta^{\mathrm{i}}_{+-}$.

(c) With β as in part (b), $\beta = \sum_{j \in I} \kappa_j^{\beta} \alpha_j$ with $\kappa_j^{\beta} \geqslant 0$ for all $j \in I$. Then $\langle \beta, \beta \rangle = \sum_{j \in I} \kappa_j^{\beta} \langle \beta, \alpha_j \rangle$, so $\langle \beta, \beta \rangle \leqslant 0$ by part (b). As there exists an $S \in \mathscr{W}$ such that $\beta = S\alpha$, (26.55) implies that $\langle \alpha, \alpha \rangle \leqslant 0$. Conversely if $\langle \alpha, \alpha \rangle \leqslant 0$ then Theorem IV shows that α cannot be real.

It will be shown later by consideration of examples that neither of the results in Theorem VIII is true for imaginary roots.

As part (b) of Theorem IX shows that $\tilde{\mathscr{L}}$ possesses imaginary roots only if the set Δ^{i}_{+-} is non-empty, it is worthwhile examining this criterion in more detail. Suppose that $\beta \in \Delta^{\mathrm{i}}_{+-}$. Then

(i) $\beta = \sum_{j \in I} \kappa_j^{\beta} \alpha_j$, where $\kappa_j^{\beta} \geqslant 0$ for all $j \in I$,

(ii) $\langle \beta, \alpha_j \rangle \leqslant 0$ for all $j \in I$, and

(iii) $\beta \in \Delta^{\mathrm{i}}$.

Conditions (i) and (ii) can be cast in matrix form by adopting the convention that for any $d_{\mathrm{A}} \times 1$ column vector $\boldsymbol{\kappa}$ with real entries κ_j ($j \in I$) the statement $\boldsymbol{\kappa} > \mathbf{0}$ means $\kappa_j > 0$ for all $j \in I$, with similar meanings being attached to $\boldsymbol{\kappa} \geqslant \mathbf{0}$, $\boldsymbol{\kappa} < \mathbf{0}$ and $\boldsymbol{\kappa} \leqslant \mathbf{0}$. The requirements (i) and (ii) can then be rewritten as

$$\boldsymbol{\kappa}^{\beta} \geqslant \mathbf{0} \tag{26.67}$$

and

$$\mathbf{A}\boldsymbol{\kappa}^{\beta} \leqslant \mathbf{0} \tag{26.68}$$

respectively. It will be seen shortly that these criteria play a significant role in the classification of Kac–Moody algebras.

4 Types of complex Kac–Moody algebras

In this section it will be assumed that $\mathbf{A}$ is a symmetrizable generalized Cartan matrix that is indecomposable. Close examination of the various possibilities shows that there are only three types of complex Kac–Moody algebra, which are called the "finite", "affine" and "indefinite" types. They can be characterized in several equivalent ways.

(*a*) *Finite Kac–Moody algebra*

This may be defined to be a complex Kac–Moody algebra $\tilde{\mathscr{L}}$ whose generalized Cartan matrix $\mathbf{A}$ is such that

(F1) $\det \mathbf{A} \neq 0$,

(F2) there exists a vector $\mathbf{u} > \mathbf{0}$ such that $\mathbf{Au} > \mathbf{0}$, and

(F3) $\mathbf{Av} \geqslant \mathbf{0}$ implies $\mathbf{v} \geqslant \mathbf{0}$.

Condition (F3) implies that $\mathbf{A}\boldsymbol{\kappa} \leqslant \mathbf{0}$ only if $\boldsymbol{\kappa} \leqslant \mathbf{0}$, so (26.67) and (26.68) together show that the subset of roots $\Delta^{i}_{+\,-}$ of Theorem IX of the previous section must be empty, thereby demonstrating that $\tilde{\mathscr{L}}$ has *no* imaginary roots in this case.

The following theorem provides an alternative but equivalent set of conditions.

Theorem I A complex Kac–Moody algebra $\tilde{\mathscr{L}}$ is "finite" if and only if *all* the principal minors of its generalized Cartan matrix $\mathbf{A}$ are *positive*.

Proof See Kac (1985a).

As mentioned earlier in Section 2, the Cartan matrices of *all* the *finite*-dimensional complex *simple* Lie algebras satisfy the conditions of Theorem I, so all of these are of this type. Moreover, by an analysis that will be outlined shortly, it can be shown that there are *no* other algebras of this type. This is the reason why these algebras are called "finite".

(*b*) *Affine Kac–Moody algebra*

This may be defined to be a complex Kac–Moody algebra $\tilde{\mathscr{L}}$ whose generalized Cartan matrix $\mathbf{A}$ is such that

(A1) $\det \mathbf{A} = 0$ but $l = d_{\mathbf{A}} - 1$,

(A2) there exists a vector $\mathbf{u} > \mathbf{0}$ such that $\mathbf{Au} = \mathbf{0}$, and

(A3) $\mathbf{Av} \geqslant \mathbf{0}$ implies $\mathbf{Av} = \mathbf{0}$.

An alternative set of conditions is embodied in the following theorem (cf. Kac 1985a).

Theorem II A complex Kac–Moody algebra $\tilde{\mathscr{L}}$ is "affine" if and only if its generalized Cartan matrix $\mathbf{A}$ is such that $\det \mathbf{A} = 0$ and all the *proper* principal minors of $\mathbf{A}$ are *positive*.

Condition (A3) implies that if $\mathbf{A}\boldsymbol{\kappa} \leqslant \mathbf{0}$ then $\mathbf{A}\boldsymbol{\kappa} = \mathbf{0}$, and condition (A2) shows that there does exist a $\boldsymbol{\kappa} \geqslant \mathbf{0}$ with this property. Equations (26.67) and (26.68) then imply that the set Δ^{i}_{+-} could be non-empty. It will be shown explicitly in the next section that imaginary roots *do* appear in this way for affine algebras.

The term "affine" is applied to this type of Kac–Moody algebra because this is an appropriate description for the special structure of its Weyl group, as will be shown in the next section. It will also be shown in that section that each complex affine Kac–Moody algebra can be constructed from a complex simple Lie algebra and so is labelled by the symbols for that algebra (such as A_l or E_6), together with a superscript q that takes values 1, 2 or 3. Those with $q = 2$ or 3 are known as "twisted" affine Kac–Moody algebras because of the rotations involved in their construction, whereas those with $q = 1$ are described as "untwisted". For example, $A_2^{(1)}$ is untwisted but $E_6^{(2)}$ and $D_4^{(3)}$ are twisted affine Kac–Moody algebras.

(*c*) *Indefinite Kac–Moody algebra*

This may be defined to be a complex Kac–Moody algebra $\tilde{\mathscr{L}}$ whose generalized Cartan matrix $\mathbf{A}$ is such that

(I1) there exists a $\mathbf{u} > \mathbf{0}$ such that $\mathbf{Au} < \mathbf{0}$, and

(I2) $\mathbf{Av} > \mathbf{0}$ and $\mathbf{v} \geqslant \mathbf{0}$ together imply that $\mathbf{v} = \mathbf{0}$.

Equivalently an indefinite Kac–Moody algebra can be characterized by the statement that its generalized Cartan matrix satisfies neither the conditions of Theorem I nor those of Theorem II. Condition (I1) taken with (26.67), (26.68) and Theorem IX of Section 3 suggest the existence of imaginary roots in this case, which is confirmed by further analysis.

The enumeration of the various possibilities for each of the above types is facilitated by the introduction for each $\tilde{\mathscr{L}}$ of its *generalized Dynkin diagram*. (These are the analogues of the Dynkin diagrams for the simple complex Lie algebras that appear in Chapter 13, Section 7, but even when the Kac–Moody algebra coincides with a simple Lie algebra the present construction is slightly different from that given in Chapter 13, Section 7.) The prescription for setting up a generalized Dynkin diagram is as follows.

(i) Assign to each simple root α_j $(j\in I)$ a "vertex", which is drawn as a small circle, and consider all pairs of vertices labelled by α_j and α_k $(j, k\in I)$ in turn.

(ii) Draw L_{jk} lines from the α_j vertex to the α_k vertex, where

$$L_{jk} = \max\{|A_{jk}|, |A_{kj}|\}.$$

(iii) Add an arrow pointing from the α_j vertex to the α_k vertex if $|A_{kj}| > 1$. (If $|A_{kj}| > 1$ *and* $|A_{jk}| > 1$, this implies that there will be arrows pointing towards both vertices.)

(iv) If $|A_{jk}A_{kj}| > 4$, draw a thick solid line from the α_j vertex to the α_k vertex and label it by the ordered pair $\{|A_{jk}|, |A_{kj}|\}$.

The procedure can be reversed to allow the deduction of the generalized Cartan matrix $\mathbf{A}$ from its generalized Dynkin diagram. Recalling that A_{jk} and A_{kj} are either zero or are negative integers, the deduction for the case of a thick solid line from the α_j vertex to the α_k vertex is trivial. Turning to the other cases, there are four possibilities.

(i) there is no line from the α_j vertex to the α_k vertex—in this case $A_{jk} = A_{kj} = 0$;

(ii) there is a single line from the α_j vertex to the α_k vertex, and this line has no arrows on it—in this case $A_{jk} = A_{kj} = -1$;

(iii) there is an arrow on the L_{jk} lines from the α_j vertex to the α_k vertex pointing to the α_k vertex, but no arrow pointing the other way—in this case $A_{kj} = -L_{jk}$, $A_{jk} = -1$.

(iv) there are arrows pointing both ways on the lines from the α_j vertex to the α_k vertex—in this case the assumption $|A_{jk}A_{kj}| \leqslant 4$ implies that $A_{kj} = A_{jk} = -2$.

Example I *The generalized Cartan matrices and generalized Dynkin diagrams for $d_{\mathbf{A}} = 2$*

It is instructive to classify the 2×2 indecomposable generalized Cartan matrices, which have the form

$$A = \begin{bmatrix} 2 & -r \\ -s & 2 \end{bmatrix},$$

where r and s are positive integers. Clearly $\det \mathbf{A} = 4 - rs$ and both proper principal submatrices are just 1×1 matrices with entry 2, so that both proper principal minors are positive. The different types are then as follows.

(a) *Finite*: in this case Theorem I requires that $\det \mathbf{A} > 0$, so $rs < 4$. There are only three possibilities giving non-isomorphic algebras:

(i) $r = 1$, $s = 1$, which corresponds to A_2;

(ii) $r = 1$, $s = 2$, which corresponds to B_2; and

(iii) $r = 1$, $s = 3$, which corresponds to G_2.

(The only other possibilities $r = 2$, $s = 1$ and $r = 3$, $s = 1$ merely correspond to reorderings of this index set I, and consequently give algebras isomorphic to B_2 and G_2 respectively.)

(b) *Affine*: in this case Theorem II requires that $\det \mathbf{A} = 0$, so $rs = 4$. There are only two possibilities giving non-isomorphic algebras:

(i) $r = 1$, $s = 4$, which corresponds to $A_2^{(2)}$; and

(ii) $r = 2$, $s = 2$, which corresponds to $A_1^{(1)}$.

(The combination $r = 4$, $s = 1$ is just a reordering of the index set I for case (i), and gives an algebra isomorphic to $A_2^{(2)}$.)

(c) *Indefinite*: the only possibility left open is $\det \mathbf{A} < 0$, so $rs > 4$, for which there is an infinite number of choices of r and s that will give non-isomorphic algebras.

The corresponding generalized Dynkin diagrams are given in Figure 26.1.

Clearly a generalized Dynkin diagram is connected if and only if the corresponding Cartan matrix $\mathbf{A}$ is indecomposable. Also, if a *proper subdiagram* is defined to be that part of a generalized Dynkin diagram that is obtained by removing one or more vertices (and the lines attached to these vertices) then each proper subdiagram corresponds to a proper principal submatrix of $\mathbf{A}$. Theorems I and II then show that for any *finite* or *affine* Kac–Moody algebra every proper subdiagram is a collection of generalized Dynkin diagrams corresponding to *finite* Kac–Moody algebras. This

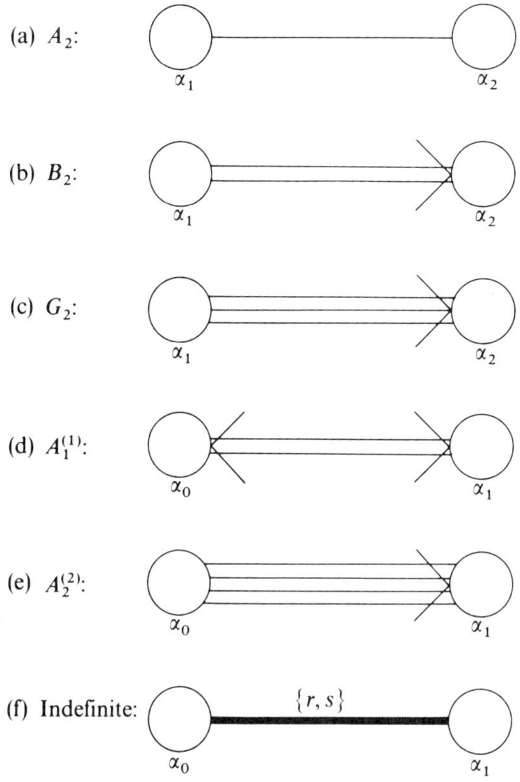

Figure 26.1 The generalized Dynkin diagrams with two connected vertices.

observation can be used to systematically construct *all* the generalized Dynkin diagrams of *finite* and *affine* type. To obtain all those of finite type, one notes first that there is only one with one vertex (which corresponds to A_1), and (as shown in Example I) there are only three with two vertices. The set of diagrams with three vertices can be found by starting with each non-equivalent two-vertex diagram, and adding a single vertex in such a way that all proper subdiagrams are of finite type and $\det \mathbf{A} > 0$. This can then be repeated to obtain all finite-type four-vertex diagrams from the three-vertex finite diagrams, and so on. For the diagrams of affine type the construction is exactly the same, except that the condition $\det \mathbf{A} = 0$ is imposed at each final stage.

A complete analysis along these lines appears in the original papers of both Kac (1968) and Moody (1968) (although the notations there are slightly different). From the full set of finite-type diagrams that is given in Figure 26.2 it will be seen that *every* complex Kac–Moody algebra of finite type is

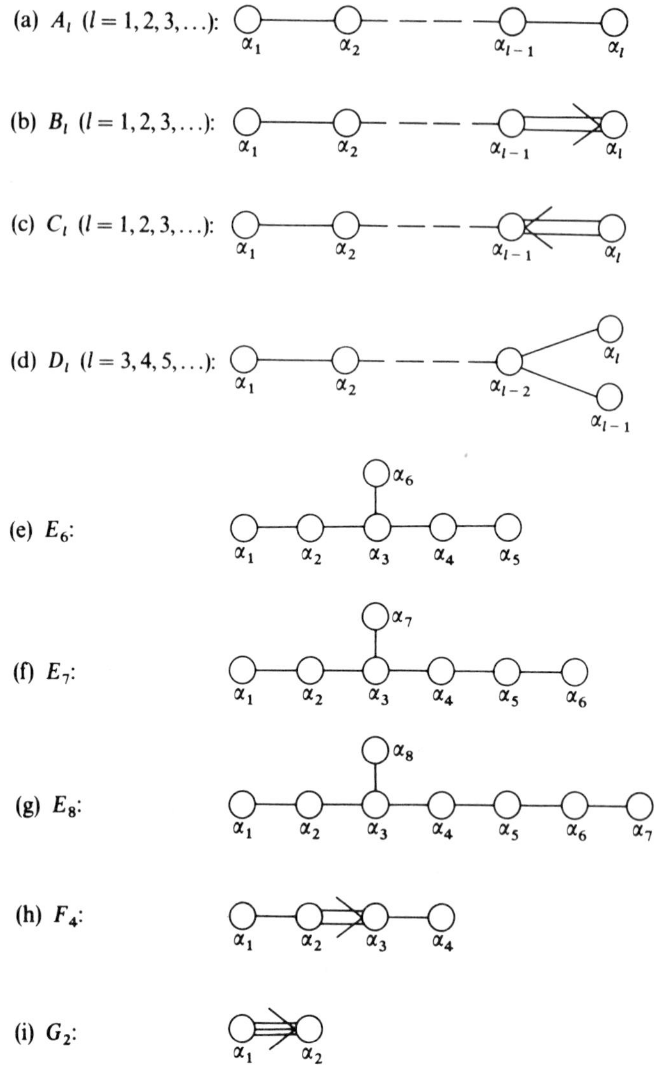

Figure 26.2 Generalized Dynkin diagrams of the complex finite Kac–Moody algebras (i.e. of the complex simple Lie algebras).

a complex *simple* Lie algebra. The generalized Dynkin diagrams of the complex untwisted and twisted affine Kac–Moody algebras are given in Figures 26.3 and 26.4. (In some cases other choices of the labelling of the vertices appear in the literature.)

It is interesting to note that in the finite and affine cases it is *not* necessary

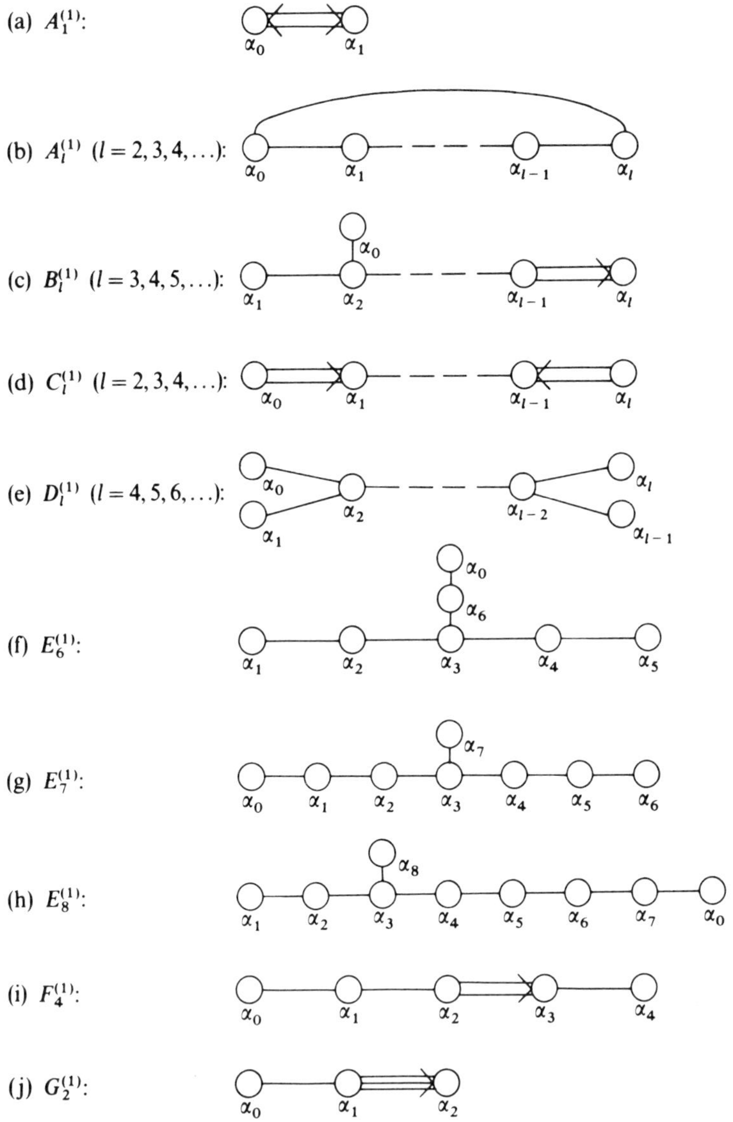

Figure 26.3 Generalized Dynkin diagrams of the complex untwisted affine Kac–Moody algebras.

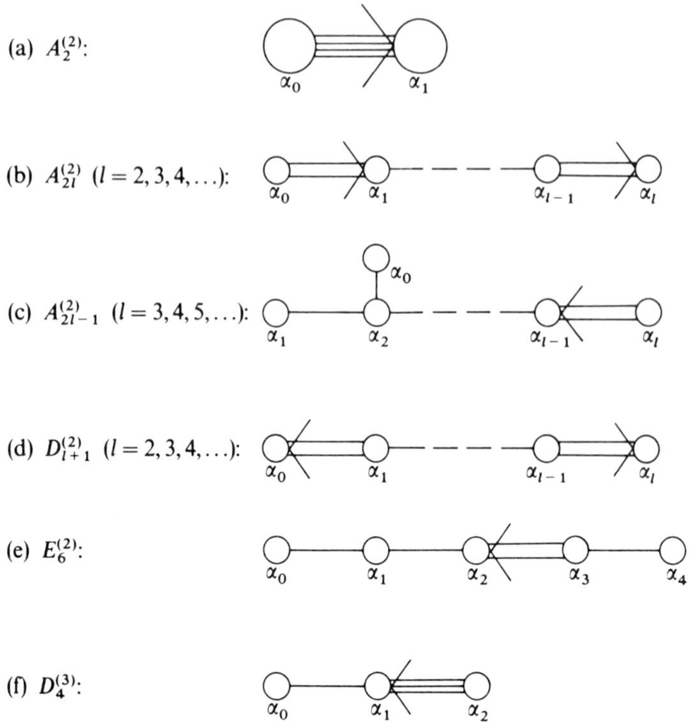

Figure 26.4 Generalized Dynkin diagrams of the complex twisted affine Kac–Moody algebras.

to *impose* the condition that **A** be symmetrizable, since in these cases it can be shown that this is *implied* by the other conditions.

The *length* of a *real* root α may be defined as $\langle \alpha, \alpha \rangle^{1/2}$, which is real by Theorem IV of Section 3. Moreover, (26.55) and (26.57) show that the number of different lengths of real roots cannot exceed the number of simple roots. If the α_j and α_k vertices are directly connected then $\langle \alpha_j, \alpha_j \rangle / \langle \alpha_k, \alpha_k \rangle = A_{kj}/A_{jk}$ (by (26.50)), and so $\langle \alpha_j, \alpha_j \rangle > \langle \alpha_k, \alpha_k \rangle$ if and only if there is a single arrow on the lines from the α_j to the α_k vertex and it points to the α_k vertex (or these vertices are connected by a thick solid line). If all the real roots have the same length then the algebra is said to be *simply laced*. Inspection of Figures 26.2–26.4 show that the only simply laced Kac–Moody algebras of finite type are A_l $(l \geqslant 1)$, D_l $(l \geqslant 3)$, E_6, E_7 and E_8, and the only affine simply laced algebras are $A_l^{(1)}$ $(l \geqslant 1)$, $D_l^{(1)}$ $(l \geqslant 4)$, $E_6^{(1)}$, $E_7^{(1)}$ and $E_8^{(1)}$. These figures also reveal that for each of the remaining finite and affine Kac–Moody algebras, with the exception of $A_{2l}^{(2)}$ $(l \geqslant 2)$, there are only *two* different real root lengths, and so for these algebras it is possible to describe each real root as being either "short" or "long".

It is obvious that if a generalized Cartan matrix **A** satisfies the conditions of Theorem I or of Theorem II then its transpose $\tilde{\mathbf{A}}$ satisfies the same conditions. Consequently the Kac–Moody algebras corresponding to **A** and $\tilde{\mathbf{A}}$ are of the *same* type. The Kac–Moody algebra associated with $\tilde{\mathbf{A}}$ is known as the *dual* of that belonging to **A**.

5 Affine Kac–Moody algebras

(*a*) *General deductions*

Complex affine Kac–Moody algebras are sometimes known as *Euclidean Lie algebras* (cf. Moody 1968, 1969) or just *affine Lie algebras*. This subsection will be devoted to the presentation of various general properties of these algebras that follow almost immediately from the definition of the previous section. However, this line of argument will not be developed very far because, as shown in subsections (*b*) and (*c*), every complex affine Kac-Moody algebra can be explicitly *constructed* from a corresponding simple complex Lie algebra, and this construction enables all their properties to be deduced quite simply and directly. Details of all of these algebras may be found in Appendix N.

As noted in Section 4, for an affine Kac–Moody algebra $d_{\mathbf{A}} = l + 1$, so that the index set I may be taken to be $\{0, 1 \ldots, l\}$, and (by (26.20)) the dimension $n_{\mathscr{H}}$ of the Cartan subalgebra $\mathscr{H}$ is given by

$$n_{\mathscr{H}} = l + 2. \tag{26.69}$$

The $l + 1$ basis elements of $\mathscr{H}$ may be taken to be H_{α_j} $(j \in I)$ or equivalently h_{α_j} $(j \in I)$. The remaining basis element of $\mathscr{H}$ may be denoted by d and may be defined to be such that

$$\alpha_j(d) = \begin{cases} 1 & \text{if } j = 0, \\ 0 & \text{if } j = 1, 2, \ldots, l. \end{cases} \tag{26.70}$$

This element d is known as the *scaling element*, and is a member of the subset $\mathscr{H}''$ of $\mathscr{H}$ (since if $d \in \mathscr{H}'$ then

$$d = \sum_{j=0}^{l} \kappa_j H_{\alpha_j},$$

which implies (by (26.21) and (26.70)) that $\tilde{\mathbf{A}}\boldsymbol{\kappa} \geqslant \mathbf{0}$ but $\tilde{\mathbf{A}}\boldsymbol{\kappa} \neq \mathbf{0}$, which is contrary to the defining condition (A3) with **A** replaced by $\tilde{\mathbf{A}}$.) By (26.39),

$$B(d, d) = 0, \tag{26.71}$$

and similarly (26.38), (26.70) and (26.49) imply that

$$B(d, H_{\alpha_k}) = \begin{cases} 2/\langle \alpha_0, \alpha_0 \rangle & \text{if } k = 0 \\ 0 & \text{if } k = 1, 2, \ldots, l. \end{cases} \tag{26.72}$$

Let $\mathbf{A}^s$ be the principal submatrix obtained from $\mathbf{A}$ by removing its 0th row and column. This submatrix $\mathbf{A}^s$ has dimension $l \times l$, and is the Cartan matrix of a simple complex Lie algebra that will be denoted by $\tilde{\mathscr{L}}^s$. $\tilde{\mathscr{L}}^s$ is isomorphic to the subalgebra of the affine algebra $\tilde{\mathscr{L}}$ that is generated by the following elements:

$$H_{\alpha_k}, \quad E_{\alpha_k}, \quad E_{-\alpha_k} \quad \text{for } k = 1, 2, \ldots, l,$$

and may be identified by this subalgebra. Then $\alpha_1, \alpha_2, \ldots, \alpha_l$ are the simple roots of $\tilde{\mathscr{L}}^s$. The set of all non-zero roots of $\tilde{\mathscr{L}}^s$ will be denoted by Δ^s, and its positive and negative subsets by Δ^s_+ and Δ^s_- respectively. Similarly the Cartan subalgebra of $\tilde{\mathscr{L}}^s$ will be denoted by $\mathscr{H}^s$.

(Clearly the above construction of $\mathbf{A}^s$ depends on the choice of the labelling of the vertices of the generalized Dynkin diagram of $\tilde{\mathscr{L}}$, and in particular on the position of the α_0 vertex. For all the statements that follow it is assumed that the choices are as in Figures 26.3 and 26.4 (which are repeated in Appendix N). Other choices can give different results; for example, for $\tilde{\mathscr{L}} = A^{(2)}_{2l}$ $(l \geqslant 2)$ the labellings of Figure 26.4 (and Figure N.12) correspond to $\tilde{\mathscr{L}}^s = B_l$, whereas a relabelling $\alpha_j \to \alpha_{l-j}$ (for $j = 0, 1, \ldots, l$) would lead to $\tilde{\mathscr{L}}^s = C_l$.)

The symmetric invariant bilinear form $B(\ ,\ )$ of $\tilde{\mathscr{L}}$ when restricted to the subalgebra $\tilde{\mathscr{L}}^s$ provides a symmetric invariant bilinear form for $\tilde{\mathscr{L}}^s$. By Theorem I of Chapter 16, Section 3, $B(a, b) = \gamma B^s(a, b)$ for all $a, b \in \tilde{\mathscr{L}}^s$, where $B^s(\ ,\ )$ is the Killing form of $\tilde{\mathscr{L}}^s$ and γ is a constant. However, $B(\ ,\ )$ has so far been specified only up to a constant multiplicative factor. The choice will henceforth be made that this factor is such as to make $\gamma = 1$, so that

$$B(a, b) = B^s(a, b) \tag{26.73}$$

for all $a, b \in \tilde{\mathscr{L}}^s$. This implies that if α is any linear functional defined on $\mathscr{H}$ such that $\alpha(h) = 0$ for $\alpha \in \mathscr{H}^{s\perp}$ (the orthogonal complement of $\mathscr{H}^s$ with respect to $B(\ ,\ )$, (cf. (21.88)) and if $h^s_\alpha \in \mathscr{H}^s$ is such that

$$B^s(h^s, h^s_\alpha) = \alpha(h^s) \tag{26.74}$$

for all $h^s \in \mathscr{H}^s$ (cf. (26.4)), then

$$h_\alpha = h^s_\alpha. \tag{26.75}$$

Moreover, if α and β are any two such linear functionals, and $\langle \alpha, \beta \rangle^s$ is defined by

$$\langle \alpha, \beta \rangle^s = B^s(h^s_\alpha, h^s_\beta) \tag{26.76}$$

(cf. (26.5)), then (26.75) implies that

$$\langle \alpha, \beta \rangle = \langle \alpha, \beta \rangle^{s}. \tag{26.77}$$

The defining condition (A2) of Section 4 indicates that there exists a vector, which will now be denoted by $\boldsymbol{\kappa}^{\delta}$, such that $\boldsymbol{\kappa}^{\delta} > \mathbf{0}$ and

$$\mathbf{A}\boldsymbol{\kappa}^{\delta} = \mathbf{0}. \tag{26.78}$$

As the elements of $\mathbf{A}$ are all integers, by an appropriate scaling the elements $\kappa_j^{\delta}(j \in I)$ may all be taken to be positive integers, with the minimum value of the set $\{\kappa_0^{\delta}, \kappa_1^{\delta}, \ldots, \kappa_l^{\delta}\}$ being 1. Detailed examination of the generalized Cartan matrices listed in Appendix N shows that for *every* affine Kac–Moody algebra.

$$\kappa_0^{\delta} = 1. \tag{26.79}$$

The linear functional δ on $\mathscr{H}$ defined by

$$\delta = \sum_{j=0}^{l} \kappa_j^{\delta} \alpha_j \tag{26.80}$$

plays a very important role. It has the interesting property (which follows immediately from (26.21) and (26.78)) that

$$\delta(H_{\alpha_k}) = 0 \tag{26.81}$$

for $k = 0, 1, \ldots, l$, and hence

$$\delta(h) = 0 \tag{26.82}$$

for all $h \in \mathscr{H}'$. However, by (26.80), (26.70) and (26.79),

$$\delta(d) = 1. \tag{26.83}$$

Also, (26.44) and (26.80) show that the corresponding element h_{δ} of $\mathscr{H}$ is given by

$$h_{\delta} = \sum_{j=0}^{l} \kappa_j^{\delta} h_{\alpha_j}, \tag{26.84}$$

and hence

$$\alpha_k(h_{\delta}) = 0 \quad \text{for} \quad k = 0, 1, \ldots, l. \tag{26.85}$$

This follows from (26.46), (26.47), (26.19) and (26.78), since

$$\alpha_k(h_{\delta}) = \sum_{j=0}^{l} \kappa_j^{\delta} \langle \alpha_k, \alpha_j \rangle = \sum_{j=0}^{l} \kappa_j^{\delta} B_{kj} = \sum_{j=0}^{l} \frac{\kappa_j^{\delta} A_{kj}}{\varepsilon_k} = 0.$$

Equations (26.34) and (26.85) together imply that $h_{\delta} \in \mathscr{C}$, the centre of $\tilde{\mathscr{L}}$. As Theorem I of Section 3 shows that $\mathscr{C}$ is one-dimensional in the case of an *affine* Kac–Moody algebra, every member of $\mathscr{C}$ is a multiple of h_δ. Moroever, (26.85) and (26.46) imply that

$$\langle \delta, \alpha_k \rangle = 0 \quad \text{for} \quad k = 0, 1, \dots, l, \tag{26.86}$$

and hence, by (26.80),

$$\langle \delta, \delta \rangle = 0. \tag{26.87}$$

The constructions of Sections 5(*b*) and 5(*c*) will demonstrate that $j\delta$ is actually a root of $\tilde{\mathscr{L}}$ for every non-zero integer j. Theorem IX of Section 3 and (26.87) then imply that $j\delta$ is an *imaginary* root for each non-zero integer j. Further analysis in (*b*) and (*c*) will prove that $\tilde{\mathscr{L}}$ has no other imaginary roots.

(*b*) *Construction of the complex untwisted affine Kac–Moody algebras*

It will now be shown how, starting from any *simple* complex Lie algebra $\tilde{\mathscr{L}}^0$, a complex untwisted affine Kac–Moody algebra may be constructed. This latter algebra will be denoted by $\tilde{\mathscr{L}}^{(1)}$. The procedure will be developed further in subsection (*c*) to give the *twisted* affine Kac–Moody algebras, which can all be constructed as subalgebras of certain of these untwisted algebras $\tilde{\mathscr{L}}^{(1)}$. (For the untwisted algebra $\tilde{\mathscr{L}}^{(1)}$ the subalgebra $\tilde{\mathscr{L}}^{\mathrm{s}}$ of the previous subsection will be identified with $\tilde{\mathscr{L}}^0$, but for the twisted algebras it will become apparent that $\tilde{\mathscr{L}}^{\mathrm{s}}$ is a proper subalgebra of $\tilde{\mathscr{L}}^0$.)

Suppose that the simple complex Lie algebra $\tilde{\mathscr{L}}^0$ has rank l^0, that $\mathscr{H}^0$ is its Cartan subalgebra, that α_j (for $j = 1, 2, \dots, l^0$) are its simple roots, that Δ^0, Δ^0_+ and Δ^0_- are its non-zero, positive and negative root systems respectively, that $B^0(\ ,\ )$ is its Killing form and that $\langle \alpha, \beta \rangle^0$ is the corresponding bilinear form for linear functionals defined on $\mathscr{H}^0$. A Weyl canonical basis may be chosen for $\tilde{\mathscr{L}}^0$. It consists of $h^0_{\alpha j}$ (for $j = 1, 2, \dots, l^0$), together with e^0_α (for all $\alpha \in \Delta^0$), and these are assumed to satisfy the relations (13.64)–(13.72) (with a superscript 0 being added where appropriate).

The first stage is to consider the *loop algebra*, which consists of all complex linear combinations of the products $t^j \otimes a^0_p$, where j takes any integer value, a^0_p are the basis elements of $\tilde{\mathscr{L}}^0$ and t is a real number. The commutator of this loop algebra may be defined by

$$[t^j \otimes a^0, t^k \otimes b^0]_- = t^{j+k} \otimes [a^0, b^0]_- \tag{26.88}$$

for all integers j and k and all a^0, $b^0 \in \tilde{\mathscr{L}}^0$, where the commutator of the right-handed side is that of $\tilde{\mathscr{L}}^0$.

This algebra may be *extended* by introducing an additional element c (so that it then consists of all complex linear combinations of the $t^j \otimes a^0_p$ and of

c), with the commutator (26.88) modified to be

$$[t^j \otimes a^0, t^k \otimes b^0]_- = t^{j+k} \otimes [a^0, b^0]_- + j\delta^{j+k,0} B^0(a^0, b^0)c \qquad (26.89)$$

for all integers j and k and all a^0, $b^0 \in \tilde{\mathscr{L}}^0$, and where it is assumed that

$$[t^j \otimes a^0, c]_- = 0 \qquad (26.90)$$

for all integers j and all $a^0 \in \tilde{\mathscr{L}}^0$. Again the commutator on the right-hand side of (26.89) is that of $\tilde{\mathscr{L}}^0$. It is easily verified that the commutator defined by (26.89) is antisymmetric and satisfies the Jacobi identity. This Lie algebra may be enlarged by adding a further element d, for which it is assumed that

$$[d, t^j \otimes a^0]_- = j t^j \otimes a^0, \qquad (26.91)$$

for all integers j and all $a^0 \in \tilde{\mathscr{L}}^0$, and that

$$[d, c]_- = 0. \qquad (26.92)$$

(Clearly (26.91) shows that d acts on the elements of the loop algebra in the same way as the differential operator $t\,\partial/\partial t$.) The resulting Lie algebra, which will be denoted by $\tilde{\mathscr{L}}^{(1)}$, will now be shown to be a complex untwisted affine Kac–Moody algebra. Its composition may be summarized by the statement

$$\tilde{\mathscr{L}}^{(1)} = (\mathbb{C}c) \oplus (\mathbb{C}d) \oplus \sum_{j=-\infty}^{\infty} (t^j \otimes \tilde{\mathscr{L}}^0). \qquad (26.93)$$

Equation (26.89) shows that the set of elements $t^0 \otimes a^0$, where $a^0 \in \tilde{\mathscr{L}}^0$, form a subalgebra of $\tilde{\mathscr{L}}^{(1)}$ that is isomorphic to $\tilde{\mathscr{L}}^0$. Also, (26.89)–(26.92) together indicate that a linearly independent and mutually commuting set of $l^0 + 2$ elements is formed by c, d and $t^0 \otimes h^0_{\alpha_k}$ (for $k = 1, 2, \ldots, l^0$). Let $\mathscr{H}^{(1)}$ be the $(l^0 + 2)$-dimensional complex vector space that has these elements as its basis, so that

$$\mathscr{H}^{(1)} = (\mathbb{C}c) \oplus (\mathbb{C}d) \oplus \sum_{k=1}^{l^0} \mathbb{C}(t^0 \otimes h^0_{\alpha_k}). \qquad (26.94)$$

It will become clear that $\mathscr{H}^{(1)}$ is the Cartan subalgebra of $\tilde{\mathscr{L}}^{(1)}$. Every linear functional α that is defined on $\mathscr{H}^0$ can be *extended* to become a linear functional on $\mathscr{H}^{(1)}$ by the definitions:

$$\alpha(t^0 \otimes h^0_{\alpha_k}) = \alpha(h^0_{\alpha_k}) \quad \text{for } k = 1, 2, \ldots, l^0, \qquad (26.95a)$$

$$\alpha(c) = 0, \qquad (26.95b)$$

$$\alpha(d) = 0. \qquad (26.95c)$$

No confusion will be caused by denoting (as here) both α and its extension by the same symbol.

Let δ be the linear functional on $\mathscr{H}^{(1)}$ defined by

$$\delta(t^0 \otimes h^0_{\alpha_k}) = 0 \quad \text{for } k = 1, 2, \ldots, l^0, \tag{26.96a}$$

$$\delta(c) = 0, \tag{26.96b}$$

$$\delta(d) = 1. \tag{26.96c}$$

Then, since (26.89)–(26.91) give

$$[t^0 \otimes h^0_{\alpha_k}, t^j \otimes e^0_\alpha]_- = \alpha(h^0_{\alpha_k})(t^j \otimes e^0_\alpha), \tag{26.97a}$$

$$[c, t^j \otimes e^0_\alpha]_- = 0, \tag{26.97b}$$

$$[d, t^j \otimes e^0_\alpha]_- = j(t^j \otimes e^0_\alpha) \tag{26.97c}$$

for any $\alpha \in \Delta^0$ and for any integer j (and for $k = 1, \ldots, l^0$), it follows from (26.95) and (26.96) that

$$[h, t^j \otimes e^0_\alpha]_- = \{j\delta(h) + \alpha(h)\}(t^j \otimes e^0_\alpha) \tag{26.98}$$

for *all* $h \in \mathscr{H}^{(1)}$. Thus $t^j \otimes e^0_\alpha$ corresponds to a root $j\delta + \alpha$ of $\tilde{\mathscr{L}}^{(1)}$; that is, $t^j \otimes e^0_\alpha \in \tilde{\mathscr{L}}^{(1)}_{j\delta+\alpha}$. Similarly for any $\beta \in \Delta^0$ and any non-zero integer j

$$[t^0 \otimes h^0_{\alpha_k}, t^j \otimes h^0_\beta]_- = 0, \tag{26.99a}$$

$$[c, t^j \otimes h^0_\beta]_- = 0, \tag{26.99b}$$

$$[d, t^j \otimes h^0_\beta]_- = j(t^j \otimes h^0_\beta), \tag{26.99c}$$

so, by (26.96),

$$[h, t^j \otimes h^0_\beta]_- = j\delta(h)(t^j \otimes h^0_\beta) \tag{26.100}$$

for all $h \in \mathscr{H}^{(1)}$. Thus $t^j \otimes h^0_\beta$ corresponds to a root $j\delta$ of $\tilde{\mathscr{L}}^{(1)}$. Moreover, there are l^0 linearly independent elements with this property, namely $t^j \otimes h^0_{\alpha_k}$ (for $k = 1, 2, \ldots, l^0$), and, since there are no further elements of $\tilde{\mathscr{L}}^{(1)}$ to consider, the root subspace $\tilde{\mathscr{L}}_{j\delta}$ must have dimension l^0.

A bilinear symmetric form $B^{(1)}(\ ,\ )$ can be set up on $\tilde{\mathscr{L}}^{(1)}$ through the definitions

$$B^{(1)}(t^j \otimes a^0, t^k \otimes b^0) = \delta^{j+k,0} B^0(a^0, b^0), \tag{26.101a}$$

$$B^{(1)}(t^j \otimes a^0, c) = 0, \tag{26.101b}$$

$$B^{(1)}(t^j \otimes a^0, d) = 0, \tag{26.101c}$$

$$B^{(1)}(c, c) = 0, \tag{26.101d}$$

$$B^{(1)}(c, d) = 1, \tag{26.101e}$$

$$B^{(1)}(d, d) = 0 \tag{26.101f}$$

for all a^0, $b^0 \in \tilde{\mathscr{L}}^0$ and all integers j and k. It is easily checked using the commutators (26.89)–(26.92) and the fact that $B^0(\ ,\ )$ is an invariant form on $\tilde{\mathscr{L}}^0$ that $B^{(1)}(\ ,\ )$ is *invariant* on $\tilde{\mathscr{L}}^{(1)}$ (i.e. that (26.37) is satisfied). Clearly $B^{(1)}(\ ,\ )$ coincides with $B^0(\ ,\ )$ on the subalgebra of $\tilde{\mathscr{L}}^{(1)}$ that is isomorphic to $\tilde{\mathscr{L}}^0$.

For any linear functional α defined on $\mathscr{H}^{(1)}$ an element $h_\alpha^{(1)}$ of $\mathscr{H}^{(1)}$ can be defined (cf. (26.42)) by

$$B^{(1)}(h, h_\alpha^{(1)}) = \alpha(h) \tag{26.102}$$

for all $h \in \mathscr{H}^{(1)}$, and for any two linear functionals α and β on $\mathscr{H}^{(1)}$ the quantity $\langle \alpha, \beta \rangle^{(1)}$ may be defined (cf. (26.45)) by

$$\langle \alpha, \beta \rangle^{(1)} = B^{(1)}(h_\alpha^{(1)}, h_\beta^{(1)}). \tag{26.103}$$

In particular, if α and β are extensions of linear functionals on $\mathscr{H}^0$ (denoted by the same symbols) then (26.101) and (26.102) imply that

$$h_\alpha^{(1)} = t^0 \otimes h_\alpha^0, \tag{26.104}$$

and hence, by (26.101) and (26.103),

$$\langle \alpha, \beta \rangle^{(1)} = \langle \alpha, \beta \rangle^0. \tag{26.105}$$

Similarly (26.101) and (26.102) show that

$$h_\delta^{(1)} = c. \tag{26.106}$$

If α_k ($k = 1, 2, \ldots, l^0$) is an extension of a simple root of $\tilde{\mathscr{L}}^0$, (26.101), (26.103), (26.104) and (26.106) show that

$$\langle \delta, \alpha_k \rangle^{(1)} = 0, \quad k = 1, 2, \ldots, l^0, \tag{26.107}$$

and

$$\langle \delta, \delta \rangle^{(1)} = 0. \tag{26.108}$$

Thus

$$\langle j\delta, j\delta \rangle^{(1)} = 0, \tag{26.109}$$

but

$$\langle j\delta + \alpha, j\delta + \alpha \rangle^{(1)} = \langle \alpha, \alpha \rangle^0 > 0 \tag{26.110}$$

for every root α that is an extension of a non-zero root of $\tilde{\mathscr{L}}^0$. Consequently the roots $j\delta$ are all imaginary, but the roots $j\delta + \alpha$ are all real.

The rest of this subsection will be devoted to establishing that $\tilde{\mathscr{L}}^{(1)}$ is isomorphic to an affine Kac–Moody algebra as defined in Section 4 and discussed in subsection (*a*) above, so for the rest of this subsection it will be assumed that $l^0 = l$. The main outstanding point is the identification of the simple roots, and in particular of α_0. In the general formulation of Kac–Moody algebras every root $\alpha \in \Delta$ can be expressed in the form (26.32), where for the affine case $I = \{0, 1, \ldots, l\}$, so that

$$\alpha = \sum_{j=0}^{l} \kappa_j^\alpha \alpha_j, \tag{26.111}$$

where the members of the set $\{\kappa_0^\alpha, \kappa_1^\alpha, \ldots, \kappa_l^\alpha\}$ are either all non-negative integers or are all non-positive integers. Clearly the roots of $\tilde{\mathscr{L}}^{(1)}$ that are extensions of those of $\tilde{\mathscr{L}}^0$ have this form, provided that $\alpha_1, \alpha_2, \ldots, \alpha_l$ are taken as l of the $l+1$ simple roots of $\tilde{\mathscr{L}}^{(1)}$. It remains to determine α_0. As the roots of $\tilde{\mathscr{L}}^{(1)}$ are $j\delta$ (for $j = \pm 1, \pm 2, \ldots$) and $j\delta + \alpha$ (for $j = 0, \pm 1, \pm 2, \ldots$, and α an extension of a root of $\tilde{\mathscr{L}}^0$) one can *define* the linear functional α_0 on $\mathscr{H}^{(1)}$ by

$$\alpha_0 = \delta - \alpha_{\rm H}, \tag{26.112}$$

where $\alpha_{\rm H}$ is the extension to $\mathscr{H}^{(1)}$ of the *highest* root of $\tilde{\mathscr{L}}^0$. (This highest root $\alpha_{\rm H}$ has the property that $\alpha_{\rm H} > \alpha$ for every other root α of $\tilde{\mathscr{L}}^0$ (cf. Chapter 13, Section 7).) It is clear then that, for any positive integer j, $j\delta$ is of the form (26.111) with $\{\kappa_0^{j\delta}, \kappa_1^{j\delta}, \ldots, \kappa_l^{j\delta}\}$ all non-negative integers. Moreover, for any positive integer j, $j\delta + \alpha$ is of the form (26.111) with $\{\kappa_0^{j\delta+\alpha}, \kappa_1^{j\delta+\alpha}, \ldots, \kappa_l^{j\delta+\alpha}\}$ all non-negative integers for *any* non-zero root α (positive *or* negative) that is an extension of a root of $\tilde{\mathscr{L}}^0$. (Closer examination shows that (26.112) is the only possible choice of α_0.)

From (26.112), (26.107) and (26.105)

$$\langle \alpha_0, \alpha_k \rangle^{(1)} = -\langle \alpha_{\rm H}, \alpha_k \rangle^0, \quad k = 1, 2, \ldots, l, \tag{26.113}$$

and

$$\langle \alpha_0, \alpha_0 \rangle^{(1)} = \langle \alpha_{\rm H}, \alpha_{\rm H} \rangle^0. \tag{26.114}$$

Calculating the matrix $\mathbf{A}^{(1)}$ using (26.50) (but with a superscript (1) added)

$$A_{jk}^{(1)} = A_{jk}^0 \quad \text{for } j, k = 1, 2, \ldots, l, \tag{26.115}$$

and, by (26.113) and (26.114), for $k = 1, 2, \ldots, l$,

$$A_{0k}^{(1)} = -\frac{2\langle \alpha_{\rm H}, \alpha_k \rangle^0}{\langle \alpha_{\rm H}, \alpha_{\rm H} \rangle^0}, \tag{26.116a}$$

$$A_{k0}^{(1)} = -\frac{2\langle \alpha_{\rm H}, \alpha_k \rangle^0}{\langle \alpha_k, \alpha_k \rangle^0}, \tag{26.116b}$$

and, of course, $A_{00}^{(1)} = 2$. As $\mathbf{A}^0$ is a Cartan matrix, all that has to be done to demonstrate that $\mathbf{A}^{(1)}$ is a generalized Cartan matrix is to establish that the $A_{0k}^{(1)}$ and $A_{k0}^{(1)}$ of (26.116) are non-positive integers for each $k = 1, 2, \ldots, l$.

Consider first $A_{k0}^{(1)}$. As $\alpha_{\rm H}$ is the highest weight of the adjoint representation of $\tilde{\mathscr{L}}^0$, by (15.12),

$$\alpha_{\mathrm{H}} = \sum_{j=1}^{l} n_j \Lambda_j,$$

where $\{n_1, n_2, \ldots, n_l\}$ is a set of non-negative integers and $\Lambda_1, \Lambda_2, \ldots, \Lambda_l$ are the fundamental weights of $\tilde{\mathscr{L}}^0$. By (15.11), for $k = 1, 2, \ldots, l$,

$$2\langle \Lambda_j, \alpha_k \rangle^0 / \langle \alpha_k, \alpha_k \rangle^0 = \delta_{jk},$$

so $A_{k0}^{(1)} = -n_k$, thereby demonstrating that $A_{k0}^{(1)}$ is a non-positive integer. Turning to $A_{0k}^{(1)}$, since

$$A_{0k}^{(1)} = \frac{A_{k0}^{(1)} \langle \alpha_k, \alpha_k \rangle^0}{\langle \alpha_{\mathrm{H}}, \alpha_{\mathrm{H}} \rangle^0},$$

it follows that

$$A_{0k}^{(1)} = -\frac{n_k \langle \alpha_k, \alpha_k \rangle^0}{\langle \alpha_{\mathrm{H}}, \alpha_{\mathrm{H}} \rangle^0}.$$

For the simply laced algebras A_l $(l \geqslant 1)$, D_l $(l \geqslant 3)$, E_6, E_7 and E_8 $\langle \alpha_k, \alpha_k \rangle^0 / \langle \alpha_{\mathrm{H}}, \alpha_{\mathrm{H}} \rangle^0 = 1$ for all $k = 1, 2, \ldots, l$. For the remaining simple Lie algebras the results of Appendix F show that for the adjoint representations when $n_k \neq 0$ then $n_k \langle \alpha_k, \alpha_k \rangle^0 / \langle \alpha_{\mathrm{H}}, \alpha_{\mathrm{H}} \rangle^0 = 1$. Thus in all cases $A_{0k}^{(1)}$ is a non-positive integer (for $k = 1, 2, \ldots, l$) and hence $\mathbf{A}^{(1)}$ is a generalized Cartan matrix.

These results are all consistent with the general theory of affine Kac–Moody algebras of subsection (*a*), with $\tilde{\mathscr{L}}$ being identified with $\tilde{\mathscr{L}}^{(1)}$, $\mathscr{H}$ with $\mathscr{H}^{(1)}$, $\mathbf{A}$ with $\mathbf{A}^{(1)}$, $B(\ ,\ )$ with $B^{(1)}(\ ,\ )$, and $\langle \alpha, \beta \rangle$ with $\langle \alpha, \beta \rangle^{(1)}$. Moreover, (26.115) shows that $\mathbf{A}^0$ is the Cartan matrix obtained by removing the 0th row and column from $\mathbf{A}^{(1)}$. Consequently in this untwisted case $\tilde{\mathscr{L}}^0$ is the simple Lie algebra $\tilde{\mathscr{L}}^{\mathrm{s}}$ that was introduced in subsection (*a*), and $B^{(1)}(\ ,\ )(= B(\ ,\ ))$ coincides with the Killing form $B^{\mathrm{s}}(\ ,\ )$ on $\tilde{\mathscr{L}}^{\mathrm{s}}$. Likewise the scaling element d of subsection (*a*) may be identified with the element d defined by (26.91) and (26.92), and the linear functional δ of (26.80) coincides with the δ defined by (26.96).

For future convenience the major results that have been established above are summarized in the following theorem.

Theorem I To each simple complex Lie algebra $\tilde{\mathscr{L}}^0$ of rank l there corresponds an "untwisted" complex affine Kac–Moody algebra $\tilde{\mathscr{L}}$ whose Cartan subalgebra is of dimension $l + 2$. The set of positive real roots of $\tilde{\mathscr{L}}$ consists of

(i) extensions of the positive roots of $\tilde{\mathscr{L}}^0$, and

(ii) the roots $j\delta + \alpha$, where $j = 1, 2, \ldots$ and α is the extension of *any* non-zero root of $\tilde{\mathscr{L}}^0$.

(The corresponding root subspaces for these are all one-dimensional.) The set of positive imaginary roots are given by $j\delta$, for $j = 1, 2, \ldots$, and for these

$$\dim \tilde{\mathscr{L}}_{j\delta} = l. \tag{26.117}$$

The basis elements of $\tilde{\mathscr{L}}^{(1)}_{j\delta+\alpha}$ ($\alpha \in \Delta^0$) may be taken to be

$$e_{j\delta+\alpha} = t^j \otimes e^0_\alpha \quad (j = 0, \pm 1, \pm 2, \ldots) \tag{26.118}$$

(cf. (26.98)), and the l basis elements of $\tilde{\mathscr{L}}^{(1)}_{j\delta}$ can be taken as

$$e^k_{j\delta} = it^j \otimes h^0_{\alpha_k} \quad (j = \pm 1, \pm 2, \ldots; k = 1, 2, \ldots, l) \tag{26.119}$$

(cf. (26.100)). Then, if $e^\beta_{j\delta}$ is defined by

$$e^\beta_{j\delta} = it^j \otimes h^0_\beta, \tag{26.120}$$

where

$$\beta = \sum_{k=1}^{l} \kappa^\beta_k \alpha_k, \tag{26.121}$$

it follows that

$$e^\beta_{j\delta} = \sum_{k=1}^{l} \kappa^\beta_k e^k_{j\delta}. \tag{26.122}$$

(The factor i has been inserted in (26.119) and (26.120) to provide a neat-looking basis for the "compact" real form of $\tilde{\mathscr{L}}^{(1)}$ (cf. (26.195)).) Then (26.98) and (26.100) can be written as

$$[h, e_{j\delta+\alpha}]_- = \{j\delta(h) + \alpha(h)\} e_{j\delta+\alpha} \tag{26.123}$$

and

$$[h, e^\beta_{j\delta}]_- = j\delta(h) e^\beta_{j\delta} \tag{26.124}$$

for all $h \in \mathscr{H}^{(1)}$.

The remaining commutation relations of the basis elements of $\tilde{\mathscr{L}}^{(1)}$ can be deduced from (26.89) and (13.64)–(13.71) and are as follows.

(i) For $\beta \in \Delta^0$ and α an extension of a root in Δ^0 and for $j = \pm 1, \pm 2, \ldots$ and $j' = 0, \pm 1, \pm 2, \ldots$

$$[e^\beta_{j\delta}, e_{j'\delta+\alpha}]_- = i\langle \beta, \alpha \rangle^0 e_{(j+j')\delta+\alpha}. \tag{26.125}$$

(ii) For $\beta, \beta' \in \Delta^0$ and $j, j' = \pm 1, \pm 2, \ldots$

$$[e^\beta_{j\delta}, e^{\beta'}_{j'\delta}]_- = -\delta_{j+j',0} \langle \beta, \beta' \rangle^0 h_{j\delta}. \tag{26.126}$$

(iii) For α and α' extensions of roots in Δ^0 and $j, j' = 0, \pm 1, \pm 2, \ldots$

$$[e_{j\delta+\alpha}, e_{j'\delta+\alpha'}]_- = N^0_{\alpha\alpha'} e_{(j+j')\delta+(\alpha+\alpha')}, \tag{26.127}$$

provided that $\alpha + \alpha' \in \Delta^0$ and $\alpha + \alpha' \neq 0$. Here $N^0_{\alpha\alpha'}$ are the structure constants of $\tilde{\mathscr{L}}^0$ that are defined (cf. (13.31) and (13.71)) by

$$[e^0_\alpha, e^0_{\alpha'}]_- = N^0_{\alpha\alpha'} e^0_{\alpha+\alpha'}. \tag{26.128}$$

If $\alpha + \alpha' \neq 0$ but $\alpha + \alpha' \notin \Delta^0$,

$$[e_{j\delta+\alpha}, e_{j'\delta+\alpha'}]_- = 0. \tag{26.129}$$

(iv) For α an extension of a root in Δ^0 and $j, j' = 0, \pm 1, \pm 2, \ldots$

$$[e_{j\delta+\alpha}, e_{j'\delta-\alpha}]_- = \begin{cases} ie_{(j+j')\delta} & \text{if } j+j' \neq 0, \\ -h_{j\delta+\alpha} & \text{if } j+j' = 0, \end{cases} \tag{26.130}$$

where it has been assumed that

$$B^0(e^0_\alpha, e^0_{-\alpha}) = -1 \tag{26.131}$$

(cf. (13.37) and (13.64)).

It is sometimes convenient to take the basis elements a^0_r $(r = 1, 2, \ldots, n^0)$ of $\tilde{\mathscr{L}}^0$ to be those of its compact real form $\mathscr{L}^0_c$ and to require that

$$B^0(a^0_r, a^0_s) = -\delta_{rs} \tag{26.131}$$

for $r, s = 1, 2, \ldots, n^0$ (n^0 being the order of $\tilde{\mathscr{L}}^0$ and of $\mathscr{L}^0_c$) (cf. (14.2)). Then

$$[a^0_r, a^0_s]_- = \sum_{u=1}^{n^0} c^u_{rs} a^0_u, \quad r, s = 1, 2, \ldots, n^0, \tag{26.132}$$

where Theorem I of Chapter 14, Section 2 shows that the structure constants c^u_{rs} are not only real but are also antisymmetric with respect to the interchange of *any* pair of indices. With this choice, (26.89) becomes

$$[t^j \otimes a^0_r, t^k \otimes a^0_s]_- = \sum_{u=1}^{n^0} c^u_{rs}(t^{j+k} \otimes a^0_u) - j\delta^{j+k,0}\delta_{rs}c. \tag{26.133}$$

Introducing the notation

$$T^r_j = it^j \otimes a^0_r \tag{26.134}$$

for $j = 0, \pm 1, \pm 2, \ldots$ and $r = 1, 2, \ldots, n^0$, and

$$f^{rst} = c^t_{rs} \tag{26.135}$$

for $r, s, t = 1, 2, \ldots, n^0$, (26.133) can be rewritten as

$$[T^r_j, T^s_k]_- = i \sum_{t=1}^{n^0} f^{rst} T^t_{j+k} + j\delta_{j+k,0}\delta^{rs}c, \tag{26.136}$$

which is in a form that is often found in the theoretical physics literature (cf.

Goddard and Olive 1986, 1988). In this notation (26.90) and (26.91) become

$$[c, T^r_j]_- = 0 \qquad (26.137a)$$

and

$$[d, T^r_j]_- = jT^r_j \qquad (26.137b)$$

for $j = 0, \pm 1, \pm 2, \dots$ and $r = 1, 2, \dots, n^0$.

It is not uncommon to find *any* infinite-dimensional Lie algebra for which (26.136) and (26.137a) provide the commutation relations between the basis elements being described as a "Kac–Moody algebra". Such an algebra does *not* necessarily have all the properties of the "genuine" Kac–Moody algebras that were defined in Section 2. One particularly important example occurs when $f^{rst} = 0$ for all $r, s, t = 1, 2, \dots, n^0$, in which case (26.136) becomes

$$[T^r_j, T^s_k]_- = j\delta_{j+k,0}\delta^{rs}c. \qquad (26.138)$$

The resulting algebra can be thought of as being built on a finite-dimensional Lie algebra $\tilde{\mathscr{L}}^0$ that is *Abelian*. This type of algebra is sometimes called an *infinite-dimensional Heisenberg algebra.*

(c) *Construction of the complex twisted affine Kac–Moody algebras*

Now let $\tilde{\mathscr{L}}^0$ be a simple complex Lie algebra of rank l^0 that possesses an *outer* automorphism, and let $\tilde{\mathscr{L}}^{(1)}$ be the untwisted affine Kac–Moody algebra constructed from $\tilde{\mathscr{L}}^0$ by the procedure of subsection (*b*). As shown in Chapter 14, Section 5, the only possible choices of $\tilde{\mathscr{L}}^0$ are A_{l^0} (for $l^0 \geqslant 2$), D_{l^0} (for $l^0 \geqslant 3$) and E_6.

Let τ be a "rotation" of the set of roots Δ^0 of $\tilde{\mathscr{L}}^0$, and let ψ_τ be the corresponding outer automorphism of $\tilde{\mathscr{L}}^0$ (cf. Theorem I of Chapter 14, Section 5). For all of these $\tilde{\mathscr{L}}^0$ (except $\tilde{\mathscr{L}}^0 = D_4$) there is only one non-trivial τ, and this satisfies

$$\tau(\tau(\alpha_k)) = \alpha_k \quad \text{for } k = 1, 2, \dots, l^0. \qquad (26.139)$$

For example (cf. (14.25)), for $\tilde{\mathscr{L}}^0 = A_{l^0}$ ($l^0 \geqslant 2$)

$$\tau(\alpha_k) = \alpha_{l^0+1-k}, \quad k = 1, 2, \dots, l^0. \qquad (26.140)$$

For $\tilde{\mathscr{L}}^0 = D_4$ there are a number of rotations satisfying (26.139), but there are also rotations such as

$$\tau(\alpha_1) = \alpha_3, \quad \tau(\alpha_2) = \alpha_2, \quad \tau(\alpha_3) = \alpha_4, \quad \tau(\alpha_4) = \alpha_1$$

for which

$$\tau(\tau(\tau(\alpha_k))) = \alpha_k \quad \text{for } k = 1, \dots, 4. \qquad (26.141)$$

The situation may be summarized by the statement that

$$\tau^q = 1,$$

where $q = 2$ or 3, the latter occurring only for the three-fold rotations of D_4.

For the corresponding outer automorphisms of $\tilde{\mathscr{L}}^0$

$$(\psi_\tau)^q = 1,$$

implying that ψ_τ has q different eigenvalues, namely $e^{2\pi ip/q}$ for $p = 0, \ldots, q-1$. The elements of $\tilde{\mathscr{L}}^0$ may be arranged as eigenvectors of ψ_τ. Let $\tilde{\mathscr{L}}_p^{0(q)}$ be the eigenspace corresponding to eigenvalue $e^{2\pi ip/q}$. Then $a \in \tilde{\mathscr{L}}_p^{0(q)}$ if $a \in \tilde{\mathscr{L}}^0$ and

$$\psi_\tau(a) = e^{2\pi ip/q} a. \tag{26.142}$$

In particular $\tilde{\mathscr{L}}_0^{0(q)}$ is the subset of elements a of $\tilde{\mathscr{L}}^0$ such that

$$\psi_\tau(a) = a. \tag{26.143}$$

Consider the subalgebra $\tilde{\mathscr{L}}^{(q)}$ of $\tilde{\mathscr{L}}^{(1)}$ that consists of all complex linear combinations of c, d and, for each $p = 0, \ldots, q-1$, of $t^j \otimes a_{pr}^0$ for all basis elements a_{pr}^0 of $\tilde{\mathscr{L}}_p^{0(q)}$ and all integers j such that $j \bmod q = p$; that is, such that $j = j'q + p$, where $j' = 0, \pm 1, \pm 2, \ldots$. This may be succinctly expressed as

$$\tilde{\mathscr{L}}^{(q)} = (\mathbb{C}c) \oplus (\mathbb{C}d) \oplus \sum_{p=0}^{q-1} \sum_{j,\, j \bmod q = p} t^j \otimes \tilde{\mathscr{L}}_p^{0(q)}. \tag{26.144}$$

It will be shown that $\tilde{\mathscr{L}}^{(q)}$ (for $q = 2$ or 3) is an affine Kac–Moody algebra, and because its construction involves a non-trivial rotation, it is said to be **twisted**. By contrast, $\tilde{\mathscr{L}}^{(1)}$ does not involve such a rotation, so it is natural to say that $\tilde{\mathscr{L}}^{(1)}$ is **untwisted**.

As the following analysis will show, there are *three series* of twisted affine Kac–Moody algebras corresponding to $q = 2$ and $\tilde{\mathscr{L}}^0 = A_{l^0}$ (l^0 even, $l^0 \geqslant 2$), $\tilde{\mathscr{L}}^0 = A_{l^0}$ (l^0 odd, $l^0 \geqslant 5$), and $\tilde{\mathscr{L}}^0 = D_{l^0}$ ($l^0 \geqslant 3$) respectively. (As noted in Chapter 13, Section 7, the algebras A_3 and D_3 are isomorphic. For convenience they are both included here in the D_{l^0} series.) The appropriate values of l^0 that give twisted algebras $\tilde{\mathscr{L}}^{(2)}$ whose generalized Cartan matrices are of dimension $(l+1) \times (l+1)$ are $l^0 = 2l$, $2l - 1$ and $l + 1$ for these three series. The resulting twisted affine Kac–Moody algebras are accordingly denoted by $A_{2l}^{(2)}$ ($l \geqslant 1$), $A_{2l-1}^{(2)}$ ($l \geqslant 3$) and $D_{l+1}^{(2)}$ ($l \geqslant 2$). There are also two special cases $E_6^{(2)}$ and $D_4^{(3)}$, which, as the notation suggests, correspond to $\tilde{\mathscr{L}}^0 = E_6$, $q = 2$ and to $\tilde{\mathscr{L}}^0 = D_4$, $q = 3$ respectively.

It is clearly necessary to examine the subspaces $\tilde{\mathscr{L}}_p^{0(q)}$ in more detail—both to establish that $\tilde{\mathscr{L}}^{(q)}$ is actually an affine Kac–Moody algebra and to determine its root structure.

The subspace $\tilde{\mathscr{L}}_0^{0(q)}$ is actually a Lie *subalgebra* of $\tilde{\mathscr{L}}^0$, because, if

$a, b \in \tilde{\mathscr{L}}_0^{0(q)}$ then, by (11.2) and (26.143),

$$\psi_\tau([a,b]_-) = [\psi_\tau(a), \psi_\tau(b)]_- = [a,b]_-,$$

so $[a,b]_- \in \tilde{\mathscr{L}}_0^{0(q)}$. Closer examination shows that $\tilde{\mathscr{L}}_0^{0(q)}$ is always a *simple* complex Lie algebra (provided that ψ_τ is an *outer* automorphism, as is being assumed here), and for $A_{2l}^{(2)}$, $A_{2l-1}^{(2)}$ and $D_{l+1}^{(2)}$ its rank is l, while for $E_6^{(2)}$ and $D_4^{(3)}$ its rank is 4 and 2 respectively. The Cartan subalgebra $\mathscr{H}^{0(q)}$ of $\tilde{\mathscr{L}}_0^{0(q)}$ consists of the elements that are common to $\mathscr{H}^0$ and $\tilde{\mathscr{L}}_0^{0(q)}$.

For $p \neq 0$ the subspace $\tilde{\mathscr{L}}_p^{0(q)}$ of $\tilde{\mathscr{L}}^0$ is not a subalgebra of $\tilde{\mathscr{L}}^0$, but is the carrier space of a representation of $\tilde{\mathscr{L}}_0^{0(q)}$. This follows because if $a \in \tilde{\mathscr{L}}_0^{0(q)}$ and $b \in \tilde{\mathscr{L}}_p^{0(q)}$ then (11.2) and (26.142) imply that $[a,b]_- \in \tilde{\mathscr{L}}_p^{0(q)}$, so a representation $\boldsymbol{\Gamma}^p$ of $\tilde{\mathscr{L}}_0^{0(q)}$ may be defined by

$$[a, a_{pr}^0]_- = \sum_{s=1}^{n_p} \Gamma^p(a)_{sr} a_{ps}^0, \quad r = 1, 2, \ldots, n_p, \tag{26.145}$$

for all $a \in \tilde{\mathscr{L}}_0^{0(q)}$, where n_p is the dimension of $\tilde{\mathscr{L}}_p^{0(q)}$ and a_{pr}^0 $(r = 1, \ldots, n_p)$ are the basis elements of $\tilde{\mathscr{L}}_p^{0(q)}$ (cf. (11.36)).

Let $\mathscr{H}^{(q)}$ be the subalgebra of $\tilde{\mathscr{L}}^{(q)}$ that consists of elements of the form $t^0 \otimes h^0$, where $h^0 \in \mathscr{H}^{0(q)}$, and linear combinations of these and of c and d. $\mathscr{H}^{(q)}$ will be identified with the Cartan subalgebra of $\tilde{\mathscr{L}}^{(q)}$. Its dimension is $l+2$ for $A_{2l}^{(2)}$, $A_{2l-1}^{(2)}$ and $D_{l+1}^{(2)}$ respectively, and is 6 and 4 for $E_6^{(2)}$ and $D_4^{(3)}$ respectively.

The following example provides both an explicit demonstration of the construction so far and an indication of the next stages.

Example I *Construction of the twisted affine Kac–Moody algebra* $A_2^{(2)}$

For $\tilde{\mathscr{L}}^0 = A_2$ (26.140) is

$$\tau(\alpha_1) = \alpha_2, \quad \tau(\alpha_2) = \alpha_1.$$

Then, by (14.32) and (14.33),

$$\psi_\tau(h_{\alpha_1}^0) = h_{\alpha_2}^0, \quad \psi_\tau(h_{\alpha_2}^0) = h_{\alpha_1}^0$$

and

$$\psi_\tau(e_{\alpha_1}^0) = e_{\alpha_2}^0, \quad \psi_\tau(e_{\alpha_2}^0) = e_{\alpha_1}^0, \quad \psi_\tau(e_{\alpha_1+\alpha_2}^0) = -e_{\alpha_1+\alpha_2}^0.$$

(Equation (14.33) shows that $\psi_\tau(e_\alpha^0) = \chi_\alpha e_{\tau(\alpha)}^0$, where $\chi_\alpha = 1$ for the simple roots and where (cf. (14.34))

$$\chi_{\alpha_1+\alpha_2} = \frac{N_{\tau(\alpha_1),\tau(\alpha_2)}^0}{N_{\alpha_1,\alpha_2}^0} \chi_{\alpha_1} \chi_{\alpha_2} = \frac{N_{\alpha_2,\alpha_1}^0}{N_{\alpha_1,\alpha_2}^0} = -1;$$

Similarly (since $\chi_{-\alpha} = \chi_\alpha$)

$$\psi_\tau(e_{-\alpha_1}^0) = e_{-\alpha_2}^0, \quad \psi_\tau(e_{-\alpha_2}^0) = e_{-\alpha_1}^0, \quad \psi_\tau(e_{-(\alpha_1+\alpha_2)}^0) = -e_{-(\alpha_1+\alpha_2)}^0.$$

Consequently $\tilde{\mathscr{L}}_0^{0(2)}$ has basis elements

$$h^0_{\alpha_1} + h^0_{\alpha_2}, \quad e^0_{\alpha_1} + e^0_{\alpha_2}, \quad e^0_{-\alpha_1} + e^0_{-\alpha_2},$$

whereas the subspace $\tilde{\mathscr{L}}_1^{0(2)}$ has basis elements

$$h^0_{\alpha_1} - h^0_{\alpha_2}, \quad e^0_{\alpha_1} - e^0_{\alpha_2}, \quad e^0_{-\alpha_1} - e^0_{-\alpha_2}, \quad e^0_{\alpha_1+\alpha_2}, \quad e^0_{-(\alpha_1+\alpha_2)}.$$

$\tilde{\mathscr{L}}_0^{0(2)}$ is isomorphic to the simple Lie algebra A_1.

Then, by (26.144), $A_2^{(2)}$ has basis elements

$$c, d;$$
$$t^j \otimes (h^0_{\alpha_1} + h^0_{\alpha_2}), \quad t^j \otimes (e^0_{\alpha_1} + e^0_{\alpha_2}), \quad t^j \otimes (e^0_{-\alpha_1} + e^0_{-\alpha_2}) \quad \text{for } j \text{ even;}$$

and

$$t^j \otimes (h^0_{\alpha_1} - h^0_{\alpha_2}), \quad t^j \otimes (e^0_{\alpha_1} - e^0_{\alpha_2}), \quad t^j \otimes (e^0_{-\alpha_1} - e^0_{-\alpha_2}),$$
$$t^j \otimes e^0_{\alpha_1+\alpha_2}, \quad t^j \otimes e^0_{-(\alpha_1+\alpha_2)} \quad \text{for } j \text{ odd.}$$

Clearly $\mathscr{H}^{(2)}$ has dimension 3, its basis elements being:

$$c, \quad d, \quad t^0 \otimes (h^0_{\alpha_1} + h^0_{\alpha_2}).$$

The most general element of $\mathscr{H}^{0(2)}$ has the form

$$h^0 = \mu(h^0_{\alpha_1} + h^0_{\alpha_2}),$$

where μ is an arbitrary complex number. Then, since

$$\alpha_1(h^0_{\alpha_1} + h^0_{\alpha_2}) = \alpha_2(h^0_{\alpha_1} + h^0_{\alpha_2}),$$

it follows that

$$\alpha_2(h^0) = \alpha_1(h^0) \tag{26.146}$$

for all $h^0 \in \mathscr{H}^{0(2)}$, and so, for all $h^0 \in \mathscr{H}^{0(2)}$

$$[h^0, h^0_{\alpha_1} + h^0_{\alpha_2}]_- = 0, \tag{26.147a}$$
$$[h^0, e^0_{\alpha_1} + e^0_{\alpha_2}]_- = \alpha_1(h^0)(e^0_{\alpha_1} + e^0_{\alpha_2}), \tag{26.147b}$$
$$[h^0, e^0_{-\alpha_1} + e^0_{-\alpha_2}]_- = -\alpha_1(h^0)(e^0_{-\alpha_1} + e^0_{-\alpha_2}), \tag{26.147c}$$

and

$$[h^0, h^0_{\alpha_1} - h^0_{\alpha_2}]_- = 0, \tag{26.148a}$$
$$[h^0, e^0_{\alpha_1} - e^0_{\alpha_2}]_- = \alpha_1(h^0)(e^0_{\alpha_1} - e^0_{\alpha_2}), \tag{26.148b}$$
$$[h^0, e^0_{-\alpha_1} - e^0_{-\alpha_2}]_- = -\alpha_1(h^0)(e^0_{-\alpha_1} - e^0_{-\alpha_2}), \tag{26.148c}$$
$$[h^0, e^0_{\alpha_1+\alpha_2}]_- = 2\alpha_1(h^0)e^0_{\alpha_1+\alpha_2}, \tag{26.148d}$$
$$[h^0, e^0_{-(\alpha_1+\alpha_2)}]_- = -2\alpha_1(h^0)e^0_{-(\alpha_1+\alpha_2)}. \tag{26.148e}$$

Thus (by 25.145)) the elements of $\tilde{\mathscr{L}}_1^{0(2)}$ provide a basis for the carrier space of a representation of $\tilde{\mathscr{L}}_0^{0(2)}$ whose highest weight is $2\alpha_1$. It then follows from

(26.97), (26.99), (26.147) and (26.148) that for all $h \in \mathscr{H}^{(2)}$

$$[h, t^j \otimes (h^0_{\alpha_1} + h^0_{\alpha_2})]_- = j\delta(h)\, t^j \otimes (h^0_{\alpha_1} + h^0_{\alpha_2}),$$
$$[h, t^j \otimes (e^0_{\alpha_1} + e^0_{\alpha_2})]_- = (j\delta + \alpha_1)(h)\, t^j \otimes (e^0_{\alpha_1} + e^0_{\alpha_2}),$$
$$[h, t^j \otimes (e^0_{-\alpha_1} + e^0_{-\alpha_2})]_- = (j\delta - \alpha_1)(h)\, t^j \otimes (e^0_{-\alpha_1} + e^0_{-\alpha_2}),$$

and

$$[h, t^j \otimes (h^0_{\alpha_1} - h^0_{\alpha_2})]_- = j\delta(h)\, t^j \otimes (h^0_{\alpha_1} - h^0_{\alpha_2}),$$
$$[h, t^j \otimes (e^0_{\alpha_1} - e^0_{\alpha_2})]_- = (j\delta + \alpha_1)(h)\, t^j \otimes (e^0_{\alpha_1} - e^0_{\alpha_2}),$$
$$[h, t^j \otimes (e^0_{-\alpha_1} - e^0_{-\alpha_2})]_- = (j\delta - \alpha_1)(h)\, t^j \otimes (e^0_{-\alpha_1} - e^0_{-\alpha_2}),$$
$$[h, t^j \otimes e^0_{\alpha_1+\alpha_2}]_- = (j\delta + 2\alpha_1)(h)\, t^j \otimes e^0_{\alpha_1+\alpha_2},$$
$$[h, t^j \otimes e^0_{-(\alpha_1+\alpha_2)}]_- = (j\delta - 2\alpha_1)(h)\, t^j \otimes e^0_{-(\alpha_1+\alpha_2)}.$$

Consequently the roots of $A_2^{(2)}$ are

(i) α_1 and $-\alpha_1$,

(ii) $j\delta, j\delta + \alpha_1$ and $j\delta - \alpha_1$ for all non-zero even j, and

(iii) $j\delta, j\delta + \alpha_1, j\delta - \alpha_1, j\delta + 2\alpha_1$ and $j\delta - 2\alpha_1$ for all odd values of j.

It is worth noting the different roles played by the simple roots α_1 and α_2 of $\tilde{\mathscr{L}}^0 = A_2$. Both are extended to become linear functionals on $\mathscr{H}^{(1)}$, but they coincide on $\mathscr{H}^{(2)}$. Consequently only one of these (chosen above to be α_1) appears in the list of roots of $\tilde{\mathscr{L}}^{(2)} = A_2^{(2)}$ (although α_2 still labels the basis vectors of $A_2^{(2)}$).

The analysis of Example I can easily be generalized to all the other cases. Let $\Delta_0^{0(q)}$ be the set of non-zero roots of $\tilde{\mathscr{L}}_0^{0(q)}$, and let $\Delta_p^{0(q)}$ be the set of non-zero weights of the representation Γ^p of $\tilde{\mathscr{L}}_0^{0(q)}$ that is defined by (26.145) for $p = 1, \ldots, q-1$. For $\tilde{\mathscr{L}}^0 = A_2$ (26.147) shows that $\Delta_0^{0(2)} = \{\alpha_1, -\alpha_1\}$ and (26.148) shows that $\Delta_1^{0(q)} = \{\alpha_1, -\alpha_1, 2\alpha_1, -2\alpha_1\}$. Let $\Pi^{(q)}$ be the set of simple roots of $\tilde{\mathscr{L}}_0^{0(q)}$. Then $\Pi^{(q)}$ may be taken to be the set of simple roots of $\tilde{\mathscr{L}}^0$ that remain distinct when considered as linear functionals on $\mathscr{H}^{0(q)}$. (Equivalently they are the set of extensions of the simple roots of $\tilde{\mathscr{L}}^0$ that are distinct when $\mathscr{H}^{(1)}$ is restricted to $\mathscr{H}^{(2)}$.) For $\tilde{\mathscr{L}}^0 = A_2$ Example I shows that $\Pi^{(q)} = \{\alpha_1\}$.

Every element α of $\Delta_p^{0(q)}$ for $p = 0, \ldots, q-1$ can be written in the form

$$\alpha = \sum_k \kappa_k^\alpha \alpha_k, \tag{26.149}$$

where the sum is over the distinct roots of $\Pi^{(q)}$ and where the coefficients κ_k^α are all integers of the same sign. Every such element of $\Delta_0^{0(q)}$, being a linear functional on $\mathscr{H}^{0(q)}$, extends to become a linear functional on $\mathscr{H}^{(q)}$.

As the set of roots of all the simple Lie algebras are known (cf. Appendix F), $\Delta_0^{0(q)}$ is determined completely once $\tilde{\mathscr{L}}_0^{0(q)}$ is identified (which can be done by evaluating the Cartan matrix of $\tilde{\mathscr{L}}_0^{0(q)}$). For $q=2$ the set $\Delta_1^{0(q)}$ can be found most simply by writing down the non-zero roots of $\tilde{\mathscr{L}}^0$ and reexpressing them in the form (26.149); that is, in terms of the simple roots of $\tilde{\mathscr{L}}_0^{0(q)}$. (For example, for $\tilde{\mathscr{L}}^0 = A_{2l}$ (26.140) implies that the generalization of (26.146) is

$$\alpha_{2l+1-k}(h^0) = \alpha_k(h^0)$$

for all $h^0 \in \mathscr{H}^{0(q)}$ and for $k = 1, 2, \ldots, l$, which expresses the roots α_{l+1}, $\alpha_{l+2}, \ldots, \alpha_{2l}$ of $\tilde{\mathscr{L}}^0$ (when regarded as linear functionals on $\mathscr{H}^{0(q)}$) in terms of the simple roots $\alpha_1, \alpha_2, \ldots, \alpha_l$ of $\tilde{\mathscr{L}}_0^{0(q)}$. Then, for instance, the highest root

$$\sum_{j=1}^{2l} \alpha_j$$

of $\tilde{\mathscr{L}}^0$ reduces on $\mathscr{H}^{0(q)}$ to

$$2 \sum_{j=1}^{l} \alpha_j.$$

On removing from this set the roots of $\tilde{\mathscr{L}}_0^{0(q)}$ (each of which has multiplicity 1), the set that is left is $\Delta_1^{0(q)}$. For $q=3$ this procedure requires only a slight modification. There is only one case, namely $D_4^{(3)}$, for which detailed examination shows that the basis vectors of $\tilde{\mathscr{L}}_1^{0(3)}$ and $\tilde{\mathscr{L}}_2^{0(3)}$ may be chosen so that $\mathbf{\Gamma}^1$ and $\mathbf{\Gamma}^2$ of $\tilde{\mathscr{L}}_0^{0(3)}$ are identical. Consequently $\Delta_1^{0(3)} = \Delta_2^{0(3)}$, and both sets can be found together by the above procedure. In all cases the representation $\mathbf{\Gamma}^1$ of $\tilde{\mathscr{L}}_0^{0(q)}$ is irreducible. Let λ_{H} be its highest weight.

It is clear from this analysis that the set of non-zero roots of $\tilde{\mathscr{L}}^{(q)}$ consists of two subsets:

(i) $j\delta$, for all non-zero integers j; and

(ii) $j\delta + \alpha$, for all j such that $j \bmod q = p$ and all α that are extensions of elements of $\Delta_p^{0(q)}$ for each $p = 0, \ldots, q-1$.

All of these can be cast in the form (26.111) if α_0 is defined for all $h \in \mathscr{H}^{(q)}$ by

$$\alpha_0(h) = \delta(h) - \lambda_{\mathrm{H}}(h), \tag{26.150}$$

where λ_{H} is the extension of $\mathscr{H}^{(q)}$ of the highest weight on $\mathscr{H}^{0(q)}$. The *positive* roots of $\tilde{\mathscr{L}}^{(q)}$ then consist of three subsets:

(i) extensions of the *positive* roots α of $\tilde{\mathscr{L}}_0^{0(q)}$;

(ii) $j\delta$, for positive integers j; and

(iii) $j\delta+\alpha$, for all positive integers j such that $j \bmod q = p$ and *all* α that are extensions of elements of $\Delta_p^{0(q)}$ for each $p=0,\ldots,q-1$.

The most general invariant symmetric bilinear form $B^{(q)}(\ ,\ )$ for the twisted algebra $\tilde{\mathscr{L}}^{(q)}$ is such that

$$B^{(q)}(a,b)=\mu^{(q)}B^{(1)}(a,b) \tag{26.151}$$

for all $a, b\in\tilde{\mathscr{L}}^{(q)}$, where $B^{(1)}(a,b)$ is the corresponding form for the untwisted algebra $\tilde{\mathscr{L}}^{(1)}$ (of which $\tilde{\mathscr{L}}^{(q)}$ is a subalgebra (cf. (26.101)), and where $\mu^{(q)}$ is any constant. A natural choice for $\mu^{(q)}$ is that which ensures that $B^{(q)}(a,b)$ coincides with the Killing form of the simple Lie algebra $\tilde{\mathscr{L}}_0^{0(q)}$ when a and b are restricted to the subalgebra that is isomorphic to $\tilde{\mathscr{L}}_0^{0(q)}$. (This subalgebra is the set of elements $t^0\otimes a^0$, where $a^0\in\tilde{\mathscr{L}}_0^{0(q)}$.) By analogy with (26.42), for any linear functional α defined on $\mathscr{H}^{(q)}$ an element $h_\alpha^{(q)}$ may be defined by

$$B^{(q)}(h,h_\alpha^{(q)})=\alpha(h) \tag{26.152}$$

for all $h\in\mathscr{H}^{(q)}$, and, for any two such functionals α and β, $\langle\alpha,\beta\rangle^{(q)}$ may be defined (cf. (26.45)) by

$$\langle\alpha,\beta\rangle^{(q)}=B^{(q)}(h_\alpha^{(q)},h_\beta^{(q)}). \tag{26.153}$$

Then if α and β are extensions of linear functionals defined on $\mathscr{H}^{0(q)}$,

$$\langle\alpha,\beta\rangle^{(q)}=\langle\alpha,\beta\rangle^{0(q)}, \tag{26.154}$$

where the latter quantity is defined with respect to the Killing form of $\tilde{\mathscr{L}}_0^{0(q)}$.

In particular, (26.152) implies that

$$h_\delta^{(q)}=\frac{c}{\mu^{(q)}},$$

and hence, by (26.153), (26.151) and (26.101),

$$\langle\delta,\alpha_k\rangle^{(q)}=0 \quad \text{for } k=1,2,\ldots,l, \tag{26.155}$$

and

$$\langle\delta,\delta\rangle^{(q)}=0.$$

Consequently

$$\langle j\delta,j\delta\rangle^{(q)}=0$$

but

$$\langle j\delta+\alpha,j\delta+\alpha\rangle^{(q)}=\langle\alpha,\alpha\rangle^{(q)}=\langle\alpha,\alpha\rangle^{0(q)}>0$$

for every α that is an extension of a non-zero linear functional on $\mathscr{H}^{0(q)}$. Consequently the roots $j\delta$ of $\tilde{\mathscr{L}}^{(q)}$ are imaginary, but the roots $j\delta+\alpha$ are all real (for $\alpha\neq 0$).

As the root $j\delta$ corresponds to basis elements of $\tilde{\mathscr{L}}^{(q)}$ of the form $t^j\otimes h^0$, where $h^0\in\mathscr{H}^0$, dim $\tilde{\mathscr{L}}_{j\delta}^{(q)}$ for j such that $j \bmod q=p$ is equal to the number of linearly independent elements of $\mathscr{H}^0$ in $\tilde{\mathscr{L}}_p^{0(q)}$. For $p=0$ this is equal to l, the rank of $\tilde{\mathscr{L}}_0^{0(q)}$. For $q=2$ and $p=1$ this set consists of the remaining l^0-l elements of $\mathscr{H}^0$, and for $q=3$ these l^0-l elements are equally shared between $\tilde{\mathscr{L}}_1^{0(q)}$ and $\tilde{\mathscr{L}}_2^{0(q)}$. Thus

$$\dim \tilde{\mathscr{L}}_{j\delta}^{(q)} = \begin{cases} l & \text{if } j \bmod q = 0, \\ \dfrac{l^0 - l}{q-1} & \text{if } j \bmod q = p, p = 1, \ldots, q-1. \end{cases} \tag{26.156}$$

It may be verified easily that for $\alpha \neq 0$

$$\dim \tilde{\mathscr{L}}_{j\delta+\alpha}^{(q)} = 1,$$

in accordance with Theorem VIII of Section 3.

The case of $A_4^{(2)}$ will now be treated in detail as an example. Details of the other twisted affine Kac–Moody algebras may be found in Appendix N.

Example II *Bilinear form and generalized Cartan matrix for* $A_4^{(2)}$

In this case $\tilde{\mathscr{L}}^0 = A_4$, and $\tilde{\mathscr{L}}_0^{0(2)}$ has 2 simple roots α_1 and α_2, for which, by (26.101), (26.151) and (26.152),

$$h_{\alpha_k}^{(2)} = \frac{1}{2\mu^{(2)}} t^0 \otimes (h_{\alpha_k}^0 + h_{\alpha_{5-k}}^0), \quad k = 1, 2.$$

Then, by (26.105), (26.151), (26.153) and Appendix F, Section 1, part (*e*),

$$\langle \alpha_1, \alpha_1 \rangle^{(2)} = \frac{1}{4\mu^{(2)}} \langle \alpha_1 + \alpha_4, \alpha_1 + \alpha_4 \rangle^0 = \frac{1}{10\mu^{(2)}}, \tag{26.157a}$$

$$\langle \alpha_2, \alpha_2 \rangle^{(2)} = \frac{1}{4\mu^{(2)}} \langle \alpha_2 + \alpha_3, \alpha_2 + \alpha_3 \rangle^0 = \frac{1}{20\mu^{(2)}}, \tag{26.157b}$$

$$\langle \alpha_1, \alpha_2 \rangle^{(2)} = \frac{1}{4\mu^{(2)}} \langle \alpha_1 + \alpha_4, \alpha_2 + \alpha_3 \rangle^0 = -\frac{1}{20\mu^{(2)}}, \tag{26.157c}$$

where $\langle \alpha, \beta \rangle^0$ indicates that the quantity is evaluated with respect to the Killing form of $\tilde{\mathscr{L}}^0 = A_4$. Thus if $\mathbf{A}^{(2)}$ denotes the generalized Cartan matrix of $\tilde{\mathscr{L}}^{(2)} = A_4^{(2)}$ then, by (26.50),

$$A_{12}^{(2)} = -1, \quad A_{21}^{(2)} = -2,$$

so the submatrix of $\mathbf{A}^{(2)}$ obtaining by removing the 0th row and column coincides with the Cartan matrix for B_2. Thus $\tilde{\mathscr{L}}_0^{0(2)} = B_2$. For B_2 the quantity $\langle \alpha_1, \alpha_1 \rangle$ has the value $\frac{1}{3}$ (cf. Appendix F, Section 2, part (*e*)), so the choice

(26.154) for $\mu^{(2)}$ implies

$$\mu^{(2)} = \frac{3}{10}. \tag{26.158}$$

Also, $\lambda_H = 2\alpha_1 + 2\alpha_2$, so $\alpha_0 = \delta - 2\alpha_1 - 2\alpha_2$. Hence, by (26.155), (26.157) and (26.158),

$$\langle \alpha_0, \alpha_1 \rangle^{(2)} = -2\langle \alpha_1 + \alpha_2, \alpha_1 \rangle^{(2)} = -\frac{1}{10\mu^{(2)}} = -\frac{1}{3},$$

$$\langle \alpha_0, \alpha_2 \rangle^{(2)} = -2\langle \alpha_1 + \alpha_2, \alpha_2 \rangle^{(2)} = 0,$$

$$\langle \alpha_0, \alpha_0 \rangle^{(2)} = 4\langle \alpha_1 + \alpha_2, \alpha_1 + \alpha_2 \rangle^{(2)} = \frac{1}{5\mu^{(2)}} = \frac{2}{3}.$$

Hence, by (26.50),

$$A_{01}^{(2)} = -1, \quad A_{02}^{(2)} = 0, \quad A_{10}^{(2)} = -2, \quad A_{20}^{(2)} = 0,$$

and so

$$\mathbf{A}^{(2)} = \begin{bmatrix} 2 & -1 & 0 \\ -2 & 2 & -1 \\ 0 & -2 & 2 \end{bmatrix}.$$

With this general construction, the submatrix of the generalized Cartan matrix for $\tilde{\mathscr{L}}^{(q)}$ obtained by removing the 0th row and column coincides with the Cartan matrix for $\tilde{\mathscr{L}}_0^{0(q)}$, so the simple Lie algebra $\tilde{\mathscr{L}}^s$ of subsection (*a*) may be identified with $\tilde{\mathscr{L}}_0^{0(q)}$ in the twisted case.

For convenience the main results on roots are gathered together in the following theorem.

Theorem II The twisted affine Kac–Moody algebra $\tilde{\mathscr{L}}^{(q)}$ has the following positive real roots:

(i) extensions of the positive roots of $\tilde{\mathscr{L}}_0^{0(q)}$; and

(ii) the roots $j\delta + \alpha$, where α is any member of $\Delta_p^{0(q)}$ and j is a positive integer such that $j \bmod q = p$ (for each $p = 0, \ldots, q-1$).

(The corresponding root spaces for all of these are one-dimensional.) The set of positive imaginary roots of $\tilde{\mathscr{L}}^{(q)}$ consists of $j\delta$, where j is any positive integer, $\dim \tilde{\mathscr{L}}_{j\delta}^{(q)}$ being given by (26.156).

The commutation relations for a twisted affine Kac–Moody algebra can be cast in a form similar to that given by (26.136)–(26.138) for the untwisted case. To this end, let a^0_{pr} $(r = 1, 2, \ldots, n_p)$ be a set of basis elements of $\tilde{\mathscr{L}}^{0(q)}_p$. Then part (*d*) of Theorem I of Chapter 13, Section 2 and (24.142) together imply that

$$B^0(a^0_{pr}, a^0_{p'r'}) = 0 \quad \text{if } (p + p') \bmod q = 0. \tag{26.159}$$

Moreover, the basis elements may be chosen so that when $(p + p') \bmod q = 0$

$$B^0(a^0_{pr}, a^0_{p's}) = -\delta_{rs} \tag{26.160}$$

for $r = 1, 2, \ldots, n_{p'}$. Also, since (26.142) and (11.2) imply that $[a^0_{pr}, a^0_{p'r'}]_- \in \mathscr{L}^{0(q)}_{p''}$, where $p'' = (p + p') \bmod q$,

$$[a^0_{pr}, a^0_{p's}]_- = \sum_{t=1}^{n_{p''}} c^{p''t}_{pr,p's} a^0_{p''t}, \tag{26.161}$$

where the basis elements may be chosen so that the coefficients in (26.161) are all real. Indeed the coefficients $c^{0t}_{0r,0s}$ are structure constants of the compact form of $\tilde{\mathscr{L}}^{0(q)}_0$ and can be taken to be antisymmetric with respect to the interchange of *any* pair of indices (cf. Chapter 14, Section 2). Also, by (26.145), $c^{pt}_{0r,ps} = \Gamma^p(a^0_{0r})_{st}$ for $p \neq 0$.

Defining T^{pr}_j by analogy with (26.134) by

$$T^{pr}_j = it^j \otimes a^0_{pr}, \tag{26.162}$$

where necessarily

$$j \bmod q = p, \tag{26.163}$$

(26.89) and (26.159)–(26.161) show that

$$[T^{pr}_j, T^{p's}_k]_- = i \sum_{t=1}^{n_{p''}} c^{p''t}_{pr,p's} T^{p''t}_{j+k} + j\delta_{j+k,0}\delta^{rs}\delta^{p''}c \tag{26.164}$$

for $p, p' = 0, \ldots, q-1$; $r = 1, 2, \ldots, n_p$; $s = 1, 2, \ldots, n_{p'}$ and $j, k = 0, \pm 1, \pm 2, \ldots$ (subject to the constraints that $j \bmod q = p$ and $k \bmod q = p'$). Also $p'' = (p + p') \bmod q$. The analogues of (26.137) and (26.138) are

$$[c, T^{pr}_j]_- = 0 \tag{26.165}$$

and

$$[d, T^{pr}_j]_- = jT^{pr}_j \tag{26.166}$$

for $j \bmod q = p$ and $r = 1, 2, \ldots, n_p$.

A very interesting situation arises if in the above construction the *outer* automorphism ψ_τ of $\tilde{\mathscr{L}}^0$ is replaced by an *inner* automorphism ψ of $\tilde{\mathscr{L}}^0$. Assuming still that $\psi^q = 1$, the analysis leading to the subalgebra $\tilde{\mathscr{L}}^{(q)}$ of (26.144) remains valid. Indeed, this construction is allowed for *any* complex simple Lie algebra $\tilde{\mathscr{L}}^0$ (and not just the ones possessing outer automorphisms).

It remains true that $\tilde{\mathscr{L}}_0^{0(q)}$ is a subalgebra of $\tilde{\mathscr{L}}^0$ and that $\tilde{\mathscr{L}}_p^{0(q)}$ for $p=1,\ldots,q-1$ are carrier spaces for representations of $\tilde{\mathscr{L}}_0^{0(q)}$, but $\tilde{\mathscr{L}}_0^{0(q)}$ might not be simple.

It is very remarkable that when ψ is an *inner* automorphism *$\tilde{\mathscr{L}}^{(q)}$ is actually isomorphic to $\tilde{\mathscr{L}}^{(1)}$ itself, even though $\tilde{\mathscr{L}}^{(q)}$ is a* **proper** *subalgebra of $\tilde{\mathscr{L}}^{(1)}$ when* $q>1$. This can be verified easily for the case in which ψ is a chief inner automorphism of $\tilde{\mathscr{L}}^0$, where $\psi=\exp\{\mathrm{ad}(h_\psi^0)\}$ for some $h_\psi^0\in\mathscr{H}^0$ (cf. Theorem III of Chapter 14, Section 4). In this case $\psi(h^0)=h^0$ for all $h^0\in\mathscr{H}^0$ and $\psi(e_\alpha^0)=\exp\{\alpha(h_\psi^0)\}\,e_\alpha^0$ for each $\alpha\in\Delta^0$. Thus $\psi^q=1$ if $q\alpha(h_\psi^0)=2\pi i r_\alpha$ for every $\alpha\in\Delta^0$, where r_α is an integer that depends on α. Let $r_\alpha=s_\alpha q+p_\alpha$, where $p_\alpha=0,1,\ldots,q-1$ and s_α is some integer, so that $q\alpha(h_\psi^0)/2\pi i=s_\alpha q+p_\alpha$ and $\exp\{\alpha(h_\psi^0)\}=\exp\{2\pi i p_\alpha/q\}$. Thus $e_\alpha\in\tilde{\mathscr{L}}_{p_\alpha}^{0(q)}$. By (26.144), the corresponding elements of $\tilde{\mathscr{L}}^{(q)}$ are $t^j\otimes e_\alpha^0$, where $j \bmod q=p_\alpha$; that is, where $j=rq+p_\alpha$, with r running over all the integers. Thus $j=(r-s_\alpha)q+q\alpha(h_\psi^0)/2\pi i$, so the allowed values of j are given by $j=q\{s+\alpha(h_\psi^0)/2\pi i\}$, where s runs over all the integers, and so the corresponding elements of $\tilde{\mathscr{L}}^{(q)}$ are

$$t^{q[s+\{\alpha(h_\psi^0)/2\pi i\}]}\otimes e_\alpha^0,$$

with $s=0,\pm1,\pm2,\ldots$. The basis elements h_α^0 of $\mathscr{H}^0$ with $\alpha=\alpha_k$ and $k=1,2,\ldots,l^0$ are all members of $\tilde{\mathscr{L}}_0^{0(q)}$, and so by (26.144) they contribute to $\tilde{\mathscr{L}}^{(q)}$ a set of elements of the form

$$t^{qs}\otimes h_{\alpha_k}^0 \quad \text{for } s=0,\pm1,\pm2,\ldots \;.$$

Consider the mapping ϕ of $\tilde{\mathscr{L}}^{(q)}$ into $\tilde{\mathscr{L}}^{(1)}$ defined by

$$\phi(c)=c/q, \tag{26.167a}$$

$$\phi(d)=qd+\frac{q}{2\pi i}\,t^0\otimes h_\psi^0, \tag{26.167b}$$

$$\phi(t^{q[s+\{\alpha(h_\psi^0)/2\pi i\}]}\otimes e_\alpha^0)=t^s\otimes e_\alpha^0, \tag{26.167c}$$

$$\phi(t^{qs}\otimes h_{\alpha_k}^0)=\begin{cases} t^s\otimes h_{\alpha_k}^0 & \text{if } s\neq 0,\\ t^0\otimes h_{\alpha_k}^0-\dfrac{\alpha(h_\psi^0)}{2pi}c & \text{if } s=0,\end{cases} \tag{26.167d}$$

where $s=0,\pm1,\pm2,\ldots$; $\alpha\in\Delta^0$ and $k=1,2,\ldots,l^0$. It is easily verified using (26.89)–(26.92) that ϕ is a homomorphic mapping. Moreover, it is clearly one-to-one, and *every* element of $\tilde{\mathscr{L}}^{(1)}$ is the image of some element of $\tilde{\mathscr{L}}^{(q)}$. Consequently ϕ is an *isomorphic mapping* of *$\tilde{\mathscr{L}}^{(q)}$ into $\tilde{\mathscr{L}}^{(1)}$*.

As *every* inner automorphism of $\tilde{\mathscr{L}}^0$ is conjugate to such a chief inner automorphism (cf. Theorem I of Chapter 14, Section 4), the above proof could be extended to *any* inner automorphism of $\tilde{\mathscr{L}}^0$.

The conclusion of this argument is that the subalgebra $\tilde{\mathscr{L}}^{(q)}$ of $\tilde{\mathscr{L}}^{(1)}$ is *not* isomorphic to $\tilde{\mathscr{L}}^{(1)}$ *only* when the twisting is performed by an *outer* automorphism of $\tilde{\mathscr{L}}^0$. Consequently the description "twisted affine Kac–Moody algebra" will be reserved *solely* for such an $\tilde{\mathscr{L}}^{(q)}$.

Both types of $\tilde{\mathscr{L}}^{(q)}$ appear in connection with "twisted" string theories (cf. Nepomechie 1986).

(*d*) *Root lattices and the Weyl group*

Returning to the general affine case, let $\tilde{\mathscr{L}}$ be any affine Kac–Moody algebra with Cartan subalgebra $\mathscr{H}$. As its dual $\mathscr{H}^*$ has the same dimension $l+2$ as $\mathscr{H}$, the set of $l+1$ linear functionals $\alpha_0, \alpha_1, \ldots, \alpha_l$ does not form a basis for $\mathscr{H}^*$ on its own. To complete this basis, let Λ_0 be the linear functional defined in $\mathscr{H}$ by

$$\Lambda_0(H_{\alpha_k}) = \begin{cases} 1 & \text{if } k=0, \\ 0 & \text{if } k=1,2,\ldots,l, \end{cases} \tag{26.168}$$

and

$$\Lambda_0(d) = 0. \tag{26.169}$$

Then (26.42), (26.71) and (26.72) imply that the corresponding element of $\mathscr{H}$ is given by

$$h_{\Lambda_0} = \tfrac{1}{2}\langle \alpha_0, \alpha_0 \rangle d. \tag{26.170}$$

Thus, by (26.45) and (26.70),

$$\langle \Lambda_0, \alpha_k \rangle = \begin{cases} \tfrac{1}{2}\langle \alpha_0, \alpha_0 \rangle & \text{if } k=0 \\ 0 & \text{if } k=1,2,\ldots,l, \end{cases} \tag{26.171}$$

and hence, by (26.80) and (26.79),

$$\langle \Lambda_0, \delta \rangle = \tfrac{1}{2}\langle \alpha_0, \alpha_0 \rangle. \tag{26.172}$$

Also, by (26.45) and (26.170),

$$\langle \Lambda_0, \Lambda_0 \rangle = 0. \tag{26.173}$$

Definition *Root lattice* Q *of* $\tilde{\mathscr{L}}$

The *root lattice* Q of the affine Kac–Moody algebra $\tilde{\mathscr{L}}$ consists of *all* linear functionals on $\mathscr{H}$ that have the form

$$\alpha = \sum_{j=0}^{l} \kappa_j^\alpha \alpha_j, \tag{26.174}$$

where each κ_j^α $(j = 0, 1, \ldots, l)$ is allowed to take *any* integer value and $\alpha_0, \alpha_1, \ldots, \alpha_l$ are the simple roots of $\tilde{\mathscr{L}}$.

If α is a root of $\tilde{\mathscr{L}}$ then $\alpha \in \mathrm{Q}$ (by (26.32)), but the converse is not necessarily true. The corresponding concept may also be defined for the simple subalgebra $\tilde{\mathscr{L}}^{\mathrm{s}}$ of $\tilde{\mathscr{L}}$ that was defined in subsection (*a*).

Definition *Root lattice* Q^{s} *of* $\tilde{\mathscr{L}}^{\mathrm{s}}$

The *root lattice* Q^{s} of the simple Lie algebra $\tilde{\mathscr{L}}^{\mathrm{s}}$ consists of *all* linear functionals defined on $\mathscr{H}^{\mathrm{s}}$ that have the form

$$\alpha = \sum_{j=1}^{l} \kappa_j^\alpha \alpha_j, \tag{26.175}$$

where each κ_j^α $(j = 1, 2, \ldots, l)$ is allowed to take *any* integer value (and $\alpha_1, \alpha_2, \ldots, \alpha_l$ are the simple roots of $\tilde{\mathscr{L}}^{\mathrm{s}}$).

Clearly Q^{s} is a sublattice of Q.

Definition *Scaled root* α^{V}

For any real root α of $\tilde{\mathscr{L}}$ the corresponding *scaled root* α^{V} may be defined by

$$\alpha^{\mathrm{V}} = \frac{2}{\langle \alpha, \alpha \rangle} \alpha. \tag{26.176}$$

Definition *Scaled root lattice* Q^{V} *of* $\tilde{\mathscr{L}}$

The *scaled root lattice* Q^{V} of the affine Kac–Moody algebra $\tilde{\mathscr{L}}$ consists of *all* linear functionals defined on $\mathscr{H}$ that have the form

$$\omega = \sum_{j=0}^{l} \mu_j^\omega \alpha_j^{\mathrm{V}}, \tag{26.177}$$

where each μ_j^ω $(j = 0, 1, \ldots, l)$ is allowed to take *any* integral value.

(The "scaled roots" and "scaled root lattice" are sometimes referred to as the "dual roots" and "dual root lattice" respectively (cf. Frenkel and Kac 1980), but this latter terminology will be avoided here in order to avoid confusion with a rather different definition of a dual root lattice that will be introduced in Chapter 29, Section 5(*a*).)

For every real $\alpha \in \mathrm{Q}$ the scaled root α^{V} is a member of Q^{V}, but not every element of Q^{V} can be obtained by "scaling" some element of Q. Moreover, in general

$$(\alpha + \beta)^{\vee} \neq \alpha^{\vee} + \beta^{\vee}. \tag{26.178}$$

These points are illustrated in the following example.

Example III *Scaled roots of $C_2^{(1)}$*

As shown in Appendix N, Section 4, for $C_2^{(1)}$, $\langle \alpha_0, \alpha_0 \rangle = \langle \alpha_2, \alpha_2 \rangle = \frac{1}{3}$ but $\langle \alpha_1, \alpha_1 \rangle = \frac{1}{6}$, so that $C_2^{(1)}$ is not simply laced. Then, by (26.175),

$$\alpha_0^{\vee} = 6\alpha_0, \quad \alpha_1^{\vee} = 12\alpha_1, \quad \alpha_2^{\vee} = 6\alpha_2.$$

Also, since $\langle \alpha_0, \alpha_1 \rangle = -\frac{1}{6}$, it follows that $\langle \alpha_0 + \alpha_1, \alpha_0 + \alpha_1 \rangle = \frac{1}{6}$, and hence $(\alpha_0 + \alpha_1)^{\vee} = 12(\alpha_0 + \alpha_1)$. Then

$$(\alpha_0 + \alpha_1)^{\vee} = 2\alpha_0^{\vee} + \alpha_1^{\vee} \neq \alpha_0^{\vee} + \alpha_1^{\vee},$$

confirming (26.178), and clearly $\alpha_0^{\vee} + \alpha_1^{\vee}$ cannot be written in the form $\alpha^{\vee}$ for any α of Q.

It will be obvious that $(\alpha + \beta)^{\vee} = \alpha^{\vee} + \beta^{\vee}$ for all $\alpha, \beta \in$ Q if and only if $\tilde{\mathscr{L}}$ is simply laced, for in this case (since $\langle \alpha_j, \alpha_j \rangle = \langle \alpha_0, \alpha_0 \rangle$ for $j = 1, 2, \ldots, l$),

$$\alpha^{\vee} = \frac{2}{\langle \alpha_0, \alpha_0 \rangle} \alpha \tag{26.179}$$

for all $\alpha \in$ Q. It should be noted that with these definitions the elements of the root lattice Q *and* of the scaled root lattice $Q^{\vee}$ are all members of $\mathscr{H}^*$, the dual space of $\mathscr{H}$.

Definition *Scaled root lattice $Q^{s\vee}$ of $\tilde{\mathscr{L}}^s$*

The *scaled root lattice* $Q^{s\vee}$ of $\tilde{\mathscr{L}}^s$ consists of *all* linear functions defined on $\mathscr{H}^s$ that have the form

$$\omega = \sum_{j=1}^{l} \mu_j^{\omega} \alpha_j^{\vee}, \tag{26.180}$$

where each μ_j^{ω} $(j = 1, 2, \ldots, l)$ is allowed to take *any* integral value.

The comments concerning the relationship of Q to $Q^{\vee}$ apply equally to Q^s and $Q^{s\vee}$.

The structure of the Weyl group $\mathscr{W}$ of the complex affine Kac–Moody algebra $\tilde{\mathscr{L}}$ can now be established.

Definition *Translation operator T_ω*

For each $\omega \in Q^{sV}$ a linear operator T_ω acting on $\mathscr{H}^*$ may be defined by

$$T_\omega(\lambda) = \lambda + \langle \lambda, \delta \rangle \omega - (\langle \lambda, \omega \rangle + \tfrac{1}{2}\langle \omega, \omega \rangle \langle \lambda, \delta \rangle)\delta \qquad (26.181)$$

for every linear functional λ defined on $\mathscr{H}$. (Here δ is the imaginary root of (26.80).)

In particular, if $\omega = \alpha^V$ for some $\alpha \in Q^s$ then (26.176) and (26.181) together imply that

$$T_{\alpha^V}(\lambda) = \lambda + \frac{2\langle \lambda, \delta \rangle}{\langle \alpha, \alpha \rangle}\alpha - \frac{2\langle \lambda, \alpha + \delta \rangle}{\langle \alpha, \alpha \rangle}\delta. \qquad (26.182)$$

If α is a *root* of $\tilde{\mathscr{L}}$ then (26.86) shows that $\langle \alpha, \delta \rangle = 0$, and so putting $\lambda = \alpha$ in (26.181) gives

$$T_\omega(\alpha) = \alpha - \langle \alpha, \omega \rangle \delta. \qquad (26.183)$$

Similarly

$$T_\omega(\delta) = \delta. \qquad (26.184)$$

Using these results, it is easily verified by evaluating $T_\omega(T_{\omega'}(\lambda))$ and $T_{\omega+\omega'}(\lambda)$ for all $\lambda \in \mathscr{H}^*$ that

$$T_\omega T_{\omega'} = T_{\omega+\omega'} \qquad (26.185)$$

for all ω and ω' of Q^{sV}, a property that suggests the description *translation* for T_ω.

Direct evaluation using (26.54) and (26.182) also shows that

$$T_{\alpha^V} = S_{\delta-\alpha} S_\alpha \qquad (26.186)$$

for any root of $\tilde{\mathscr{L}}^s$ for which $\delta - \alpha$ is a real root of $\tilde{\mathscr{L}}$. In particular, since $\delta - \alpha_j$ is a real root of $\tilde{\mathscr{L}}$ for $j = 1, 2, \ldots, l$,

$$T_{\alpha_j^V} = S_{\delta-\alpha_j} S_{\alpha_j}, \qquad (26.187)$$

showing that T_ω with $\omega = \alpha_j^V$ is a member of the Weyl group $\mathscr{W}$ of $\tilde{\mathscr{L}}$ for $j = 1, 2, \ldots, l$. However, (26.185) implies that if ω is given by (26.180) then

$$T_\omega = \prod_{j=1}^{l} (T_{\alpha_j^V})^{\mu_j^\omega}, \qquad (26.188)$$

so $T_\omega \in \mathscr{W}$ for all $\omega \in Q^{sV}$.

It is also easily shown that

$$S T_\omega S^{-1} = T_{S(\omega)} \qquad (26.189)$$

for any $S \in \mathscr{W}$, demonstrating (cf. Chapter 2, Section 3) that the set $\mathscr{T}$ of translations T_ω forms an *invariant* Abelian subgroup of $\mathscr{W}$.

The Weyl group $\mathscr{W}^s$ of $\tilde{\mathscr{L}}^s$ may be identified with the subgroup of $\mathscr{W}$ that is generated by the Weyl reflections S_{α_j} $(j = 1, 2, \ldots, l)$.

Theorem III If $\mathscr{W}$ and $\mathscr{W}^s$ are the Weyl groups of $\tilde{\mathscr{L}}$ and $\tilde{\mathscr{L}}^s$ respectively, and $\mathscr{T}$ is the Abelian invariant subgroup of translations T_ω $(\omega \in Q^{sV})$, then $\mathscr{W}$ has the semi-direct product structure:

$$\mathscr{W} = \mathscr{T} \circledS \mathscr{W}^s. \tag{26.190}$$

Proof It has just been demonstrated that $\mathscr{T}$ is an invariant subgroup of $\mathscr{W}$, so (cf. Chapter 2, Section 7) it remains only to show (i) that $\mathscr{T}$ and $\mathscr{W}^s$ have only the identity element of $\mathscr{W}$ in common and (ii) that every element of $\mathscr{W}$ can be written as a product of an element of $\mathscr{T}$ and an element of $\mathscr{W}^s$. To prove (i), it is sufficient to note that, by (26.54) and (26.171), $S_\alpha(\Lambda_0) = \Lambda_0$ for all $\alpha = \alpha_k$ with $k = 1, 2, \ldots, l$, and hence $S(\Lambda_0) = \Lambda_0$ for all $S \in \mathscr{W}^s$, whereas by (26.181), (26.180) and (26.171)

$$T_\omega(\Lambda_0) = \Lambda_0 + \langle \Lambda_0, \delta \rangle (\omega - \tfrac{1}{2}\langle \omega, \omega \rangle \delta), \tag{26.191}$$

which (by (26.172)) is equal to Λ_0 only when $\omega = 0$. To establish (ii), it needs only to show that every S_α with $\alpha = \alpha_j$ and $j = 0, 1, \ldots, l$ can be written in the desired product form, the general result then following from (26.189). As this is trivially true for $j = 1, 2, \ldots, l$ (the corresponding element of $\mathscr{T}$ being the identity), it remains only to consider the case of S_α with $\alpha = \alpha_0$. But, by (26.187), (26.112) and (26.150),

$$S_{\alpha_0} = T_{\alpha^\vee}(S_\beta)^{-1},$$

where $\beta = \alpha_H$ in the untwisted case and $\beta = \lambda_H$ in the twisted case, so the result is true for S_α with $\alpha = \alpha_0$ as well.

The structure exhibited in (26.190), which is very similar to that of a symmorphic space group with $\mathscr{W}^s$ playing the role of the rotation group $\mathscr{G}_0$ (cf. Theorem I of Chapter 9, Section 2), is responsible for the appellation *affine* being attached to this type of Kac–Moody algebra.

One useful property of $\mathscr{W}$ is that

$$S(\delta) = \delta \tag{26.192}$$

for every $S \in \mathscr{W}$. (This follows immediately from (26.54) and (26.86) when $S = S_\alpha$ with $\alpha = \alpha_j$ and $j = 0, 1, \ldots, l$, and is true in general because these fundamental reflections generate the whole of $\mathscr{W}$.)

(e) *The compact real form of a complex affine Kac–Moody algebra*

For convenience consider first the case in which the complex affine Kac–Moody algebra $\tilde{\mathscr{L}}$ is *untwisted*. Let $\tilde{\mathscr{L}}^0$ $(= \tilde{\mathscr{L}}^s)$ be the corresponding complex simple Lie algebra, and let h_α^0 (for $\alpha = \alpha_j$ with $j = 1, \dots, l$) and e_α^0 (for $\alpha \in \Delta^0$) be the elements of the Weyl canonical basis of $\tilde{\mathscr{L}}^0$. Then, as shown in Theorem IV of Chapter 14, Section 2, the set of elements ih_α^0 (for $\alpha = \alpha_j$ with $j = 1, 2, \dots, l$) together with $e_\alpha^0 + e_{-\alpha}^0$ and $i(e_\alpha^0 - e_{-\alpha}^0)$ for all $\alpha \in \Delta_+^0$ form a basis for the *compact real form* $\mathscr{L}_c^0$ of $\tilde{\mathscr{L}}^0$.

Although various criteria for defining the compact real form were discussed in Chapter 14, Section 2, in order to generalize the concept to affine Kac–Moody algebras it is necessary to introduce a different specification. To this end, let ϕ_0 be the *conjugation* defined on the complex Lie algebra $\tilde{\mathscr{L}}^0$ by

$$\phi_0\left(\sum_{p=1}^{n_0} \mu_p a_p^0\right) = \sum_{p=1}^{n_0} \mu_p^* a_p^0,$$

where a_p^0 $(p = 1, 2, \dots, n_0)$ are basis elements of $\mathscr{L}_c^0$ and μ_p $(p = 1, 2, \dots, n_0)$ are a set of arbitrary complex numbers. Clearly

$$\phi_0\left(\phi_0\left(\sum_{p=1}^{n_0} \mu_p a_p^0\right)\right) = \sum_{p=1}^{n_0} \mu_p a_p^0,$$

so ϕ_0 is *involutive*, although ϕ_0 is clearly *not* a *linear* operator. Then $\mathscr{L}_c^0$ is the subset of elements a of $\tilde{\mathscr{L}}^0$ such that

$$\phi_0(a) = a. \tag{26.193}$$

This may be used to specify $\mathscr{L}_c^0$.

Every element of the untwisted complex affine Kac–Moody algebra $\tilde{\mathscr{L}}$ $(= \tilde{\mathscr{L}}^{(1)})$ constructed in subsection (*b*) has the form

$$a = \sum_{j=-\infty}^{\infty} \sum_{p=1}^{n_0} \mu_{jp}(t^j \otimes a_p^0) + \mu_c c + \mu_d d,$$

where the coefficients μ_{jp}, μ_c and μ_d are complex numbers, and where it may be assumed (as above) that the a_p^0 are basis elements of the compact real form $\mathscr{L}_c^0$ of $\tilde{\mathscr{L}}^0$. Define the conjugation ϕ of $\tilde{\mathscr{L}}$ by

$$\phi\left(\sum_{j=-\infty}^{\infty} \sum_{p=1}^{n_0} \mu_{jp}(t^j \otimes a_p^0) + \mu_c c + \mu_d d\right)$$
$$= \sum_{j=-\infty}^{\infty} \sum_{p=1}^{n_0} \mu_{jp}^*(t^{-j} \otimes a_p^0) - \mu_c c - \mu_d d,$$

so that ϕ also is involutive. By analogy with (26.193), the *compact real form* $\mathscr{L}_c$ of the *untwisted* affine Kac–Moody algebra $\tilde{\mathscr{L}}$ is then defined to be the subset of elements a of $\tilde{\mathscr{L}}$ such that

$$\phi(a) = a.$$

Clearly the basis elements of $\mathscr{L}_c$ are

$$ic, \quad id, \tag{26.194a}$$

$$t^0 \otimes a_p^0 \quad \text{for } p = 1, 2, \ldots, n_0, \tag{26.194b}$$

$$t^j \otimes a_p^0 + t^{-j} \otimes a_p^0 \quad \text{for } p = 1, 2, \ldots, n_0 \text{ and } j = 1, 2, \ldots, \tag{26.194c}$$

$$i(t^j \otimes a_p^0 - t^{-j} \otimes a_p^0) \quad \text{for } p = 1, 2, \ldots, n_0 \text{ and } j = 1, 2, \ldots . \tag{26.194d}$$

Thus $\mathscr{L}_c^0$ is a subalgebra of $\mathscr{L}_c$. In terms of the canonical basis elements of $\mathscr{L}_c^0$, the basis elements of $\mathscr{L}_c$ are

$$ic, \quad id, \tag{26.195a}$$

$$ih_{\alpha_k} \ (= it^0 \otimes h_{\alpha_k}^0) \quad \text{for } k = 1, 2, \ldots, l, \tag{26.195b}$$

$$e_\alpha + e_{-\alpha} \quad \text{for } \alpha \in \Delta_+^0, \tag{26.195c}$$

$$i(e_\alpha - e_{-\alpha}) \quad \text{for } \alpha \in \Delta_+^0, \tag{26.195d}$$

$$e_{j\delta}^k + e_{-j\delta}^k \quad \text{for } k = 1, 2, \ldots, l \text{ and } j = 1, 2, \ldots, \tag{26.195e}$$

$$i(e_{j\delta}^k - e_{-j\delta}^k) \quad \text{for } k = 1, 2, \ldots, l \text{ and } j = 1, 2, \ldots, \tag{26.195e}$$

$$e_{j\delta+\alpha} + e_{-(j\delta+\alpha)} \quad \text{for } \alpha \in \Delta^0 \text{ and } j = 1, 2, \ldots, \tag{26.195f}$$

$$i(e_{j\delta+\alpha} - e_{-(j\delta+\alpha)}) \quad \text{for } \alpha \in \Delta^0 \text{ and } j = 1, 2, \ldots . \tag{26.195g}$$

(Here the definitions (26.118) and (26.119) have been invoked.) It is easy to check that with this basis all the structure constants are *real*, so that $\mathscr{L}_c$ is a *real* Lie algebra.

For a *twisted* complex affine Kac–Moody algebra $\tilde{\mathscr{L}}^{(q)}$ the compact real form consists of the set of elements common to $\tilde{\mathscr{L}}^{(q)}$ and to $\mathscr{L}_c^{(1)}$ (the compact real form of $\tilde{\mathscr{L}}^{(1)}$).

The classification of the involutive automorphisms of affine Kac–Moody algebras has been investigated by Levstein (1988).

6 Kac–Moody superalgebras

The general line of argument that is given for Lie algebras in Sections 2–5 of this chapter can also, with minor modification, be applied to Lie superalgebras. The resulting structures are now often referred to as *Kac–Moody superalgebras*, although in the original investigations by Kac

(1977a, 1978a) they were called "contragredient Lie superalgebras". The argument will not be developed in detail here, but instead the present account will be confined to indicating the main points where the theory for Kac–Moody superalgebras has features that are not possessed by Kac–Moody algebras. For brevity only *complex* Kac–Moody superalgebras will be considered.

The starting point in the superalgebra case is again the generalized Cartan matrix **A**, but the definition given in Section 2 requires a slight modification. As before, each row (and corresponding column) of **A** is associated with a simple root, but in the superalgebra case the set of simple roots is divided into two classes, the "even" and the "odd" simple roots.

Definition *Generalized Cartan matrix* **A**

A *generalized Cartan matrix* **A** is defined to be a $d_{\mathbf{A}} \times d_{\mathbf{A}}$ matrix (with an index set I (with $d_{\mathbf{A}}$ entries) labelling its rows and columns) such that

(i) either

(a) $A_{jj} = 2$ for all $j \in \mathrm{I}$,

or

(b) $\det \mathbf{A} \neq 0$ and

$$A_{jj} = \begin{cases} 2 & \text{if } \alpha_j \text{ is even,} \\ 0 \text{ or } 2 & \text{if } \alpha_j \text{ is odd;} \end{cases}$$

(ii) for $j \neq k$ ($j, k \in \mathrm{I}$), A_{jk} is either zero or a negative integer; and

(iii) for $j \neq k$ ($j, k \in \mathrm{I}$), $A_{jk} = 0$ if and only if $A_{kj} = 0$.

Again it is convenient to assume also that **A** is both "symmetrizable" and "indecomposable" (in the same senses as those defined in Section 2). If the set of simple odd roots is *empty* then the conditions (i) (a) and (ii) (b) together reduce to the single condition (i) of the definition of **A** given for Kac–Moody algebras in Section 2.

Assuming that **A** has rank L, a Cartan subalgebra $\mathscr{H}$ can be introduced as complex vector space of dimension $n_{\mathscr{H}}$ where $n_{\mathscr{H}}$ is defined by

$$n_{\mathscr{H}} = 2d_{\mathbf{A}} - L. \tag{26.196}$$

(If the set of simple odd roots is *empty*, one may write $L = l$, and then (26.196) reduces to (26.20).) Introducing a set of $d_{\mathbf{A}}$ linearly independent elements h_j of $\mathscr{H}$, a set of $d_{\mathbf{A}}$ simple roots α_j may be defined (as in (26.21)) to be the linear functionals on $\mathscr{H}$ that are such that

$$\alpha_k(h_j) = A_{jk}.$$

To each α_j ($j \in I$) there may be assigned the pair of generators E_{α_j} and $E_{-\alpha_j}$, it being assumed that these are both even if α_j is even and both odd if α_j is odd. A complex Lie superalgebra $\tilde{\mathscr{L}}'_s$ can be generated by these $2d_A$ elements and the elements of $\mathscr{H}$, these generators being assumed to have the commutation and anticommutation relations that are obtained by just replacing the Lie brackets in (26.22)–(26.25) by generalized Lie brackets (all the elements of $\mathscr{H}$ being assumed to be even). The corresponding Kac–Moody superalgebra $\tilde{\mathscr{L}}_s$ associated with the generalized Cartan matrix $\mathbf{A}$ is then defined to be the factor superalgebra $\tilde{\mathscr{L}}'_s/\mathscr{R}$, where $\mathscr{R}$ is the maximal invariant subalgebra of $\tilde{\mathscr{L}}'_s$ whose intersection with $\mathscr{H}$ is trivial.

It is obvious that if the set of simple odd roots is *empty* then the Kac–Moody superalgebra is actually just a Kac–Moody *algebra*. Consequently attention will be confined for the rest of this section to the case in which there exists at least one simple odd root.

Kac (1977a) has shown that those Kac–Moody superalgebras for $\det \mathbf{A} \neq 0$ are *finite-dimensional* Lie superalgebras, and that this process produces the basic classical simple Lie superalgebras $A(r/s)$ (for $r > s \geqslant 0$), $B(r/s)$ (for $r \geqslant 0$ and $s \geqslant 1$), $C(s)$ (for $s \geqslant 2$), $D(r/s)$ (for $r \geqslant 2$ and $s \geqslant 1$), $D(2/1;\alpha)$ (for any complex α except 0, -1 or ∞), $F(4)$ and $G(3)$, as well as $\mathrm{sl}(r+1/r+1;\mathbb{C})$) for $r \geqslant 1$), whose factor algebra with its invariant Abelian subalgebra is $A(r/r)$. That is, *all* the basic classical simple complex Lie superalgebras of Chapter 25 can be constructed in this way. Moreover, if l is the rank of the simple Lie superalgebra then $L = l$ except for $A(r/r)$ (with $r \geqslant 1$), where $L = l + 1$ (cf. (25.62)).

If $\det \mathbf{A} = 0$ but all proper principal minors of $\mathbf{A}$ are non-zero, a class of *infinite-dimensional* superalgebras emerge that are known as *affine Kac–Moody superalgebras* or as *affine Lie superalgebras*. Each generalized Cartan matrix may again be described by a generalized Dynkin diagram, the rules being exactly as specified in Section 3, the only new feature of the superalgebra case being that even simple roots are indicated by "white" vertices and odd simple roots by "black" vertices (as in Chapter 25, Section 3(*c*)). (The "grey" vertices of Chapter 25, Section 3(*c*) are effectively excluded by the requirement in the above definition of $\mathbf{A}$ that if $\det \mathbf{A} = 0$ then $A_{ji} = 2$ for *all* $j \in \mathrm{I}$). Kac (1978a) has given a complete enumeration of these affine superalgebras, and has shown that they can all be constructed by the method described in Sections 5(*b*) and 5(*c*), the only feature that is essentially new being that, whereas in the affine algebra case the process starts with a simple Lie algebra $\tilde{\mathscr{L}}^0$, in the superalgebra case it must start with a simple Lie superalgebra. The representations of the affine Kac–Moody superalgebras and of the affine Kac–Moody algebras have a number of properties in common. For details see Kac (1978a), Dobrev (1986d, 1987b) and Golitzin (1988).

It is interesting to note that the only Kac–Moody superalgebras that possess an odd simple root and have a generalized Cartan matrix that satisfies *both* of the conditions (i) (a) and (i) (b) are the simple Lie superalgebras $B(0/S)$ (for $S \geqslant 1$).

Another class of infinite-dimensional Lie superalgebras that involves Kac–Moody algebras is discussed in Chapter 28, Section 4.

27

Representations of Kac–Moody Algebras

1 Highest weight representations of general Kac–Moody algebras

Many of the general features of the representation theory of semi-simple Lie algebras that were presented in Chapter 15 carry over to non-finite Kac–Moody algebras. However, since the resulting representations are always *infinite*-dimensional (apart from certain one-dimensional representations), they are more difficult to construct explicitly, and our present knowledge of them is less complete, even in the affine case.

The discussion must necessarily be in terms of modules (cf. Chapter 11, Section 4), each of which consists of a carrier space V and a set of linear operators $\Phi(a)$ (one for each element a of the Kac–Moody algebra $\tilde{\mathscr{L}}$) that are assumed to act on V and for which

$$\Phi([a,b]_-) = [\Phi(a), \Phi(b)]_- \tag{27.1}$$

for all $a, b \in \tilde{\mathscr{L}}$.

The present account will be confined largely to *integrable highest weight representations.* For these it is assumed that the carrier space V is a direct sum of *finite*-dimensional *weight subspaces.* Each weight subspace is associated with a *weight* λ, which is a linear functional on the Cartan subalgebra $\mathscr{H}$ of $\tilde{\mathscr{L}}$, the elements of the weight subspace all being eigenvectors of the operators $\Phi(h)$ with eigenvalue $\lambda(h)$. That is (cf. (15.3))

$$\Phi(h)\psi(\lambda) = \lambda(h)\psi(\lambda) \tag{27.2}$$

for all $h \in \mathscr{H}$. The dimension of this subspace is called the *multiplicity* of the weight λ and is denoted by $m(\lambda)$. It is also assumed that one of these weights, which is called the *highest weight* and is denoted by Λ, is such that

$$\Phi(e_\alpha)\psi(\Lambda) = 0 \tag{27.3}$$

for *every positive root* α of $\tilde{\mathscr{L}}$. The final assumption (for which the description "integrable" is attached) is that for every simple root α_k ($k \in \mathrm{I}$, I being the index set of $\tilde{\mathscr{L}}$) and for every $\psi \in \mathrm{V}$ there exist positive integers N and N' such that

$$\Phi(e_{\alpha_k})^N \psi = 0 \tag{27.4}$$

and

$$\Phi(e_{-\alpha_k})^{N'} \psi = 0. \tag{27.5}$$

With these assumptions, Theorems I, II, IV and V of Chapter 15, Section 2 apply with at most minor modifications, as do Theorems I and II of Chapter 15, Section 3 (although the latter requires a more significant modification). For convenience they will be restated explicitly, but their proofs, which are essentially as before, will be omitted.

Theorem I If λ is a weight of an integrable highest weight representation of a Kac–Moody algebra $\tilde{\mathscr{L}}$ then $\lambda + \alpha$ is also a weight of the same representation for each $\alpha \in \Delta$ (the set of non-zero roots of $\tilde{\mathscr{L}}$) such that $\Phi(e_\alpha)\psi(\lambda) \neq 0$.

Theorem II For any weight λ of an integrable highest weight representation of Kac–Moody algebra $\tilde{\mathscr{L}}$ and for any *real* root α of Δ, $2\langle \lambda, \alpha \rangle / \langle \alpha, \alpha \rangle$ is an *integer*.

Theorem III For every weight λ of an integrable highest weight representation of a Kac–Moody algebra $\tilde{\mathscr{L}}$ and for every element S of the Weyl group $\mathscr{W}$ of $\tilde{\mathscr{L}}$, $S\lambda$ is a weight of the same representation with the same multiplicity, i.e.

$$m(S\lambda) = m(\lambda) \tag{27.6}$$

for all $S \in \mathscr{W}$.

Theorem IV Let α be a non-zero *real* root of a Kac–Moody algebra $\tilde{\mathscr{L}}$ and let λ be a weight of an integrable highest weight representation of $\tilde{\mathscr{L}}$. Then there exist two non-negative integers p and q (which depend on α and λ) such that $\lambda + k\alpha$ is in the α-string of weights containing λ for *every* integer k that satisfies the relations $-p \leqslant k \leqslant q$. Moreover, p and q are such that

$$p - q = \frac{2\langle \lambda, \alpha \rangle}{\langle \alpha, \alpha \rangle}. \tag{27.7}$$

Theorem V If Λ is the highest weight of an *irreducible* integrable highest weight representation of a Kac–Moody algebra $\tilde{\mathscr{L}}$ then

(a) $m(\Lambda) = 1$, and

(b) every other weight λ of the representation has the form

$$\lambda = \Lambda - \sum_{k \in \mathrm{I}} q_k \alpha_k, \tag{27.8}$$

where q_k ($k \in \mathrm{I}$) are a set of *non-negative integers.*

Theorem V allows a lexicographic ordering to be put on the set of weights of an irreducible integrable highest weight representation (cf. Chapter 13, Section 7). Suppose that λ and λ' are two weights of such a representation, λ being given by (27.8) and λ' by

$$\lambda' = \Lambda - \sum_{k \in \mathrm{I}} q'_k \alpha_k.$$

Then

$$\lambda - \lambda' = -\sum_{k \in \mathrm{I}} (q_k - q'_k)\alpha_k,$$

and one may say that $\lambda > \lambda'$ if the first non-vanishing member of the set $-(q_k - q'_k)$ ($k \in \mathrm{I}$) is positive. With this definition, $\Lambda > \lambda$ for every other weight λ of the representation, so Λ is a highest weight in this usual sense.

Theorem VI For every *irreducible* integrable highest weight representation of a Kac–Moody algebra $\tilde{\mathscr{L}}$ the highest weight Λ is such that

$$\frac{2\langle \Lambda, \alpha_k \rangle}{\langle \alpha_k, \alpha_k \rangle} = n_k \tag{27.9}$$

for all $k \in \mathrm{I}$, where the n_k are set of *non-negative integers.* Moreover, every linear functional Λ on $\mathscr{H}$ satisfying these conditions is the highest weight for a unique irreducible integrable highest weight representation of $\tilde{\mathscr{L}}$.

Proof The first part is proved as in the corresponding part of Theorem II of Chapter 15, Section 3. For the second part see Kac (1985a).

It should be noted that in general (27.9) does *not* imply a result similar to (15.12), since it can happen, as in the case of affine Kac–Moody algebras, that $\mathbf{A}^{-1}$ does not exist, and so the latter part of the proof of Theorem II of Chapter 15, Section 3 cannot be applied. This point will be discussed further in the next subsection.

It should also be noted that Theorem III of Chapter 15, Section 2 does *not* generalize to all Kac–Moody algebras, since it need not be true that the α_k ($k \in \mathrm{I}$) form a basis for $\mathscr{H}^*$ *on their own.* Indeed, in the affine case they have to be supplemented by the Λ_0 of (26.168) and (26.169).

The following theorem gives the major results on reducibility.

Theorem VII (a) Every reducible integrable highest weight representation of a Kac–Moody algebra $\tilde{\mathscr{L}}$ is *completely* reducible; that is, it decomposes into a direct sum of irreducible representations of $\tilde{\mathscr{L}}$.

(b) The direct product of any two integrable highest weight irreducible representations of a Kac–Moody algebra $\tilde{\mathscr{L}}$ is a direct sum integrable highest weight irreducible representations of $\tilde{\mathscr{L}}$.

(c) Every integrable highest weight irreducible representation of a Kac–Moody algebra $\tilde{\mathscr{L}}$ remains *irreducible* when restricted to the "derived" subalgebra $\tilde{\mathscr{L}}'' = [\tilde{\mathscr{L}}, \tilde{\mathscr{L}}]_-$.

Proof See Kac (1985a).

As discussed in Chapter 26, Section 3, the derived subalgebra $\tilde{\mathscr{L}}'' = [\tilde{\mathscr{L}}, \tilde{\mathscr{L}}]_-$ has as its basis elements all the basis elements of $\tilde{\mathscr{L}}$ except those of $\mathscr{H}''$. In the case of *affine* Lie algebras this means that the only basis elements of $\tilde{\mathscr{L}}$ that is not in $\tilde{\mathscr{L}}''$ is the scaling element d.

The theory can be developed further for *general* Kac–Moody algebras (cf. Kac 1985a), but for ease of exposition the rest of the presentation here will be for the *affine* case.

2 Highest weight representations of affine Kac–Moody algebras

It can be verified by inspection of the commutation relations (26.89)–(26.92) that both the untwisted and twisted affine Kac–Moody algebras $\tilde{\mathscr{L}}$ possess a *family* of *one-dimensional* irreducible representations that are specified by an *arbitrary complex number* μ and are defined by

$$\mathbf{\Gamma}^{\mu}(d) = [\mu] \tag{27.10}$$

and

$$\mathbf{\Gamma}^{\mu}(a) = [0] \tag{27.11}$$

for any other basis element a of $\tilde{\mathscr{L}}$. Being one-dimensional, $\mathbf{\Gamma}^{\mu}$ has only one weight, its highest weight, which is given by $\Lambda = \mu\delta$. By (26.86) for $\mathbf{\Gamma}^{\mu}$

$$\frac{2\langle \Lambda, \alpha_k \rangle}{\langle \alpha_k, \alpha_k \rangle} = 0$$

for all $k = 0, 1, \ldots, l$, so that for *every* irreducible representation of this family the integers n_k of (27.9) are such that $n_k = 0$ for $k = 0, 1, \ldots, l$.

All the other integrable highest weight representations are associated with highest weights formed from the "fundamental weights".

Definition *Fundamental weights of a complex affine Kac–Moody algebra* $\tilde{\mathscr{L}}$

The $l+1$ *fundamental weights* $\Lambda_0(h), \Lambda_1(h), \ldots, \Lambda_l(h)$ of $\tilde{\mathscr{L}}$ are the $l+1$ linear functionals defined on $\mathscr{H}$ by

$$\frac{2\langle \Lambda_j, \alpha_k \rangle}{\langle \alpha_k, \alpha_k \rangle} = \delta_{jk} \tag{27.12}$$

for $j, k = 0, 1, \ldots, l$, and

$$\Lambda_j(d) = 0 \tag{27.13}$$

for $j = 0, 1, \ldots, l$.

Equations (27.12) are the analogues of (15.11) for the simple Lie algebra situation, but (15.10) cannot be generalized, since $\mathbf{A}^{-1}$ does not exist in the affine case because det $\mathbf{A} = 0$. The linear functional Λ_0 introduced in (26.168) and (26.169) is one of these fundamental weights.

If Λ_j^s $(j = 1, 2, \ldots, l)$ are the extensions to $\mathscr{H}$ from $\mathscr{H}^s$ of the fundamental weights of the simple Lie algebra $\tilde{\mathscr{L}}^s$ of Chapter 26, Section 5(*a*), it is easily verified that for $j = 1, 2, \ldots, l$ the relations (27.12) and (27.13) are satisfied with

$$\Lambda_j = \Lambda_j^s + \mu_j \Lambda_0, \tag{27.14a}$$

where

$$\mu_j = -\sum_{k=1}^{l} A_{0k}((\mathbf{A}^s)^{-1})_{kj}, \tag{27.14b}$$

$\mathbf{A}^s$ being the Cartan matrix of $\tilde{\mathscr{L}}^s$.

Example I *Fundamental weights of* $\tilde{\mathscr{L}} = A_2^{(1)}$

As shown in Example I of Chapter 15, Section 3, for $\tilde{\mathscr{L}}^s = A_2$

$$\Lambda_1^s = \tfrac{2}{3}\alpha_1 + \tfrac{1}{3}\alpha_2, \quad \Lambda_2^s = \tfrac{1}{3}\alpha_1 + \tfrac{2}{3}\alpha_2.$$

Thus, by (27.14a, b), the fundamental weights of $\tilde{\mathscr{L}} = A_2^{(1)}$ are Λ_0 together with

$$\Lambda_1 = \tfrac{2}{3}\alpha_1 + \tfrac{1}{3}\alpha_2 + \Lambda_0$$

and

$$\Lambda_2 = \tfrac{1}{3}\alpha_1 + \tfrac{2}{3}\alpha_2 + \Lambda_0.$$

Definition *Standard irreducible representations of a complex affine Kac–Moody algebra* $\tilde{\mathscr{L}}$

A *standard* irreducible representation of a complex affine Kac–Moody algebra $\tilde{\mathscr{L}}$ is an irreducible integrable highest weight representation whose highest weight Λ is given by

$$\Lambda = \sum_{k=0}^{l} n_k \Lambda_k, \tag{27.15}$$

where $n_1, n_2, \ldots, n_l$ are a set of *non-negative integers*, of which at least one is non-zero.

The most general solution of the equations (27.9) (for $\mathrm{I} = \{0, 1, \ldots, l\}$) is

$$\Lambda = \sum_{k=0}^{l} n_k \Lambda_k + \mu\delta, \tag{27.16}$$

where μ is an arbitrary complex number. The corresponding irreducible representation can be constructed as the direct product of the standard irreducible representation with highest weight given by (27.15) and the one-dimensional irreducible representation Γ^{μ} of (27.10) and (27.11). Equation (16.42) shows that to each weight λ of the former representation there corresponds a weight $\lambda + \mu\delta$ of the direct product representation, and these have the same multiplicity. Clearly the whole structure of the irreducible representation with highest weight (27.16) is determined by that of the standard irreducible representation with the highest weight (27.15), and so attention can be confined to these standard irreducible representations.

It should be noted that the *adjoint* representation of $\tilde{\mathscr{L}}$ is *not* a highest weight representation because its set of weights (being the roots of $\tilde{\mathscr{L}}$—cf. Example I of Chapter 15, Section 2) contains the set $\{j\delta\}$, where j takes *all* integer values, both positive and negative.

Although it is possible to develop the theory further for general affine Kac–Moody algebras, for convenience *the rest of the discussion will be confined to the* **untwisted** *case alone.*

As (26.90) and (26.92) show that the element c of $\tilde{\mathscr{L}}$ commutes with *every* element of $\tilde{\mathscr{L}}$, it follows that in any irreducible representation with highest weight Λ

$$\Phi(c) = c_{\Lambda} 1, \tag{27.17}$$

where 1 is the identity operator and c_{Λ} is a number that depends on Λ. Consequently

$$\Phi(c)\psi(\lambda) = c_{\Lambda}\psi(\lambda) \tag{27.18}$$

for every weight λ of the representation. Consideration of (27.2) for the special case $\lambda = \Lambda$ shows that $c_\Lambda = \Lambda(c)$, which, together with (26.46) and (26.106), implies that

$$c_\Lambda = \langle \Lambda, \delta \rangle. \tag{27.19}$$

Thus, by (26.80) and (27.12), if Λ is given by (27.15) then

$$c_\Lambda = \tfrac{1}{2} \sum_{k=0}^{l} n_k \kappa_k^{\delta} \langle \alpha_k, \alpha_k \rangle. \tag{27.20}$$

Clearly $c_\Lambda = 0$ for the one-dimensional irreducible representations $\boldsymbol{\Gamma}^\mu$ of (27.10) and (27.11), and $c_\Lambda > 0$ for every standard irreducible representation.

Definition *The level of a standard irreducible representation of an untwisted Kac–Moody algebra $\tilde{\mathscr{L}}$*

The *level*, level(Λ), of the standard irreducible representation of an untwisted Kac–Moody algebra $\tilde{\mathscr{L}}$ with highest weight Λ is defined by

$$\text{level}(\Lambda) = \frac{2\langle \Lambda, \delta \rangle}{\langle \alpha_{\mathrm{H}}, \alpha_{\mathrm{H}} \rangle}, \tag{27.21}$$

where α_{H} is the highest root of the simple Lie algebra $\tilde{\mathscr{L}}^0$ associated with $\tilde{\mathscr{L}}$ (cf. (26.112)).

By (27.19) and (27.20),

$$\text{level}(\Lambda) = \frac{2c_\Lambda}{\langle \alpha_{\mathrm{H}}, \alpha_{\mathrm{H}} \rangle} = \sum_{k=0}^{l} n_k \kappa_k^{\delta} \frac{\langle \alpha_k, \alpha_k \rangle}{\langle \alpha_{\mathrm{H}}, \alpha_{\mathrm{H}} \rangle}. \tag{27.22}$$

For the simply laced algebras $A_l^{(1)}$ $(l \geq 1)$, $D_l^{(1)}$ $(l \geq 4)$, $E_6^{(1)}$, $E_7^{(1)}$ and $E_8^{(1)}$ (27.22) implies that

$$\text{level}(\Lambda) = \sum_{k=0}^{l} n_k \kappa_k^{\delta}. \tag{27.23}$$

For the remaining untwisted affine algebras $B_l^{(1)}$ $(l \geq 3)$, $C_l^{(1)}$ $(l \geq 2)$, $F_4^{(1)}$ and $G_2^{(1)}$ inspection of the results given in Appendix N shows that α_{H} is always a "long" root, and every simple root α_k $(k = 0, 1, \ldots, l)$ is such that

$$m_k \langle \alpha_k, \alpha_k \rangle = \langle \alpha_{\mathrm{H}}, \alpha_{\mathrm{H}} \rangle,$$

where m_k is either 1 or 2 (depending on whether α_k is "short" or "long"), so (27.22) reduces to

$$\text{level}(\Lambda) = \sum_{k=0}^{l} n_k \frac{\kappa_k^{\delta}}{m_k}. \tag{27.24}$$

However, in all cases κ_k^δ/m_k is always a positive integer. Thus for *every* untwisted affine Kac–Moody algebra level(Λ) is a *positive integer*.

Special interest attaches to the standard irreducible representations of level 1. Inspection of the coefficients κ_k^δ and the simple root lengths given in Appendix N shows that the highest weights for these are as follows:

$$A_l^{(1)} (l \geq 1): \quad \Lambda = \Lambda_0, \Lambda_1, \ldots, \Lambda_l; \tag{27.25a}$$

$$B_l^{(1)} (l \geq 3): \quad \Lambda = \Lambda_0, \Lambda_1, \Lambda_l; \tag{27.25b}$$

$$C_l^{(1)} (l \geq 2): \quad \Lambda = \Lambda_0, \Lambda_1, \ldots, \Lambda_l; \tag{27.25c}$$

$$D_l^{(1)} (l \geq 4): \quad \Lambda = \Lambda_0, \Lambda_1, \Lambda_{l-1}, \Lambda_l; \tag{27.25d}$$

$$E_6^{(1)}: \quad \Lambda = \Lambda_0, \Lambda_1, \Lambda_5; \tag{27.25e}$$

$$E_7^{(1)}: \quad \Lambda = \Lambda_0, \Lambda_6; \tag{27.25f}$$

$$E_8^{(1)}: \quad \Lambda = \Lambda_0; \tag{27.25g}$$

$$F_4^{(1)}: \quad \Lambda = \Lambda_0, \Lambda_4; \tag{27.25h}$$

$$G_2^{(1)}: \quad \Lambda = \Lambda_0, \Lambda_1. \tag{27.25i}$$

In every case level$(\Lambda_0) = 1$.

For the *simply laced* algebras the Λ_k associated with level 1 representations correspond to vertices of the generalized Dynkin diagrams that can be mapped into each other by diagram rotations τ belonging to outer automorphisms of $\tilde{\mathscr{L}}$. For example, for $\tilde{\mathscr{L}} = A_1^{(1)}$ the appropriate rotation τ is given by $\tau(\alpha_0) = \alpha_1$ and $\tau(\alpha_1) = \alpha_0$. Consequently, once the standard irreducible representation with highest weight Λ_0 has been found, *all* the other standard irreducible representations of level 1 of the same algebra $\tilde{\mathscr{L}}$ can be determined using the appropriate automorphisms. Thus attention can be concentrated on the standard irreducible representation with highest weight Λ_0, which is known as *the basic representation* of $\tilde{\mathscr{L}}$. The structure of these basic representations for the simply laced algebras was first determined by the *vertex operator construction* by Frenkel and Kac (1980), and will be described in Section 4, where further references will be given.

A generalization of the vertex operator construction to level 1 representations of the *non-simply-laced* untwisted algebras $B_l^{(1)}$ $(l \geq 3)$, $C_l^{(1)}$ $(l \geq 2)$, $F_4^{(1)}$ and $G_2^{(1)}$ has been given by Goddard *et al.* (1986b).

The standard irreducible representations of level m, where $m > 1$, can be found from the level 1 representations by considering the proper subalgebra $\tilde{\mathscr{L}}^{(1)}(m)$ of the untwisted algebra $\tilde{\mathscr{L}}^{(1)}$ $(= \tilde{\mathscr{L}})$ of Chapter 26, Section 5(*b*), which is defined to have basis elements c, d and $t^{mj} \otimes a_p^0$, where j runs over all integral values and a_p^0 are the basis elements of the simple Lie algebra $\tilde{\mathscr{L}}^0$.

The mapping ϕ from $\tilde{\mathscr{L}}^{(1)}(m)$ to $\tilde{\mathscr{L}}^{(1)}(=\tilde{\mathscr{L}})$ defined by

$$\phi(t^{mj}\otimes a_p^0)=t^j\otimes a_p^0, \tag{27.26a}$$

$$\phi(c)=c/m, \tag{27.26b}$$

$$\phi(d)=md \tag{27.26c}$$

is an *isomorphic* mapping of $\tilde{\mathscr{L}}^{(1)}(m)$ *onto* $\tilde{\mathscr{L}}^{(1)}$ (for it is easily checked that the commutation relations (26.89)–(26.92) are preserved, and that *every* element of $\tilde{\mathscr{L}}^{(1)}$ is the image of some element of $\tilde{\mathscr{L}}^{(1)}(m)$ under this mapping). Thus, if the operators $\Phi(a)$ for $a\in\tilde{\mathscr{L}}^{(1)}(=\tilde{\mathscr{L}})$ belong to a standard irreducible representation of level 1, and $\psi\in\mathrm{V}$ is any member of its carrier space, and if the operators $\Psi(a)$ for all $a\in\tilde{\mathscr{L}}^{(1)}(=\tilde{\mathscr{L}})$ are defined by

$$\Psi(a)=\Phi(\phi^{-1}(a)),$$

then the operators $\Psi(a)$ also provide a representation of $\tilde{\mathscr{L}}^{(1)}(=\tilde{\mathscr{L}})$ on V. Moreover, if $\Phi(h)\psi(\Lambda)=\Lambda(h)\psi(\Lambda)$ then

$$\Psi(c)\psi(\Lambda)=(mc_\Lambda)\psi(\Lambda),$$

implying that the level of this representation is m. Similarly, by (27.26a) and (26.104),

$$\Psi(h_{\alpha_k})\psi(\Lambda)=\Lambda(h_{\alpha_k})\psi(\Lambda),$$

so this representation has the same n_k $(k=1,2,\dots,l)$ as the one from which it is formed.

Turning to the simple subalgebra $\tilde{\mathscr{L}}^0(=\tilde{\mathscr{L}}^{\mathrm{s}})$ of $\tilde{\mathscr{L}}(=\tilde{\mathscr{L}}^{(1)})$, it is obvious that the highest weight Λ of (27.15) provides a linear functional on $\mathscr{H}^0$. As

$$\Lambda(h_{\alpha_j})=\sum_{k=0}^{l} n_k\langle\Lambda_k,\alpha_j\rangle=\tfrac{1}{2}n_j\langle\alpha_j,\alpha_j\rangle=\sum_{k=1}^{l} n_k\langle\Lambda_k^0,\alpha_j\rangle,$$

where $\Lambda_k^0(=\Lambda_k^{\mathrm{s}})$ are the fundamental weights of $\tilde{\mathscr{L}}^0(=\tilde{\mathscr{L}}^{\mathrm{s}})$, the corresponding eigenvector $\psi(\Lambda)$ belongs to the irreducible representation of $\tilde{\mathscr{L}}^0$ whose highest weight Λ^0 is given by

$$\Lambda^0=\sum_{k=1}^{l} n_k\Lambda_k^0.$$

Consequently the standard irreducible representation of $\tilde{\mathscr{L}}(=\tilde{\mathscr{L}}^{(1)})$ with Λ given by (27.15) can be regarded as being specified by its level together with this irreducible representation of $\tilde{\mathscr{L}}^0$ with highest weight Λ^0. In particular, for the *basic* representation, since $\Lambda=\Lambda_0$, the corresponding irreducible representation of $\tilde{\mathscr{L}}^0$ is the trivial one-dimensional representation with highest weight 0.

Theorem IV of the previous section dealt with strings of weights involving *real* roots. The corresponding result for the *imaginary* roots $j\delta$ is provided in the following theorem.

Theorem I Let λ be any weight of a standard irreducible representation of an untwisted affine Kac–Moody algebra $\tilde{\mathscr{L}}$. Then there exists a non-negative integer p (which depends on λ) such $\lambda - k\delta$ is in the δ-string of weights containing λ for *every* integer $k \geqslant -p$.

Proof See Appendix O, Section 1.

One implication of this theorem is that every δ-string has *infinite* length, so every standard irreducible representation must be *infinite-dimensional.*

Definition *Maximal weights of a standard irreducible representation of an untwisted affine Kac–Moody algebra*

A weight λ of a standard irreducible representation of an untwisted affine Kac–Moody algebra $\tilde{\mathscr{L}}$ is said to be *maximal* if $\lambda + \delta$ is *not* a weight of the representation.

The properties of maximal weights are summarized in the next theorem.

Theorem II For a standard irreducible representation of an untwisted affine Kac–Moody algebra $\tilde{\mathscr{L}}$

(a) the highest weight Λ is a maximal weight;

(b) each weight λ is associated with a unique maxiınal weight $\lambda_{\max}$ and a non-negative integer j by the relation

$$\lambda = \lambda_{\max} - j\delta, \tag{27.27}$$

and the set of *all* weights of the representation is obtained by letting j take all the values $0, 1, 2, \ldots$ for each $\lambda_{\max}$ in turn; and

(c) if $S \in \mathscr{W}$, the Weyl group of $\tilde{\mathscr{L}}$, and $\lambda_{\max}$ is a maximal weight, then $S\lambda_{\max}$ is also a maximal weight.

Proof (a) This follows immediately from the definitions of maximal weight and highest weight.

(b) This is an immediate consequence of Theorem I.

(c) Theorem III of Section 1 shows that $S\lambda_{\max}$ is a weight of the representation, so all that has to be demonstrated is that $S\lambda_{\max}$ is maximal.

Suppose to the contrary that $S\lambda_{\max} = \lambda'_{\max} - j\delta$, where $\lambda'_{\max}$ is a maximal weight and j is a positive integer. Then $\lambda_{\max} = S^{-1}\lambda'_{\max} - jS^{-1}\delta = S^{-1}\lambda'_{\max} - j\delta$ by (26.192), which is a contradiction, since it has been assumed that $\lambda_{\max}$ is maximal.

Definition *Maximal dominant weights of a standard irreducible representation of an untwisted Kac–Moody algebra*

A maximal weight $\lambda_{\max}$ of a standard irreducible representation of an untwisted Kac–Moody algebra $\tilde{\mathscr{L}}$ is said to be *dominant* if

$$\frac{2\langle \lambda_{\max}, \alpha_k \rangle}{\langle \alpha_k, \alpha_k \rangle} = m_k, \tag{27.28}$$

where m_k $(k = 0, 1, \ldots, l)$ are a set of *non-negative integers*.

That the m_k of (27.28) are integers is ensured by Theorem II of Section 1, so the only *new* requirement here is that they be *non-negative*.

Theorem III For a standard irreducible representation of an untwisted affine Kac–Moody algebra $\tilde{\mathscr{L}}$

(a) the highest weight Λ is a maximal dominant weight;

(b) at least one member of the set m_k $(k = 0, 1, \ldots, l)$ of (27.28) is non-zero;

(c) for every maximal weight $\lambda_{\max}$ there exists an element S of the Weyl group $\mathscr{W}$ of $\tilde{\mathscr{L}}$ and a unique dominant maximal weight $\lambda^+_{\max}$ such that $\lambda_{\max} = S\lambda^+_{\max}$; and

(d) the set of maximal dominant weights is *finite*.

Proof See Appendix O, Section 1.

Theorem IV For the basic representation (with highest weight Λ_0) of a simply laced untwisted affine Kac–Moody algebra $A^{(1)}_l$ $(l \geq 1)$, $D^{(1)}_l$ $(l \geq 4)$, $E^{(1)}_6$, $E^{(1)}_7$ or $E^{(1)}_8$

(a) the *only* maximal dominant weight is Λ_0;

(b) the set of *maximal* weights is given by

$$\lambda_{\max} = \Lambda_0 + \alpha - \frac{\langle \alpha, \alpha \rangle}{\langle \alpha_0, \alpha_0 \rangle}\delta, \tag{27.29}$$

where α runs over all members of the root lattice Q^0 of $\tilde{\mathscr{L}}^0 (= \tilde{\mathscr{L}}^s)$;

(c) the set of *all* weights is given by

$$\lambda = \Lambda_0 + \alpha - \left(j + \frac{\langle \alpha, \alpha \rangle}{\langle \alpha_0, \alpha_0 \rangle} \right) \delta, \tag{27.30}$$

where α runs over all members of the root lattice Q^0 of $\tilde{\mathscr{L}}^0 (= \tilde{\mathscr{L}}^s)$ and j takes all non-negative integer values.

Proof See Appendix O, Section 1.

The last theorem of this section gives some information on the inner product on the carrier space V.

Theorem V The carrier space V of an integrable highest weight representation of an affine Kac–Moody algebra $\tilde{\mathscr{L}}$ is an inner product space, the inner product (,) being such that for any two weight vectors $\psi(\lambda)$ and $\psi'(\lambda')$

$$(\psi(\lambda), \psi'(\lambda')) = 0 \quad \text{unless } \lambda = \lambda'$$

and

$$(\Phi(a)\psi, \phi) = -(\psi, \Phi(a)\phi)$$

for all $\psi, \phi \in V$ and for every element a of the compact real form $\mathscr{L}_c$ of $\tilde{\mathscr{L}}$.

Proof See Kac (1985a).

Such a representation is said to be *unitary*, because in the analogous simple case a representation with such a property would give rise (on exponentiation) to a *unitary* representation of the corresponding compact universal linear group $\hat{\mathscr{G}}$.

The next section is devoted to a short account of the character formulae and their implications, and is followed in Section 4 by a description of the vertex construction of the basic representation. The discussion of the representations of Kac–Moody algebras will be concluded in Section 5 with a construction in terms of fermion creation and annihilation operators.

Other aspects of the representation theory of Kac–Moody algebras have been discussed by Jantzen (1979), Kac and Kazhdan (1979), Rocha–Coridi and Wallach (1983), Feingold (1984), Goodman and Wallach (1984a, b), Misra (1984), Jakobsen and Kac (1985), Patera (1986), Mickelsson (1985), Altschuler (1986), Dobrev (1986b), Wakimoto (1986), Goddard (1987), Kass and Patera (1987), Mazouni (1987), Wallach (1987), Chari and Pressley (1988), Kac and Raina (1988), Kac and Sanielevici (1988), Kac and Wakimoto (1988), Miki (1988) and Wang (1988). In particular, Mandia (1987) has given a detailed account of the structure of the level 1 standard representations of the affine Lie algebras $B_l^{(1)}$, $F_4^{(1)}$ and $G_2^{(1)}$, and Bernard and Thierry–Mieg (1987) have discussed at length the level 1 representations of all the non-simply-laced Kac–Moody algebras.

3 Character formulae

A number of interesting results follow from the generalization by Kac (1974) of the character formula of Weyl (1925, 1926a, b) for a semi-simple Lie algebra $\tilde{\mathscr{L}}$. In this latter case one can consider the representation of the universal linear group $\hat{\mathscr{G}}$ of the compact form $\mathscr{L}_c$ of $\tilde{\mathscr{L}}$ corresponding to the irreducible representation $\mathbf{\Gamma}$ with highest weight Λ, and for an element $\exp h$ of $\hat{\mathscr{G}}$, where $h \in \mathscr{H} \cap \mathscr{L}_c$, its character is

$$\chi_\Lambda(h) = \operatorname{tr}\{\mathbf{\Gamma}(\exp h)\} = \operatorname{tr}(\exp \mathbf{\Gamma}(h)),$$

(cf. Chapter 4, Section 6 and (11.13)). As $\mathbf{\Gamma}(h)$ is diagonal, with diagonal entries given by the weights $\lambda(h)$,

$$\chi_\Lambda(h) = \sum_\lambda m(\lambda) \exp\{\lambda(h)\}, \tag{27.31}$$

where $m(\lambda)$ is the multiplicity of λ and the sum is over all the distinct weights of the representation. *Weyl's character formula* for semi-simple Lie algebras is

$$\chi_\Lambda(h) = \frac{\sum_{S \in \mathscr{W}} (\det S) \exp\{S(\Lambda + \rho)(h)\}}{\exp\{\rho(h)\} \prod_{\alpha \in \Delta_+} \{1 - \exp\{-\alpha(h)\}\}}, \tag{27.32}$$

where the sum is over all the rotations S of the Weyl group $\mathscr{W}$ of $\tilde{\mathscr{L}}$, and ρ is the linear functional defined on $\mathscr{H}$ by

$$\frac{2\langle \rho, \alpha_j \rangle}{\langle \alpha_j, \alpha_j \rangle} = 1 \tag{27.33}$$

(for $j = 1, 2, \ldots, l$), so that

$$\rho(h) = \tfrac{1}{2} \sum_{\alpha \in \Delta_+} \alpha(h).$$

(In Chapter 13, 15 and 16 ρ was denoted by δ (cf. (13.77).) Here S has an associated $l \times l$ matrix $\mathbf{S}$ defined by

$$S\alpha_j = \sum_{k=1}^{l} S_{kj} \alpha_k,$$

and $\det S$ is defined to be the determinant of $\mathbf{S}$. It follows from the definition (13.73) that for the fundamental reflections (i.e. the Weyl reflections associated with the simple roots)

$$\det S_{\alpha_j} = -1$$

(for $j = 1, 2, \ldots, l$), so if S is the product of r fundamental reflections then

$$\det S = (-1)^r.$$

For example, for $\tilde{\mathscr{L}} = A_1$, since $\rho = \frac{1}{2}\alpha_1$ and since $\mathscr{W}$ consists only of the identity and the fundamental reflection S_α for which $\alpha = \alpha_1$, for the irreducible representation with highest weight $\Lambda = n_1\Lambda_1 = \frac{1}{2}n_1\alpha_1$ and for

$$h = 2i\omega h_{\alpha_1}$$

(with $n_1 = 2j + 1$) Weyl's character formula (27.32) reduces to

$$\chi_\Lambda(h) = \frac{\sin(j + \frac{1}{2})\omega}{\sin \frac{1}{2}\omega}.$$

This agrees with (12.33).

The irreducible representation with highest weight $\Lambda = 0$ is the trivial one-dimensional representation, so $\chi_0(h) = 1$ for all $h \in \mathscr{H}$. Equation (27.32) then gives the *denominator formula* for *semi-simple* algebras:

$$\exp\{\rho(h)\} \prod_{\alpha \in \Delta_+} \{1 - \exp\{-\alpha(h)\}\} = \sum_{S \in \mathscr{W}} (\det S) \exp\{S\rho(h)\}. \quad (27.34)$$

Weyl's dimensionality formula is obtained from (27.32) by considering the limit $h \to 0$.

Retaining the definition (27.31), Kac (1974) obtained as the generalization of (27.32) to an *affine Kac–Moody algebra* $\tilde{\mathscr{L}}$ the *character formula*

$$\chi_\Lambda(h) = \frac{\sum_{S \in \mathscr{W}} (\det S) \exp\{S(\Lambda + \rho)(h)\}}{\exp\{\rho(h)\} \prod_{\alpha \in \Delta_+} \{1 - \exp\{-\alpha(h)\}\}^{m(\alpha)}}, \quad (27.35)$$

where $m(\alpha)$ is the multiplicity of the root α, ρ is defined (cf. (27.33)) by

$$\frac{2\langle \rho, \alpha_j \rangle}{\langle \alpha_j, \alpha_j \rangle} = 0$$

(for $j = 0, 1, \ldots, l$) and

$$\rho(d) = 0,$$

and $\det S$ is defined by the obvious extension of the semi-simple Lie algebra discussion. The *denominator formula* for *affine Kac–Moody* algebras that generalizes (27.34) is then

$$\exp\{\rho(h)\} \prod_{\alpha \in \Delta_+} \{1 - \exp\{-\alpha(h)\}\}^{m(\alpha)} = \sum_{S \in \mathscr{W}} (\det S) \exp\{S\rho(h)\}. \quad (27.36)$$

The really significant new feature of (27.35) and (27.36) is the presence of the *imaginary* roots in Δ_+.

It has been shown by Kac (1978a), Feingold and Lepowsky (1978) and Kac and Peterson (1980) that for the *basic* representation of Theorem IV of Section 2 Kac's character formula (27.35) simplifies to

$$\chi_{\Lambda_0}(h) = \frac{\exp\{\Lambda_0(h)\} \sum_{\alpha \in Q^0} \exp\{\alpha(h) - (\langle \alpha, \alpha \rangle / \langle \alpha_0, \alpha_0 \rangle)\delta(h)\}}{\phi(\exp\{-\delta(h)\})^l},$$

where

$$\phi(q) = \prod_{k=1}^{\infty} (1 - q^k).$$

From this it may be inferred that the multiplicity $m(\lambda)$ of the weight λ of (27.30) is given by

$$m(\lambda) = p^{(l)}(j), \tag{27.37}$$

where

$$\sum_{j=0}^{\infty} p^{(l)}(j) q^j = \phi(q)^{-l}. \tag{27.38}$$

Kac's character formula (27.35) has many interesting applications in pure mathematics, particularly to such topics as Dedekind's η-function, theta functions and Hecke modular functions (cf. Kac (1974, 1978a, 1980, 1985a), Moody (1975), Feingold and Lepowsky (1978), Kac and Peterson (1980), MacDonald (1981), Frenkel (1982), Feingold and Frenkel (1983) and Frenkel *et al.* (1984)).

4 The vertex construction of the basic representation of a simply laced untwisted affine Kac–Moody algebra

Let $\tilde{\mathscr{L}}$ be a *simply laced* untwisted complex affine Kac–Moody algebra (that is, let $\tilde{\mathscr{L}} = A_l^{(1)} (l \geq 1)$, $D_l^{(1)} (l \geq 4)$, $E_6^{(1)}$, $E_7^{(1)}$ or $E_8^{(1)}$, and let $\tilde{\mathscr{L}}^0 (= \tilde{\mathscr{L}}^s)$ be its complex simple Lie subalgebra that is defined as in Chapter 26, Sections 5(*a*) and 5(*b*). The objective of the present section is to construct an explicit form for the operators of the basic representation, the weight structure of which was deduced in Theorem IV of Section 2.

As $\langle \alpha, \alpha \rangle^0$ has the same value for *every* root α of $\tilde{\mathscr{L}}^0$ (and for their extensions to $\tilde{\mathscr{L}}$), a scaling factor η can be defined by

$$\eta = \left(\frac{2}{\langle \alpha, \alpha \rangle^0} \right)^{1/2} \tag{27.39}$$

for any $\alpha \in \Delta^0$. Consideration of Appendix F shows that η takes the value $\{2(l+1)\}^{1/2}$ for $\tilde{\mathscr{L}}^0 = A_l$ $(l \geqslant 1)$, $(l-1)^{1/2}$ for D_l $(l \geqslant 4)$, $2\sqrt{3}$ for E_6, 6 for E_7, and $2\sqrt{15}$ for E_8.

By (13.25) and (13.51), the Chevelley basis elements H_α^0, E_α^0 and $E_{-\alpha}^0$ of $\tilde{\mathscr{L}}^0$ are related to the corresponding Weyl elements by

$$H_\alpha^0 = \eta^2 h_\alpha^0, \tag{27.40a}$$

$$E_\alpha^0 = \eta e_\alpha^0, \tag{27.40b}$$

$$E_{-\alpha}^0 = -\eta e_{-\alpha}^0 \tag{27.40c}$$

for all $\alpha \in \Delta_+^0$, and, by (13.26),

$$B^0(E_\alpha^0, E_{-\alpha}^0) = \eta^2 \tag{27.41}$$

for all $\alpha \in \Delta_+^0$. It is convenient also to "scale" the "central charge" c of (26.89) to given an element C such that

$$C = \eta^2 c. \tag{27.42}$$

Then, for the *basic* representation, by (27.18) and (27.22),

$$\Phi(C)\psi = \psi \tag{27.43}$$

for every $\psi \in \mathrm{V}$.

Let H_r^0 $(r = 1, 2, \ldots, l)$ be an orthonormal basis of the Cartan subalgebra $\mathscr{H}^0$ of $\tilde{\mathscr{L}}^0$ with respect to the Killing form $B^0(\ ,\ )$ of $\tilde{\mathscr{L}}^0$, so that

$$B^0(H_r^0, H_s^0) = \delta_{rs} \tag{27.44}$$

for $r, s = 1, 2, \ldots, l$ (cf. (13.56)). Then, as was shown in Chapter 13, Section 6, every linear functional α defined on $\mathscr{H}^0$ can be associated with an l-component vector $\boldsymbol{\alpha}$ by the prescription

$$\boldsymbol{\alpha} = (\alpha(H_1^0), \alpha(H_2^0), \ldots, \alpha(H_l^0)). \tag{27.45}$$

Moreover (cf. (13.59)),

$$\boldsymbol{\alpha}\cdot\boldsymbol{\beta} = \langle \alpha, \beta \rangle^0, \tag{27.46}$$

and, if h_α^0 is defined as in (13.11) (with a superscript 0 being inserted where appropriate), (13.57) shows that

$$h_\alpha^0 = \sum_{r=1}^{l} \alpha(H_r^0) H_r^0. \tag{27.47}$$

In constructing the basic representation of $\tilde{\mathscr{L}}$ it is convenient to introduce for every linear functional α on $\mathscr{H}^0$ a "scaled" l-component vector $\hat{\boldsymbol{\alpha}}$ by the definition

$$\hat{\boldsymbol{\alpha}} = \eta \boldsymbol{\alpha}, \tag{27.48}$$

so that

$$\hat{\boldsymbol{\alpha}} = (\eta\alpha(H_1^0), \eta\alpha(H_2^0), \ldots, \eta\alpha(H_l^0)). \tag{27.49}$$

Equations (27.48), (27.46) and (27.39) then imply that

$$\hat{\boldsymbol{\alpha}}\cdot\hat{\boldsymbol{\alpha}} = 2 \tag{27.50}$$

for *every* root α of Δ^0 and that for any two simple roots α_j and α_k

$$\hat{\boldsymbol{\alpha}}_j\cdot\hat{\boldsymbol{\alpha}}_k = A_{jk}. \tag{27.51}$$

Moreover, if α and β are any two roots of Δ^0, Theorem X of Chapter 13, Section 6 shows that $2\langle\alpha,\beta\rangle/\langle\alpha,\alpha\rangle = 0, \pm 1, \pm 2$ or ± 3, and, since detailed examination demonstrates that the values ± 3 cannot occur for simply laced algebras,

$$\hat{\boldsymbol{\alpha}}\cdot\hat{\boldsymbol{\beta}} = 0, \ \pm 1 \text{ or } \pm 2. \tag{27.52}$$

If $\alpha + \beta$ is also a non-zero root, (27.50) shows that

$$2 = (\hat{\boldsymbol{\alpha}} + \hat{\boldsymbol{\beta}})\cdot(\hat{\boldsymbol{\alpha}} + \hat{\boldsymbol{\beta}}) = \hat{\boldsymbol{\alpha}}\cdot\hat{\boldsymbol{\alpha}} + 2\hat{\boldsymbol{\alpha}}\cdot\hat{\boldsymbol{\beta}} + \hat{\boldsymbol{\beta}}\cdot\hat{\boldsymbol{\beta}} = 4 + 2\hat{\boldsymbol{\alpha}}\cdot\hat{\boldsymbol{\beta}},$$

so that

$$\hat{\boldsymbol{\alpha}}\cdot\hat{\boldsymbol{\beta}} = -1. \tag{27.53}$$

It can be shown that the converse is also true; that is, if α, $\beta\in\Delta^0$ and (27.53) holds then $\alpha + \beta\in\Delta^0$. Also,

$$\hat{\boldsymbol{\alpha}}\cdot\hat{\boldsymbol{\beta}} = -2 \tag{27.54}$$

if and only if $\beta = -\alpha$. Finally it follows from (27.52) and (27.50) that if α and β are any two members of the root lattice Q^0 of $\tilde{\mathscr{L}}^0$ then

$$\hat{\boldsymbol{\alpha}}\cdot\hat{\boldsymbol{\beta}} = \text{integer} \tag{27.55}$$

and

$$\hat{\boldsymbol{\alpha}}\cdot\hat{\boldsymbol{\alpha}} = \text{even integer}, \tag{27.56}$$

and a root lattice vector α of Q^0 is itself a root of $\tilde{\mathscr{L}}^0$ if and only if $\hat{\boldsymbol{\alpha}}\cdot\hat{\boldsymbol{\alpha}} = 2$.

Example I *Scaled root vectors for* $\tilde{\mathscr{L}}^0 = A_2$

By Example IV of Chapter 13, Section 6, $\boldsymbol{\alpha}_1 = \frac{1}{6}(2\sqrt{3}, 0)$ and $\boldsymbol{\alpha}_2 = \frac{1}{6}(-\sqrt{3}, 3)$ for $\tilde{\mathscr{L}}^0 = A_2$. As $\eta = \sqrt{6}$, it follows from (27.48) that

$$\hat{\boldsymbol{\alpha}}_1 = \sqrt{2}\,(1, 0),$$

$$\hat{\boldsymbol{\alpha}}_2 = \sqrt{2}\,(-\tfrac{1}{2}, \tfrac{1}{2}\sqrt{3}),$$

and hence

$$\hat{\boldsymbol{\alpha}}_1 + \hat{\boldsymbol{\alpha}}_2 = \sqrt{2}\,(\tfrac{1}{2}, \tfrac{1}{2}\sqrt{3}).$$

Clearly

$$\hat{\boldsymbol{\alpha}}_1 \cdot \hat{\boldsymbol{\alpha}}_1 = \hat{\boldsymbol{\alpha}}_2 \cdot \hat{\boldsymbol{\alpha}}_2 = (\hat{\boldsymbol{\alpha}}_1 + \hat{\boldsymbol{\alpha}}_2) \cdot (\hat{\boldsymbol{\alpha}}_1 + \hat{\boldsymbol{\alpha}}_2) = 2,$$

in accordance with (27.50), and

$$\hat{\boldsymbol{\alpha}}_1 \cdot \hat{\boldsymbol{\alpha}}_2 = -1,$$

in agreement with (27.53). Also,

$$\hat{\boldsymbol{\alpha}}_1 \cdot (\hat{\boldsymbol{\alpha}}_1 + \hat{\boldsymbol{\alpha}}_2) = \hat{\boldsymbol{\alpha}}_2 \cdot (\hat{\boldsymbol{\alpha}}_1 + \hat{\boldsymbol{\alpha}}_2) = 1,$$

the fact that these are not equal to -1 indicating that $2\alpha_1 + \alpha_2$ and $\alpha_1 + 2\alpha_2$ are not roots of A_2. Also, if $\alpha = \kappa_1^\alpha \alpha_1 + \kappa_2^\alpha \alpha_2$ then $\hat{\boldsymbol{\alpha}} \cdot \hat{\boldsymbol{\alpha}} = 2$ if and only if $(\kappa_1^\alpha)^2 + (\kappa_2^\alpha)^2 - \kappa_1^\alpha \kappa_2^\alpha = 1$, for which the only integer solutions are $\{\kappa_1^\alpha, \kappa_2^\alpha\} = \{\pm 1, 0\}$, $\{0, \pm 1\}$ and $\{\pm 1, \pm 1\}$, which correspond to the roots $\pm\alpha_1$, $\pm\alpha_2$ and $\pm(\alpha_1 + \alpha_2)$ respectively.

As was noted in Chapter 13, Section 8, with the *Chevalley basis* of $\tilde{\mathscr{L}}^0$ consisting of H_α^0 for $\alpha = \alpha_k$ $(k = 1, 2, \ldots, l)$ together with E_α^0 and $E_{-\alpha}^0$ for all $\alpha \in \Delta_+^0$, *all* the structure constants of $\mathscr{L}^0$ are *integers*. In the special case in which $\tilde{\mathscr{L}}^0$ is *simply laced* it can be verified that if α, β and $\alpha + \beta \in \Delta^0$ then

$$[E_\alpha^0, E_\beta^0]_- = \varepsilon(\alpha, \beta) E_{\alpha+\beta}^0, \tag{27.57}$$

where the *only* possible values of $\varepsilon(\alpha, \beta)$ are

$$\varepsilon(\alpha, \beta) = \pm 1. \tag{27.58}$$

Comparison with (13.31) shows that (by (27.40))

$$\varepsilon(\alpha, \beta) = \eta N_{\alpha,\beta}^0,$$

where η is defined by (27.39) and

$$[e_\alpha^0, e_\beta^0]_- = N_{\alpha,\beta}^0 e_{\alpha+\beta}^0.$$

The requirement $N_{\beta,\alpha}^0 = -N_{\alpha,\beta}^0$ implies that $\varepsilon(\beta, \alpha) = -\varepsilon(\alpha, \beta)$, which, by (27.58), is equivalent to

$$\varepsilon(\alpha, \beta)\varepsilon(\beta, \alpha) = -1 \tag{27.59}$$

for all $\alpha, \beta \in \Delta^0$ such that $\alpha + \beta \in \Delta^0$.

In terms of the Chevalley basis of $\tilde{\mathscr{L}}^0$ and the C of (27.42), the commutation relations of $\tilde{\mathscr{L}}$ are

$$[t^j \otimes H_\alpha^0, t^k \otimes H_\beta^0]_- = j\delta^{j+k,0} \hat{\boldsymbol{\alpha}} \cdot \hat{\boldsymbol{\beta}} C, \tag{27.60}$$

$$[t^j \otimes H_\alpha^0, t^k \otimes E_\beta^0]_- = \hat{\boldsymbol{\alpha}} \cdot \hat{\boldsymbol{\beta}} (t^{j+k} \otimes E_\beta^0), \tag{27.61}$$

$$[t^j \otimes E^0_\alpha, t^k \otimes E^0_\beta]_- = \begin{cases} 0 & \text{if } \alpha + \beta \notin \Delta^0 \text{ and } \alpha + \beta \neq 0, \\ \varepsilon(\alpha, \beta) t^{j+k} \otimes E^0_{\alpha+\beta} & \text{if } \alpha + \beta \in \Delta^0, \\ t^{j+k} \otimes H^0_\alpha + j\delta^{j+k,0} C & \text{if } \alpha + \beta = 0, \end{cases} \tag{27.62}$$

for $j, k = 0, \pm 1, \pm 2, \ldots$ and $\alpha, \beta \in \Delta^0$ (by (27.57), (26.133), (27.41), (27.42) and (13.9)). Also, by (26.90) and (26.91),

$$[C, t^j \otimes H^0_\alpha]_- = 0, \tag{27.63a}$$

$$[C, t^j \otimes E^0_\alpha]_- = 0 \tag{27.63b}$$

and

$$[d, t^j \otimes H^0_\alpha]_- = j t^j \otimes H^0_\alpha, \tag{27.64a}$$

$$[d, t^j \otimes E^0_\alpha]_- = j t^j \otimes E^0_\alpha \tag{27.64b}$$

for $j = 0, \pm 1, \pm 2, \ldots$ and $\alpha \in \Delta^0$, (26.92) being unchanged.

It has been shown by Frenkel and Kac (1980) that the domain of the quantities $\varepsilon(\alpha, \beta)$ introduced in (27.57) for $\alpha, \beta \in \Delta^0$ and $\alpha + \beta \in \Delta^0$ can be extended so that $\varepsilon(\alpha, \beta)$ becomes defined for *all* α and β of the root lattice Q^0 ($= \mathrm{Q}^s$) of $\tilde{\mathscr{L}}^0$ in such a way that

$$\varepsilon(\alpha, \beta) = \pm 1, \tag{27.65}$$

$$\varepsilon(\alpha, \beta)\varepsilon(\alpha + \beta, \gamma) = \varepsilon(\beta, \gamma)\varepsilon(\alpha, \beta + \gamma), \tag{27.66}$$

$$\varepsilon(\alpha, \beta)\varepsilon(\beta, \alpha) = \exp(i\pi \hat{\boldsymbol{\alpha}} \cdot \hat{\boldsymbol{\beta}}), \tag{27.67}$$

$$\varepsilon(\alpha, 0) = 1, \tag{27.68}$$

$$\varepsilon(\alpha, -\alpha) = 1, \tag{27.69}$$

for all $\alpha, \beta, \gamma \in \mathrm{Q}^0$. In the special case in which $\alpha, \beta \in \Delta^0$ and $\alpha + \beta \in \Delta^0$ (27.53) shows that (27.67) reduces to (27.59).

The carrier space V of the basic representation of $\tilde{\mathscr{L}}$ is given by

$$\mathrm{V} = \mathrm{V}_\mathrm{F} \otimes \mathrm{V}_\mathrm{L}, \tag{27.70}$$

where V_F is a "Fock space" and V_L is a space associated with the root lattice Q^0 of $\tilde{\mathscr{L}}^0$. These will now be considered in turn.

Let $A^{r\dagger}_j$ and A^r_j be a set of *boson creation and annihilation operators* that are defined for all positive integer values of j and for $r = 1, 2, \ldots, l$ and that are assumed to satisfy the commutation relations

$$[A^r_j, A^{s\dagger}_k]_- = j\delta_{jk}\delta^{rs}1, \tag{27.71a}$$

$$[A^r_j, A^s_k]_- = 0, \tag{27.71b}$$

$$[A^{r\dagger}_j, A^{s\dagger}_k]_- = 0, \tag{27.71c}$$

for $j, k = 1, 2, \ldots$ and $r, s = 1, 2, \ldots, l$. The *vacuum state* $|0\rangle$ is defined by

$$A_j^r|0\rangle = 0 \tag{27.72}$$

for all $j = 1, 2, \ldots$ and $r = 1, 2, \ldots, l$. The Fock space V_F is then defined to be the infinite-dimensional complex vector space that has as its basis vectors the vacuum state $|0\rangle$ together with the *state vectors*

$$\Psi(n_1^1, \ldots, n_j^r, \ldots) = \prod_{j=1}^{\infty} \prod_{r=1}^{l} \{(n_j^r)!\, j^{n_j^r}\}^{-1/2} (A_j^{r\dagger})^{n_j^r}|0\rangle, \tag{27.73}$$

where the *occupation numbers* n_j^r can take any non-negative integer values. The vacuum state $|0\rangle$ corresponds to all the occupation numbers n_j^r being zero, i.e.

$$\Psi(0, 0, \ldots) = |0\rangle.$$

With these definitions,

$$A_j^{r\dagger}\Psi(n_1^1, \ldots, n_j^r, \ldots) = \{j(n_j^r + 1)\}^{1/2}\Psi(n_1^1, \ldots, n_j^r + 1, \ldots); \tag{27.74}$$

that is, the occupation number of the state corresponding to the pair j and r is increased by 1, all the others being unchanged, while

$$A_j^r\Psi(n_1^1, \ldots, n_j^r, \ldots) = \begin{cases} (jn_j^r)^{1/2}\Psi(n_1^1, \ldots, n_j^r - 1, \ldots) & \text{if } n_j^r > 0, \\ 0 & \text{if } n_j^r = 0. \end{cases} \tag{27.75}$$

If $n_j^r > 0$, this shows that A_j^r reduces the occupation number of the state corresponding to j and r by 1, all others being unchanged, whereas if $n_j^r = 0$ the state vector is completely annihilated.

An inner product can be defined on V_F by the assumption that on its basis vectors

$$(\Psi(n_1^1, \ldots, n_j^r, \ldots), \Psi(n_1'^1, \ldots, n_j'^r, \ldots))_F = \begin{cases} 1 & \text{if } n_j^r = n_j'^r \text{ for all } j = 1, 2, \ldots \text{ and } r = 1, 2, \ldots, l, \\ 0 & \text{otherwise,} \end{cases} \tag{27.76}$$

so the state vectors $\Psi(n_1^1, \ldots, n_j^r, \ldots)$ form an orthonormal set. The operator $A_j^{r\dagger}$ is then the adjoint of A_j^r for every $j = 1, 2, \ldots$ and $r = 1, 2, \ldots, l$ (cf. Appendix B, Section 4).

For notational convenience one may define the l-component operators $\mathbf{A}_j^\dagger$ and $\mathbf{A}_j$ by

$$\mathbf{A}_j^\dagger = (A_j^{1\dagger}, A_j^{2\dagger}, \ldots, A_j^{l\dagger}) \tag{27.77a}$$

and

$$\mathbf{A}_j = (A_j^1, A_j^2, \ldots, A_j^l), \tag{27.77b}$$

for each $j = 1, 2, \ldots$. Then, by (27.49),

$$\hat{\boldsymbol{\alpha}} \cdot \mathbf{A}_j^\dagger = \eta \sum_{r=1}^{l} \alpha(H_r^0) A_j^{r\dagger} \tag{27.78a}$$

and

$$\hat{\boldsymbol{\alpha}} \cdot \mathbf{A}_j = \eta \sum_{r=1}^{l} \alpha(H_r^0) A_j^r. \tag{27.78b}$$

The vector space V_L is defined to be the set of all complex linear combinations of the formal exponentials e^α that may be set up for each α of the root lattice Q^0 of $\tilde{\mathscr{L}}^0$. The inner product of two such basis elements e^α and e^β of V_L may be defined by

$$(e^\alpha, e^\beta)_L = \delta^{\alpha,\beta} \tag{27.79}$$

for all $\alpha, \beta \in Q^0$, so the basis elements e^α form an orthonormal set.

With the usual definition of an inner product of a direct product space (cf. Appendix 13, Section 7), the vectors $\Psi(n_1^1, \ldots, n_j^r, \ldots) \otimes e^\gamma$ form an orthonormal basis for $V = V_F \otimes V_L$. The creation and annihilation operators $A_j^{r\dagger}$ and A_j^r may be considered to act on V, with the understanding that they affect only members of V_F. Henceforth, for brevity a typical basis element of V will be denoted by $\psi_F \otimes e^\gamma$ (where ψ_F is a basis vector of V_F and $\gamma \in Q^0$), unless a more explicit expression is required. In order to preserve the signs in the commutation relations (27.62), it is necessary to introduce for each $\alpha \in Q^0$ an operator c_α by the definition

$$c_\alpha(\psi_F \otimes e^\gamma) = \varepsilon(\alpha, \gamma)(\psi_F \otimes e^\gamma) \tag{27.80}$$

for all $\gamma \in Q^0$.

The **vertex operator** $U(\alpha, z)$ is defined for each $\alpha \in Q^0$ by

$$U(\alpha, z) = \exp\{i\hat{\boldsymbol{\alpha}} \cdot \mathbf{Q}_<(z)\} M_\alpha(z) \exp\{i\hat{\boldsymbol{\alpha}} \cdot \mathbf{Q}_>(z)\}, \tag{27.81}$$

where $\mathbf{Q}_<(z)$ and $\mathbf{Q}_>(z)$ are l-component operators whose components are given by

$$Q_<^r(z) = -i \sum_{j=1}^{\infty} \frac{z^j}{j} A_j^{r\dagger} \tag{27.82}$$

and

$$Q_>^r = i \sum_{j=1}^{\infty} \frac{z^{-j}}{j} A_j^r \tag{27.83}$$

for $r = 1, 2, \ldots, l$, and where $M_\alpha(z)$ is the operator defined by

$$M_\alpha(z)(\psi_F \otimes e^\gamma) = z^{\hat{\boldsymbol{\alpha}} \cdot \hat{\boldsymbol{\gamma}} + \hat{\boldsymbol{\alpha}} \cdot \hat{\boldsymbol{\alpha}}/2}(\psi_F \otimes e^{\alpha+\gamma}) \tag{27.84}$$

for all $\gamma \in Q^0$. In (27.81)–(27.84) z is an arbitrary complex number. This operator $U(\alpha, z)$ is an adaptation of the vertex operator of the dual resonance models of hadron physics (cf. Fubini *et al.* 1969, Fubini and Veneziano 1970, Nambu 1970, Gervais 1970, Schwarz 1973, Jacob 1974, Mandelstam 1974, Scherk 1975).

Equations (27.84), (27.55) and (27.56) show that the power of z in $M_\alpha(z)$ is always an *integer*, and so $M_\alpha(z)$ and $U(\alpha, z)$ are both single-valued about $z = 0$. Consequently $U(\alpha, z)$ has a Laurent expansion

$$U(\alpha, z) = \sum_{j=-\infty}^{\infty} U_j(\alpha) z^{-j}, \tag{27.85}$$

where the Laurent coefficients $U_j(\alpha)$, which are operators acting on V, are given by

$$U_j(\alpha) = \frac{1}{2\pi i} \oint z^{j-1} U(\alpha, z)\, \mathrm{d}z \tag{27.86}$$

for $j = 0, \pm 1, \pm 2, \ldots$, the integration being over any closed contour enclosing the point $z = 0$ (cf. Copson 1935).

Theorem I (a) The operators of the basic representation (with highest weight Λ_0) of a simply-laced untwisted affine Kac–Moody algebra $\tilde{\mathscr{L}}$ act on the carrier space V defined by (27.70) and are given by

(i)
$$\Phi(C) = 1, \tag{27.87}$$

where 1 is the identity operator;

(ii)
$$\Phi(d) = -N_{\mathrm{F}} - N_{\mathrm{L}}, \tag{27.88}$$

where

$$N_{\mathrm{F}} = \sum_{j=1}^{\infty} \sum_{r=1}^{l} A_j^{r\dagger} A_j^r \tag{27.89}$$

and

$$N_{\mathrm{L}}(\psi_{\mathrm{F}} \otimes \mathrm{e}^{\gamma}) = \tfrac{1}{2} \hat{\boldsymbol{\gamma}} \cdot \hat{\boldsymbol{\gamma}} (\psi_{\mathrm{F}} \otimes \mathrm{e}^{\gamma}) \tag{27.90}$$

for all $\gamma \in \mathrm{Q}^0$;

(iii) for $\alpha \in \Delta^0$

$$\Phi(t^0 \otimes H_\alpha^0)(\psi_{\mathrm{F}} \otimes \mathrm{e}^{\gamma}) = \hat{\boldsymbol{\alpha}} \cdot \hat{\boldsymbol{\gamma}} (\psi_{\mathrm{F}} \otimes \mathrm{e}^{\gamma}) \tag{27.91}$$

for all $\gamma \in \mathrm{Q}^0$;

(iv) for $j = 1, 2, \ldots$ and $\alpha \in \Delta^0$

$$\Phi(t^j \otimes H_\alpha^0) = \hat{\boldsymbol{\alpha}} \cdot \mathbf{A}_j \tag{27.92}$$

and

$$\Phi(t^{-j} \otimes H_\alpha^0) = \hat{\boldsymbol{\alpha}} \cdot \mathbf{A}_j^\dagger; \tag{27.93}$$

(v) for $j = 0, \pm 1, \pm 2, \ldots$ and $\alpha \in \Delta^0$

$$\Phi(t^j \otimes E^0_\alpha) = U_j(\alpha) c_\alpha. \tag{27.94}$$

(b) The weights of the basic representation are given by

$$\lambda = \Lambda_0 + \gamma - (m + \tfrac{1}{2}\hat{\gamma}\cdot\hat{\gamma})\delta, \tag{27.95}$$

where γ runs over all members of the root lattice Q^0 of $\tilde{\mathscr{L}}^0$ and $m = 0, 1, 2, \ldots,$ with every vector

$$\Psi(n^1_1, \ldots, n^r_j, \ldots) \otimes e^\gamma \tag{27.96}$$

such that

$$m = \sum_{j=1}^{\infty} \sum_{r=1}^{l} j n^r_j \tag{27.97}$$

being a weight vector $\psi(\lambda)$ with this weight as eigenvalue.

(c) For all $a \in \mathscr{L}_c$, the compact real form of $\tilde{\mathscr{L}}$, and for all $\psi, \phi \in V$

$$(\Phi(a)\psi, \phi) = -(\psi, \Phi(a)\phi), \tag{27.98}$$

so the basic representation is "unitary".

Proof See Appendix O, Section 2.

The results of this theorem were first established by Frenkel and Kac (1980), a related analysis also being given by Segal (1981). For the modifications requires for the level 1 standard irreducible representations of the twisted affine Kac–Moody algebras $A^{(2)}_{2l}$ ($l \geqslant 1$), $A^{(2)}_{2l-1}$ ($l \geqslant 3$) and $D^{(2)}_{l+1}$ ($l \geqslant 2$) see Frenkel (1985). Goddard *et al.* (1986b, 1987a) and Olive (1987) have given the generalization to the level 1 standard irreducible representations of the non-simply-laced untwisted affine Kac–Moody algebras. For other developments involving the vertex construction see Lepowski and Wilson (1978, 1984), Kac *et al.* (1981), Lepowski *et al.* (1984), Kaplansky and Santharoubane (1984), Lepowski and Primc (1984a, b, 1985), Kac and Peterson (1985), Lepowski (1985a, b), Alvarez *et al.* (1986), Goddard and Olive (1984, 1986, 1988), Corrigan (1986), Goddard (1987), Goddard *et al.* (1987b), Golitzin (1988) and Sorba and Torresani (1988).

It will be noted that part (b) of the above theorem is consistent with part (c) of Theorem VI of Section 2, and implies the multiplicity formulae (27.37) and (27.38). For example, for $m = 0$ in (27.95), (27.97) implies that $n^r_j = 0$ for all j and r, so $\psi(\lambda)$ is given by $|0\rangle \otimes e^\gamma$, which, being unique, corresponds to $m(\lambda) = 1$. Similarly, for $m = 1$ (27.97) implies that $n^r_1 = 1$ for just one r, and all other n^r_j are zero, so in this case $m(\lambda) = l$, the weight vectors $\psi(\lambda)$ being $A^{r\dagger}_1|0\rangle \otimes e^\gamma$ for $r = 1, 2, \ldots, l$. Likewise, for $m = 2$ the possible independent weight vectors are

$A_2^{r\dagger}|0\rangle \otimes e^{\gamma}$ (for $r = 1, 2, \ldots, l$) *and* $A_1^{r\dagger} A_1^{s\dagger}|0\rangle \otimes e^{\gamma}$ (for $r, s = 1, 2, \ldots, l$, with $r \leqslant s$). Consequently for $m = 2$ the multiplicity is $m(\lambda) = l + \frac{1}{2}l(l+1)$.

As noted in Chapter 26, Section 5(*b*), the subset of elements of the Kac–Moody algebra $\tilde{\mathscr{L}}$ that have the form $t^0 \otimes a^0$, where a^0 is any element of the complex simple Lie algebra $\tilde{\mathscr{L}}^0$, form a subalgebra of $\tilde{\mathscr{L}}$ that is isomorphic to $\tilde{\mathscr{L}}^0$. The basic representation of $\tilde{\mathscr{L}}$ therefore provides a representation of $\tilde{\mathscr{L}}^0$, but since the basic representation is infinite-dimensional and the highest weight representations of $\tilde{\mathscr{L}}^0$ are all finite-dimensional, this representation of $\tilde{\mathscr{L}}^0$ will be highly reducible. The following theorem shows that one of the irreducible representations of $\tilde{\mathscr{L}}^0$ that appear in this reduction is the *adjoint* representation of $\tilde{\mathscr{L}}^0$ (cf. Chapter 11, Section 5), a fact that plays a crucial role in the construction of the heterotic string (cf. Chapter 29, Section 5(*b*)).

Theorem II The two sets of element:

(i) $|0\rangle \otimes e^{\gamma}$, for every non-zero root γ of $\tilde{\mathscr{L}}^0$, and

(ii) $A_1^{s\dagger}|0\rangle \otimes e^0$, for $s = 1, 2, \ldots, l$,

together form a basis for the *adjoint* representation of $\tilde{\mathscr{L}}^0$ (the operators $\Phi(t^0 \otimes H_\alpha^0)$ and $\Phi(t^0 \otimes E_\alpha^0)$ being as defined in (27.91) and (27.94) respectively).

Two comments are worth making here. First, it should be recalled that a root lattice vector γ of $\tilde{\mathscr{L}}^0$ is a root of $\tilde{\mathscr{L}}^0$ if and only if $\hat{\gamma} \cdot \hat{\gamma} = 2$. Secondly, the basis vectors of the adjoint representation of $\tilde{\mathscr{L}}^0$ in its usual definition (11.36) are the basis elements of $\tilde{\mathscr{L}}^0$ itself, and the proof of this theorem will demonstrate explicitly that the identification of this carrier space with that of the basic representation is given by

(i)
$$|0\rangle \otimes e^{\gamma} \Leftrightarrow t^0 \otimes E_\gamma^0 \tag{27.99}$$

for every non-zero root γ of $\tilde{\mathscr{L}}^0$, and

(ii)
$$A_1^{s\dagger}|0\rangle \otimes e^0 \Leftrightarrow t^0 \otimes H_s \tag{27.100}$$

for $s = 1, 2, \ldots, l$.

Proof See Appendix O, Section 2.

5 Representations of untwisted affine Kac–Moody algebras in terms of fermion creation and annihilation operators

Suppose that the untwisted affine Kac–Moody algebra $\tilde{\mathscr{L}}(=\tilde{\mathscr{L}}^{(1)})$ is constructed from a simple Lie algebra $\tilde{\mathscr{L}}^0$ exactly as in Chapter 26, Section 3(*b*) and that the basis elements a_m^0 ($m = 1, 2, \ldots, n^0$) of $\tilde{\mathscr{L}}^0$ are those of its compact real form $\mathscr{L}_c^0$. Let $\boldsymbol{\Gamma}$ be a representation of $\mathscr{L}_c^0$ of dimension D^0. As noted in Chapter 26, Section 2, after an appropriate similarity transformation the matrices $\boldsymbol{\Gamma}(a_m^0)$ may be taken to be anti-Hermitian. The additional assumption will now be made that they are *real*, implying that they are also *antisymmetric*. (The extension to non-real representations will be considered briefly at the end of this section.)

Let B_p^r be a set of operators satisfying the anticommutation relations

$$[B_p^r, B_q^s]_+ = \delta^{rs}\delta_{p+q,0}\,1, \tag{27.101}$$

where r and s take values $1, 2, \ldots, D_0$, and where p and q either take all integer values or they take all "half-integer" values (i.e. $p, q = \pm\frac{1}{2}, \pm\frac{3}{2}, \ldots$). As such operators appear in the theory of the fermionic string, where they correspond to the Ramond and Neveu–Schwarz models respectively (cf. Chapter 29, Section 3), the integer index case will be referred to as the *Ramond case* and the half-integer case as the *Neveu–Schwarz* case.

With the operators $B_p^{r\dagger}$ defined for $p > 0$ by

$$B_p^{r\dagger} = B_{-p}^r, \tag{27.102}$$

(27.101) can be rewritten for $p > 0$ and $q > 0$ as

$$[B_p^{r\dagger}, B_q^s]_+ = \delta^{rs}\delta_{pq}1, \tag{27.103a}$$

$$[B_p^{r\dagger}, B_q^{s\dagger}]_+ = 0, \tag{27.103b}$$

$$[B_p^r, B_q^s]_+ = 0, \tag{27.103c}$$

so that for $p > 0$ the operators $B_p^{r\dagger}$ and B_p^r can be interpreted as fermion creation and annihilation operators respectively.

In the *Neveu–Schwarz* case there is a unique *vacuum* state, which may be denoted by $|0\rangle$, such that

$$B_p^r|0\rangle = 0 \tag{27.104}$$

for $r = 1, 2, \ldots, D_0$ and for all $p > 0$. The corresponding Fock space is similar to that introduced in Section 4 in connection with bosonic creation and annihilation operators in that it is defined to be the infinite-dimensional complex vector space that has as its basis vectors the vacuum state $|0\rangle$ together with the state vectors

$$\Psi(n_1^1, \ldots, n_p^r, \ldots) = \prod_{p>0}\prod_{r=1}^{D_0}(B_p^{r\dagger})^{n_p^r}|0\rangle \tag{27.105}$$

(cf. (27.73)), but the new feature of the fermion case is that with $r = s$ and $p = q$ (27.103) implies that $B_p^{r\dagger}B_p^{r\dagger} = 0$, so the *only* possible values of the occupation numbers n_p^r are 0 and 1. The vacuum state corresponds to $n_p^r = 0$ for all

$r = 1, 2, \ldots, D_0$ and all $p > 0$; that is,

$$\Psi(0, 0, \ldots) = |0\rangle. \tag{27.106}$$

As (27.103) also implies that $B_p^{r\dagger} B_q^{s\dagger} = -B_q^{s\dagger} B_p^{r\dagger}$, it is clear that for the definition (27.105) to be unambiguous it is necessary to order the pairs of state labels (p, r), in a specific way. The choice that will be adopted here is that in (27.105) $B_p^{r\dagger}$ lies to the *left* of $B_q^{s\dagger}$ when either $p < q$ or when $p = q$ and $r < s$. With this choice,

$$B_p^{r\dagger}\Psi(n_1^1, \ldots, n_p^r, \ldots) = \begin{cases} (-1)^{\Sigma}\Psi(n_1^1, \ldots, n_p^r = 1, \ldots) & \text{if } n_p^r = 0, \\ 0 & \text{if } n_p^r = 1. \end{cases} \tag{27.107}$$

That is, if $n_p^r = 0$, the occupation number of the state corresponding to the pair (p, r) is increased to 1, all other occupation numbers being unchanged, whereas if $n_p^r = 1$ the state vector is annihilated by $B_p^{r\dagger}$. In (27.107) Σ is defined as the sum of the occupation numbers n_q^s corresponding to all pairs (q, s) for which either $q < p$ or alternatively $q = p$ and $s < r$. Similarly

$$B_p^r\Psi(n_1^1, \ldots, n_p^r, \ldots) = \begin{cases} (-1)^{\Sigma}\Psi(n_1^1, \ldots, n_p^r = 0, \ldots) & \text{if } n_p^r = 1, \\ 0 & \text{if } n_p^r = 0, \end{cases} \tag{27.108}$$

with Σ as before. Thus if $n_p^r = 1$, the occupation number of the pair (p, r) is decreased to 0, all the other occupation numbers being unchanged, but if $n_p^r = 0$, the state vector is annihilated by B_p^r. An inner product can be defined in the Fock space by the assumption that the state vectors $\Psi(n_1^1, \ldots, n_j^r, \ldots)$ form an orthonormal set (cf. (27.76)). Then $B_p^{r\dagger}$ is the adjoint of B_p^r for all $p > 0$ and $r = 1, 2, \ldots, D_0$.

The situation for the *Ramond* case is a little more complicated, since in this case (27.101) with $p = q = 0$ gives

$$[B_0^r, B_0^s]_+ = \delta^{rs} 1 \tag{27.109}$$

for $r, s = 1, 2, \ldots, D_0$. Consider the set of $d_0 \times d_0$ matrices γ_E^r that satisfy the anticommutation relations

$$[\gamma_E^r, \gamma_E^s]_+ = 2\delta^{rs}\mathbf{1} \tag{27.110}$$

for $r, s = 1, 2, \ldots, D_0$. As they form a representation of the D_0-dimensional Euclidean Clifford algebra, the analysis of Appendix L shows that

$$d_0 = \begin{cases} 2^{D_0/2} & \text{if } D_0 \text{ is even,} \\ 2^{(D_0-1)/2} & \text{if } D_0 \text{ is odd.} \end{cases} \tag{27.111}$$

Let $|\mu; 0\rangle$ for $\mu = 1, 2, \ldots, d_0$ be d_0 vacuum state vectors chosen so that

$$B_0^r|\mu; 0\rangle = 2^{-1/2} \sum_{\kappa=1}^{d_0} (\gamma_E^r)_{\kappa\mu}|\kappa; 0\rangle \tag{27.112}$$

and

$$B^r_p|\mu;0\rangle = 0 \quad \text{for } p>0. \tag{27.113}$$

Then clearly (27.109) is satisfied. Each state vector of the corresponding Fock space requires an additional label μ ($=1,2,\ldots,d_0$) to identify the vacuum state on which it is based. That is, the generalization of (27.105) is

$$\Psi(\mu;n^1_1,\ldots,n^r_p,\ldots) = \prod_{p>0}\prod_{r=1}^{D_0}(B^{r+}_p)^{n^r_p}|\mu;0\rangle, \tag{27.114}$$

and then (27.106)–(27.108) generalize in the obvious way.

In *both* the Neveu–Schwarz and Ramond cases the *normal product* of a pair of operators B^r_p and B^s_q may be denoted by $:B^r_pB^s_q:$ and defined by

$$:B^r_pB^s_q: = \begin{cases} B^r_pB^s_q & \text{if } p<q, \\ \tfrac{1}{2}(B^r_pB^s_q - B^s_qB^r_p) & \text{if } p=q, \\ -B^s_qB^r_p & \text{if } p>q. \end{cases} \tag{27.115}$$

(This contains as a special case the usual normal product for fermionic operators in which creation operators appear to the left of annihilation operators, with a factor -1 being inserted if the order of factors is changed.) Taken with (27.101), (27.115) implies that

$$:B^r_pB^s_q: = \begin{cases} B^r_pB^s_q & \text{if } p<q, \\ B^r_pB^s_q - \delta^{rs}\delta_{p+q,0}1 & \text{if } p\geq q, \end{cases}$$

and hence that

$$:B^r_pB^s_q: = \begin{cases} B^r_pB^s_q - \delta^{rs}1 & \text{if } p+q=0 \text{ and } p\geq 0, \\ B^r_pB^s_q & \text{otherwise.} \end{cases} \tag{27.116}$$

Theorem I The set of operators defined by

$$\Phi(t^j\otimes a^0_m) = \frac{1}{2}\sum_{r,s=1}^{D_0}\Gamma(a^0_m)_{rs}\sum_p :B^r_{j-p}B^s_p: \tag{27.117}$$

for $j=0,\pm 1,\pm 2,\ldots$ and $m=1,2,\ldots,n^0$, and by

$$\Phi(c) = \tfrac{1}{2}\gamma 1, \tag{27.118}$$

where γ is the Dynkin index of the representation Γ of $\tilde{\mathscr{L}}^0$ (cf. (16.12)), form a representation of the derived algebra $[\tilde{\mathscr{L}},\tilde{\mathscr{L}}]_-$ of the untwisted affine Kac–Moody algebra $\tilde{\mathscr{L}}$. (In (27.117) the index p runs over all integer values in the Ramond case and over all half-integer values in the Neveu–Schwarz case, the carrier space of the representation being the corresponding Fock space.)

Proof See Appendix O, Section 3.

Equations (27.117) imply that in the Neveu–Schwarz case

$$\Phi(t^j \otimes a_m^0)|0\rangle = 0 \tag{27.119}$$

for all $j \geqslant 0$ (and $m = 1, 2, \ldots, n^0$), so the vacuum state is the highest weight vector, and this is associated with the one-dimensional trivial representation of the simple Lie algebra $\tilde{\mathscr{L}}^0$. In the Ramond case

$$\Phi(t^j \otimes a_m^0)|\mu; 0\rangle = 0 \tag{27.120}$$

for $j > 0$ (and $m = 1, 2, \ldots, n^0$ *and* $\mu = 1, 2, \ldots, d_0$), and the vacuum vectors $|\mu; 0\rangle$ form the basis for a d_0-dimensional representation $\boldsymbol{\Gamma}'$ of the simple Lie algebra $\mathscr{L}^0$ that is given by

$$\boldsymbol{\Gamma}'(a_m^0) = \frac{1}{4}\sum_{r,s=1}^{D_0} \Gamma(a_m^0)_{rs}(\gamma_{\mathrm{E}}^r \gamma_{\mathrm{E}}^s - \gamma_{\mathrm{E}}^s \gamma_{\mathrm{E}}^r).$$

As the antisymmetry property of $\boldsymbol{\Gamma}$ implies that $\Gamma(a_m^0)_{rr} = 0$ for each r, (27.110) shows that

$$\boldsymbol{\Gamma}'(a_m^0) = \frac{1}{2}\sum_{r,s=1}^{D_0} \Gamma(a_m^0)_{rs}\gamma_{\mathrm{E}}^r \gamma_{\mathrm{E}}^s. \tag{27.121}$$

Thus in both cases the representation of $[\tilde{\mathscr{L}}, \tilde{\mathscr{L}}]_-$ is a highest weight representation, and (27.118) and (27.22) together show that its level is $\gamma/\langle \alpha_{\mathrm{H}}, \alpha_{\mathrm{H}}\rangle$. In particular, if $\boldsymbol{\Gamma}$ is the adjoint representation of $\mathscr{L}_{\mathrm{c}}^0$ (which is real and for which $\gamma = 1$), then the level is $(\langle \alpha_{\mathrm{H}}, \alpha_{\mathrm{H}}\rangle)^{-1}$. Also, in the special case in which $\mathscr{L}_{\mathrm{c}}^0 = \mathrm{so}(N)$ for $N \geqslant 5$ it is easily shown that $\gamma = \langle \alpha_{\mathrm{H}}, \alpha_{\mathrm{H}}\rangle$ for the N-dimensional defining "vector" representation, and so the resulting level is 1.

If the representation $\boldsymbol{\Gamma}$ of the compact real form $\mathscr{L}_{\mathrm{c}}^0$ is anti-Hermitian but is *not* real then a real antisymmetric representation $\boldsymbol{\Gamma}''$ of dimension $2D_0$ can be defined for all $a \in \mathscr{L}_{\mathrm{c}}^0$ by

$$\boldsymbol{\Gamma}''(a) = \begin{bmatrix} \frac{1}{2}\{\boldsymbol{\Gamma}(a)^* + \boldsymbol{\Gamma}(a)\} & \frac{1}{2}i\{\boldsymbol{\Gamma}(a)^* - \boldsymbol{\Gamma}(a)\} \\ -\frac{1}{2}i\{\boldsymbol{\Gamma}(a)^* - \boldsymbol{\Gamma}(a)\} & \frac{1}{2}\{\boldsymbol{\Gamma}(a)^* + \boldsymbol{\Gamma}(a)\} \end{bmatrix},$$

which is such that

$$\boldsymbol{\Gamma}''(a) = \mathbf{S}^{-1}\begin{bmatrix} \boldsymbol{\Gamma}(a) & \mathbf{0} \\ \mathbf{0} & \boldsymbol{\Gamma}(a)^* \end{bmatrix}\mathbf{S}$$

with

$$\mathbf{S} = 2^{-1/2}\begin{bmatrix} \mathbf{1} & \mathbf{1} \\ i\mathbf{1} & -i\mathbf{1} \end{bmatrix}.$$

The above analysis can then be repeated with $\mathbf{\Gamma}$ replaced by $\mathbf{\Gamma}''$ (and D_0 replaced by $2D_0$).

This type of representation is sometimes called a *spinor representation*. For further details see Frenkel (1980, 1981), Kac and Pearson (1981), Goddard and Olive (1986, 1988) and Alvarez *et al.* (1986).

28

The Virasoro Algebra and Superalgebras

1 The conformal algebras

The Virasoro algebra was first introduced by Virasoro (1970) in the dual resonance model for hadrons. As will be discussed in Chapter 29, both this algebra and its supersymmetric extensions have played very important roles in the development of string and superstring theories. The purpose of the present chapter is to describe the main properties of these algebras, including some interesting relationships with the affine Kac–Moody algebras.

Because string theory is a field theory based on two real independent variables σ and τ, the so-called "world-sheet" coordinates, which can be combined to form a single complex variable z, meromorphic functions of z appear naturally in the theory. As conformal invariance plays a significant part in the developments, the first matter that will be discussed is the general theory of conformal groups and their associated Lie algebras.

The conformal group of a real D-dimensional Euclidean or pseudo-Euclidean space with metric tensor $\eta_{\mu\lambda}$ (where $\eta_{\mu\lambda} = \pm\delta_{\mu\lambda}$ for $\mu, \lambda = 1, 2, \ldots, D$) is *defined* as the set of coordinate transformations

$$x^{\mu} \rightarrow x'^{\mu}(x^1, x^2, \ldots, x^D)$$

that are such that

$$\sum_{\mu,\lambda=1}^{D} \eta_{\mu\lambda} \frac{\partial x^{\mu}}{\partial x'^{\kappa}} \frac{\partial x^{\lambda}}{\partial x'^{\sigma}} = \rho(x^1, x^2, \ldots, x^D)\eta_{\kappa\sigma}, \tag{28.1}$$

where ρ is some function of $x^1, x^2, \ldots, x^D$ that depends on the transformation. It is obvious that the set of such transformations do indeed form a group, and it is also clear that they leave invariant the "angle" whose cosine is

$$\frac{\sum_{\mu,\lambda=1}^{D} \eta_{\mu\lambda}\,\mathrm{d}x^{\mu}\,\mathrm{d}y^{\lambda}}{\left(\sum_{\kappa,\sigma=1}^{D} \eta_{\kappa\sigma}\,\mathrm{d}x^{\kappa}\,\mathrm{d}x^{\sigma}\right)^{1/2}\left(\sum_{\tau,\nu=1}^{D} \eta_{\tau\nu}\,\mathrm{d}y^{\tau}\,\mathrm{d}y^{\nu}\right)^{1/2}}$$

between any two infinitesimal vectors $\mathrm{d}x^1, \mathrm{d}x^2, \ldots, \mathrm{d}x^D$ and $\mathrm{d}y^1, \mathrm{d}y^2, \ldots, \mathrm{d}y^D$. For an infinitesimal transformation of the form

$$x'^{\mu} = x^{\mu} + \delta x^{\mu}$$

the above condition implies that

$$\partial_{\mu}(\delta x_{\lambda}) + \partial_{\lambda}(\delta x_{\mu}) = \eta_{\mu\lambda}\,\delta\rho,$$

where

$$\rho = 1 + \delta\rho,$$

which gives, on taking the trace,

$$\delta\rho = \frac{2}{D}\sum_{\mu=1}^{D} \partial_{\mu}(\delta x^{\mu}).$$

For $D \geqslant 3$ the conformal groups are all *finite*-dimensional Lie groups. The one of this type that is most commonly encountered is the conformal group of 4-dimensional Minkowski space–time, which of course corresponds to $D = 4$ and $\eta_{\mu\lambda} = \mathrm{diag}(-1, -1, -1, 1)$. Clearly the set of all 4-dimensional Poincaré transformations forms a 10-parameter subgroup of this group (with $\rho = 1$). This conformal group also includes the 1-parameter subgroup of "scale" transformations (otherwise known as "dilatations"), for which $x'^{\mu} = \Lambda x^{\mu}$, where Λ is any non-zero real constant (independent of μ), as well as the 4-parameter subgroup of "special conformal transformations", which have the form

$$x'^{\mu} = \left(x^{\mu} - c^{\mu}\sum_{\kappa,\sigma=1}^{4} \eta_{\kappa\sigma}x^{\kappa}x^{\sigma}\right)\Big/\phi(x),$$

where

$$\phi(x) = 1 - \sum_{\kappa,\sigma=1}^{4} \eta_{\kappa\sigma}c^{\kappa}x^{\sigma} + \left(\sum_{\kappa,\sigma=1}^{4} \eta_{\kappa\sigma}c^{\kappa}c^{\sigma}\right)\left(\sum_{\lambda,\nu=1}^{4} \eta_{\lambda\nu}x^{\lambda}x^{\nu}\right)$$

and where c^1, c^2, c^3 and c^4 are any four real numbers. The associated real Lie algebra has dimension 15 and is isomorphic to su(2, 2) and to so(4, 2) (cf. Mack and Salam 1969).

In contrast, the conformal groups for $D = 2$ are *infinite*-dimensional. Consider first the conformal group of 2-dimensional real *Euclidean* space (which corresponds to $\eta_{\mu\lambda} = \mathrm{diag}(1, 1)$). The defining condition (28.1) is satisfied in this case if and only if the coordinate transformation is such that

$$\frac{\partial x'^1}{\partial x^1} = \frac{\partial x'^2}{\partial x^2} \quad \text{and} \quad \frac{\partial x'^2}{\partial x^1} = -\frac{\partial x'^1}{\partial x^2}$$

and then

$$\rho(x^1, x^2) = \left(\frac{\partial x^1}{\partial x'^1}\right)^2 + \left(\frac{\partial x^2}{\partial x'^1}\right)^2.$$

With the complex variables z and z' defined by

$$z = x^1 + ix^2$$

and

$$z' = x'^1 + ix'^2,$$

these will be recognized as the Cauchy–Riemann relations (cf. Copson 1935). Thus every transformation of the conformal group of the 2-dimensional real Euclidean space is associated with an analytic function $F(z)$ by the prescription

$$z' = F(z),$$

the product of transformations corresponding to the analytic functions $F(z)$ and $G(z)$ being $F(G(z))$. Consequently the 2-dimensional Euclidean conformal group is isomorphic to the group of all functions of z that are analytic in some simply connected domain of the z plane (which will be assumed to contain the point $z = 0$), the group product of such functions being as just defined. Henceforth this latter group will also be referred to as a conformal group.

The identity element of this group corresponds to the function $F(z) = z$. Each element of this conformal group that is close to the identity corresponds to a function of the form

$$F(z) = z + \varepsilon(z),$$

where $\varepsilon(z)$ is a function of z that can be written in the form of a Laurent expansion

$$\varepsilon(z) = \sum_{J=-\infty}^{\infty} \varepsilon_J z^{J+1}, \tag{28.2}$$

where the ε_J are a set of small parameters. (At first sight it might seem surprising that negative powers of z could appear in (28.2), given the assumption that $F(z)$ is analytic at $z = 0$. To see how these can arise, consider the family of functions

$$F(z) = \left(\frac{az^n + b}{b^*z^n + a^*}\right)^{1/n},$$

where a and b are complex numbers such that $|a|^2 - |b|^2 = 1$, and where n is a positive integer. Each of these functions is analytic and one-to-one for $|z| < |a|/|b|$, *but*, for small b, expanding $F(z)$ in powers of b and b^* by the binomial theorem,

$$F(z) = \left(\frac{a}{a^*}\right)^{1/n}\left(1 + \frac{b}{naz^n} + \cdots\right)\left(1 - \frac{b^* z^n}{na^*} + \cdots\right),$$

in which (for $n > 1$) negative powers of z appear.) As an infinite number of independent parameters ε_J are involved in (28.2), this conformal group of one complex dimension is like the Lie groups introduced in Chapter 3, but each element depends on a countable *infinity* of parameters.

By analogy with (1.17), for each element $F(z)$ of this conformal group an operator P_F acting on functions $f(z)$ may be defined by

$$P_F f(z) = f(F^{-1}(z)),$$

(where F^{-1} is the inverse function defined by $z = F^{-1}(z')$ when $z' = F(z)$). Introducing the Lie algebra operator P_J corresponding to the parameter ε_J (by analogy with the definition of $P(\mathbf{a})$ in Chapter 10, Section 4) by the prescription

$$P_J f(z) = \lim_{\varepsilon_J \to 0} \frac{(P_{F_J} - 1) f(z)}{\varepsilon_J},$$

where $F_J(z) = z + \varepsilon_J z^{J+1}$ (all the other parameters being taken to be zero), it follows that

$$P_J = -z^{J+1}\frac{d}{dz}.$$

Clearly the commutation relations of the basis elements P_J of this infinite-dimensional "conformal algebra" are

$$[P_J, P_K]_- = (J - K)P_{J+K} \tag{28.3}$$

for $J, K = 0, \pm 1, \pm 2, \ldots$.

It may be noted that for each positive integer J the elements P_J, P_{-J} and P_0 form the basis of a Lie algebra of dimension 3, which can easily be shown to be isomorphic to the simple non-compact real Lie algebra su(1, 1).

For the conformal group of 2-dimensional *Minkowski* space–time the situation is very similar. With $D = 2$ and $\eta_{\mu\lambda} = \mathrm{diag}(-1, 1)$ the conditions (28.1) are satisfied if and only if

$$\frac{\partial x'^1}{\partial x^1} = \frac{\partial x'^2}{\partial x^2} \quad \text{and} \quad \frac{\partial x'^2}{\partial x^1} = \frac{\partial x'^1}{\partial x^2}.$$

The analysis could be continued as above, but instead these relations will be recast in the form of the Cauchy–Riemann relations by formally replacing x^2 and x'^2 by $-ix^2$ and $-ix'^2$ respectively. This enables all the machinery of complex variable theory to be applied. (Because the 2-dimensional Euclidean and Minkowski conformal groups are so intimately related, there is a tendency in the literature to omit any mention of the distinction and to refer to both as "the" 2-dimensional conformal group.) For a review of their representations see Jackiw (1987).

Limitations of space do not permit any detailed *general* discussion of conformal and superconformal covariant two-dimensional field theories, and of the role played in them by the Virasoro algebra and superalgebras. However, special mention must be made of the work of Belavin *et al.* (1984a, b), which is a development of ideas on such matters as operator-product expansions that date from the early days of dual models (cf. Fubini and Veneziano 1970, Di Vecchia *et al.* 1972, Brink *et al.* 1973), and of the supersymmetric extension of this analysis by Friedan *et al.* (1985) and Bershadsky *et al.* (1985).

2 Representations of the Virasoro algebra

Definition *Virasoro algebra*

The *Virasoro algebra* is an infinite-dimensional Lie algebra with basis elements L_J ($J = 0, \pm 1, \pm 2, \ldots$) and C_V that satisfy the commutation relations

$$[L_J, L_K]_- = (J - K)L_{J+K} + \tfrac{1}{12}J(J^2 - 1)\delta_{J+K,0}C_V \tag{28.4}$$

(for all $J, K = 0, \pm 1, \pm 2, \ldots$) and

$$[L_J, C_V]_- = 0 \tag{28.5}$$

(for all $J = 0, \pm 1, \pm 2, \ldots$).

It is easy to check that (28.4) and (28.5) satisfy the Jacobi identity. As the structure constants are all real, the algebra can be regarded as a real Lie algebra, or, when convenient, as a complex Lie algebra.

It is obvious from the discussion of the previous section that the Virasoro algebra is an extension of the Lie algebra of the conformal group in two dimensions, for which the commutation relations (28.3) are those of (28.4) but with the second term on the right-hand side omitted.

Henceforth it will be assumed that the elements L_J and C_V of the Virasoro algebra are operators acting on a carrier space V. V will be assumed to be a vector space that is provided with an inner product that satisfies the conditions (a)–(c) of the definition of an inner product space given in Appendix B, Section 2, but the conditions (d) and (e) will not be an immediate requirement. If V has no proper invariant subspace then the operators form an irreducible representation of the algebra, and in this case

$$C_{\mathrm{V}} = c_{\mathrm{V}} 1, \tag{28.6}$$

where c_{V} is a constant, which is known as the *central charge* of the representation. Also, if

$$L_J^{\dagger} = L_{-J} \tag{28.7}$$

(for all $J = 0, \pm 1, \pm 2, \ldots$) then (27.4) implies that

$$C_{\mathrm{V}}^{\dagger} = C_{\mathrm{V}}. \tag{28.8}$$

Such a set of operators are said to form a *unitary representation* of the Virasoro algebra. If the representation is both irreducible and unitary then (28.6) and (28.8) together imply that the constant c_{V} must be *real*.

Definition *Highest weight representation of the Virasoro algebra*

The set of Virasoro operators L_J are said to form a *highest weight representation* of the Virasoro algebra if there exists a vector $\Psi(h)$ of V such that

$$L_J \Psi(h) = 0 \quad \text{for all } J > 0 \tag{28.9}$$

and

$$L_0 \Psi(h) = h \Psi(h). \tag{28.10}$$

Then $\Psi(h)$ is called the *highest weight vector* of the representation.

With this assumption, the other basis elements of V are of the form

$$L_{J_1} L_{J_2} \cdots L_{J_n} \Psi(h), \tag{28.11}$$

where n is any positive integer and $\{J_1, J_2, \ldots, J_n\}$ is any set of n negative integers satisfying

$$J_1 \leqslant J_2 \leqslant \cdots \leqslant J_n.$$

Theorem I In a unitary irreducible highest weight representation of the Virasoro algebra for which the inner product (ψ, ψ) is non-negative for every ψ of the carrier space V the *only* allowed pairs of values of the central charge c_{V} and the eigenvalue h of (28.10) are such that *either*

(a) $c_{\mathrm{V}} \geqslant 1$ and $h \geqslant 0$

or

(b) $c_V = 1 - \dfrac{6}{(m+2)(m+3)}$

and

$$h = \frac{\{(m+3)p - (m+2)q\}^2 - 1}{4(m+2)(m+3)},$$

where m can take any non-negative integer value, p can take values $1, 2, \ldots, m+1$, and $q = 1, 2, \ldots, p$.

Proof Only a brief indication of the basis for the argument will be given here. For the complete proof see Friedan *et al.* (1984a, b, 1986), who have shown (using the work of Kac (1979) and Feigin and Fuchs (1982)) that no other values of c_V and h are possible, and see also Goddard and Olive (1985, 1988) and Goddard *et al.* (1985, 1986a), who have produced explicit constructions for each of the above pairs.

An inner product can be induced on V by assuming that

$$(\Psi(h), \Psi(h)) = 1 \tag{28.12}$$

and then invoking (28.4), (28.6), (28.7), (28.9) and (28.10) for all the other basis vectors of the form (28.11). For example, by (28.7),

$$(L_{-J}\Psi(h), L_{-J}\Psi(h)) = (\Psi(h), L_J L_{-J}\Psi(h)).$$

However, by (28.4) and (28.6),

$$\begin{aligned} L_J L_{-J} &= [L_J, L_{-J}]_- + L_{-J}L_J \\ &= 2JL_0 + \tfrac{1}{12}J(J^2-1)c_V 1 + L_{-J}L_J \end{aligned}$$

so, for $J > 0$, by (28.9), (28.10) and (28.12),

$$(L_{-J}\Psi(h), L_{-J}\Psi(h)) = 2Jh + \tfrac{1}{12}J(J^2-1)c_V. \tag{28.13}$$

In particular, for $J = 1$

$$(L_{-1}\Psi(h), L_{-1}\Psi(h)) = 2h.$$

Thus if it is required that $(L_{-1}\Psi(h), L_{-1}\Psi(h)) \geqslant 0$ then $h \geqslant 0$. Similarly if it is required that $(L_{-J}\Psi(h), L_{-J}\Psi(h)) \geqslant 0$ for all $J > 0$ then (28.13) shows that it is necessary to have $c_V \geqslant 0$.

The possibility that the carrier space V contains *null vectors* ψ such that $(\psi, \psi) = 0$ even though $\psi \neq 0$ is not excluded, although it can be shown that this cannot occur if $c_V > 1$ and $h > 0$.

The following theorem is very useful in applications.

Theorem II Suppose that the operators L_J^p (for $p = 1, 2$) provide two unitary irreducible highest weight representations of the Virasoro algebra (28.4), the corresponding central charges being c_V^p and the carrier spaces being V^p. Suppose that a set of operators L_J are defined (for $J = 0, \pm 1, \pm 2, \ldots$) on the direct product space $V^1 \otimes V^2$ by

$$L_J(\psi^1 \otimes \psi^2) = (L_J^1 \psi^1 \otimes \psi^2) + (\psi^1 \otimes L_J^2 \psi^2)$$

for all $\psi^1 \in V^1$ and $\psi^2 = V^2$. Then the operators L_J form a unitary highest weight representation of the Virasoro algebra with central charge c_V given by

$$c_V = c_V^1 + c_V^2.$$

Moreover, if $\Psi^p(h^p)$ is the highest weight vector of V^p and

$$L_0^p \Psi^p(h^p) = h^p \Psi^p(h^p)$$

for $p = 1$ and 2 then $\Psi^1(h^1) \otimes \Psi^2(h^2)$ is the highest weight vector of this representation, and its eigenvalue with L_0 is h, where

$$h = h^1 + h^2.$$

Proof As the argument is completely straightforward, the details will be omitted.

In the next section two constructions of the representations of the Virasoro algebra will be presented. Various aspects of the representations of the Virasoro algebra have been investigated by Feigin and Fuchs (1982, 1983), Kaplansky (1982), Rocha-Caridi and Wallach (1983, 1984), Rocha-Caridi (1984), Thorn (1984a, b), Corrigan (1986), Kaplansky and Santharoubane (1984), Goodman (1985), Kent (1985), Dobrev (1986b), Gomes (1986), Sasaki and Yamanaka (1986), Rittenberg (1986), Bowcock and Goddard (1987), Goddard *et al.* (1987b), Jain *et al.* (1987), Kato and Matsuda (1987b), Zhang (1987), Baake (1988), Bagger *et al.* (1988), Fairlie *et al.* (1988b), Kac and Raina (1988) and Meurman and Santharoubane (1988). A construction of the Virasoro algebra in terms of a finite number of generators satisfying a small number of conditions has been presented by Fairlie *et al.* (1988a).

3 Some constructions of highest weight representations of the Virasoro algebra

The simplest construction of a unitary irreducible highest weight representation of the Virasoro algebra is in terms of bosonic operators. Let

A^r_j be a set of such operators, where $r = 1, 2, \ldots, D_0$ and where j takes *all* integer values, it being assumed that these operators satisfy the commutation relations

$$[A^r_j, A^s_k]_- = j\delta^{rs}\delta_{j+k,0}1 \tag{28.14}$$

for r, $s = 1, 2, \ldots, D_0$ and j, $k = 0, \pm 1, \pm 2, \ldots$. With the operators $A^{r\dagger}_j$ defined for $j > 0$ by

$$A^{r\dagger}_j = A^r_{-j}, \tag{28.15}$$

(28.14) can be rewritten for $j > 0$ and $k > 0$ as

$$[A^r_j, A^{s\dagger}_k]_- = j\delta^{rs}\delta_{jk}1, \tag{28.16a}$$

$$[A^r_j, A^s_k]_- = 0, \tag{28.16b}$$

$$[A^{r\dagger}_j, A^{s\dagger}_k]_- = 0, \tag{28.16c}$$

which have the same form as (27.71) (with l replaced by D_0). Consequently for $j > 0$ the operator $A^{r\dagger}_j$ $(= A^r_{-j})$ can be interpreted as a bosonic creation operator and A^r_j can be considered as a bosonic annihilation operator. The bosonic Fock space can then be set up exactly as in Chapter 27, Section 4 (but with l replaced by D_0). The only new feature here is that the operators A^r_0 (for $r = 1, 2, \ldots, D_0$) commute with each other and with all the creation and annihilation operators. Consequently the vacuum state $|0\rangle$ can be chosen to be a simultaneous eigenvector of all the A^r_0. Denoting the corresponding eigenvalues by α^r, this implies that for $r = 1, 2, \ldots, D_0$

$$A^r_0|0\rangle = \alpha^r|0\rangle \tag{28.17}$$

and (as in (27.72))

$$A^r_j|0\rangle = 0 \quad \text{for } j > 0. \tag{28.18}$$

(Comparison with (26.138) shows that the A^r_j form a Heisenberg algebra.)

The *normal product* of a pair A^r_j and A^s_k may be defined to be

$$:A^r_jA^s_k: = \begin{cases} A^r_jA^s_k & \text{if } j < k, \\ \frac{1}{2}(A^r_jA^s_k + A^s_kA^r_j) & \text{if } p = q, \\ A^s_kA^r_j & \textit{if } j > k. \end{cases} \tag{28.19}$$

(This contains as a special case the usual normal product for bosonic operators in which creation operators appear to the left of annihilation operators.) With the definition

$$L_J = \frac{1}{2}\sum_{j=-\infty}^{\infty}\sum_{r=1}^{D_0} :A^r_{J+j}A^r_{-j}: \tag{28.20}$$

for $J = 0, \pm 1, \pm 2, \ldots$, it follows that for *odd* J

$$L_J = \sum_{j=(J+1)/2}^{\infty} \sum_{r=1}^{D_0} A^r_{J-j} A^r_j, \tag{28.21}$$

and for *even* J

$$L_J = \sum_{j=J/2+1}^{\infty} \sum_{r=1}^{D_0} A^r_{J-j} A^r_j + \frac{1}{2} \sum_{r=1}^{D_0} A^r_{J/2} A^r_{J/2}. \tag{28.22}$$

Theorem I With the operators L_J defined by (28.20), the following hold.

(a) The commutation relations with the A^s_k are

$$[L_J, A^s_k]_- = -k A^s_{J+k} \tag{28.23}$$

for $s = 1, 2, \ldots, D_0$ and for all integers k and J.

(b) The operators L_J provide a unitary irreducible highest weight representation of the Virasoro algebra (28.4) in which the carrier space is the bosonic Fock space V_F, and for which the central charge is given by

$$c_V = D_0. \tag{28.24}$$

The vacuum state $|0\rangle$ of V_F is the highest weight vector of the representation; that is,

$$L_J|0\rangle = 0 \quad \text{for } J > 0. \tag{28.25}$$

Moreover,

$$L_0|0\rangle = \left\{ \frac{1}{2} \sum_{r=1}^{D_0} (\alpha^r)^2 \right\} |0\rangle, \tag{28.26}$$

so that

$$h = \frac{1}{2} \sum_{r=1}^{D_0} (\alpha^r)^2. \tag{28.27}$$

Proof See Appendix O, Section 4.

As the α^r are arbitrary real numbers, (28.27) shows that in this construction h can take any non-negative real value.

It will now be shown that a representation of the Virasoro algebra can be constructed from any standard representation of any untwisted affine Kac–Moody algebra in a way that generalizes many of the features of Theorem I. This also provides an explicit construction of the semi-direct sum of the Virasoro algebra and the extended loop algebra of Chapter 26, Section

5(*b*) (whose basis elements are $t^j \otimes a_r^0$ and c) in which

$$[L_J, t^j \otimes a_r^0]_- = -jt^{J+j} \otimes a_r^0 \tag{28.28}$$

for all integers j and J and for $r = 1, 2, \ldots, n^0$, and for which all commutators with c and C_V are zero.

Let $\tilde{\mathscr{L}}^0$ $(= \tilde{\mathscr{L}}^s)$ be the complex simple Lie algebra of the untwisted affine Kac–Moody algebra $\tilde{\mathscr{L}}$ $(= \tilde{\mathscr{L}}^{(1)})$ (defined as in Chapter 26, Section 5(*b*)), $\tilde{\mathscr{L}}^0$ being assumed (as before) to have dimension n^0. Let a_r^0 $(r = 1, 2, \ldots, n^0)$ be the basis elements of $\tilde{\mathscr{L}}^0$, and suppose that they are chosen so that

$$B^0(a_r^0, a_s^0) = -\delta_{rs} \tag{28.29}$$

for r, $s = 1, 2, \ldots, n^0$ (cf. (26.131)), so that the commutation relations of the corresponding basis elements of $\tilde{\mathscr{L}}$ are given by

$$[t^j \otimes a_r^0, t^k \otimes a_s^0]_- = \sum_{p=1}^{n^0} c_{rs}^p (t^{j+k} \otimes a_p^0) - j\delta^{j+k,0}\delta_{rs} c \tag{28.30}$$

(cf. (26.133)), where the structure constants c_{rs}^p are antisymmetric with respect to the interchange of *any* pair of indices.

Suppose that the operators $\Phi(t^j \otimes a_r^0)$, $\Phi(c)$ and $\Phi(d)$ provide an integrable highest weight irreducible representation of $\tilde{\mathscr{L}}$ with highest weight

$$\Lambda = \sum_{k=0}^{l} n_k \Lambda_k + \mu\delta.$$

As $t^j \otimes a_r^0$ for $j > 0$ corresponds to a *positive* root of $\tilde{\mathscr{L}}$ (cf. (26.118) and (26.119) and Theorem I of Chapter 26, Section 5(*b*)), (27.3) shows that

$$\Phi(t^j \otimes a_r^0)\psi(\Lambda) = 0 \tag{28.31}$$

for all $j > 0$ and $r = 1, 2, \ldots, n^0$. The highest weight vector $\psi(\Lambda)$ can therefore be thought of as being a "vacuum" state vector, with the operators $\Phi(t^j \otimes a_r^0)$ for $j > 0$ playing the role of annihilation operators. The corresponding *normal product* for a pair of operators formed from elements of $\tilde{\mathscr{L}}^0$ may, by analogy with (28.19), be defined by

$$:\Phi(t^j \otimes a_r^0)\Phi(t^k \otimes a_s^0):$$

$$= \begin{cases} \Phi(t^j \otimes a_r^0)\Phi(t^k \otimes a_s^0) & \text{if } j < k \\ \tfrac{1}{2}\{\Phi(t^j \otimes a_r^0)\Phi(t^k \otimes a_s^0) + \Phi(t^k \otimes a_s^0)\Phi(t^j \otimes a_r^0)\} & \text{if } j = k, \\ \Phi(t^k \otimes a_s^0)\Phi(t^j \otimes a_r^0) & \text{if } j > k. \end{cases} \tag{28.32}$$

The carrier space of this representation of $\tilde{\mathscr{L}}$ will be denoted by V, and it may be assumed that on V

$$\Phi(c) = c_\Lambda 1 \tag{28.33}$$

(cf. (27.17)), an explicit expression for c_Λ being given by (27.20).

Now define the operators L_J (for $J = 0, \pm 1, \pm 2, \ldots$) acting on V by

$$L_J\psi = -\frac{1}{\kappa}\sum_{j=-\infty}^{\infty}\sum_{r=1}^{n^0}\{:\Phi(t^{J+j}\otimes a_r^0)\Phi(t^{-j}\otimes a_r^0):\psi\} \tag{28.34}$$

for every $\psi \in \mathrm{V}$, where

$$\kappa = 1 + 2c_\Lambda \tag{28.35}$$

(cf. (28.20)). By (28.32), the analogue of (28.21) for *odd* J is

$$L_J\psi = -\frac{2}{\kappa}\sum_{j=(J+1)/2}^{\infty}\sum_{r=1}^{n^0}\Phi(t^{J-j}\otimes a_r^0)\Phi(t^j\otimes a_r^0)\psi, \tag{28.36}$$

and for *even* J the analogue of (28.22) is

$$\begin{aligned} L_J\psi = &-\frac{2}{\kappa}\sum_{j=J/2+1}^{\infty}\sum_{r=1}^{n^0}\Phi(t^{J-j}\otimes a_r^0)\Phi(t^j\otimes a_r^0)\psi \\ &-\frac{1}{\kappa}\sum_{r=1}^{n^0}\Phi(t^{J/2}\otimes a_r^0)\Phi(t^{J/2}\otimes a_r^0)\psi. \end{aligned} \tag{28.37}$$

Thus, for example, by (28.31) and (28.36),

$$L_{-1}\psi(\Lambda) = -\frac{2}{\kappa}\sum_{r=1}^{n^0}\Phi(t^{-1}\otimes a_r^0)\Phi(t^0\otimes a_r^0)\psi(\Lambda),$$

where only a *finite* number of terms appear on the right-hand side. As every weight vector $\psi(\lambda)$ can be obtained from the highest weight vector $\psi(\Lambda)$ by acting with a linear combination of products of *finite* numbers of operators of the form $\Phi(t^k\otimes a_s^0)$ with $k \leqslant 0$, it follows that when L_J acts as any $\psi(\lambda)$ only a *finite* number of non-zero contributions appear (although the non-zero terms that appear do depend on λ). Thus the infinite sum in (28.34) is effectively only a *finite* sum, and so no problems of convergence arise.

As

$$[\Phi(t^{J+j}\otimes a_r^0), \Phi(t^{-j}\otimes a_r^0)]_- = -(J+j)\delta_{J,0}\Phi(c)$$

(by (28.30) and the fact that $c_{rr}^p = 0$), it follows from (28.32) that

$$\begin{aligned} &:\Phi(t^{J+j}\otimes a_r^0)\Phi(t^{-j}\otimes a_r^0): \\ &= \begin{cases} \Phi(t^{J+j}\otimes a_r^0)\Phi(t^{-j}\otimes a_r^0) & \text{if } J \leqslant -2j, \\ \Phi(t^{J+j}\otimes a_r^0)\Phi(t^{-j}\otimes a_r^0) - (J+j)\delta_{J,0}\Phi(c) & \text{if } J > -2j, \end{cases} \end{aligned} \tag{28.38}$$

so the "normal ordering" only plays a significant role in the definition of L_J for the case $J = 0$.

Theorem II (a) The operators L_J defined by (28.34) (with κ given by (28.35)) satisfy the Virasoro algebra (28.4) and (28.5) provided that C_V has the form (28.6) and the eigenvalue c_V is given by

$$c_V = \frac{2c_\Lambda n^0}{2c_\Lambda + 1}. \tag{28.39}$$

(b) Moreover,

$$[L_J, \Phi(t^k \otimes a_s^0)]_- = -k\Phi(t^{J+k} \otimes a_s^0) \tag{28.40}$$

for all $J, k = 0, \pm 1, \pm 2, \ldots$ and $s = 1, 2, \ldots, n^0$, and

$$[L_J, \Phi(c)]_- = 0 \tag{28.41}$$

for $J = 0, \pm 1, \pm 2, \ldots$.

(c) The operator L_0 is related to $\Phi(d)$ by

$$L_0 = -\Phi(d) + \left\{ \Lambda(d) + \frac{C_2(\Lambda^0)}{\kappa} \right\} 1, \tag{28.42}$$

where $C_2(\Lambda^0)$ is the eigenvalue of the second-order Casimir operator of the irreducible representation of the complex simple Lie algebra $\tilde{\mathscr{L}}^0$ with highest weight $\Lambda^0 = \sum_{k=1}^{L} n_k \Lambda_k^0$, where Λ_k^0 are the fundamental weights of $\tilde{\mathscr{L}}^0$.

(d) If $\psi(\Lambda)$ is the highest weight vector of the Kac–Moody algebra representation then

$$L_J \psi(\Lambda) = 0 \quad \text{for } J > 0 \tag{28.43}$$

and

$$L_0 \psi(\Lambda) = \frac{C_2(\Lambda^0)}{\kappa} \psi(\Lambda) \tag{28.44}$$

($C_2(\Lambda^0)$ being as in part (c)).

(e) With the definition (28.34),

$$L_J^\dagger = L_{-J} \tag{28.45}$$

for all $J = 0, \pm 1, \pm 2, \ldots$.

Proof See Appendix O, Section 4.

The essential ideas in this theorem can be traced back to some work of Sugawara (1968), so (28.34) is often described as *Sugawara's construction* of the Virasoro algebra. The argument that gives the value κ of (28.35) is rather subtle, since a rather plausible line of reasoning gives the incorrect value $\kappa = 2c_\Lambda$. The correct expression (28.35) was first derived for an arbitrary untwisted affine Kac–Moody algebra by Knizhnik and Zamolodchikov (1984), Goodman and Wallach (1984a, b), Goddard and Olive (1985) and Todorov (1985). The origin of the discrepancy is explained in Appendix O, Section 4. For the generalization of Theorem II to twisted affine Kac–Moody algebras see Nepomechie (1986).

In terms of the level of the irreducible representation of $\tilde{\mathscr{L}}$, (28.39) can be rewritten (using (27.21)) as

$$c_{\mathrm{V}} = n^0 \left\{ 1 + \frac{1}{\mathrm{level}(\Lambda) \langle \alpha_{\mathrm{H}}, \alpha_{\mathrm{H}} \rangle} \right\}^{-1} \tag{28.46}$$

where α_{H} is the highest root of $\tilde{\mathscr{L}}^0$. (The values of $\langle \alpha_{\mathrm{H}}, \alpha_{\mathrm{H}} \rangle$ are given in Appendix N.) By (16.3) and (16.4), the eigenvalue $C_2(\Lambda^0)$ of the second-order Casimir operator of (28.42) and (28.44) is given by

$$C_2(\Lambda^0) = \left\langle \sum_{k=1}^{l} n_k \Lambda_k, \left\{ \sum_{j=1}^{l} n_j \Lambda_j + \sum_{\alpha \in \Delta_+^0} \alpha \right\} \right\rangle. \tag{28.47}$$

For an extensive account of the development of the ideas of this section see Goddard and Olive (1986).

4 Virasoro superalgebras

The *Virasoro superalgebras* are Lie superalgebras that contain the Virasoro algebra in their even parts. Because the Virasoro algebra is the extension of the Lie algebra of conformal transformations in two dimensions, these algebras are often known as the *superconformal algebras in two dimensions.*

There are two "$N = 1$" Virasoro superalgebras. They are called the *Ramond superalgebra* and the *Neveu–Schwarz superalgebra* because they originally appeared in the fermionic string theories of Ramond (1971) and Neveu and Schwarz (1971a, b). The commutation and anticommutation relations of their basis elements both have the same form, namely

$$[L_J, L_K]_- = (J-K)L_{J+K} + \tfrac{1}{12}J(J^2-1)\delta_{J+K,0}C_{\mathrm{V}}, \tag{28.48a}$$

$$[L_J, C_{\mathrm{V}}]_- = 0, \tag{28.48b}$$

$$[L_J, G_R]_- = (\tfrac{1}{2}J - R)G_{J+R}, \tag{28.48c}$$

$$[C_{\mathrm{V}}, G_R]_- = 0, \tag{28.48d}$$

$$[G_R, G_S]_+ = 2L_{R+S} + \tfrac{1}{3}(R^2 - \tfrac{1}{4})\delta_{R+S,0}C_{\mathrm{V}}, \tag{28.48e}$$

but for the *Ramond* superalgebra the index set for the odd basis elements G_R is $R = 0, \pm 1, \pm 2, \ldots$, i.e. R takes all *integer* values, whereas for the *Neveu–*

Schwarz superalgebra $R = \pm\frac{1}{2}, \pm\frac{3}{2}, \pm\frac{5}{2}, \ldots$, i.e. R takes all *half-integer* values. (In (28.48) J and K take all integer values.)

Assuming that these are algebras of operators acting in an inner product space V, the discussion proceeds along the same lines as for the Virasoro algebra. In particular, for a unitary irreducible representation (28.6)–(28.8) hold, together with the additional requirement that

$$G_R^\dagger = G_{-R}$$

for all R of the appropriate index set. Moreover, for a highest weight representation, in addition to (28.9) and (28.10), it is now required that also

$$G_R \Psi(h) = 0 \tag{28.49}$$

for all $R > 0$.

For the *Ramond* superalgebra (28.48) shows that $[G_0, L_0]_- = 0$, so that if $L_0 \Psi(h) = h\Psi(h)$ then $G_0 \Psi(h)$ is also an eigenvector of L_0 with the same eigenvalue h, unless $G_0\Psi(h) = 0$.

These Virasoro superalgebras each have an explicit representation in terms of bosonic and fermionic creation and annihilation operators that generalizes the representation for the Virasoro algebra that was given in Theorem I of Section 3. (Indeed, it was essentially in this form that these superalgebras first appeared (cf. Ramond 1971, Neveu and Schwarz 1971a, b).) Let A_j^r (for $j = 0, \pm 1, \pm 2, \ldots$ and for $r = 1, 2, \ldots, D_0$) be a set of bosonic operators with commutation relations

$$[A_j^r, A_k^s]_- = j\delta^{rs}\delta_{j+k,0}1 \tag{28.50}$$

(as in (28.14)), and let B_p^r (for $r = 1, 2, \ldots, D_0$) be a set of fermionic operators with anticommutation relations

$$[B_p^r, B_q^s]_+ = \delta^{rs}\delta_{p+q,0}1 \tag{28.51}$$

(as in (27.99)), it being understood that for the Ramond superalgebra p and q take all integer values, whereas for the Neveu–Schwarz superalgebra they take all half-integer values. Moreover,

$$[A_j^r, B_q^s]_- = 0 \tag{28.52}$$

for all the allowed values of r, s, j and q. (It should be noted that for each value of j and p the number of bosonic operators A_j^r and fermionic operators B_p^r is assumed to be the *same*, since in both cases r takes D_0 values.) Define the normal products by (28.19) and (27.113), and define L_J and G_R by

$$\begin{aligned} L_J = \frac{1}{2}\sum_{j=-\infty}^{\infty}\sum_{r=1}^{D_0} :A_{J+j}^r A_{-j}^r: - \frac{1}{2}\sum_p \sum_{r=1}^{D_0} (p + \tfrac{1}{2}J):B_{J+p}^r B_{-p}^r: \\ + \tfrac{1}{16}\delta_{J,0}\eta D_0 1 \end{aligned} \tag{28.53}$$

and

$$G_R = \sum_{j=-\infty}^{\infty} \sum_{r=1}^{D_0} A^r_{-j} B^r_{R+j}, \tag{28.54}$$

where in (28.53) the sum over p is over all integer values for the Ramond superalgebra and all half-integer values for the Neveu–Schwarz superalgebra, and where η takes the values 1 and 0 for the Ramond and Neveu–Schwarz cases respectively.

Then for *odd J*

$$L_J = \sum_{j=(J+1)/2}^{\infty} \sum_{r=1}^{D_0} A^r_{J-j} A^r_j - \sum_{p>J/2} \sum_{r=1}^{D_0} (\tfrac{1}{2}J - p) B^r_{J-p} B^r_p, \tag{28.55}$$

and for *even J*

$$\begin{aligned} L_J = & \sum_{j=J/2+1}^{\infty} \sum_{r=1}^{D_0} A^r_{J-j} A^r_j + \frac{1}{2} \sum_{r=1}^{D_0} A_{J/2} A_{J/2} \\ & - \sum_{p>J/2} \sum_{r=1}^{D_0} (\tfrac{1}{2}J - p) B^r_{J-p} B^r_p \\ & + \tfrac{1}{16} \delta_{J,0} \eta D_0 1. \end{aligned} \tag{28.56}$$

For the Neveu–Schwarz algebra there is a vacuum state vector $|0\rangle$ (which is unique to an arbitrary multiplicative factor) such that (for $r = 1, 2, \ldots, D_0$)

$$A^r_j |0\rangle = 0 \quad \text{for } j = 1, 2, 3, \ldots \tag{28.57}$$

$$A^r_0 |0\rangle = \alpha^r |0\rangle, \tag{28.58}$$

$$B^r_p |0\rangle = 0 \quad \text{for } p = \tfrac{1}{2}, \tfrac{3}{2}, \tfrac{5}{2}, \ldots \quad . \tag{28.59}$$

(This vacuum state vector $|0\rangle$ is the direct product of the Neveu–Schwarz vacuum state of (27.102) and the bosonic vacuum state vector of (28.18).)

For the Ramond superalgebra the situation is essentially as described in Chapter 27, Section 5, the vacuum state being d_0-fold degenerate, d_0 being given by (27.109). The corresponding vacuum state vectors $|\mu; 0\rangle$ are then such that for $r = 1, 2, \ldots, D_0$ and $\mu = 1, 2, \ldots, d_0$

$$A^r_j |\mu; 0\rangle = 0 \quad \text{for } j = 1, 2, 3, \ldots, \tag{28.60}$$

$$A^r_0 |\mu; 0\rangle = \alpha^r |\mu; 0\rangle, \tag{28.61}$$

$$B^r_p |\mu; 0\rangle = 0 \quad \text{for } p = 1, 2, 3, \ldots, \tag{28.62}$$

$$B^r_0 |\mu; 0\rangle = 2^{-1/2} \sum_{\kappa=1}^{d_0} (\gamma^r_{\mathrm{E}})_{\kappa\mu} |\kappa; 0\rangle \tag{28.63}$$

(cf. (28.17), (28.18), (27.110) and (27.111)). Here γ^r_{E} are the $d_0 \times d_0$ matrices satisfying the D_0-dimensional Euclidean Clifford algebra (27.108).

Theorem I With the definitions (28.53) and (28.54), the operators L_J and G_R satisfy the commutation and anticommutation relations (28.48) provided that C_{V} is given by (28.6) and its eigenvalue c_{V} is such that

$$c_{\mathrm{V}} = \tfrac{3}{2} D_0. \tag{28.64}$$

Moreover, for the Neveu–Schwarz superalgebra, with the vacuum state $|0\rangle$ defined by (28.57)–(28.59),

$$L_J|0\rangle = 0 \quad \text{for } J = 1, 2, 3, \ldots, \tag{28.65}$$

$$G_R|0\rangle = 0 \quad \text{for } R = \tfrac{1}{2}, \tfrac{3}{2}, \tfrac{5}{2}, \ldots, \tag{28.66}$$

$$L_0|0\rangle = \frac{1}{2}\sum_{r=1}^{D_0} (\alpha^r)^2 |0\rangle. \tag{28.67}$$

Similarly, for the Ramond superalgebra, with the vacuum states $|\mu; 0\rangle$ (for $\mu = 1, 2, \ldots, d_0$) defined by (28.60)–(28.63),

$$L_J|\mu; 0\rangle = 0 \quad \text{for } J = 1, 2, 3, \ldots, \tag{28.68}$$

$$G_R|\mu; 0\rangle = 0 \quad \text{for } R = 1, 2, 3, \ldots, \tag{28.69}$$

$$L_0|\mu; 0\rangle = \left\{\frac{1}{2}\sum_{r=1}^{D_0} (\alpha^r)^2 + \tfrac{1}{16} D_0\right\}|\mu; 0\rangle, \tag{28.70}$$

$$G_0|\mu; 0\rangle = 2^{-1/2} \sum_{r=1}^{D_0} \sum_{\kappa=1}^{d_0} (\alpha^r \gamma_{\mathrm{E}}^r)_{\kappa\mu} |\kappa; 0\rangle. \tag{28.71}$$

Proof The first term in the expression (28.53) for L_J is the corresponding operator (28.20) for the bosonic operators alone, and this may be denoted by L_J^b. As shown in Theorem I of Section 3, these satisfy the Virasoro algebra with the central charge eigenvalue (now denoted by c_{V}^b) given by $c_{\mathrm{V}}^b = D_0$. Denoting the sum of the remaining two terms of (28.53) by L_J^{f}, it can be shown by a similar argument that these operators also satisfy the Virasoro algebra, but with a central charge eigenvalue $c_{\mathrm{V}}^{\mathrm{f}}$ given by $c_{\mathrm{V}}^{\mathrm{f}} = \frac{1}{2}D_0$. Theorem II of Section 2 then shows that, since the operators L_J of (28.53) are given by $L_J = L_J^{\mathrm{b}} + L_J^{\mathrm{f}}$, they too satisfy the Virasoro algebra with central charge eigenvalue $c_{\mathrm{V}} = c_{\mathrm{V}}^{\mathrm{b}} + c_{\mathrm{V}}^{\mathrm{f}} = \frac{3}{2}D_0$. The remaining expressions are easily checked in the obvious way.

It may be noted that if the additional requirement is imposed that

$$G_0|\mu; 0\rangle = 0 \tag{28.72}$$

(for $\mu = 1, 2, \ldots, d_0$), (28.71) requires that if $\boldsymbol{\psi}_0$ is the $d_0 \times 1$ matrix with entries $|\mu; 0\rangle$ (for $\mu = 1, 2, \ldots, d_0$) then

$$\sum_{r=1}^{D_0} (\alpha^r \tilde{\gamma}_{\mathrm{E}}^r)\boldsymbol{\psi}_0 = \mathbf{0}.$$

Application of (27.110) then shows that

$$\sum_{r=1}^{D_0} (\alpha^r)^2 \boldsymbol{\psi}_0 = \mathbf{0},$$

so, since $\boldsymbol{\psi}_0$ is necessarily non-zero,

$$\sum_{r=1}^{D_0} (\alpha^r)^2 = 0, \tag{28.73}$$

Hence, by (28.70), with this additional requirement,

$$L_0 \,|\, \mu; 0\rangle = \tfrac{1}{16} D_0 \,|\, \mu; 0\rangle \tag{28.74}$$

for $\mu = 1, 2, \ldots, d_0$.

The unitary irreducible highest weight representations of the Ramond and Neveu–Schwarz superalgebras have been investigated in detail by Friedan *et al.* (1984b, 1985) and Goddard *et al.* (1986a), who obtained results corresponding to those of Theorem I of Section 2, and by Meurman and Rocha–Caridi (1986). Other interesting results on this subject have been found by Horsley (1978), Kac (1979), Kaplansky (1982), Bershadsky *et al.* (1985), Eichenherr (1985), Kuperschmidt (1985), Dobrev (1986a, c), Kac and Wakimoto (1986), Nam (1986, 1987), Baake (1988), Fairlie *et al.* (1988b) and Kiritsis (1988a, b). Extensions of these superalgebras to higher spin have been introduced by Apikyan (1987).

At the "$N = 2$" level there are again two non-trivial Virasoro superalgebras, one whose odd part consists of two copies of the odd part of the Ramond superalgebra, and the other whose odd part is two copies of the Neveu–Schwarz superalgebra. These two copies will be indicated by putting superscripts 1 or 2 on the G_R. Close study (cf. Di Vecchia *et al.* 1985a, Aratyn and Damgaard 1986, Nam 1986) shows that $[G_R^1, G_S^2]_+$ cannot be written in terms of the elements of the Virasoro algebra alone, and that it is necessary to extend the even part of the Lie superalgebra so that it includes not only the Virasoro algebra but a Heisenberg algebra as well. Denoting the basis elements of this Heisenberg algebra by T_J (where $J = 0, \pm 1, \pm 2, \ldots$), the full set of commutation and anticommutation relations is

$$[L_J, L_K]_- = (J - K) L_{J+K} + \tfrac{1}{12} J(J^2 - 1)\delta_{J+K,0} C_V, \tag{28.75a}$$

$$[L_J, C_V]_- = 0, \tag{28.75b}$$

$$[L_J, T_K]_- = -K T_{J+K}, \tag{28.75c}$$

$$[C_V, T_K]_- = 0, \tag{28.75d}$$

$$[T_J, T_K]_- = \tfrac{1}{12} J \delta_{J+K,0} C_V, \tag{28.75e}$$

$$[L_J, G_R^p]_- = (\tfrac{1}{2}J - R) G_{J+R}^p, \tag{28.75f}$$

$$[C_V, G^p_R]_- = 0, \tag{28.75g}$$

$$[T_J, G^1_R]_- = \tfrac{1}{2} i G^2_{J+R}, \tag{28.75h}$$

$$[T_J, G^2_R]_- = -\tfrac{1}{2} i G^1_{J+R}, \tag{28.75i}$$

$$[G^1_R, G^2_S]_+ = 2i(R-S)T_{R+S}, \tag{28.75j}$$

$$[G^p_R, G^p_S]_+ = 2L_{R+S} + \tfrac{1}{3}(R^2 - \tfrac{1}{4})\delta_{R+S,0} C_V, \tag{28.75k}$$

where $J, K = 0, \pm 1, \pm 2, \ldots$ and $p = 1$ or 2, and R and S take integer values in the "Ramond" case and half-integer values in the "Neveu–Schwarz" case. This algebra can be traced back to the work of Ademollo *et al.* (1976a, b, c). The representations of these algebras have also been discussed by Boucher *et al.* (1986), Di Vecchia *et al.* (1986a, b), Kiritsis (1986, 1987), Matsuo (1987), Nam (1986, 1987), Schwimmer and Seiberg (1987), Waterson (1986), Zamolodchikov and Fateev (1986), Dobrev (1987a), Kato and Matsuda (1987a), Qiu (1987a, b), Rittenberg and Schwimmer (1987), Rocha-Caridi (1987), Dobrev and Ganchev (1988), Eguchi and Taormina (1988b), Li and Warner (1988), Lynker and Schimmrigk (1988) and Yang (1988). The uniqueness of these algebras has been discussed by Ramond and Schwarz (1976), Gastmans *et al.* (1987) and Van Proeyen (1987). Virasoro superalgebras with $N > 2$ have also been discussed by these authors and by Schwimmer and Seiberg (1987), Chang and Kumar (1987), Kent and Riggs (1987), Yu (1987a, b), Eguchi and Taormina (1988a) and Sevrin *et al.* (1988).

Di Vecchia *et al.* (1985b) and Kac and Todorov (1985, 1986) have shown how to construct *supersymmetric extensions of affine Kac–Moody algebras.* These are Lie superalgebras that contain the derived algebra $[\tilde{\mathscr{L}}, \tilde{\mathscr{L}}]_-$ of any chosen untwisted affine Kac–Moody algebra $\tilde{\mathscr{L}}$ in their even parts. The *even* basis elements may be taken to be the element c and the T^r_j defined in (26.134), with the structure constants f^{rst} given by (26.135), so the "even" commutation relations are those of (26.136) and (26.137a). For the *odd* part there are *two choices*, which can be labelled by a parameter κ taking the values 0 and $\frac{1}{2}$. For each choice the odd basis elements may be denoted by $S^r_{j+\kappa}$ (for $r = 1, 2, \ldots, n^0$, n^0 being the order of the simple Lie algebra $\tilde{\mathscr{L}}^0$ from which $\tilde{\mathscr{L}}$ is constructed, and $j = 0, \pm 1, \pm 2, \ldots$), with commutation and anticommutation relations

$$[T^r_j, S^s_{k+\kappa}]_- = i \sum_{t=1}^{n^0} f^{rst} S^t_{j+\kappa+k}, \tag{28.76a}$$

$$[c, S^r_{j+\kappa}]_- = 0, \tag{28.76b}$$

$$[S^r_{j+\kappa}, S^s_{k-\kappa}]_+ = \delta_{j+k,0}\delta^{rs} c \tag{28.76c}$$

(for $r, s = 1, 2, \ldots, n^0$ and $j, k = 0, \pm 1, \pm 2, \ldots$).

The superalgebra with $\kappa = 0$ can be combined in a semi-direct sum with the Ramond superalgebra in the same way that a Kac–Moody and a Virasoro algebra can be put together. Similarly the superalgebra with $\kappa = \frac{1}{2}$ can be combined with the Neveu–Schwarz superalgebra. The generalization of (28.28) for both cases is

$$[T^r_j, G_{k+\kappa}]_- = jS^r_{j+k+\kappa}, \tag{28.77a}$$

$$[c, G_{k+\kappa}]_- = 0, \tag{28.77b}$$

$$[S^r_{j+\kappa}, G_{k-\kappa}]_+ = T^r_{j+k}, \tag{28.77c}$$

$$[L_J, T^r_j]_- = -jT^r_{J+j}, \tag{28.77d}$$

$$[L_J, S^r_{j+\kappa}]_- = -(j+\kappa+\tfrac{1}{2})S^r_{J+j+\kappa}, \tag{28.77e}$$

where $r = 1, 2, \ldots, n^0$ and $j, k, J = 0, \pm 1, \pm 2, \ldots$. The commutation and anticommutation relations for the Ramond and Neveu–Schwarz basis elements are given in (28.48). These semi-direct sums have been called *superconformal current algebras* by Kac and Todorov (1985, 1986), who have studied their unitary irreducible highest weight representations. Other aspects of the representation theory of these superalgebras have been investigated by Windey (1986) and Nam (1987).

29

Algebraic Aspects of the Theory of Strings and Superstrings

1 Introduction

The theory of strings has experienced an explosion of interest since the successful formulation by Green and Schwarz (1981, 1982a, b, c) of its supersymmetric counterpart, the theory of "superstrings". The focus here will be on the "free" theory, with the algebraic aspect being given prominence, particular emphasis being placed on the role of the infinite-dimensional Lie algebras and superalgebras of the previous three chapters.

In Section 2 the "bosonic string" will be described in detail. Not only was this the first to be discussed, but many of its features reoccur in discussions of its successors. The first of these was the "spinning string" of Ramond (1971) and Neveu and Schwarz (1971a, b), an account of which is given in Section 3. The Green–Schwarz superstring is introduced in Section 4, and a development of it, the "heterotic string" of Gross *et al.* (1985a, b, 1986) forms the subject of Section 5. In Section 6 the account is concluded with a few lines on further developments.

The literature is enormous, so for the most part references will be given only to the most important original papers. Others may be found in the reviews mentioned in the various sections.

2 The bosonic string

(*a*) *The Lagrangian density for the bosonic string*

The bosonic string will be treated in some detail not merely because it was the first theory of this type to be investigated but also because it has appeared as part of every subsequent development in this area. Its origins lie in the "dual theory" of hadrons (cf. Fubini *et al.* 1969, Fubini and Veneziano 1970, Gervais 1970, Nambu 1970), comprehensive reviews of which have been given by Schwarz (1973), Frampton (1974, 1986), Jacob (1974), Mandelstam (1974), Rebbi (1974) and Scherk (1975). However, with the realization by Scherk and Schwarz (1974a, b, 1975) that the basic structure of the string theory was such that it could be applied more promisingly to gravitation and the other fundamental interactions of the elementary particles, there has been a major shift of emphasis. This is reflected in the present treatment, where the discussions will start with certain Lagrangian densities, the original motivations for which are now mainly of historical interest. That is, the initial assumptions are to be justified by their physical consequences rather than by any inherent plausibility.

Let $x^\mu(\sigma, \tau)$ be D real variables (each with the dimensions of length) that depend on two independent real variables σ and τ (which, for convenience, will be assumed to be dimensionless). The index μ will be allowed to take values $1, 2, \ldots, D$, the first $D-1$ corresponding to space-like coordinates, and x^D being a time-like coordinate. The quantity σ will be assumed to lie in the interval $0 \leqslant \sigma \leqslant \pi$, but τ will be allowed to take any real value. The interpretation will be that each fixed value of σ specifies a particular point of the string, different values of σ corresponding to different points, and that, with σ fixed but τ varying, the D coordinates $x^\mu(\sigma, \tau)$ describe the trajectory of that point in a D-dimensional space–time. As each value of τ then corresponds to an "event" in D-dimensional space–time, for any fixed value of σ there will be a one-to-one correspondence between $x^\mu(\sigma, \tau)$ and τ, so τ is a measure of the time development of the associated point of the string. (This matter will be taken up again later.) The variables σ and τ will be called *world-sheet coordinates*, σ being described as the spatial and τ as the time coordinate, whereas the x^μ will be referred to as *space–time coordinates*. If

$$x^\mu(0, \tau) = x^\mu(\pi, \tau) \tag{29.1}$$

for $\mu = 1, 2, \ldots, D$ and all values of τ, the string is said to be *closed*, since its end points always coincide in space–time. This situation is illustrated in Figure 29.1(a). The alternative, which is that of the "*open*" string, is shown in Figure 29.1(b). Although the aim is to describe phenomena in a 4-dimensional space–time, it will become apparent that this can only be achieved by compactification from higher values of D. Consequently the value of D will not be specified initially.

It is sometimes convenient to denote the world-sheet coordinates as ζ^α, where α takes values 1 and 2, and where $\zeta^1 = \sigma$ and $\zeta^2 = \tau$. This may be taken further by writing $\partial/\partial\zeta^\alpha$ as ∂_α and putting $\partial_1 = \partial/\partial\sigma = \partial_\sigma$ and $\partial_2 = \partial/\partial\tau = \partial_\tau$. Let $g_{\alpha\beta}(\sigma, \tau)$ be the world-sheet metric tensor (for $\alpha, \beta = 1, 2$), which is assumed

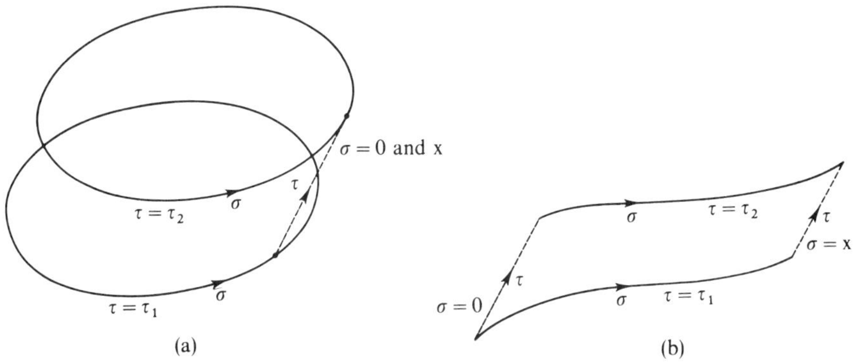

Figure 29.1 (a) Propagation of a closed string from "time" $\tau = \tau_1$ to "time" $\tau = \tau_2$. (b) Propagation of an open string from "time" $\tau = \tau_1$ to "time" $\tau = \tau_2$.

to transform as a symmetric second-rank covariant tensor under world-sheet coordinate transformations, let $g^{\alpha\beta}(\sigma, \tau)$ be its associate contravariant tensor, defined by

$$g^{\alpha\beta}(\sigma, \tau) g_{\alpha\gamma}(\sigma, \tau) = \delta^{\beta}_{\gamma} \tag{29.2}$$

(for $\beta, \gamma = 1, 2$), and let g be the determinant of $g_{\alpha\beta}(\sigma, \tau)$ considered as a 2×2 matrix. (It will be assumed that $g_{\alpha\beta}(\sigma, \tau)$ has signature 0 (cf. Appendix B, Section 5.) The Einstein summation convention will be used for repeated Greek indices throughout this chapter unless otherwise indicated.

The action S for the bosonic string may be taken to be

$$S = -\frac{1}{4\pi\alpha' \hbar c^2} \int_0^{\pi} d\sigma \int_{\tau_1}^{\tau_2} d\tau \, \eta_{\mu\lambda} (-g)^{1/2} g^{\alpha\beta} \, \partial_{\alpha} x^{\mu} \, \partial_{\beta} x^{\lambda} \tag{29.3}$$

(cf. Brink *et al.* 1976, Deser and Zumino 1976, Brink and Schwarz 1977). Here $\eta_{\mu\lambda}$ is the metric tensor for D-dimensional Minkowski space–time, which may be taken to be diagonal with the first $D - 1$ diagonal entries assuming the value -1 and the last diagonal entry having value 1, i.e.

$$\eta_{\mu\lambda} = \text{diag}(-1, -1, \ldots, -1, 1). \tag{29.4}$$

In (29.3) τ_1 and τ_2 are two fixed values of τ.

The quantity α' of (29.3) is taken to be a constant. With S assigned its usual dimensions of energy × time (i.e. mass × length^2 × time^{-1}), and with all other quantities in (29.3) except x^{μ} and x^{λ} dimensionless, it follows that α' must have the dimensions of energy^{-2} (i.e. mass^{-2} × length^{-4} × time^4). (In the old dual resonance model for hadrons α' gave the slope of the Regge trajectories.) The quantity $T = (4\pi\alpha' \hbar c^2)^{-1}$ has the dimensions of force and is often called the "string tension". Regarding α' as a fundamental constant, with the same standing as Planck's constant h (and $\hbar = h/2\pi$) and the speed of light c, quantities of any dimension can be constructed for appropriate combinations of α', $\hbar$ and c. In particular, $\alpha'\hbar^2c^2$ and $(\alpha'c^2)^{-1}$ have the dimensions of length^2 and momentum^2. It is usual in the literature to use units in which $\hbar = 1$ and $c = 1$, and either $\alpha' = 1$ (as in the important early paper of Goddard *et al.* 1973), or, more commonly, $\alpha' = \frac{1}{2}$, so that these quantities often do not appear explicitly. Of course, in principle it is easy to reconstruct the appropriate combinations (of α', $\hbar$ and c by dimensional analysis, but in doing so it has to be borne in mind that some quantities may not have the usual dimensions. For example, if $\mathscr{L}$ is the Lagrangian density associated with the action S of (29.3) by

$$S = \int_0^{\pi} \mathrm{d}\sigma \int_{\tau_1}^{\tau_2} \mathrm{d}\tau\, \mathscr{L}, \tag{29.5}$$

so that

$$\mathscr{L} = -(4\pi\alpha'\hbar c^2)^{-1}\eta_{\mu\lambda}(-g)^{1/2}g^{\alpha\beta}\,\partial_\alpha x^\mu\,\partial_\beta x^\lambda, \tag{29.6}$$

then $\mathscr{L}$ has the *same* dimensions as S, whereas in the usual 4-dimensional theories $\mathscr{L}$ has dimensions mass $\times$ length^{-1} $\times$ time^{-2}. One advantage of exhibiting the α' dependence of quantities explicitly is that it facilitates consideration of the so-called "field theory limit" in which $\alpha' \to 0$.

In applying the variational principle to (29.3), the quantities $g^{\alpha\beta}$ (for $\alpha, \beta = 1, 2$) and $x^\mu(\sigma, \tau)$ may be taken as being independent variables. The resulting Euler–Lagrange equations for the $g^{\alpha\beta}$,

$$\frac{\partial\mathscr{L}}{\partial g^{\alpha\beta}} - \partial_\sigma\left(\frac{\partial\mathscr{L}}{\partial(\partial_\sigma g^{\alpha\beta})}\right) - \partial_\tau\left(\frac{\partial\mathscr{L}}{\partial(\partial_\tau g^{\alpha\beta})}\right) = 0,$$

reduce to

$$\eta_{\mu\lambda}g_{\alpha\beta}g^{\gamma\delta}\partial_\gamma x^\mu\partial_\delta x^\lambda - 2\eta_{\mu\lambda}\partial_\alpha x^\mu\partial_\beta x^\lambda = 0 \tag{29.7}$$

on using the identity

$$2\frac{\partial(-g)^{1/2}}{\partial g^{\alpha\beta}} = -\frac{(-g)^{1/2}}{g_{\alpha\beta}}. \tag{29.8}$$

Defining $h_{\alpha\beta}$ by

$$h_{\alpha\beta} = \eta_{\mu\lambda}\,\partial_\alpha x^\mu\,\partial_\beta x^\lambda,$$

(29.7) requires that

$$g_{\alpha\beta}g^{\gamma\delta}h_{\gamma\delta} = 2h_{\alpha\beta},$$

which implies that $h_{\alpha\beta} = \kappa g_{\alpha\beta}$, where κ is some world-sheet scalar. Absorbing the κ into the x^μ, this gives

$$g_{\alpha\beta} = \eta_{\mu\lambda}\,\partial_\alpha x^\mu\,\partial_\beta x^\lambda. \tag{29.9}$$

Substituting back into (29.6) and using (29.2),

$$\mathscr{L} = -(2\pi\alpha'\hbar c^2)^{-1}(-g)^{1/2}, \tag{29.10}$$

which is the Lagrangian density written down by Nambu (1970) and Goto (1971). Written out explicitly, using (29.9), (29.10) is

$$\begin{aligned}\mathscr{L} = -(2\pi\alpha'\hbar c^2)^{-1}\{&(\eta_{\mu\lambda}\,\partial_\sigma x^\mu\,\partial_\tau x^\lambda)(\eta_{\mu'\lambda'}\,\partial_\sigma x^{\mu'}\,\partial_\tau x^{\lambda'})\\ &-(\eta_{\mu\lambda}\,\partial_\sigma x^\mu\,\partial_\sigma x^\lambda)(\eta_{\mu'\lambda'}\,\partial_\tau x^{\mu'}\,\partial_\tau x^{\lambda'})\}^{1/2}.\end{aligned} \tag{29.11}$$

As it has been assumed that the world-sheet metric tensor $g_{\alpha\beta}$ is a second-rank covariant tensor, after a reparametrization of the world-sheet coordinates of the form

$$\zeta^1 = \sigma \to \sigma' = \sigma'(\tau, \sigma) = \zeta'^1,$$
$$\zeta^2 = \tau \to \tau' = \tau'(\tau, \sigma) = \zeta'^2,$$

the new metric tensor $g'_{\alpha\beta}$ is given by

$$g'_{\alpha\beta} = g_{\gamma\delta}\frac{\partial\zeta^\gamma}{\partial\zeta'^\alpha}\frac{\partial\zeta^\delta}{\partial\zeta'^\beta}. \tag{29.12}$$

It is always possible to choose the transformation (29.12) so that

$$g'_{11} + g'_{22} = 0$$

and

$$g'_{12} = g'_{21} = 0.$$

Henceforth it will be assumed (when convenient) that this has been done, and the "new" world-sheet coordinates σ' and τ' will also be denoted by σ and τ. That is, it will be assumed that

$$g_{11} + g_{22} = 0$$

and

$$g_{12} = g_{21} = 0.$$

By (29.9), this requires that

$$\eta_{\mu\lambda}(\partial_\sigma x^\mu\,\partial_\sigma x^\lambda + \partial_\tau x^\mu\,\partial_\tau x^\lambda) = 0 \tag{29.13}$$

and

$$\eta_{\mu\lambda}\,\partial_\sigma x^\mu\,\partial_\tau x^\lambda = 0. \tag{29.14}$$

These equations imply that

$$g_{\alpha\beta} = \exp\{\phi(\sigma, \tau)\}\begin{bmatrix} -1 & 0 \\ 0 & 1 \end{bmatrix}, \tag{29.15}$$

where $\phi(\sigma, \tau)$ is some function of σ and τ. As $(-g)^{1/2} = \exp\{\phi(\sigma, \tau)\}$, and, by (29.15),

$$g^{\alpha\beta} = \exp\{-\phi(\sigma, \tau)\}\begin{bmatrix} -1 & 0 \\ 0 & 1 \end{bmatrix}, \tag{29.16}$$

it follows that

$$g^{\alpha\beta}(-g)^{1/2} = \begin{bmatrix} -1 & 0 \\ 0 & 1 \end{bmatrix} = \eta^{\alpha\beta}, \tag{29.17}$$

where the *world-sheet* Minkowski covariant tensor $\eta^{\alpha\beta}$ has been introduced. (This will be distinguished from the corresponding *space–time* Minkowski metric tensor $\eta^{\mu\lambda}$ by the convention employed for indices.) With the choice of $g_{\alpha\beta}$ given by (29.15), (29.6) reduces (by (29.17)) to

$$\mathscr{L} = -(4\pi\alpha'\hbar c^2)^{-1}\eta_{\mu\lambda}\eta^{\alpha\beta}\,\partial_\alpha x^\mu\,\partial_\beta x^\lambda. \tag{29.18}$$

Equation (28.1) shows that the set of world-sheet coordinate transformations that preserve the form (29.15) for the world-sheet metric tensor $g_{\alpha\beta}$ is precisely the conformal group of 2-dimensional Minkowski space–time, and therefore this conformal group is the symmetry group of the Lagrangian density (29.18). For this reason the form (29.18) is said to correspond to the *conformal gauge*. As noted in Chapter 28, Section 1, the conformal group of 2-dimensional real Euclidean space and the conformal group of the complex plane can be obtained by replacing τ by $-i\tau$ and considering $\tau + i\sigma$ as a complex variable. By this means, any of these conformal groups can be regarded as the symmetry group of the Lagrangian density (29.18). This observation has prompted great interest in the theory of general 2-dimensional conformally invariant field theories.

(b) *The classical open string*

For the classical open string the Euler–Lagrange equation for x^μ (for $\mu = 1, 2, \ldots, D$).

$$\frac{\partial\mathscr{L}}{\partial x^\mu} + \partial_\sigma\left(\frac{\partial\mathscr{L}}{\partial(\partial_\sigma x^\mu)}\right) + \partial_\tau\left(\frac{\partial\mathscr{L}}{\partial(\partial_\tau x^\mu)}\right) = 0, \tag{29.19}$$

has to be supplemented by the "end conditions"

$$\frac{\partial\mathscr{L}}{\partial(\partial_\sigma x^\mu)} = 0 \quad \text{for } \sigma = 0 \text{ and } \pi \tag{29.20}$$

(cf. Goddard *et al.* 1973). From (29.11), (29.19) gives

$$\frac{\partial^2 x^\mu}{\partial\sigma^2} - \frac{\partial^2 x^\mu}{\partial\tau^2} = 0, \tag{29.21}$$

and the end conditions (29.20) become

$$\partial_\sigma x^\mu = 0 \quad \text{for } \sigma = 0 \text{ and } \pi. \tag{29.22}$$

The most general solution of (29.21) satisfying the end conditions is

$$x^\mu(\sigma, \tau) = q^\mu + 2\alpha' \hbar c^2 p^\mu \tau + (2\alpha' \hbar^2 c^2)^{1/2} i \sum_{j=-\infty, j\neq 0}^{\infty} \frac{a_j^\mu}{j} \mathrm{e}^{-ij\tau} \cos j\sigma, \tag{29.23}$$

where q^μ, p^μ and the a_j^μ are all independent of σ and τ, p^μ has the dimensions of momentum, and the quantities a_j^μ are dimensionless. It is often convenient to use a modified form of (29.23) for which

$$x^\mu(\sigma, \tau) = q^\mu + (2\alpha' \hbar^2 c^2)^{1/2} \left(a_0^\mu \tau + i \sum_{j=-\infty, j\neq 0}^{\infty} \frac{a_j^\mu}{j} \mathrm{e}^{-ij\tau} \cos j\sigma \right), \tag{29.24}$$

where

$$a_0^\mu = (2\alpha' c^2)^{1/2} p^\mu, \tag{29.25}$$

so that the α_0^μ are also dimensionless. The reality requirement $x^\mu(\sigma, \tau)^* = x(\sigma, \tau)$ implies that for $\mu = 1, 2, \ldots, D$

$$q^{\mu *} = q^\mu, \tag{29.26}$$

$$p^{\mu *} = p^\mu, \tag{29.27}$$

$$a_j^{\mu *} = a_{-j}^\mu \quad (\text{for } j = 0, \pm 1, \pm 2, \ldots). \tag{29.28}$$

For the point of the string corresponding to $\sigma = \frac{1}{2}\pi$, (29.23) shows that

$$x^\mu(\tfrac{1}{2}\pi, \tau) = q^\mu + 2\alpha' \hbar c^2 p^\mu \tau. \tag{29.29}$$

Comparing this with the solution of the relativistic wave equation of motion for a free point particle, it can be seen that (apart from a constant of proportionality) τ can be identified with the proper time of the point $\sigma = \frac{1}{2}\pi$. The interpretation of (29.29) is then that q^μ are the space–time coordinates of the point $\sigma = \frac{1}{2}\pi$ at the proper time $\tau = 0$, p^μ are the components of the momentum of the string as a whole, and the last term of (29.23) describes oscillations of the string relative to the point $\sigma = \frac{1}{2}\pi$.

An energy–momentum current $p_\tau^\mu(\sigma, \tau)$ may be defined by

$$p_\tau^\mu(\sigma, \tau) = \frac{\partial \mathscr{L}}{\partial(\partial_\tau x_\mu)}, \tag{29.30}$$

where $x_\mu = \eta_{\mu\lambda} x^\lambda$, so that, by (29.11),

$$p_\tau^\mu(\sigma, \tau) = (2\pi\alpha' \hbar c^2)^{-1} \partial_\tau x^\mu. \tag{29.31}$$

Thus, by (29.23),

$$\int_0^\pi p_\tau^\mu(\sigma, \tau)\, \mathrm{d}\sigma = p^\mu, \tag{29.32}$$

which, as noted previously, is independent of σ and τ. The end condition (29.20) can then be written as

$$p_\tau^\mu(0,\tau) = p_\tau^\mu(\pi,\tau) = 0, \tag{29.33}$$

which says that no energy–momentum flows out of the ends of the string.

It remains to determine the conditions that must be satisfied by the parameters of (29.23) in order that the requirements (29.13) and (29.14) be satisfied. There are two ways of proceeding. The one that involves treating *all* the variables x^μ (for $\mu = 1, 2, \ldots, D$) on the *same* footing is known as the *covariant formulation*. The alternative, in which one direction is singled out for special treatment, is referred to as the *light-cone formulation*. These two ways of dealing with the constraints give rise to two approaches to quantization, as will be discussed in detail in the next subsection.

The first stage in the *covariant formulation* is to extend the range of σ from $[0,\pi]$ to $[-\pi,\pi]$. As $\partial_\tau x^\mu$ and $\partial_\sigma x^\mu$ as functions of σ are even and odd respectively, and as

$$\eta_{\mu\lambda}(\partial_\sigma x^\mu + \partial_\tau x^\mu)(\partial_\sigma x^\lambda + \partial_\tau x^\lambda) = \eta_{\mu\lambda}(\partial_\sigma x^\mu\, \partial_\sigma x^\lambda + \partial_\tau x^\mu\, \partial_\tau x^\lambda) + 2\eta_{\mu\lambda}\, \partial_\sigma x^\mu\, \partial_\tau x^\lambda, \tag{29.34}$$

the first two terms of the right-hand side of (29.34) are even in σ but the remaining term is odd in σ. Consequently the imposition of the *single* condition that

$$\eta_{\mu\lambda}(\partial_\sigma x^\mu + \partial_\tau x^\mu)(\partial_\sigma x^\lambda + \partial_\tau x^\lambda) = 0 \tag{29.35}$$

for all σ in $[-\pi,\pi]$ and all τ is equivalent to the *pair* of requirements (29.13) and (29.14). As (29.24) gives

$$\partial_\sigma x^\mu + \partial_\tau x^\mu = (2\alpha'\hbar^2 c^2)^{1/2} \sum_{j=-\infty}^{\infty} a_j^\mu\, \mathrm{e}^{-ij(\sigma+\tau)}, \tag{29.36}$$

(29.35) can be rewritten as

$$\sum_{J=-\infty}^{\infty} L_J \exp\{-iJ(\sigma+\tau)\} = 0 \tag{29.37}$$

where

$$L_J = -\tfrac{1}{2}\eta_{\mu\lambda} \sum_{j=-\infty}^{\infty} a_{J+j}^\mu a_{-j}^\lambda. \tag{29.38}$$

Thus if the infinite set of conditions

$$L_J = 0 \quad \text{for } J = 0, \pm 1, \pm 2, \ldots \tag{29.39}$$

are satisfied then the solutions (29.23) (or equivalently (29.24)) of the equations of motion are consistent with the requirements (29.13) and (29.14). In particular, the condition $L_0 = 0$ implies, by (29.38) and (29.25), that

$$\eta_{\mu\lambda} p^\mu p^\lambda = -\frac{1}{\alpha' c^2} \sum_{j=1}^{\infty} a_j^\mu a_{-j}^\lambda \eta_{\mu\lambda}. \tag{29.40}$$

The first step in the *light-cone formulation* is to introduce for any D-component vector A^μ ($\mu = 1, 2, \ldots, D$) the quantities A^+ and A^- by

$$A^+ = 2^{-1/2}(A^D + A^{D-1}) \tag{29.41}$$

and

$$A^- = 2^{-1/2}(A^D - A^{D-1}). \tag{29.42}$$

The remaining $D-2$ components $A^1, A^2, \ldots, A^{D-2}$ will be called the "transverse components" of the vector. If A^μ and B^μ are any two vectors in D-dimensional Minkowski space–time then

$$\eta_{\mu\lambda} A^\mu B^\lambda = A^+ B^- + A^- B^+ - \sum_{r=1}^{D-2} A^r B^r. \tag{29.43}$$

In particular, these considerations apply to the vectors $x^\mu(\sigma, \tau)$, q^μ, p^μ and a_j^μ of (29.23) and (29.24).

It can be shown (cf. Goddard *et al.* 1973) that it is possible to consider further reparametrizations of the form (29.12) that leave the conditions (29.13) and (29.14) unchanged, and that by this means it is possible to arrange that

$$x^+(\sigma, \tau) = (2\alpha' \hbar c^2)^{1/2} p^+ \tau; \tag{29.44}$$

that is (cf. (29.23)),

$$q^+ = 0 \tag{29.45}$$

and

$$a_j^+ = 0 \quad \text{for } j \neq 0. \tag{29.46}$$

The conditions (29.13) and (29.14) can then be solved to give $\partial_\tau x^-$ and $\partial_\sigma x^-$ in terms of derivatives of $x^+(\sigma, \tau)$ and the transverse components, and a detailed analysis (cf. Goddard *et al.* 1973) shows that, with

$$x^-(\sigma, \tau) = q^- + (2\alpha' \hbar^2 c^2)^{1/2} \left(a_0^- \tau + i \sum_{j=-\infty, j\neq 0}^{\infty} \frac{a_j^-}{j} e^{-ij\tau} \cos j\sigma \right) \tag{29.47}$$

(as implied by (29.24)), the quantities a_j^- are given by

$$a_J^- = (a_0^+)^{-1} L_J^{\text{tr}} \tag{29.48}$$

for $J = 0, \pm 1, \pm 2, \ldots$, where

$$L_J^{\text{tr}} = \frac{1}{2} \sum_{j=-\infty}^{\infty} \sum_{r=1}^{D-2} a_{J+j}^r a_{-j}^r, \tag{29.49}$$

which has the same form as (29.38) but involves the transverse components only. Then, by (29.38), (29.43) and (29.49),

$$L_J = L_J^{\mathrm{tr}} - \frac{1}{2}\sum_{j=-\infty}^{\infty} (a_{J+j}^{+}a_{-j}^{-} + a_{J+j}^{-}a_{-j}^{+}), \tag{29.50}$$

and so, by (29.46),

$$L_J = L_J^{\mathrm{tr}} - a_0^{+}a_J^{-}. \tag{29.51}$$

Thus it follows from (29.48) that the conditions (29.39) are satisfied. In this analysis the transverse component quantities a_j^r (for $r = 1, 2, \ldots, D-2$ and $j = 0, \pm 1, \pm 2, \ldots$) are not restricted in any way. Consequently in the light-cone formulation the constraints (29.13) and (29.14) are satisfied if the *only* independent quantities are taken to be p^+, q^- and the parameters q^r, p^r and a_j^r (for $j = \pm 1, \pm 2, \ldots$) of the transverse modes.

Defining the rest mass M by

$$M^2c^2 = \eta_{\mu\lambda}p^{\mu}p^{\lambda}, \tag{29.52}$$

(29.43) gives

$$M^2c^2 = 2p^+p^- - \sum_{r=1}^{D-2}(p^r)^2. \tag{29.53}$$

However, by (29.25) and (29.48),

$$p^+p^- = (2\alpha'c^2)^{-1}a_0^{+}a_0^{-} = (2\alpha'c^2)^{-1}L_0^{\mathrm{tr}},$$

so

$$M^2c^2 = (\alpha'c^2)^{-1}L_0^{\mathrm{tr}} - \sum_{r=1}^{D-2}(p^r)^2. \tag{29.54}$$

It may be noted that the equations (29.21) for $\mu = r = 1, 2, \ldots, D-2$ follow from the "light-cone Lagrangian density" $\mathscr{L}^{\mathrm{lc}}$ defined (cf. (29.18)) by

$$\mathscr{L}^{\mathrm{lc}} = \frac{1}{4\pi\alpha'\hbar c^2}\sum_{r=1}^{D-2}\eta^{\alpha\beta}\,\partial_{\alpha}x^r\,\partial_{\beta}x^r, \tag{29.55}$$

with the end conditions

$$\frac{\partial\mathscr{L}^{\mathrm{lc}}}{\partial(\partial_{\sigma}x^r)} = 0 \tag{29.56}$$

for $\sigma = 0$ and π and $r = 1, 2, \ldots, D-2$.

(c) *Light-cone quantization of the open bosonic string*

In the quantized version of the light-cone formulation all of the independent quantities q^-, p^+, q^r and a_j^r (for $r = 1, 2, \ldots, D-2$ and $j = 0, \pm 1, \pm 2, \ldots$) are treated as *operators* acting on a Hilbert space, and it has been convincingly argued by Goddard *et al.* (1973) that their commutation relations may be taken to be

$$[a_j^r, a_k^s]_- = j\delta^{rs}\delta_{j+k,0}1 \tag{29.57}$$

for $r, s = 1, 2, \ldots, D-2$ and $j, k = 0, \pm 1, \pm 2, \ldots,$

$$[q^r, a_0^s]_- = -i\delta^{rs}(2\alpha'\hbar^2c^2)^{1/2}1 \tag{29.58}$$

for $r, s = 1, 2, \ldots, D-2$, and

$$[q^-, p^+]_- = -i\hbar 1, \tag{29.59}$$

all other commutators being zero. The relations (29.57) have the same form as (28.14) with

$$A_j^r = a_j^r \tag{29.60}$$

and

$$D_0 = D - 2. \tag{29.61}$$

Consequently the operators a_j^r for $j > 0$ can be regarded as annihilation operators, while the a_j^r for $j < 0$ can be considered to be creation operators. It is convenient to write

$$a_j^{r\dagger} = a_{-j}^r \quad \text{for } j > 0 \tag{29.62}$$

(cf. (28.15)), which is consistent with the operator analogue of (29.28). By virtue of (29.25),

$$a_0^r = (2\alpha' c^2)^{1/2} p^r \tag{29.63}$$

for $r = 1, 2, \ldots, D-2$. Let $|p_0^1, \ldots; 0\rangle$ be the state vector such that

$$a_j^r|p_0^1, \ldots; 0\rangle = 0 \tag{29.64}$$

for all $r = 1, 2, \ldots, D-2$ and all $j > 0$, and also

$$p^r|p_0^1, \ldots; 0\rangle = p_0^r|p_0^1, \ldots; 0\rangle \tag{29.65}$$

for $r = 1, 2, \ldots, D-2$. This vector represents a state in which the string as a whole moves with transverse momentum components $p_0^1, p_0^2, \ldots, p_0^{D-2}$ but with *no* oscillations. As (29.57), (29.62) and (29.63) show that p^r commutes with $a_j^{s\dagger}$ for all $r, s = 1, 2, \ldots, D-2$ and $j > 0$, it follows that *every* excited state formed by acting on $|p_0^1, \ldots; 0\rangle$ with creation operators has the *same* transverse momentum. For example,

$$p^r(a_j^{s\dagger}|p_0^1, \ldots; 0\rangle) = p_0^r(a_j^{s\dagger}|p_0^1, \ldots; 0\rangle).$$

Thus every state in the Fock space constructed from the "vacuum state" $|p_0^1, \ldots; 0\rangle$ corresponds to the same transverse momentum.

The operators corresponding to (29.49) may be defined as

$$L_J^{\text{tr}} = \frac{1}{2} \sum_{j=-\infty}^{\infty} \sum_{r=1}^{D-2} :a^r_{J+j} a^r_{-j}: \quad , \tag{29.66}$$

where the normal product of (28.19) has been introduced. (Only for $J = 0$ is the insertion of the normal product significant, since for $J \neq 0$ all the terms on the right-hand side of (29.66) commute.) Comparison of (29.66) with (28.20) shows that the operators form a realization of a Virasoro algebra, which will be called the *transverse Virasoro algebra*. As $D_0 = D - 2$ for this algebra, its central charge c_V is shown by (28.24) to be

$$c_V = D - 2. \tag{29.67}$$

Then, by (28.4),

$$[L_J^{\text{tr}}, L_K^{\text{tr}}]_- = (J - K)L_{J+K}^{\text{tr}} + \tfrac{1}{12}(D-2)J(J^2 - 1)\delta_{J+K,0}1. \tag{29.68}$$

Also, by (28.25),

$$L_J^{\text{tr}}|p_0^1, \ldots; 0\rangle = 0 \quad \text{for } J > 0, \tag{29.69}$$

and, by (28.23) and (29.60),

$$[L_J^{\text{tr}}, a_k^s]_- = -ka^s_{J+k} \tag{29.70}$$

for $J, k = 0, \pm 1, \pm 2, \ldots$ and $s = 1, 2, \ldots, D-2$. Moreover, by (28.22), (29.63) and (29.66),

$$L_0^{\text{tr}} = \sum_{j=1}^{\infty} \sum_{r=1}^{D-2} a_j^{r\dagger} a_j^r + \alpha' c^2 \sum_{r=1}^{D-2} (p^r)^2, \tag{29.71}$$

which can be rewritten in terms of the "transverse excitation number operator"

$$N^{\text{tr}} = \sum_{j=1}^{\infty} \sum_{r=1}^{D-2} a_j^{r\dagger} a_j^r \tag{29.72}$$

as

$$L_0^{\text{tr}} = N^{\text{tr}} + \alpha' c^2 \sum_{r=1}^{D-2} (p^r)^2. \tag{29.73}$$

In the classical formulation of the light-cone approach the parameters a_J^- are related to the quantities L_J^{tr} by (29.48). For $J \neq 0$ there is clearly no difficulty in assuming that the same relationship holds for the associated operators in the quantum formulation, since the operators in L_J^{tr} for $J \neq 0$ mutually commute. However, for $J = 0$ the ordering of operators is significant, and so it is necessary to assume that the quantum analogue of (29.48) for

$J = 0$ is

$$a_0^- a_0^+ = L_0^{\text{tr}} - \alpha_0 1, \tag{29.74}$$

where α_0 is some constant that has to be determined.

With the assumption (29.74), the analogue of (29.54) for the rest mass operator M is

$$M^2 c^2 = (\alpha' c^2)^{-1}(L_0^{\text{tr}} - \alpha_0 1) - \sum_{r=1}^{D-2} (p^r)^2. \tag{29.75}$$

Thus, by (29.73),

$$(Mc^2)^2 = \alpha'^{-1}(N^{\text{tr}} - \alpha_0 1). \tag{29.76}$$

In the light-cone formulation there are no subsidiary conditions to worry about, the Hilbert space of physical states being the whole of the Fock spaces based on vacuum states $|p_0^1, \ldots; 0\rangle$. However, it is not obvious that the theory is invariant under D-dimensional space–time Lorentz transformations, since the $D-2$ transverse components have been treated differently from the two remaining components. The D-dimensional Lorentz algebra generators may be defined by

$$M^{\mu\lambda} = \frac{1}{2}\int_0^\pi \mathrm{d}\sigma\,(x^\mu p_\tau^\lambda + p_\tau^\lambda x^\mu - x^\lambda p_\tau^\mu - p_\tau^\mu x^\lambda) \tag{29.77}$$

for $\mu, \lambda = 1, 2, \ldots, D$. The expected commutation relations (cf. (17.192)) are

$$[M^{\mu\lambda}, M^{\alpha\beta}]_- = \frac{\hbar}{i}(\eta^{\lambda\alpha} M^{\mu\beta} - \eta^{\lambda\beta} M^{\mu\alpha} - \eta^{\mu\alpha} M^{\lambda\beta} + \eta^{\mu\beta} M^{\lambda\alpha}), \tag{29.78}$$

which, by (29.42), includes the relations

$$[M^{r-}, M^{s-}] = 0 \tag{29.79}$$

for $r, s = 1, 2, \ldots, D-2$. However, when the operators $M^{\mu\lambda}$ are expressed in terms of the operators $q^-, p^+, L_j^{\text{tr}}$ and a_j^r using (29.23), (29.47) and (29.64), the Virasoro relations (29.66) and (29.68) show that

$$[M^{r-}, M^{s-}]_- = -\frac{\alpha' \hbar^2 c^2}{4(p^+)^2} \sum_{j=1}^{\infty} \left\{ j\left(1 - \frac{D-2}{24}\right) + \frac{1}{j}\left(\frac{D-2}{24} - \alpha_0\right)\right\}(a_j^r a_j^s - a_j^s a_j^r) \tag{29.80}$$

(Goddard *et al.* 1973). Consequently (29.79) is satisfied only if

$$D = 26, \tag{29.81}$$

and then only if the parameter α_0 is chosen so that

$$\alpha_0 = 1. \tag{29.82}$$

(All the other commutation relations (29.78) are satisfied.) These arguments show that the theory for the bosonic string is consistent *only* for a 26-dimensional space–time. (The fact that (29.76) and (29.82) imply that the $D-2$ "transverse" states $a_1^{r\dagger}|0\rangle$ (for $r=1,2,\ldots,D-2$) are massless is compatible with the Lorentz invariance.) However, by (29.76) and (29.82),

$$(Mc^2)^2|0\rangle = -\alpha'^{-1}|0\rangle, \tag{29.83}$$

so the vacuum state $|0\rangle$ is *tachyonic*!

(d) *Covariant quantization of the open bosonic string*

In this formulation *all* of the D space–time coordinates are treated on the same footing, so that manifest D-dimensional Lorentz invariance is preserved. This means that for each $\mu = 1,2,\ldots,D$ *all* of the quantities q^μ, p^μ and a_j^μ (for $j=\pm 1, \pm 2,\ldots$) of (29.23) are now to be treated as being *independent* operators, the appropriate covariant commutation relations for which are

$$[a_j^\mu, a_k^\lambda]_- = -j\eta^{\mu\lambda}\delta_{j+k,0}1 \tag{29.84}$$

and

$$[q^\mu, p^\lambda]_- = -i\hbar\eta^{\mu\lambda}1, \tag{29.85}$$

with all other commutators being zero.

A set of Fock spaces may again be constructed in the same way as in Chapter 27, Section 4 and Chapter 28, Section 3. That is, one may assume that there exists a set of vacuum state vectors $|p_0^1,\ldots;0\rangle$ such that

$$a_j^\mu|p_0^1,\ldots;0\rangle = 0 \quad \text{for } j>0 \text{ and } \mu=1,2,\ldots,D, \tag{29.86}$$

and for which

$$p^\mu|p_0^1,\ldots;0\rangle = p_0^\mu|p_0^1,\ldots;0\rangle \quad \text{for } \mu=1,2,\ldots,D, \tag{29.87}$$

so that, by (29.25),

$$a_0^\mu|p_0^1,\ldots;0\rangle = (2\alpha'c^2)^{1/2}p_0^\mu|p_0^1,\ldots;0\rangle \quad \text{for } \mu=1,2,\ldots,D. \tag{29.88}$$

The operators a_j^μ for $j>0$ act as annihilation operators and

$$a_j^{\mu\dagger} = a_{-j}^\mu \quad \text{for } j>0 \tag{29.89}$$

are the corresponding creation operators. The Fock space built on $|p_0^1,\ldots;0\rangle$ then consists of all complex linear combinations of the state vectors

$$\Psi(n_1^1,\ldots,n_j^\mu,\ldots) = \prod_{j=1}^{\infty}\prod_{\mu=1}^{D}\{(n_j^\mu)!j^{n_j^\mu}\}^{-1/2}(a_j^{\mu\dagger})^{n_j^\mu}|p_0^1,\ldots;0\rangle \tag{29.90}$$

(cf. (27.73)), where the occupation numbers n_j^μ can take any non-negative integer values. (*Note*: in (29.90) the Einstein summation convention for Greek indices does not apply.)

Not all of these vectors can correspond to physical states, since some have "negative norm". For example, for $j > 0$, by (29.84), (29.86) and (29.89),

$$\| a_j^{D\dagger} | p_0^1, \ldots ; 0 \rangle \|^2 = \langle p_0^1, \ldots ; 0 | a_j^D a_{-j}^D | p_0^1, \ldots ; 0 \rangle$$
$$= -j \langle p_0^1, \ldots ; 0 | p_0^1, \ldots ; 0 \rangle,$$

so that if $|p_0^1, \ldots ; 0\rangle$ is assumed to have positive norm then $a_j^{D\dagger} | p_0^1, \ldots ; 0\rangle$ has negative norm. These negative-norm states can be eliminated by introducing the *covariant Virasoro algebra*, whose basis elements are defined by

$$L_J = -\frac{1}{2} \sum_{j=-\infty}^{\infty} \eta_{\mu\lambda} :a_{J+j}^\mu a_{-j}^\lambda: \quad . \tag{29.91}$$

With A_j^μ chosen so that

$$A_j^\mu = a_j^\mu \quad \text{for } \mu = 1, 2, \ldots, D-1 \tag{29.92}$$

and

$$A_j^\mu = i a_j^\mu \quad \text{for } \mu = D, \tag{29.93}$$

it follows that

$$\sum_{r=1}^{D} A_j^r A_k^r = -\eta_{\mu\lambda} a_j^\mu a_k^\lambda \tag{29.94}$$

for all j and k, and the commutation relations (29.84) imply that

$$[A_j^\mu, A_k^\lambda]_- = j \delta^{\mu\lambda} \delta_{j+k,0} 1 \tag{29.95}$$

for $\mu, \lambda = 1, 2, \ldots, D$ and $j, k = 0, \pm 1, \pm 2, \ldots$, which have the same form as (28.14) with

$$D_0 = D. \tag{29.96}$$

Although (29.89) and (29.92) show that

$$A_j^{\mu\dagger} = A_{-j}^\mu \quad \text{for } j > 0 \text{ and } \mu = 1, 2, \ldots, D-1, \tag{29.97}$$

(29.89) and (29.93) together give

$$A_j^{\mu\dagger} = -A_{-j}^\mu \quad \text{for } j > 0 \text{ and } \mu = D. \tag{29.98}$$

However, the analysis leading to Theorem I of Chapter 28, Section 3 is based on the commutation relations (28.14) (i.e. (29.95)) alone, so that, by (29.91) and (29.94),

$$L_J = \frac{1}{2} \sum_{j=-\infty}^{\infty} \sum_{r=1}^{D} :A_{J+j}^r A_{-j}^r: \quad , \tag{29.99}$$

exactly as in (28.20), and

$$[L_J, L_K]_- = (J-K)L_{J+K} + \tfrac{1}{12}DJ(J^2-1)\delta_{J+K,0}1,$$

since (28.24) and (29.96) imply that the central charge c_V is given by

$$c_V = D. \tag{29.100}$$

Moreover, by (28.23), (29.92) and (29.93),

$$[L_J, a_k^\mu]_- = -ka_{J+k}^\mu \tag{29.101}$$

for $\mu = 1, 2, \ldots, D$ and $J, k = 0, \pm 1, \pm 2, \ldots$. Also, (28.25) shows that

$$L_J|p_0^1, \ldots; 0\rangle = 0 \quad \text{for } J > 0, \tag{29.102}$$

and (28.26) and (29.88) imply that

$$L_0|p_0^1, \ldots; 0\rangle = -(\alpha' c^2)\eta_{\mu\lambda}p_0^\mu p_0^\lambda |p_0^1, \ldots; 0\rangle. \tag{29.103}$$

The "no-ghost theorem" of Brower (1972) and Goddard and Thorn (1972) then states that the set of states Ψ that satisfy the subsidiary conditions

$$L_J\Psi = 0 \quad \text{for } J > 0 \tag{29.104}$$

and

$$L_0\Psi = \alpha_0\Psi \tag{29.105}$$

does not contain any negative-norm states if either

$$D = 26 \text{ and } \alpha_0 = 1, \tag{29.106}$$

or

$$D < 26 \text{ and } \alpha_0 < 1. \tag{29.107}$$

However, the latter possibility gives rise to various difficulties that make it untenable (cf. Brower 1972, Goddard and Thorn 1972, Goddard *et al.* 1973, Scherk 1975, Polyakov 1981a). (The conditions (29.104) and (29.105) are operator analogues of the classical constraints (29.39).)

The general mass formula is easily obtained, since, by (28.22) (with $J = 0$) and (29.94),

$$L_0 = -\sum_{j=1}^{\infty} \eta_{\mu\lambda}a_{-j}^\mu a_j^\lambda - \tfrac{1}{2}\eta_{\mu\lambda}a_0^\mu a_0^\lambda, \tag{29.108}$$

so that, by (29.25),

$$L_0 = -\sum_{j=1}^{\infty} \eta_{\mu\lambda}a_j^{\mu\dagger} a_j^\lambda - \alpha' c^2 \eta_{\mu\lambda}p^\mu p^\lambda. \tag{29.109}$$

Thus, with the covariant excitation number operator defined by

$$N = -\sum_{j=1}^{\infty} \eta_{\mu\lambda}a_j^{\mu\dagger} a_j^\lambda, \tag{29.110}$$

and with the mass operator M given by

$$M^2c^2 = \eta_{\mu\lambda}p^\mu p^\lambda, \tag{29.111}$$

it follows from (29.109) that

$$(Mc^2)^2 = \alpha'^{-1}(N - L_0).$$

Hence, on the physical states, the condition (29.105) with $\alpha_0 = 1$ implies that

$$(Mc^2)^2\Psi = \alpha'^{-1}(N-1)\Psi.$$

In particular, since

$$N|p_0^1,\ldots;0\rangle = 0,$$

it follows that

$$(Mc^2)^2|p_0^1,\ldots;0\rangle = -\alpha'^{-1}|p_0^1,\ldots;0\rangle$$

(cf. (29.83)).

Thus both of the major conclusions of the light-cone formulation of quantization are confirmed in this covariant approach; that is, the theory is only consistent for a 26-dimensional space–time, and it contains a tachyonic state.

The Virasoro operators L_J are actually related to the energy–momentum tensor of the system. The Lagrangian density (29.18) can be regarded as corresponding to a set of D world-sheet scalar fields $x^\mu(\sigma,\tau)$ ($\mu = 1,2,\ldots,D$), and so the standard 4-dimensional Minkowski space–time scalar field energy–momentum tensor (cf. Schweber 1961) can be modified to produce the world-sheet energy–momentum tensor

$$T_{\alpha\beta} = (2\alpha'\hbar c^2)^{-1}\eta_{\mu\lambda}(\partial_\alpha x^\mu\,\partial_\beta x^\lambda - \tfrac{1}{2}\eta_{\alpha\beta}\eta^{\gamma\delta}\,\partial_\gamma x^\mu\,\partial_\delta x^\lambda), \tag{29.112}$$

where $\alpha,\beta,\gamma,\delta$ take the values 1 and 2, and $\mu,\lambda = 1,2,\ldots,D$. This tensor is traceless, symmetric and conserved; that is, $\eta^{\alpha\beta}T_{\alpha\beta} = 0$, $T_{\alpha\beta} = T_{\beta\alpha}$ and $\partial_\alpha T^{\alpha\beta} = 0$. Clearly

$$T_{11} = T_{22} = (4\alpha'\hbar c^2)^{-1}\eta_{\mu\lambda}(\partial_\sigma x^\mu\partial_\sigma x^\lambda + \partial_\tau x^\mu\partial_\tau x^\lambda) \tag{29.113}$$

and

$$T_{12} = T_{21} = (2\alpha'\hbar c^2)^{-1}\eta_{\mu\lambda}\partial_\sigma x^\mu\partial_\tau x^\lambda. \tag{29.114}$$

Thus for the combination

$$T = -\hbar^{-1}(T_{11} + T_{12})$$

it follows on substituting the expressions for $x^\mu(\sigma,\tau)$ given in (29.24) that

$$T = -\tfrac{1}{2}\eta_{\mu\lambda}\sum_{j,k=-\infty}^{\infty} a_j^\mu a_k^\lambda\,\mathrm{e}^{-i(j+k)(\tau+\sigma)}.$$

On replacing τ by $-i\tau$ and defining z by

$$z = \mathrm{e}^{\tau + i\sigma},$$

it is clear that T is a function of z alone, and so can be written henceforth as $T(z)$, and that (cf. (29.38))

$$T(z) = \sum_{J=-\infty}^{\infty} L_J z^{-J}.$$

At the classical level this is all trivial, since the equations (29.7), (29.9) and (29.112) imply that all the components $T_{\alpha\beta}$ are identically zero, and hence $T(z)$ is zero, which is consistent with the classical condition (29.39) that $L_J = 0$ for all J. However, in the covariant quantization scheme these conditions are relaxed and replaced by (29.104) and (29.105), so that $T(z)$ is no longer identically zero. Taking $T(z)$ to be the operator obtained from the corresponding classical expression by normal-ordering, so that

$$T(z) = -\tfrac{1}{2}\eta_{\mu\lambda} \sum_{j,k=-\infty}^{\infty} :a_j^{\mu} a_k^{\lambda}: \mathrm{e}^{-i(j+k)(\tau+\sigma)},$$

then

$$T(z) = \sum_{J=-\infty}^{\infty} L_J z^{-J},$$

where now the L_J are the Virasoro operators (29.91). In fact, for any physical state vector Ψ (29.104) implies that the quantity $T(z)\Psi$ is an *analytic* function of z.

(e) *The closed bosonic string*

For the closed bosonic string the equation of motion is again (29.21), but the boundary conditions (29.22) have to be replaced by (29.1). The most general solution is then of the form

$$x^{\mu}(\sigma,\tau) = q^{\mu} + 2\alpha'\hbar c^2 p^{\mu}\tau + (2\alpha'\hbar^2 c^2)^{1/2} \times \frac{i}{2} \sum_{j=-\infty, j\neq 0}^{\infty} \left(\frac{a_{\mathrm{R}j}^{\mu}}{j} \mathrm{e}^{-2ij(\tau-\sigma)} + \frac{a_{\mathrm{L}j}^{\mu}}{j} \mathrm{e}^{-2ij(\tau+\sigma)} \right), \tag{29.115}$$

where q^{μ}, p^{μ} and the $a_{\mathrm{R}j}^{\mu}$ and $a_{\mathrm{L}j}^{\mu}$ are all independent of σ and τ, p^{μ} has the dimensions of momentum and the quantities $a_{\mathrm{R}j}^{\mu}$ and $a_{\mathrm{L}j}^{\mu}$ are dimensionless. It is convenient to write

$$a_{\mathrm{R}0}^{\mu} = a_{\mathrm{L}0}^{\mu} = (\tfrac{1}{2}\alpha' c^2)^{1/2} p^{\mu}, \tag{29.116}$$

so that (29.115) can be rewritten as

$$x^{\mu}(\sigma,\tau) = q^{\mu} + (2\alpha'\hbar^2 c^2)^{1/2} \left\{ a_{\mathrm{R}0}^{\mu}(\tau-\sigma) + \frac{i}{2} \sum_{j=-\infty, j\neq 0}^{\infty} \frac{a_{\mathrm{R}j}^{\mu}}{j} \mathrm{e}^{-2ij(\tau-\sigma)} + a_{\mathrm{L}0}^{\mu}(\tau+\sigma) + \frac{i}{2} \sum_{j=-\infty, j\neq 0}^{\infty} \frac{a_{\mathrm{L}j}^{\mu}}{j} \mathrm{e}^{-2ij(\tau+\sigma)} \right\}. \tag{29.117}$$

The terms involving the combination $\tau-\sigma$ are described as corresponding to *right-moving modes*, since if $\sigma-\tau=$ constant then $\sigma=$ constant $+\tau$, and so σ moves to the right along the σ-axis as τ increases. Similarly the terms involving the combination $\sigma+\tau$ are said to belong to *left-moving modes*. The subscripts "R" and "L" on the quantities $a^{\mu}_{\mathrm{R}j}$ and $a^{\mu}_{\mathrm{L}j}$ are inserted to give an indication of this terminology.

Most of the theory of the closed string is obtained by replacing the quantities a^{μ}_{j} of the open string by $a^{\mu}_{\mathrm{R}j}$ and by $a^{\mu}_{\mathrm{L}j}$. For example, the classical reality conditions (29.26) and (29.27) are unchanged, and (29.28) is replaced by

$$a^{\mu *}_{\mathrm{R}j}=a^{\mu}_{\mathrm{R}(-j)},\quad a^{\mu *}_{\mathrm{L}j}=a^{\mu}_{\mathrm{L}(-j)}\quad \text{for } j=0,\pm 1,\pm 2,\ldots\;. \tag{29.118}$$

Moreover, the conditions (29.39) are replaced by the pair

$$L_{\mathrm{R}J}=0,\quad L_{\mathrm{L}J}=0\quad \text{for } J=0,\pm 1,\pm 2,\ldots, \tag{29.119}$$

where

$$L_{\mathrm{R}J}=-\tfrac{1}{2}\eta_{\mu\lambda}\sum_{j=-\infty}^{\infty} a^{\mu}_{\mathrm{R}(J+j)}a^{\lambda}_{\mathrm{R}(-j)} \tag{29.120}$$

and

$$L_{\mathrm{L}J}=-\tfrac{1}{2}\eta_{\mu\lambda}\sum_{j=-\infty}^{\infty} a^{\mu}_{\mathrm{L}(J+j)}a^{\lambda}_{\mathrm{L}(-j)}. \tag{29.121}$$

Similarly in the classical light-cone formulation of the closed string the independent quantities are p^{+}, q^{-}, and the parameters q^{r}, p^{r}, $a^{r}_{\mathrm{R}j}$ and $a^{r}_{\mathrm{L}j}$ (for $j=\pm 1,\pm 2,\ldots$) of the transverse modes, with

$$a^{-}_{\mathrm{R}J}=(a^{+}_{\mathrm{R}J})^{-1}L^{\mathrm{tr}}_{\mathrm{R}J} \tag{29.122}$$

for $J=0,\pm 1,\pm 2,\ldots$ (cf. (29.48)), where

$$L^{\mathrm{tr}}_{\mathrm{R}J}=\frac{1}{2}\sum_{j=-\infty}^{\infty}\sum_{r=1}^{D-2} a^{r}_{\mathrm{R}(J+j)}a^{r}_{\mathrm{R}(-j)}, \tag{29.123}$$

$a^{-}_{\mathrm{L}J}$ being given by a similar expression involving left-moving quantities only.

In the *light-cone quantization* of the closed bosonic string the independent operators are $p^{+},q^{-},q^{r},a^{r}_{\mathrm{R}0}(=a^{r}_{\mathrm{L}0}),a^{r}_{\mathrm{R}j}$ and $a^{r}_{\mathrm{L}j}$ (for $r=1,2,\ldots,D-2$ and $j=\pm 1,\pm 2,\ldots$), with the commutation relations (29.57) replaced by

$$[a^{r}_{\mathrm{R}j},a^{s}_{\mathrm{R}k}]_{-}=j\delta^{rs}\delta_{j+k,0}1, \tag{29.124}$$

$$[a^{r}_{\mathrm{L}j},a^{s}_{\mathrm{L}k}]_{-}=j\delta^{rs}\delta_{j+k,0}1, \tag{29.125}$$

$$[a^{r}_{\mathrm{R}j},a^{s}_{\mathrm{L}k}]_{-}=0 \tag{29.126}$$

for $r,s=1,2,\ldots,D-2$ and $j,k=0,\pm 1,\pm 2,\ldots$. Also, (29.58) becomes

$$[q^{r},a^{s}_{\mathrm{R}0}]_{-}=-i\delta^{rs}(\tfrac{1}{2}\alpha'\hbar^{2}c^{2})^{1/2}1, \tag{29.127}$$

(29.59) is unchanged, and all other commutators are zero. The "right-moving and left-moving transverse Virasoro algebras" have basis elements given by

$$L^{\mathrm{tr}}_{\mathrm{R}J} = \frac{1}{2} \sum_{j=-\infty}^{\infty} \sum_{r=1}^{D-2} :a^{r}_{\mathrm{R}(J+j)} a^{r}_{\mathrm{R}(-j)}: \tag{29.128}$$

and

$$L^{\mathrm{tr}}_{\mathrm{L}J} = \frac{1}{2} \sum_{j=-\infty}^{\infty} \sum_{r=1}^{D-2} :a^{r}_{\mathrm{L}(J+j)} a^{r}_{\mathrm{L}(-j)}: \tag{29.129}$$

(cf. (29.66)), so that both have the same central charge $c_{\mathrm{V}} = D - 2$ (as in (29.67)). The analogue of (29.73) is (by (29.116)) the *pair* of equations

$$L^{\mathrm{tr}}_{\mathrm{R}0} = N^{\mathrm{tr}}_{\mathrm{R}} + \tfrac{1}{4}\alpha' c^2 \sum_{r=1}^{D-2} (p^r)^2 \tag{29.130}$$

and

$$L^{\mathrm{tr}}_{\mathrm{L}0} = N^{\mathrm{tr}}_{\mathrm{L}} + \tfrac{1}{4}\alpha' c^2 \sum_{r=1}^{D-2} (p^r)^2, \tag{29.131}$$

where

$$N^{\mathrm{tr}}_{\mathrm{R}} = \sum_{j=1}^{\infty} \sum_{r=1}^{D-2} a^{r\dagger}_{\mathrm{R}j} a^{r}_{\mathrm{R}j} \tag{29.132}$$

and

$$N^{\mathrm{tr}}_{\mathrm{L}} = \sum_{j=1}^{\infty} \sum_{r=1}^{D-2} a^{r\dagger}_{\mathrm{L}j} a^{r}_{\mathrm{L}j} \tag{29.133}$$

(cf. (29.72)). Similarly the analogue of (29.74) is the set of conditions

$$a^{-}_{\mathrm{R}0} a^{+}_{\mathrm{R}0} = a^{-}_{\mathrm{L}0} a^{+}_{\mathrm{L}0} = L^{\mathrm{tr}}_{\mathrm{R}0} - \alpha_0 1 = L^{\mathrm{tr}}_{\mathrm{L}0} - \alpha_0 1, \tag{29.134}$$

where the constant α_0 is again determined by the requirement of Lorentz invariance in D-dimensional space–time and is again given by $\alpha_0 = 1$. Moreover, the theory is again only consistent for $D = 26$. As (29.111) and (29.116) give

$$M^2 c^2 = \frac{4}{\alpha' c^2} a^{+}_{\mathrm{R}0} a^{-}_{\mathrm{R}0} - \sum_{r=1}^{D-2} (p^r)^2, \tag{29.135}$$

it follows from (29.130) and (29.134) that

$$(Mc^2)^2 = \frac{4}{\alpha'}(N^{\mathrm{tr}}_{\mathrm{R}} - 1). \tag{29.136}$$

However, since (29.135) can be written as

$$M^2 c^2 = \frac{4}{\alpha' c^2} a^{+}_{\mathrm{L}0} a^{-}_{\mathrm{L}0} - \sum_{r=1}^{D-2} (p^r)^2, \tag{29.137}$$

it is also true that

$$(Mc^2)^2 = \frac{4}{\alpha'}(N_{\mathrm{L}}^{\mathrm{tr}} - 1). \tag{29.138}$$

Taken together, (29.136) and (29.138) imply that for every physical state Ψ

$$N_{\mathrm{R}}^{\mathrm{tr}}\Psi = N_{\mathrm{L}}^{\mathrm{tr}}\Psi. \tag{29.139}$$

As vacuum states have zero eigenvalues with $N_{\mathrm{R}}^{\mathrm{tr}}$ and $N_{\mathrm{L}}^{\mathrm{tr}}$, (29.136) and (29.137) show that they again correspond to tachyonic modes. Equations (29.136), (29.138) and (29.139) also show that the massless states are obtained by acting with the operators $a_{\mathrm{R}1}^{r\dagger}a_{\mathrm{L}1}^{s\dagger}$ on the vacuum states (for all $r, s = 1, 2, \ldots, D-2$). These include the states that Scherk and Schwarz (1974a, b, 1975) interpreted as corresponding to the graviton. Indeed, their consideration of the series expansion in powers of α' for the combined open and closed string model shows that to first order the result is the same as for a system consisting of interacting vector bosons and gravitons, provided that

$$\alpha' = \pi G/\alpha^2\hbar c^5, \tag{29.140}$$

where G is Newton's gravitational constant and α ($\approx \frac{1}{137}$) is the fine structure constant. (Here it is assumed that the gauge coupling constant of the vector bosons is given by the electron's charge.) This suggests that

$$\alpha' \approx 4 \times 10^{-34}\,\mathrm{GeV}^{-2}.$$

Then, for example, the first excited state with positive rest mass has, by (29.139), a rest mass M that is given by

$$M = 4\alpha\pi^{-1/2}M_{\mathrm{P}},$$

where

$$M_{\mathrm{P}} = (\hbar c/G)^{1/2} \approx 2.2 \times 10^{-5}\,\mathrm{g}$$

is the Planck mass, so that

$$M_{\mathrm{P}}c^2 \approx 1.2 \times 10^{19}\,\mathrm{GeV}.$$

Thus, with this sort of value of α', only the massless states have any chance of physical relevance. This is also true of the theories involving fermions that follow.

The generalization of the *covariant quantization* treatment to the closed bosonic string is similar. For $\mu = 1, 2, \ldots, D$ all the quantities q^μ, p^μ, $a_{\mathrm{R}j}^\mu$ and $a_{\mathrm{L}j}^\mu$ (for $j = \pm 1, \pm 2, \ldots$) are treated as being independent operators, so, with $a_{\mathrm{R}0}^\mu$ ($= a_{\mathrm{L}0}^\mu$) defined by (29.116), the commutation relations (29.84) are replaced by

$$[a_{\mathrm{R}j}^\mu, a_{\mathrm{R}k}^\lambda]_- = -j\eta^{\mu\lambda}\delta_{j+k,0}1, \tag{29.141}$$

$$[a^{\mu}_{\mathrm{L}j}, a^{\lambda}_{\mathrm{L}k}]_- = -j\eta^{\mu\lambda}\delta_{j+k,0}1, \tag{29.142}$$

$$[a^{\mu}_{\mathrm{R}j}, a^{\lambda}_{\mathrm{L}k}]_- = 0, \tag{29.143}$$

(29.85) being unchanged, and all the other commutators being zero. The "right-moving and left-moving covariant Virasoro algebras" have basis elements defined by

$$L_{\mathrm{R}J} = -\frac{1}{2}\sum_{j=-\infty}^{\infty} \eta_{\mu\lambda} :a^{\mu}_{\mathrm{R}(J+j)}a^{\lambda}_{\mathrm{R}(-j)}: \tag{29.144}$$

and

$$L_{\mathrm{L}J} = -\frac{1}{2}\sum_{j=-\infty}^{\infty} \eta_{\mu\lambda} :a^{\mu}_{\mathrm{L}(J+j)}a^{\lambda}_{\mathrm{L}(-j)}: \tag{29.145}$$

(cf. (29.91)), and both have central charge $c_{\mathrm{V}} = D$. The "no-ghost theorem" then states that the set of states Ψ that satisfy the subsidiary conditions

$$L_{\mathrm{R}J}\Psi = 0 \text{ and } L_{\mathrm{L}J}\Psi = 0 \quad \text{for} \quad J > 0 \tag{29.146}$$

and

$$L_{\mathrm{R}0}\Psi = \alpha_0\Psi, \quad L_{\mathrm{L}0}\Psi = \alpha_0\Psi \tag{29.147}$$

do not contain any negative-norm states if $D = 26$ and $\alpha_0 = 2$.

The connection between the quantized world-sheet energy–momentum tensor (29.112) and the Virasoro operators that was discussed for the open string at the end of the last subsection generalizes to the case of the closed string. On substituting (29.117) into (29.113) and (29.114) and normal-ordering, the combinations

$$T_{\mathrm{R}} = -(4\hbar)^{-1}(T_{11} - T_{12})$$

and

$$T_{\mathrm{L}} = -(4\hbar)^{-1}(T_{11} + T_{12})$$

become

$$T_{\mathrm{R}} = -\tfrac{1}{2}\eta_{\mu\lambda}\sum_{j,k=-\infty}^{\infty} :a^{\mu}_{\mathrm{R}j}a^{\lambda}_{\mathrm{R}k}: \mathrm{e}^{-2i(j+k)(\tau-\sigma)}$$

and

$$T_{\mathrm{L}} = -\tfrac{1}{2}\eta_{\mu\lambda}\sum_{j,k=-\infty}^{\infty} :a^{\mu}_{\mathrm{L}j}a^{\lambda}_{\mathrm{L}k}: \mathrm{e}^{-2i(j+k)(\tau+\sigma)}.$$

On replacing τ by $-i\tau$ and defining z by

$$z = \mathrm{e}^{2(\tau - i\sigma)},$$

it follows that T_{R} is a function of z alone, and is given by

$$T_{\mathrm{R}}(z) = \sum_{J=-\infty}^{\infty} L_{\mathrm{R}J} z^{-J},$$

where now the $L_{\mathrm{R}J}$ are the Virasoro operators (29.144). Similarly T_{L} is a function of the complex conjugate $\bar{z}$ of z alone, and is given by

$$T_{\mathrm{L}}(\bar{z}) = \sum_{J=-\infty}^{\infty} L_{\mathrm{L}J} \bar{z}^{-J},$$

where the $L_{\mathrm{L}J}$ are the Virasoro operators (29.145).

3 The spinning string of Ramond, Neveu and Schwarz

The "spinning string" was first discussed by Ramond (1971) and Neveu and Schwarz (1971a, b). Its Lagrangian density (with the same status as (29.18)) is

$$\mathscr{L} = -(4\pi\alpha'\hbar c^2)^{-1}\eta_{\mu\lambda}\eta^{\alpha\beta}\,\partial_\alpha x^\mu\,\partial_\beta x^\lambda + i\hbar\eta_{\mu\lambda}\eta^{\alpha\beta}\boldsymbol{\psi}^{\mu\dagger}\boldsymbol{\rho}^2\boldsymbol{\rho}_\alpha\,\partial_\beta\boldsymbol{\psi}^\lambda, \qquad (29.148)$$

where

$$\boldsymbol{\psi}^\mu(\sigma,\tau) = \begin{bmatrix} \psi_1^\mu(\sigma,\tau) \\ \psi_2^\mu(\sigma,\tau) \end{bmatrix} \qquad (29.149)$$

is a function of σ and τ that transforms as a 2-component spinor under world-sheet transformations but as a D-component vector under space–time transformations. In (29.148)

$$\boldsymbol{\rho}_\alpha = \eta_{\alpha\beta}\boldsymbol{\rho}^\beta, \qquad (29.150)$$

where $\boldsymbol{\rho}^1$ and $\boldsymbol{\rho}^2$ are two 2×2 matrices that satisfy the conditions

$$\boldsymbol{\rho}^\alpha\boldsymbol{\rho}^\beta + \boldsymbol{\rho}^\beta\boldsymbol{\rho}^\alpha = 2\eta^{\alpha\beta}\mathbf{1}, \qquad (29.151)$$

(for α, $\beta = 1, 2$), $\eta^{\alpha\beta}$ being defined by (29.17) and $\eta_{\alpha\beta}$ being its associated covariant tensor. That is, $\boldsymbol{\rho}^1$ and $\boldsymbol{\rho}^2$ form a realization of the Clifford algebra of a 2-dimensional Minkowski space–time. One choice of these matrices is

$$\boldsymbol{\rho}^1 = i\boldsymbol{\sigma}_1 = \begin{bmatrix} 0 & i \\ i & 0 \end{bmatrix} \qquad (29.152)$$

and

$$\boldsymbol{\rho}^2 = \boldsymbol{\sigma}_2 = \begin{bmatrix} 0 & -i \\ i & 0 \end{bmatrix}. \qquad (29.153)$$

All other quantities in (29.148) are as described in Section 2(*a*). The components of $\boldsymbol{\psi}^\mu(\sigma,\tau)$ may be taken to be dimensionless.

Although it is possible to construct an action analogous to (29.3) that incorporates world-sheet reparametrization invariance (cf. Iwasaki and Kikkawa 1973, Brink *et al.* 1976, Deser and Zumino 1976, Brink and Schwarz 1977), (29.148) will be taken as the starting point of the present treatment. Moreover, although both the open and closed spinning strings can be quantized in both the light-cone and the covariant formulations, the present account will be confined to a discussion of the open string, most emphasis being place on the covariant approach.

The equation of motion for the bosonic coordinates $x^\mu(\sigma, \tau)$ remains (29.21), and, with the open string boundary conditions (29.22) imposed, the general solution is again (29.23), or equivalently (29.24).

For the fields $\boldsymbol{\psi}^\mu(\sigma, \tau)$ the Euler–Lagrange equation is the massless 2-dimensional Dirac equation

$$\eta^{\alpha\beta}\boldsymbol{\rho}_\alpha\, \partial_\beta \boldsymbol{\psi}^\mu(\sigma, \tau) = 0 \tag{29.154}$$

(cf. (17.97)), which, with the choices (29.152) and (29.153), becomes

$$\partial_\sigma \psi_1^\mu(\sigma, \tau) + \partial_\tau \psi_1^\mu(\sigma, \tau) = 0 \tag{29.155}$$

and

$$\partial_\sigma \psi_2^\mu(\sigma, \tau) - \partial_\tau \psi_2^\mu(\sigma, \tau) = 0. \tag{29.156}$$

As the matrices $\boldsymbol{\rho}^1$ and $\boldsymbol{\rho}^2$ are both purely imaginary, i.e. they form a Majorana representation of the Clifford algebra, the coefficients in the equations (29.155) and (29.156) are all real, and so the components of $\boldsymbol{\psi}^\mu(\sigma, \tau)$ may be taken to be purely real. There are two possible inequivalent boundary conditions, namely the periodic conditions

$$\psi_1^\mu(0, \tau) = \psi_2^\mu(0, \tau), \quad \psi_1^\mu(\pi, \tau) = \psi_2^\mu(\pi, \tau), \tag{29.157}$$

which were first discussed by Ramond (1971), and the antiperiodic conditions

$$\psi_1^\mu(0, \tau) = \psi_2^\mu(0, \tau), \quad \psi_1^\mu(\pi, \tau) = -\psi_2^\mu(\pi, \tau), \tag{29.158}$$

which were investigated first by Neveu and Schwarz (1971a, b). Later work (cf. Gliozzi *et al.* 1976, 1977) has shown that it is useful to regard these as giving two "sectors" of the same theory, rather than two separate theories, and they are usually called the *Ramond sector* and the *Neveu–Schwarz sector* respectively.

For the Ramond sector the most general solution of (29.155) and (29.156) has the form

$$\psi_1^\mu(\sigma, \tau) = 2^{-1/2} \sum_{p=-\infty}^{\infty} b_p^\mu \exp\{-ip(\tau - \sigma)\} \tag{29.159}$$

and

$$\psi_2^\mu(\sigma, \tau) = 2^{-1/2} \sum_{p=-\infty}^{\infty} b_p^\mu \exp\{-ip(\tau + \sigma)\}, \tag{29.160}$$

the sum being over all integers (as indicated), whereas for the Neveu–Schwarz sector the general solution is

$$\psi_1^\mu(\sigma, \tau) = 2^{-1/2} \sum_{p \in Z + 1/2} b_p^\mu \exp\{-ip(\tau - \sigma)\} \tag{29.161}$$

and

$$\psi_2^\mu(\sigma, \tau) = 2^{-1/2} \sum_{p \in Z + 1/2} b_p^\mu \exp\{-ip(\tau + \sigma)\}, \tag{29.162}$$

where in this case the p-summations are over all the "half-integers". In both cases the assumed reality of $\psi^\mu(\sigma, \tau)$ implies that the parameters b_p^μ satisfy

$$b_p^{\mu *} = b_{-p}^\mu \tag{29.163}$$

for all allowed values of p.

In the covariant formulation of the open spinning string the bosonic part may be treated exactly as in Section 2(d). For the new part involving $\psi^\mu(\sigma, \tau)$ the covariant anticommutation relations may be taken to be

$$[b_p^\mu, b_q^\lambda]_+ = -\eta^{\mu\lambda}\delta_{p+q,0}1 \tag{29.164}$$

for μ, $\lambda = 1, 2, \ldots, D$ and for all allowed values of p and q, and it may be assumed that the b_p^μ commute with all the bosonic operators. In the Ramond sector (29.164) includes the special case:

$$[b_0^\mu, b_0^\lambda]_+ = -\eta^{\mu\lambda}1. \tag{29.165}$$

With (29.92) and (29.93) and the corresponding relations

$$B_p^\mu = b_p^\mu \quad \text{for } \mu = 1, 2, \ldots, D-1 \tag{29.166}$$

and

$$B_p^\mu = ib_p^\mu \quad \text{for } \mu = D, \tag{29.167}$$

the commutation and anticommutation relations become

$$[A_j^\mu, A_k^\lambda]_- = j\delta^{\mu\lambda}\delta_{j+k,0}1, \tag{29.168}$$

as in (29.95), and

$$[B_p^\mu, B_q^\lambda]_+ = \delta^{\mu\lambda}\delta_{p+q,0}1, \tag{29.169}$$

together with

$$[A_j^\mu, B_k^\lambda]_- = 0. \tag{29.170}$$

These are identical with (28.20), (28.51) and (28.52) with

$$D_0 = D, \tag{29.171}$$

so the basis elements of the Ramond and Neveu–Schwarz superalgebras, i.e. of the two $N = 1$ Virasoro superalgebras, may be constructed as in (28.53) and (28.54). Written in terms of the operators a_j^μ and b_p^μ, these are

$$L_J = -\frac{1}{2}\sum_{j=-\infty}^{\infty} :\eta_{\mu\lambda}a_{J+j}^\mu a_{-j}^\lambda: + \frac{1}{2}\sum_p (p + \tfrac{1}{2}J):\eta_{\mu\lambda}b_{J+p}^\mu b_{-p}^\lambda: + \tfrac{1}{16}\delta_{J,0}\eta D1 \tag{29.172}$$

and

$$G_R = -\sum_{j=-\infty}^{\infty} \eta_{\mu\lambda}a_{-j}^\mu b_{R+j}^\lambda. \tag{29.173}$$

The subscript p in (29.172) and the subscript R in (29.173) run over all integer values for the Ramond superalgebra and over all half-integer values for the Neveu – Schwarz superalgebra, and $\eta = 1$ for the Ramond superalgebra but $\eta = 0$ for the Neveu–Schwarz case.

For the *Neveu–Schwarz sector* the "no-ghost theorem" (Goddard and Thorn 1972, Schwarz 1972, Brower and Friedman 1973, Brink and Nielsen 1973, Corrigan and Goddard 1974, Iwasaki and Kikkawa 1973) states that the set of states Ψ that satisfy the conditions

$$L_J\Psi = 0 \quad \text{for } J > 0, \tag{29.174}$$

$$G_R\Psi = 0 \quad \text{for } R > 0, \tag{29.175}$$

$$(L_0 - \alpha_0)\Psi = 0 \tag{29.176}$$

does not contain any negative-norm states if

$$D = 10 \text{ and } \alpha_0 = \tfrac{1}{2} \tag{29.177}$$

or

$$D < 10 \text{ and } \alpha_0 \leqslant \tfrac{1}{2}, \tag{29.178}$$

but again the latter possibility gives rise to various difficulties that make it untenable (cf. the papers just referred to and Polyakov 1981b). (In (29.174) and (29.176) L_J is as defined in (29.172).) The general mass formula is then obtained as in the theory of the bosonic string, since (28.56) for $J = 0$ implies that

$$L_0 = -\sum_{j=1}^{\infty} \eta_{\mu\lambda}a_{-j}^\mu a_j^\lambda - \sum_{p>0} \eta_{\mu\lambda}pb_{-p}^\mu b_p^\lambda - \tfrac{1}{2}\eta_{\mu\lambda}a_0^\mu a_0^\lambda, \tag{29.179}$$

and hence, by (29.25), (29.176) and (29.177), on the physical states Ψ

$$M^2c^2\Psi = (\alpha' c^2)^{-1}(N_a + N_b - \tfrac{1}{2}1)\Psi, \tag{29.180}$$

where

$$N_a = -\sum_{j=1}^{\infty} \eta_{\mu\lambda} a_j^{\mu\dagger} a_j^{\lambda} \tag{29.181}$$

and

$$N_b = -\sum_{p>0} \eta_{\mu\lambda} p b_p^{\mu\dagger} b_p^{\lambda}, \tag{29.182}$$

$a_j^{\mu\dagger}$ being defined by (29.89) and $b_p^{\mu\dagger}$ by

$$b_p^{\mu\dagger} = b_{-p}^{\mu} \quad \text{for } p > 0 \tag{29.183}$$

(cf. (29.163)). In particular, the vacuum states, which have zero eigenvalues with the excitation number operators N_a and N_b, are tachyonic, and the states obtained from these by acting with the operators $b_{1/2}^{\mu\dagger}$ are massless. Introducing the "G-parity operator" by the definition

$$G = (-1)^{\zeta}, \tag{29.184}$$

where

$$\zeta = \sum_{p>0} \eta_{\mu\lambda} b_p^{\mu\dagger} b_p^{\lambda} - 1,$$

so that G has eigenvalues 1 and -1, it is clear that the set of Neveu–Schwarz physical states Ψ can be divided into two classes. For those for which

$$G\Psi = \Psi \tag{29.185}$$

(29.184) and (29.180) imply that the eigenvalue of $\alpha' M^2 c^4$ is a non-negative integer, so this set contains no tachyons, whereas if

$$G\Psi = -\Psi \tag{29.186}$$

then the corresponding eigenvalue of $\alpha' M^2 c^4$ can have the values $-\frac{1}{2}, \frac{1}{2}, \frac{3}{2}, \ldots$.

These conclusions are confirmed in the light-cone analysis of Iwasaki and Kikkawa (1973). The conditions (29.13) and (29.14) have to be modified for the spinning string, but the results are very similar to those presented for the bosonic string in Sections 2(*b*) and 2(*c*). In particular, the operators q^+ and a_j^+ (for $j \neq 0$) can be taken to be zero, while

$$a_J^- a_0^+ = L_J^{\text{tr}} - \alpha_0 \delta_{J,0} 1 \tag{29.187}$$

(cf. (29.48) and (29.74)), the only new feature here being that the L_J^{tr} are now the even basis elements of the "transverse Neveu–Schwarz superalgebra" that are given (cf. (29.53)) by

$$L_J^{\text{tr}} = \frac{1}{2} \sum_{j=-\infty}^{\infty} \sum_{r=1}^{D} :a_{J+j}^r a_{-j}^r: - \frac{1}{2} \sum_{p} \sum_{r=1}^{D} (p + \tfrac{1}{2}J) :b_{J+p}^r b_{-p}^r: \; . \tag{29.188}$$

Similarly one may take $b_p^+ = 0$ for all p and

$$b_R^- a_0^+ = G_R^{\text{tr}}, \tag{29.189}$$

where G_R^{tr} are the odd basis elements of the transverse Neveu–Schwarz superalgebra, and are given

$$G_R^{\text{tr}} = \sum_{j=-\infty}^{\infty} \sum_{r=1}^{D} a_{-j}^r b_{R+j}^r \tag{29.190}$$

(cf. (29.54)). The independent operators of the light-cone formulation of the open string Neveu–Schwarz sector are q^-, p^+ and (for $r = 1, 2, \ldots, D-2$) q^r, a_j^r (for $j = 0, \pm 1, \pm 2, \ldots$) and b_p^r (for $p = \pm\frac{1}{2}, \pm\frac{3}{2}, \ldots$). Closure of the D-dimensional Lorentz algebra is only possible if

$$D = 10 \text{ and } \alpha_0 = \tfrac{1}{2}, \tag{29.191}$$

so the analogue of (29.76) is

$$(Mc^2)^2 = \alpha'^{-1}(N_a^{\text{tr}} + N_b^{\text{tr}} - \tfrac{1}{2}1), \tag{29.192}$$

where

$$N_a^{\text{tr}} = \sum_{j=1}^{\infty} \sum_{r=1}^{D-2} a_j^{r\dagger} a_j^r \tag{29.193}$$

and

$$N_b^{\text{tr}} = \sum_{p>0} \sum_{r=1}^{D-2} p b_p^{r\dagger} b_p^r. \tag{29.194}$$

As expected, the tachyonic vacuum states appear again.

For the Ramond sector the corresponding "no-ghost theorem" (cf. Schwarz 1973, Rebbi 1974, Scherk 1975) states that the set of states Ψ such that

$$L_J \Psi = 0 \quad \text{for } J > 0 \tag{29.195}$$

and

$$G_R \Psi = 0 \quad \text{for } R = 0, 1, 2, \ldots \tag{29.196}$$

does not contain any negative-norm states provided again that

$$D = 10. \tag{29.197}$$

Equations (28.68) and (28.69) show that the vacuum state vectors satisfy (29.195) and (29.196) for $R > 0$. The requirement that (29.196) be satisfied for the vacuum states for $R = 0$ implies by (28.74) and (29.171) that they have eigenvalue $\frac{1}{16}D$ with L_0. Equations (28.56) and (29.25) then imply that these vacuum states are massless and that the general mass formula is

$$(Mc^2)^2 \Psi = \alpha'^{-1}(N_a + N_b)\Psi, \tag{29.198}$$

where N_a is defined by (29.181) and N_b is given by (29.182) (with p now being summed over all positive integer values). Indeed, (28.73) and the identity

$$\sum_{r=1}^{D} (\alpha^r)^2 = -2\alpha' c^2 \eta_{\mu\lambda} p_0^\mu p_0^\lambda \tag{29.199}$$

(which follows from (29.25) and (29.94)) demonstrate the zero-mass nature of the vaccum states immediately. Thus there is no tachyonic mode in the Ramond sector.

Gliozzi *et al.* (1976, 1977) showed that the G-parity operator (29.184) can be extended to the Ramond sector in such a way that the eigenspace of physical states with eigenvalue $+1$ is not only tachyon-free but is also supersymmetric in 10-dimensional space–time. The superstring theory of Green and Schwarz (1981, 1982a, b, c), which forms the subject of the next section, provides a formalism in which this supersymmetry is much more obvious.

The relationship between the energy–momentum tensor and the Virasoro algebra that was discussed for bosonic strings at the ends of Sections 2(*d*) and 2(*e*) generalizes to the Virasoro superalgebras of the spinning string. The even basis elements L_J again appear in Laurent expansions of the energy–momentum tensor, the components of which contain a world-sheet fermion contribution in addition to the bosonic part (29.112). For details see Bershadsky *et al.* (1985) and Friedan *et al.* (1985).

4 The superstring of Green and Schwarz

(*a*) *The light-cone Lagrangian density*

By constructing a manifestly supersymmetric and tachyon-free string theory with other very significant properties, Green and Schwarz (1981, 1982a, b, c) triggered off an explosion of interest in string theories. This subsection is devoted to presenting the algebraic background to the theory and setting up the "light-cone" Lagrangian density, its application to open and closed strings being left to the two subsections that follows.

Right from the start of the discussion it will be assumed that $D = 10$; that is, that the space–time is 10-dimensional. Not only is this suggested by the analysis of the previous section, but 10-dimensional Minkowski space–time has certain unique properties that make it particularly promising for the construction of supersymmetric theories. In particular, the Lie algebra of the little group of massless representations (cf. Chapter 23, Section 7(*a*), part (B)) is so(8), which is the compact form of the simple complex Lie algebra D_4. Both of these algebras have *three* inequivalent 8-dimensional irreducible representations. One is provided by the defining representation of so(8) in which each matrix of so(8) is represented by itself. This will be called the "vector" representation. The other two 8-dimensional irreducible representations are known as "spinor" representations, since both provide representations of the universal covering group Spin(8) of so(8), but not of the Lie group SO(8) itself (cf. Chapter 15, Section 6). The fact that the vector and spinor irreducible representations have the *same* dimension can be traced to the three-fold symmetry of the Dynkin diagram of D_4 (cf. Chapter 14, Section 5, particularly Figure 14.1), a property known as "triality", and clearly augers well for the construction of a supersymmetric theory.

Let γ^{μ}_{10M} (for $\mu = 1, 2, \ldots, 10$) be the Dirac matrices for 10-dimensional Minkowski space–time, so that

$$[\gamma^{\mu}_{10M}, \gamma^{\lambda}_{10M}]_{+} = 2\eta^{\mu\lambda}\mathbf{1}, \tag{29.200}$$

where $\eta^{\mu\lambda}$ is given by (29.4) with $D = 10$. As noted in Appendix L, Section 2(*a*), these matrices must have dimension 32×32, and may be chosen to be purely imaginary. A convenient realization is given by

$$\gamma^{r}_{10M} = \begin{bmatrix} i\gamma^{r}_{8E} & \mathbf{0} \\ \mathbf{0} & i\gamma^{r}_{8E} \end{bmatrix}, \quad r = 1, 2, \ldots, 8, \tag{29.201}$$

$$\gamma^{9}_{10M} = \begin{bmatrix} \mathbf{0} & -i\mathbf{1}_{16} \\ -i\mathbf{1}_{16} & \mathbf{0} \end{bmatrix}, \tag{29.202}$$

$$\gamma^{10}_{10M} = \begin{bmatrix} \mathbf{0} & -i\mathbf{1}_{16} \\ i\mathbf{1}_{16} & \mathbf{0} \end{bmatrix}, \tag{29.203}$$

where γ^{r}_{8E} (for $r = 1, 2, \ldots, 8$) are the Dirac matrices for a 8-dimensional Euclidean space–time, i.e.

$$[\gamma^{r}_{8E}, \gamma^{s}_{8E}]_{+} = 2\delta^{rs}\mathbf{1}. \tag{29.204}$$

These γ^{r}_{8E} matrices have dimension 16×16 and may be selected to be purely real.

A particularly convenient choice is to take

$$\gamma^{r}_{8E} = \begin{bmatrix} \mathbf{0} & \mathbf{c}^{r} \\ \tilde{\mathbf{c}}^{r} & \mathbf{0} \end{bmatrix}, \tag{29.205}$$

with the 8×8 matrices $\mathbf{c}^{r}$ ($r = 1, 2, \ldots, 8$) all being real. Clearly (20.204) is satisfied if and only if

$$\mathbf{c}^{r}\tilde{\mathbf{c}}^{s} + \mathbf{c}^{s}\tilde{\mathbf{c}}^{r} = 2\delta^{rs}\mathbf{1}_{8} \tag{29.206}$$

and

$$\tilde{\mathbf{c}}^{r}\mathbf{c}^{s} + \tilde{\mathbf{c}}^{s}\mathbf{c}^{r} = 2\delta^{rs}\mathbf{1}_{8} \tag{29.207}$$

for $r, s = 1, 2, \ldots, 8$. With

$$\mathbf{c}^{8} = -\mathbf{1}_{8}, \tag{29.208}$$

(29.206) and (30.207) with $r = 1, 2, \ldots, 7$ and $s = 8$ imply that $\mathbf{c}^r$-must be antisymmetric for $r = 1, 2, \ldots, 7$, and then, for $r, s = 1, 2, \ldots, 7$, (29.206) and (29.207) both reduce to

$$\mathbf{c}^r\mathbf{c}^s + \mathbf{c}^s\mathbf{c}^r = -2\delta^{rs}\mathbf{1}_8, \tag{20.209}$$

so that

$$\mathbf{c}^r = i\gamma^r_{7\mathrm{E}} \tag{29.210}$$

for $r = 1, 2, \ldots, 7$, where $\gamma^r_{7\mathrm{E}}$ satisfy the 7-dimensional Euclidean Clifford algebra. One explicit realization (cf. Green and Schwarz 1983) is

$$\left.\begin{array}{lll} \mathbf{c}^1 = i\boldsymbol{\sigma}_2 \otimes i\boldsymbol{\sigma}_2 \otimes i\boldsymbol{\sigma}_2, & \mathbf{c}^2 = \mathbf{1}_2 \otimes \boldsymbol{\sigma}_1 \otimes i\boldsymbol{\sigma}_2, & \mathbf{c}^3 = \mathbf{1}_2 \otimes \boldsymbol{\sigma}_3 \otimes i\boldsymbol{\sigma}_2, \\ \mathbf{c}^4 = \boldsymbol{\sigma}_1 \otimes i\boldsymbol{\sigma}_2 \otimes \mathbf{1}_2, & \mathbf{c}^5 = \boldsymbol{\sigma}_3 \otimes i\boldsymbol{\sigma}_2 \otimes \mathbf{1}_2, & \mathbf{c}^6 = i\boldsymbol{\sigma}_2 \otimes \mathbf{1}_2 \otimes \boldsymbol{\sigma}_1, \\ \mathbf{c}^7 = i\boldsymbol{\sigma}_2 \otimes \mathbf{1}_2 \otimes \boldsymbol{\sigma}_3, & & \end{array}\right\} \tag{29.211}$$

which are all real and antisymmetric.

In accordance with (29.41) and (29.42), (29.202) and (29.203) imply that

$$\gamma^+_{10\mathrm{M}} = 2^{-1/2}(\gamma^{10}_{10\mathrm{M}} + \gamma^9_{10\mathrm{M}}) = \begin{bmatrix} \mathbf{0} & -i2^{1/2}\mathbf{1}_{16} \\ \mathbf{0} & \mathbf{0} \end{bmatrix} \tag{29.212}$$

and

$$\gamma^-_{10\mathrm{M}} = 2^{-1/2}(\gamma^{10}_{10\mathrm{M}} - \gamma^9_{10\mathrm{M}}) = \begin{bmatrix} \mathbf{0} & \mathbf{0} \\ i2^{1/2}\mathbf{1}_{16} & \mathbf{0} \end{bmatrix}. \tag{29.213}$$

Moreover, with $\gamma^{11}_{10\mathrm{M}}$ and $\gamma^9_{8\mathrm{E}}$ defined by

$$\gamma^{11}_{10\mathrm{M}} = \prod_{\mu=1}^{10} \gamma^\mu_{10\mathrm{M}} \tag{29.214}$$

and

$$\gamma^9_{8\mathrm{E}} = \prod_{r=1}^{8} \gamma^r_{8\mathrm{E}}, \tag{29.215}$$

and, with the $\gamma^r_{8\mathrm{E}}$ being given by (29.205) and with the constituent matrices $\mathbf{c}^r$ being as in (29.208) and (29.211),

$$\gamma^9_{8\mathrm{E}} = \begin{bmatrix} -\mathbf{1}_8 & \mathbf{0} \\ \mathbf{0} & \mathbf{1}_8 \end{bmatrix}. \tag{29.216}$$

Then, by (29.201)–(29.203),

$$\gamma^{11}_{10\mathrm{M}} = \begin{bmatrix} -\mathbf{1}_8 & \mathbf{0} & \mathbf{0} & \mathbf{0} \\ \mathbf{0} & \mathbf{1}_8 & \mathbf{0} & \mathbf{0} \\ \mathbf{0} & \mathbf{0} & \mathbf{1}_8 & \mathbf{0} \\ \mathbf{0} & \mathbf{0} & \mathbf{0} & -\mathbf{1}_8 \end{bmatrix}. \tag{29.217}$$

Thus, if $\mathbf{S}$ is a 32×1 column matrix partitioned into four 8×1 column matrices $\mathbf{s}_j$ $(j = 1, \ldots, 4)$ by putting

$$\mathbf{S} = \begin{bmatrix} \mathbf{s}_1 \\ \mathbf{s}_2 \\ \mathbf{s}_3 \\ \mathbf{s}_4 \end{bmatrix}, \tag{29.218}$$

then, if

$$\gamma^{+}_{10\mathrm{M}}\mathbf{S} = \mathbf{0}, \tag{29.219}$$

(29.212) shows that

$$\mathbf{s}_3 = \mathbf{s}_4 = \mathbf{0}. \tag{29.220}$$

Moreover, if it is also required that

$$\gamma^{11}_{10\mathrm{M}}\mathbf{S} = \mathbf{S} \tag{29.221}$$

then (29.217) implies that

$$\mathbf{s}_2 = \mathbf{0} \tag{29.222}$$

as well, whereas if it is required that $\mathbf{S}$ satisfy (29.219) and

$$\gamma^{11}_{10\mathrm{M}}\mathbf{S} = -\mathbf{S} \tag{29.223}$$

then (29.220) holds and

$$\mathbf{s}_1 = \mathbf{0} \tag{29.224}$$

as well. Thus if $\mathbf{S}$ satisfies (29.219) and (29.221) then

$$\mathbf{S} = \begin{bmatrix} \mathbf{s} \\ \mathbf{0} \\ \mathbf{0} \\ \mathbf{0} \end{bmatrix}, \tag{29.225}$$

where the components of $\mathbf{S}$ are arbitrary, whereas if $\mathbf{S}$ satisfies (29.219) and (29.223) then

$$\mathbf{S} = \begin{bmatrix} \mathbf{0} \\ \mathbf{s} \\ \mathbf{0} \\ \mathbf{0} \end{bmatrix}, \tag{29.226}$$

the components of $\mathbf{S}$ again being arbitrary. In each case the result is that the 32-component column matrix $\mathbf{S}$ effectively depends on a *single* 8-component column matrix $\mathbf{s}$.

The basis elements of the Lie algebra so(8) may be denoted by L_{pq}, where $p, q = 1, 2, \ldots, 8$ and $p < q$. As in Appendix G, Section 2 (where these elements are denoted by M_{pq}), it is convenient to extend the range of values of the pairs (p, q) to cover *all* the pairs with $p, q = 1, 2, \ldots, 8$, with the assumption that

$$L_{pq} = -L_{qp},$$

and hence that $L_{pq} = 0$ if $p = q$. The commutation relations are then

$$[L_{pq}, L_{rs}]_- = \delta_{qr} L_{ps} - \delta_{qs} L_{pr} - \delta_{pr} L_{qs} + \delta_{ps} L_{qr} \tag{29.227}$$

(which is consistent with (17.13) with $g_{pq} = \eta_{pq} = -\delta_{pq}$). The 8-dimensional vector irreducible representation of so(8) is then provided by

$$\boldsymbol{\Gamma}^{\mathrm{v}}(L_{pq})_{jk} = \delta_{pj}\delta_{qk} - \delta_{pk}\delta_{qj}, \tag{29.228}$$

whose eigenvalues are i, $-i$ and 0, the last being six-fold degenerate. A 16-dimensional spinor representation of so(8) is given by

$$\boldsymbol{\Gamma}(L_{pq}) = \tfrac{1}{4}[\boldsymbol{\gamma}^p_{8\mathrm{E}}, \boldsymbol{\gamma}^q_{8\mathrm{E}}]_-$$

(the apparent sign change relative to (17.135) being caused by the choice of conventions in which the L_{pq} correspond to (17.13) with $g_{pq} = \eta_{pq} = -\delta_{pq}$ but the $\boldsymbol{\gamma}^p_{8\mathrm{E}}$ correspond (by (29.204)) to (17.98) with $g_{pq} = \eta_{pq} = \delta_{pq}$). With the choice (29.205),

$$\boldsymbol{\Gamma}(L_{pq}) = \boldsymbol{\Gamma}^{\mathrm{s}}(L_{pq}) \oplus \boldsymbol{\Gamma}^{\mathrm{s}'}(L_{pq}), \tag{29.229}$$

where

$$\boldsymbol{\Gamma}^{\mathrm{s}}(L_{pq}) = \tfrac{1}{4}(\mathbf{c}^p\tilde{\mathbf{c}}^q - \mathbf{c}^q\tilde{\mathbf{c}}^p) \tag{29.230}$$

and

$$\boldsymbol{\Gamma}^{\mathrm{s}'}(L_{pq}) = \tfrac{1}{4}(\tilde{\mathbf{c}}^p\mathbf{c}^q - \tilde{\mathbf{c}}^q\mathbf{c}^p) \tag{29.231}$$

are the two 8-dimensional spinor irreducible representations of so(8). (In terms of the notation introduced in Chapter 15, Section 3, the basis of the Cartan subalgebra of D_4 may be chosen so that $\boldsymbol{\Gamma}^{\mathrm{v}}, \boldsymbol{\Gamma}^{\mathrm{s}}$, and $\boldsymbol{\Gamma}^{\mathrm{s}'}$ are equivalent to $\boldsymbol{\Gamma}(\{1,0,0,0\})$, $\boldsymbol{\Gamma}(\{0,0,1,0\})$, and $\boldsymbol{\Gamma}(\{0,0,0,1\})$ respectively.)

The "light-cone" Lagrangian density proposed by Green and Schwarz (1982a) is

$$\mathscr{L}^{\mathrm{lc}} = -\frac{1}{4\pi\alpha'\hbar c^2}\sum_{r=1}^{8} \eta^{\alpha\beta}\partial_\alpha x^r \partial_\beta x^r + \frac{i\hbar}{4\pi}\sum_{A,B,C=1}^{2}\ \sum_{a,b,c=1}^{32} \eta^{\alpha\beta} S^{\dagger Bb}(\boldsymbol{\gamma}^{10}_{10\mathrm{M}})^{ba}(\boldsymbol{\rho}^2)^{BA}(\boldsymbol{\gamma}^{-}_{10\mathrm{M}})^{ac}(\boldsymbol{\rho}_\alpha)^{AC}\,\partial_\beta S^{Cc}. \tag{29.232}$$

Here $x^r(\sigma, \tau)$ for $r = 1, 2, \ldots, 8$ are the transverse components of the string coordinates, so that the first term of (29.232) is identical with (29.55). The quantities $S^{Aa}(\sigma, \tau)$ (for $A = 1, 2$ and $a = 1, 2, \ldots, 32$) may be taken as the components of two Majorana–Weyl spinors $\mathbf{S}^A(\sigma, \tau)$, each of which is a 32×1 matrix. The Majorana condition implies that

$$\mathbf{S}^A(\sigma, \tau)^* = \mathbf{S}^A(\sigma, \tau), \tag{29.233}$$

which is compatible with all the matrices $\boldsymbol{\gamma}^{\mu}_{10\mathrm{M}}$ (for $\mu = 1, 2, \ldots, 10$), being chosen so to be purely imaginary (as in (29.201)–(29.203)), and the Weyl condition can be expressed as

$$\boldsymbol{\gamma}^{11}_{10\mathrm{M}}\mathbf{S}^A(\sigma, \tau) = \eta^A \mathbf{S}^A(\sigma, \tau), \tag{29.234}$$

where

$$\eta^A = +1 \text{ or } -1. \tag{29.235}$$

The matrices $\boldsymbol{\rho}^2$ and $\boldsymbol{\rho}_\alpha$ (for $\alpha = 1, 2$) are as defined in (29.151)–(29.153). Imposing the further condition that

$$\boldsymbol{\gamma}^{+}_{10\mathrm{M}}\mathbf{S}^A(\sigma, \tau) = \mathbf{0} \tag{29.236}$$

for $A = 1, 2$, it follows from (29.225) that if $\eta^A = 1$ then $\mathbf{S}^A(\sigma, \tau)$ can be written as

$$\mathbf{S}^A(\sigma, \tau) = 8^{1/4}\begin{bmatrix} \mathbf{s}^A(\sigma, \tau) \\ \mathbf{0} \\ \mathbf{0} \\ \mathbf{0} \end{bmatrix}, \tag{29.237}$$

whereas if $\eta^A = -1$ then (29.226) implies that $\mathbf{S}^A(\sigma, \tau)$ has the form

$$\mathbf{S}^A(\sigma, \tau) = 8^{1/4}\begin{bmatrix} \mathbf{0} \\ \mathbf{s}^A(\sigma, \tau) \\ \mathbf{0} \\ \mathbf{0} \end{bmatrix}. \tag{29.238}$$

In both cases $\mathbf{s}^A(\sigma, \tau)$ are two 8×1 matrices (the factors $8^{1/4}$ being included for later numerical convenience). The Lagrangian density (29.232) can then be rewritten as

$$\mathscr{L}^{\mathrm{lc}} = -\frac{1}{4\pi\alpha'\hbar c^2}\sum_{r=1}^{8} \eta^{\alpha\beta}\,\partial_\alpha x^r\,\partial_\beta x^r + \frac{i\hbar}{\pi}\sum_{A,B,C=1}^{2}\sum_{p=1}^{8} \eta^{\alpha\beta}\, s^{\dagger Bp}(\boldsymbol{\rho}^2)^{BA}(\boldsymbol{\rho}_\alpha)^{AC}\,\partial_\beta s^{Cp}. \tag{29.239}$$

A covariant Lagrangian density that reduces to (29.232) in the light-cone gauge was subsequently found by Green and Schwarz (1984a), but, although it has a number of interesting symmetries, it has not been found possible to construct a covariant scheme of quantization based on it (for reasons discussed in detail by Green and Schwarz 1984b). Consequently the treatment that follows will proceed from (29.232) and its equivalent form (29.239).

(b) *Light-cone quantization of the open superstring*

The open string boundary conditions for the bosonic coordinates $x^r(\sigma, \tau)$ $(r = 1, 2, \ldots, 8)$ are

$$\partial_\sigma x^r(\sigma, \tau) = 0 \quad \text{for } \sigma = 0 \text{ and } \pi, \tag{29.240}$$

as implied by (29.22). The equation of motion that follows from (29.239) is

$$\frac{\partial^2 x^r}{\partial \sigma^2} - \frac{\partial^2 x^r}{\partial \tau^2} = 0 \tag{29.241}$$

(cf. (29.21)), so the solution can again be written as

$$x^r(\sigma, \tau) = q^r + 2\alpha' \hbar c^2 p^r \tau + (2\alpha' \hbar^2 c^2)^{1/2} i \sum_{j=-\infty, j\neq 0}^{\infty} \frac{a_j^r}{j} \mathrm{e}^{-ij\tau} \cos j\sigma, \tag{29.242}$$

(as in (29.24)), or equivalently as

$$x^r(\sigma, \tau) = q^r + (2\alpha' \hbar^2 c^2)^{1/2} \left(a_0^r \tau + i \sum_{j=-\infty, j\neq 0}^{\infty} \frac{a_j^r}{j} \mathrm{e}^{-ij\tau} \cos j\sigma \right) \tag{29.243}$$

(as in (29.26)), where

$$a_0^r = (2\alpha' c^2)^{1/2} p^r \tag{29.244}$$

(as in (29.25)).

For the fermionic coordinates $\mathbf{S}^A(\sigma, \tau)$ the open string boundary conditions proposed by Green and Schwarz (1982a) are

$$\mathbf{S}^1(0, \tau) = \mathbf{S}^2(0, \tau) \tag{29.245}$$

and

$$\mathbf{S}^1(\pi, \tau) = \mathbf{S}^2(\pi, \tau). \tag{29.246}$$

These imply immediately (by (29.234)) that

$$\eta^1 = \eta^2 \tag{29.247}$$

and, by (29.237) and (29.238), that

$$\mathbf{s}^1(0, \tau) = \mathbf{s}^2(0, \tau) \tag{29.248}$$

and

$$\mathbf{s}^1(\pi, \tau) = \mathbf{s}^2(\pi, \tau). \tag{29.249}$$

(As will be discussed shortly, these are compatible with a global 10-

dimensional supersymmetry, but no such supersymmetry is possible if these boundary conditions involve quantities with opposite signs.)

The Euler–Lagrange equations for the $\mathbf{s}^A(\sigma, \tau)$ that follow from (29.239) are

$$\partial_\sigma s^{1r}(\sigma, \tau) + \partial_\tau s^{1r}(\sigma, \tau) = 0 \tag{29.250}$$

and

$$\partial_\sigma s^{2r}(\sigma, \tau) - \partial_\tau s^{2r}(\sigma, \tau) = 0, \tag{29.251}$$

where $s^{Ar}(\sigma, \tau)$ (for $r = 1, 2, \ldots, 8$) are the components of $\mathbf{s}^A(\sigma, \tau)$.

As in (29.155) and (29.156), these are of the form of the 2-dimensional Dirac equations corresponding to *world-sheet spinors*, but the new feature is that these fields carry a *space–time spinor* index r instead of the *space–time vector* index μ of (29.155) and (29.156). The solution of (29.250) and (29.251) satisfying (29.248) and (29.249) can be written as

$$s^{1r}(\sigma, \tau) = 2^{-1/2} \sum_{j=-\infty}^{\infty} s_j^r \exp\{-ij(\tau - \sigma)\} \tag{29.252}$$

and

$$s^{2r}(\sigma, \tau) = 2^{-1/2} \sum_{j=-\infty}^{\infty} s_j^r \exp\{-ij(\tau + \sigma)\}, \tag{29.253}$$

where the s_j^r (for $r = 1, 2, \ldots, 8$ and $j = 0, \pm 1, \pm 2, \ldots$) are a set of coefficients for which the reality condition (29.245) implies

$$s_j^{r*} = s_{-j}^r. \tag{29.254}$$

In the light-cone quantization formulation all of the parameters such as a_j^r and s_j^r in (29.242), (29.243), (29.252) and (29.253) are treated as operators acting in a Hilbert space. The bosonic operators may be assumed to retain the commutation relations (29.57)–(29.59), and the fermionic operators s_j^r may be assumed to satisfy the anticommutation relations

$$[s_j^r, s_k^s]_+ = \delta^{rs}\delta_{j+k,0}1, \tag{29.255}$$

with the commutators of all other pairs of operators being zero.

As far as $x^+(\sigma, \tau)$ and $x^-(\sigma, \tau)$ are concerned, it is assumed that

$$x^+(\sigma, \tau) = (2\alpha' \hbar c^2)^{1/2} p^+ \tau \tag{29.256}$$

(as in (29.44)) and $x^-(\sigma, \tau)$ is again given by (29.47), the coefficients a_J^- again being determined by (29.48), that is (on using (29.55)), by

$$a_J^- = (2\alpha' c^2)^{1/2} (p^+)^{-1} L_J^{\mathrm{tr}},$$

but where now

$$L_J^{\mathrm{tr}} = \frac{1}{2} \sum_{j=-\infty}^{\infty} \sum_{r=1}^{D} :a_{J+j}^r a_{-j}^r: - \frac{1}{2} \sum_{j=1}^{\infty} \sum_{r=1}^{D} (j + \tfrac{1}{2}J) :s_{J+j}^r s_{-j}^r: \quad , \tag{29.257}$$

which is similar in many respects to (29.188), and which shows that the Virasoro algebra appears yet again.

Green and Schwarz (1981) showed that it is possible to write down expressions for the 10-dimensional space–time Lorentz generators in terms of all these operators that do indeed close to give this Lorentz algebra, provided that the mass-shell condition

$$(Mc^2)^2 = \alpha'^{-1} N^{\rm tr} \tag{29.258}$$

is satisfied, where $N^{\rm tr}$ is the excitation number operator defined by

$$N^{\rm tr} = \sum_{j=1}^{\infty} \sum_{r=1}^{8} a_j^{r\dagger} a_j^r + \sum_{j=1}^{\infty} \sum_{r=1}^{8} j s_j^{r\dagger} s_j^r, \tag{29.259}$$

and where, as usual (cf. (29.62)), for $j > 0$

$$a_j^{r\dagger} = a_{-j}^r$$

and the quantum analogue of (29.254) is

$$s_j^{r\dagger} = s_{-j}^r. \tag{29.260}$$

As the eigenvalues of $N^{\rm tr}$ are non-negative integers, (29.258) shows that this theory contains no tachyonic mode. Moreover, since the vacuum states have zero eigenvalue with $N^{\rm tr}$, they must be massless.

These vacuum states will now be examined in more detail. With $j = k = 0$, (29.255) contains as a special case the anticommutation relations

$$[s_0^r, s_0^s]_+ = \delta^{rs} 1. \tag{29.261}$$

As these have a matrix representation in terms of the 16×16 Euclidean Dirac matrices γ_{8E}^r in which

$$\Gamma(s_0^r) = 2^{-1/2} \gamma_{8E}^r \tag{29.262}$$

(cf. (29.204)), it follows that the vacuum state is 16-fold degenerate. From the form (29.205) it is clear that the basis vectors for the representation (29.262) may be divided into two sets of eight, which may be denoted by $|0, 1, j\rangle$ and $|0, 2, j\rangle$ (for $j = 1, 2, \ldots, 8$), and which may be chosen so that

$$s_0^p |0, 1, j\rangle = 2^{-1/2} \sum_{k=1}^{8} c_{jk}^p |0, 2, k\rangle \tag{29.263}$$

and

$$s_0^p |0, 2, j\rangle = 2^{-1/2} \sum_{k=1}^{8} c_{kj}^p |0, 1, k\rangle \tag{29.264}$$

(for $j = 1, 2, \ldots, 8$). (In these and the vacuum state vectors used subsequently the dependence on the momentum will not be displayed explicitly.) It then follows from (29.230) that

$$s_0^p s_0^q |0,1,j\rangle = \tfrac{1}{2}\delta^{pq}|0,1,j\rangle + \sum_{k=1}^{8} \Gamma^s(L_{pq})_{kj}|0,1,k\rangle \tag{29.265}$$

and from (29.231) that

$$s_0^p s_0^q |0,2,j\rangle = \tfrac{1}{2}\delta^{pq}|0,2,j\rangle + \sum_{k=1}^{8} \Gamma^{s'}(L_{pq})_{kj}|0,2,k\rangle \tag{29.266}$$

(for $p, q, j = 1, 2, \ldots, 8$). The analysis of Lorentz algebra operators given by Green and Schwarz (1981, 1982b) shows that on the vacuum state vectors the transverse rotation operators (i.e. those of so(8)) reduce to

$$\Phi_0(L_{pq}) = \frac{1}{2}\sum_{a,b=1}^{8} \Gamma^s(L_{pq})_{ab} s_0^a s_0^b \tag{29.267}$$

if $\eta^1 = \eta^2 = 1$, and to

$$\Phi_0(L_{pq}) = \frac{1}{2}\sum_{a,b=1}^{8} \Gamma^{s'}(L_{pq})_{ab} s_0^a s_0^b \tag{29.268}$$

if $\eta^1 = \eta^2 = -1$. (Actually the $\Phi_0(L_{pq})$ are related to the R_0^{pq} of Green and Schwarz by $\Phi_0(L_{pq}) = -R_0^{pq}$.)

Suppose first that $\eta^1 = \eta^2 = 1$. Then, by (29.265) and (29.267),

$$\Phi_0(L_{pq})|0,1,j\rangle = \frac{1}{2}\sum_{k,a,b=1}^{8} \Gamma^s(L_{pq})_{ab}\Gamma^s(L_{ab})_{kj}|0,1,k\rangle \tag{29.269}$$

and

$$\Phi_0(L_{pq})|0,2,j\rangle = \frac{1}{2}\sum_{k,a,b=1}^{8} \Gamma^s(L_{pq})_{ab}\Gamma^{s'}(L_{ab})_{kj}|0,2,k\rangle. \tag{29.270}$$

This shows that the $|0,1,j\rangle$ transform as a 8-dimensional representation $\mathbf{\Gamma}^{ss}$ of so(8) defined by

$$\mathbf{\Gamma}^{ss}(L_{pq}) = \frac{1}{2}\sum_{a,b=1}^{8} \Gamma^s(L_{pq})_{ab}\mathbf{\Gamma}^s(L_{ab}), \tag{29.271}$$

and the $|0,2,j\rangle$ transform as another 8-dimensional representation $\mathbf{\Gamma}^{ss'}$ of so(8) defined by

$$\mathbf{\Gamma}^{ss'}(L_{pq}) = \frac{1}{2}\sum_{a,b=1}^{8} \Gamma^s(L_{pq})_{ab}\mathbf{\Gamma}^{s'}(L_{ab}). \tag{29.272}$$

Clearly both of these representations are irreducible. The question that must now be settled is how they are related to $\mathbf{\Gamma}^v$, $\mathbf{\Gamma}^s$, and $\mathbf{\Gamma}^{s'}$.

Detailed calculation shows that the matrices $\mathbf{\Gamma}^{ss'}(L_{pq})$ have eigenvalues i,

$-i$ and 0, the last with a six-fold degeneracy, which is exactly the same as for the matrices $\mathbf{\Gamma}^{\mathrm{v}}(L_{pq})$, the sets of eigenvalues for $\mathbf{\Gamma}^{\mathrm{s}}(L_{pq})$ and $\mathbf{\Gamma}^{\mathrm{s}'}(L_{pq})$ being quite different. Consequently the representation $\mathbf{\Gamma}^{\mathrm{ss}'}$ must be equivalent to the vector representation $\mathbf{\Gamma}^{\mathrm{v}}$, and so there must exist eight vectors $|0, \mathrm{v}, j\rangle$, each a linear combination of the $|0, 2, k\rangle$ (for $k = 1, 2, \ldots, 8$), such that

$$\Phi_0(L_{pq})|0, \mathrm{v}, j\rangle = \sum_{k=1}^{8} \Gamma^{\mathrm{v}}(L_{pq})_{kj}|0, \mathrm{v}, k\rangle \tag{29.273}$$

for $p, q, j = 1, 2, \ldots, 8$. Then let $|0, \mathrm{s}', j\rangle$ be defined for $j = 1, 2, \ldots, 8$ by

$$|0, \mathrm{s}', j\rangle = \kappa \sum_{a,k=1}^{8} c_{aj}^{k} s_0^a |0, \mathrm{v}, k\rangle, \tag{29.274}$$

where κ is a normalization constant, assumed independent of j. Equation (29.264) implies that these vectors $|0, \mathrm{s}', j\rangle$ are linear combinations of the $|0, 1, k\rangle$. As

$$[\Phi_0(L_{pq}), s_0^a]_- = \sum_{b=1}^{8} \Gamma^{\mathrm{s}}(L_{pq})_{ba} s_0^b \tag{29.275}$$

(which follows immediately from (29.261) and (29.267)), (29.228), (29.206) and (29.207) show that

$$\Phi_0(L_{pq})|0, \mathrm{s}', j\rangle = \sum_{k=1}^{8} \Gamma^{\mathrm{s}'}(L_{pq})_{kj}|0, \mathrm{s}', k\rangle. \tag{29.276}$$

Thus the vectors $|0, \mathrm{s}', j\rangle$ form a basis for the spinor irreducible representation $\mathbf{\Gamma}^{\mathrm{s}'}$ (as anticipated in the notation). The conclusion is therefore that if $\eta^1 = \eta^2 = 1$ then eight of the massless ground state vectors ($|0, \mathrm{v}, j\rangle$ for $j = 1, 2, \ldots, 8$)) transform as the vector representation $\mathbf{\Gamma}^{\mathrm{v}}$ and the remaining eight massless ground state vectors ($|0, \mathrm{s}', j\rangle$ for $j = 1, 2, \ldots, 8$) transform as the spinor representation $\mathbf{\Gamma}^{\mathrm{s}'}$.

In the case in which $\eta^1 = \eta^2 = -1$ the conclusion is similar in that eight of the massless ground state vectors again transform as the vector representation $\mathbf{\Gamma}^{\mathrm{v}}$, the remaining eight massless ground state vectors transforming this time as the spinor representation $\mathbf{\Gamma}^{\mathrm{s}}$.

The action formed from the Lagrangian density (29.232) using (29.5) is invariant under the supersymmetry transformations

$$x^r(\sigma, \tau) \to x^r(\sigma, \tau) + \delta x^r(\sigma, \tau)$$

and

$$S^{Aa}(\sigma, \tau) \to S^{Aa}(\sigma, \tau) + \delta S^{Aa}(\sigma, \tau),$$

where

$$\delta x^r(\sigma, \tau) = \left(\frac{2\alpha' \hbar^2 c^2}{p^+}\right)^{1/2} \sum_{A,B=1}^{2} \sum_{a,b,c=1}^{32} \varepsilon^{Bb} (\gamma_{10\mathrm{M}}^{10})^{ba} (\boldsymbol{\rho}^2)^{BA} (\gamma_{10\mathrm{M}}^{r})^{ac} S^{Ac} \tag{29.277}$$

and

$$\delta S^{Aa}(\sigma,\tau) = (2\alpha'\hbar^2c^2p^+)^{-1/2} \sum_{B=1}^{2}\sum_{b=1}^{32} \eta_{\mu\lambda}(\gamma^-_{10M}\gamma^\mu_{10M})^{ab}\eta^{\alpha\beta}(\mathbf{\rho}_\alpha)^{AB}\varepsilon^{Bb}\partial_\beta x^\lambda, \tag{29.278}$$

where $\boldsymbol{\varepsilon}^1$ and $\boldsymbol{\varepsilon}^2$ are two 32-component Majorana–Weyl spinors that are independent of σ and τ (and are of dimension mass$^{-1/2}$ $\times$ length$^{-1/2}$ $\times$ time$^{1/2}$), provided that $\boldsymbol{\varepsilon}^1 + \boldsymbol{\varepsilon}^2 = \mathbf{0}$. Green and Schwarz (1981, 1982a) showed that they give rise to a single global 10-dimensional supersymmetry; that is, that they are associated with operators Q_a (for $a = 1, 2, \ldots, 32$) that can be written in terms of the a^r_j and s^r_j and that form the odd part of the 10-dimensional $N = 1$ Poincaré superalgebra. This theory is often described as being of type I.

(c) *Light-cone quantization of the closed superstring*

For the closed superstring the situation for the bosonic coordinates $x^r(\sigma,\tau)$ is essentially the same as in the theory of the bosonic string that was presented in Section 2(*e*). In particular, the equation of motion and the boundary conditions are given by (29.21) and (29.1), and its solution has the form (29.115), or equivalently (29.117). That is,

$$x^\mu(\sigma,\tau) = q^\mu + (2\alpha'\hbar^2c^2)^{1/2}\left\{a^\mu_{\mathrm{R}0}(\tau-\sigma) + \frac{i}{2}\sum_{j=-\infty,\, j\neq 0}^{\infty}\frac{a^\mu_{\mathrm{R}j}}{j}\mathrm{e}^{-2ij(\tau-\sigma)}\right.$$
$$\left. + a^\mu_{\mathrm{L}0}(\tau+\sigma) + \frac{i}{2}\sum_{j=-\infty,\, j\neq 0}^{\infty}\frac{a^\mu_{\mathrm{L}j}}{j}\mathrm{e}^{-2ij(\tau+\sigma)}\right\}. \tag{29.117}$$

As before, the subscripts "R" and "L" correspond to the right-moving and left-moving modes respectively.

For the fermionic coordinates $\mathbf{S}^A(\sigma,\tau)$ the closed string boundary conditions proposed by Green and Schwarz (1982a) are

$$\mathbf{S}^A(0,\tau) = \mathbf{S}^A(\pi,\tau) \quad \text{for } A = 1, 2, \tag{29.279}$$

which imply, by (29.237) and (29.238), that

$$\mathbf{s}^A(0,\tau) = \mathbf{s}^A(\pi,\tau) \quad \text{for } A = 1, 2. \tag{29.280}$$

The equations of motion are still the 2-dimensional Dirac equations (29.250) and (29.251) of the open string case, but the solutions satisfying the closed string boundary conditions (29.280) have the form

$$s^{1r}(\sigma,\tau) = 2^{-1/2}\sum_{j=-\infty}^{\infty} s^r_{\mathrm{R}j}\exp\{-ij(\tau-\sigma)\} \tag{29.281}$$

and

$$s^{2r}(\sigma, \tau) = 2^{-1/2} \sum_{j=-\infty}^{\infty} s^r_{Lj} \exp\{-ij(\tau+\sigma)\}, \tag{29.282}$$

where now the coefficients s^r_{Rj} and s^r_{Lj} (for $r = 1, 2, \ldots, 8$ and $j = 0, \pm 1, \pm 2, \ldots$) are independent of each other. (Again the subscripts "R" and "L" indicate the correspondence to right-moving and left-moving modes.) In contrast with the situation for the open string, the boundary conditions (29.279) and (29.281) do not relate $\mathbf{s}^1$ and $\mathbf{s}^2$, and consequently the eigenvalues η^1 and η^2 of (29.234) are independent of each other. Clearly there are essentially only two cases: one in which they both have the same signs, which for ease of exposition will be taken to be such that $\eta^1 = \eta^2 = 1$, and the other in which they have opposite signs, where the choice will be made that $\eta^1 = -\eta^2 = 1$.

In the light-cone formulation of quantization the bosonic operators are assumed to satisfy the commutation relations (29.124)–(29.127), and the fermion operators s^r_{Rj} and s^r_{Lj} may be assumed to satisfy the anticommutation relations

$$[s^r_{Rj}, s^s_{Rk}]_+ = \delta^{rs}\delta_{j+k,0}1, \tag{29.283}$$

$$[s^r_{Lj}, s^s_{Lk}]_+ = \delta^{rs}\delta_{j+k,0}1, \tag{29.284}$$

$$[s^r_{Rj}, s^s_{Lk}]_+ = 0 \tag{29.285}$$

(which generalize (29.255)), with the commutators of all other pairs of operators being zero.

Again it is possible to write down expressions for the 10-dimensional space–time Lorentz generators that do close to give this Lorentz algebra provided that mass-shell conditions are satisfied. In the closed string case these conditions are

$$(Mc^2)^2 = \frac{4}{\alpha'} N^{tr}_R \tag{29.286}$$

and

$$(Mc^2)^2 = \frac{4}{\alpha'} N^{tr}_L, \tag{29.287}$$

where

$$N^{tr}_R = \sum_{j=1}^{\infty}\sum_{r=1}^{8} a^{r\dagger}_{Rj} a^r_{Rj} + \sum_{j=1}^{\infty}\sum_{r=1}^{8} j s^{r\dagger}_{Rj} s^r_{Rj} \tag{29.288}$$

and

$$N^{tr}_L = \sum_{j=1}^{\infty}\sum_{r=1}^{8} a^{r\dagger}_{Lj} a^r_{Lj} + \sum_{j=1}^{\infty}\sum_{r=1}^{8} j s^{r\dagger}_{Lj} s^r_{Lj}, \tag{29.289}$$

and where (as usual)

$$a^{r\dagger}_{Rj} = a^r_{R(-j)}, \quad a^{r\dagger}_{Lj} = a^r_{L(-j)}$$

and

$$s_{\mathrm{R}j}^{r\dagger} = s_{\mathrm{R}(-j)}^{r}, \quad s_{\mathrm{L}j}^{r\dagger} = s_{\mathrm{L}(-j)}^{r}.$$

It should be noted that (29.286) and (29.287) imply that

$$N_{\mathrm{R}}^{\mathrm{tr}}\Psi = N_{\mathrm{L}}^{\mathrm{tr}}\Psi \tag{29.290}$$

for every physical state Ψ (which has the same form as (29.139)), so (29.286) can be rewritten as

$$(Mc^2)^2 = \frac{2}{\alpha'}(N_{\mathrm{L}}^{\mathrm{tr}} + N_{\mathrm{R}}^{\mathrm{tr}}). \tag{29.291}$$

Again these number operators have no negative eigenvalues, so the closed superstring has no tachyonic modes. Moreover, the vacuum states have zero eigenvalue with all of these operators, so again they must be massless.

These vacuum states may be analysed along the lines described in the previous subsection. In the closed string case the analogue of (29.261) is

$$[s_{\mathrm{RO}}^{r}, s_{\mathrm{RO}}^{s}]_{+} = \delta^{rs}1, \tag{29.292}$$

$$[s_{\mathrm{LO}}^{r}, s_{\mathrm{LO}}^{s}]_{+} = \delta^{rs}1, \tag{29.293}$$

$$[s_{\mathrm{RO}}^{r}, s_{\mathrm{LO}}^{s}]_{+} = 0. \tag{29.294}$$

As these anticommutation relations involve two sets of operators s_{RO}^{r} and s_{LO}^{r} that each have the properties possessed previously by the s_0^r, for the case in which $\eta^1 = \eta^2 = 1$ the ground state vectors have the form of direct products of either $|0, \mathrm{v}, j\rangle_{\mathrm{R}}$ or $|0, \mathrm{s}', j\rangle_{\mathrm{R}}$ (for $j = 1, 2, \ldots, 8$) with either $|0, \mathrm{v}, k\rangle_{\mathrm{L}}$ or $|0, \mathrm{s}', k\rangle_{\mathrm{L}}$ (for $k = 1, 2, \ldots, 8$), where, for example, the "right-moving" analogue of (29.274) is

$$|0, \mathrm{s}', j\rangle_{\mathrm{R}} = \kappa \sum_{a,k=1}^{8} c_{aj}^{k} s_{\mathrm{RO}}^{a} |0, \mathrm{v}, k\rangle_{\mathrm{R}}. \tag{29.295}$$

In the case $\eta^1 = -\eta^2 = 1$ the analysis is similar, but $|0, \mathrm{s}', k\rangle_{\mathrm{L}}$ is replaced by $|0, \mathrm{s}, k\rangle_{\mathrm{L}}$. In both cases the ground state has degeneracy 256 ($= 16 \times 16$).

As noted by Green and Schwarz (1982a), the spin content of this ground state multiplet is exactly the same as that of $N = 2$, $D = 10$ supergravity. If $\eta^1 = -\eta^2 = 1$ then the two gravitinos contained in $|0, \mathrm{v}, j\rangle_{\mathrm{R}} \otimes |0, \mathrm{s}', k\rangle_{\mathrm{L}}$ and $|0, \mathrm{s}', j\rangle_{\mathrm{R}} \otimes |0, \mathrm{v}, k\rangle_{\mathrm{L}}$ have opposite handedness, and, since $\Gamma(\{0,0,0,1\}) \otimes \Gamma(\{0,0,1,0\}) \simeq \Gamma(\{0,0,1,1\}) \oplus \Gamma(\{1,0,0,0\})$, in which the constituents of the direct sum are a 56-dimensional irreducible representation and the 8-dimensional vector representation, the states $|0, \mathrm{s}', j\rangle_{\mathrm{R}} \otimes |0, \mathrm{s}, k\rangle_{\mathrm{L}}$ are associated with a vector and a third-rank antisymmetric tensor field. This corresponds of the reduction of $D = 11$ supergravity to 10 dimensions. On the other hand, if $\eta^1 = \eta^2 = 1$, the gravitinos have the same handedness and, since $\Gamma(\{0,0,0,1\}) \otimes \Gamma(\{0,0,0,1\})$ is the direct sum of a 35-, a 28- and a 1-

dimensional irreducible representation, the states $|0, s', j\rangle_R \otimes |0, s', k\rangle_L$ correspond to a scalar, a second-rank antisymmetric tensor field and a fourth-rank antisymmetric self-dual tensor field. The cases where $\eta^1 = -\eta^2$ and $\eta^1 = \eta^2$ are known as the type IIA and type IIB closed string theories respectively.

For the closed string the Lagrangian density (29.232) is again invariant under the supersymmetry transformations (29.277) and (29.278), but now the quantities $\boldsymbol{\varepsilon}^1$ and $\boldsymbol{\varepsilon}^2$ are independent. This implies (cf. Green and Schwarz 1981, 1982a) that there exists a set of operators Q_{Aa} (for $A = 1, 2$ and $a = 1, 2, \ldots, 32$) that form the odd part of the $N = 2$ 10-dimensional Poincaré superalgebra.

In the case in which $\eta^1 = \eta^2$ the set of states that are obtained from the vacuum states by acting with the creation operators $a_{Rj}^{r\dagger}$, $a_{Lj}^{r\dagger}$, $s_{Rj}^{r\dagger}$ and $s_{Lj}^{r\dagger}$ and that are *symmetrical* with respect to the interchange of $a_{Rj}^{r\dagger}$ with $a_{Lj}^{r\dagger}$ and of $s_{Rj}^{r\dagger}$ with $s_{Lj}^{r\dagger}$ form a subspace of the set of physical states. (Clearly they satisfy (29.292).) This truncated theory has only $N = 1$ 10-dimensional Poincaré supersymmetry, and can describe the interaction of closed strings with open strings. It is known as the type I closed string theory. The zero-mass states of the type I open and closed superstrings together form a $N = 1$ supersymmetric Yang–Mills multiplet coupled to a $N = 1$ supergravity multiplet.

(*d*) *Torus compactification, the field theory limit and interactions*

Hitherto it has been assumed that superstrings move in a 10-dimensional Minkowski space–time. However, the real world is 4-dimensional, and the question naturally arises as to how these two spaces can be related. The most promising approach is to adopt the Kaluza–Klein philosophy, and to assume that six of these dimensions, all spatial, are compactified so that they end up so small as to be undetectable. One way of doing this is to replace the Minkowski metric tensor $\eta_{\mu\lambda}$ that appears in the Lagrangian density for the bosonic coordinates (29.6) by an appropriate metric tensor $G_{\mu\lambda}$ that is a function of the bosonic coordinates, but this has the disadvantage that the Euler–Lagrange equation that would replace (29.21) (and (29.241)) would necessarily be more complicated. A simpler alternative is to assume that six of the spatial coordinates lie on a 6-dimensional torus.

The original formulation of toroidal compactification for bosonic strings was given by Cremmer and Scherk (1976), its generalization for the superstring being developed by Green *et al.* (1982). As the idea also plays a crucial role in the theory of the heterotic string, but with a different number of coordinates being compactified, as well as in other contexts, it is worth setting up the theory in a fairly general way. Consider therefore D coordinates x^μ ($\mu = 1, 2, \ldots, D$), with x^D being time-like and all the rest space-like, and suppose that D' of these lie on a D'-dimensional torus $T^{D'}$. For definiteness these will be labelled $x^1, x^2, \ldots, x^{D'}$. The torus $T^{D'}$ is constructed by taking D' linearly independent vectors $\mathbf{f}_1, \mathbf{f}_2, \ldots, \mathbf{f}_{D'}$ in $\mathbb{R}^{D'}$. Then, with $\mathbf{r} = (x^1, x^2, \ldots, x^{D'})$, the points $\mathbf{r}$ and $\mathbf{r}'$ are deemed to be equivalent if

$$\mathbf{r}' = \mathbf{r} + \sum_{s=1}^{D'} n^s \mathbf{f}_s, \tag{29.296}$$

for *any* set of D' integers $\{n^1, n^2, \ldots, n^{D'}\}$, this equivalence being indicated by writing $\mathbf{r}' \equiv \mathbf{r}$. Let $\mathbf{f}_s = (f_s^1, f_s^2, \ldots, f_s^{D'})$ (for $s = 1, 2, \ldots, D'$) be the expressions for these vectors in term of their components in $\mathbb{R}^{D'}$.

The set of vectors $\mathbf{f}_1, \mathbf{f}_2, \ldots, \mathbf{f}_{D'}$ form the basic lattice vectors of a lattice in $\mathbb{R}^{D'}$, so the condition (29.296) can be expressed as the statement that two points are equivalent if and only if they differ by a lattice vector. The D'-dimensional torus is thus the D'-dimensional generalization of the 3-dimensional configuration considered in Chapter 8 for a crystalline solid with Born cyclic boundary conditions, and, as noted there, the momentum eigenvalues are determined by the reciprocal lattice vectors. The basic reciprocal lattice vectors $\mathbf{f}_1^*, \mathbf{f}_2^*, \ldots, \mathbf{f}_{D'}^*$ corresponding to $\mathbf{f}_1, \mathbf{f}_2, \ldots, \mathbf{f}_{D'}$ may be defined by

$$\mathbf{f}_r \cdot \mathbf{f}_s^* = \delta_{rs} \tag{29.297}$$

for $r, s = 1, 2, \ldots, D'$. These vectors $\mathbf{f}_1^*, \mathbf{f}_2^*, \ldots, \mathbf{f}_{D'}^*$ are often called in this field the "basic dual lattice vectors", and the reciprocal lattice based upon them is known as the "dual lattice". The allowed values of the compactified momentum components p^s (for $s = 1, 2, \ldots, D'$) are then given by

$$p^s = 2\pi\hbar \sum_{r=1}^{D'} m^r f_r^{s*}, \tag{29.298}$$

for every set of D' integers $\{m^1, m^2, \ldots, m^{D'}\}$.

For *open* strings it is only through this effect that toroidal compactification modifies the mass spectrum. However, for *closed* strings it is natural to replace the boundary condition (29.1) by the weaker condition that

$$x^s(0, \tau) \equiv x^s(\pi, \tau) \tag{29.299}$$

for $s = 1, 2, \ldots, D'$. By (29.296), this implies that

$$x^s(\pi, \tau) = x^s(0, \tau) + \sum_{s=1}^{D'} n^r f_r^s \tag{29.300}$$

for $s = 1, 2, \ldots, D'$ and any set of D' integers $\{n^1, n^2, \ldots, n^{D'}\}$. The most general solution (29.115) for $\mu = s = 1, 2, \ldots, D'$ is therefore slightly modified to become

$$x^s(\sigma,\tau) = q^s + 2\alpha'\hbar c^2 p^s\tau + 2L^s\sigma + (2\alpha'\hbar^2c^2)^{1/2}\tfrac{1}{2}i \sum_{j=-\infty,j\neq 0}^{\infty}\left(\frac{a^s_{\mathrm{R}j}}{j}\mathrm{e}^{-2ij(\tau-\sigma)} + \frac{a^s_{\mathrm{L}j}}{j}\mathrm{e}^{-2ij(\tau+\sigma)}\right), \quad (29.301)$$

where (29.300) requires that

$$L^s = (2\pi)^{-1}\sum_{r=1}^{D'} n^r f^s_r. \quad (29.302)$$

Equation (29.301) can then be recast in the form of (29.117) provided that (29.116) is now replaced by

$$a^s_{\mathrm{R}0} = (2\alpha'\hbar^2c^2)^{-1/2}(\alpha'\hbar c^2 p^s - L^s) \quad (29.303)$$

and

$$a^s_{\mathrm{L}0} = (2\alpha'\hbar^2c^2)^{-1/2}(\alpha'\hbar c^2 p^s + L^s). \quad (29.304)$$

As it is clear from (29.296) that all the components f^s_r have the dimensions of length, it is convenient to introduce for each vector $\mathbf{f}_r$ a parameter R_r with the dimensions of length, and to define the dimensionless basic lattice vectors $\mathbf{e}_1, \mathbf{e}_2, \ldots, \mathbf{e}_{D'}$ by

$$\mathbf{e}_r = R_r^{-1}\mathbf{f}_r \quad (29.305)$$

for $r = 1, 2, \ldots, D'$. The basic reciprocal lattice vectors $\mathbf{e}^*_1, \mathbf{e}^*_2, \ldots, \mathbf{e}^*_{D'}$, defined in accordance with (29.297) by

$$\mathbf{e}_r \cdot \mathbf{e}^*_s = \delta_{rs}, \quad (29.306)$$

are also dimensionless and are such that

$$\mathbf{e}^*_r = R_r\mathbf{f}^*_r. \quad (29.307)$$

In terms of these vectors, (29.298) can be rewritten as

$$p^s = 2\pi\hbar\sum_{r=1}^{D'}\frac{m^r e^{s*}_r}{R_r}, \quad (29.308)$$

and (29.302) becomes

$$L^s = (2\pi)^{-1}\sum_{r=1}^{D'} n^r e^s_r R_r, \quad (29.309)$$

where $\mathbf{e}_s = (e^1_s, e^2_s, \ldots, e^{D'}_s)$ and $\mathbf{e}^*_s = (e^{*1}_s, e^{*2}_s, \ldots, e^{*D'}_s)$.

The simplest choice of basic lattice vectors, which is the one adopted in the compactification of the superstring, is to take

$$R_r = 2\pi R \quad (29.310)$$

for $r = 1, 2, \ldots, D'$ $(=6)$, and to assume that

$$e^s_r = \delta^s_r \quad (29.311)$$

for $r, s = 1, 2, \ldots, D'$ $(= 6)$. That is, $\mathbf{f}_1 = (2\pi R, 0, 0, \ldots, 0)$, $\mathbf{f}_2 = (0, 2\pi R, 0, \ldots, 0)$, and so on. Then

$$e_r^{*s} = \delta_r^s, \tag{29.312}$$

and so $\mathbf{f}_1^* = ((2\pi R)^{-1}, 0, 0, \ldots, 0)$, $\mathbf{f}_2^* = (0, (2\pi R)^{-1}, 0, \ldots, 0)$, and so on. With this choice, (29.308) simplifies to give

$$p^s = \hbar m^s / R, \tag{29.313}$$

and (29.309) becomes

$$L^s = R n^s. \tag{29.314}$$

Thus, from (29.303) and (29.304),

$$a_{\mathrm{R}0}^s = (2\alpha' \hbar^2 c^2)^{-1/2} \left(\frac{\alpha' \hbar^2 c^2}{R} m^s - R n^s \right) \tag{29.315}$$

and

$$a_{\mathrm{L}0}^s = (2\alpha' \hbar^2 c^2)^{-1/2} \left(\frac{\alpha' \hbar^2 c^2}{R} m^s + R n^s \right). \tag{29.316}$$

For the *open* superstring Green *et al.* (1982) showed that the restriction (29.298) on the allowed values of the compactified momentum components implies that the mass formula (29.258) is modified to give

$$(Mc^2)^2 = \frac{c^2 \hbar^2}{R^2} \sum_{r=1}^{6} (m^r)^2 + \frac{1}{\alpha'} N^{\mathrm{tr}}. \tag{29.317}$$

Similarly, for the *closed* superstrings the modification of (29.286) implied by the replacement of (29.116) by (29.315) and (29.316) is

$$(Mc^2)^2 = \frac{c^2 \hbar^2}{R^2} \sum_{r=1}^{6} (m^r)^2 + \frac{R^2}{\alpha'^2 c^2 \hbar^2} \sum_{r=1}^{6} (n^r)^2 + \frac{2}{\alpha'} (N_{\mathrm{R}}^{\mathrm{tr}} + N_{\mathrm{L}}^{\mathrm{tr}}), \tag{29.318}$$

where the physical states Ψ must still satisfy the condition (29.292).

In the limit of the open and closed type I superstring theory in which $\alpha' \to 0$ in such a way that $\alpha'/R \to 0$ and $R \to 0$ Green *et al.* (1982) showed that the resulting massless states have both the particle content and the interactions of $N = 4$, $D = 4$ supersymmetric Yang–Mills theory. They also showed that the corresponding limit for the type II closed superstring theories is that of $N = 8$, $D = 4$ supergravity.

Apart from some passing comments, nearly all of the previous discussion has been concerned with *free* strings and superstrings. However, the study of their *interactions* is clearly of the greatest importance. Although an immense amount of effort has been applied to examining the interactions of the bosonic and spinning strings, comprehensive reviews of which have been given by Schwarz (1973), Frampton (1974, 1986), Jacob (1974), Mandelstam (1974),

Rebbi (1974) and Scherk (1975), the superstring theory of Green and Schwarz was the first to be free of the complications of unphysical tachyonic modes. As this naturally facilitates the treatment of interactions, their discussion has been deferred until this point, and even now will be confined to a brief introductory sketch of some of the main points.

The investigation of superstring interactions by Green and Schwarz (1982b, c, 1983), and its generalization to the heterotic string by Gross *et al.* (1986), has been within the light-cone formulation of vertex operators, particular use being made of the vertex operators for the emission of massless particles. In this formulation the amplitude $A(1, 2, \ldots, N)$ corresponding to the N-particle tree diagram of Figure 29.2 is given by

$$A(1,2,\ldots,N) = \langle 1 | V(2)\, \Delta\, V(3)\, \Delta \cdots \Delta\, V(N-1) | N \rangle,$$

where V indicates a vertex operator and Δ a propagator. For example, the vertex operator for the emission of a massless vector particle with momentum k^{μ} and polarization vector $\zeta^{\mu}(k)$ in the simplest case in which $\zeta^{-}(k) = 0$ and the only non-vanishing component of momentum is k^{-} is given for the open string by

$$V_{\mathrm{B}}(\zeta, k^{-}, \tau) = g \sum_{r=1}^{D-2} \zeta^{r} p^{r}(\tau) \exp\left\{ -\frac{ik^{-}x^{+}(0,\tau)}{\hbar} \right\},$$

where

$$p^{r}(\tau) = (2\alpha' c^{2})^{-1/2} \sum_{j=-\infty}^{\infty} a_{j}^{r} \mathrm{e}^{-ij\tau}$$

and where g is another coupling constant (which is independent of α'). For the open superstring the propagator is given by

$$\Delta = \alpha' / L_{0}^{\mathrm{tr}},$$

where L_{0}^{tr} is as defined in (29.257). The important point to observe is that the vertex operator and the propagator both involve the annihilation and creation operators of the free theory. This expression for the vertex operator is valid

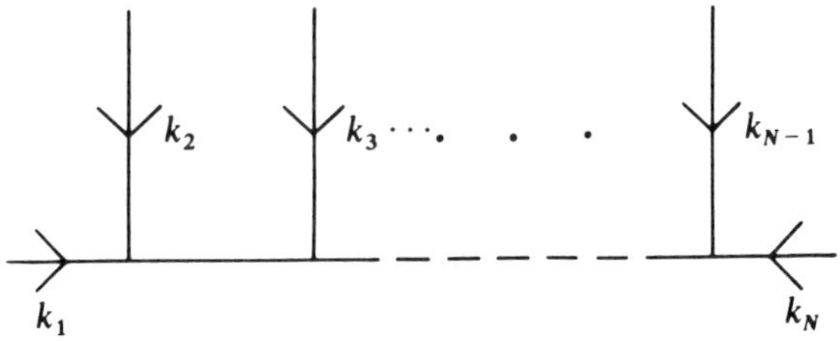

Figure 29.2 The N-particle tree diagram for superstring interactions.

both for the bosonic string and the superstring, but to get the corresponding expressions for other momenta it is necessary to apply the appropriate Lorentz transformation operators, which are different in the two cases. For details of the resulting expressions for both the open and the closed superstrings, and their generalizations to the emission of other types of massless particle, see Green and Schwarz (1982b, c, 1983) and Schwarz (1982b).

In the case of the open superstring the coupling constant g is known as the Yang–Mills coupling constant because it is involved in the interactions of massless vector bosons and reduces to the coupling constant of the Yang–Mills theory in the field theory limit in which $\alpha' \to 0$. For the closed superstring the corresponding coupling constant is usually denoted by κ, and, since it is involved in the three-graviton interaction, it is referred to as a gravitational coupling constant. In a type I theory incorporating both open and closed strings, since the open strings can join at their ends to form closed strings, it follows that κ is proportional to g^2.

The prescription developed earlier by Paton and Chan (1969) for inserting internal symmetries into the bosonic string can be applied to the type I superstring theories. In this procedure the complete N-particle tree amplitude is given by

$$T = \sum \operatorname{tr}\{\boldsymbol{\lambda}_{r_1}\boldsymbol{\lambda}_{r_2}\cdots\boldsymbol{\lambda}_{r_N}\}\, A(r_1, r_2, \ldots, r_N),$$

where the sum is over all permutations $\{r_1, r_2, \ldots, r_N\}$ of $\{1, 2, \ldots, N\}$, two permutations that are cyclic (such as $\{1, 2, 3\}$ and $\{2, 3, 1\}$ for $N = 3$) being counted only once. In this expression $\boldsymbol{\lambda}_r$ are matrices representing the basis elements of some Lie algebra. It was conjectured by Schwarz (1982a) and proved by Marcus and Sagnotti (1982) that the only possible candidates are the *classical* compact Lie algebras so(n), sp($\frac{1}{2}n$) and u(n), and that the $\boldsymbol{\lambda}_r$ must be in the defining representation of dimension n in each case.

Alvarez-Gaumé and Witten (1983) showed that type IIB supergravity in 10 dimensions is anomaly-free, and since this theory appears as the field theory limit of the type IIB superstring, it follows that the same result must be true for the type IIB superstring. Green and Schwarz (1984d, 1985a, b) then showed that a type I superstring with so(32) symmetry introduced by the Chan–Paton method is both anomaly- and infinity-free. It was also demonstrated by Green and Schwarz (1984d) that with supersymmetric $D = 10$ Yang–Mills fields coupled to those of $N = 1, D = 10$ supergravity the anomalies cancel if the gauge algebra is either so(32) or the compact real form of $E_8 \oplus E_8$. As far as the so(32) is concerned, these two statements are consistent, since this latter field theory multiplet appears as the $\alpha' \to 0$ limit of the type I superstring.

The question that now arose was how the $E_8 \oplus E_8$ algebra could be incorporated into a string theory, since, as it is not a classical Lie algebra, it could not be introduced by the Chan–Paton method. The heterotic string theory of Gross *et al.* (1985a, b, 1986) that will be described in the next section provided an answer.

5 The heterotic string

(a) *The right-moving and left-moving modes*

The heterotic string as proposed by Gross *et al.* (1985a, b, 1986) involves a combination of a bosonic string in 26 dimensions and a superstring based on 10-dimensional space–time that is nevertheless consistent and that continues to be supersymmetric.

The *bosonic* part of this heterotic string is assumed to consist of 26 bosonic coordinates $x^{\mu}(\sigma, \tau)$ for $\mu = 1, 2, \ldots, 26$, with $x^{26}(\sigma, \tau)$ being time-like and all the others being space-like. Of these, $x^{17}(\sigma, \tau), \ldots, x^{26}(\sigma, \tau)$ are assumed to describe a 10-dimensional Minkowski space–time, so that they satisfy the closed string boundary conditions (29.1). However, $x^{1}(\sigma, \tau), \ldots, x^{16}(\sigma, \tau)$ are assumed to lie on a 16-dimensional torus T^{16}, and so are required to satisfy the toroidal boundary conditions (20.299) (or equivalently (29.300)). Moreover, it is also assumed that the compactified coordinates $x^{s}(\sigma, \tau)$ (for $s = 1, 2, \ldots, 16$) are *entirely left-moving*. Gross *et al.* (1985a, b) have argued that this latter requirement results in a set of commutation relations

$$[q^{r}, p^{s}] = \tfrac{1}{2} i \delta^{rs} \tag{29.319}$$

(for $r, s = 1, 2, \ldots, 16$) in which an extra factor of $\frac{1}{2}$ appears relative to the usual Heisenberg commutation relations. This has the consequence that a factor of $\frac{1}{2}$ has to be inserted in the expression for the compactified momentum components p^{s} of (29.308), which now becomes

$$p^{s} = \pi \hbar \sum_{r=1}^{16} \frac{m^{r} e_{r}^{s*}}{R_{r}} \tag{29.320}$$

(for $s = 1, 2, \ldots, 16$). Applying further the toroidal compactification theory developed in the previous subsection with $D' = 16$, the requirement that the compactified coordinates be entirely left-moving implies that $a_{\mathrm{R}0}^{s} = 0$, and hence, by (29.303), that

$$\alpha' \hbar c^{2} p^{s} = L^{s} \tag{29.321}$$

for $s = 1, 2, \ldots, 16$, and consequently, by (29.305), that

$$a_{\mathrm{L}0}^{s} = (2\alpha' c^{2})^{1/2} p^{s} \tag{29.322}$$

for $s = 1, 2, \ldots, 16$. Substituting (29.309) and (29.320) into (29.321) leads to the requirement that for every set of integers $\{m_{1}, m_{2}, \ldots, m_{D'}\}$ there exists a set of integers $\{n_1, n_2, \ldots, n_{D'}\}$ that satisfy

$$\sum_{r=1}^{16} n^r e_r^s R_r = 2\pi^2 \alpha' \hbar^2 c^2 \sum_{r=1}^{16} \frac{m^r e_r^{*s}}{R_r} \tag{29.323}$$

and vice versa. Clearly this is possible only if

$$(R_r)^2 = 2\pi^2 \alpha' \hbar^2 c^2 \tag{29.324}$$

for *all* $r = 1, 2, \ldots, 16$. A preliminary estimate of the value of R_r obtained by using the expression (29.140) for α' gives

$$R_r = (2\pi^3/\alpha^2)^{1/2} L_{\mathrm{P}}, \tag{29.325}$$

where

$$L_{\mathrm{P}} = (\hbar G/c^3)^{1/2} \approx 1.6 \times 10^{-33}\,\mathrm{cm} \tag{29.326}$$

is the Planck length and α is the fine structure constant, and so, as is required by the Kaluza–Klein philosophy, the compactified torus is certainly very small! (Gross *et al.* (1986) have shown by consideration of the low energy limit of the interacting heterotic string that within this theory (29.140) is slightly modified to become

$$\alpha' = 16\pi G/g^2 \hbar c^5,$$

where g is a Yang–Mills coupling constant. This gives

$$R_r = (32\pi^3/g^2)^{1/2} L_{\mathrm{P}},$$

which is still of the order of the Planck length.) With the choice (29.324), the condition (29.323) reduces to

$$\sum_{r=1}^{16} n^r e_r^s = \sum_{r=1}^{16} m^r e_r^{*s}, \tag{29.327}$$

which will be satisfied if every basic dual lattice vector $\mathbf{e}_r^*$ is a linear combination of the basic lattice vectors $\mathbf{e}_1, \mathbf{e}_2, \ldots, \mathbf{e}_{16}$ with integer coefficients, and vice versa, and indeed the study of interactions (Gross *et al.* (1985b, 1986)) shows that these conditions are absolutely necessary. Such a lattice is said to be *self-dual.* On quantization, the quantities $a_{\mathrm{R}0}^s$ and p^s (for $s = 1, 2, \ldots, 16$) become operators, which are related to each other by (29.322) and which commute with all the other bosonic and fermionic operators. No confusion will arise if the operators p^s (for $s = 1, 2, \ldots, 16$) and their eigenvalues are denoted by the same symbols, since it will always be clear from the context which is intended. With each set of these eigenvalues being gathered into a 16-component vector $\mathbf{p} = (p^1, p^2, \ldots, p^{16})$, (29.321), (29.309) and (29.324) show that there is a one-to-one correspondence between the vectors and the lattice torus vectors of the form

$$(2\alpha' c^2)^{1/2} \mathbf{p} = \sum_{r=1}^{16} n^r \mathbf{e}_r. \tag{29.328}$$

The *fermionic* part of the heterotic string is taken to be as for the closed superstring, but is assumed to entirely to be *entirely right-moving*. Thus the components $s^{1r}(\sigma, \tau)$ continue to be given by (29.281), but the components $s^{2r}(\sigma, \tau)$ of (29.282) are assumed to be such that

$$s^{2r}(\sigma, \tau) = 0 \tag{29.329}$$

for $r = 1, 2, \ldots, 8$. Consequently, in the light-cone quantization, only the right-moving operators $s^r_{\mathbf{R}j}$ appear, and they are assumed to continue to satisfy the anticommutation relations (29.283). For convenience it will also be assumed that $\eta^1 = 1$.

Gross *et al.* (1985a, b) have shown that the analogue of (29.290) for the heterotic string is

$$N^{\mathrm{tr}}_{\mathbf{R}}\Psi = \left\{N^{\mathrm{tr}}_{\mathrm{L}} - 1 + \alpha' c^2 \sum_{r=1}^{16} (p^r)^2\right\}\Psi \tag{29.330}$$

for every physical state Ψ, where

$$N^{\mathrm{tr}}_{\mathbf{R}} = \sum_{j=1}^{\infty} \sum_{r=17}^{24} a^{r\dagger}_{\mathbf{R}j} a^r_{\mathbf{R}j} + \sum_{j=1}^{\infty} \sum_{r=1}^{8} j s^{r\dagger}_{\mathbf{R}j} s^r_{\mathbf{R}j} \tag{29.331}$$

and

$$N^{\mathrm{tr}}_{\mathrm{L}} = \sum_{j=1}^{\infty} \sum_{r=1}^{24} a^{r\dagger}_{\mathrm{L}j} a^r_{\mathrm{L}j}. \tag{29.332}$$

However, by (29.328), the eigenvalues of the third term on the right-hand side of (29.330) have the form

$$\frac{1}{2} \sum_{r,s=1}^{16} n^r n^s \mathbf{e}_r \cdot \mathbf{e}_s \quad \left(= \frac{1}{2}\left|\sum_{r=1}^{16} n^r \mathbf{e}_r\right|^2\right). \tag{29.333}$$

As all the other quantities on both sides of (29.330) have integer eigenvalues, the quantity (29.333) must be an integer (for all sets of integers $\{n_1, n_2, \ldots, n_{D'}\}$), which implies that for each pair $r, s = 1, 2, \ldots, 16$ the scalar product $\mathbf{e}_r \cdot \mathbf{e}_s$ must be an *integer*, and, moreover, for each $r = 1, 2, \ldots, 16$ the product $\mathbf{e}_r \cdot \mathbf{e}_r$ must be an *even integer*. Such a lattice is called an *even integral lattice*. Gross *et al.* (1985a, b) have also shown that the mass-shell condition that generalizes (29.291) is

$$(Mc^2)^2 = \frac{2}{\alpha'}\left\{\alpha' c^2 \sum_{r=1}^{16} (p^r)^2 + N^{\mathrm{tr}}_{\mathbf{R}} + N^{\mathrm{tr}}_{\mathrm{L}} - 1\right\}, \tag{29.334}$$

so that, on any physical state Ψ, by (29.330),

$$(Mc^2)^2\Psi = \frac{4}{\alpha'} N^{\mathrm{tr}}_{\mathbf{R}}\Psi. \tag{29.335}$$

As the operator $N^{\mathrm{tr}}_{\mathbf{R}}$ has only non-negative eigenvalues, (29.335) shows that there are *no* tachyonic modes for the heterotic string.

The state vectors for this system are obtained by acting on the vacuum state vectors with the creation operators $a_{\mathrm{R}j}^{r\dagger}$ (for $r = 1, 2, \ldots, 24$), $a_{\mathrm{L}j}^{r\dagger}$ (for $r = 17, 18, \ldots, 24$) and $s_{\mathrm{R}j}^{r\dagger}$ (for $r = 1, 2, \ldots, 8$), j being allowed to take any positive integer value in each case. These vacuum state vectors can be taken to be direct products of "right-handed" and "left-handed" vacuum state vectors .

The "right-handed" vacuum state vectors have the properties described in Section 4(*b*); that is, there are eight vectors corresponding to the vector representation $\boldsymbol{\Gamma}^{\mathrm{v}}$ of so(8), which may be denoted by $|0, \mathrm{v}, k\rangle_{\mathrm{R}}$ for $k = 1, 2, \ldots, 8$, and there are eight vectors corresponding to the spinor representation $\boldsymbol{\Gamma}^{\mathrm{s}'}$, which may be denoted by $|0, \mathrm{s}', k\rangle_{\mathrm{R}}$ for $k = 1, 2, \ldots, 8$. (To apply the theory of Section 4(*b*), all that has to be done is to append a subscript "R" in all the obvious places, as was done, for example, in (29.295).)

As the fermionic part of the heterotic string has been assumed to be purely left-handed, there are no operators $s_{\mathrm{L}0}^{r}$ in this theory. Consequently there is no "spin" degeneracy for the "left-handed" vacuum states. As always, the vacuum state vectors are eigenvectors of the momentum operators, although hitherto it has normally been convenient to not display the momentum dependence. However, it is now important to exhibit the dependence on the compactified momentum components explicitly, and as these involve only left-moving components, these may be incorporated as part of the "left-moving" vacuum state vectors. Consequently such a state vector may be denoted by $|0, \Sigma_r n^r \mathbf{e}_r\rangle_{\mathrm{L}}$, where, by (29.328),

$$p^s|0, \Sigma_r n^r \mathbf{e}_r\rangle_{\mathrm{L}} = (2\alpha' c^2)^{-1/2}\left(\sum_{r=1}^{16} n^p e_p^s\right)|0, \Sigma_r n^r \mathbf{e}_r\rangle_{\mathrm{L}} \tag{29.336}$$

for $s = 1, 2, \ldots, 16$, and so (cf. (29.333))

$$\alpha' c^2 \sum_{s=1}^{16} (p^s)^2 |0, \Sigma_r n^r \mathbf{e}_r\rangle_{\mathrm{L}} = \frac{1}{2}\left|\sum_{p=1}^{16} n^p \mathbf{e}_p\right|^2 |0, \Sigma_r n^r \mathbf{e}_r\rangle_{\mathrm{L}}. \tag{29.337}$$

With this notation, the vacuum state vectors of the system have the form

$$|0, \mathrm{v}, k\rangle_{\mathrm{R}} \otimes |0, \Sigma_r n^r \mathbf{e}_r\rangle_{\mathrm{L}} \quad \text{and} \quad |0, \mathrm{s}', k\rangle_{\mathrm{R}} \otimes |0, \Sigma_r n^r \mathbf{e}_r\rangle_{\mathrm{L}}$$

for $k = 1, 2, \ldots, 8$.

The subset of these with $n^1 = n^2 = \cdots = n^{16} = 0$ do *not* satisfy the condition (29.330), since they have zero eigenvalues with $N_{\mathrm{R}}^{\mathrm{tr}}$, $N_{\mathrm{L}}^{\mathrm{tr}}$ and the third term on the right-hand side of (29.330) (by (29.337)). Consequently they do *not* correspond to *physical* states.

Equation (29.334) shows that there are *three* types of state vectors corresponding to *zero-mass* states:

(i) vectors of the form

$$|0, \mathrm{v}, k\rangle_{\mathrm{R}} \otimes a^{r\dagger}_{\mathrm{L}1} |0, \mathbf{0}\rangle_{\mathrm{L}} \quad \text{and} \quad |0, \mathrm{s}', k\rangle_{\mathrm{R}} \otimes a^{r\dagger}_{\mathrm{L}1} |0, \mathbf{0}\rangle_{\mathrm{L}}$$

for $k = 1, 2, \ldots, 8$ and $r = 17, 18, \ldots, 24$;

(ii) vectors of the form

$$|0, \mathrm{v}, k\rangle_{\mathrm{R}} \otimes a^{r\dagger}_{\mathrm{L}1} |0, \mathbf{0}\rangle_{\mathrm{L}} \quad \text{and} \quad |0, \mathrm{s}', k\rangle_{\mathrm{R}} \otimes a^{r\dagger}_{\mathrm{L}1} |0, \mathbf{0}\rangle_{\mathrm{L}}$$

for $k = 1, 2, \ldots, 8$ and $r = 1, 2, \ldots, 16$;

(iii) vectors of the form

$$|0, \mathrm{v}, k\rangle_{\mathrm{R}} \otimes |0, \Sigma_r n^r \mathbf{e}_r\rangle_{\mathrm{L}} \quad \text{and} \quad |0, \mathrm{s}', k\rangle_{\mathrm{R}} \otimes |0, \Sigma_r n^r \mathbf{e}_r\rangle_{\mathrm{L}}$$

for $k = 1, 2, \ldots, 8$ and lattice vectors such that

$$\left| \sum_{r=1}^{16} n^r \mathbf{e}_r \right|^2 = 2. \tag{29.338}$$

It can be shown that the 128 states of type (i) form an irreducible $N = 1$, $D = 10$ supergravity multiplet.

As will be discussed in detail in the next subsection, there are only *two* 16-dimensional self-dual even integral lattices, and for both of these the number of vectors satisfying the condition (29.338) is 480. Consequently the total number of states of types (ii) and (iii) together is $16 \times (480 + 16) = 16 \times 496$. As will become apparent, the two self-dual even integral lattices in the next subsection are associated with semi-simple Lie algebras whose dimension is 496 in both cases, so that in both cases 496 is the dimension of the adjoint representation. This allows the states (ii) and (iii) together to form bases for $N = 1$, $D = 10$ Yang–Mills supermultiplets.

(*b*) *The appearance of the* $E_8 \oplus E_8$ *algebra and the* Spin (32)/Z_2 *group*

The general theory of lattices (cf. Serre 1973, Goddard and Olive 1984) shows that self-dual even integral lattices occur only in spaces whose dimensions are integral multiples of 8, that in 8 dimensions there is only one such lattice, that for 16 dimensions (the situation that is of interest in connection with heterotic strings) there are two distinct lattices with these properties, and that for all three of these lattices

$$\mathbf{e}_r \cdot \mathbf{e}_r = 2 \quad \text{for } r = 1, 2, \ldots \quad . \tag{29.339}$$

The question of whether any of these is a root lattice of a complex semi-simple Lie algebra $\tilde{\mathscr{L}}^0$ of rank l $(= 8$ or $16)$ will now be examined.

Clearly the condition (29.339) implies that such an algebra would have to be simply laced; that is, $\langle \alpha_j, \alpha_j \rangle$ must have the same value for all $j = 1, 2, \ldots, l$ (cf. Chapter 26, Section 4). It is therefore convenient to use the scaling introduced in (27.48), whereby every linear functional α defined on the Cartan subalgebra of a complex semi-simple Lie algebra $\tilde{\mathscr{L}}^0$ has associated with it an l-component vector $\hat{\boldsymbol{\alpha}}$, and if α is a root then (cf. (27.50))

$$\hat{\boldsymbol{\alpha}} \cdot \hat{\boldsymbol{\alpha}} = 2.$$

Thus, in particular, for the vectors $\hat{\boldsymbol{\alpha}}_1, \hat{\boldsymbol{\alpha}}_2, \ldots, \hat{\boldsymbol{\alpha}}_l$ corresponding to the simple roots $\alpha_1, \alpha_2, \ldots, \alpha_l$

$$\hat{\boldsymbol{\alpha}}_j \cdot \hat{\boldsymbol{\alpha}}_j = 2 \tag{29.340}$$

for $j = 1, 2, \ldots, l$. Also, (27.39), (27.46), (27.48) and (13.61) imply that

$$\hat{\boldsymbol{\alpha}}_j \cdot \hat{\boldsymbol{\alpha}}_k = A_{jk} \tag{29.341}$$

for $j, k = 1, 2, \ldots, l$, where A_{jk} are the elements of the Cartan matrix of $\tilde{\mathscr{L}}^0$. As these elements are integers, the lattice with basic lattice vectors defined by

$$\mathbf{e}_j = \hat{\boldsymbol{\alpha}}_j \tag{29.342}$$

(for $j = 1, 2, \ldots, l$) is certainly an even integral lattice. Moreover, if $\hat{\boldsymbol{\Lambda}}_1, \hat{\boldsymbol{\Lambda}}_2, \ldots, \hat{\boldsymbol{\Lambda}}_l$ are the vectors corresponding to the fundamental weights $\Lambda_1, \Lambda_2, \ldots, \Lambda_l$ of $\tilde{\mathscr{L}}^0$, which are such that

$$\frac{2\langle \Lambda_j, \alpha_k \rangle}{\langle \alpha_k, \alpha_k \rangle} = \delta_{jk}$$

for $j, k = 1, 2, \ldots, l$ (cf. (15.11)), then (27.39), (27.46) and (27.48) imply that

$$\hat{\boldsymbol{\Lambda}}_j \cdot \hat{\boldsymbol{\alpha}}_k = \delta_{jk} \tag{29.343}$$

for $j, k = 1, 2, \ldots, l$. Consequently $\hat{\boldsymbol{\Lambda}}_1, \hat{\boldsymbol{\Lambda}}_2, \ldots, \hat{\boldsymbol{\Lambda}}_l$ are the basic dual lattice vectors. Furthermore, (15.10) implies that

$$\hat{\boldsymbol{\Lambda}}_j = \sum_{k=1}^{l} (\mathbf{A}^{-1})_{kj} \hat{\boldsymbol{\alpha}}_k, \tag{29.344}$$

for $j, k = 1, 2, \ldots, l$, whose inverse is

$$\hat{\boldsymbol{\alpha}}_j = \sum_{k=1}^{l} A_{kj} \hat{\boldsymbol{\Lambda}}_k \tag{29.345}$$

for $j = 1, 2, \ldots, l$. As the elements A_{kj} are all integers, (29.345) shows that every basic root lattice vector $\hat{\boldsymbol{\alpha}}_j$ is an integral linear combination of the basic dual lattice vectors, but (29.344) shows that the converse is true only if the elements of $\mathbf{A}^{-1}$ are integers. Thus the root lattice of a simply laced complex semi-simple Lie algebra $\tilde{\mathscr{L}}^0$ is a self-dual integral lattice if and only if the elements of $\mathbf{A}^{-1}$ are integers.

Inspection of the matrices $\mathbf{A}^{-1}$ given in Appendix F shows that this condition is satisfied only for $\tilde{\mathscr{L}}^0 = E_8$, which gives the one 8-dimensional self-dual even integral lattice, for $\tilde{\mathscr{L}}^0 = E_8 \oplus E_8$, which gives one of the two 16-dimensional self-dual even integral lattices, for $\tilde{\mathscr{L}}^0 = E_8 \oplus E_8 \oplus E_8$, and so on.

With $\tilde{\mathscr{L}}^0 = E_8 \oplus E_8$ and the identification (29.342) (with $l = 16$), the vertex construction of the basic representations of the affine Kac–Moody algebras developed in Chapter 27, Section 4 is easily applied. Let

$$A_j^{r\dagger} = a_{\mathrm{L}j}^{r\dagger} \tag{29.346}$$

and

$$A_j^r = a_{\mathrm{L}j}^r, \tag{29.347}$$

where $A_j^{r\dagger}$ and A_j^r are the creation and annihilation operators of (27.71) and $a_{\mathrm{L}j}^{r\dagger}$ and $a_{\mathrm{L}j}^r$ are the corresponding operators for the left-moving modes of the heterotic string that were used in the previous subsection. Identify the basis elements of the carrier space of the representation with the state vectors of the previous subsection by the prescription

$$|0, \Sigma_r n^r \mathbf{e}_r\rangle_{\mathrm{L}} = |0\rangle \otimes \mathrm{e}^{\gamma}, \tag{29.348}$$

$$a_{\mathrm{L}j}^{r\dagger}|0, \Sigma_r n^r \mathbf{e}_r\rangle_{\mathrm{L}} = A_j^{r\dagger}|0\rangle \otimes \mathrm{e}^{\gamma}, \tag{29.349}$$

and so on, where

$$\hat{\gamma} = \sum_{r=1}^{16} n^r \mathbf{e}_r. \tag{29.350}$$

By (29.342), this latter equation can be expressed as

$$\hat{\gamma} = \sum_{r=1}^{16} n^r \hat{\boldsymbol{\alpha}}_r. \tag{29.351}$$

It follows that the set of lattice vectors satisfying (29.338) is precisely the set of root vectors of $E_8 \oplus E_8$. Moreover, by (29.349),

$$a_{\mathrm{L}1}^{r\dagger}|0, \mathbf{0}\rangle_{\mathrm{L}} = A_1^{r\dagger}|0\rangle \otimes \mathrm{e}^0. \tag{29.352}$$

Thus Theorem II of Chapter 27, Section 4 shows that the left-moving parts of the sets (ii) and (iii) of the zero-mass state vectors of the previous subsection transform as the adjoint representation of the complex semi-simple Lie algebra $\tilde{\mathscr{L}}^0 = E_8 \oplus E_8$. As the direct products of these left-moving state vectors with $|0, \mathrm{v}, k\rangle_{\mathrm{R}}$ (for $k = 1, 2, \ldots, 8$) transform as the vector representation $\boldsymbol{\Gamma}^{\mathrm{v}}$ under so(8) transformations (and hence transform as vectors under the whole of the 10-dimensional Poincaré algebra), and since in a gauge theory the massless vector particles transform according to the adjoint representation of the gauge algebra, it is natural with this construction to interpret $E_8 \oplus E_8$ as the gauge algebra.

Although the other 16-dimensional self-dual even integral lattice cannot be a root lattice of a complex semi-simple Lie algebra, it is very closely related to such a lattice. Indeed, if $\alpha_1, \alpha_2, \ldots, \alpha_{16}$ are the simple roots of $\tilde{\mathscr{L}}^0 = D_{16}$, which is the complexification of the compact real Lie algebra so(32), then it is easily checked that

$$\mathbf{e}_j = \begin{cases} \frac{1}{2}(\hat{\boldsymbol{\alpha}}_1 + \hat{\boldsymbol{\alpha}}_3 + \hat{\boldsymbol{\alpha}}_5 + \cdots + \hat{\boldsymbol{\alpha}}_{13} + \hat{\boldsymbol{\alpha}}_{15}) & \text{for } j = 1, \\ \hat{\boldsymbol{\alpha}}_j & \text{for } j = 2, 3, \ldots, 16 \end{cases} \tag{29.353}$$

defines a self-dual even integral lattice whose basic bual lattice vectors are given by

$$\mathbf{e}_j^* = \begin{cases} 2\hat{\boldsymbol{\Lambda}}_1 & \text{for } j = 1, \\ \hat{\boldsymbol{\Lambda}}_j & \text{for } j = 2, 4, 6, 8, 10, 12, 14, \\ \hat{\boldsymbol{\Lambda}}_j - \hat{\boldsymbol{\Lambda}}_1 & \text{for } j = 3, 5, 7, 9, 11, 13, 15, \\ \hat{\boldsymbol{\Lambda}}_{16} & \text{for } j = 16. \end{cases} \tag{29.354}$$

As

$$\hat{\boldsymbol{\alpha}}_1 = 2\mathbf{e}_1 - \mathbf{e}_3 - \mathbf{e}_5 - \cdots - \mathbf{e}_{13} - \mathbf{e}_{15}, \tag{29.355}$$

(29.353) shows that every root lattice vector of D_{16} is a member of this lattice (although the converse is not true). That is, the root lattice of D_{16} is a sublattice of the "toroidal" lattice.

As was shown in Chapter 15, Section 6 and Appendix H, the universal linear group of the compact real form so(32) of D_{16} is $\tilde{\mathscr{G}} = \text{Spin}(32)$, and its centre $Z(\tilde{\mathscr{G}})$ consists of elements of the form $\exp(h)$, where (cf. (15.25))

$$h = \sum_{k=1}^{l} p_k \hat{h}_k$$

and where

$$\hat{h}_k = \frac{4\pi i h^0_{\Lambda_k}}{\langle \alpha_k, \alpha_k \rangle}$$

for $k = 1, 2, \ldots, l$ $(= 16)$, the superscript 0 being added here to conform with the usage in Chapters 26 and 27. Moreover, (15.31), (15.10), (27.39) and (27.40) show that for a simply laced algebra, such as D_{16}, $\exp(h)$ and $\exp(h')$ are equivalent if

$$h - h' = 2\pi i \sum_{k=1}^{l} M_k H^0_{\alpha_k}, \tag{29.356}$$

where $\{M_1, M_2, \ldots, M_l\}$ is any set of integers. As $h' = 0$ corresponds to the identity element of $\hat{\mathscr{G}}$, (29.356) shows that for every root lattice vector

$$\sum_{k=1}^{l} M_k \alpha_k$$

of $\tilde{\mathscr{L}}^0$ the corresponding element of $\tilde{\mathscr{G}}$ of the form

$$\exp\left(2\pi i \sum_{k=1}^{l} M_k \alpha_k\right)$$

is equivalent to the identity element. For $\hat{\mathscr{G}} = \mathrm{Spin}(32)$, it was noted in Appendix H, Section 4 that

$$Z(\hat{G}) = Z_2(\exp(h)) \otimes Z_2(\exp(h')), \tag{29.357}$$

where

$$h = 4\pi i(l-1)h_{\alpha_1+\alpha_3+\cdots+\alpha_{15}}$$

and

$$h' = 4\pi i(l-1)(h_{\alpha_{15}} + h_{\alpha_{16}}).$$

These expressions can be rewritten using (27.39) and (27.40) to give

$$h = 2\pi i H^0_{(\alpha_1+\alpha_3+\alpha_5+\cdots+\alpha_{15})/2} \tag{29.358}$$

and

$$h' = \pi i H^0_{\alpha_{15}+\alpha_{16}}. \tag{29.359}$$

It follows from (29.358) and (29.353) that every element of the toroidal lattice is associated with an element of $Z_2(\exp(h))$ under the exponential mapping, all the elements of the form

$$\sum_{r=1}^{16} n^r \mathbf{e}_r$$

with n^1 odd being associated with $\exp(h)$ (with h given by (29.358)), and those with n^1 even being associated with the identity. That is, the toroidal lattice corresponds to the group $\mathrm{Spin}(32)/Z_2(\exp(h))$. (It should be noted that there is a small typographical error in earlier printings of Appendix H, Section 4, part B(i), since $\mathrm{SO}(32) = \hat{\mathscr{G}}/Z_2(\exp(h'))$ (and not $\hat{\mathscr{G}}/Z_2(\exp(h))$ as was stated there).)

There is no difficulty in generalizing the vertex construction of the basic representation of the Kac–Moody algebra $\tilde{\mathscr{L}} = D_{16}^{(1)}$ by allowing the lattice vectors γ of V_L to run over all the members of the lattice defined by the basic lattice vectors (29.353), and not just over the sublattice of root lattice vectors. (The proof given in Appendix O, Section 2 requires no modification.) The representation of so(32) that emerges can be exponentiated to provided a representation of its universal linear group $\hat{\mathscr{G}} = \text{Spin}(32)$. It is easily checked that the kernel of this representation is $Z_2(\exp(h))$. (This follows because, by (27.91),

$$\Phi(t^0 \otimes 2\pi i H_\alpha^0)(\psi_F \otimes e^\gamma) = 2\pi i \hat{\alpha} \cdot \hat{\gamma}(\psi_F \otimes e^\gamma)$$

and with $\alpha = \frac{1}{2}(\alpha_1 + \alpha_3 + \alpha_5 + \cdots + \alpha_{15})$ the quantity $2\pi i \hat{\alpha} \cdot \hat{\gamma}$ is always an integer multiple of $2\pi i$. Consequently $\Phi(\exp(h)) = 1$, the identity operator, but a similar argument shows that $\Phi(\exp(h')) \neq 1$.) Thus the vertex representation provides a *faithful* representation of $\text{Spin}(32)/Z_2(\exp(h))$.

The argument presented earlier in this subsection concerning the massless states of $E_8 \oplus E_8$ can be repeated for this toroidal lattice. It shows that the massless states transform as the adjoint representation of the algebra so(32), and hence, by the discussion just given, $\text{Spin}(32)/Z_2(\exp(h))$ is the gauge group for this torus.

For both toroidal lattices it was shown by Gross *et al.* (1985a, b, 1986) that the theory is $D = 10$ Lorentz-invariant, is free of anomalies, and possesses a single set of operators Q_a that act non-trivially on the right-moving states and that form the odd part of the $N = 1$ 10-dimensional Poincaré superalgebra. The detailed study of interactions for the heterotic string is contained in the paper by Gross *et al.* (1986).

6 Further developments

The preceding sections contain a treatment of most of the basic ideas on strings and superstrings, with particular emphasis being placed on its algebraic aspects, but the development of these concepts has been the subject of several thousand papers in the last few years alone. As it is clearly impossible to attempt even a brief summary of this work, the present section has the much more modest aim of merely indicating a few starting points for some further reading.

The most comprehensive account is the two volume monograph of Green *et al.* (1987a, b), and there is an introductory text by Kaku (1988), while other useful shorter general articles have been produced by Brink (1985), Clavelli (1986), Green (1983, 1986), Mandelstam (1984), Nicolai (1987) and Schwarz (1982b, 1984a, b, 1985b, 1986, 1987). The articles by Ellis (1986), Nilles (1986), Ibáñez (1987), Binétruy (1988) and Zwirner (1988) have concentrated more on phenomenological aspects. Introductory reviews of string field theory have been written by Jevicki (1988), Kazama (1987), Kaku (1986, 1987, 1988), West (1986b) and Siegel (1988), and the connection with the non-linear sigma model has been reviewed by Hull (1987).

On the pressing question of the reduction of the theory to a 4-dimensional space–time, one active line of approach involving compactification on a Calibi–Yau manifold has been initiated by Candelas *et al.* (1985), Strominger and Witten (1985) and Witten (1985, 1986). The investigation of orbifold compactification as an alternative has been started by Dixon *et al.* (1985, 1986), and Narain (1986) and Narain *et al.* (1987) have made significant developments in the method of toroidal compactification. For a general review of compactification see Dolan (1986).

Many of the original articles and reviews on strings and superstrings have been gathered together in the useful two-volume compilation of Schwarz (1985a).

Appendices

Appendix K

Proofs of Certain Theorems on Supermatrices and Lie Superalgebras

1 Proofs of Theorems I and IV of Chapter 20, Section 4

Theorem I (a) Suppose that $\mathbf{A}$ is a $(p/0) \times (p/0)$ even supermatrix (that is, $\mathbf{A}$ is a $p \times p$ matrix all of whose elements are members of $\mathbb{C}\mathbf{B}_{L0}$). Let

$$\mathbf{A} = \sum_{\mu} \mathbf{A}_{\mu} \mathcal{E}_{\mu} \tag{20.50}$$

be the expansion corresponding to (20.25), where for each index set μ the matrix $\mathbf{A}_{\mu}$ is a $p \times p$ matrix with complex entries. Then $\mathbf{A}$ is invertible if and only if the matrix $\mathbf{A}_{\varnothing}$ corresponding to the Grassmann identity $\mathcal{E}_{\varnothing}\,(=1)$ is invertible.

(b) If $\mathbf{M}$ is a $(p/q) \times (p/q)$ even supermatrix, partitioned as in (20.40), then $\mathbf{M}$ is invertible if and only if its submatrices $\mathbf{A}$ and $\mathbf{D}$ are invertible.

(c) If $\mathbf{M}$ and $\mathbf{N}$ are $(p/q) \times (p/q)$ even invertible supermatrices then $\mathbf{MN}$ is also a $(p/q) \times (p/q)$ even invertible supermatrix.

Proof (a) If $\mathbf{B}$ is another $(p/0) \times (p/0)$ even supermatrix, with expansion $\mathbf{B} = \sum_{\mu} \mathbf{B}_{\mu} \mathcal{E}_{\mu}$, it follows (cf. (20.28)) that

$$\mathbf{AB} = \sum_{\mu,\mu'} \mathbf{A}_{\mu} \mathbf{B}_{\mu'} (\mathcal{E}_{\mu} \mathcal{E}_{\mu'}).$$

In particular, $\mathbf{A}_{\varnothing}\mathbf{B}_{\varnothing}$ is the matrix of coefficients of the Grassmann identity $\mathcal{E}_{\varnothing}$ $(=1)$ in $\mathbf{AB}$. Thus if $\mathbf{A}$ is invertible and has inverse $\mathbf{B}$ $(=\mathbf{A}^{-1})$, it is necessary that $\mathbf{A}_{\varnothing}\mathbf{B}_{\varnothing} = \mathbf{1}$, implying that $\mathbf{A}_{\varnothing}$ is invertible and $(\mathbf{A}_{\varnothing})^{-1} = \mathbf{B}_{\varnothing}$.

Conversely, if $\mathbf{A}_\varnothing$ is invertible, on putting $\mathbf{B}_\varnothing = (\mathbf{A}_\varnothing)^{-1}$, the product $\mathbf{AB}$ differs from $\mathbf{1}$ in terms of level 2 and above. However, with the further choice $\mathbf{B}_{\{j,k\}} = -(\mathbf{A}_\varnothing)^{-1}\mathbf{A}_{\{j,k\}}(\mathbf{A}_\varnothing)^{-1}$ for all $j, k = 1, 2, \ldots, L$, $\mathbf{AB}$ differs from $\mathbf{1}$ only in terms of level 4 and above. Clearly the higher-level coefficients $\mathbf{B}_\mu$ can be chosen successively so as to get $\mathbf{AB}$ to be equal to $\mathbf{1}$ at all levels. The resulting matrix $\mathbf{B}$ is then the inverse of $\mathbf{A}$.

(b) Let

$$\mathbf{N} = \begin{bmatrix} \mathbf{E} & \mathbf{F} \\ \mathbf{G} & \mathbf{H} \end{bmatrix} \tag{K.1}$$

be another $(p/q) \times (p/q)$ even supermatrix (with the same partitioning as (20.40)). If $\mathbf{MN} = \mathbf{1}$ then $\mathbf{AE} + \mathbf{BG} = \mathbf{1}$ and $\mathbf{CF} + \mathbf{DH} = \mathbf{1}$. As the elements of $\mathbf{B}$, $\mathbf{C}$, $\mathbf{F}$ and $\mathbf{G}$ are members of $\mathbb{C}B_{L1}$, no elements in $\mathbf{BG}$ or $\mathbf{CF}$ contain the Grassmann identity component. This implies that $\mathbf{A}_\varnothing\mathbf{E}_\varnothing = \mathbf{1}$ and $\mathbf{D}_\varnothing\mathbf{H}_\varnothing = \mathbf{1}$. Thus if $\mathbf{M}$ is invertible and $\mathbf{N} = \mathbf{M}^{-1}$ then $\mathbf{A}_\varnothing$ and $\mathbf{D}_\varnothing$ must be invertible (with inverses $\mathbf{E}_\varnothing$ and $\mathbf{H}_\varnothing$ respectively), and hence by part (a), $\mathbf{A}$ and $\mathbf{D}$ must be invertible.

Conversely if $\mathbf{A}$ and $\mathbf{D}$ are invertible then by part (a) $\mathbf{A}_\varnothing$ and $\mathbf{D}_\varnothing$ must also be invertible. Let $\mathbf{E}_\varnothing = (\mathbf{A}_\varnothing)^{-1}$ and $\mathbf{H}_\varnothing = (\mathbf{D}_\varnothing)^{-1}$. It will be shown first that $\mathbf{MN}$ is invertible. Clearly $\mathbf{MN} = \mathbf{1} - \mathbf{R}$, where $\mathbf{R}$ contains no non-zero elements with a Grassmann identity component. Hence $\mathbf{R}$ is nilpotent; that is, there exists a positive integer r such that $\mathbf{R}^{r+1} = \mathbf{0}$. Then

$$(\mathbf{1} + \mathbf{R} + \mathbf{R}^2 + \cdots + \mathbf{R}^r)(\mathbf{1} - \mathbf{R}) = (\mathbf{1} - \mathbf{R})(\mathbf{1} + \mathbf{R} + \mathbf{R}^2 + \cdots + \mathbf{R}^r) = \mathbf{1},$$

showing that $\mathbf{1} + \mathbf{R} + \mathbf{R}^2 + \cdots + \mathbf{R}^r$ is the inverse of $\mathbf{1} - \mathbf{R}$, i.e. of $\mathbf{MN}$. A similar argument shows that $\mathbf{NM}$ is invertible. However, by (20.43), $\mathbf{N}(\mathbf{MN}) = (\mathbf{NM})\mathbf{N}$, so

$$(\mathbf{NM})^{-1}\mathbf{N} = \mathbf{N}(\mathbf{MN})^{-1}. \tag{K.2}$$

Define $\mathbf{M}'$ by $\mathbf{M}' = \mathbf{N}(\mathbf{MN})^{-1}$. Then $\mathbf{MM}' = \mathbf{1}$ and, by (K.2), $\mathbf{M}'\mathbf{M} = \mathbf{1}$, so $\mathbf{M}'$ is the inverse of $\mathbf{M}$.

(c) With $\mathbf{M}$ and $\mathbf{N}$ being *any* two $(p/q) \times (p/q)$ even supermatrices, partitioned as in (20.40) and (K.1), the Grassmann identity component of $\mathbf{MN}$ is

$$\begin{bmatrix} \mathbf{A}_\varnothing\mathbf{E}_\varnothing & \mathbf{0} \\ \mathbf{0} & \mathbf{D}_\varnothing\mathbf{H}_\varnothing \end{bmatrix}.$$

If $\mathbf{M}$ and $\mathbf{N}$ are invertible then part (b) implies that $\mathbf{A}_\varnothing$, $\mathbf{E}_\varnothing$, $\mathbf{D}_\varnothing$ and $\mathbf{H}_\varnothing$ are invertible, and hence so are $\mathbf{A}_\varnothing\mathbf{E}_\varnothing$ and $\mathbf{D}_\varnothing\mathbf{H}_\varnothing$. A further application of part (b) then shows that $\mathbf{MN}$ is invertible. Clearly $\mathbf{MN}$ is a $(p/q) \times (p/q)$ *even* supermatrix.

Theorem IV (a) If $\mathbf{M}$ and $\mathbf{N}$ are any two members of $GL(p/q; \mathbb{C}B_L)$ then

$$\text{sdet}\,(\mathbf{MN}) = (\text{sdet}\,\mathbf{M})(\text{sdet}\,\mathbf{N}). \tag{20.75}$$

(b) If $\mathbf{M} \in GL(p/q; \mathbb{C}B_L)$ then

$$\text{sdet}\,(\mathbf{M}^{st}) = \text{sdet}\,\mathbf{M}. \tag{20.76}$$

(c) If $\mathbf{M}$ is an even $(p/q) \times (p/q)$ supermatrix then $\exp \mathbf{M}$ is an even invertible $(p/q) \times (p/q)$ supermatrix and

$$\text{sdet}\,(\exp \mathbf{M}) = \exp\,(\text{str}\,\mathbf{M}). \tag{20.77}$$

(d) An expression equivalent to (20.74) is

$$\text{sdet}\,\mathbf{M} = \frac{\det \mathbf{A}}{\det\,(\mathbf{D} - \mathbf{C}\mathbf{A}^{-1}\mathbf{B})}. \tag{20.78}$$

Proof (a) Let G_+, G_0 and G_- be the subgroups of $GL(p/q; \mathbb{C}B_L)$ consisting of supermatrices of the forms:

$$G_+ = \left\{\begin{bmatrix} \mathbf{1} & \mathbf{B} \\ \mathbf{0} & \mathbf{1} \end{bmatrix}\right\}, \quad G_0 = \left\{\begin{bmatrix} \mathbf{A} & \mathbf{0} \\ \mathbf{0} & \mathbf{D} \end{bmatrix}\right\}, \quad G_- = \left\{\begin{bmatrix} \mathbf{1} & \mathbf{0} \\ \mathbf{C} & \mathbf{1} \end{bmatrix}\right\}.$$

Any $\mathbf{M} \in GL(p/q; \mathbb{C}B_L)$ can be written as $\mathbf{M} = \mathbf{M}_+\mathbf{M}_0\mathbf{M}_-$, where $\mathbf{M}_+ \in G_+$, $\mathbf{M}_0 \in G_0$ and $\mathbf{M}_- \in G_-$, for

$$\mathbf{M} = \begin{bmatrix} \mathbf{A} & \mathbf{B} \\ \mathbf{C} & \mathbf{D} \end{bmatrix} = \begin{bmatrix} \mathbf{1} & \mathbf{B}\mathbf{D}^{-1} \\ \mathbf{0} & \mathbf{1} \end{bmatrix} \begin{bmatrix} \mathbf{A} - \mathbf{B}\mathbf{D}^{-1}\mathbf{C} & \mathbf{0} \\ \mathbf{0} & \mathbf{D} \end{bmatrix} \begin{bmatrix} \mathbf{1} & \mathbf{0} \\ \mathbf{D}^{-1}\mathbf{C} & \mathbf{1} \end{bmatrix}.$$

The first stage in the proof is to show that (20.75) is true

(i) for any $\mathbf{M} \in GL(p/q; \mathbb{C}B_L)$ but $\mathbf{N} \in G_0$ or $\mathbf{N} \in G_-$, and

(ii) for any $\mathbf{N} \in GL(p/q; \mathbb{C}B_L)$ but $\mathbf{M} \in G_+$ or $\mathbf{M} \in G_0$.

For example, if $\mathbf{N} \in G_-$, so that

$$\mathbf{N} = \begin{bmatrix} \mathbf{1} & \mathbf{0} \\ \mathbf{G} & \mathbf{1} \end{bmatrix},$$

(20.74) gives

$$\begin{aligned} \text{sdet}(\mathbf{MN}) &= \text{sdet}\begin{bmatrix} \mathbf{A} + \mathbf{BG} & \mathbf{B} \\ \mathbf{C} + \mathbf{DG} & \mathbf{D} \end{bmatrix} \\ &= \frac{\det\{(\mathbf{A} + \mathbf{BG}) - \mathbf{B}\mathbf{D}^{-1}(\mathbf{C} + \mathbf{DG})\}}{\det \mathbf{D}} \end{aligned}$$

$$= \frac{\det(\mathbf{A} - \mathbf{B}\mathbf{D}^{-1}\mathbf{C})}{\det \mathbf{D}}$$

$$= \operatorname{sdet} \mathbf{M} = (\operatorname{sdet} \mathbf{M})(\operatorname{sdet} \mathbf{N}),$$

the proofs for the other parts of cases (i) and (ii) being similar.

Then, with $\mathbf{M} = \mathbf{M}_+\mathbf{M}_0\mathbf{M}_-$ (as above) and, correspondingly, with $\mathbf{N} = \mathbf{N}_+\mathbf{N}_0\mathbf{N}_-$, application of (i) and (ii) implies that

$$\begin{aligned}\operatorname{sdet}(\mathbf{MN}) &= (\operatorname{sdet} \mathbf{M}_+)(\operatorname{sdet} \mathbf{M}_0)(\operatorname{sdet}(\mathbf{M}_-\mathbf{N}_+))(\operatorname{sdet} \mathbf{N}_0)(\operatorname{sdet} \mathbf{N}_-)\\ &= (\operatorname{sdet} \mathbf{M}_0)(\operatorname{sdet}(\mathbf{M}_-\mathbf{N}_+))(\operatorname{sdet} \mathbf{N}_0),\end{aligned}$$

and since (i) and (ii) also imply that

$$\operatorname{sdet} \mathbf{M} = (\operatorname{sdet} \mathbf{M}_+)(\operatorname{sdet} \mathbf{M}_0)(\operatorname{sdet} \mathbf{M}_-) = \operatorname{sdet} \mathbf{M}_0$$

and the corresponding equation for $\mathbf{N}$, to prove that (20.75) is valid for *all* $\mathbf{M}$, $\mathbf{N} \in \mathrm{GL}(p/q; \mathbb{C}B_L)$ it remains only to demonstrate that

$$\operatorname{sdet}(\mathbf{M}_-\mathbf{N}_+) = 1. \tag{K.3}$$

Defining an "elementary" member of G_+ to be a supermatrix of the form

$$\begin{bmatrix} \mathbf{1} & \mathbf{F} \\ \mathbf{0} & \mathbf{1} \end{bmatrix} \tag{K.4}$$

in which $\mathbf{F}$ contains *only one* non-zero element (which is necessarily a member of $\mathbb{C}B_{L1}$), since

$$\begin{bmatrix} \mathbf{1} & \mathbf{F} \\ \mathbf{0} & \mathbf{1} \end{bmatrix}\begin{bmatrix} \mathbf{1} & \mathbf{F}' \\ \mathbf{0} & \mathbf{1} \end{bmatrix} = \begin{bmatrix} \mathbf{1} & \mathbf{F}+\mathbf{F}' \\ \mathbf{0} & \mathbf{1} \end{bmatrix},$$

it follows that every member of G_+ is a product of elementary members of G_+. It will first be shown that (K.3) is true for all $\mathbf{M}_- \in G_-$ provided that $\mathbf{N}_+$ is an elementary member of G_+. Suppose that $\mathbf{N}_+$ has the form (K.4) with $\mathbf{F}$ containing only one non-zero element f ($\in \mathbb{C}B_{L1}$). As

$$\operatorname{sdet}(\mathbf{M}_-\mathbf{N}_+) = \operatorname{sdet}\begin{bmatrix} \mathbf{1} & \mathbf{F} \\ \mathbf{C} & \mathbf{1}+\mathbf{CF} \end{bmatrix} = \frac{\det\{\mathbf{1} - \mathbf{F}(\mathbf{1}+\mathbf{CF})^{-1}\mathbf{C}\}}{\det(\mathbf{1}+\mathbf{CF})},$$

it is necessary to evaluate $(\mathbf{1}+\mathbf{CF})^{-1}$. As $\mathbf{CF}$ is a multiple of f and $f^2 = 0$, it follows that $(\mathbf{CF})^2 = \mathbf{0}$ and hence $(\mathbf{1}+\mathbf{CF})^{-1} = \mathbf{1} - \mathbf{CF}$. Moreover, since $(\mathbf{FC})^2 = \mathbf{0}$ for the same reason,

$$\operatorname{sdet}(\mathbf{M}_-\mathbf{N}_+) = \frac{\det(\mathbf{1} - \mathbf{FC})}{\det(\mathbf{1}+\mathbf{CF})}.$$

However, in these circumstances $\det(\mathbf{1} - \mathbf{FC}) = 1 - \operatorname{tr}(\mathbf{FC})$ and $\det(\mathbf{1}+\mathbf{CF}) = 1 + \operatorname{tr}(\mathbf{CF}) = 1 - \operatorname{tr}(\mathbf{FC})$ (since the entries of $\mathbf{F}$ and $\mathbf{C}$ are all in $\mathbb{C}B_{L1}$). This establishes (K.3) for the special case in which $\mathbf{N}_+$ is an *elementary* member of G_+.

The proof of (K.3) for the general case follows by successive application of this result for the special case just established. For example, if $\mathbf{N}_+ = \mathbf{N}_+^1\mathbf{N}_+^2$, where $\mathbf{N}_+^1$ and $\mathbf{N}_+^2$ are elementary members of G_+, then on writing $\mathbf{M}_-\mathbf{N}_+^1 = (\mathbf{M}_-\mathbf{N}_+^1)_+(\mathbf{M}_-\mathbf{N}_+^1)_0(\mathbf{M}_-\mathbf{N}_+^1)_-$, application of previously established results shows that

$$\begin{aligned}\operatorname{sdet}(\mathbf{M}_-\mathbf{N}_+) &= \operatorname{sdet}(\mathbf{M}_-\mathbf{N}_+^1)_0 \operatorname{sdet}((\mathbf{M}_-\mathbf{N}_+^1)_-\mathbf{N}_+^1)\\ &= \operatorname{sdet}(\mathbf{M}_-\mathbf{N}_+^1)_0\\ &= \operatorname{sdet}(\mathbf{M}_-\mathbf{N}_+^1) = 1.\end{aligned}$$

This completes the proof of (20.75).

(b) By (20.58) and (20.74),

$$\begin{aligned}\operatorname{sdet}(\mathbf{M}^{st}) &= \frac{\det\{\tilde{\mathbf{A}} - \tilde{\mathbf{C}}(\tilde{\mathbf{D}}^{-1})(-\tilde{\mathbf{B}})\}}{\det \tilde{\mathbf{D}}}\\ &= \frac{\det(\mathbf{A} - \mathbf{B}\mathbf{D}^{-1}\mathbf{C})}{\det \mathbf{D}}\end{aligned}$$

(since the elements of $\mathbf{B}$ and $\mathbf{C}$ are in $\mathbb{C}\mathbf{B}_{L1}$).

(c) With $\mathbf{M}$ as in (20.40),

$$\mathbf{M}^2 = \begin{bmatrix} \mathbf{A}^2 + \mathbf{BC} & \mathbf{AB} + \mathbf{BD} \\ \mathbf{CA} + \mathbf{DC} & \mathbf{CB} + \mathbf{D}^2 \end{bmatrix},$$

and, since the elements of $\mathbf{B}$ and $\mathbf{C}$ are odd, the Grassmann identity component of $\mathbf{M}^2$ is

$$\begin{bmatrix} (\mathbf{A}_\varnothing)^2 & \mathbf{0} \\ \mathbf{0} & (\mathbf{D}_\varnothing)^2 \end{bmatrix}$$

(cf. (20.50)). Similarly the Grassmann identity component of $\mathbf{M}^n$ for any $n \geqslant 1$ is

$$\begin{bmatrix} (\mathbf{A}_\varnothing)^n & \mathbf{0} \\ \mathbf{0} & (\mathbf{D}_\varnothing)^n \end{bmatrix},$$

and hence, by (10.1), the Grassmann identity component of $\exp \mathbf{M}$ is

$$\begin{bmatrix} \exp \mathbf{A}_\varnothing & \mathbf{0} \\ \mathbf{0} & \exp \mathbf{D}_\varnothing \end{bmatrix}.$$

By part (g) of Theorem III of Chapter 10, Section 2, $\exp \mathbf{A}_\varnothing$ and $\exp \mathbf{D}_\varnothing$ are invertible, so Theorem I of Chapter 20, Section 4 shows that $\exp \mathbf{M}$ must be invertible.

To establish (20.77), suppose first that $\mathbf{M}$ has the special form

$$\mathbf{M} = \begin{bmatrix} \mathbf{A} & \mathbf{B} \\ \mathbf{0} & \mathbf{D} \end{bmatrix}. \tag{K.5}$$

Then, by (10.1),

$$\exp \mathbf{M} = \begin{bmatrix} \exp \mathbf{A} & \mathbf{B}' \\ \mathbf{0} & \exp \mathbf{D} \end{bmatrix},$$

where $\mathbf{B}'$ is some fairly complicated expression, which need not be determined explicitly, since by (20.74),

$$\text{sdet}\, \mathbf{M} = \frac{\det(\exp \mathbf{A})}{\det(\exp \mathbf{D})}.$$

Application of part (f) of Theorem III of Chapter 10, Section 2 reduces this to

$$\text{sdet}\, \mathbf{M} = \frac{\exp(\operatorname{tr} \mathbf{A})}{\exp(\operatorname{tr} \mathbf{D})},$$

which shows, by (20.67), that (20.77) is valid for $\mathbf{M}$ of the form (K.4). A similar argument shows that (20.77) is valid for $\mathbf{M}$ of the form

$$\mathbf{M} = \begin{bmatrix} \mathbf{A} & \mathbf{0} \\ \mathbf{C} & \mathbf{D} \end{bmatrix}. \tag{K.6}$$

To demonstrate that (20.77) is valid for *any* $(p/q) \times (p/q)$ even supermatrix $\mathbf{M}$, first make use of the results just established and (20.75) to deduce that

$$\begin{aligned}
&\text{sdet}\left\{\exp\begin{bmatrix} \mathbf{A} & \mathbf{B} \\ \mathbf{C} & \mathbf{D} \end{bmatrix} \exp\begin{bmatrix} -\mathbf{A} & -\mathbf{B} \\ \mathbf{0} & -\mathbf{D} \end{bmatrix}\right\} \\
&= \text{sdet}\left\{\exp\begin{bmatrix} \mathbf{A} & \mathbf{B} \\ \mathbf{C} & \mathbf{D} \end{bmatrix}\right\} \text{sdet}\left\{\exp\begin{bmatrix} -\mathbf{A} & -\mathbf{B} \\ \mathbf{0} & -\mathbf{D} \end{bmatrix}\right\} \\
&= \text{sdet}\left\{\exp\begin{bmatrix} \mathbf{A} & \mathbf{B} \\ \mathbf{C} & \mathbf{D} \end{bmatrix}\right\} \exp\left\{-\operatorname{str}\begin{bmatrix} \mathbf{A} & \mathbf{B} \\ \mathbf{0} & \mathbf{D} \end{bmatrix}\right\} \\
&= \text{sdet}\left\{\exp\begin{bmatrix} \mathbf{A} & \mathbf{B} \\ \mathbf{C} & \mathbf{D} \end{bmatrix}\right\} \exp\left\{-\operatorname{str}\begin{bmatrix} \mathbf{A} & \mathbf{B} \\ \mathbf{C} & \mathbf{D} \end{bmatrix}\right\}.
\end{aligned} \tag{K.7}$$

However, by the Campbell–Baker–Hausdorff formula (cf. Theorem II of Chapter 10, Section 2),

$$\exp\begin{bmatrix} \mathbf{A} & \mathbf{B} \\ \mathbf{C} & \mathbf{D} \end{bmatrix} \exp\begin{bmatrix} -\mathbf{A} & -\mathbf{B} \\ \mathbf{0} & -\mathbf{D} \end{bmatrix} = \exp\left\{\begin{bmatrix} \mathbf{0} & \mathbf{0} \\ \mathbf{C} & \mathbf{0} \end{bmatrix} + \mathbf{M}'\right\}, \tag{K.8}$$

where $\mathbf{M}'$ is a sum of commutators, and, moreover, with the usual partitioning, the form of $\mathbf{M}'$ is

$$\mathbf{M}' = \begin{bmatrix} \mathbf{A}' & \mathbf{0} \\ \mathbf{C}' & \mathbf{D}' \end{bmatrix}.$$

As the matrix in the exponent on the right-hand-side of (K.8) is of the type (K.6), for which it has already been established that (20.77) is valid, it follows that

$$\operatorname{sdet}\left\{\begin{bmatrix} \mathbf{0} & \mathbf{0} \\ \mathbf{C} & \mathbf{0} \end{bmatrix} + \mathbf{M}'\right\} = \exp\left\{\operatorname{str}\left(\begin{bmatrix} \mathbf{0} & \mathbf{0} \\ \mathbf{C} & \mathbf{0} \end{bmatrix} + \mathbf{M}'\right)\right\}.$$

However, as $\mathbf{M}'$ is a sum of commutators, (20.69) implies that $\operatorname{str}\mathbf{M}' = 0$, and so it follows from (K.7) and (K.8) that

$$\operatorname{sdet}\left\{\exp\begin{bmatrix} \mathbf{A} & \mathbf{B} \\ \mathbf{C} & \mathbf{D} \end{bmatrix}\right\}\exp\left\{-\operatorname{str}\begin{bmatrix} \mathbf{A} & \mathbf{B} \\ \mathbf{C} & \mathbf{D} \end{bmatrix}\right\} = 1.$$

Thus (20.77) is valid for any $(p/q) \times (p/q)$ even supermatrix $\mathbf{M}$.

(d) As

$$\mathbf{M} = \begin{bmatrix} \mathbf{A} & \mathbf{B} \\ \mathbf{C} & \mathbf{D} \end{bmatrix} = \begin{bmatrix} \mathbf{1} & \mathbf{0} \\ \mathbf{C}\mathbf{A}^{-1} & \mathbf{1} \end{bmatrix}\begin{bmatrix} \mathbf{A} & \mathbf{B} \\ \mathbf{0} & \mathbf{D} - \mathbf{C}\mathbf{A}^{-1}\mathbf{B} \end{bmatrix},$$

it follows from (20.75) that

$$\begin{aligned}\operatorname{sdet}\mathbf{M} &= \operatorname{sdet}\begin{bmatrix} \mathbf{1} & \mathbf{0} \\ \mathbf{C}\mathbf{A}^{-1} & \mathbf{1} \end{bmatrix}\operatorname{sdet}\begin{bmatrix} \mathbf{A} & \mathbf{B} \\ \mathbf{0} & \mathbf{D} - \mathbf{C}\mathbf{A}^{-1}\mathbf{B} \end{bmatrix} \\ &= \frac{\det\mathbf{A}}{\det(\mathbf{D} - \mathbf{C}\mathbf{A}^{-1}\mathbf{B})}, \quad \text{by (20.74).}\end{aligned}$$

2 Proof of Theorem I of Chapter 21, Section 4

Theorem I Suppose that the matrices $\boldsymbol{\Gamma}(a)$ form a (d_0/d_1)-dimensional *irreducible* graded representation of a Lie superalgebra $\mathscr{L}_s$, and that $\mathbf{M}$ is a matrix of $M(d_0/d_1; \mathbb{C})$ such that

$$[\mathbf{M}, \boldsymbol{\Gamma}(a)] = \mathbf{0} \tag{21.62}$$

for all $a \in \mathscr{L}_s$. We then have the following.

(a) If $\mathbf{M}$ is *even* then $\mathbf{M}$ must have the form

$$\mathbf{M} = \kappa\mathbf{1}, \tag{21.63}$$

where κ is any constant and $\mathbf{1}$ is the $(d_0 + d_1)$-dimensional unit matrix.

(b) If $\mathbf{M}$ is *odd* and

$$d_0 = d_1 \tag{21.64}$$

and there exists a non-singular $d_0 \times d_0$ matrix $\mathbf{B}$ such that

$$\mathbf{\Gamma}_{11}(a) = \mathbf{B}^{-1}\mathbf{\Gamma}_{00}(a)\mathbf{B} \tag{21.65}$$

for all $a \in \mathscr{L}_0$, and

$$\mathbf{\Gamma}_{01}(a) = -\mathbf{B}\mathbf{\Gamma}_{10}(a)\mathbf{B} \tag{21.66}$$

for all $a \in \mathscr{L}_1$, then $\mathbf{M}$ must have the form

$$\mathbf{M} = \mu \begin{bmatrix} \mathbf{0} & \mathbf{B} \\ \mathbf{B}^{-1} & \mathbf{0} \end{bmatrix}, \tag{21.67}$$

where μ is any constant. However, if any of the conditions (21.64), (21.65) or (21.66) are *not* satisfied then there is *no* odd $\mathbf{M}$ of $M(d_0/d_1;\mathbb{C})$ satisfying (21.62). (In (21.65) and (21.66) the submatrices $\mathbf{\Gamma}_{00}(a)$, $\mathbf{\Gamma}_{11}(a)$, $\mathbf{\Gamma}_{01}(a)$ and $\mathbf{\Gamma}_{10}(a)$ are as defined in (21.52) and (21.53).)

Proof (a) Let ψ_j $(j = 1, 2, \ldots, d_0 + d_1)$ be a homogeneous basis for the carrier space V of the irreducible graded representation $\mathbf{\Gamma}$. Let

$$\phi_k = \sum_{j=1}^{d_0+d_1} M_{jk}\psi_j$$

for $k = 1, 2, \ldots, d_0 + d_1$, and let $\mathscr{S}$ be the subspace of V spanned by the ϕ_k $(k = 1, 2, \ldots, d_0 + d_1)$. Then, for any $a \in \mathscr{L}_s$, with $\mathbf{M}$ assumed to be *even*, (21.62) and (21.21) imply that $\mathbf{M}\mathbf{\Gamma}(a) = \mathbf{\Gamma}(a)\mathbf{M}$ for all $a \in \mathscr{L}_s$, and hence

$$\Phi(a)\phi_j = \sum_{k=1}^{d_0+d_1} \Gamma(a)_{kj}\phi_k,$$

so $\Phi(a)\phi_j \in \mathscr{S}$. As $\phi_j \in V_0$ for $j = 1, 2, \ldots, d_0$, and $\phi_j \in V_1$ for $j = d_0 + 1, \ldots, d_0 + d_1$, it follows that $\mathscr{S}$ is an invariant graded subspace of V. However, as $\mathbf{\Gamma}$ is assumed to be *irreducible*, V has no invariant graded subspaces of dimension less than its own. Thus either $\phi_j = 0$ for $j = 1, 2, \ldots, d_0 + d_1$, implying that $\mathbf{M} = \mathbf{0}$, or else $\mathscr{S}$ coincides with V. In this latter case the ϕ_j $(j = 1, 2, \ldots, d_0 + d_1)$ must be linearly independent, implying that $\det \mathbf{M} \neq 0$. Now let $\mathbf{M}' = \mathbf{M} - \kappa\mathbf{1}$, where κ is a constant chosen so that $\mathbf{M}' = 0$. Then $\mathbf{M}'$ is even and $\mathbf{M}'\mathbf{\Gamma}(a) = \mathbf{\Gamma}(a)\mathbf{M}'$ for all $a \in \mathscr{L}_s$, so repeating the argument with $\mathbf{M}$ replaced by $\mathbf{M}'$ leads to the conclusion that either $\det \mathbf{M}' \neq 0$ (which is forbidden) or $\mathbf{M}' = \mathbf{0}$. Thus the only possibility not excluded is $\mathbf{M}' = \kappa\mathbf{1}$.

(b) Suppose now that $\mathbf{M}$ is *odd* and has the form

$$\mathbf{M} = \begin{bmatrix} \mathbf{0} & \mathbf{B}' \\ \mathbf{C}' & \mathbf{0} \end{bmatrix}.$$

Then $\mathbf{M}^2$ is even and applying the argument in (a) with $\mathbf{M}$ replaced by $\mathbf{M}^2$ shows that

$$\sum_{k=1}^{d_0+d_1} (\mathbf{M}^2)_{kj}\psi_k$$

(with $j = 1, 2, \ldots, d_0 + d_1$) must be a basis for V. However, as

$$\sum_{k=1}^{d_0+d_1} M_{kj}\psi_k$$

is odd when $j = 1, 2, \ldots, d_0$ and is even when $j = d_0 + 1, \ldots, d_0 + d_1$, this is only possible if $d_0 = d_1$ (i.e. if (21.64) is satisfied). Moreover, the condition (21.63) now becomes $\mathbf{M}^2 = \mu^2\mathbf{1}$, where μ is any constant, which implies that $\mathbf{B}'\mathbf{C}' = \mathbf{C}'\mathbf{B}' = \mu^2\mathbf{1}$, so that $\mathbf{B}'$ is invertible and $\mathbf{C}' = \mu^2(\mathbf{B}')^{-1}$. With $\mathbf{B}$ defined by $\mathbf{B} = \mu^{-1}\mathbf{B}'$, $\mathbf{C}' = \mu\mathbf{B}^{-1}$ and so

$$\mathbf{M} = \mu\begin{bmatrix} \mathbf{0} & \mathbf{B} \\ \mathbf{B}^{-1} & \mathbf{0} \end{bmatrix}.$$

Applying (21.62) for $a \in \mathscr{L}_0$ then requires that (21.65) be satisfied, whereas applying (21.62) for $a \in \mathscr{L}_1$ implies that (21.66) must hold.

3 Proof of Theorem III of Chapter 21, Section 5

Theorem III Let $\mathscr{L}'_s$ be an invariant graded subalgebra of $\mathscr{L}_s$, and let $\mathscr{L}'^{\perp}_s$ be the "orthogonal complement of $\mathscr{L}'_s$ in $\mathscr{L}_s$ with respect to the Killing form $B(\ ,\ )$ of $\mathscr{L}_s$", which is defined to be the subset of elements of $\mathscr{L}_s$ such that $a \in \mathscr{L}'^{\perp}_s$ if

$$B(a, b) = 0 \tag{21.88}$$

for all $b \in \mathscr{L}'_s$. Then $\mathscr{L}'^{\perp}_s$ is also an invariant graded subalgebra of $\mathscr{L}_s$.

Proof It will first be demonstrated that $\mathscr{L}'^{\perp}_s$ is an invariant subalgebra of $\mathscr{L}_s$, and then it will be shown that it is graded. Suppose that $a \in \mathscr{L}'^{\perp}_s$, $b \in \mathscr{L}'_s$ and $c \in \mathscr{L}_s$. Then, by (21.86), $B([a, c], b) = B(a, [c, b])$. But $[c, b] \in \mathscr{L}'_s$, so by the definition (21.88), $B(a, [c, b]) = 0$, and hence $B([a, c], b) = 0$, implying that $[a, c] \in \mathscr{L}'^{\perp}_s$ and thus showing that $\mathscr{L}'^{\perp}_s$ is an invariant subalgebra of $\mathscr{L}_s$.

To show that $\mathscr{L}'^{\perp}_s$ is graded, suppose that $a \in \mathscr{L}'^{\perp}_s$ and $a = a_{\text{even}} + a_{\text{odd}}$, where $a_{\text{even}} \in \mathscr{L}_0$ and $a_{\text{odd}} \in \mathscr{L}_1$. Then, for any $b \in \mathscr{L}'_s$, with the corresponding decomposition $b = b_{\text{even}} + b_{\text{odd}}$, by (21.88), (21.84) and the bilinearity property of $B(\ ,\ )$

$$0 = B(a, b) = B(a_{\text{even}}, b_{\text{even}}) + B(a_{\text{odd}}, b_{\text{odd}}).$$

In particular, this is true when $b_{\text{odd}} = 0$, in which case $B(a_{\text{even}}, b_{\text{even}}) = 0$ for all $b_{\text{even}} \in \mathscr{L}'_s$. But as $B(a_{\text{even}}, b_{\text{odd}}) = 0$ by (21.84) for all $b_{\text{odd}} \in \mathscr{L}'_s$, it follows that $B(a_{\text{even}}, b) = 0$ for all $b \in \mathscr{L}'_s$. Thus $a_{\text{even}} \in \mathscr{L}'^{\perp}_s$. Similarly $a_{\text{odd}} \in \mathscr{L}'^{\perp}_s$, and hence $\mathscr{L}'^{\perp}_s$ is graded.

4 Proofs of Theorems II, III, IV and V of Chapter 25, Section 2

Theorem II If $\mathscr{L}_s$ is a Lie superalgebra whose Killing form is non-degenerate then $\mathscr{L}_s$ is either a *simple* Lie superalgebra with non-degenerate Killing form or else it is a direct sum of such superalgebras.

Proof By generalizing the line of argument in the first paragraph of the proof of Theorem V of Chapter 13, Section 2 to make it apply to Lie superalgebras, it is easily shown that if $\mathscr{L}_s$ possesses an Abelian invariant subalgebra then the Killing form of $\mathscr{L}_s$ must be degenerate. Thus if $\mathscr{L}_s$ has non-degenerate Killing form then $\mathscr{L}_s$ has no Abelian invariant graded subalgebras, proper or otherwise.

If $\mathscr{L}_s$ has *no* proper invariant graded subalgebra then, since $\mathscr{L}_s$ is not Abelian, $\mathscr{L}_s$ must be simple. It therefore remains to consider the case in which $\mathscr{L}_s$ does possess at least one proper invariant graded subalgebra.

Let $\mathscr{L}'_s$ be a minimal invariant graded subalgebra of $\mathscr{L}_s$. ("Minimal" means that $\mathscr{L}'_s$ does not possess any proper graded subalgebra that is an invariant graded subalgebra of $\mathscr{L}_s$.) By Theorem III of Chapter 21, Section 5, the orthogonal complement $\mathscr{L}'^{\perp}_s$ of $\mathscr{L}'_s$ with respect to the Killing form of $\mathscr{L}_s$ is also an invariant graded subalgebra of $\mathscr{L}_s$. Consequently the intersection $\mathscr{L}'_s \cap \mathscr{L}'^{\perp}_s$ (the set of elements common to both $\mathscr{L}'_s$ and $\mathscr{L}'^{\perp}_s$) is also an invariant subalgebra of $\mathscr{L}_s$, but, since $\mathscr{L}'_s$ is minimal, $\mathscr{L}'_s \cap \mathscr{L}'^{\perp}_s$ is either the trivial set $\{0\}$ or $\mathscr{L}'_s$ itself. In the latter case $\mathscr{L}'_s$ must be a subset of $\mathscr{L}'^{\perp}_s$, and so for *any* $a \in \mathscr{L}_s$ and any $b, c \in \mathscr{L}'_s$, since $B([a,[b,c]]) = B([a,b],c)$ (by (21.86)) and $[a,b] \in \mathscr{L}'_s$, so that $[a,b] \in \mathscr{L}'^{\perp}_s$, it follows that $B([a,[b,c]]) = 0$. As the Killing form of $\mathscr{L}_s$ is assumed to be non-degenerate, this implies that $[b,c] = 0$ for all $b, c \in \mathscr{L}'_s$, so that $\mathscr{L}'_s$ has to be an Abelian invariant graded subalgebra of $\mathscr{L}_s$. However, it has already been shown that $\mathscr{L}_s$ has no subalgebras of this type, so $\mathscr{L}'_s \cap \mathscr{L}'^{\perp}_s$ cannot be equal to $\mathscr{L}'_s$.

The only possibility that remains is that $\mathscr{L}'_s \cap \mathscr{L}'^{\perp}_s = \{0\}$, which implies that $\mathscr{L}_s$ is the vector space direct sum of $\mathscr{L}'_s$ and $\mathscr{L}'^{\perp}_s$. However, if $a' \in \mathscr{L}'_s$ and $b' \in \mathscr{L}'^{\perp}_s$, since both are invariant subalgebras of $\mathscr{L}_s$, it follows that $[a', b'] \in \mathscr{L}'_s \cap \mathscr{L}'^{\perp}_s (= \{0\})$, and hence $[a', b'] = 0$. Thus

$$\mathscr{L}_s = \mathscr{L}'_s \oplus \mathscr{L}'^{\perp}_s,$$

which is a Lie superalgebra direct sum. This shows that any invariant graded subalgebra of $\mathscr{L}'_s$ or $\mathscr{L}'^{\perp}_s$ would also be an invariant graded subalgebra of $\mathscr{L}_s$. Thus $\mathscr{L}'_s$ possesses no proper invariant graded subalgebras, and, since $\mathscr{L}'_s$ is not Abelian, it follows that $\mathscr{L}'_s$ *must be simple*. By part (g) of Theorem III of Chapter 21, Section 5, the Killing form of $\mathscr{L}'_s$ is equal to the Killing form of $\mathscr{L}_s$ restricted to $\mathscr{L}'_s$, and this is clearly non-degenerate.

The argument can then be repeated with $\mathscr{L}_s$ replaced by $\mathscr{L}'^{\perp}_s$, and so on.

Theorem III If $B'(\ , \ )$ and $B''(\ , \ )$ are any two non-degenerate bilinear supersymmetric consistent invariant forms on a simple Lie superalgebra $\mathscr{L}_s$ then

$$B''(a,b) = \lambda B'(a,b) \tag{25.2}$$

for all $a, b \in \mathscr{L}_s$, where λ is a constant independent of a and b.

Proof For all $a, b \in \mathscr{L}_s$ let $\psi(a) \in \mathscr{L}_s$ be defined by

$$B''(a,b) = B'(\psi(a), b). \tag{K.9}$$

Then ψ is a *linear* operator acting on $\mathscr{L}_s$. Moreover, if $a \in \mathscr{L}_0$ then, writing $\psi(a) = \psi(a)_{\text{even}} + \psi(a)_{\text{odd}}$ (where $\psi(a)_{\text{even}} \in \mathscr{L}_0$ and $\psi(a)_{\text{odd}} \in \mathscr{L}_1$), it follows from (21.84) (with $B(\ , \ )$ replaced by $B''(\ , \ )$) that if $b \in \mathscr{L}_1$ then

$$0 = B''(a,b) = B'(\psi(a)_{\text{odd}}, b).$$

But $B'(\psi(a)_{\text{odd}}, b) = 0$ if $b \in \mathscr{L}_0$, so $B'(\psi(a)_{\text{odd}}, b) = 0$ for *all* $b \in \mathscr{L}_s$. However, $B'(\ , \ )$ is assumed to be non-degenerate, so $\psi(a)_{\text{odd}} = 0$. This shows that if $a \in \mathscr{L}_0$ then $\psi(a) \in \mathscr{L}_0$. Similarly if $a \in \mathscr{L}_1$ then $\psi(a) \in \mathscr{L}_1$.

As $B''(\ , \ )$ is assumed to be invariant,

$$B''([a,b],c) = B''(a,[b,c])$$

for all $a, b, c \in \mathscr{L}_s$. Thus, by (K.9),

$$B'(\psi([a,b]),c) = B'(\psi(a),[b,c]),$$

and hence, by (21.86),

$$B'(\psi([a,b]),c) = B'([\psi(a),b],c).$$

As $B'(\ , \ )$ is assumed to be non-degenerate, this implies that

$$\psi([a,b]) = [\psi(a),b] \tag{K.10}$$

for all $a, b \in \mathscr{L}_s$. Let λ be an eigenvalue of ψ and let S_λ be the corresponding subspace of eigenvectors, so that $a \in S_\lambda$ if $\psi(a) = \lambda a$. Then, by (K.10), if $a \in S_\lambda$ then $[a,b] \in S_\lambda$ for all $b \in \mathscr{L}_s$. Thus S_λ is an invariant subalgebra of $\mathscr{L}_s$, and, since $\psi(a) \in \mathscr{L}_0$ if $a \in \mathscr{L}_0$ and $\psi(a) \in \mathscr{L}_1$ if $a \in \mathscr{L}_1$, S_λ must be an invariant graded subalgebra of $\mathscr{L}_s$. As $\mathscr{L}_s$ is simple, either $S_\lambda = \{0\}$ or $S_\lambda = \mathscr{L}_s$. But since ψ is a linear operator acting in a finite-dimensional space, ψ must possess at least one eigenvalue λ with $S_\lambda \neq \{0\}$. For this eigenvalue $S_\lambda = \mathscr{L}_s$ (which indicates that there is no other eigenvalue of ψ), so that, by (K.9),

$$B''(a, b) = \lambda B'(a, b)$$

for all $a, b \in \mathscr{L}_s$.

Theorem IV If $\mathscr{L}_s$ is a simple Lie superalgebra with even subspace $\mathscr{L}_0$ and non-trivial odd subspace $\mathscr{L}_1$ then

(a) the representation of $\mathscr{L}_0$ on $\mathscr{L}_1$ is faithful,

(b) $[\mathscr{L}_1, \mathscr{L}_1] = \mathscr{L}_0$, and

(c) $[\mathscr{L}_0, \mathscr{L}_1] = \mathscr{L}_1$.

Proof (a) For each $a \in \mathscr{L}_0$ let $\mathbf{D}(a)$ be the representation of $\mathscr{L}_0$ on $\mathscr{L}_1$ as defined in (21.16). Suppose that this is *not* faithful, so that there exist two elements a and a' of $\mathscr{L}_0$ with $a \neq a'$ such that $\mathbf{D}(a) = \mathbf{D}(a')$. Then $\mathbf{D}(a - a') = \mathbf{0}$ with $a - a' \neq 0$. Let $\mathscr{L}'_0$ be the subset of elements $b \in \mathscr{L}_0$ such that $\mathbf{D}(b) = \mathbf{0}$. Clearly $\mathscr{L}'_0$ is an invariant subalgebra of $\mathscr{L}_0$, and since (21.16) shows that $[b, c] = 0$ for all $b \in \mathscr{L}'_0$ and $c \in \mathscr{L}_1$, it follows that $\mathscr{L}'_0$ is an invariant graded subalgebra of $\mathscr{L}_s$, which is impossible if $\mathscr{L}_s$ is simple. Thus the representation of $\mathscr{L}_0$ on $\mathscr{L}_1$ must be faithful.

(b) The fourth statement of (21.6) shows that $[\mathscr{L}_1, \mathscr{L}_1]$ is a subset of $\mathscr{L}_0$. With $\mathscr{L}_1$ non-trivial, Theorem II of Chapter 21, Section 3 demonstrates that $[\mathscr{L}_1, \mathscr{L}_1] \oplus \mathscr{L}_1$ is an invariant graded subalgebra of $\mathscr{L}_s$. With $\mathscr{L}_s$ assumed to be simple, this subalgebra must coincide with $\mathscr{L}_s$, which is only possible if $[\mathscr{L}_1, \mathscr{L}_1] = \mathscr{L}_0$.

(c) The proof is similar to that of (b).

Theorem V If $\mathscr{L}_s$ is a Lie superalgebra with non-trivial even subspace $\mathscr{L}_0$ and non-trivial odd subspace $\mathscr{L}_1$ and if

(a) the representation of $\mathscr{L}_0$ on $\mathscr{L}_1$ is faithful and irreducible, and
(b) $[\mathscr{L}_1, \mathscr{L}_1] = \mathscr{L}_0$,

then $\mathscr{L}_s$ is simple.

Proof It has to be shown that if $\mathscr{L}_s$ is not simple then at least one of the conditions (a) and (b) is violated. Suppose first that $\mathscr{L}_s$ is Abelian. Then $[\mathscr{L}_1, \mathscr{L}_1] = \{0\}$ and the representation of $\mathscr{L}_0$ on $\mathscr{L}_1$ is such that $\mathbf{D}(a) = \mathbf{0}$ for all $a \in \mathscr{L}_0$ (cf. (21.16)), so both (a) and (b) are violated. The other possibility is that $\mathscr{L}_s$ possesses a proper invariant graded subalgebra $\mathscr{L}'_s$. Suppose that this is so, and that $\mathscr{L}'_0$ and $\mathscr{L}'_1$ are the even and odd subspaces of $\mathscr{L}'_s$. Then if $\mathscr{L}'_1$ is a *non-trivial proper* subspace of $\mathscr{L}_1$, the representation of $\mathscr{L}_0$ on $\mathscr{L}_1$ must be reducible, which is contrary to proposition (a). On the other hand, if $\mathscr{L}'_1 = \{0\}$ then $\mathscr{L}'_0$ must be a non-trivial subspace of $\mathscr{L}_0$ for which $[a', b] \in \mathscr{L}'_0$ for all $a' \in \mathscr{L}'_0$ and $b \in \mathscr{L}_s$, implying that $[a', b] = 0$ for all $a' \in \mathscr{L}'_0$ and $b \in \mathscr{L}_1$, so that $\mathbf{D}(a') = \mathbf{0}$ for all $a' \in \mathscr{L}'_0$, and so that the proposition (a) is again violated. Finally, if $\mathscr{L}'_1 = \mathscr{L}_1$ then $\mathscr{L}'_0$ must be a *proper* subspace of $\mathscr{L}_0$; but if $a, b \in \mathscr{L}_1$ then $a, b \in \mathscr{L}'_1$ and so $[a, b] \in \mathscr{L}'_0$, which contradicts the part of proposition (b) that asserts that every $c \in \mathscr{L}_0$ can be written in the form $c = [a, b]$ for some $a, b \in \mathscr{L}_1$.

5 Proofs of Theorems VI, VII, VIII, IX, XVI, XX, XXII and XXIII of Chapter 25, Sections 3(*a*), 3(*b*) and 3(*c*)

Theorem VI If $\mathscr{L}_s$ is a classical simple Lie superalgebra and if the representation of $\mathscr{L}_0$ on $\mathscr{L}_1$ is *irreducible* then its even part $\mathscr{L}_0$ is *semi-simple.*

Proof As $\mathscr{L}_s$ is classical its even part $\mathscr{L}_0$ is given by $\mathscr{L}_0^{\mathrm{A}} \oplus \mathscr{L}_0^{\mathrm{ss}}$ (cf. (25.8)). Suppose that $\mathscr{L}_0$ is *not* semi-simple, so that $\mathscr{L}_0^{\mathrm{A}} \neq \{0\}$, and that the representation of $\mathscr{L}_0$ on $\mathscr{L}_1$ is *irreducible.* It will be shown that a contradiction ensues. Let $c \in \mathscr{L}_0^{\mathrm{A}}$ and let $\mathbf{D}$ be the representation of $\mathscr{L}_0$ on $\mathscr{L}_1$ (as defined in (21.16)). Thus for any $a \in \mathscr{L}_0$, since $[a, c]_- = 0$, it follows that $\mathbf{D}(a)\mathbf{D}(c) = \mathbf{D}(c)\mathbf{D}(a)$, and hence, by Schur's lemma for Lie algebra representations, $\mathbf{D}(c) = \gamma \mathbf{1}$, where γ is some constant. Then, by (21.16),

$$[c, b] = \gamma b \tag{K.11}$$

for all $b \in \mathscr{L}_1$. Part (a) of Theorem IV of Chapter 25, Section 2 then shows that $\gamma \neq \{0\}$ (since if $\gamma = 0$ the representation of $\mathscr{L}_0$ on $\mathscr{L}_1$ would not be faithful).

As $[\mathscr{L}_1, \mathscr{L}_1] = \mathscr{L}_0$ (by part (b) of Theorem IV of Chapter 25, Section 2), it follows that for any $a \in \mathscr{L}_0^{\mathrm{ss}}$ there exist elements b' and b'' of $\mathscr{L}_1$ such that $a = [b', b'']$. Then, by (21.3), (21.8) and (K.11),

$$[c, a] = -[b', [b'', c]] + [b'', [c, b']] = 2\gamma a,$$

which conflicts with the commutation relation $[c, a] = 0$.

Theorem VII If $\mathscr{L}_s$ is a classical simple Lie superalgebra then the representation of $\mathscr{L}_0$ on $\mathscr{L}_1$ is *either*

(i) irreducible

or

(ii) equivalent to the direct sum of *two* irreducible representations of $\mathscr{L}_0$.

Moreover, in case (ii), if $\mathscr{L}_1^1$ and $\mathscr{L}_1^2$ are the invariant subspaces of $\mathscr{L}_1$ corresponding to these two irreducible representations then

$$[\mathscr{L}_1^1, \mathscr{L}_1^1] = [\mathscr{L}_1^2, \mathscr{L}_1^2] = 0 \tag{25.6}$$

and

$$[\mathscr{L}_1^1, \mathscr{L}_1^2] = \mathscr{L}_0. \tag{25.7}$$

Proof It will be shown that if the representation of $\mathscr{L}_0$ on $\mathscr{L}_1$ is reducible then it is equivalent to the direct sum of only *two* irreducible representations. The relations (25.6) and (25.7) will appear in the course of the argument.

Suppose that the representation of $\mathscr{L}_0$ on $\mathscr{L}_1$ is equivalent to the direct sum of r irreducible representations of $\mathscr{L}_0$, where $r \geqslant 2$. This implies that $\mathscr{L}_1$ contains r subspaces $\mathscr{L}_1^p$ ($p = 1, 2, \ldots, r$) such that if $b^p \in \mathscr{L}_1^p$ then $[a, b^p] \in \mathscr{L}_1^p$ for all $a \in \mathscr{L}_0$. It is easily shown that for each p ($= 1, 2, \ldots, r$) the subspace $[\mathscr{L}_1^p, \mathscr{L}_1^p] \oplus [\mathscr{L}_1^p, [\mathscr{L}_1^p, \mathscr{L}_1^p]]$ is a proper invariant graded subalgebra of $\mathscr{L}_s$, and, since $\mathscr{L}_s$ is simple, this implies that

$$[\mathscr{L}_1^p, \mathscr{L}_1^p] = \{0\} \tag{K.12}$$

for all $p = 1, 2, \ldots, r$. The argument can be repeated with $\mathscr{L}_1^p$ replaced by

$$\mathscr{L}_1'^p = \sum_{q=1, q \neq p}^{r} \oplus \mathscr{L}_1^q$$

(which is the direct sum of all the subspaces in $\mathscr{L}_1$ except $\mathscr{L}_1^p$), giving

$$\sum_{q,q'=1; q,q' \neq p}^{r} [\mathscr{L}_1'^q, \mathscr{L}_1'^{q'}] = \{0\} \tag{K.13}$$

for all $p = 1, 2, \ldots, r$.

For the case $r = 2$ (K.12) gives

$$[\mathscr{L}_1^1, \mathscr{L}_1^1] = [\mathscr{L}_1^2, \mathscr{L}_1^2] = \{0\}$$

(i.e. (25.6)), and (K.13) merely repeats these results. As $[\mathscr{L}_1, \mathscr{L}_1] = \mathscr{L}_0$ (part (b) of Theorem IV of Chapter 25, Section 2) and $\mathscr{L}_1 = \mathscr{L}_1^1 \oplus \mathscr{L}_1^2$, it follows from (25.6) that

$$[\mathscr{L}_1^1, \mathscr{L}_1^2] = \mathscr{L}_0,$$

which is (25.7).

However, for the case $r = 3$, (K.12) gives

$$[\mathscr{L}_1^1, \mathscr{L}_1^1] = [\mathscr{L}_1^2, \mathscr{L}_1^2] = [\mathscr{L}_1^3, \mathscr{L}_1^3] = \{0\}, \tag{K.14}$$

while for $p = 1$ (K.13) gives

$$[\mathscr{L}_1^2 \oplus \mathscr{L}_1^3, \mathscr{L}_1^2 \oplus \mathscr{L}_1^3] = \{0\},$$

so that, by (K.14),

$$[\mathscr{L}_1^2, \mathscr{L}_1^3] = \{0\}.$$

Similarly (K.13) for $p = 2$ and $p = 3$ taken with (K.14) gives

$$[\mathscr{L}_1^1, \mathscr{L}_1^3] = [\mathscr{L}_1^1, \mathscr{L}_1^2] = \{0\}.$$

Thus for $r = 3$

$$[\mathscr{L}_1^p, \mathscr{L}_1^q] = \{0\}$$

for all $p, q = 1, 2, 3$. But $\mathscr{L}_1 = \mathscr{L}_1^1 \oplus \mathscr{L}_1^2 \oplus \mathscr{L}_1^3$, so that it is impossible to satisfy the relation $[\mathscr{L}_1, \mathscr{L}_1] = \mathscr{L}_0$ (cf. part (b) of Theorem IV of Chapter 25, Section 2) for $r = 3$. A similar argument applies for any $r > 3$. Thus the representation of $\mathscr{L}_0$ on $\mathscr{L}_1$ can contain at most *two* irreducible representations.

Theorem VIII If $\mathscr{L}_s$ is a complex classical simple Lie superalgebra and the centre of its even part $\mathscr{L}_0$ is *non-trivial* then

(i) $\mathscr{L}_0^{\mathrm{A}}$ is *one*-dimensional, and

(ii) the representation of $\mathscr{L}_0$ on $\mathscr{L}_1$ is equivalent to the direct sum of *two* irreducible representations of $\mathscr{L}_0$.

Moreover, if $\mathscr{L}_1^1$ and $\mathscr{L}_1^2$ are the two invariant subspaces of $\mathscr{L}_1$ corresponding to the two irreducible representations then there exists an element c of $\mathscr{L}_0^{\mathrm{A}}$ such that

$$[c, b^1] = b^1 \tag{25.8}$$

for all $b^1 \in \mathscr{L}_1^1$, and

$$[c, b^2] = -b^2 \tag{25.9}$$

for all $b^2 \in \mathscr{L}_1^2$.

Proof If $\mathscr{L}_0^{\mathrm{A}}$ is non-trivial (i.e. $\mathscr{L}_0^{\mathrm{A}} \neq \{0\}$) then Theorem VI shows that the representation of $\mathscr{L}_0$ on $\mathscr{L}_1$ must be reducible, and then Theorem VII demonstrates that it must be equivalent to the direct sum of *two* irreducible representations of $\mathscr{L}_0$. Let $\mathbf{D}^1$ and $\mathbf{D}^2$ be these two irreducible representations, and suppose that the basis elements of $\mathscr{L}_1$ are chosen so that

$$\mathbf{D}(a) = \begin{bmatrix} \mathbf{D}^1(a) & \mathbf{0} \\ \mathbf{0} & \mathbf{D}^2(a) \end{bmatrix}$$

for all $a \in \mathscr{L}_0$ (cf. (21.16)). Then Schur's Lemma for Lie algebras (Theorem II of Chapter 11, Section 4) implies that if $c' \in \mathscr{L}_0^{\mathrm{A}}$ then

$$\mathbf{D}(c') = \begin{bmatrix} \kappa_1' \mathbf{1} & \mathbf{0} \\ \mathbf{0} & \kappa_2' \mathbf{1} \end{bmatrix}, \tag{K.15}$$

where κ_1' and κ_2' are two complex numbers. Thus if $\mathscr{L}_1^1$ and $\mathscr{L}_1^2$ are the two invariant subspaces of $\mathscr{L}_1$ corresponding to $\mathbf{D}^1$ and $\mathbf{D}^2$ then (21.16) and (K.15) together imply that

$$[c', b^1] = \kappa_1' b^1 \tag{K.16}$$

for all $b^1 \in \mathscr{L}_1^1$, and

$$[c', b^2] = \kappa_2' b^2 \tag{K.17}$$

for all $b^2 \in \mathscr{L}_1^2$. However, by (25.7), for *any* $a \in \mathscr{L}_0^{\mathrm{ss}}$ there exist elements b^1 and b^2 of $\mathscr{L}_1^1$ and $\mathscr{L}_1^2$ respectively such that $a = [b^1, b^2]$. Then, by (21.3), (21.8), (K.16) and (K.17),

$$[c', a] = -[b^1, [b^2, c']] + [b^2, [c', b^1]] = (\kappa_1' + \kappa_2')a.$$

However, by (25.8), $[c', a] = 0$ for $c' \in \mathscr{L}_0^{\mathrm{A}}$ and $a \in \mathscr{L}_0^{\mathrm{ss}}$, so $\kappa_1' + \kappa_2' = 0$, and hence, from (K.15),

$$\mathbf{D}(c') = \begin{bmatrix} \kappa_1' \mathbf{1} & \mathbf{0} \\ \mathbf{0} & -\kappa_1' \mathbf{1} \end{bmatrix}. \tag{K.18}$$

Thus, defining $c (\in \mathscr{L}_0^{\mathrm{A}})$ by $c = c'/\kappa_1'$, (K.18) gives

$$\mathbf{D}(c) = \begin{bmatrix} \mathbf{1} & \mathbf{0} \\ \mathbf{0} & -\mathbf{1} \end{bmatrix},$$

which (by (21.16)) implies (25.9) and (25.10).

Now suppose that $\mathscr{L}_0^{\mathrm{A}}$ is of dimension greater than one and c'' is an element of $\mathscr{L}_0^{\mathrm{A}}$ that is linearly independent of c. The above argument can then be repeated to give

$$\mathbf{D}(c'') = \begin{bmatrix} \kappa_1'' \mathbf{1} & \mathbf{0} \\ \mathbf{0} & -\kappa_1'' \mathbf{1} \end{bmatrix}$$

(cf. (K.18)), where κ_1'' is some complex number, and hence that

$$\mathbf{D}(c''/\kappa_1'') = \begin{bmatrix} \mathbf{1} & \mathbf{0} \\ \mathbf{0} & -\mathbf{1} \end{bmatrix}.$$

Thus in the representation of $\mathscr{L}_0$ on $\mathscr{L}_1$ both c''/κ_1'' and c are represented by the same matrix. This conflicts with part (a) of Theorem IV of Chapter 21,

Section 2, which requires that this representation be faithful, so $\mathscr{L}_0^A$ *cannot* have dimension greater than one.

Theorem IX If $\mathscr{L}_s$ is a simple Lie superalgebra and the Killing form of $\mathscr{L}_s$ is *non-degenerate* then $\mathscr{L}_s$ is *classical.*

Proof For every $a \in \mathscr{L}_s$ the adjoint representation **ad** of $\mathscr{L}_s$ has the partitioned form

$$\mathbf{ad}(a) = \begin{bmatrix} \mathbf{ad}_0(a) & \mathbf{0} \\ \mathbf{0} & \mathbf{ad}_1(a) \end{bmatrix},$$

where $\mathbf{ad}_0$ is the adjoint representation of $\mathscr{L}_0$ (cf. (11.36)) and $\mathbf{ad}_1$ is the representation of $\mathscr{L}_0$ on $\mathscr{L}_1$ (denoted in (21.16) and elsewhere by **D**). By (21.78) and (21.80), if $B(\ ,\ )$ is the Killing form of $\mathscr{L}_s$ then for all $a, b \in \mathscr{L}_0$

$$B(a, b) = B_0(a, b) - B_1(a, b), \tag{K.19}$$

where

$$\mathrm{B}_0(a, b) = \mathrm{tr}\,\{\mathbf{ad}_0(a)\,\mathbf{ad}_0(b)\}$$

and

$$B_1(a, b) = \mathrm{tr}\,\{\mathbf{ad}_1(a)\,\mathbf{ad}_1(b)\}$$

(cf. (25.3)). Let $\mathscr{L}_{00}^{\perp}$ and $\mathscr{L}_{01}^{\perp}$ be the orthogonal complements of $\mathscr{L}_0$ with respect to $B_0(\ ,\ )$ and $B_1(\ ,\ )$ respectively (cf. (25.4)). Then if $a \in \mathscr{L}_{00}^{\perp} \cap \mathscr{L}_{01}^{\perp}$, it follows that $B_0(a, b) = B_1(a, b) = 0$ for all $b \in \mathscr{L}_0$, and hence (by (K.19)) that $B(a, b) = 0$ for all $b \in \mathscr{L}_0$. As $B(a, b) = 0$ for all $b \in \mathscr{L}_1$ by (21.84) (since $a \in \mathscr{L}_0$), it follows that $B(a, b) = 0$ for *all* $b \in \mathscr{L}_s$. With $B(\ ,\ )$ assumed to be non-degenerate, this implies that $a = 0$, and so $\mathscr{L}_{00}^{\perp} \cap \mathscr{L}_{01}^{\perp} = \{0\}$. Theorem IV of Chapter 25, Section 3(*a*) then shows that the nilpotent radical of $\mathscr{L}_0$ is $\{0\}$, and so, by Theorem III of that section, $\mathscr{L}_0$ must be reductive. Theorem V of that section then shows that $\mathscr{L}_s$ must be classical.

Theorem XVI The Killing form of a basic classical simple complex Lie superalgebra $\tilde{\mathscr{L}}_s$ is *non-degenerate* if and only if

$$\gamma_{\mathrm{D}}^{j} \neq 1$$

for every simple Lie algebra $\tilde{\mathscr{L}}_0^{sj}$ in $\tilde{\mathscr{L}}_0$ (cf. (25.16) and (25.17)), γ_{D}^{j} being the Dynkin index of (25.20) and (25.21).

Proof Suppose that $\gamma_{\mathrm{D}}^{j} = 1$ for some $\tilde{\mathscr{L}}_0^{sj}$ in $\tilde{\mathscr{L}}_0$. Then, by (25.21), $B(a, b) = 0$ for all $a, b \in \tilde{\mathscr{L}}_0^{sj}$, and, in particular, $B(h, h') = 0$ for all h, h' of the Cartan subalgebra $\mathscr{H}_0^{j}$ of $\tilde{\mathscr{L}}_0^{sj}$. However, as $B(\ ,\ )$ provides a bilinear symmetric invariant form on the whole of $\tilde{\mathscr{L}}_0^{ss}$, $B(h, h') = 0$ if $h \in \mathscr{H}_0^{j}$ and $h' \in \mathscr{H}_0^{k}$, where $j \neq k$. Moreover, if $\tilde{\mathscr{L}}_0^{A} \neq \{0\}$ then the fact that $[c, a] = 0$ for all $a \in \tilde{\mathscr{L}}_0^{ss}$ taken with the results of Theorem VIII of Chapter 25, Section 3(a) show that

$$\mathbf{ad}(c) = \left[\begin{array}{c|cc} \mathbf{0} & \mathbf{0} & \mathbf{0} \\ \hline \mathbf{0} & \mathbf{1} & \mathbf{0} \\ \mathbf{0} & \mathbf{0} & -\mathbf{1} \end{array}\right],$$

where the two unit submatrices have the dimensions of the subspaces $\tilde{\mathscr{L}}_1^1$ and $\tilde{\mathscr{L}}_1^2$ of that theorem. Then, by (21.78) and (21.80), for all $h \in \mathscr{H}_0^j$,

$$B(h, c) = -\operatorname{tr}\{\mathbf{D}^1(h) - \mathbf{D}^2(h)\},$$

where $\mathbf{D}^1$ and $\mathbf{D}^2$ are the two irreducible representations of $\tilde{\mathscr{L}}_0$ corresponding to these subspaces. However, Theorem VI of Chapter 15, Section 2 shows that $\operatorname{tr}\mathbf{D}^1(h) = \operatorname{tr}\mathbf{D}^2(h) = 0$. Thus if $h \in \mathscr{H}_0^j$ then $B(h, h') = 0$ for *all* $h' \in \mathscr{H}_s$, implying that $B(\ ,\ )$ is degenerate on $\mathscr{H}_s$, and thereby conflicting with the result of Theorem XV of Chapter 25, Section 3(b). Thus $\gamma_{\mathbf{D}}^j \neq 1$ for every $\tilde{\mathscr{L}}_0^{sj}$ in $\tilde{\mathscr{L}}_0$. The converse is obvious.

Theorem XX If $\tilde{\mathscr{L}}_s$ is a basic classical simple complex Lie superalgebra and α is an *odd* root of $\tilde{\mathscr{L}}_s$ such that $\langle\alpha, \alpha\rangle \neq 0$ then 2α is an *even* root of $\tilde{\mathscr{L}}_s$ and

$$\dim \tilde{\mathscr{L}}_{s\alpha} = \dim \tilde{\mathscr{L}}_{s(-\alpha)} = 1. \tag{25.49}$$

However, 3α is never a root of $\tilde{\mathscr{L}}_s$.

Proof Suppose that $a_\alpha \in \tilde{\mathscr{L}}_{s\alpha}$ and $a_\alpha \in \tilde{\mathscr{L}}_1$. Theorem XVIII of Chapter 25, Section 3(b) shows that $-\alpha$ is also a root, which must necessarily be odd (although it could be even as well). Let $a_{-\alpha} \in \tilde{\mathscr{L}}_{s(-\alpha)}$ and $a_{-\alpha} \in \tilde{\mathscr{L}}_1$. Then, by (21.15),

$$[a_\alpha, [a_\alpha, a_{-\alpha}]] + [a_\alpha, [a_{-\alpha}, a_\alpha]] + [a_{-\alpha}, [a_\alpha, a_\alpha]] = 0.$$

However, by (25.30), this can be rewritten as

$$[a_{-\alpha}, [a_\alpha, a_\alpha]] + 2B'(a_\alpha, a_{-\alpha})[a_\alpha, h_\alpha] = 0.$$

But $[a_\alpha, h_\alpha] = \langle\alpha, \alpha\rangle a_\alpha$, so

$$[a_{-\alpha}, [a_\alpha, a_\alpha]] = -2B'(a_\alpha, a_{-\alpha})\langle\alpha, \alpha\rangle a_\alpha.$$

Thus if $\langle\alpha, \alpha\rangle \neq 0$ then it follows that $[a_\alpha, a_\alpha] \neq 0$, and hence, since $[a_\alpha, a_\alpha] \in \tilde{\mathscr{L}}_0$ and must correspond to root 2α, 2α must be an even root of $\tilde{\mathscr{L}}_s$.

By Theorem X of Chapter 25, Section 3(a), $\dim \tilde{\mathscr{L}}_{s(2\alpha)} = 1$, which implies that $\dim \tilde{\mathscr{L}}_{s\alpha} = 1$, and hence that $\dim \tilde{\mathscr{L}}_{s(-\alpha)} = 1$. 3α is not a root because the generalized Jacobi identity (21.15) shows that $[a_\alpha, [a_\alpha, a_\alpha]] = 0$.

Theorem XXII The Weyl group of $\tilde{\mathscr{L}}_s$ is isomorphic to the Weyl group of its semi-simple subalgebra $\tilde{\mathscr{L}}_0^{ss}$.

Proof Suppose that α is the extension of a root α^j of the simple complex Lie algebra $\tilde{\mathscr{L}}_0^{sj}$ (cf. (25.17)). Then, since $B'(\ ,\ )$ provides a bilinear symmetric invariant form on $\tilde{\mathscr{L}}_0^{sj}$, Theorem I of Chapter 16, Section 3 shows that there exists a complex number μ^j such that

$$B'(a, b) = \mu^j B_0^j(a, b) \tag{K.20}$$

for all $a, b \in \tilde{\mathscr{L}}_0^{sj}$, where $B_0^j(\ ,\ )$ is the Killing form of $\tilde{\mathscr{L}}_0^{sj}$. (If the Killing form $B(\ ,\ )$ of $\tilde{\mathscr{L}}_s$ is not identically zero and $B'(a,b) = B(a,b)$ for all $a, b \in \tilde{\mathscr{L}}_s$ then (25.21) implies that $\mu^j = 1 - \gamma_{\mathrm{D}}^j$. If $B(\ ,\ )$ is identically zero then (K.20) is still valid, but $\mu^j \neq 1 - \gamma_{\mathrm{D}}^j$.) Then

$$\langle \alpha, \alpha \rangle = \langle \alpha^j, \alpha^j \rangle_0^j / \mu^j, \tag{K.21}$$

and if λ^j is the restriction of λ to $\mathscr{H}_0^j$, the Cartan subalgebra of $\tilde{\mathscr{L}}_0^{sj}$, then

$$\langle \alpha, \lambda \rangle = \langle \alpha^j, \lambda^j \rangle_0^j / \mu^j. \tag{K.22}$$

(In (K.21) and (K.22) the quantities $\langle\ ,\ \rangle_0^j$ are evaluated from $B_0^j(\ ,\ )$ using (13.14).) Thus, by (25.72),

$$(S_\alpha \lambda)(h) = \lambda(h) - \frac{2\langle \lambda^j, \alpha^j \rangle_0^j}{\langle \alpha^j, \alpha^j \rangle_0^j} \alpha(h),$$

for all $h \in \mathscr{H}_s$. Thus

$$(S_\alpha \lambda)(h) = (S_{\alpha^j} \lambda^j)(h). \tag{K.23}$$

for all $h \in \mathscr{H}_0^j$, where the Weyl reflection on the right-hand side of (K.23) is the Weyl reflection of $\tilde{\mathscr{L}}_0^{ss}$ corresponding to α^j. This implies that the Weyl groups of $\tilde{\mathscr{L}}_0$ and $\tilde{\mathscr{L}}_0^{ss}$ are isomorphic.

Theorem XXIII For any element S of the Weyl group $\mathscr{W}$ of $\mathscr{L}_s$, if $\gamma \in \Delta_0$ then $S\gamma \in \Delta_0$, and if $\gamma \in \Delta_1$ then $S\gamma \in \Delta_1$.

Proof It is sufficient to prove these resuts for the case $S = S_\alpha$, where α is any member of Δ_0.

As the non-zero even roots of $\mathscr{L}_s$ are extensions of the non-zero roots of $\mathscr{L}_0^{ss}$, the quoted result for the *even* roots follow immediately from (K.23) and the corresponding result for non-zero roots of semi-simple complex Lie algebras that was presented in Chapter 13, Section 9. If $\gamma \in \Delta_1$ then its restriction γ^j to $\mathscr{H}_0^j$ is a weight of some irreducible representation of $\tilde{\mathscr{L}}_0^{sj}$, and so, by Theorem IV of Chapter 15, Section 2, $S_{\alpha^j}\gamma^j$ is a weight of the *same* irreducible representation of $\tilde{\mathscr{L}}_0^{sj}$. The stated result then follows from (K.23).

6 Proofs of Theorems III and IV of Chapter 25, Section 4(*a*)

Theorem III If $\tilde{\mathscr{L}}_s$ is a type I basic classical simple complex Lie superalgebra $A(r/s)$ (for $r > s \geqslant 0$) or $C(s)$ (for $s \geqslant 2$), or if $\tilde{\mathscr{L}}_s = \mathrm{sl}(r+1/r+1;\mathbb{C})$ (for $r \geqslant 1$), then the graded representation with highest weight Λ is of *finite* dimension if and only if the numerical mark n_j is a *non-negative integer* for every distinguished simple root α_j of $\tilde{\mathscr{L}}_s$ such that $\langle \alpha_j, \alpha_j \rangle \neq 0$.

Proof That the conditions stated are *necessary* for the graded representation to be finite-dimensional is easily demonstrated. It is clear that the graded representation of $\tilde{\mathscr{L}}_s$ provides a representation of its semi-simple Lie subalgebra $\tilde{\mathscr{L}}_0^{ss}$ and Λ restricted to $\mathscr{H}_0^{ss}$ is a highest weight of this representation of $\tilde{\mathscr{L}}_0^{ss}$. If α_j is a distinguished simple root such that $\langle \alpha_j, \alpha_j \rangle \neq 0$ then α_j is an *even* root of $\tilde{\mathscr{L}}_s$.

However, if α is any *even* root of $\tilde{\mathscr{L}}_s$, so that $\langle \alpha, \alpha \rangle \neq 0$, and if $H_\alpha (\in \mathscr{H}_s)$ is defined (cf. (13.25) and (25.63)) by $H_\alpha = \{2/\langle \alpha, \alpha \rangle\} h_\alpha$ (where h_α is defined by (25.25)), and if E_α and $E_{-\alpha}$ are elements of $\tilde{\mathscr{L}}_{s\alpha}$ and $\tilde{\mathscr{L}}_{s(-\alpha)}$ respectively such that $B'(E_\alpha, E_{-\alpha}) = 2/\langle \alpha, \alpha \rangle$ (cf. (13.26)), then it follows (as in (13.27)) that $[H_\alpha, E_\alpha]_- = 2E_\alpha$, $[H_\alpha, E_{-\alpha}] = -2E_{-\alpha}$ and $[E_\alpha, E_{-\alpha}]_- = H_\alpha$. Thus in a representation of $\tilde{\mathscr{L}}_s$ in which $\Phi(a)$ is the operator corresponding to a for each $a \in \tilde{\mathscr{L}}_s$ the operators A_3, A_+ and A_- defined by $A_3 = \frac{1}{2}\Phi(H_\alpha)$, $A_+ = \Phi(E_\alpha)$ and $A_- = \Phi(E_{-\alpha})$ (cf. (13.28)) satisfy the commutation relations (12.14), (12.15) and (12.16), and hence form the basis of a subalgebra of $\tilde{\mathscr{L}}_s$ that is isomorphic to an A_1 simple complex Lie algebra. As shown in Theorem I of Chapter 12, Section 3, each finite-dimensional irreducible representation of this Lie algebra is specified by a non-negative integer or half-integer j (i.e. $j = 0, \frac{1}{2}, 1, \frac{3}{2}, \ldots$). Such a representation has dimension $d = 2j + 1$, the $2j + 1$ eigenvalues of A_3 being $m = j, j-1, \ldots, -j$. Thus j is the maximum eigenvalue of A_3, and hence the corresponding maximum eigenvalue of $\Phi(H_\alpha)$ is $2j$, which is a *non-negative integer*. Thus, putting $\alpha = \alpha_j$ and invoking (25.77), it follows immediately that n_j must be a *non-negative integer* for every simple root α_j of $\tilde{\mathscr{L}}_s$ such that $\langle \alpha_j, \alpha_j \rangle \neq 0$.

For a proof of the *sufficiency* of these conditions see Kac (1977a).

It is worth noting that the arguments given in this proof do *not* apply to any of the *odd* roots α of $\tilde{\mathscr{L}}_s$, since even if $\langle \alpha, \alpha \rangle \neq 0$ the equation for $[E_\alpha, E_{-\alpha}]$ would involve an anticommutator, so the resulting subalgebra would not be a Lie algebra. Indeed, if α is an odd root of $\tilde{\mathscr{L}}_s$ such that $\langle \alpha, \alpha \rangle = 0$, and if $h_\alpha (\in \mathscr{H}_s)$ is defined by (25.25) and a_α and $a_{-\alpha}$ are elements of $\tilde{\mathscr{L}}_{s\alpha}$ and $\tilde{\mathscr{L}}_{s(-\alpha)}$ respectively, then, by (25.15) and (25.29), $[h_\alpha, a_\alpha]_- = 0$, $[h_\alpha, a_{-\alpha}]_- = 0$ and, by (25.30), $[a_\alpha, a_{-\alpha}]_+ = B'(a_\alpha, a_{-\alpha})h_\alpha$, so h_α, a_α and $a_{-\alpha}$ form the basis of a 3-

dimensional Lie superalgebra. This has a set of irreducible graded representations with $d_0 = d_1 = 1$ that are such that

$$\Gamma(a_\alpha) = \begin{bmatrix} 0 & \eta_+ \\ 0 & 0 \end{bmatrix}, \quad \Gamma(a_{-\alpha}) = \begin{bmatrix} 0 & 0 \\ \eta_- & 0 \end{bmatrix}$$

and

$$\Gamma(h_\alpha) = \frac{2\eta_+\eta_-}{B'(a_\alpha, a_{-\alpha})} \begin{bmatrix} 1 & 0 \\ 0 & 1 \end{bmatrix},$$

where η_+ and η_- are *any* complex numbers.

Theorem IV If $\tilde{\mathscr{L}}_s$ is a type II basic classical simple complex Lie superalgebra then the graded representation with highest weight Λ is of *finite* dimension if and only if *all* the following conditions are satisfied.

(1) For *every* distinguished simple root α_j such that $\langle \alpha_j, \alpha_j \rangle \neq 0$ the numerical mark n_j must be a *non-negative integer*.

(2) If β is the extension of a simple root of $\tilde{\mathscr{L}}_0^{ss}$ that is not a simple root of $\tilde{\mathscr{L}}_s$ and if N_β is defined by

$$N_\beta = \frac{2\langle \Lambda, \beta \rangle}{\langle \beta, \beta \rangle} \tag{25.80}$$

then N_β must be a *non-negative integer*.

(3) (a) If $\tilde{\mathscr{L}}_s = B(r/s)$ (for $r \geqslant 1$ and $s \geqslant 1$) and if $N_\beta < r$ then

$$n_{s+N_\beta+1} = n_{s+N_\beta+2} = \cdots = n_{s+r} = 0.$$

(b) If $\tilde{\mathscr{L}}_s = D(r/s)$ (for $r \geqslant 2$ and $s \geqslant 1$), $D(2/1;\alpha)$, $F(4)$ or $G(3)$ then the consistency conditions are as detailed in the statment of the theorem that is given in Chapter 25, Section 4(*a*).

Proof Only the demonstration that (1)–(3) provide a set of *necessary* conditions will be presented here. For a proof of the *sufficiency* of these conditions see Kac (1977a). Conditions (1) and (2) arise in exactly the same way as the conditions of Theorem III above, so attention will be concentrated on the condition (3).

The essential ideas are best seen by considering the simplest non-trivial example $\tilde{\mathscr{L}}_s = B(2/1)$ (details of which can be found in Appendix M, Section 3). For this superalgebra $\beta = 2(\alpha_1 + \alpha_2 + \alpha_3)$ is the extension of the simple root of $\tilde{\mathscr{L}}_0^{s1} (= C_1)$ that is not a simple root of $\tilde{\mathscr{L}}_s$.

Consider first the linear functional $\Lambda - \alpha_1$. If this is a weight of $\tilde{\mathscr{L}}_s$ then it is a *highest* weight of $\tilde{\mathscr{L}}_0^{s1} (= C_1)$ (since otherwise $\Lambda - \alpha_1 + \beta (= \Lambda + \alpha_1 + 2\alpha_2 + 2\alpha_3)$ would be a weight of $\tilde{\mathscr{L}}_s$, which is not possible because Λ is assumed to be the highest weight of the irreducible representation of $\tilde{\mathscr{L}}_s$.) *If* this is a highest weight of $\tilde{\mathscr{L}}_0^{s1}(=C_1)$ then $2\langle \Lambda-\alpha,\beta\rangle/\langle\beta,\beta\rangle$ must be a non-negative integer. However, using the results of Appendix M, Section 3,

$$\frac{2\langle \Lambda-\alpha_1,\beta\rangle}{\langle\beta,\beta\rangle}=N_\beta-1,$$

which is non-negative only if $N_\beta \geqslant 1$. Thus if $N_\beta=0$, the linear functional $\Lambda-\alpha_1$ cannot be a weight of $\tilde{\mathscr{L}}_s$, which implies, by Theorem II of Chapter 25, Section 4(*a*), that

$$\Phi(e_{-\alpha_1})\psi(\Lambda)=0.$$

Then

$$\Phi(e_{\alpha_1})\Phi(e_{-\alpha_1})\psi(\Lambda)=0,$$

and since

$$[e_{\alpha_1},e_{-\alpha_1}]_+=B'(e_{\alpha_1},e_{-\alpha_1})h_{\alpha_1}$$

(by (25.30)) and

$$\Phi(e_{\alpha_1})\psi(\Lambda)=0,$$

it follows that

$$\Lambda(h_{\alpha_1})=0.$$

But, by (25.76),

$$\Lambda(h_{\alpha_1})=\langle\Lambda,\alpha_1\rangle=n_1\langle\alpha_1,\alpha_2\rangle,$$

so $N_\beta=0$ implies that $n_1=0$. However, using the results of Appendix M, Section 3, $N_\beta=n_1-n_2-\frac{1}{2}n_3$, so with $N_\beta=0$ and $n_1=0$, and with n_2 and n_3 non-negative, it follows that $n_2=0$ and $n_3=0$, which is the part of the consistency condition for $B(2/1)$ that is quoted for the case $N_\beta=0$.

Now consider the linear functional $\Lambda-2\alpha_1-\alpha_2(=\Lambda-\alpha_1-(\alpha_1+\alpha_2))$ and apply the same argument. (Here α_1 and $\alpha_1+\alpha_2$ are the only two odd roots of $B(2/1)$ of the form $\varepsilon_p^1-\varepsilon_q^2$ (cf. Appendix M, Section 3).) If $\Lambda-2\alpha_1-\alpha_2$ is a highest weight of $\tilde{\mathscr{L}}_0^{s1}$ $(=C_1)$ then $2\langle\Lambda-2\alpha_1-\alpha_2,\beta\rangle/\langle\beta,\beta\rangle$ must be a non-negative integer. However,

$$\frac{2\langle\Lambda-2\alpha_1-\alpha_2,\beta\rangle}{\langle\beta,\beta\rangle}=N_\beta-2,$$

which is non-negative only if $N_\beta\geqslant 2$. Thus if $N_\beta=0$ or 1, the linear functional $\Lambda-\alpha_1-\alpha_2$ cannot be a weight of $\tilde{\mathscr{L}}_s$. If $N_\beta=0$ then the linear functional $\Lambda-\alpha_1-\alpha_2$ is not a weight because $\Lambda-\alpha_1$ is not a weight of $\tilde{\mathscr{L}}_s$ (this being the situation just considered above), but if $N_\beta=1$ then the quantity $\Lambda-\alpha_1-\alpha_2$ is not a weight of $\tilde{\mathscr{L}}_s$, even though $\Lambda-\alpha_1$ is a weight. This means that

$$\Phi(e_{-(\alpha_1+\alpha_2)})\Phi(e_{-\alpha_1})\psi(\Lambda)=0$$

but

$$\Phi(e_{-\alpha_1})\psi(\Lambda)\neq 0.$$

The latter condition indicates that $n_1 \neq 0$, while the former implies by an extension of the argument given above that

$$\{\langle \Lambda, \alpha_1 + \alpha_2 \rangle - \langle \alpha_1, \alpha_1 + \alpha_2 \rangle\} \langle \Lambda, \alpha_1 \rangle = 0,$$

from which it may be deduced, using the results of Appendix M, Section 3, that $n_1 - n_2 = 1$ if $N_\beta = 1$. Thus, since $N_\beta = n_1 - n_2 - \frac{1}{2}n_3$, it follows that $n_3 = 0$, which is the part of the consistency condition for $B(2/1)$ quoted above for $N_\beta = 1$.

The origin of the consistency conditions for all the other type II superalgebras is similar.

Appendix L

Clifford Algebras

1 The Clifford algebras of D-dimensional space–times

Consider a D-dimensional Euclidean or pseudo-Euclidean space with a covariant metric tensor η_{pq}, where

$$\eta_{pq} = \pm\,\delta_{pq} \tag{L.1}$$

for $p, q = 1, 2, \ldots, D$. Although the theory can be developed for any such space, the ones of most interest are the D-dimensional Euclidean space, for which

$$\eta_{pq} = \delta_{pq} \tag{L.2}$$

for $p, q = 1, 2, \ldots, D$, and D-dimensional Minkowski space–time, for which it will be assumed that

$$\eta_{pq} = \begin{cases} -1 & \text{for } p = q = 1, 2, \ldots, D-1, \\ +1 & \text{for } p = q = D, \\ 0 & \text{for } p \neq q. \end{cases} \tag{L.3}$$

The Clifford algebra of such a space is defined to have a basis consisting of the identity 1, D generators γ_p (for $p = 1, 2, \ldots, D$), which are assumed to satisfy the conditions

$$\gamma_p\gamma_q + \gamma_q\gamma_p = 2\eta_{pq}1 \tag{L.4}$$

(for $p, q = 1, 2, \ldots, D$), and all independent products of these generators. Because of the condition (L.4), there is only a finite number of such products, and indeed the Clifford algebra has the finite dimension 2^D. In particular (L.4) imply that

$$\gamma_p\gamma_q = -\gamma_q\gamma_p \quad (\text{if } p \neq q) \tag{L.5}$$

and

$$(\gamma_p)^2 = \eta_{pq}1, \tag{L.6}$$

(for $p = 1, 2, \ldots, D$). The "highest" product is $\gamma_1\gamma_2\gamma_3\cdots\gamma_{D-1}\gamma_D$.

Example I *Clifford algebras for $D = 4$*

For $D = 4$ the 16 ($= 2^4$) linearly independent basis elements are

$$1;$$
$$\gamma_1, \gamma_2, \gamma_3, \gamma_4;$$
$$\gamma_1\gamma_2, \gamma_1\gamma_3, \gamma_1\gamma_4, \gamma_2\gamma_3, \gamma_2\gamma_4, \gamma_3\gamma_4;$$
$$\gamma_1\gamma_2\gamma_3, \gamma_1\gamma_2\gamma_4, \gamma_1\gamma_3\gamma_4, \gamma_2\gamma_3\gamma_4;$$
$$\gamma_1\gamma_2\gamma_3\gamma_4.$$

Any other product of generators is a multiple of one of these. For example $\gamma_2\gamma_1 = -\gamma_1\gamma_2$ and $\gamma_1\gamma_2\gamma_1\gamma_4 = -\gamma_1\gamma_1\gamma_2\gamma_4 = -(\eta_{11}1)\gamma_2\gamma_4 = -\eta_{11}\gamma_2\gamma_4$.

Let $\boldsymbol{\gamma}_p$ (for $p = 1, 2, \ldots, D$) be a set of $d \times d$ matrices that form a matrix representation of the algebra, so that

$$\boldsymbol{\gamma}_p\boldsymbol{\gamma}_q + \boldsymbol{\gamma}_q\boldsymbol{\gamma}_p = 2\eta_{pq}\mathbf{1}_d \tag{L.7}$$

(for $p, q = 1, 2, \ldots, D$). In the case of $D = 4$ Minkowski space–time this is the condition satisfied by the Dirac matrices (cf. (17.114)). This appendix will be largely devoted to describing those aspects of the representation theory of Clifford algebras that are relevant to the study of Poincaré superalgebras and supergroups.

Clearly if the matrices $\boldsymbol{\gamma}_p$ form a d-dimensional representation of a Clifford algebra then so too do the $d \times d$ matrices

$$\boldsymbol{\gamma}'_p = \mathbf{S}\boldsymbol{\gamma}_p\mathbf{S}^{-1} \tag{L.8}$$

for any non-singular $d \times d$ matrix $\mathbf{S}$, and these two representations are said to be equivalent. It is also obvious that if the matrices $\boldsymbol{\gamma}_p$ form a d-dimensional representation of a Clifford algebra then so too do their negatives $-\boldsymbol{\gamma}_p$, their transposes $\tilde{\boldsymbol{\gamma}}_p$, their complex conjugates $\boldsymbol{\gamma}_p^*$, and all sets of matrices formed by taking any combination of these operations. Moreover in the Euclidean case the indices p may be permuted in any way.

If $\boldsymbol{\gamma}_p^{\text{Euclidean}}$ are a set of $d \times d$ matrices satisfying the D-dimensional

Euclidean Clifford algebra then the set of $d \times d$ matrices defined by

$$\gamma_p^{\text{Minkowski}} = (\eta_{pp})^{1/2} \gamma_p^{\text{Euclidean}} \tag{L.9}$$

(for $p = 1, 2, \ldots, D$) form a d-dimensional representation of the D-dimensional Minkowski Clifford algebra. Consequently all questions concerning representations are equivalent for the Euclidean and Minkowski cases, *except* for those relating to the *reality* of the representations. As the D-dimensional Minkowski case is the one that is of greatest interest in connection with supersymmetries, most of the rest of this appendix will be devoted to the study of its Clifford algebra.

It is easily checked that the set of $d \times d$ matrices $\Gamma^{\text{spin}}(L_{pq})$ defined by

$$\Gamma^{\text{spin}}(L_{pq}) = -\tfrac{1}{4}[\gamma_p, \gamma_q]_- \tag{L.10}$$

(for $p, q = 1, 2, \ldots, D$) form a d-dimensional representation of the real Lie algebra so$(D-1, 1)$, which is of course the D-dimensional homogeneous Lorentz algebra. Here L_{pq} (for $p, q = 1, 2, \ldots, D$, with $p < q$) are the basis elements of so$(D-1, 1)$, and, with the assumption that

$$L_{qp} = -L_{pq}, \tag{L.11}$$

they satisfy the commutation relations

$$[L_{pq}, L_{rs}]_- = \eta_{pr} L_{qs} - \eta_{ps} L_{qr} - \eta_{qr} L_{ps} + \eta_{qs} L_{pr} \tag{L.12}$$

(cf. (17.13) and (23.30)). It then follows from (L.7) and (L.10) that

$$\begin{aligned} \Gamma^{\text{spin}}([L_{pq}, L_{rs}]_-) = {} & \eta_{pr} \Gamma^{\text{spin}}(L_{qs}) - \eta_{ps} \Gamma^{\text{spin}}(L_{qr}) \\ & - \eta_{qr} \Gamma^{\text{spin}}(L_{ps}) + \eta_{qs} \Gamma^{\text{spin}}(L_{pr}). \end{aligned} \tag{L.13}$$

It should be noted that, by virtue of (L.7), (L.10) can be rewritten as

$$\Gamma^{\text{spin}}(L_{pq}) = -\tfrac{1}{2} \gamma_p \gamma_q \tag{L.14}$$

(for $p, q = 1, 2, \ldots, D$, with $p \neq q$).

2 Irreducible representations for the case in which D is even

(a) *Explicit expressions for the matrices*

Theorem I If D is *even* then there exists (up to equivalence) only *one* irreducible representation of the D-dimensional Minkowski Clifford algebra. Its dimension d is given by

$$d = 2^{D/2}, \tag{L.15}$$

and a convenient explicit form for it is given (cf. Brauer and Weyl 1935) by

(i) for $1 \leqslant p \leqslant \frac{1}{2}D$

$$\boldsymbol{\gamma}_p^{\mathrm{B}} = (\eta_{pp})^{1/2}\boldsymbol{\sigma}_3 \otimes \boldsymbol{\sigma}_3 \otimes \boldsymbol{\sigma}_3 \otimes \cdots \otimes \boldsymbol{\sigma}_3 \otimes \boldsymbol{\sigma}_1 \otimes \mathbf{1}_2 \otimes \mathbf{1}_2 \otimes \cdots \otimes \mathbf{1}_2 \otimes \mathbf{1}_2, \tag{L.16}$$

which is a direct product of $\frac{1}{2}D$ 2×2 matrices in which the Pauli spin matrix $\boldsymbol{\sigma}_1$ appears in the pth position, the matrices in the first $p-1$ positions are all equal to $\boldsymbol{\sigma}_3$, and those in the last $\frac{1}{2}D - p$ positions are all equal to the 2×2 identity matrix $\mathbf{1}_2$;

(ii) for $\frac{1}{2}D + 1 \leqslant p \leqslant D$

$$\boldsymbol{\gamma}_p^{\mathrm{B}} = (\eta_{pp})^{1/2}\boldsymbol{\sigma}_3 \otimes \boldsymbol{\sigma}_3 \otimes \boldsymbol{\sigma}_3 \otimes \cdots \otimes \boldsymbol{\sigma}_3 \otimes \boldsymbol{\sigma}_2 \otimes \mathbf{1}_2 \otimes \mathbf{1}_2 \otimes \cdots \otimes \mathbf{1}_2 \otimes \mathbf{1}_2, \tag{L.17}$$

which is a direct product of $\frac{1}{2}D$ 2×2 matrices in which the matrix $\boldsymbol{\sigma}_2$ appears in the $p - \frac{1}{2}D$ position, the matrices in the first $p - \frac{1}{2}D - 1$ positions are all equal to $\boldsymbol{\sigma}_3$, and those in the last $\frac{3}{2}D - p$ positions are all equal to $\mathbf{1}_2$.

Proof It is easily checked that the matrices given in (L.16) and (L.17) do form a d-dimensional representation of the D-dimensional Minkowski Clifford algebra. It remains to show that it is irreducible, and that there are no irreducible representations that are not equivalent to it.

This can be done by introducing the corresponding *Clifford group*, which is defined to consist of all the basis elements of the Clifford algebra together with their negatives, so, for example, for $D = 2$ the Clifford group consists of $1, \gamma_1, \gamma_2, \gamma_1\gamma_2, -1, -\gamma_1, -\gamma_2$ and $-\gamma_1\gamma_2$. The Clifford group has order 2^{D+1}, this being twice the dimension of the Clifford algebra. It is clear that every representation of the Clifford algebra provides a representation of the corresponding Clifford group, but, as will be seen, the converse is not necessarily true. It is easily checked that the Clifford group possesses $2^D + 1$ classes, with 1 and -1 each being in classes of their own but every other generator or product of generators being in a class of two with its negative. (For example, for $D = 2$ the classes are $\{1\}, \{-1\}, \{\gamma_1, -\gamma_1\}, \{\gamma_2, -\gamma_2\}$ and $\{\gamma_1\gamma_2, -\gamma_1\gamma_2\}$.) It is clear that the Clifford group has 2^D *one*-dimensional irreducible representations, which are specified by $\boldsymbol{\Gamma}(\boldsymbol{\gamma}_p) = [\pm(\eta_{pp})^{1/2}]$ for $p = 1, 2, \ldots, D$ (the 2^D different irreducible representations corresponding to the 2^D different combinations of signs). Theorem VIII of Chapter 4, Section 6 then implies that there is only *one* further irreducible representation (which is unique up to equivalence), and Theorem VII of the same section indicates that its dimension d is such that $d^2 + 2^D = 2^{D+1}$. Thus $d^2 = 2^D$, and so $d = 2^{D/2}$. It is obvious that there are *no one*-dimensional representations of the D-dimensional

Minkowski Clifford *algebra*, so the only irreducible representation of the Clifford group that provides a representation of the corresponding algebra is the one of dimension $2^{D/2}$.

The superscript B is inserted in (L.16) and (L.17) to distinguish this specific form from its equivalent representations. For the case $D = 4$ four equivalent representations are given in (17.117), (17.118), (23.14) and (23.25).

It should be noted that

$$(\gamma_p^{\mathrm{B}})^{\dagger} = \begin{cases} -\gamma_p^{\mathrm{B}} & \text{for } p = 1, 2, \ldots, D-1 \\ \gamma_p^{\mathrm{B}} & \text{for } p = D, \end{cases} \tag{L.18}$$

which can be expressed alternatively as

$$(\gamma_p^{\mathrm{B}})^{\dagger} = \gamma_D^{\mathrm{B}} \gamma_p^{\mathrm{B}} \gamma_D^{\mathrm{B}} \tag{L.19}$$

for $p = 1, 2, \ldots, D$. These properties (L.18) and (L.19) are preserved in any similarity transformation of the form (L.8), provided that a *unitary* matrix **S** is used, and only such similarity transformations will be considered henceforth.

It is interesting to note that if D is *even* and the space is *Euclidean*, and moreover if the γ_p are assumed to be *self-adjoint* operators acting on a Hilbert space, then, with the combinations b_p defined by

$$b_p = 2^{-1/2}(\gamma_p + i\gamma_{p+D/2})$$

for $p = 1, 2, \ldots, \frac{1}{2}D$, it follows that

$$b_p^{\dagger} = 2^{-1/2}(\gamma_p - i\gamma_{p+D/2});$$

and (L.2) and (L.4) then imply that

$$[b_p, b_q]_+ = 0,$$
$$[b_p^{\dagger}, b_q^{\dagger}]_+ = 0,$$
$$[b_p^{\dagger}, b_q]_+ = \delta_{pq} 1$$

(for $p, q = 1, 2, \ldots, \frac{1}{2}D$). That is, the D-dimensional Euclidean Clifford algebra for even D is isomorphic to that of $\frac{1}{2}D$ fermion creation operators $b_p^{\dagger}$ and $\frac{1}{2}D$ fermion annihilation operators b_p. The Fock space obtained by taking a "vacuum state" vector $|0\rangle$ such that

$$b_q|0\rangle = 0$$

for $p = 1, 2, \ldots, D$ and acting on $|0\rangle$ with all linearly independent products of the creation operators $b_p^{\dagger}$ has $d = 2^{D/2}$ basis elements. It provides an alternative basis for the d-dimensional representation of the D-dimensional Euclidean Clifford algebra. (As noted in Example I of Chapter 21, Section 2, this Euclidean Clifford algebra is also known as a "Heisenberg superalgebra".)

(b) *Chirality of the representation*

The generalization of the γ_5 matrix of the $D = 4$ case (cf. (23.7)) is provided for general *even* values of D by

$$\gamma_{D+1} = i^{(D-2)/2}\gamma_1\gamma_2\gamma_3\cdots\gamma_{D-1}\gamma_D. \tag{L.20}$$

This coefficient is chosen so that with the D-dimensional Minkowski metric (L.3) the relations (L.4) give

$$(\gamma_{D+1})^2 = \mathbf{1}_d,$$

which indicates that the only possible eigenvalues of γ_{D+1} are 1 and -1. Indeed, substituting (L.16) and (L.17) (with the D-dimensional Minkowski metric (L.3)) into (L.20) gives

$$\gamma^{\mathrm{B}}_{D+1} = (-1)^{(D-2)/2}\, \boldsymbol{\sigma}_3 \otimes \boldsymbol{\sigma}_3 \otimes \boldsymbol{\sigma}_3 \otimes \cdots \boldsymbol{\sigma}_3 \otimes \boldsymbol{\sigma}_3,$$

showing that (apart from a factor ± 1) $\gamma^{\mathrm{B}}_{D+1}$ is equal to the direct product of $\frac{1}{2}D$ $\boldsymbol{\sigma}_3$ matrices, which implies that $\gamma^{\mathrm{B}}_{D+1}$ is diagonal, with $\frac{1}{2}d$ diagonal elements taking value 1 and $\frac{1}{2}d$ diagonal elements taking value -1. Consequently in *any* representation γ_{D+1} has $\frac{1}{2}d$ eigenvalues with value 1 and $\frac{1}{2}d$ eigenvalues with value -1. Thus if D is *even*, there always exists a representation, which will be called a *chiral representation* and in which the matrices will be denoted by γ^{C}_p, for which

$$\gamma^{\mathrm{C}}_{D+1} = \begin{bmatrix} \mathbf{1}_{d/2} & \mathbf{0} \\ \mathbf{0} & -\mathbf{1}_{d/2} \end{bmatrix}. \tag{L.21}$$

(Chiral representations are sometimes known alternatively as "Weyl representations".) The chiral representations are not unique, since if a similarity transformation is applied to a chiral representation with a $d \times d$ non-singular matrix $\mathbf{S}$ of the form

$$\mathbf{S} = \begin{bmatrix} \mathbf{s}_1 & \mathbf{0} \\ \mathbf{0} & \mathbf{s}_2 \end{bmatrix},$$

where $\mathbf{s}_1$ and $\mathbf{s}_2$ both have dimension $\frac{1}{2}d \times \frac{1}{2}d$, then the resulting representation is also a chiral representation. The above considerations show that if D is *even*, there exists a similarity transformation with a real $d \times d$ non-singular matrix $\mathbf{T}$ that merely interchanges rows and interchanges columns such that

$$\mathbf{T}\gamma^{\mathrm{B}}_{D+1}\mathbf{T}^{-1} = \gamma^{\mathrm{C}}_{D+1}. \tag{L.22}$$

Thus, apart from this relabelling of rows and columns, the representation of the Clifford algebra presented in (L.16) and (L.17) is a chiral representation.

It should be noted that the relations (L.4) also imply that

$$\gamma_p\gamma_{D+1} + \gamma_{D+1}\gamma_p = \mathbf{0} \tag{L.23}$$

for $p = 1, 2, \ldots, D$.

Turning to the associated representation $\boldsymbol{\Gamma}^{\text{spin}}$ of so$(D-1,1)$ defined in (L.10), the relations (L.14) and (L.23) imply that in *every* representation

$$\gamma_{D+1}\boldsymbol{\Gamma}^{\text{spin}}(L_{pq}) = \boldsymbol{\Gamma}^{\text{spin}}(L_{pq})\gamma_{D+1} \qquad \text{(L.24)}$$

for all $p, q = 1, 2, \ldots, D$ such that $p \neq q$. As γ_{D+1} is not a multiple of the unit matrix $\mathbf{1}_d$, Schur's Lemma for Lie algebras (Theorem II of Chapter 11, Section 4) implies that the representation $\boldsymbol{\Gamma}^{\text{spin}}$ of so$(D-1,1)$ must be *reducible*. Closer investigation shows that $\boldsymbol{\Gamma}^{\text{spin}}$ is equivalent to the direct sum of *two* irreducible representations of so$(D-1,1)$ that both have dimension $\frac{1}{2}d$ $(=2^{D/2-1})$, and which may be denoted by $\boldsymbol{\Gamma}_{\text{L}}$ and $\boldsymbol{\Gamma}_{\text{R}}$. These correspond to the irreducible representations $\boldsymbol{\Gamma}(\{0,0,\ldots,0,0,1\})$ and $\boldsymbol{\Gamma}(\{0,0,\ldots,0,1,0\})$ of the associated simple complex Lie algebra $D_{D/2}$, and so are "spinor representations" in that they provide faithful representations for the universal linear group Spin$(D-1,1)$ of so$(D-1,1)$.

Theorem II The spinor representation $\boldsymbol{\Gamma}^{\text{spin}}$ of so$(D-1,1)$ is the *direct sum* of the irreducible representations $\boldsymbol{\Gamma}_{\text{L}}$ and $\boldsymbol{\Gamma}_{\text{R}}$ if and only if the matrices representing the Minkowski Clifford algebra are in a *chiral* representation.

Proof If $\gamma_{D+1} = \gamma^{\text{C}}_{D+1}$ then (L.21) and (L.24) together imply immediately that $\boldsymbol{\Gamma}^{\text{spin}}$ must have a direct sum form. Conversely, if $\boldsymbol{\Gamma}^{\text{spin}}$ is the direct sum of two $\frac{1}{2}d$-dimensional irreducible representations then (L.24) and Schur's Lemma for Lie algebras require that γ_{D+1} must have the form

$$\gamma_{D+1} = \begin{bmatrix} \lambda\mathbf{1}_{d/2} & \mathbf{0} \\ \mathbf{0} & \mu\mathbf{1}_{d/2} \end{bmatrix},$$

where λ and μ are two complex numbers. As γ_{D+1} has $\frac{1}{2}d$ eigenvalues with value 1 and $\frac{1}{2}d$ eigenvalues with value -1, it follows that $\lambda = -\mu = \pm 1$. The choice $\lambda = -\mu = 1$ immediately gives the stated result, while with the other choice $\lambda = -\mu = -1$ a similarity transformation can be applied that merely interchanges the two constituent irreducible representations and the two non-zero blocks in γ_{D+1}, giving again the form $\gamma_{D+1} = \gamma^{\text{C}}_{D+1}$.

The "chiral" spinor representation of so$(D-1,1)$ will henceforth be denoted by $\boldsymbol{\Gamma}^{\text{spin}}_{\text{C}}$, and it will be assumed that the labels "L" and "R" of $\boldsymbol{\Gamma}_{\text{L}}$ and $\boldsymbol{\Gamma}_{\text{R}}$ are chosen so that

$$\boldsymbol{\Gamma}^{\text{spin}}_{\text{C}}(L_{pq}) = \begin{bmatrix} \boldsymbol{\Gamma}_{\text{L}}(L_{pq}) & \mathbf{0} \\ \mathbf{0} & \boldsymbol{\Gamma}_{\text{R}}(L_{pq}) \end{bmatrix} \qquad \text{(L.25)}$$

when γ^{C}_{D+1} is defined as in (L.21).

With the choice of matrices (L.16) and (L.17), it only requires a similarity transformation with the reordering matrix $\mathbf{T}$ of (L.22) to put $\boldsymbol{\Gamma}^{\text{spin}}_{\text{B}}$ into the direct sum form. That is, if D is *even*,

$$\mathbf{T}\boldsymbol{\Gamma}_{\mathrm{B}}^{\mathrm{spin}}(L_{pq})\mathbf{T}^{-1} = \begin{bmatrix} \boldsymbol{\Gamma}_{\mathrm{L}}(L_{pq}) & \mathbf{0} \\ \mathbf{0} & \boldsymbol{\Gamma}_{\mathrm{R}}(L_{pq}) \end{bmatrix}$$

for all $p, q = 1, 2, \ldots, D$ such that $p \neq q$.

(c) *Reality of the spinor representation of* so$(D-1, 1)$

The spinor representation $\boldsymbol{\Gamma}^{\mathrm{spin}}$ of so$(D-1, 1)$ defined by (L.10) is *real* if either

(i) the matrices $\boldsymbol{\gamma}_p$ are a set of *real* matrices (for $p = 1, 2, \ldots, D$),

or

(ii) the matrices $\boldsymbol{\gamma}_p$ are a set of *imaginary* matrices (for $p = 1, 2, \ldots, D$).

In both cases the set of matrices $\boldsymbol{\gamma}_p$ are said to form a *Majorana representation* of the D-dimensional Minkowski Clifford algebra, and may be denoted by $\boldsymbol{\gamma}_p^{\mathrm{M}}$. When such a representation exists it is not unique, since applying to it a similarity transformation of the form (L.8) with a real orthogonal matrix $\mathbf{S}$ will produce another Majorana representation.

Theorem III (a) For $D \bmod 8 = 0, 2$ or 4, that is, if

$$D = \begin{cases} 8, 16, 24, \ldots, \text{ or} \\ 2, 10, 18, 26, \ldots, \text{ or} \\ 4, 12, 20, 28, \ldots, \end{cases}$$

the irreducible representation of the D-dimensional Minkowski Clifford algebra consisting of the matrices $\boldsymbol{\gamma}_p^{\mathrm{B}}$ of (L.16) and (L.17) is equivalent to a Majorana representation. In such a case let $\mathbf{S}_{\mathrm{B}}^{\mathrm{M}}$ be the $d \times d$ non-singular matrix with the property that

$$\boldsymbol{\gamma}_p^{\mathrm{M}} = \mathbf{S}_{\mathrm{B}}^{\mathrm{M}} \boldsymbol{\gamma}_p^{\mathrm{B}} (\mathbf{S}_{\mathrm{B}}^{\mathrm{M}})^{-1} \tag{L.26}$$

for $p = 1, 2, \ldots, D$.

(b) For $D \bmod 8 = 0$ the matrices $\boldsymbol{\gamma}_p^{\mathrm{M}}$ are all *real*, and $\mathbf{S}_{\mathrm{B}}^{\mathrm{M}}$ may be chosen so that

$$\mathbf{S}_{\mathrm{B}}^{\mathrm{M}} = 2^{-1/2}(\mathbf{1}_d + i\boldsymbol{\gamma}_{D/2+1}^{\mathrm{B}} \boldsymbol{\gamma}_{D/2+2}^{\mathrm{B}} \cdots \boldsymbol{\gamma}_{D-1}^{\mathrm{B}}), \tag{L.27}$$

in which case

$$\boldsymbol{\gamma}_p^{\mathrm{M}} = \begin{cases} \boldsymbol{\gamma}_p^{\mathrm{B}} & \text{if } p = \tfrac{1}{2}D + 1, \ldots, D-1, \\ -i\boldsymbol{\gamma}_p^{\mathrm{B}} \boldsymbol{\gamma}_{D/2+1}^{\mathrm{B}} \boldsymbol{\gamma}_{D/2+2}^{\mathrm{B}} \cdots \boldsymbol{\gamma}_{D-1}^{\mathrm{B}} & \text{if } p = 1, \ldots, \tfrac{1}{2}D \text{ or } p = D. \end{cases} \tag{L.28}$$

(c) For $D \bmod 8 = 2$ there are two subcases:

(i) the matrices γ_p^{M} can be chosen so that they are all *imaginary*, and then $\mathbf{S}_{\mathrm{B}}^{\mathrm{M}}$ may be chosen so that

$$\mathbf{S}_{\mathrm{B}}^{\mathrm{M}} = \mathbf{1}_2 \quad (\text{for } D = 2) \tag{L.29}$$

and

$$\mathbf{S}_{\mathrm{B}}^{\mathrm{M}} = 2^{-1/2}(\mathbf{1}_d + i\gamma_{D/2+1}^{\mathrm{B}}\gamma_{D/2+2}^{\mathrm{B}}\cdots\gamma_{D-1}^{\mathrm{B}}) \quad (\text{for } D \geqslant 10), \tag{L.30}$$

in which case for $D = 2$

$$\gamma_p^{\mathrm{M}} = \gamma_p^{\mathrm{B}} \tag{L.31}$$

(for $p = 1, 2$) and for $D \geqslant 10$

$$\gamma_p^{\mathrm{M}} = \begin{cases} \gamma_p^{\mathrm{B}} & \text{if } p = 1, \ldots, \tfrac{1}{2}D \text{ or } p = D, \\ -i\gamma_p^{\mathrm{B}}\gamma_{D/2+1}^{\mathrm{B}}\gamma_{D/2+2}^{\mathrm{B}}\cdots\gamma_{D-1}^{\mathrm{B}} & \text{if } p = \tfrac{1}{2}D+1, \tfrac{1}{2}D+2, \ldots, D-1; \end{cases} \tag{L.32}$$

(ii) the matrices γ_p^{M} can alternatively be chosen so that they are all *real*, and then $\mathbf{S}_{\mathrm{B}}^{\mathrm{M}}$ may be chosen so that

$$\mathbf{S}_{\mathrm{B}}^{\mathrm{M}} = 2^{-1/2}(\mathbf{1}_d + i\gamma_1^{\mathrm{B}}\gamma_2^{\mathrm{B}}\cdots\gamma_{D/2}^{\mathrm{B}}\gamma_D^{\mathrm{B}}), \tag{L.33}$$

in which case

$$\gamma_p^{\mathrm{M}} = \begin{cases} \gamma_p^{\mathrm{B}} & \text{if } p = \tfrac{1}{2}D+1, \tfrac{1}{2}D+2, \ldots, D-1, \\ -i\gamma_p^{\mathrm{B}}\gamma_1^{\mathrm{B}}\gamma_2^{\mathrm{B}}\cdots\gamma_{D/2}^{\mathrm{B}}\gamma_{D-1}^{\mathrm{B}} & \text{if } p = 1, 2, \ldots, \tfrac{1}{2}D \text{ or } p = D. \end{cases} \tag{L.34}$$

(d) For $D \bmod 8 = 4$ the matrices γ_p^{M} are all *imaginary*, and $\mathbf{S}_{\mathrm{B}}^{\mathrm{M}}$ may be chosen so that

$$\mathbf{S}_{\mathrm{B}}^{\mathrm{M}} = 2^{-1/2}(\mathbf{1}_d - \gamma_1^{\mathrm{B}}\gamma_2^{\mathrm{B}}\cdots\gamma_{D/2}^{\mathrm{B}}\gamma_{\mathrm{D}}^{\mathrm{B}}), \tag{L.35}$$

in which case

$$\gamma_p^{\mathrm{M}} = \begin{cases} \gamma_p^{\mathrm{B}} & \text{if } p = 1, 2, \ldots, \tfrac{1}{2}D \text{ or } p = D, \\ \gamma_p^{\mathrm{B}}\gamma_1^{\mathrm{B}}\gamma_2^{\mathrm{B}}\cdots\gamma_{D/2}^{\mathrm{B}}\gamma_{D-1}^{\mathrm{B}} & \text{if } p = \tfrac{1}{2}D+1, \tfrac{1}{2}D+2, \ldots, D-1. \end{cases} \tag{L.36}$$

Proof In the case $D = 2$, by (L.16) and (L.17),

$$\gamma_1^{\mathrm{B}} = i\boldsymbol{\sigma}_1, \quad \gamma_2^{\mathrm{B}} = \boldsymbol{\sigma}_2, \tag{L.37}$$

which are both imaginary, so this irreducible representation of the 2-dimensional Minkowski Clifford algebra is itself a Majorana representation. Equations (L.29) and (L.31) then follow immediately.

Now consider the general case in which D is *even* and $D > 2$. Suppose first that the γ_p^{M} are all *imaginary* matrices, so that by (L.26)

$$(\mathbf{S}_{\mathrm{B}}^{\mathrm{M}}\gamma_p^{\mathrm{B}}(\mathbf{S}_{\mathrm{B}}^{\mathrm{M}})^{-1})^* = -\mathbf{S}_{\mathrm{B}}^{\mathrm{M}}\gamma_p^{\mathrm{B}}(\mathbf{S}_{\mathrm{B}}^{\mathrm{M}})^{-1}$$

for $p = 1, \ldots, D$, which is so if and only if

$$((\mathrm{S}_{\mathrm{B}}^{\mathrm{M}})^{-1}\mathrm{S}_{\mathrm{B}}^{\mathrm{M}*})\gamma_p^{\mathrm{B}*}((\mathrm{S}_{\mathrm{B}}^{\mathrm{M}})^{-1}\mathrm{S}_{\mathrm{B}}^{\mathrm{M}*})^{-1} = -\gamma_p^{\mathrm{B}} \tag{L.38}$$

for $p = 1, \ldots, D$. However, with the metric tensor of (L.3), if D is even then the matrices $\gamma_1^{\mathrm{B}}, \gamma_2^{\mathrm{B}}, \ldots, \gamma_{D/2-1}^{\mathrm{B}}, \gamma_{D/2}^{\mathrm{B}}$ and γ_D^{B} of (L.16) and (L.17) are imaginary and the matrices $\gamma_{D/2+1}^{\mathrm{B}}, \gamma_{D/2+2}^{\mathrm{B}}, \ldots, \gamma_{D-2}^{\mathrm{B}}$ and $\gamma_{D-1}^{\mathrm{B}}$ are real, so the condition (L.38) reduces to the pair of conditions

$$((\mathrm{S}_{\mathrm{B}}^{\mathrm{M}})^{-1}\mathrm{S}_{\mathrm{B}}^{\mathrm{M}*})\gamma_p^{\mathrm{B}}((\mathrm{S}_{\mathrm{B}}^{\mathrm{M}})^{-1}\mathrm{S}_{\mathrm{B}}^{\mathrm{M}*})^{-1} = \gamma_p^{\mathrm{B}} \quad \text{for } p = 1, 2, \ldots, \tfrac{1}{2}D \text{ and } p = D, \tag{L.39}$$

and

$$((\mathrm{S}_{\mathrm{B}}^{\mathrm{M}})^{-1}\mathrm{S}_{\mathrm{B}}^{\mathrm{M}*})\gamma_p^{\mathrm{B}}((\mathrm{S}_{\mathrm{B}}^{\mathrm{M}})^{-1}\mathrm{S}_{\mathrm{B}}^{\mathrm{M}*})^{-1} = -\gamma_p^{\mathrm{B}}$$
$$\text{for } p = \tfrac{1}{2}D + 1, \tfrac{1}{2}D + 2, \ldots, D - 2 \text{ and } p = D - 1. \tag{L.40}$$

As $(\mathrm{S}_{\mathrm{B}}^{\mathrm{M}})^{-1}\mathrm{S}_{\mathrm{B}}^{\mathrm{M}*}$ must be a linear combination of $\mathbf{1}_d$ and the products of the matrices γ_p^{B}, use of (L.7) shows that the only candidate is

$$(\mathrm{S}_{\mathrm{B}}^{\mathrm{M}})^{-1}\mathrm{S}_{\mathrm{B}}^{\mathrm{M}*} = \begin{cases} \kappa\gamma_{D/2+1}^{\mathrm{B}}\gamma_{D/2+2}^{\mathrm{B}}\cdots\gamma_{D-2}^{\mathrm{B}}\gamma_{D-1}^{\mathrm{B}} & \text{for } \tfrac{1}{2}D \text{ odd,} \\ \kappa\gamma_1^{\mathrm{B}}\gamma_2^{\mathrm{B}}\cdots\gamma_{D/2}^{\mathrm{B}}\gamma_D^{\mathrm{B}} & \text{for } \tfrac{1}{2}D \text{ even,} \end{cases} \tag{L.41}$$

where κ is some complex number.

Suppose first that $\frac{1}{2}D$ is *odd*. Then

$$\mathrm{S}_{\mathrm{B}}^{\mathrm{M}*} = \mathrm{S}_{\mathrm{B}}^{\mathrm{M}}\kappa\gamma_{D/2+1}^{\mathrm{B}}\gamma_{D/2+2}^{\mathrm{B}}\cdots\gamma_{D-2}^{\mathrm{B}}\gamma_{D-1}^{\mathrm{B}}, \tag{L.42}$$

which, on taking the complex conjugate and substituting back in (L.42), gives the condition

$$\mathrm{S}_{\mathrm{B}}^{\mathrm{M}} = \mathrm{S}_{\mathrm{B}}^{\mathrm{M}}|\kappa|^2(-1)^K, \tag{L.43}$$

which is only satisfied if

$$K = \text{even integer}, \tag{L.44}$$

where

$$\mathrm{K} = \tfrac{1}{4}D(\tfrac{1}{2}D - 1). \tag{L.45}$$

The only solutions of (L.45) for $\frac{1}{2}D$ odd are given by $\frac{1}{2}D = 4m + 1$, where m is any integer; that is, the γ_p^{B} of (L.16) and (L.17) can only all be equivalent to *imaginary* matrices for $\frac{1}{2}D$ *odd* if $D \bmod 8 = 2$.

In the case in which $\frac{1}{2}D$ is *even* the consistency conditions for the γ_p^{B} of (L.16) and (L.17) to be equivalent to *imaginary* matrices again reduce to the form (L.44), but in this case (L.45) is replaced by

$$K = \tfrac{1}{2}D + \tfrac{1}{2}(\tfrac{1}{2}D + 1)(\tfrac{1}{2}D + 2), \tag{L.46}$$

whose only solutions are of the form $\frac{1}{2}D = 4m + 2$, where m is any integer. Thus the $\gamma_p^{\rm B}$ of (L.16) and (L.17) can only all be equivalent to *imaginary* matrices for $\frac{1}{2}D$ *even* if $D \bmod 8 = 4$.

It is easily shown that the unitary matrices $\mathbf{S}_{\rm B}^{\rm M}$ of (L.30) and (L.35) satisfy (L.41) for $D \bmod 8 = 2$ ($D \geqslant 10$) and $D \bmod 8 = 4$ respectively.

This analysis is easily repeated for the case in which the $\gamma_p^{\rm B}$ of (L.16) and (L.17) are assumed to be equivalent to *real* matrices. In this case the right-hand sides of (L.38)–(L.40) have to be multiplied by -1, so that (L.41) is replaced by

$$(\mathbf{S}_{\rm B}^{\rm M})^{-1}\mathbf{S}_{\rm B}^{\rm M*} = \begin{cases} \kappa\gamma_{D/2+1}^{\rm B}\gamma_{D/2+2}^{\rm B}\cdots\gamma_{D-2}^{\rm B}\gamma_{D-1}^{\rm B} & \text{for } \frac{1}{2}D \text{ even}, \\ \kappa\gamma_1^{\rm B}\gamma_2^{\rm B}\cdots\gamma_{D/2}^{\rm B}\gamma_D^{\rm B} & \text{for } \frac{1}{2}D \text{ odd}, \end{cases} \tag{L.47}$$

where κ is some complex number, the only change from (L.29) being that the expressions for $\frac{1}{2}D$ even and $\frac{1}{2}D$ odd are interchanged. Consequently (L.44) and (L.45) must hold for $\frac{1}{2}D$ even, which implies that with D even the only solution is $D \bmod 8 = 0$, and (L.44) and (L.46) must hold for $\frac{1}{2}D$ odd, for which the only solution for D even is $D \bmod 8 = 2$. Thus the $\gamma_p^{\rm B}$ of (L.16) and (L.17) can only all be equivalent to *real* matrices if $D \bmod 8 = 0$ or 2. It is easily shown that the unitary matrices $\mathbf{S}_{\rm B}^{\rm M}$ of (L.27) and (L.33) satisfy (L.47) for $D \bmod 8 = 0$ and 2 respectively.

It should be noted that *no* Majorana representations exist for $D \bmod 8 = 6$, that is, for $D = 6, 14, 22, \ldots$. Further information on Majorana representations appears in Section 4.

Theorem IV For $D \bmod 8 = 2$, that is, for $D = 2, 10, 18, 26, \ldots$, and for these cases *alone*, there exists a spinor representation $\mathbf{\Gamma}^{\rm spin}$ of so$(D-1, 1)$ defined by (L.10) that *both* has the *chiral* form (L.26) *and* is composed of *real* matrices, so the two constituent irreducible representations $\mathbf{\Gamma}_{\rm L}$ and $\mathbf{\Gamma}_{\rm R}$ of (L.25) are both *real*. The associated representation of the D-dimensional Minkowski Clifford algebra is called a *Majorana–Weyl* representation.

Proof For the case $D = 2$ (L.14) and (L.37) together give

$$\mathbf{\Gamma}_{\rm B}^{\rm spin}(L_{12}) = -\tfrac{1}{2}i\boldsymbol{\sigma}_1\boldsymbol{\sigma}_2 = \begin{bmatrix} -\frac{1}{2} & 0 \\ 0 & \frac{1}{2} \end{bmatrix}, \tag{L.48}$$

so $\mathbf{\Gamma}_{\rm B}^{\rm spin}$ is clearly both real and chiral. Clearly, since $\mathbf{\Gamma}_{\rm C}^{\rm spin}(L_{12}) = \mathbf{\Gamma}_{\rm B}^{\rm spin}(L_{12})$ in this case,

$$\mathbf{\Gamma}_{\rm L}(L_{12}) = [-\tfrac{1}{2}], \quad \mathbf{\Gamma}_{\rm R}(L_{12}) = [\tfrac{1}{2}]. \tag{L.49}$$

For D even and $D > 2$ the argument proceeds as follows. Let $\mathbf{\Gamma}_{\rm M'}^{\rm spin}$ be a Majorana representation. The $\mathbf{\Gamma}_{\rm M'}^{\rm spin}$ is chiral if and only if

$$\gamma_{D+1}^{\rm C}\mathbf{\Gamma}_{\rm M'}^{\rm spin}(L_{pq}) = \mathbf{\Gamma}_{\rm M'}^{\rm spin}(L_{pq})\gamma_{D+1}^{\rm C} \tag{L.50}$$

for all $p, q = 1, \ldots, D$ (with $p \neq q$). Let $\boldsymbol{\Gamma}_{\mathrm{M}}^{\mathrm{spin}}$ be the Majorana representation defined by

$$\boldsymbol{\Gamma}_{\mathrm{M'}}^{\mathrm{spin}}(L_{pq}) = \mathbf{T}^{-1}\boldsymbol{\Gamma}_{\mathrm{M}}^{\mathrm{spin}}(L_{pq})\mathbf{T},$$

where $\mathbf{T}$ is the real non-singular matrix of (L.22). Let $\mathbf{S}_{\mathrm{B}}^{\mathrm{M}}$ be the $d \times d$ non-singular matrix of (L.26) that transforms the matrices $\boldsymbol{\gamma}_p^{\mathrm{B}}$ into the Majorana representation $\boldsymbol{\gamma}_p^{\mathrm{M}}$ so that

$$\boldsymbol{\Gamma}_{\mathrm{M}}^{\mathrm{spin}}(L_{pq}) = \mathbf{S}_{\mathrm{B}}^{\mathrm{M}}\boldsymbol{\Gamma}_{\mathrm{B}}^{\mathrm{spin}}(L_{pq})(\mathbf{S}_{\mathrm{B}}^{\mathrm{M}})^{-1},$$

where $\boldsymbol{\Gamma}_{\mathrm{B}}^{\mathrm{spin}}$ denotes the spinor representation constructed from the matrices $\boldsymbol{\gamma}_p^{\mathrm{B}}$ of (L.16) and (L.17). Then (L.22) and (L.50) imply that representation $\boldsymbol{\Gamma}_{\mathrm{M}}^{\mathrm{spin}}$ is chiral if and only if

$$\boldsymbol{\gamma}_{D+1}^{\mathrm{B}}\boldsymbol{\Gamma}_{\mathrm{M}}^{\mathrm{spin}}(L_{pq}) = \boldsymbol{\Gamma}_{\mathrm{M}}^{\mathrm{spin}}(L_{pq})\boldsymbol{\gamma}_{D+1}^{\mathrm{B}} \tag{L.51}$$

for all $p, q = 1, \ldots, D$ such that $p \neq q$, which, by (L.24) (with an index "B" inserted), is satisfied if and only if

$$\mathbf{S}_{\mathrm{B}}^{\mathrm{M}}\boldsymbol{\gamma}_{D+1}^{\mathrm{B}}(\mathbf{S}_{\mathrm{B}}^{\mathrm{M}})^{-1} = \boldsymbol{\gamma}_{D+1}^{\mathrm{B}}.$$

It is easily checked that this is satisfied for the expressions (L.30) and (L.33) of the case $D \bmod 8 = 2$, but is not valid for the expressions (L.27) and (L.35) that correspond to $D \bmod 8 = 0$ and 4.

Further information on these Majorana–Weyl representations may be found in Section 4.

(d) *The generalized charge conjugation matrices*

If D is *even* then the representation of the D-dimensional Minkowski Clifford algebra by the matrices $\tilde{\boldsymbol{\gamma}}_p^{\mathrm{B}}$ is equivalent to that by the matrices $\boldsymbol{\gamma}_p^{\mathrm{B}}$ *and* to that by the matrices $-\boldsymbol{\gamma}_p^{\mathrm{B}}$. Consequently there are *two* inequivalent ways of generalizing the charge conjugation matrix of (23.17).

(i) With $\mathbf{C}^{\mathrm{B}}$ defined by

$$\mathbf{C}^{\mathrm{B}} = \eta\boldsymbol{\sigma}_2 \otimes \boldsymbol{\sigma}_1 \otimes \boldsymbol{\sigma}_2 \otimes \boldsymbol{\sigma}_1 \otimes \ldots, \tag{L.52}$$

which is the direct product of $\frac{1}{2}D$ Pauli spin matrices and in which η is some complex number, and with the $\boldsymbol{\gamma}_p^{\mathrm{B}}$ defined in (L.16) and (L.17),

$$\tilde{\boldsymbol{\gamma}}_p^{\mathrm{B}} = -(\mathbf{C}^{\mathrm{B}})^{-1}\boldsymbol{\gamma}_p^{\mathrm{B}}\mathbf{C}^{\mathrm{B}} \tag{L.53}$$

for $p = 1, 2, \ldots, D$. Moreover every $d \times d$ matrix $\mathbf{C}^{\mathrm{B}}$ satisfying (L.53) must have the form (L.52).

For $D \bmod 8 = 2$ and 4

$$\tilde{\mathbf{C}}^{\mathrm{B}} = -\mathbf{C}^{\mathrm{B}}, \tag{L.54}$$

but for $D \bmod 8 = 0$ and 6

$$\tilde{\mathbf{C}}^{\mathrm{B}} = \mathbf{C}^{\mathrm{B}}. \tag{L.55}$$

For all even values of D

$$\mathbf{C}^{\mathrm{B}}\mathbf{C}^{\mathrm{B}\dagger} = \mathbf{C}^{\mathrm{B}\dagger}\mathbf{C}^{\mathrm{B}} = \mathbf{1}_d, \tag{L.56}$$

provided that η is chosen so that $|\eta|^2 = 1$.

If $\mathbf{S}_{\mathrm{B}}^{\mathrm{M}}$ is the $d \times d$ non-singular matrix of (L.26) that transforms the $\boldsymbol{\gamma}_p^{\mathrm{B}}$ defined in (L.16) and (L.17) into a Majorana representation $\boldsymbol{\gamma}_p^{\mathrm{M}}$ (which, as noted in Theorem III above can only exist for $D \bmod 8 = 0$, 2 and 4) then the corresponding charge conjugation matrix $\mathbf{C}^{\mathrm{M}}$ of the Majorana representation is defined by

$$\mathbf{C}^{\mathrm{M}} = \mathbf{S}_{\mathrm{B}}^{\mathrm{M}}\mathbf{C}^{\mathrm{B}}\tilde{\mathbf{S}}_{\mathrm{B}}^{\mathrm{M}} \tag{L.57}$$

(cf. (17.156)). Then

$$\tilde{\boldsymbol{\gamma}}_p^{\mathrm{M}} = -(\mathbf{C}^{\mathrm{M}})^{-1}\boldsymbol{\gamma}_p^{\mathrm{M}}\mathbf{C}^{\mathrm{M}} \tag{L.58}$$

for $p = 1, 2, \ldots, D$, and $\mathbf{C}^{\mathrm{M}}$ is antisymmetric if $D \bmod 8 = 2$ or 4, but is symmetric if $D \bmod 8 = 0$ or 6. Moreover, by an appropriate choice of the factor η, $\mathbf{C}^{\mathrm{M}}$ can be chosen to be real or imaginary (whichever is required).

Thus for $D \bmod 8 = 2$ and 4 the products $\boldsymbol{\gamma}_{\mathrm{M}}^p\mathbf{C}^{\mathrm{M}}$ can be taken to be real and symmetric for all $p = 1, 2, \ldots, D$. Although (L.57) could be used to obtain an explicit expression for $\mathbf{C}^{\mathrm{M}}$ in these cases, the following alternative argument is simpler. As $\mathbf{S}_{\mathrm{B}}^{\mathrm{M}}$ is a unitary matrix, the identity (L.19) retains its form in the Majorana representation, so that

$$\boldsymbol{\gamma}_p^{\mathrm{M}\dagger} = \boldsymbol{\gamma}_D^{\mathrm{M}}\boldsymbol{\gamma}_p^{\mathrm{M}}\boldsymbol{\gamma}_D^{\mathrm{M}} \tag{L.59}$$

for $p = 1, 2, \ldots, D$. For a Majorana representation by *imaginary* matrices this reduces to

$$\boldsymbol{\gamma}_p^{\mathrm{M}} = -\boldsymbol{\gamma}_D^{\mathrm{M}}\boldsymbol{\gamma}_p^{\mathrm{M}}\boldsymbol{\gamma}_D^{\mathrm{M}} \tag{L.60}$$

for $p = 1, 2, \ldots, D$, and hence $\mathbf{C}^{\mathrm{M}} = \xi\boldsymbol{\gamma}_D^{\mathrm{M}}$, where ξ is some complex number. With the choice $\xi = -1$, i.e.

$$\mathbf{C}^{\mathrm{M}} = -\boldsymbol{\gamma}_D^{\mathrm{M}}, \tag{L.61}$$

not only are the products $\boldsymbol{\gamma}_{\mathrm{M}}^p\mathbf{C}^{\mathrm{M}}$ real and symmetric but also the convention (17.167) is generalized in a natural manner. With the choice (L.61), equations (L.3), (L.5) and (L.6) imply that

$$\operatorname{tr}(\boldsymbol{\gamma}_{\mathrm{M}}^p\mathbf{C}^{\mathrm{M}}) = \begin{cases} 0 & \text{for } p = 1, 2, \ldots, D-1, \\ -D & \text{for } p = D. \end{cases} \tag{L.62}$$

(ii) With $\mathbf{C}^{\mathrm{B}}$ defined by

$$\mathbf{C}^{\mathrm{B}} = \eta\boldsymbol{\sigma}_1 \otimes \boldsymbol{\sigma}_2 \otimes \boldsymbol{\sigma}_1 \otimes \boldsymbol{\sigma}_2 \otimes \cdots, \tag{L.63}$$

which is the direct product of $\frac{1}{2}D$ Pauli spin matrices and in which η is some complex number, and with the $\boldsymbol{\gamma}_p^{\mathrm{B}}$ defined in (L.16) and (L.17),

$$\tilde{\boldsymbol{\gamma}}_p^{\mathrm{B}} = (\mathbf{C}^{\mathrm{B}})^{-1}\boldsymbol{\gamma}_p^{\mathrm{B}}\mathbf{C}^{\mathrm{B}} \tag{L.64}$$

for $p = 1, 2, \ldots, D$. Moreover every $d \times d$ matrix $\mathbf{C}^{\mathrm{B}}$ satisfying (L.64) must have the form (L.63).

For $D \bmod 8 = 0$ and 2

$$\tilde{\mathbf{C}}^{\mathrm{B}} = \mathbf{C}^{\mathrm{B}}, \tag{L.65}$$

but for $D \bmod 8 = 4$ and 6

$$\tilde{\mathbf{C}}^{\mathrm{B}} = -\mathbf{C}^{\mathrm{B}}. \tag{L.66}$$

For all even values of D

$$\mathbf{C}^{\mathrm{B}}\mathbf{C}^{\mathrm{B}\dagger} = \mathbf{C}^{\mathrm{B}\dagger}\mathbf{C}^{\mathrm{B}} = \mathbf{1}_d, \tag{L.67}$$

provided that η is chosen so that $|\eta|^2 = 1$.

If $\mathbf{S}_{\mathrm{B}}^{\mathrm{M}}$ is the $d \times d$ non-singular matrix of (L.26) then the corresponding generalized charge conjugation matrix $\mathbf{C}^{\mathrm{M}}$ of the Majorana representation is defined by

$$\mathbf{C}^{\mathrm{M}} = \mathbf{S}_{\mathrm{B}}^{\mathrm{M}}\mathbf{C}^{\mathrm{B}}\tilde{\mathbf{S}}_{\mathrm{B}}^{\mathrm{M}}. \tag{L.68}$$

Then

$$\tilde{\boldsymbol{\gamma}}_p^{\mathrm{M}} = (\mathbf{C}^{\mathrm{M}})^{-1}\boldsymbol{\gamma}_p^{\mathrm{M}}\mathbf{C}^{\mathrm{M}} \tag{L.69}$$

for $p = 1, 2, \ldots, D$, and $\mathbf{C}^{\mathrm{M}}$ is symmetric if $D \bmod 8 = 0$ or 2, but is antisymmetric if $D \bmod 8 = 4$ or 6. Moreover, by an appropriate choice of the factor η, $\mathbf{C}^{\mathrm{M}}$ can be chosen to be real or imaginary (whichever is required).

Thus for $D \bmod 8 = 0$ and 2 the products $\boldsymbol{\gamma}_{\mathrm{M}}^{p}\mathbf{C}^{\mathrm{M}}$ can be taken to be real and symmetric for all $p = 1, 2, \ldots, D$. Although (L.68) could be used to obtain an explicit expression for $\mathbf{C}^{\mathrm{M}}$ in these cases, a modification of the alternative argument used in (i) is simpler. As $\mathbf{S}_{\mathrm{B}}^{\mathrm{M}}$ is again a unitary matrix, the identity (L.69) is again valid. For a Majorana representation by *real* matrices this reduces to

$$\tilde{\boldsymbol{\gamma}}_p^{\mathrm{M}} = \boldsymbol{\gamma}_D^{\mathrm{M}}\boldsymbol{\gamma}_p^{\mathrm{M}}\boldsymbol{\gamma}_D^{\mathrm{M}} \tag{L.70}$$

for $p = 1, 2, \ldots, D$, and hence $\mathbf{C}^{\mathrm{M}} = \xi\boldsymbol{\gamma}_D^{\mathrm{M}}$, where ξ is some complex number. Again, with the choice $\xi = -1$. i.e.

$$\mathbf{C}^{\mathrm{M}} = -\boldsymbol{\gamma}_D^{\mathrm{M}}, \tag{L.71}$$

not only are the products $\boldsymbol{\gamma}_{\mathrm{M}}^{p}\mathbf{C}^{\mathrm{M}}$ real and symmetric but (L.62) holds as well.

It should be noted that with *both* of the definitions (L.52) and (L.63) for $\mathbf{C}^{\mathrm{B}}$ it is true that in every representation

$$\tilde{\gamma}_q\tilde{\gamma}_p = \mathbf{C}^{-1}\gamma_q\gamma_p\mathbf{C} \tag{L.72}$$

for all p, $q = 1, 2, \ldots, D$, and all even D, *but* that *only* for $D \bmod 8 = 0$, 2 and 4 can the generalized charge conjugation matrix be chosen in such a way that the products $\gamma^p\mathbf{C}$ are *symmetric* for $p = 1, 2, \ldots, D$. Moreover, for $D \bmod 8 = 0, 2$ and 4 the $\mathbf{C}^{\mathrm{M}}$ can be chosen so that the Majorana matrix products $\gamma^p_{\mathrm{M}}\mathbf{C}^{\mathrm{M}}$ are all *real*. However, for $D \bmod 8 = 6$ *neither* of these two latter properties can be achieved.

3 Irreducible representations for the case in which D is odd

(a) Explicit expressions for the matrices

Theorem I If D is *odd* then there exist up to equivalence only *two* irreducible representations of the D-dimensional Minkowski Clifford algebra. Both have the same dimension d, which is given by

$$d = 2^{(D-1)/2}, \tag{L.73}$$

and they merely differ in their overall sign. A convenient explicit form for them is given for $D = 1$ by

$$\gamma^{\mathrm{B}}_1 = [\pm(\eta_{11})^{1/2}]$$

and for $D \geqslant 3$ as follows:

(i) for $1 \leqslant p \leqslant \frac{1}{2}(D-1)$

$$\gamma^{\mathrm{B}}_p = \pm(\eta_{pp})^{1/2}\boldsymbol{\sigma}_3 \otimes \boldsymbol{\sigma}_3 \otimes \boldsymbol{\sigma}_3 \otimes \cdots \otimes \boldsymbol{\sigma}_3 \otimes \boldsymbol{\sigma}_1 \otimes \mathbf{1}_2 \otimes \mathbf{1}_2 \otimes \cdots \otimes \mathbf{1}_2 \otimes \mathbf{1}_2, \tag{L.74}$$

which is a direct product of $\frac{1}{2}(D-1)$ 2×2 matrices in which the Pauli spin matrix $\boldsymbol{\sigma}_1$ appears in the pth position, the matrices in the first $p-1$ positions are all equal to $\boldsymbol{\sigma}_3$, and those in the last $\frac{1}{2}(D-1) - p$ positions are all equal the 2×2 identity matrix $\mathbf{1}_2$;

(ii) for $\frac{1}{2}(D-1) + 1 \leqslant p \leqslant D-1$

$$\gamma^{\mathrm{B}}_p = \pm(\eta_{pp})^{1/2}\boldsymbol{\sigma}_3 \otimes \boldsymbol{\sigma}_3 \otimes \boldsymbol{\sigma}_3 \otimes \cdots \otimes \boldsymbol{\sigma}_3 \otimes \boldsymbol{\sigma}_2 \otimes \mathbf{1}_2 \otimes \mathbf{1}_2 \otimes \cdots \otimes \mathbf{1}_2 \otimes \mathbf{1}_2, \tag{L.75}$$

which is a direct product of $\frac{1}{2}(D-1)$ 2×2 matrices in which the matrix $\boldsymbol{\sigma}_2$ appears in the $[p - \frac{1}{2}(D-1)]$th position, the matrices in the first

$p - \frac{1}{2}(D-1) - 1$ positions are all equal to $\boldsymbol{\sigma}_3$, and those in the last $\frac{3}{2}(D-1) - p$ positions are all equal to $\mathbf{1}_2$;

(iii) for $p = D$

$$\boldsymbol{\gamma}_p^{\mathrm{B}} = \pm(\eta_{pp})^{1/2}\boldsymbol{\sigma}_3 \otimes \boldsymbol{\sigma}_3 \otimes \sigma_3 \otimes \cdots \otimes \boldsymbol{\sigma}_3, \tag{L.76}$$

which is a direct product of $\frac{1}{2}(D-1)$ $\boldsymbol{\sigma}_3$ matrices.

One of the irreducible representations corresponds to taking the positive signs in (L.59)–(L.61), and the other to taking the negative signs (these explicit forms appear in Brauer and Weyl 1935). The superscript "B" in (L.74)–(L.76) is inserted to distinguish these representations from those equivalent to them.

Proof The case $D = 1$ is trivial and obvious, so attention will be devoted to the case $D \geqslant 3$. The Clifford group can be set up in the same way as in Theorem I of Section 2, its order again being 2^{D+1}, but for odd D it contains $2^D + 2$ classes, every element of the group being in the same class as its negative *except* for 1, -1, $\gamma_1\gamma_2\cdots\gamma_{D-1}\gamma_D$ and $-\gamma_1\gamma_2\cdots\gamma_{D-1}\gamma_D$, which are each in classes of their own. As the Clifford group still has 2^D one-dimensional irreducible representations, Theorem VIII of Chapter 4, Section 6 implies for the case of odd D that there are *two* further inequivalent irreducible representations. If they have dimensions d_1 and d_2 then Theorem VII of Chapter 4, Section 6 shows that $(d_1)^2 + (d_2)^2 + 2^D = 2^{D+1}$, implying that $(d_1)^2 + (d_2)^2 = 2^D$, for which the only possible solution is $d_1 = d_2 = 2^{(D-1)/2}$. None of the one-dimensional representations of the Clifford group provide representations of the Clifford algebra, but it is easily checked that the matrices of (L.74)–(L.76) do provide two inequivalent representations.

It should be noted that for odd D it remains true that

$$\boldsymbol{\gamma}_p^{\mathrm{B}\dagger} = \begin{cases} -\boldsymbol{\gamma}_p^{\mathrm{B}} & \text{for } p = 1, 2, \ldots, D-1 \\ \boldsymbol{\gamma}_p^{\mathrm{B}} & \text{for } p = D, \end{cases} \tag{L.77}$$

which can again be expressed alternatively as

$$\boldsymbol{\gamma}_p^{\mathrm{B}\dagger} = \boldsymbol{\gamma}_D^{\mathrm{B}}\boldsymbol{\gamma}_p^{\mathrm{B}}\boldsymbol{\gamma}_D^{\mathrm{B}} \tag{L.78}$$

for $p = 1, 2, \ldots, D$. These properties (L.77) and (L.78) are preserved in any similarity transformation of the form (L.8) provided that a *unitary* matrix $\mathbf{S}$ is used, and again only such similarity transformations will be considered.

(b) *Non-chirality of the representations*

The generalization of the $\boldsymbol{\gamma}_5$ matrix of the $D = 4$ cases (cf. (23.7)) is provided for general *odd* values of D by

$$\gamma_{D+1} = i^{(D+1)/2}\gamma_1\gamma_2\gamma_3\cdots\gamma_{D-1}\gamma_D.$$

Again this coefficient is chosen so that with the D-dimensional Minkowski metric (L.3) the relations (L.4) give

$$(\gamma_{D+1})^2 = \mathbf{1}_d,$$

which indicates that the only possible eigenvalues of γ_{D+1} are 1 and -1. However, substitution of (L.74)–(L.76) (with the D-dimensional Minkowski metric (L.3)) shows that $\gamma^{\mathrm{B}}_{D+1} = \pm\mathbf{1}_d$, which implies that in *every* irreducible representation of the D-dimensional Minkowski Clifford algebra

$$\gamma_{D+1} = \pm\mathbf{1}_d,$$

the two signs corresponding to the two different irreducible representations. Thus if D is *odd*, a representation in which γ_{D+1} has the form (L.21) *never* exists; that is, there is *no* chiral representation.

For the case in which D is *odd*, since the two irreducible representations of the D-dimensional Minkowski Clifford algebra merely differ in sign, on using the definition (L.10), they both produce the *same* representation $\mathbf{\Gamma}^{\mathrm{spin}}$ of so$(D-1,1)$, whose dimension is $d = 2^{(D-1)/2}$. Moreover, when D is *odd* this spinor representation $\mathbf{\Gamma}^{\mathrm{spin}}$ is an *irreducible* representation of so$(D-1,1)$. (Although (L.24) remains valid, the fact that γ_{D+1} is merely a multiple of the unit matrix means that a Schur's Lemma argument does not imply reducibility if D is odd). Consequently $\mathbf{\Gamma}^{\mathrm{spin}}$ *cannot* be reduced into left- and right-handed parts, indicating again that the notion of chirality is inapplicable. With D *odd*, $\mathbf{\Gamma}^{\mathrm{spin}}$ corresponds to the irreducible representation $\mathbf{\Gamma}(\{0,0,\ldots,0,1\})$ of the associated simple complex Lie algebra $B_{(D-1)/2}$, and is called a "spinor representation" because it provides a faithful representation for the universal linear group Spin$(D-1,1)$ of SO$(D-1,1)$.

(c) *Reality of the spinor representation of* so$(D-1,1)$

When D is odd it is still true that the spinor representation $\mathbf{\Gamma}^{\mathrm{spin}}$ of so$(D-1,1)$ defined by (L.10) is *real* if either

(i) the matrices γ_p are a set of *real* matrices (for $p = 1,2,\ldots,D$),

or

(ii) the matrices γ_p are a set of *imaginary* matrices (for $p = 1,2,\ldots,D$).

In both cases the set of matrices γ_p are again said to form a *Majorana representation* of the D-dimensional Minkowski Clifford algebra, and may be denoted by γ^{M}_p. When such a representation exists it is not unique, for applying to it a similarity transformation of the form (L.8) with a real orthogonal matrix $\mathbf{S}$ will produce another Majorana representation.

Theorem II (a) For $D \bmod 8 = 1$ or 3, that is, if

$$D = \begin{cases} 9, 17, \ldots, \text{or} \\ 3, 11, \ldots, \end{cases}$$

the irreducible representation of the D-dimensional Minkowski Clifford algebra consisting of the matrices γ_p^{B} of (L.74)–(L.76) is equivalent to a Majorana representation. In such a case let $\mathbf{S}_{\mathrm{B}}^{\mathrm{M}}$ be the $d \times d$ non-singular matrix with the property that

$$\gamma_p^{\mathrm{M}} = \mathbf{S}_{\mathrm{B}}^{\mathrm{M}} \gamma_p^{\mathrm{B}} (\mathbf{S}_{\mathrm{B}}^{\mathrm{M}})^{-1} \tag{L.79}$$

for $p = 1, 2, \ldots, D$.

(b) For $D \bmod 8 = 1$ the matrices γ_p^{M} are all *real*. For $D = 1$ the matrix γ_1^{M} may be taken to be γ_1^{B}. For $D \geqslant 9$ the matrices $\mathbf{S}_{\mathrm{B}}^{\mathrm{M}}$ may be chosen so that

$$\mathbf{S}_{\mathrm{B}}^{\mathrm{M}} = 2^{-1/2}(\mathbf{1}_d + i\gamma_1^{\mathrm{B}}\gamma_2^{\mathrm{B}} \cdots \gamma_{(D-1)/2}^{\mathrm{B}}), \tag{L.80}$$

in which case

$$\gamma_p^{\mathrm{M}} = \begin{cases} \gamma_p^{\mathrm{B}} & \text{if } p = \tfrac{1}{2}(D-1)+1, \ldots, D, \\ -i\gamma_p^{\mathrm{B}}\gamma_1^{\mathrm{B}}\gamma_2^{\mathrm{B}} \cdots \gamma_{(D-1)/2}^{\mathrm{B}} & \text{if } p = 1, 2, \ldots, \tfrac{1}{2}(D-1). \end{cases} \tag{L.81}$$

(c) For $D \bmod 8 = 3$ the matrices γ_p^{M} are all *imaginary*, and $\mathbf{S}_{\mathrm{B}}^{\mathrm{M}}$ may be chosen so that

$$\mathbf{S}_{\mathrm{B}}^{\mathrm{M}} = 2^{-1/2}(\mathbf{1}_d - \gamma_1^{\mathrm{B}}\gamma_2^{\mathrm{B}} \cdots \gamma_{(D-1)/2}^{\mathrm{B}}), \tag{L.82}$$

in which case

$$\gamma_p^{\mathrm{M}} = \begin{cases} \gamma_p^{\mathrm{B}} & \text{if } p = 1, 2, \ldots, \tfrac{1}{2}(D-1), \\ \gamma_p^{\mathrm{B}}\gamma_1^{\mathrm{B}}\gamma_2^{\mathrm{B}} \cdots \gamma_{(D-1)/2}^{\mathrm{B}} & \text{if } p = \tfrac{1}{2}(D-1)+1, \ldots, \mathrm{D} \end{cases} \tag{L.83}$$

Proof The case $D = 1$ is trivial and obvious, so attention will be devoted to the case $D \geqslant 3$. The line of argument is essentially the same as for the proof of Theorem III of Section 2, and so only the new features will be mentioned.

Suppose first that the γ_p^{M} are all *imaginary* matrices. With the metric tensor of (L.3), the matrices $\gamma_1^{\mathrm{B}}, \gamma_2^{\mathrm{B}}, \ldots, \gamma_{(D-1)/2}^{\mathrm{B}}$ are imaginary and the matrices $\gamma_{(D-1)/2+1}^{\mathrm{B}}, \gamma_{(D-1)/2+2}^{\mathrm{B}}, \ldots, \gamma_{D-1}^{\mathrm{B}}$, and γ_D^{B} are real, so the condition (L.38) reduces to the pair of conditions

$$((\mathbf{S}_{\mathrm{B}}^{\mathrm{M}})^{-1}\mathbf{S}_{\mathrm{B}}^{\mathrm{M}*})\gamma_p^{\mathrm{B}}((\mathbf{S}_{\mathrm{B}}^{\mathrm{M}})^{-1}\mathbf{S}_{\mathrm{B}}^{\mathrm{M}*})^{-1} = \gamma_p^{\mathrm{B}} \quad \text{for } p = 1, 2, \ldots, \tfrac{1}{2}(D-1), \tag{L.84}$$

and

$$((\mathbf{S}_{\mathrm{B}}^{\mathrm{M}})^{-1}\mathbf{S}_{\mathrm{B}}^{\mathrm{M}*})\gamma_p^{\mathrm{B}}((\mathbf{S}_{\mathrm{B}}^{\mathrm{M}})^{-1}\mathbf{S}_{\mathrm{B}}^{\mathrm{M}*})^{-1} = -\gamma_p^{\mathrm{B}} \quad \text{for } p = \tfrac{1}{2}(D-1)+1, \tfrac{1}{2}(D-1)+2, \ldots, D. \tag{L.85}$$

Again $(\mathbf{S}^{\mathrm{M}}_{\mathrm{B}})^{-1}\mathbf{S}^{\mathrm{M}*}_{\mathrm{B}}$ has to be a linear combination of $\mathbf{1}_d$ and the products of the matrices $\boldsymbol{\gamma}^{\mathrm{B}}_p$. Use of (L.7) shows that there are *no* solutions of (L.84) and (L.85) if $D \bmod 8 = 1$ or 5, but for $D \bmod 8 = 3$ there is the possibility

$$(\mathbf{S}^{\mathrm{M}}_{\mathrm{B}})^{-1}\mathbf{S}^{\mathrm{M}*}_{\mathrm{B}} = \kappa\boldsymbol{\gamma}^{\mathrm{B}}_1\boldsymbol{\gamma}^{\mathrm{B}}_2\cdots\boldsymbol{\gamma}^{\mathrm{B}}_{(D-1)/2}, \tag{L.86}$$

and for $D \bmod 8 = 7$ there are the two possibilities

$$(\mathbf{S}^{\mathrm{M}}_{\mathrm{B}})^{-1}\mathbf{S}^{\mathrm{M}*}_{\mathrm{B}} = \kappa\boldsymbol{\gamma}^{\mathrm{B}}_1\boldsymbol{\gamma}^{\mathrm{B}}_2\cdots\boldsymbol{\gamma}^{\mathrm{B}}_{(D-1)/2} \tag{L.87}$$

and

$$(\mathbf{S}^{\mathrm{M}}_{\mathrm{B}})^{-1}\mathbf{S}^{\mathrm{M}*}_{\mathrm{B}} = \kappa\boldsymbol{\gamma}^{\mathrm{B}}_{(D-1)/2+1}\boldsymbol{\gamma}^{\mathrm{B}}_{(D-1)/2+2}\cdots\boldsymbol{\gamma}^{\mathrm{B}}_{D}. \tag{L.88}$$

However, for $D \bmod 8 = 7$ both of these lead to an inconsistency when the complex conjugate is taken and substituted back in, so there are no solutions to (L.85) and (L.85) if $D \bmod 8 = 7$. For $D \bmod 8 = 3$ (L.86) has no inconsistencies, and indeed always has a solution of the form (L.82).

This analysis is easily repeated for the case in which the $\boldsymbol{\gamma}^{\mathrm{B}}_p$ of (L.74)–(L.76) are assumed to be equivalent to *real* matrices. In this case the right-hand sides of (L.84) and (L.85) have to be multiplied by -1. There are no consistent solutions for $D \bmod 8 = 3, 5$ and 7, but there are solutions for $D \bmod 8 = 1$, of which one is given by (L.80).

It should be noted that *no* Majorana representations exist for $D \bmod 8 = 5$ and for $D \bmod 8 = 7$; that is, for $D = 5, 13, \ldots$, and $7, 15, \ldots$. Also as there is no chiral representation for odd D, there cannot be any Majorana–Weyl representations for odd D.

(*d*) *The generalized charge conjugation matrices*

If $\boldsymbol{\gamma}^{\mathrm{B}}_p$ are as defined in (L.74)–(L.76) then for *odd D* and *odd* $\frac{1}{2}(D-1)$ the representation of the D-dimensional Minkowski Clifford algebra given by the set of matrices $\tilde{\boldsymbol{\gamma}}^{\mathrm{B}}_p$ is equivalent to that given by the set $-\boldsymbol{\gamma}^{\mathrm{B}}_p$, whereas for *odd D* and *even* $\frac{1}{2}(D-1)$ the set of matrices $\tilde{\boldsymbol{\gamma}}^{\mathrm{B}}_p$ is equivalent to the set $\boldsymbol{\gamma}^{\mathrm{B}}_p$. Each gives rise to a generalized charge conjugation matrix $\mathbf{C}^{\mathrm{B}}$.

(i) For *D odd* and $\frac{1}{2}(D-1)$ *odd* the matrix $\mathbf{C}$ defined by

$$\mathbf{C}^{\mathrm{B}} = \eta\boldsymbol{\sigma}_2 \otimes \boldsymbol{\sigma}_1 \otimes \boldsymbol{\sigma}_2 \otimes \boldsymbol{\sigma}_1 \otimes \cdots, \tag{L.89}$$

which is the direct product of $\frac{1}{2}(D-1)$ Pauli spin matrices and for which η is some complex number, has the property that

$$\tilde{\boldsymbol{\gamma}}^{\mathrm{B}}_p = -(\mathbf{C}^{\mathrm{B}})^{-1}\boldsymbol{\gamma}^{\mathrm{B}}_p\mathbf{C}^{\mathrm{B}} \tag{L.90}$$

for $p = 1, 2, \ldots, D$. Moreover, every $d \times d$ matrix $\mathbf{C}^{\mathrm{B}}$ satisfying (L.90) must have the form (L.89). For $D \bmod 8 = 3$

$$\tilde{\mathbf{C}}^{\mathrm{B}} = -\mathbf{C}^{\mathrm{B}}, \tag{L.91}$$

but for $D \bmod 8 = 7$

$$\tilde{\mathbf{C}}^{\mathrm{B}} = \mathbf{C}^{\mathrm{B}}, \tag{L.92}$$

and in both cases

$$\mathbf{C}^{\mathrm{B}}\mathbf{C}^{\mathrm{B}\dagger} = \mathbf{C}^{\mathrm{B}\dagger}\mathbf{C}^{\mathrm{B}} = \mathbf{1}_d, \tag{L.93}$$

provided that η is chosen so that $|\eta|^2 = 1$.

If $D \bmod 8 = 3$ and $\mathbf{S}_{\mathrm{B}}^{\mathrm{M}}$ is the $d \times d$ non-singular matrix of (L.79) that transforms the $\boldsymbol{\gamma}_p^{\mathrm{B}}$ defined in (L.74)–(L.76) into a Majorana representation $\boldsymbol{\gamma}_p^{\mathrm{M}}$, the corresponding charge conjugation matrix $\mathbf{C}^{\mathrm{M}}$ of the Majorana representation is defined by

$$\mathbf{C}^{\mathrm{M}} = \mathbf{S}_{\mathrm{B}}^{\mathrm{M}}\mathbf{C}^{\mathrm{B}}\tilde{\mathbf{S}}_{\mathrm{B}}^{\mathrm{M}}. \tag{L.94}$$

Then

$$\tilde{\boldsymbol{\gamma}}_p^{\mathrm{M}} = -(\mathbf{C}^{\mathrm{M}})^{-1}\boldsymbol{\gamma}_p^{\mathrm{M}}\mathbf{C}^{\mathrm{M}} \tag{L.95}$$

for $p = 1, 2, \ldots, D$, and $\mathbf{C}^{\mathrm{M}}$ is antisymmetric if $D \bmod 8 = 3$. Moreover, by an appropriate choice of the factor η, $\mathbf{C}^{\mathrm{M}}$ can be chosen to be real or imaginary (whichever is required). Thus for $D \bmod 8 = 3$ the products $\boldsymbol{\gamma}_{\mathrm{M}}^{p}\mathbf{C}^{\mathrm{M}}$ can be taken to be real and symmetric for all $p = 1, 2, \ldots, D$. Although (L.94) could be used to obtain an explicit expression for $\mathbf{C}^{\mathrm{M}}$ in this case, the simpler alternative argument of Section 2 can be employed again. As $\mathbf{S}_{\mathrm{B}}^{\mathrm{M}}$ is a unitary matrix, the identity (L.78) retains its form in the Majorana representation, so that

$$(\boldsymbol{\gamma}_p^{\mathrm{M}})^{\dagger} = \boldsymbol{\gamma}_D^{\mathrm{M}}\boldsymbol{\gamma}_p^{\mathrm{M}}\boldsymbol{\gamma}_D^{\mathrm{M}} \tag{L.96}$$

for $p = 1, 2, \ldots, D$. For a Majorana representation by *imaginary* matrices this reduces to

$$\tilde{\boldsymbol{\gamma}}_p^{\mathrm{M}} = -\boldsymbol{\gamma}_D^{\mathrm{M}}\boldsymbol{\gamma}_p^{\mathrm{M}}\boldsymbol{\gamma}_D^{\mathrm{M}} \tag{L.97}$$

for $p = 1, 2, \ldots, D$, and hence $\mathbf{C}^{\mathrm{M}} = \xi\boldsymbol{\gamma}_D^{\mathrm{M}}$, where ξ is some complex number. With the choice $\xi = -1$, i.e.

$$\mathbf{C}^{\mathrm{M}} = -\boldsymbol{\gamma}_D^{\mathrm{M}}, \tag{L.98}$$

the products $\boldsymbol{\gamma}_{\mathrm{M}}^{p}\mathbf{C}^{\mathrm{M}}$ are all real and symmetric. With the choice (L.61), equations (L.3), (L.5) and (L.6) again imply that

$$\operatorname{tr}(\boldsymbol{\gamma}_{\mathrm{M}}^{p}\mathbf{C}^{\mathrm{M}}) = \begin{cases} 0 & \text{for } p = 1, 2, \ldots, D-1, \\ -D & \text{for } p = D. \end{cases} \tag{L.99}$$

(ii) For D *odd* and $\frac{1}{2}(D-1)$ *even* the matrix $\mathbf{C}^{\mathrm{B}}$ defined by

$$\mathbf{C}^{\mathrm{B}} = \eta \boldsymbol{\sigma}_1 \otimes \boldsymbol{\sigma}_2 \otimes \boldsymbol{\sigma}_1 \otimes \boldsymbol{\sigma}_2 \otimes \cdots, \tag{L.100}$$

which is the direct product of $\frac{1}{2}(D-1)$ Pauli spin matrices and where η is some complex number, has the property that

$$\tilde{\boldsymbol{\gamma}}_p^{\mathrm{B}} = (\mathbf{C}^{\mathrm{B}})^{-1} \boldsymbol{\gamma}_p^{\mathrm{B}} \mathbf{C}^{\mathrm{B}} \tag{L.101}$$

for $p = 1, 2, \ldots, D$. Moreover, every $d \times d$ matrix $\mathbf{C}^{\mathrm{B}}$ satisfying (L.101) must have the form (L.100). For $D \bmod 8 = 1$

$$\tilde{\mathbf{C}}^{\mathrm{B}} = \mathbf{C}^{\mathrm{B}}, \tag{L.102}$$

but for $D \bmod 8 = 5$

$$\tilde{\mathbf{C}}^{\mathrm{B}} = -\mathbf{C}^{\mathrm{B}}, \tag{L.103}$$

and in both cases

$$\mathbf{C}^{\mathrm{B}} \mathbf{C}^{\mathrm{B}\dagger} = \mathbf{C}^{\mathrm{B}\dagger} \mathbf{C}^{\mathrm{B}} = \mathbf{1}_d, \tag{L.104}$$

provided that η is chosen so that $|\eta|^2 = 1$.

If $D \bmod 8 = 1$ and $\mathbf{S}_{\mathrm{B}}^{\mathrm{M}}$ is the $d \times d$ non-singular matrix of (L.79) then the corresponding modified charge conjugation matrix $\mathbf{C}^{\mathrm{M}}$ of the Majorana representation is defined by

$$\mathbf{C}^{\mathrm{M}} = \mathbf{S}_{\mathrm{B}}^{\mathrm{M}} \mathbf{C}^{\mathrm{B}} \tilde{\mathbf{S}}_{\mathrm{B}}^{\mathrm{M}}. \tag{L.105}$$

Then

$$\tilde{\boldsymbol{\gamma}}_p^{\mathrm{M}} = (\mathbf{C}^{\mathrm{M}})^{-1} \boldsymbol{\gamma}_p^{\mathrm{M}} \mathbf{C}^{\mathrm{M}} \tag{L.106}$$

for $p = 1, 2, \ldots, D$, and $\mathbf{C}^{\mathrm{M}}$ is symmetric. Moreover, by an appropriate choice of the factor η, $\mathbf{C}^{\mathrm{M}}$ can be chosen to be real or imaginary (whichever is required). Thus for $D \bmod 8 = 1$ the products $\boldsymbol{\gamma}_{\mathrm{M}}^{p} \mathbf{C}^{\mathrm{M}}$ can be taken to be real and symmetric for all $p = 1, 2, \ldots, D$. Although (L.105) could be used to obtain an explicit expression for $\mathbf{C}^{\mathrm{M}}$, in this case the simpler alternative argument of (i) can be employed again. As $\mathbf{S}_{\mathrm{B}}^{\mathrm{M}}$ is a unitary matrix, the identity (L.78) retains its form (L.96) in the Majorana representation, so that for a Majorana representation by *real* matrices (L.96) reduces to

$$\tilde{\boldsymbol{\gamma}}_p^{\mathrm{M}} = \boldsymbol{\gamma}_D^{\mathrm{M}} \boldsymbol{\gamma}_p^{\mathrm{M}} \boldsymbol{\gamma}_D^{\mathrm{M}} \tag{L.107}$$

for $p = 1, 2, \ldots, D$, and hence $\mathbf{C}^{\mathrm{M}} = \xi \boldsymbol{\gamma}_D^{\mathrm{M}}$, where ξ is some complex number. With the choice $\xi = -1$, i.e.

$$\mathbf{C}^{\mathrm{M}} = -\boldsymbol{\gamma}_D^{\mathrm{M}}, \tag{L.108}$$

the products $\boldsymbol{\gamma}_{\mathrm{M}}^{p} \mathbf{C}^{\mathrm{M}}$ are all real and symmetric, and again (L.99) holds.

It should be noted that with *both* of the definitions (L.89) and (L.100) for

C it is true that

$$\tilde{\gamma}_q \tilde{\gamma}_p = \mathbf{C}^{-1} \gamma_q \gamma_p \mathbf{C} \tag{L.109}$$

for all $p, q = 1, 2, \ldots, D$, but that *only* for $D \bmod 8 = 1$ and 3 can **C** be chosen in such a way that the products $\gamma^p \mathbf{C}$ are *symmetric* for $p = 1, 2, \ldots, D$ (a property that is unchanged by similarity transformations) and such that it is also true that their Majorana forms $\gamma_{\mathrm{M}}^p \mathbf{C}^{\mathrm{M}}$ are all *real*. For $D \bmod 8 = 5$ and 7 *neither* of these two latter properties can be achieved.

4 Connections between representations of the D-dimensional Minkowski Clifford algebra and those of the $(D-2)$-dimensional Euclidean Clifford algebra

Some very useful information on the representations of the D-dimensional Minkowski Clifford algebra can be found from considering those of the $(D-2)$-dimensional Euclidean Clifford algebra

Theorem I For the D-dimensional *Euclidean* Clifford algebra with metric tensor (L.2) each irreducible representation of dimension d, where d is given by (L.15) or (L.73) as appropriate, can be chosen to be *real* if $D \bmod 8 = 0, 1$ or 2, and can be chosen to be *imaginary* if $D \bmod 8 = 0, 6$ or 7.

Proof The same line of argument as that given for Theorem III of Section 2 and Theorem II of Section 3 can be applied, since the explicit matrices quoted in (L.16), (L.17) and (L.74)–(L.76) are also valid for the D-dimensional *Euclidean* Clifford algebra provided that the metric tensor (L.2) is inserted. The details will be omitted.

As noted previously, if D is *even* then there is (up to equivalence) only *one* irreducible representation, whereas if D is *odd* then there are *two* inequivalent irreducible representations that merely differ in sign.

The connections between the real and imaginary irreducible representations of the D-dimensional Minkowski Clifford algebra and those of the $(D-2)$-dimensional Euclidean Clifford algebra are given by the following theorem.

Theorem II (a) If $D > 2$ and $D \bmod 8 = 0, 1$ or 2 then the D-dimensional Minkowski Clifford algebra with metric tensor (L.3) has a *real* irreducible representation of dimension d (where d is given by (L.15) or (L.73) as appropriate) whose matrices $\gamma_p^{(M,D)}$ are given in terms of the matrices $\gamma_p^{(E,D-2)}$ of an *imaginary* $\frac{1}{2}d$-dimensional irreducible representation of the $(D-2)$-dimensional Euclidean Clifford algebra with metric tensor (L.2) by

$$\gamma_p^{(M,D)} = \begin{bmatrix} i\gamma_p^{(E,D-2)} & \mathbf{0} \\ \mathbf{0} & -i\gamma_p^{(E,D-2)} \end{bmatrix} \quad \text{for } p = 1, 2, \ldots, D-2, \tag{L.110}$$

$$\gamma_{D-1}^{(M,D)} = \begin{bmatrix} \mathbf{0} & \mathbf{1} \\ -\mathbf{1} & \mathbf{0} \end{bmatrix}, \tag{L.111}$$

$$\gamma_D^{(M,D)} = \begin{bmatrix} \mathbf{0} & \mathbf{1} \\ \mathbf{1} & \mathbf{0} \end{bmatrix}. \tag{L.112}$$

(b) If $D > 2$ and $D \bmod 8 = 2, 3$ or 4 then the D-dimensional Minkowski Clifford algebra with metric tensor (L.3) has an *imaginary* irreducible representation of dimension d (where d is given by (L.15) or (L.73) as appropriate) whose matrices $\gamma_p^{(M,D)}$ are given in terms of the matrices $\gamma_p^{(E,D-2)}$ of a *real* $\frac{1}{2}d$-dimensional irreducible representation of the $(D-2)$-dimensional Euclidean Clifford algebra with metric tensor (L.2) by

$$\gamma_p^{(M,D)} = \begin{bmatrix} i\gamma_p^{(E,D-2)} & \mathbf{0} \\ \mathbf{0} & -i\gamma_p^{(E,D-2)} \end{bmatrix} \quad \text{for } p = 1, 2, \ldots, D-2, \tag{L.113}$$

$$\gamma_{D-1}^{(M,D)} = \begin{bmatrix} \mathbf{0} & i\mathbf{1} \\ i\mathbf{1} & \mathbf{0} \end{bmatrix}, \tag{L.114}$$

$$\gamma_D^{(M,D)} = \begin{bmatrix} \mathbf{0} & i\mathbf{1} \\ -i\mathbf{1} & \mathbf{0} \end{bmatrix}. \tag{L.115}$$

Proof It is easy to check that the above assertions are true.

Of course all the matrices of the D-dimensional Minkowski Clifford algebras that are mentioned are necessarily in a Majorana representation (although in the case $D = 4$ this representation is different (but equivalent) to that presented in (17.119) and (23.14)). With the corresponding generalized charge conjugation matrices $\mathbf{C}^{(M,D)}$ being defined by

$$\mathbf{C}^{(M,D)} = -\gamma_D^{(M,D)} \tag{L.116}$$

(cf. (L.61), (L.71), (L.98) and (L.108)), it follows from (L.111), (L.112), (L.114), and (L.115) that

$$(\gamma_{D-1}^{(M,D)} + \gamma_D^{(M,D)})\mathbf{C}^{(M,D)} = \begin{bmatrix} -2\mathbf{1}_{d/2} & \mathbf{0} \\ \mathbf{0} & \mathbf{0} \end{bmatrix}. \tag{L.117}$$

In (L.110)–(L.115) and (L.117) all submatrices are of dimension $\frac{1}{2}d \times \frac{1}{2}d$. It should be noted that this representation of the D-dimensional Minkowski Clifford algebra is *not chiral.*

Theorem III (a) If D is *even* and > 2 then the D-dimensional Minkowski Clifford algebra with metric tensor (L.3) has a *chiral* irreducible representation of dimension d (when d is given by (L.15)) whose matrices $\gamma_p^{(M,D)}$ are given in terms of the matrices $\gamma_p^{(E,D-2)}$ of a $\frac{1}{2}d$-dimensional irreducible representation of the $(D-2)$-dimensional Euclidean Clifford algebra with metric tensor (L.2) by

$$\gamma_p^{(M,D)} = \begin{bmatrix} \mathbf{0} & -\gamma_p^{(E,D-2)} \\ \gamma_p^{(E,D-2)} & \mathbf{0} \end{bmatrix} \quad \text{for } p = 1, 2, \ldots, D-1 \tag{L.118}$$

and

$$\gamma_D^{(M,D)} = \begin{bmatrix} \mathbf{0} & -\mathbf{1} \\ -\mathbf{1} & \mathbf{0} \end{bmatrix}, \tag{L.119}$$

where $\gamma_{D-1}^{(E,D-2)}$ is defined by

$$\gamma_{D-1}^{(E,D-2)} = -i^{(D-2)/2}\gamma_1^{(E,D-2)}\gamma_2^{(E,D-2)}\cdots\gamma_{D-3}^{(E,D-2)}\gamma_{D-2}^{(E,D-2)}. \tag{L.120}$$

(in (L.118)–(L.120) all the submatrices are of dimension $\frac{1}{2}d \times \frac{1}{2}d$.)

(b) For $D > 2$ and $D \bmod 8 = 2$ (i.e. for $D = 10, 18, 26, \ldots$) if the matrices $\gamma_p^{(E,D-2)}$ of the $(D-2)$-dimensional Euclidean Clifford algebra are chosen to be *real* for $p = 1, 2, \ldots, D-2$ then the matrices $\gamma_p^{(M,D)}$ of the D-dimensional Minkowski Clifford algebra that are defined by (L.118) and (L.119) are also *real.* Thus (L.118) and (L.119) provide a *Majorana–Weyl* representation of the D-dimensional Minkowski Clifford algebra for $D = 10, 18, 26, \ldots$.

(c) For $D > 2$ and $D \bmod 8 = 2$ (i.e. for $D = 10, 18, 26, \ldots$) the $\frac{1}{2}d \times \frac{1}{2}d$ matrices $\gamma_p^{(E,D-2)}$ of the $(D-2)$-dimensional Euclidean Clifford algebra can not only be chosen to be *real* for $p = 1, 2, \ldots, D-2$ but can *also* be chosen to be *chiral* in the sense that

$$\gamma_{D-1}^{(E,D-2)} = \begin{bmatrix} \mathbf{1}_{d/4} & \mathbf{0} \\ \mathbf{0} & -\mathbf{1}_{d/4} \end{bmatrix}; \tag{L.121}$$

that is, they can be chosen themselves to form a *Majorana–Weyl* representation.

Proof (a) With the phase choice in (L.120), the defining relations (L.4) for the Euclidean metric tensor (L.2) imply that

$$(\gamma_{D-1}^{(\mathrm{E},D-2)})^2 = \mathbf{1}_{d/2} \tag{L.122}$$

and

$$\gamma_{D-1}^{(\mathrm{E},D-2)}\gamma_p^{(\mathrm{E},D-2)} + \gamma_p^{(\mathrm{E},D-2)}\gamma_{D-1}^{(\mathrm{E},D-2)} = \mathbf{0} \tag{L.123}$$

for $p = 1, 2, \ldots, D-2$. It can then be verified that the matrices of (L.118) and (L.119) form a representation of the D-dimensional Minkowski Clifford algebra. Moreover, by (L.20),

$$\gamma_{D+1}^{(\mathrm{M},D)} = \begin{bmatrix} \mathbf{1}_{d/2} & \mathbf{0} \\ 0 & -\mathbf{1}_{d/2} \end{bmatrix}, \tag{L.124}$$

which shows that this representation is *chiral*.

(b) As $(D-2) \bmod 8 = 0$ when $D \bmod 8 = 2$, Theorem I above shows that the $(D-2)$-dimensional Euclidean Clifford algebra does possess a *real* representation of dimension $\frac{1}{2}d$. By (L.120), the matrix $\gamma_{D-1}^{(\mathrm{E},D-2)}$ is also real, and the reality of the matrices of (L.118) and (L.119) is then obvious.

(c) The argument is similar to that for (a) and (b) and just involves defining the *real* $\frac{1}{2}d \times \frac{1}{2}d$ matrices $\gamma_p^{(\mathrm{E},D-2)}$ of the $(D-2)$-dimensional Euclidean Clifford algebra in terms of the *imaginary* $\frac{1}{4}d \times \frac{1}{4}d$ matrices $\gamma_p^{(\mathrm{E},D-4)}$ of the $(D-4)$-dimensional Euclidean Clifford algebra (which Theorem I above shows to exist) by the prescription that

$$\gamma_p^{(\mathrm{E},D-2)} = \begin{bmatrix} \mathbf{0} & i\gamma_p^{(\mathrm{E},D-4)} \\ -i\gamma_p^{(\mathrm{E},D-4)} & \mathbf{0} \end{bmatrix} \quad \text{for } p = 1, 2, \ldots, D-3 \tag{L.125}$$

and

$$\gamma_{D-2}^{(\mathrm{E},D-2)} = \begin{bmatrix} \mathbf{0} & -\mathbf{1} \\ -\mathbf{1} & \mathbf{0} \end{bmatrix}. \tag{L.126}$$

(In conformity with (L.120), $\gamma_{D-3}^{(\mathrm{E},D-4)}$ is defined by

$$\gamma_{D-3}^{(\mathrm{E},D-4)} = -i^{(D-4)/2}\gamma_1^{(\mathrm{E},D-4)}\gamma_2^{(\mathrm{E},D-4)}\cdots\gamma_{D-5}^{(\mathrm{E},D-4)}\gamma_{D-4}^{(\mathrm{E},D-4)},$$

which is also imaginary.) The stated properties of the matrices $\gamma_p^{(\mathrm{E},D-2)}$ for $p = 1, 2, \ldots, D-1$ then follow immediately. (In (L.125) and (L.126) all the submatrices are of dimension $\frac{1}{4}d \times \frac{1}{4}d$.)

With the *Majorana–Weyl* representation of the $(D-2)$-dimensional Euclidean Clifford algebra substituted into (L.118) and (L.119) to give a *Majorana–Weyl* representation of the D-dimensional Minkowski Clifford algebra, and in particular with (L.121) substituted into (L.118), and with the charge conjugation again being chosen so that $\mathbf{C}^{(\mathrm{M},D)} = -\gamma_D^{(\mathrm{M},D)}$ (as in (L.116)), it follows from (L.118) and (L.119) that

$$(\gamma_{D-1}^{(\mathrm{M},D)} + \gamma_D^{(\mathrm{M},D)})\mathbf{C}^{(\mathrm{M},D)} = \begin{bmatrix} -2\mathbf{1}_{d/4} & \mathbf{0} & \mathbf{0} & \mathbf{0} \\ \mathbf{0} & \mathbf{0} & \mathbf{0} & \mathbf{0} \\ \mathbf{0} & \mathbf{0} & \mathbf{0} & \mathbf{0} \\ \mathbf{0} & \mathbf{0} & \mathbf{0} & -2\mathbf{1}_{d/4} \end{bmatrix} \tag{L.127}$$

(In (L.127) all the submatrices are of dimension $\frac{1}{4}d \times \frac{1}{4}d$.)

Moreover, with this *Majorana–Weyl* representation of the $(D-2)$-dimensional Euclidean Clifford algebra,

(i) the d-dimensional spinor representations $\mathbf{\Gamma}_{\mathrm{C}}^{\mathrm{spin}}$ of so$(D-1,1)$ reduces to the two $\frac{1}{2}d$-dimensional representations $\mathbf{\Gamma}_{\mathrm{L}}$ and $\mathbf{\Gamma}_{\mathrm{R}}$ of so$(D-1,1)$ (as in (L.25)); that is,

$$\mathbf{\Gamma}_{\mathrm{C}}^{\mathrm{spin}}(L_{pq}) = \begin{bmatrix} \mathbf{\Gamma}_{\mathrm{L}}(L_{pq}) & \mathbf{0} \\ \mathbf{0} & \mathbf{\Gamma}_{\mathrm{R}}(L_{pq}) \end{bmatrix},$$

(ii) on the so$(D-2)$ subalgebra with basis elements L_{pq} with $p, q = 1, 2, \ldots, D-2$ (and $p < q$) (L.118) shows that these two representations $\mathbf{\Gamma}_{\mathrm{L}}$ and $\mathbf{\Gamma}_{\mathrm{R}}$ are *identical*; and

(iii) on this so$(D-2)$ subalgebra (L.125) and (L.126) imply that

$$\mathbf{\Gamma}_{\mathrm{L}}(L_{pq}) = \mathbf{\Gamma}_{\mathrm{R}}(L_{pq}) = \begin{bmatrix} \mathbf{\Gamma}_{\mathrm{L}}^{\mathrm{so}(D-2)}(L_{pq}) & \mathbf{0} \\ \mathbf{0} & \mathbf{\Gamma}_{\mathrm{R}}^{\mathrm{so}(D-2)}(L_{pq}) \end{bmatrix}, \tag{L.128}$$

where $\mathbf{\Gamma}_{\mathrm{L}}^{\mathrm{so}(D-2)}$ and $\mathbf{\Gamma}_{\mathrm{R}}^{\mathrm{so}(D-2)}$ are the two inequivalent left- and right-handed $\frac{1}{4}d$-dimensional irreducible representations of so$(D-2)$, which are given by

$$\mathbf{\Gamma}_{\mathrm{L}}^{\mathrm{so}(D-2)}(L_{pq}) = \mathbf{\Gamma}_{\mathrm{R}}^{\mathrm{so}(D-2)}(L_{pq}) = \tfrac{1}{2}\gamma_p^{(\mathrm{E},D-4)}\gamma_q^{(\mathrm{E},D-4)} \tag{L.129}$$

for $p, q = 1, 2, \ldots, D-3$ (with $p \neq q$), but

$$\mathbf{\Gamma}_{\mathrm{L}}^{\mathrm{so}(D-2)}(L_{pq}) = \tfrac{1}{2}i\gamma_p^{(\mathrm{E},D-4)} \tag{L.130}$$

and

$$\mathbf{\Gamma}_{\mathrm{R}}^{\mathrm{so}(D-2)}(L_{pq}) = -\tfrac{1}{2}i\gamma_p^{(\mathrm{E},D-4)} \tag{L.131}$$

for $p = 1, 2, \ldots, D-3$ and $q = D-2$.

In the interesting case in which $D = 10$, since $d = 32$, the representation $\mathbf{\Gamma}_C^{spin}$ of so(9, 1) is 32-dimensional, the irreducible representations $\mathbf{\Gamma}_L$ and $\mathbf{\Gamma}_R$ of so(9, 1) are 16-dimensional, and the irreducible representations $\mathbf{\Gamma}_L^{so(8)}$ and $\mathbf{\Gamma}_R^{so(8)}$ of so(8) are 8-dimensional.

5 A matrix identity for the $D = 4$ Minkowski Clifford algebra

The 4-dimensional Minkowski Clifford algebra possesses 16 ($=2^4$) basis elements. These may be taken to consist of the identity 1, the 4 generators γ_r (for $r = 1, \ldots, 4$), the 6 products $\gamma_r\gamma_s$ (for $r, s = 1, \ldots, 4$, with $r < s$), γ_5 and the 4 products $\gamma_5\gamma_r$ (for $r = 1, \ldots, 4$). Let γ_Λ denote a typical element, and $\boldsymbol{\gamma}_\Lambda$ its corresponding 4-dimensional matrix representation. In every case

$$(\gamma_\Lambda)^2 = \xi_\Lambda 1,$$

where $\xi_\Lambda = 1$ or -1. For example, $\xi_\Lambda = -1$ for $\gamma_\Lambda = \gamma_1$ but $\xi_\Lambda = 1$ for $\gamma_\Lambda = \gamma_5\gamma_1$. If $\gamma_\Lambda = \gamma_{\Lambda'}$ it follows that

$$\mathrm{tr}\,(\boldsymbol{\gamma}_\Lambda \boldsymbol{\gamma}_{\Lambda'}) = 4\xi_\Lambda, \tag{L.132}$$

whereas if $\gamma_\Lambda \neq \gamma_{\Lambda'}$ then

$$\mathrm{tr}\,(\boldsymbol{\gamma}_\Lambda \boldsymbol{\gamma}_{\Lambda'}) = 0. \tag{L.133}$$

Now let $\mathbf{M}$ and $\mathbf{N}$ be any 4×4 matrices (with real or complex entries) and consider the set of products $M_{aa'}N_{bb'}$ in which a' and b are fixed but a and b' are allowed to vary over all values from 1 to 4. These can be regarded as forming a 4×4 matrix with rows specified by a and columns by b'. As the complete set of the $\boldsymbol{\gamma}_\Lambda$ form a basis for the set of all 4×4 matrices, one can write

$$M_{aa'}N_{bb'} = \sum_\Lambda c(\Lambda)_{a'b}(\boldsymbol{\gamma}_\Lambda)_{ab'} \tag{L.134}$$

(for $a, a', b, b' = 1, \ldots, 4$), where the $c(\Lambda)_{a'b}$ are a set of real or complex numbers. Multiplying both sides of (L.134) by $(\boldsymbol{\gamma}_{\Lambda'})_{b'a}$, summing over a and b' from 1 to 4 and invoking (L.132) and (L.133) gives $c(\Lambda)_{a'b} = \frac{1}{4}(\mathbf{N}\boldsymbol{\gamma}_\Lambda\mathbf{M})_{ba'}/\xi_\Lambda$. Substituting back into (L.134) then gives

$$M_{aa'}N_{bb'} = \frac{1}{4}\sum_\Lambda \frac{(\mathbf{N}\boldsymbol{\gamma}_\Lambda\mathbf{M})_{ba'}(\boldsymbol{\gamma}_\Lambda)_{ab'}}{\xi_\Lambda}. \tag{L.135}$$

Let $\boldsymbol{\psi}^1, \boldsymbol{\psi}^2, \boldsymbol{\psi}^3$, and $\boldsymbol{\psi}^4$ be any four 4×1 column vectors whose component quantities mutually *anticommute*; that is, suppose that $[(\boldsymbol{\psi}^r)_a, (\boldsymbol{\psi}^s)_b]_+ = 0$ if

$r \neq s$ (for all a and $b = 1, \ldots, 4$). Then multiplying both sides of (L.135) by $(\boldsymbol{\psi}^1)_a(\boldsymbol{\psi}^2)_{a'}(\boldsymbol{\psi}^3)_b(\boldsymbol{\psi}^4)_{b'}$ and summing over all values of a, a', b and b' from 1 to 4 gives

$$(\tilde{\boldsymbol{\psi}}^1 \mathbf{M} \boldsymbol{\psi}^2)(\tilde{\boldsymbol{\psi}}^3 \mathbf{N} \boldsymbol{\psi}^4) = -\frac{1}{4} \sum_{\Lambda} \frac{(\tilde{\boldsymbol{\psi}}^3 \mathbf{N} \boldsymbol{\gamma}_\Lambda \mathbf{M} \boldsymbol{\psi}^2)(\tilde{\boldsymbol{\psi}} \boldsymbol{\gamma}_\Lambda \boldsymbol{\psi}^4)}{\xi_\Lambda} \tag{L.136}$$

(Wess and Zumino 1974a).

Appendix M

Properties of the Classical Simple Complex Lie Superalgebras

In this appendix frequent use will be made of the square matrices $\mathbf{e}_{pq}$ defined by

$$(\mathbf{e}_{pq})_{jk} = \delta_{jp}\delta_{kq}$$

(i.e. the only non-zero element of $\mathbf{e}_{pq}$ is a 1 in the (p, q) position). This matrix will sometimes be denoted alternatively by $\mathbf{e}_{p,q}$. Its dimension will always be clear from the context.

1 The basic type 1 classical simple complex Lie superalgebras $A(r/s)$, $r > s \geq 0$

(a) A realization of $\tilde{\mathscr{L}}_s = A(r/s)$ (for $r > s \geq 0$) is provided by $\mathrm{sl}(r+1/s+1; \mathbb{C})$, considered as a complex Lie superalgebra. (See Example II of Chapter 21, Section 2 for the definition and details of $\mathrm{sl}(r+1/s+1; \mathbb{C})$.) It follows from (21.30) and (21.31) that the dimension m of the even part $\tilde{\mathscr{L}}_0$ of $\tilde{\mathscr{L}}_s$ is given by $m = (r+1)^2 + (s+1)^2 - 1$, and the dimension n of the odd part $\tilde{\mathscr{L}}_1$ of $\tilde{\mathscr{L}}_s$ is given by $n = 2(r+1)(s+1)$.

(b) The *even subalgebra* $\tilde{\mathscr{L}}_0$ of $\tilde{\mathscr{L}}_s = A(r/s)$ (for $r > s \geq 0$) is $\tilde{\mathscr{L}}_0 = \tilde{\mathscr{L}}_0^{\mathrm{A}} \oplus \tilde{\mathscr{L}}_0^{\mathrm{ss}}$, where $\tilde{\mathscr{L}}_0^{\mathrm{A}}$ is a one-dimensional Abelian Lie algebra, and $\tilde{\mathscr{L}}_0^{\mathrm{ss}} = \tilde{\mathscr{L}}_0^{\mathrm{s}1} \oplus \tilde{\mathscr{L}}_0^{\mathrm{s}2}$ with $\tilde{\mathscr{L}}_0^{\mathrm{s}1} = A_r$ and $\tilde{\mathscr{L}}_0^{\mathrm{s}2} = A_s$ if $s > 0$, but $\tilde{\mathscr{L}}_0^{\mathrm{ss}} = \tilde{\mathscr{L}}_0^{\mathrm{s}1} = A_r$ if $s = 0$. The *rank* l of $\tilde{\mathscr{L}}_s$ is given by $l = r + s + 1$.

(c) A convenient basis of the *Cartan subalgebra* $\mathscr{H}_s$ of $\tilde{\mathscr{L}}_s = A(r/s)$ (for

$r > s \geqslant 0$) is provided by

$$\mathbf{c} = -\begin{bmatrix} \frac{s+1}{r-s}\mathbf{1}_{r+1} & \mathbf{0} \\ \mathbf{0} & \frac{r+1}{r-s}\mathbf{1}_{s+1} \end{bmatrix}$$

(which provides a basis for $\tilde{\mathscr{L}}_0^{\mathrm{A}}$),

$$\mathbf{h}_j^1 = \mathbf{e}_{j,j} - \mathbf{e}_{j+1,j+1}, \quad j = 1, 2, \ldots, r$$

(which provides a basis for the Cartan subalgebra $\mathscr{H}_0^1$ of $\tilde{\mathscr{L}}_0^{s1} = A_r$) and, for $s > 0$,

$$\mathbf{h}_j^2 = \mathbf{e}_{j+r+1,j+r+1} - \mathbf{e}_{j+r+2,j+r+2}, \quad j = 1, 2, \ldots, s$$

(which provides a basis for the Cartan subalgebra $\mathscr{H}_0^2$ of $\tilde{\mathscr{L}}_0^{s2} = A_s$).

(d) Two useful sets of *linear functionals* of $\mathscr{H}_s$ are defined for $s > 0$ by

$$\varepsilon_p^1(\mathbf{h}) = \begin{cases} 1 & \text{if } \mathbf{h} = \mathbf{c}, \\ \delta_{jp} - \delta_{j+1,p} & \text{if } \mathbf{h} = \mathbf{h}_j^1, j = 1, 2, \ldots, r, \\ 0 & \text{if } \mathbf{h} = \mathbf{h}_j^2, j = 1, 2, \ldots, s, \end{cases}$$

for $p = 1, 2, \ldots, r+1$, and

$$\varepsilon_p^2(\mathbf{h}) = \begin{cases} 0 & \text{if } \mathbf{h} = \mathbf{c}, \\ \delta_{jp} - \delta_{j+1,p} & \text{if } \mathbf{h} = \mathbf{h}_j^2, j = 1, 2, \ldots, s, \\ 0 & \text{if } \mathbf{h} = \mathbf{h}_j^1, j = 1, 2, \ldots, r, \end{cases}$$

for $p = 1, 2, \ldots, s+1$. The set $\varepsilon_1^1, \varepsilon_2^1, \ldots, \varepsilon_{r+1}^1$ are linearly independent, but the set $\varepsilon_1^2, \varepsilon_2^2, \ldots, \varepsilon_{s+1}^2$ are not, since

$$\varepsilon_{s+1}^2 = -\sum_{p=1}^{s} \varepsilon_p^2.$$

For $s = 0$ the definitions are

$$\varepsilon_p^1(\mathbf{h}) = \begin{cases} 1 & \text{if } \mathbf{h} = \mathbf{c}, \\ \delta_{jp} - \delta_{j+1,p} & \text{if } \mathbf{h} = \mathbf{h}_j^1, \quad j = 1, 2, \ldots, r, \end{cases}$$

and

$$\varepsilon_1^2(\mathbf{h}) = 0$$

for all $\mathbf{h} \in \mathscr{H}_s$.

(e) $\tilde{\mathscr{L}}_s = A(r/s)$ (for $r > s \geqslant 0$) has $r(r+1) + s(s+1)$ *even roots*, which fall into two classes:

(1) $\alpha = \varepsilon_p^1 - \varepsilon_q^1$ for $p, q = 1, 2, \ldots, r+1$ $(p \neq q)$ (for which the corresponding root subspace $\tilde{\mathscr{L}}_{s\alpha}$ is one-dimensional and has basis $\mathbf{e}_{p,q}$), which is an extension of a root of $\tilde{\mathscr{L}}_0^{s1} = A_r$;

(2) $\alpha = \varepsilon_p^2 - \varepsilon_q^2$ for $p, q = 1, 2, \ldots, s+1$ $(p \neq q)$ (for which the corresponding root subspace $\tilde{\mathscr{L}}_{s\alpha}$ is one-dimensional and has basis $\mathbf{e}_{p+r+1, q+r+1}$), which is an extension of a root of $\tilde{\mathscr{L}}_0^{s2} = A_s$, provided that $s > 0$. If $s = 0$, this latter class of even roots is empty.

(f) $\tilde{\mathscr{L}}_s = A(r/s)$ (for $r > s \geqslant 0$) has $2(r+1)(s+1)$ *odd roots*, which fall into two classes:

(1) $\alpha = \varepsilon_q^1 - \varepsilon_q^2$ for $p = 1, 2, \ldots, r+1$ and $q = 1, 2, \ldots, s+1$ (for which the corresponding root subspace $\tilde{\mathscr{L}}_{s\alpha}$ is one-dimensional and has basis $\mathbf{e}_{p,q+r+1}$);

(2) $\alpha = -(\varepsilon_p^1 - \varepsilon_q^2)$ for $p = 1, 2, \ldots, r+1$ and $q = 1, 2, \ldots, s+1$ (for which the corresponding root subspace $\tilde{\mathscr{L}}_{s\alpha}$ is one-dimensional and has basis $\mathbf{e}_{q+r+1,p}$).

(g) The *distinguished* set of *simple roots* of $\tilde{\mathscr{L}}_s = A(r/s)$ (for $r > s \geqslant 0$) may be taken to be

(1) simple even roots

$$\alpha_j = \varepsilon_j^1 - \varepsilon_{j+1}^1 \quad \text{for } j = 1, 2, \ldots, r,$$

and, for $s > 0$,

$$\alpha_{j+r+1} = \varepsilon_j^2 - \varepsilon_{j+1}^2 \quad \text{for } j = 1, 2, \ldots, s;$$

(2) simple odd root

$$\alpha_{r+1} = \varepsilon_{r+1}^1 - \varepsilon_1^2.$$

The number of simple roots is $L = l = r + s + 1$.

(h) In terms of these simple roots, the expressions for the *positive even* roots of $\tilde{\mathscr{L}}_s = A(r/s)$ (for $r > s \geqslant 0$) are

(1) for $p, q = 1, 2, \ldots, r+1$ with $p < q$

$$\alpha = \varepsilon_p^1 - \varepsilon_q^1 = \sum_{j=p}^{q-1} \alpha_j;$$

(2) provided that $s > 0$, for $p, q = 1, 2, \ldots, s+1$ with $p < q$

$$\alpha = \varepsilon_p^2 - \varepsilon_q^2 = \sum_{j=p}^{q-1} \alpha_{j+r+1}.$$

In particular, the extensions of the *simple* roots of $\tilde{\mathscr{L}}_0^{s1} = A_r$ are

$$\alpha_j^1 = \varepsilon_j^1 - \varepsilon_{j+1}^1 = \alpha_j \quad \text{for } j = 1, 2, \ldots, r,$$

and the extensions of the *simple* roots of $\tilde{\mathscr{L}}_0^{s2} = A_s$ are (for $s > 0$)

$$\alpha_j^2 = \varepsilon_j^2 - \varepsilon_{j+1}^2 = \alpha_{j+r+1} \quad \text{for } j = 1, 2, \ldots, s.$$

That is, for $A(r/s)$ (with $r > s \geqslant 0$) *every* extension of a *simple* root of $\tilde{\mathscr{L}}_0^{ss}$ is a *simple* root of $\tilde{\mathscr{L}}_s$.

(i) In terms of these simple roots $\alpha_1, \alpha_2, \ldots, \alpha_{r+s+1}$, the expressions for the *positive odd* roots of $\tilde{\mathscr{L}}_s = A(r/s)$ (for $r > s \geqslant 0$) are

$$\alpha = \varepsilon_p^1 - \varepsilon_q^2 = \begin{cases} \alpha_{r+1} & \text{for } p = r+1 \text{ and } q = 1, \\ \alpha_{r+1} + \sum_{j=p}^{r} \alpha_j & \text{for } p = 1, 2, \ldots, r \text{ and } q = 1, \\ \alpha_{r+1} + \sum_{k=2}^{q} \alpha_{k+r} & \text{for } p = r+1 \text{ and } q = 2, 3, \ldots, s+1, \\ \alpha_{r+1} + \sum_{j=p}^{r} \alpha_j + \sum_{k=2}^{q} \alpha_{k+r} & \text{for } p = 1, 2, \ldots, r \\ & \text{and } q = 2, 3, \ldots, s+1. \end{cases} \tag{M.1}$$

(The latter two sets do not occur if $s = 0$.)

(j) The *Killing form* of $\tilde{\mathscr{L}}_s = A(r/s)$ (for $r > s \geqslant 0$) is given by

$$B(\mathbf{M}, \mathbf{N}) = 2(r-s)\,\mathrm{str}\,(\mathbf{MN}) \tag{M.2}$$

for all $\mathbf{M}, \mathbf{N} \in A(r/s)$, and is *not identically zero*.

(k) For $j = 1, 2, \ldots, r$

$$\mathbf{h}_{\alpha_j} = \frac{1}{2(r-s)} \mathbf{h}_j^1,$$

and, if $s > 0$, for $j = 1, 2, \ldots, s$

$$\mathbf{h}_{\alpha_{j+r+1}} = -\frac{1}{2(r-s)} \mathbf{h}_j^2.$$

Also

$$\mathbf{h}_{\alpha_{r+1}} = -\frac{1}{2(r+1)(s+1)}\mathbf{c} - \frac{1}{2(r+1)(r-s)}\sum_{j=1}^{r} j\mathbf{h}_j^1 + \frac{1}{2(s+1)(r-s)}\sum_{j=1}^{s}(s+1-j)\mathbf{h}_j^2$$

(with the last term being absent if $s = 0$).

(l) For $s > 0$ the quantities $\langle \alpha_j, \alpha_k \rangle$ of $\tilde{\mathscr{L}}_s = A(r/s)$ (for $r > s \geqslant 0$) are given by

$$\langle \alpha_j, \alpha_k \rangle = \begin{cases} (r-s)^{-1} & \text{if } j = k \quad (j = 1, 2, \ldots, r), \\ -(r-s)^{-1} & \text{if } j = k \quad (j = r+2, r+3, \ldots, r+s+1), \\ \{2(r-s)\}^{-1} & \text{if } j = k \pm 1 \quad (j, k = r+1, r+2, \ldots, r+s+1), \\ -\{2(r-s)\}^{-1} & \text{if } j = k \pm 1 \quad (j, k = 1, 2, \ldots, r+1), \\ 0 & \text{for all other values of } j \text{ and } k. \end{cases}$$

For $s = 0$ the corresponding quantities are

$$\langle \alpha_j, \alpha_k \rangle = \begin{cases} r^{-1} & \text{if } j = \quad (j = 1, 2, \ldots, r), \\ -(2)^{-1} & \text{if } j = k \pm 1 \quad (j = 1, 2, \ldots, r+1), \\ 0 & \text{for all other values of } j \text{ and } k. \end{cases}$$

It should be noted that for both cases

$$\langle \alpha_{r+1}, \alpha_{r+1} \rangle = 0.$$

(m) For $s > 0$, with the index $(r+1)'$ of (24.64) being chosen to be $r+2$, the *Cartan matrix* of $A(r/s)$ is the $(r+s+1) \times (r+s+1)$ matrix

$$\mathbf{A} = \left[\begin{array}{c|c|c} & 0 & \\ & \vdots & \\ \mathbf{A}(A_r) & & \mathbf{0} \\ & 0 & \\ & -1 & \\ \hline 0 \cdots 0 \; -1 & 0 & 1 \quad 0 \cdots 0 \\ \hline & -1 & \\ & \vdots & \\ \mathbf{0} & & \mathbf{A}(A_s) \\ & 0 & \\ & 0 & \end{array}\right], \tag{M.3}$$

where the $(r+1)$th row and column have been partitioned off for extra clarity, where $\mathbf{A}(A_r)$ and $\mathbf{A}(A_s)$ are the Cartan matrices of the simple Lie algebras A_r and A_s respectively (cf. Appendix F, Section 1), and $\mathbf{0}$ denotes a submatrix whose entries are all zero.

Similarly, for $s = 0$, with the index $(r+1)'$ being chosen to be r, the Cartan matrix of $A(r/s)$ is the $(r+1) \times (r+1)$ matrix

$$\mathbf{A} = \left[\begin{array}{c|c} & 0 \\ \mathbf{A}(A_r) & \vdots \\ & 0 \\ & -1 \\ \hline 0 \cdots 0 \quad 1 & 0 \end{array}\right]$$

(n) The *generalized Dynkin diagrams* corresponding to $A(r/s)$ for $r > s > 0$ and $A(r/0)$ are shown in Figures M.1 and M.2 respectively.

(o) For *every odd root* α of $\tilde{\mathscr{L}}_s = A(r/s)$ (for $r > s \geqslant 0$)

$$\langle \alpha, \alpha \rangle = 0.$$

(p) As $\tilde{\mathscr{L}}_0^A \neq \{0\}$, the *representation of* $\tilde{\mathscr{L}}_0$ *on* $\tilde{\mathscr{L}}_1$ is the direct sum of two irreducible representations of $\tilde{\mathscr{L}}_0$ (cf. Theorem VIII of Chapter 25, Section 3(*a*)).

(1) One of these has as its carrier space the union of all root subspaces $\tilde{\mathscr{L}}_{s\alpha}$ corresponding to all the positive roots. On $\tilde{\mathscr{L}}_0^{ss}$ this representation is the direct product of the $(r+1)$-dimensional irreducible representation of $\tilde{\mathscr{L}}_0^{s1} = A_r$ whose highest weight is $\Lambda = \Lambda_1$ with the $(s+1)$-dimensional irreducible representation of $\tilde{\mathscr{L}}_0^{s2} = A_s$ whose highest weight is $\Lambda = \Lambda_s$. The Dynkin index of the first of these irreducible representations is $\{2(r+1)\}^{-1}$, and as the direct product

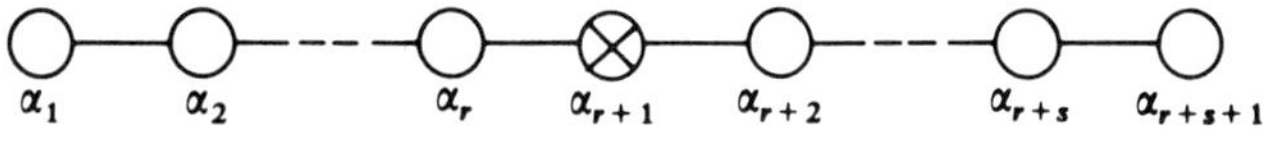

Figure M.1 Generalized Dynkin diagram for $A(r/s)$ (for $r > s > 0$) with the "distinguished" choice of simple roots.

Figure M.2 Generalized Dynkin diagram for $A(r/0)$ (for $r > 0$) with the "distinguished" choice of simple roots.

representation reduces to $s+1$ copies of this irreducible representation on $\tilde{\mathscr{L}}_0^{s1}$, the Dynkin index on $\tilde{\mathscr{L}}_0^{s1}$ of this direct product representation is $(s+1)/2(r+1)$. Similarly the Dynkin index of this direct product representation on $\tilde{\mathscr{L}}_0^{s2}$ is $(r+1)/2(s+1)$.

(2) The other irreducible representation of $\tilde{\mathscr{L}}_0$ corresponds to the negative odd roots. On $\tilde{\mathscr{L}}_0^{ss}$ it is again a direct product of an $(r+1)$-dimensional irreducible representation of $\tilde{\mathscr{L}}_0^{s1}$ (this time with highest weight $\Lambda = \Lambda_r$) with an $(s+1)$-dimensional irreducible representation of $\tilde{\mathscr{L}}_0^{s2}$ (whose highest weight is $\Lambda = \Lambda_1$). The Dynkin indices are as for the representation of (1).

It follows from these considerations that the Dynkin indices of (24.20) are given by

$$\gamma_{\mathbf{D}}^1 = \frac{s+1}{r+1}, \quad \gamma_{\mathbf{D}}^2 = \frac{r+1}{s+1},$$

so

$$1 - \gamma_{\mathbf{D}}^1 = \frac{r-s}{r+1}, \quad 1 - \gamma_{\mathbf{D}}^2 = -\frac{r-s}{s+1}. \tag{M.4}$$

In the discussion just given it has been assumed that $s > 0$, but the modifications required for the case $s = 0$ are obvious and trivial.

2 The basic type I classical simple complex Lie superalgebras $A(r/r), r \geqslant 1$

(a) Consider $sl(r+1/r+1;\mathbb{C})$ regarded as a complex Lie superalgebra. (See Example II of Chapter 21, Section 2 for the definition and details of $sl(r+1/r+1;\mathbb{C})$). The one-dimensional subspace with basis element $\mathbf{1}_{2r+2}$ forms a graded invariant subalgebra, so $sl(r+1/r+1;\mathbb{C})$ is not simple. For $r \geqslant 0$ the Lie superalgebra $A(r/r)$ may be *defined* as $sl(r+1/r+1;\mathbb{C})$ *factored* by this subspace (cf. Chapter 25, Section 2 for a general treatment of this process). This implies that if $\mathbf{M}$ and $\mathbf{N}$ are any two matrices of $sl(r+1/r+1;\mathbb{C})$ that are such that

$$\mathbf{M} - \mathbf{N} = \kappa \mathbf{1}_{2r+2}, \tag{M.5}$$

where κ is some complex number, then $\mathbf{M}$ and $\mathbf{N}$ are to be regarded as corresponding to the *same* element of $A(r/r)$. With this proviso, no confusion will arise if the elements of $A(r/r)$ are denoted by the corresponding elements of $sl(r+1/r+1;\mathbb{C})$. It follows from (21.30) and (21.31) that the dimension m of the even part $\tilde{\mathscr{L}}_0$ of $\tilde{\mathscr{L}}_s = A(r/r)$ is given by $m = 2r(r+2)$, and the dimension n of the odd part $\tilde{\mathscr{L}}_1$ of $\tilde{\mathscr{L}}_s$ is given by $n = 2(r+1)^2$.

For the special case in which $r=0$ the even part $\tilde{\mathscr{L}}_0$ is trivial, and the basis elements $\mathbf{e}_{12}$ and $\mathbf{e}_{21}$ of $\tilde{\mathscr{L}}_1$ are such that (by (M.5))

$$[\mathbf{e}_{12},\mathbf{e}_{12}]=[\mathbf{e}_{21},\mathbf{e}_{21}]=[\mathbf{e}_{12},\mathbf{e}_{21}]=\mathbf{0},$$

implying that $\mathbf{e}_{12}$ and $\mathbf{e}_{21}$ are both basis elements of one-dimensional graded invariant subalgebras. Thus $A(0/0)$ is not simple, but it can be shown that $A(r/r)$ is simple for $r \geqslant 1$.

(b) The *even subalgebra* $\tilde{\mathscr{L}}_0$ of $\tilde{\mathscr{L}}_s = A(r/r)$ (for $r \geqslant 1$) is $\tilde{\mathscr{L}}_0 = \tilde{\mathscr{L}}_0^{ss} = \tilde{\mathscr{L}}_0^{s1} \oplus \tilde{\mathscr{L}}_0^{s2}$, with $\tilde{\mathscr{L}}_0^{s1}$ and $\tilde{\mathscr{L}}_0^{s2}$ both being isomorphic to A_r. The *rank* l of $\tilde{\mathscr{L}}_s$ is given by $l=2r$.

(c) A convenient basis of the *Cartan subalgebra* $\mathscr{H}_s$ of $\tilde{\mathscr{L}}_s = A(r/r)$ (for $r \geqslant 1$) is provided by

$$\mathbf{h}_j^1 = \mathbf{e}_{j,j} - \mathbf{e}_{j+1,j+1}, \quad j=1,2,\ldots,r,$$

and

$$\mathbf{h}_j^2 = \mathbf{e}_{j+r+1,j+r+1} - \mathbf{e}_{j+r+2,j+r+2}, \quad j=1,2,\ldots,r$$

(which provide bases for the Cartan subalgebras $\mathscr{H}_0^1$ and $\mathscr{H}_0^2$ of $\tilde{\mathscr{L}}_0^{s1}$ and $\tilde{\mathscr{L}}_0^{s2}$ respectively).

(d) Two useful sets of *linear functionals* on $\mathscr{H}_s$ are defined by

$$\varepsilon_p^a(\mathbf{h}_j^b) = \delta^{ab}(\delta_{jp} - \delta_{j+1,p})$$

for $a,b=1,2$ and $j,p=1,2,\ldots,r+1$. Neither set is linearly independent, since

$$\varepsilon_{r+1}^a = -\sum_{p=1}^{r} \varepsilon_p^a$$

for $a=1,2$. In particular, for $A(1/1)$

$$\varepsilon_2^1 = -\varepsilon_1^1, \quad \varepsilon_2^2 = -\varepsilon_1^2. \tag{M.6}$$

(e) $\tilde{\mathscr{L}}_s = A(r/r)$ (for $r \geqslant 1$) has $2r(r+1)$ *even roots*:

$$\alpha = \varepsilon_p^a - \varepsilon_q^a \quad \text{for } a=1,2 \quad \text{and} \quad p,q=1,2,\ldots,r+1 \quad (p \neq q)$$

(for which the corresponding root subspaces $\tilde{\mathscr{L}}_{s\alpha}$ are one-dimensional with basis element $\mathbf{e}_{p,q}$ for $a=1$ and $\mathbf{e}_{p+r+1,q+r+1}$ for $a=2$). These are extensions of roots of $\tilde{\mathscr{L}}_0^{sa}$ $(a=1,2)$.

(f) $\tilde{\mathscr{L}}_s = A(r/r)$ (for $r \geqslant 1$) has $2(r+1)^2$ *odd roots*, which fall into two classes:

(1) $\alpha = \varepsilon_p^1 - \varepsilon_q^2$ for $p,q=1,2,\ldots,r+1$ (for which $\mathbf{e}_{p,q+r+1}$ is a basis element of $\tilde{\mathscr{L}}_{s\alpha}$);

(2) $\alpha = -(\varepsilon^1_p - \varepsilon^2_p)$ for $p, q = 1, 2, \ldots, r+1$ (for which $\mathbf{e}_{q+r+1,p}$ is a basis elemement of $\tilde{\mathscr{L}}_{s\alpha}$).

For $r > 1$ these are all linearly independent, so $\dim \tilde{\mathscr{L}}_{s\alpha} = 1$ for every odd root, but for $r = 1$ (i.e. for $A(1/1)$), by (M.6), $\varepsilon^1_2 - \varepsilon^2_2 = -(\varepsilon^1_1 - \varepsilon^2_1)$, $\varepsilon^1_1 - \varepsilon^2_2 = \varepsilon^1_1 + \varepsilon^2_1$ and $\varepsilon^1_2 - \varepsilon^2_1 = -(\varepsilon^1_1 + \varepsilon^2_1)$, so there are four distinct roots $\pm(\varepsilon^1_1 + \varepsilon^2_1)$ and $\pm(\varepsilon^1_1 - \varepsilon^2_1)$, for which $\tilde{\mathscr{L}}_{s\alpha}$ is two-dimensional in each case. Moreover, with the subspace $\mathcal{N}^+$ of (24.50) being taken to have basis $\mathbf{e}_{p,q}$ (for $p, q = 1, 2, \ldots, 2r+2$ with $p < q$), it follows that each of these odd roots is *both* positive *and* negative for $A(1/1)$. (See also subsection (i) below.)

(g) The *distinguished* set of *simple roots* of $\tilde{\mathscr{L}}_s = A(r/r)$ (for $r \geqslant 1$) may be taken to be

(1) simple even roots

$$\alpha_j = \varepsilon^1_j - \varepsilon^1_{j+1} \quad \text{for } j = 1, 2, \ldots, r,$$

and

$$\alpha_{j+r+1} = \varepsilon^2_j - \varepsilon^2_{j+1} \quad \text{for } j = 1, 2, \ldots, r;$$

(2) simple odd root

$$\alpha_{r+1} = \varepsilon^1_{r+1} - \varepsilon^2_1.$$

The number of simple roots is $L = 2r + 1 \neq l$. They are *not* linearly independent, since

$$\sum_{j=1}^{r+1} j\alpha_j + \sum_{j=r+2}^{2r+1} (2r + 2 - j)\alpha_j = 0. \tag{M.7}$$

(h) In terms of these simple roots, the expressions for the *positive even* roots of $\tilde{\mathscr{L}}_s = A(r/r)$ (for $r \geqslant 1$) are

(1) for $p, q = 1, 2, \ldots, r+1$, with $p < q$,

$$\alpha = \varepsilon^1_p - \varepsilon^1_q = \sum_{j=p}^{q-1} \alpha_j,$$

(2) for $p, q = 1, 2, \ldots, r+1$, with $p < q$,

$$\alpha = \varepsilon^2_p - \varepsilon^2_q = \sum_{j=p}^{q-1} \alpha_{j+r+1}.$$

In particular, the extensions of the *simple* roots of $\tilde{\mathscr{L}}^{s1}_0$ are

$$\alpha^1_j = \varepsilon^1_j - \varepsilon^1_{j+1} = \alpha_j \quad \text{for } j = 1, 2, \ldots, r,$$

and the extensions of the *simple* roots of $\tilde{\mathscr{L}}^{s2}_0$ are

$$\alpha^2_j = \varepsilon^2_j - \varepsilon^2_{j+1} = \alpha_{j+r+1} \quad \text{for } j = 1, 2, \ldots, r.$$

That is, for $A(r/r)$ (with $r \geqslant 1$) *every* extension of a *simple* root of $\tilde{\mathscr{L}}_0^{ss}$ is a *simple* root of $\tilde{\mathscr{L}}_s$.

(i) In terms of these simple roots $\alpha_1, \alpha_2, \ldots, \alpha_{2r+1}$, the expressions for the *positive odd* roots of $\tilde{\mathscr{L}}_s = A(r/r)$ (for $r \geqslant 1$) are as given in (M.1) (with s being put equal to r).

However, in the special case when $r = 1$ (i.e. for $A(1/1)$), (M.7) becomes $\alpha_1 + 2\alpha_2 + \alpha_3 = 0$, and so

$$\begin{aligned}\varepsilon_2^1 - \varepsilon_1^2 &= \alpha_2 = -(\alpha_1 + \alpha_2 + \alpha_3),\\ \varepsilon_1^1 - \varepsilon_1^2 &= \alpha_1 + \alpha_2 = -(\alpha_2 + \alpha_3),\\ \varepsilon_2^1 - \varepsilon_2^2 &= \alpha_2 + \alpha_3 = -(\alpha_1 + \alpha_2),\\ \varepsilon_1^1 - \varepsilon_2^2 &= \alpha_1 + \alpha_2 + \alpha_3 = -\alpha_2,\end{aligned}$$

demonstrating (as in (f) above) that each of the odd roots of $A(1/1)$ is *both* positive *and* negative.

(j) The *Killing form* of $\mathrm{sl}(r+1/r+1;\mathbb{C})$ is given by (M.2), and this is identically zero when $r = s$. Thus, by (M.5), the *Killing form* of $\tilde{\mathscr{L}}_s = A(r/r)$ (for $r \geqslant 1$) is *identically zero*. A convenient non-degenerate bilinear supersymmetric consistent invariant form that can be used in its place is provided by

$$B'(\mathbf{M}, \mathbf{N}) = \mathrm{str}\,(\mathbf{MN})$$

for all $\mathbf{M}, \mathbf{N} \in A(r/r)$. This is not identically zero and is compatible with (M.5).

(k) For $j = 1, 2, \ldots, r$

$$\mathbf{h}_{\alpha_j} = \mathbf{h}_j^1,$$

and

$$\mathbf{h}_{\alpha_{j+r+1}} = -\mathbf{h}_j^2.$$

Also

$$\mathbf{h}_{\alpha_{r+1}} = \frac{1}{r+1}\sum_{j=1}^{r}\{-j\mathbf{h}_j^1 + (r+1-j)\mathbf{h}_j^2\}.$$

(l) The quantities $\langle \alpha_j, \alpha_k \rangle$ of $\tilde{\mathscr{L}}_s = A(r/r)$ (for $r \geqslant 1$) are given by

$$\langle \alpha_j, \alpha_k \rangle = \begin{cases} 2 & \text{if } j = k \quad (j = 1, 2, \ldots, r),\\ -2 & \text{if } j = k \quad (j = r+2, r+3, \ldots, 2r+1),\\ 1 & \text{if } j = k \pm 1 \quad (j, k = r+1, r+2, \ldots, 2r+1),\\ -1 & \text{if } j = k \pm 1 \quad (j, k = 1, 2, \ldots, r+1),\\ 0 & \text{for all other values of } j \text{ and } k.\end{cases}$$

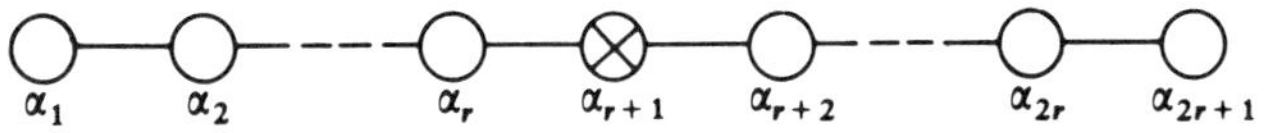

Figure M.3 Generalized Dynkin diagram for $A(r/r)$ (for $r \geqslant 1$) with the "distinguished" choice of simple roots.

It should be noted that

$$\langle \alpha_{r+1}, \alpha_{r+1} \rangle = 0.$$

(m) The *Cartan matrix* of $A(r/r)$ is the $(2r+1) \times (2r+1)$ matrix given in (M.3) (with s put equal to r).

(n) The *generalized Dynkin diagram* corresponding to $A(r/r)$ (for $r \geqslant 1$) is displayed in Figure M.3.

(o) For *every odd root* α of $\tilde{\mathscr{L}}_s = A(r/r)$ (for $r \geqslant 1$)

$$\langle \alpha, \alpha \rangle = 0.$$

(p) The discussion of the *representation of* $\tilde{\mathscr{L}}_0$ *on* $\tilde{\mathscr{L}}_1$ for $\tilde{\mathscr{L}}_s = A(r/r)$ (for $r \geqslant 1$) is exactly as given for $\tilde{\mathscr{L}}_s = A(r/s)$ (with s put equal to r) in part (p) of Section 1 above. Equations (M.4) show that for $A(r/r)$

$$1 - \gamma_{\mathrm{D}}^1 = 1 - \gamma_{\mathrm{D}}^2 = 0,$$

implying, by Theorem XIII of Chapter 25, Section 3(*a*), that the Killing form of $A(r/r)$ (for $r \geqslant 1$) is identically zero (as was noted above in subsection (j)).

3 The basic type II classical simple complex Lie superalgebras $B(r/s)$, $r \geqslant 0, s \geqslant 1$

(a) A realization of $\tilde{\mathscr{L}}_s = B(r/s)$ (for $r \geqslant 0$ and $s \geqslant 1$) is provided by $\mathrm{osp}(2r+1/2s; \mathbb{C})$, considered as a complex Lie superalgebra. (See Example III of Chapter 21, Section 2 for the definition and details of $\mathrm{osp}(2r+1/2s; \mathbb{C})$). In the discussion that follows, the matrix $\mathbf{G}$ of (21.32) will be taken to have the form (21.43). It follows from (21.40) and (21.41) that the dimension m of the even part $\tilde{\mathscr{L}}_0$ of $\tilde{\mathscr{L}}_s$ is given by $m = r(2r+1) + s(2s+1)$, and the dimension n of the odd part $\tilde{\mathscr{L}}_1$ of $\tilde{\mathscr{L}}_s$ is given by $n = 2s(2r+1)$.

(b) For $r \geqslant 1$ and $s \geqslant 1$ the *even subalgebra* $\tilde{\mathscr{L}}_0$ of $\tilde{\mathscr{L}}_s = B(r/s)$ is $\tilde{\mathscr{L}}_0 = \tilde{\mathscr{L}}_0^{\mathrm{ss}}$, where $\tilde{\mathscr{L}}_0^{\mathrm{ss}} = \tilde{\mathscr{L}}_0^{\mathrm{s}1} \oplus \tilde{\mathscr{L}}_0^{\mathrm{s}2}$ with $\tilde{\mathscr{L}}_0^{\mathrm{s}1} = C_s$ and $\tilde{\mathscr{L}}_0^{\mathrm{s}2} = B_r$. The *rank* l of $\tilde{\mathscr{L}}_s$ is given by $l = r + s$.

For $r=0$ and $s \geqslant 1$ (i.e. for $B(0/s)$) $\tilde{\mathscr{L}}_0 = \tilde{\mathscr{L}}_0^{ss} = \tilde{\mathscr{L}}_0^{s1}$, with $\tilde{\mathscr{L}}_0^{s1} = C_s$, and so $l = s$.

In both cases it is the set of $2s \times 2s$ matrices $\mathbf{D}$ of (21.35) that is isomorphic to $\tilde{\mathscr{L}}_0^{s1} = C_s$. Equations (21.33) and (21.37) imply that they can be partitioned in the form

$$\mathbf{D} = \begin{bmatrix} \mathbf{a} & \mathbf{b} \\ \mathbf{c} & -\tilde{\mathbf{a}} \end{bmatrix}, \tag{M.8}$$

where $\mathbf{a}$ is an arbitrary complex $s \times s$ matrix and $\mathbf{b}$ and $\mathbf{c}$ are symmetric complex $s \times s$ matrices.

For $r \geqslant 1$ it is the set of $(2r+1) \times (2r+1)$ matrices $\mathbf{A}$ of (21.35) that is isomorphic to $\tilde{\mathscr{L}}_0^{s2} = B_r$. Equations (21.36) and (21.43) imply that they can be partitioned in the form

$$\mathbf{A} = \begin{bmatrix} \mathbf{a} & \mathbf{b} & \mathbf{c} \\ \mathbf{d} & -\tilde{\mathbf{a}} & \mathbf{f} \\ -\tilde{\mathbf{f}} & -\tilde{\mathbf{c}} & \mathbf{0} \end{bmatrix},$$

where $\mathbf{a}$ is an arbitrary complex $r \times r$ matrix, $\mathbf{c}$ and $\mathbf{f}$ are arbitrary complex $r \times 1$ matrices, and $\mathbf{b}$ and $\mathbf{d}$ are antisymmetric complex $r \times r$ matrices.

(c) A convenient basis of the *Cartan subalgebra* $\mathscr{H}_s$ of $\tilde{\mathscr{L}}_s = B(r/s)$ (for $r \geqslant 0$ and $s \geqslant 1$) is provided by

$$\mathbf{h}_j^1 = \mathbf{e}_{j+2r+1,j+2r+1} - \mathbf{e}_{j+2r+s+1,j+2r+s+1}, \quad j = 1, 2, \ldots, s$$

(which provides a basis for the Cartan subalgebra $\mathscr{H}_0^1$ of $\tilde{\mathscr{L}}_0^{s1} = C_s$), and, for $r \geqslant 1$

$$\mathbf{h}_j^2 = \mathbf{e}_{j,j} - \mathbf{e}_{j+r,j+r}, \quad j = 1, 2, \ldots, r$$

(which provides a basis for the Cartan subalgebra $\mathscr{H}_0^2$ of $\tilde{\mathscr{L}}_0^{s2} = B_r$).

(d) For $r \geqslant 1$ two useful sets of *linear functionals* $\varepsilon_1^1, \varepsilon_2^1, \ldots, \varepsilon_s^1$ and $\varepsilon_1^2, \varepsilon_2^2, \ldots, \varepsilon_r^2$ on $\mathscr{H}_s$ are defined by

$$\varepsilon_p^1(\mathbf{h}) = \begin{cases} \delta_{jp} & \text{if } \mathbf{h} = \mathbf{h}_j^1 \quad \text{for } j = 1, 2, \ldots, s, \\ 0 & \text{if } \mathbf{h} \in \mathscr{H}_0^2, \end{cases}$$

for $p = 1, 2, \ldots, s$, and

$$\varepsilon_p^2(\mathbf{h}) = \begin{cases} \delta_{jp} & \text{if } \mathbf{h} = \mathbf{h}_j^2 \quad \text{for } j = 1, 2, \ldots, r, \\ 0 & \text{if } \mathbf{h} \in \mathscr{H}_0^1, \end{cases}$$

for $p = 1, 2, \ldots, r$.

For $r = 0$, only the set $\varepsilon_1^1, \varepsilon_2^1, \ldots, \varepsilon_s^1$ defined by

$$\varepsilon_p^1(\mathbf{h}_j^1) = \delta_{jp} \quad \text{for } j, p = 1, 2, \ldots, s$$

is needed.

(e) $\tilde{\mathscr{L}}_s = B(r/s)$ has $2(r^2 + s^2)$ *even roots*, which for $r \geqslant 1$ and $s \geqslant 1$ fall into eight classes:

(1) $\alpha = \varepsilon_p^1 - \varepsilon_q^1$ for $p, q = 1, 2, \ldots, s$ $(p \neq q)$ (for which the corresponding root subspace $\tilde{\mathscr{L}}_{s\alpha}$ is one-dimensional and has basis $\mathbf{e}_{2r+1+p, 2r+1+q} - \mathbf{e}_{2r+1+s+q, 2r+1+s+p}$);

(2) $\alpha = \varepsilon_p^1 + \varepsilon_q^1$ for $p, q = 1, 2, \ldots, s$ $(p \leqslant q)$ (for which $\tilde{\mathscr{L}}_{s\alpha}$ is one-dimensional and has basis $\mathbf{e}_{2r+1+p, 2r+1+s+q} + \mathbf{e}_{2r+1+q, 2r+1+s+p}$);

(3) $\alpha = -(\varepsilon_p^1 + \varepsilon_q^1)$ for $p, q = 1, 2, \ldots, s$ $(p \leqslant q)$ (for which $\tilde{\mathscr{L}}_{s\alpha}$ is one-dimensional and has basis $\mathbf{e}_{2r+1+s+p, 2r+1+q} + \mathbf{e}_{2r+1+s+q, 2r+1+p}$);

(4) $\alpha = \varepsilon_p^2 - \varepsilon_q^2$ for $p, q = 1, 2, \ldots, r$ $(p \neq q)$ (for which $\tilde{\mathscr{L}}_{s\alpha}$ is one-dimensional and has basis $\mathbf{e}_{p,q} - \mathbf{e}_{q+r, p+r}$);

(5) $\alpha = \varepsilon_p^2 + \varepsilon_q^2$ for $p, q = 1, 2, \ldots, r$ $(p < q)$ (for which $\tilde{\mathscr{L}}_{s\alpha}$ is one-dimensional and has basis $\mathbf{e}_{p, q+r} - \mathbf{e}_{q, p+r}$);

(6) $\alpha = -(\varepsilon_p^2 + \varepsilon_q^2)$ for $p, q = 1, 2, \ldots, r$ $(p < q)$ (for which $\tilde{\mathscr{L}}_{s\alpha}$ is one-dimensional and has basis $\mathbf{e}_{p+r, q} - \mathbf{e}_{q+r, p}$);

(7) $\alpha = \varepsilon_p^2$ for $p = 1, 2, \ldots, r$ (for which $\tilde{\mathscr{L}}_{s\alpha}$ is one-dimensional and has basis $\mathbf{e}_{p, 2r+1} - \mathbf{e}_{2r+1, p+r}$);

(8) $\alpha = -\varepsilon_p^2$ for $p = 1, 2, \ldots, r$ (for which $\tilde{\mathscr{L}}_{s\alpha}$ is one-dimensional and has basis $\mathbf{e}_{p+r, 2r+1} - \mathbf{e}_{2r+1, p}$);

The first three classes are extensions of roots of $\tilde{\mathscr{L}}_0^{s1} = C_s$, and the last five are extensions of roots of $\tilde{\mathscr{L}}_0^{s2} = B_r$. Of course, if $r = 0$, the latter set of roots do not appear. (If $s = 1$, the set (1) is empty, as are the sets (4), (5) and (6) if $r = 1$.)

(f) $\tilde{\mathscr{L}}_s = B(r/s)$ has $2s(2r + 1)$ *odd roots*, which for $r \geqslant 1$ and $s \geqslant 1$ fall into six classes:

(1) $\alpha = \varepsilon_q^1$ for $q = 1, 2, \ldots, s$ (for which $\tilde{\mathscr{L}}_{s\alpha}$ is one-dimensional with **B** basis element $\mathbf{e}_{2r+1, 2r+1+s+q}$);

(2) $\alpha = -\varepsilon_q^1$ for $q = 1, 2, \ldots, s$ (for which $\tilde{\mathscr{L}}_{s\alpha}$ is one-dimensional with **B** basis element $\mathbf{e}_{2r+1, 2r+1+q}$);

(3) $\alpha = \varepsilon_p^1 + \varepsilon_q^2$ for $p = 1, 2, \ldots, r$ and $q = 1, 2, \ldots, s$ (for which $\tilde{\mathscr{L}}_{s\alpha}$ is one-dimensional with **B** basis element $\mathbf{e}_{p,2r+1+s+q}$);

(4) $\alpha = -(\varepsilon_p^1 + \varepsilon_q^2)$ for $p = 1, 2, \ldots, r$ and $q = 1, 2, \ldots, s$ (for which $\tilde{\mathscr{L}}_{s\alpha}$ is one-dimensional with **B** basis element $\mathbf{e}_{p+r,2r+1+q}$);

(5) $\alpha = \varepsilon_p^1 - \varepsilon_q^2$ for $p = 1, 2, \ldots, r$ and $q = 1, 2, \ldots, s$ (for which $\tilde{\mathscr{L}}_{s\alpha}$ is one-dimensional with **B** basis element $\mathbf{e}_{p+r,2r+1+s+q}$);

(6) $\alpha = -(\varepsilon_p^1 - \varepsilon_q^2)$ for $p = 1, 2, \ldots, r$ and $q = 1, 2, \ldots, s$ (for which $\tilde{\mathscr{L}}_{s\alpha}$ is one-dimensional with **B** basis element $\mathbf{e}_{p,2r+1+q}$).

(Here the "**B** basis element" is that part of the basis element whose non-zero matrix elements lie in the submatrix **B** of (21.38), the corresponding "**C** basis element" being given by $\mathbf{C} = \mathbf{J}^{-1}\tilde{\mathbf{B}}\mathbf{G}$ (cf. (21.39).)

If $r = 0$ (i.e. for $B(0/s)$), the last four classes of roots do not appear.

(f) The *distinguished* set of *simple roots* of $\tilde{\mathscr{L}}_s = B(r/s)$ (for $r \geqslant 1$ and $s \geqslant 1$) may be taken to be

(1) simple even roots

$$\begin{aligned} \alpha_j &= \varepsilon_j^1 - \varepsilon_{j+1}^1 \quad \text{for } j = 1, 2, \ldots, s-1 \text{ (if } s > 1), \\ \alpha_{j+s} &= \varepsilon_j^2 - \varepsilon_{j+1}^2 \quad \text{for } j = 1, 2, \ldots, r-1 \text{ (if } r > 1), \\ \alpha_{r+s} &= \varepsilon_r^2; \end{aligned}$$

(2) simple odd root

$$\alpha_s = \varepsilon_s^1 - \varepsilon_1^2.$$

For $r = 0$ (i.e. for $B(0/s)$) (with $s \geqslant 1$) the "distinguished" set of simple roots may be taken to be

(1) simple even roots

$$\alpha_j = \varepsilon_j^1 - \varepsilon_{j+1}^1 \quad \text{for } j = 1, 2, \ldots, s-1 \text{ (if } s > 1);$$

(2) simple odd root

$$\alpha_s = \varepsilon_s^1.$$

In both cases the number of simple roots is $L = l = r + s$.

(h) In terms of these simple roots, the expressions for the *positive even* roots of $\tilde{\mathscr{L}}_s = B(r/s)$ (for $r \geqslant 1$ and $s \geqslant 1$) are

(1) for $p, q = 1, 2, \ldots, s$, with $p < q$,

$$\alpha = \varepsilon_p^1 - \varepsilon_q^1 = \sum_{j=p}^{q-1} \alpha_j;$$

(2) for $p, q = 1, 2, \ldots, s$, with $p < q$,

$$\alpha = \varepsilon_p^1 + \varepsilon_q^1 = \sum_{j=p}^{q-1} \alpha_j + 2 \sum_{j=q}^{r+s} \alpha_j;$$

(3) for $p = 1, 2, \ldots, s$,

$$\alpha = 2\varepsilon_p^1 = 2 \sum_{j=p}^{r+s} \alpha_j;$$

(4) for $p, q = 1, 2, \ldots, r$, with $p < q$,

$$\alpha = \varepsilon_p^2 - \varepsilon_q^2 = \sum_{j=p}^{q-1} \alpha_{j+s};$$

(5) for $p, q = 1, 2, \ldots, r$, with $p < q$,

$$\alpha = \varepsilon_p^2 + \varepsilon_q^2 = \sum_{j=p}^{q-1} \alpha_{j+s} + 2 \sum_{j=q}^{r} \alpha_{j+s};$$

(6) for $p = 1, 2, \ldots, r$

$$\alpha = \varepsilon_p^2 = \sum_{j=p}^{r} \alpha_{j+s}.$$

In particular, the extensions of the *simple* roots of $\tilde{\mathscr{L}}_0^{s1} = C_s$ are (for $s > 1$)

$$\alpha_j^1 = \varepsilon_j^1 - \varepsilon_{j+1}^1 = \alpha_j \quad \text{for } j = 1, 2, \ldots, s-1, \tag{M.9}$$

and

$$\alpha_s^1 = 2\varepsilon_s^1 = 2 \sum_{j=s}^{r+s} \alpha_j, \tag{M.10}$$

and the extensions of the *simple* roots of $\tilde{\mathscr{L}}_0^{s2} = B_r$ are

$$\alpha_j^2 = \alpha_{j+s} \quad \text{for } j = 1, 2, \ldots, r.$$

It should be noted that

$$\beta = \alpha_s^1 = 2 \sum_{j=s}^{r+s} \alpha_j \tag{M.11}$$

is an extension of a simple root of $\tilde{\mathscr{L}}_0^{s1}$ but it is *not* a simple root of $\tilde{\mathscr{L}}_s$.

In the case $r = 0$ (i.e. for $B(0/s)$) the extensions of the positive roots of $\tilde{\mathscr{L}}_0^{s1} = C_s$ are as above (in (1)–(3) (with $r = 0$)), but the sets (4)–(6) do not appear. Equations (M.9)–(M.11) remain valid, but when $r = 0$ (M.11) reduces to

$$\beta = \alpha_s^1 = 2\alpha_s.$$

(i) In terms of these simple roots $\alpha_1, \alpha_2, \ldots, \alpha_{r+s}$, the expressions for the *positive odd* roots of $\tilde{\mathscr{L}}_s = B(r/s)$ (for $r \geqslant 1$ and $s \geqslant 1$) are

(1) for $p = 1, 2, \ldots, s$

$$\alpha = \varepsilon_p^1 = \sum_{j=p}^{r+s} \alpha_j;$$

(2) for $p = 1, 2, \ldots, s$ and $q = 1, 2, \ldots, r$

$$\alpha = \varepsilon_p^1 + \varepsilon_q^2 = \sum_{j=p}^{q+s-1} \alpha_j + 2 \sum_{j=q+s}^{r+s} \alpha_j;$$

(3) for $p = 1, 2, \ldots, s$ and $q = 1, 2, \ldots, r$

$$\alpha = \varepsilon_p^1 - \varepsilon_q^2 = \sum_{j=p}^{q+s-1} \alpha_j.$$

In the case $r = 0$ (i.e. for $B(0/s)$) the last two classes do not appear, and for the first class

$$\alpha = \varepsilon_p^1 = \sum_{j=p}^{s} \alpha_j$$

for $p = 1, 2, \ldots, s$.

(j) The *Killing form* of $\tilde{\mathscr{L}}_s = B(r/s)$ (for $r \geqslant 0$ and $s \geqslant 1$) is given by

$$B(\mathbf{M}, \mathbf{N}) = (2r - 2s - 1)\,\mathrm{str}\,(\mathbf{MN})$$

for all $\mathbf{M}, \mathbf{N} \in B(r/s)$ and is *not identically zero.*

(k) For $j = 1, 2, \ldots, s$

$$\mathbf{h}_{\varepsilon_j^1} = -\frac{1}{2(2r - 2s - 1)} \mathbf{h}_j^1,$$

and, if $r \geqslant 1$, for $j = 1, 2, \ldots, r$

$$\mathbf{h}_{\varepsilon_j^2} = \frac{1}{2(2r - 2s - 1)} \mathbf{h}_j^2.$$

(l) The quantities $\langle \alpha_j, \alpha_k \rangle$ of $\tilde{\mathscr{L}}_s = B(r/s)$ (for $r \geqslant 1$ and $s \geqslant 1$) are given by

$$(2r - 2s - 1)\langle \alpha_j, \alpha_k \rangle = \begin{cases} 1 & \text{if } j = k\,(j = s+1, s+2, \ldots, s+r-1), \\ -1 & \text{if } j = k\,(j = 1, 2, \ldots, s-1)\,(\text{for } s > 1), \\ \tfrac{1}{2} & \text{if } j = k \pm 1\,(j, k = 1, 2, \ldots, s) \text{ and } j = k = s + r, \\ -\tfrac{1}{2} & \text{if } j = k \pm 1\,(j, k = s, s+1, \ldots, s+r), \\ 0 & \text{for all other values of } j \text{ and } k. \end{cases}$$

It should be noted that for $r \geqslant 1$ and $s \geqslant 1$

$$\langle \alpha_s, \alpha_s \rangle = 0. \tag{M.12}$$

For $r = 0$ (and $s \geqslant 1$) the corresponding quantities are

$$(2s+1)\langle \alpha_j, \alpha_k \rangle = \begin{cases} 1 & \text{if } j = k \; (j = 1, 2, \ldots, s-1) \, (\text{for } s > 1), \\ \frac{1}{2} & \text{if } j = k = s, \\ -\frac{1}{2} & \text{if } j = k \pm 1 \; (j, k = 1, 2, \ldots, s), \\ 0 & \text{for all other values of } j \text{ and } k. \end{cases}$$

and so, for $r = 0$ and $s \geqslant 1$,

$$\langle \alpha_s, \alpha_s \rangle = \frac{1}{2(2s+1)} \neq 0$$

(in contrast with (M.12)).

(m) For $r \geqslant 1$ and $s \geqslant 2$, with the index s' of (24.64) chosen to be $s+1$, the *Cartan matrix* of $B(r/s)$ is the $(r+s) \times (r+s)$ matrix

$$\mathbf{A} = \left[\begin{array}{cc|c|cc} \multicolumn{2}{c|}{\mathbf{A}(A_{s-1})} & \begin{matrix} 0 \\ \vdots \\ 0 \\ -1 \end{matrix} & \multicolumn{2}{c}{\mathbf{0}} \\ \hline 0 \cdots 0 & -1 & 0 & 1 & 0 \cdots 0 \\ \hline \multicolumn{2}{c|}{\mathbf{0}} & \begin{matrix} -1 \\ 0 \\ \vdots \\ 0 \end{matrix} & \multicolumn{2}{c}{\mathbf{A}(B_r)} \end{array}\right],$$

where the sth row and column have been partitioned off for extra clarity, and where $\mathbf{A}(A_{s-1})$ and $\mathbf{A}(B_r)$ are the Cartan matrices of the complex simple Lie algebras A_{s-1} and B_r respectively (cf. Appendix F, Sections 1 and 2).

Similarly, for $r \geqslant 1$ and $s = 1$ (with $s' = 1' = 2$) the Cartan matrix of $B(r/s)$ is the $(r+1) \times (r+1)$ matrix

$$\mathbf{A} = \left[\begin{array}{c|cc} 0 & 1 & 0 \cdots 0 \\ \hline -1 & & \\ 0 & & \\ \vdots & \multicolumn{2}{c}{\mathbf{A}(B_r)} \\ 0 & & \end{array}\right].$$

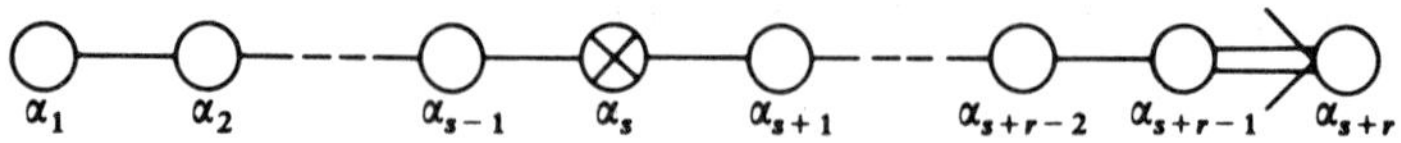

Figure M.4 Generalized Dynkin diagram for $B(r/s)$ (for $r \geqslant 1$ and $s \geqslant 1$) with the "distinguished" choice of simple roots.

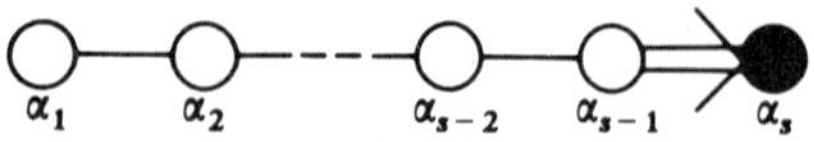

Figure M.5 Generalized Dynkin diagram for $B(0/s)$ (for $s \geqslant 1$) with the "distinguished" choice of simple roots.

However, for $r = 0$ and $s \geqslant 1$ the *Cartan matrix* of $B(0/s)$ is the $s \times s$ matrix

$$\mathbf{A} = \mathbf{A}(B_s),$$

where $\mathbf{A}(B_s)$ is the Cartan matrix of B_s (cf. Appendix F, Section 2).

(n) The *generalized Dynkin diagrams* corresponding to $B(r/s)$ (for $r \geqslant 1$ and $s \geqslant 1$) and $B(0/s)$ (for $s \geqslant 1$) are shown in Figures M.4 and M.5 respectively.

(o) For $r \geqslant 1$ and $s \geqslant 1$

$$\langle \alpha, \alpha \rangle = 0$$

for every odd root α of $B(r/s)$ of the form

$$\alpha = \pm \varepsilon_p^1 \pm \varepsilon_q^2$$

($p = 1, 2, \ldots, s$ and $q = 1, 2, \ldots, r$), but

$$\langle \alpha, \alpha \rangle = -\frac{1}{2(2r - 2s - 1)} \neq 0$$

for every odd root α of $B(r/s)$ of the form

$$\alpha = \pm \varepsilon_p^1$$

($p = 1, 2, \ldots, s$).

For $r = 0$ and $s \geqslant 1$ (i.e. for $B(0/s)$)

$$\langle \alpha, \alpha \rangle = \frac{1}{2(2s + 1)} \neq 0$$

for *every* odd root α.

(p) For $r \geqslant 1$ and $s \geqslant 1$ the *representation of* $\tilde{\mathscr{L}}_0$ *on* $\tilde{\mathscr{L}}_1$ is an *irreducible* representation of $\tilde{\mathscr{L}}_0$ that is the direct product of the $2s$-dimensional irreducible representation of $\tilde{\mathscr{L}}_0^{s1}$ $(=C_s)$ whose highest weight is $\Lambda = \Lambda_1$ with the $(2r+1)$-dimensional irreducible representation of $\tilde{\mathscr{L}}_0^{s2}$ $(=B_r)$ whose highest weight is $\Lambda = \Lambda_1$. The Dynkin index of the first of these irreducible representations is $\{2(s+1)\}^{-1}$, and, since the direct product representation reduces to $2r+1$ copies of this irreducible representation on $\tilde{\mathscr{L}}_0^{s1}$, the Dynkin index $\gamma_{\mathbf{D}}^1$ is $(2r+1)/2(s+1)$. Similarly $\gamma_{\mathbf{D}}^2 = 2s/(2r-1)$.

Thus the coefficients of (24.21) are

$$1 - \gamma_{\mathbf{D}}^1 = -\frac{2r-2s-1}{2(s+1)}$$

and

$$1 - \gamma_{\mathbf{D}}^2 = \frac{2r-2s-1}{2r-1},$$

neither of which is zero.

For $r = 0$ and $s \geqslant 1$ the representation of $\tilde{\mathscr{L}}_0$ $(=C_s)$ on $\tilde{\mathscr{L}}_1$ is the $2s$-dimensional irreducible representation of C_s whose highest weight is $\Lambda = \Lambda_1$. In this case

$$1 - \gamma_{\mathbf{D}}^1 = \frac{2s+1}{2(s+1)}.$$

4 The basic type I classical simple complex Lie superalgebras $C(s)$, $s \geqslant 2$

(a) A realization of $\tilde{\mathscr{L}}_s = C(s)$ (for $s \geqslant 2$) is provided by $\mathrm{osp}(2/2s-2;\mathbb{C})$, considered as a complex Lie superalgebra. (See Example III of Chapter 21, Section 2 for the definition and details of $\mathrm{osp}(2/2s-2;\mathbb{C})$). In the discussion that follows, the matrix $\mathbf{G}$ of (21.32) will be taken to have the form (21.42) (with $r = 1$). It follows from (21.40) and (21.41) that the dimension m of the even part $\tilde{\mathscr{L}}_0$ of $\tilde{\mathscr{L}}_s$ is given by $m = 2s^2 - 3s + 2$, and the dimension n of the odd part $\tilde{\mathscr{L}}_1$ of $\tilde{\mathscr{L}}_s$ is given by $n = 4(s-1)$. $C(2)$ is isomorphic to $A(2/1)$.

(b) The *even subalgebra* $\tilde{\mathscr{L}}_0$ of $\tilde{\mathscr{L}}_s = C(s)$ (for $s \geqslant 2$) is $\tilde{\mathscr{L}}_0 = \tilde{\mathscr{L}}_0^{\mathrm{A}} \oplus \tilde{\mathscr{L}}_0^{\mathrm{ss}}$, where $\tilde{\mathscr{L}}_0^{\mathrm{A}}$ is a one-dimensional Abelian Lie algebra and $\tilde{\mathscr{L}}_0^{\mathrm{ss}} = \tilde{\mathscr{L}}_0^{\mathrm{s1}} = C_{s-1}$, this being isomorphic to the set of $(2s-2) \times (2s-2)$ matrices $\mathbf{D}$ of (21.35). Equations (21.33) and (21.37) imply that they can be partitioned in the form

$$\mathbf{D} = \begin{bmatrix} \mathbf{a} & \mathbf{b} \\ \mathbf{c} & -\tilde{\mathbf{a}} \end{bmatrix},$$

where **a** is an arbitrary complex $(s-1)\times(s-1)$ matrix and **b** and **c** are symmetric complex $(s-1)\times(s-1)$ matrices. The *rank* l of $\tilde{\mathscr{L}}_s$ is given by $l=s$.

(c) A convenient basis of the *Cartan subalgebra* $\mathscr{H}_s$ of $\tilde{\mathscr{L}}_s = C(s)$ (for $s \geqslant 2$) is provided by

$$\mathbf{c} = \mathbf{e}_{1,1} - \mathbf{e}_{2,2}$$

(which provides a basis for $\tilde{\mathscr{L}}_0^{\mathrm{A}}$), and

$$\mathbf{h}_j^1 = \mathbf{e}_{j+2,j+2} - \mathbf{e}_{j+s+1,j+s+1}, \quad j=1,2,\ldots,s-1$$

(which provides a basis for the Cartan subalgebra $\mathscr{H}_0^1$ of $\tilde{\mathscr{L}}_0^{\mathrm{s}1} = C_{s-1}$).

(d) Two useful sets of *linear functionals* $\varepsilon_1^1, \varepsilon_2^1, \ldots, \varepsilon_{s-1}^1$ and ε_1^2 on $\mathscr{H}_s$ are defined by

$$\varepsilon_p^1(\mathbf{h}) = \begin{cases} \delta_{jp} & \text{if } \mathbf{h} = \mathbf{h}_j^1 \quad \text{for } j=1,2,\ldots,s-1, \\ 0 & \text{if } \mathbf{h} = \mathbf{c}, \end{cases}$$

for $p=1,2,\ldots,s-1$, and

$$\varepsilon_1^2(\mathbf{h}) = \begin{cases} 1 & \text{if } \mathbf{h} = \mathbf{c}, \\ 0 & \text{if } \mathbf{h} = \mathscr{H}_0^1. \end{cases}$$

(e) $\tilde{\mathscr{L}}_s = C(s)$ (for $s \geqslant 2$) has $2(s-1)^2$ *even roots*, which fall into three classes:

(1) $\alpha = \varepsilon_p^1 - \varepsilon_q^1$ for $p,q = 1,2,\ldots,s-1$ $(p \neq q)$ (for $s \geqslant 3$) (for which the corresponding roots subspace $\tilde{\mathscr{L}}_{s\alpha}$ is one-dimensional and has basis $\mathbf{e}_{p+2,q+2} - \mathbf{e}_{q+1+s,p+1+s}$);

(2) $\alpha = \varepsilon_p^1 + \varepsilon_q^1$ for $p,q=1,2,\ldots,s-1$ $(p \leqslant q)$ (for $s \geqslant 3$) (for which $\tilde{\mathscr{L}}_{s\alpha}$ one-dimensional and has basis $\mathbf{e}_{p+2,q+1+s} + \mathbf{e}_{q+2,p+1+s}$);

(3) $\alpha = -(\varepsilon_p^1 + \varepsilon_q^1)$ for $p,q=1,2,\ldots,s-1$ $(p \leqslant q)$ (for which $\tilde{\mathscr{L}}_{s\alpha}$ is one-dimensional and has basis $\mathbf{e}_{p+1+s,q+2} + \mathbf{e}_{q+1+s,p+2}$).

All of these are extensions of roots of $\tilde{\mathscr{L}}_0^{\mathrm{s}1} = C_{s-1}$.

(f) $\tilde{\mathscr{L}}_s = C(s)$ (for $s \geqslant 2$) has $4(s-1)$ *odd roots*, which fall into four classes:

(1) $\alpha = \varepsilon_p^1 + \varepsilon_1^2$ for $p=1,2,\ldots,s-1$ (for which $\tilde{\mathscr{L}}_{s\alpha}$ is one-dimensional with **B** basis element $\mathbf{e}_{1,p+1+s}$);

(2) $\alpha = -(\varepsilon_p^1 + \varepsilon_1^2)$ for $p=1,2,\ldots,s-1$ (for which $\tilde{\mathscr{L}}_{s\alpha}$ is one-dimensional with **B** basis element $\mathbf{e}_{2,p+2}$);

(3) $\alpha = \varepsilon_p^1 - \varepsilon_1^2$ for $p=1,2,\ldots,s-1$ (for which $\tilde{\mathscr{L}}_{s\alpha}$ is one-dimensional with **B** basis element $\mathbf{e}_{2,p+1+s}$);

(4) $\alpha = -(\varepsilon_p^1 - \varepsilon_1^2)$ for $p = 1, 2, \ldots, s-1$ (for which $\tilde{\mathscr{L}}_{s\alpha}$ is one-dimensional with **B** basis element $\mathbf{e}_{1,p+2}$).

(Here the "**B** basis element" is that part of the basis element whose non-zero matrix elements lie in the submatrix **B** of (21.38), the corresponding "**C** basis element" being given by $\mathbf{C} = \mathbf{J}^{-1}\tilde{\mathbf{B}}\mathbf{G}$ (cf. (21.39).)

(g) The *distinguished* set of *simple roots* of $\tilde{\mathscr{L}}_s = C(s)$ (for $s \geqslant 2$) may be taken to be

(1) simple even roots

$$\alpha_j = \varepsilon_{j-1}^1 - \varepsilon_j^1 \quad \text{for } j = 2, 3, \ldots, s-1 \text{ (for } s \geqslant 3),$$
$$\alpha_s = 2\varepsilon_{s-1}^1;$$

(2) simple odd root

$$\alpha_1 = \varepsilon_1^2 - \varepsilon_1^1.$$

The number of simple roots is $L = l = s$.

(h) In terms of these simple roots, the expressions for the *positive even* roots of $\tilde{\mathscr{L}}_s = C(s)$ (for $s \geqslant 2$) are

(1) for $p, q = 1, 2, \ldots, s-1$, with $p < q$, (for $s \geqslant 3$)

$$\alpha = \varepsilon_p^1 - \varepsilon_q^1 = \sum_{j=p+1}^{q} \alpha_j;$$

(2) for $p, q = 1, 2, \ldots, s-2$, with $p < q$, (for $s \geqslant 4$)

$$\alpha = \varepsilon_p^1 + \varepsilon_q^1 = \sum_{j=p+1}^{q} \alpha_j + 2 \sum_{j=q+1}^{s-1} \alpha_j + \alpha_s;$$

(3) for $p = 1, 2, \ldots, s-2$ (for $s \geqslant 3$)

$$\alpha = \varepsilon_p^1 + \varepsilon_{s-1}^1 = \sum_{j=p+1}^{s-1} \alpha_j + \alpha_s;$$

(4) for $p = 1, 2, \ldots, s-2$ (for $s \geqslant 3$)

$$\alpha = 2\varepsilon_p^1 = 2 \sum_{j=p+1}^{s-1} \alpha_j + \alpha_s;$$

(5) the single root

$$\alpha = 2\varepsilon_{s-1}^1 = \alpha_s.$$

In particular, the extensions of the *simple* roots of $\tilde{\mathscr{L}}_0^{s1} = C_{s-1}$ (for $s \geqslant 2$) are

$$\alpha_j^1 = \alpha_{j+1} \quad \text{for } j = 1, 2, \ldots, s-1,$$

so *every* extension of a *simple* root of $\tilde{\mathscr{L}}_0^{ss}$ is a *simple* root of $\tilde{\mathscr{L}}_s$.

(i) In terms of these simple roots $\alpha_1, \alpha_2, \ldots, \alpha_s$, the expressions for the *positive odd* roots of $\tilde{\mathscr{L}}_s = C(s)$ (for $s \geqslant 2$) are

(1) for $p = 1, 2, \ldots, s-2$ (for $s \geqslant 3$)

$$\alpha = \varepsilon_1^2 + \varepsilon_p^1 = \sum_{j=1}^{p} \alpha_j + 2 \sum_{j=p+1}^{s-1} \alpha_j + \alpha_s;$$

(2) for $p = 1, 2, \ldots, s-1$

$$\alpha = \varepsilon_1^2 - \varepsilon_p^1 = \sum_{j=1}^{p} \alpha_j;$$

(3) the single root

$$\alpha = \varepsilon_1^2 + \varepsilon_{s-1}^1 = \sum_{j=1}^{s} \alpha_j.$$

(j) The *Killing form* of $\tilde{\mathscr{L}}_s = C(s)$ (for $s \geqslant 2$) is given by

$$B(\mathbf{M}, \mathbf{N}) = -2(s-1)\,\mathrm{str}(\mathbf{MN})$$

for all $\mathbf{M}, \mathbf{N} \in C(s)$, and is *not identically zero.*

(k) For $j = 1, 2, \ldots, s-1$

$$\mathbf{h}_{\varepsilon_j^2} = \frac{1}{4(s-1)} \mathbf{h}_j^1,$$

and

$$\mathbf{h}_{\varepsilon_1^2} = \frac{1}{4(s-1)} \mathbf{c}.$$

(l) The quantities $\langle \alpha_j, \alpha_k \rangle$ of $\tilde{\mathscr{L}}_s = C(s)$ (for $s \geqslant 2$) are given by

$$\langle \alpha_j, \alpha_k \rangle = \begin{cases} \{2(s-1)\}^{-1} & \text{if } j = k \ (j = 2, 3, \ldots, s-1) \ (\text{for } s \geqslant 3), \\ (s-1)^{-1} & \text{if } j = k = s, \\ -\{4(s-1)\}^{-1} & \text{if } j = k \pm 1 \ (j, k = 1, 2, \ldots, s-1), \\ -\{2(s-1)\}^{-1} & \text{if } j = s-1, k = s \text{ and } j = s, k = s-1, \\ 0 & \text{for all other values of } j \text{ and } k. \end{cases}$$

It should be noted that

$$\langle \alpha_1, \alpha_1 \rangle = 0.$$

(m) With the index 1′ of (24.64) being chosen to be 2, the *Cartan matrix* of $C(s)$ (for $s \geqslant 2$) is the $s \times s$ matrix

$$\mathbf{A} = \left[\begin{array}{c|cccc} 0 & 1 & 0 & \cdots & 0 \\ \hline -1 & & & & \\ 0 & & \mathbf{A}(C_{s-1}) & & \\ \vdots & & & & \\ 0 & & & & \end{array}\right],$$

where the first row and column have been partitioned off for extra clarity, and where $\mathbf{A}(C_{s-1})$ is the Cartan matrix of the simple complex Lie algebra C_{s-1} (cf. Appendix F, Section 3).

(n) The *generalized Dynkin diagram* corresponding to $C(s)$ (for $s \geqslant 2$) is given in Figure M.6.

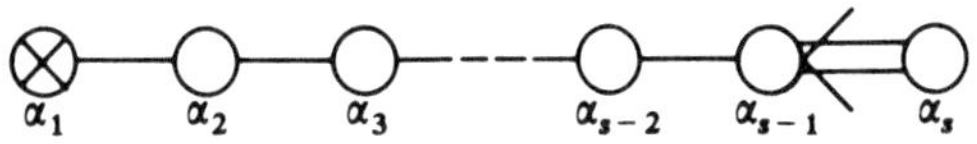

Figure M.6 Generalized Dynkin diagram for $C(s)$ (for $s \geqslant 2$) with the "distinguished" choice of simple roots.

(o) For *every odd root* α of $C(s)$ (for $s \geqslant 2$)

$$\langle \alpha, \alpha \rangle = 0.$$

(p) As $\tilde{\mathscr{L}}_0^{\mathrm{A}} \neq \{0\}$, the *representation of* $\tilde{\mathscr{L}}_0$ *on* $\tilde{\mathscr{L}}_1$ is the direct sum of two *irreducible* representations of $\tilde{\mathscr{L}}_0$ (cf. Theorem VIII of Chapter 25, Section 3(a)). On $\tilde{\mathscr{L}}_0^{\mathrm{ss}} = \tilde{\mathscr{L}}_0^{\mathrm{s}1} = C_{s-1}$ these are equivalent irreducible representations of dimension $2(s-1)$, highest weight $\Lambda = \Lambda_1$ and Dynkin index $(2s)^{-1}$. Thus $\gamma_{\mathrm{D}}^1 = s^{-1}$ (which is not equal to 1 for $s \geqslant 2$).

5 The basic type II classical simple complex Lie superalgebras $D(r/s)$, $r \geqslant 2, s \geqslant 1$

(a) A realization of $\tilde{\mathscr{L}}_s = D(r/s)$ (for $r \geqslant 2$ and $s \geqslant 1$) is provided by $\mathrm{osp}(2r/2s; \mathbb{C})$, considered as a complex Lie superalgebra. (See Example III of

Chapter 21, Section 2 for the definition and details of osp($2r/2s$; $\mathbb{C}$).) In the discussion that follows, the matrix **G** of (21.32) will be taken to have the form (21.42). Equations (21.40) and (21.41) imply that the dimension m of the even part $\tilde{\mathscr{L}}_0$ of $\tilde{\mathscr{L}}_s$ is given by $m = r(2r-1) + s(2s+1)$, and the dimension n of the odd part $\tilde{\mathscr{L}}_1$ of $\tilde{\mathscr{L}}_s$ is given by $n = 4rs$.

(b) The *even subalgebra* $\tilde{\mathscr{L}}_0$ of $\tilde{\mathscr{L}}_s = D(r/s)$ (for $r \geqslant 2$ and $s \geqslant 1$) is $\tilde{\mathscr{L}}_0 = \tilde{\mathscr{L}}_0^{ss}$, where $\tilde{\mathscr{L}}_0^{ss} = \tilde{\mathscr{L}}_0^{s1} \oplus \tilde{\mathscr{L}}_0^{s2}$ with $\tilde{\mathscr{L}}_0^{s1} = C_s$ and $\tilde{\mathscr{L}}_0^{s2} = D_r$. The *rank* l of $\tilde{\mathscr{L}}_s$ is given by $l = r + s$.

The Lie algebra $\tilde{\mathscr{L}}_0^{s1} = C_s$ is isomorphic to the set of $2s \times 2s$ matrices **D** of (21.35), which may be partitioned in the form (M.8).

Similarly the Lie algebra $\tilde{\mathscr{L}}_0^{s2} = D_r$ is isomorphic to the set of matrices **A** of (21.35). Equations (21.36) and (21.43) imply that they can be partitioned in the form

$$\mathbf{A} = \begin{bmatrix} \mathbf{a} & \mathbf{b} \\ \mathbf{c} & -\tilde{\mathbf{a}} \end{bmatrix},$$

where **a** is an arbitrary complex $r \times r$ matrix and **b** and **c** are antisymmetric complex $r \times r$ matrices.

(c) A convenient basis of the *Cartan subalgebra* $\mathscr{H}_s$ of $\tilde{\mathscr{L}}_s = D(r/s)$ (for $r \geqslant 2$ and $s \geqslant 1$) is provided by

$$\mathbf{h}_j^1 = \mathbf{e}_{j+2r,j+2r} - \mathbf{e}_{j+2r+s,j+2r+s}, \quad j = 1, 2, \ldots, s$$

(which provides a basis for the Cartan subalgebra $\mathscr{H}_0^1$ of $\tilde{\mathscr{L}}_0^{s1} = C_s$) and

$$\mathbf{h}_j^2 = \mathbf{e}_{j,j} - \mathbf{e}_{j+r,j+r}, \quad j = 1, 2, \ldots, r$$

(which provides a basis for the Cartan subalgebra $\mathscr{H}_0^2$ of $\tilde{\mathscr{L}}_0^{s2} = D_r$).

(d) Two useful sets of *linear functionals* $\varepsilon_1^1, \varepsilon_2^1, \ldots, \varepsilon_s^1$ and $\varepsilon_1^2, \varepsilon_2^2, \ldots, \varepsilon_r^2$ on $\mathscr{H}_s$ are defined by

$$\varepsilon_p^1(\mathbf{h}) = \begin{cases} \delta_{jp} & \text{if } \mathbf{h} = \mathbf{h}_j^1 \quad \text{for } j = 1, 2, \ldots, s, \\ 0 & \text{if } \mathbf{h} \in \mathscr{H}_0^2, \end{cases}$$

for $p = 1, 2, \ldots, s$, and

$$\varepsilon_p^2(\mathbf{h}) = \begin{cases} \delta_{jp} & \text{if } \mathbf{h} = \mathbf{h}_j^2 \quad \text{for } j = 1, 2, \ldots, r, \\ 0 & \text{if } \mathbf{h} \in \mathscr{H}_0^1, \end{cases}$$

for $p = 1, 2, \ldots, r$.

(e) $\tilde{\mathscr{L}}_s = D(r/s)$ (for $r \geqslant 2$ and $s \geqslant 1$) has $2(s^2 + r^2 - r)$ *even roots*, which fall into six classes:

(1) $\alpha = \varepsilon_p^1 - \varepsilon_q^1$ for $p, q = 1, 2, \ldots, s$ $(p \neq q)$ (for $s \geqslant 2$) (for which the corresponding root subspace $\tilde{\mathscr{L}}_{s\alpha}$ is one-dimensional and has basis $\mathbf{e}_{2r+p,2r+q} - \mathbf{e}_{2r+s+q,2r+s+p}$);

(2) $\alpha = \varepsilon_p^1 + \varepsilon_q^1$ for $p, q = 1, 2, \ldots, s$ $(p \leqslant q)$ (for which $\tilde{\mathscr{L}}_{s\alpha}$ is one-dimensional and has basis $\mathbf{e}_{2r+p,2r+s+q} + \mathbf{e}_{2r+q,2r+s+p}$);

(3) $\alpha = -(\varepsilon_p^1 + \varepsilon_q^1)$ for $p, q = 1, 2, \ldots, s$ $(p \leqslant q)$ (for which $\tilde{\mathscr{L}}_{s\alpha}$ is one-dimensional and has basis $\mathbf{e}_{2r+s+p,2r+q} + \mathbf{e}_{2r+s+q,2r+p}$);

(4) $\alpha = \varepsilon_p^2 - \varepsilon_q^2$ for $p, q = 1, 2, \ldots, r$ $(p \neq q)$ (for which $\tilde{\mathscr{L}}_{s\alpha}$ is one-dimensional and has basis $\mathbf{e}_{p,q} - \mathbf{e}_{q+r,p+r}$);

(5) $\alpha = \varepsilon_p^2 + \varepsilon_q^2$ for $p, q = 1, 2, \ldots, r$ $(p < q)$ (for which $\tilde{\mathscr{L}}_{s\alpha}$ is one-dimensional and has basis $\mathbf{e}_{p,q+r} - \mathbf{e}_{q,p+r}$);

(6) $\alpha = -(\varepsilon_p^2 + \varepsilon_q^2)$ for $p, q = 1, 2, \ldots, r$ $(p < q)$ (for which $\tilde{\mathscr{L}}_{s\alpha}$ is one-dimensional and has basis $\mathbf{e}_{p+r,q} - \mathbf{e}_{q+r,p}$);

The first three classes are extensions of a root of $\tilde{\mathscr{L}}_0^{s1} = C_s$, and the last three are extensions of roots $\tilde{\mathscr{L}}_0^{s2} = D_r$.

(f) $\tilde{\mathscr{L}}_s = D(r/s)$ (for $r \geqslant 2$ and $s \geqslant 1$) has $4rs$ *odd roots*, which fall into four classes:

(1) $\alpha = \varepsilon_p^1 + \varepsilon_q^2$ for $p = 1, 2, \ldots, s$ and $q = 1, 2, \ldots, r$ (for which $\tilde{\mathscr{L}}_{s\alpha}$ is one-dimensional with **B** basis element $\mathbf{e}_{p,q+2r+s}$);

(2) $\alpha = -(\varepsilon_p^1 + \varepsilon_q^2)$ for $p = 1, 2, \ldots, s$ and $q = 1, 2, \ldots, r$ (for which $\tilde{\mathscr{L}}_{s\alpha}$ is one-dimensional with **B** basis element $\mathbf{e}_{p+r,q+2r}$);

(3) $\alpha = \varepsilon_p^1 - \varepsilon_q^2$ for $p = 1, 2, \ldots, s$ and $q = 1, 2, \ldots, r$ (for which $\tilde{\mathscr{L}}_{s\alpha}$ is one-dimensional with **B** basis element $\mathbf{e}_{p+r,q+2r+s}$);

(4) $\alpha = -(\varepsilon_p^1 - \varepsilon_q^2)$ for $p = 1, 2, \ldots, s$ and $q = 1, 2, \ldots, r$ (for which $\tilde{\mathscr{L}}_{s\alpha}$ is one-dimensional with **B** basis element $\mathbf{e}_{p,q+2r}$).

(Here the "**B** basis element" is that part of the basis element whose non-zero matrix elements lie in the submatrix **B** of (21.38), the corresponding "**C** basis element" being given by $\mathbf{C} = \mathbf{J}^{-1}\tilde{\mathbf{B}}\mathbf{G}$ (cf. (21.39).)

(g) The *distinguished* set of *simple roots* of $\tilde{\mathscr{L}}_s = D(r/s)$ (for $r \geqslant 2$ and $s \geqslant 1$) may be taken to be

(1) simple even roots

$$\begin{aligned} \alpha_j &= \varepsilon_j^1 - \varepsilon_{j+1}^1 \quad \text{for } j = 1, 2, \ldots, s-1, \\ \alpha_{j+s} &= \varepsilon_j^2 - \varepsilon_{j+1}^2 \quad \text{for } j = 1, 2, \ldots, r-1, \\ \alpha_{r+s} &= \varepsilon_{r-1}^2 + \varepsilon_r^2; \end{aligned}$$

(2) simple odd root

$$\alpha_s = \varepsilon_s^1 - \varepsilon_1^2.$$

The number of simple roots is $L = l = r + s$.

(h) In terms of these simple roots the expressions for the *positive even* roots of $\tilde{\mathscr{L}}_s = D(r/s)$ (for $r \geqslant 2$ and $s \geqslant 1$) are

(1) if $s \geqslant 2$, for $p, q = 1, 2, \ldots, s$, with $p < q$,

$$\alpha = \varepsilon_p^1 - \varepsilon_q^1 = \sum_{j=p}^{q-1} \alpha_j;$$

(2) if $s \geqslant 2$, for $p, q = 1, 2, \ldots, s$, with $p < q$,

$$\alpha = \varepsilon_p^1 + \varepsilon_q^1 = \sum_{j=p}^{q-1} \alpha_j + 2 \sum_{j=q}^{r+s-2} \alpha_j + \sum_{j=r+s-1}^{r+s} \alpha_j;$$

(3) for $p = 1, 2, \ldots, s$,

$$\alpha = 2\varepsilon_p^1 = 2 \sum_{j=p}^{r+s-2} \alpha_j + \sum_{j=r+s-1}^{r+s} \alpha_j;$$

(4) for $p, q = 1, 2, \ldots, r$, with $p < q$,

$$\alpha = \varepsilon_p^2 - \varepsilon_q^2 = \sum_{j=p+s}^{q+s-1} \alpha_j;$$

(5) if $r \geqslant 4$, for $p, q = 1, 2, \ldots, r-2$, with $p < q$,

$$\alpha = \varepsilon_p^2 + \varepsilon_q^2 = \sum_{j=p+s}^{q+s-1} \alpha_j + 2 \sum_{j=q+s}^{r+s-2} \alpha_j + \sum_{j=r+s-1}^{r+s} \alpha_j;$$

(6) if $r \geqslant 3$, for $p = 1, 2, \ldots, r-2$,

$$\alpha = \varepsilon_p^2 + \varepsilon_{r-1}^2 = \sum_{j=p+s}^{r+s} \alpha_j;$$

(7) if $r \geqslant 3$, for $p = 1, 2, \ldots, r-2$,

$$\alpha = \varepsilon_p^2 + \varepsilon_r^2 = 2 \sum_{j=p+s}^{r+s-2} \alpha_j + \alpha_{r+s};$$

(8) the single root

$$\alpha = \varepsilon_{r-1}^2 + \varepsilon_r^2 = \alpha_{r+s}.$$

In particular, the extensions of the *simple* roots of $\tilde{\mathscr{L}}_0^{s1} = C_s$ are (for $s > 1$)

$$\alpha_j^1 = \varepsilon_j^1 - \varepsilon_{j+1}^1 = \alpha_j \quad \text{for } j = 1, 2, \ldots, s-1,$$

and

$$\alpha_s^1 = 2\varepsilon_s^1 = 2\sum_{j=s}^{r+s-2} \alpha_j + \sum_{j=r+s-1}^{r+s} \alpha_j,$$

and the extensions of the *simple* roots of $\tilde{\mathscr{L}}_0^{s2} = D_r$ are

$$\alpha_j^2 = \alpha_{j+s} \quad \text{for } j = 1, 2, \ldots, r.$$

It should be noted that

$$\beta = \alpha_s^1 = 2\sum_{j=s}^{r+s-2} \alpha_j + \sum_{j=r+s-1}^{r+s} \alpha_j$$

is an extension of a simple root of $\tilde{\mathscr{L}}_0^{s1}$, but it is *not* a simple root of $\tilde{\mathscr{L}}_s$.

(i) In terms of these simple roots $\alpha_1, \alpha_2, \ldots, \alpha_{r+s}$, the expressions for the *positive odd* roots of $\tilde{\mathscr{L}}_s = D(r/s)$ (for $r \geqslant 2$ and $s \geqslant 1$) are

(1) for $p = 1, 2, \ldots, s$ and $q = 1, 2, \ldots, r$

$$\alpha = \varepsilon_p^1 - \varepsilon_q^2 = \sum_{j=p}^{q+s-1} \alpha_j;$$

(2) if $r \geqslant 3$, for $p = 1, 2, \ldots, s$ and $q = 1, 2, \ldots, r-2$

$$\alpha = \varepsilon_p^1 + \varepsilon_q^2 = \sum_{j=p}^{q+s-1} \alpha_j + 2\sum_{j=q+s}^{r+s-2} \alpha_j + \sum_{j=r+s-1}^{r+s} \alpha_j;$$

(3) for $p = 1, 2, \ldots, s$

$$\alpha = \varepsilon_p^1 + \varepsilon_{r-1}^2 = \sum_{j=p}^{r+s} \alpha_j;$$

(4) for $p = 1, 2, \ldots, s$

$$\alpha = \varepsilon_p^1 + \varepsilon_r^2 = \sum_{j=p}^{r+s-2} \alpha_j + \alpha_{r+s}.$$

(j) The *Killing form* of $\tilde{\mathscr{L}}_s = D(r/s)$ (for $r \geqslant 2$ and $s \geqslant 1$) is given by

$$B(\mathbf{M}, \mathbf{N}) = 2(r - s - 1)\,\mathrm{str}\,(\mathbf{MN})$$

for all $\mathbf{M}, \mathbf{N} \in D(r/s)$. Thus the Killing form is identically zero if $r = s + 1$, but not otherwise. For $r = s + 1$ a convenient non-degenerate bilinear supersymmetric consistent form $B'(\ ,\ )$ that can be used in its place is provided by

$$B'(\mathbf{M}, \mathbf{N}) = \mathrm{str}\,(\mathbf{MN})$$

for all $\mathbf{M}, \mathbf{N} \in D(s + 1/s)$. The cases can be unified by taking

$$B'(\mathbf{M}, \mathbf{N}) = \kappa_B\,\mathrm{str}\,(\mathbf{MN}),$$

where

$$\kappa_B = \begin{cases} 2(r-s-1) & \text{if } r \neq s+1, \\ 1 & \text{if } r = s+1. \end{cases}$$

(k) For $j = 1, 2, \ldots, s$

$$\mathbf{h}_{\varepsilon_j^1} = -\frac{1}{2\kappa_B}\mathbf{h}_j^1,$$

and for $j = 1, 2, \ldots, r$

$$\mathbf{h}_{\varepsilon_j^2} = \frac{1}{2\kappa_B}\mathbf{h}_j^2.$$

(l) The quantities $\langle \alpha_j, \alpha_k \rangle$ of $\tilde{\mathscr{L}}_s = D(r/s)$ (for $r \geqslant 2$ and $s \geqslant 1$) are given by

$$\kappa_B\langle \alpha_j, \alpha_k \rangle = \begin{cases} 1 & \text{if } j = k\,(j = s+1, s+2, \ldots, r+s), \\ -1 & \text{if } j = k\,(j = 1, 2, \ldots, s-1), \\ \frac{1}{2} & \text{if } j = k \pm 1\,(j, k = 1, 2, \ldots, s), \\ -\frac{1}{2} & \text{if } j = k \pm 1\,(j, k = s, s+1, \ldots, r+s-2), \text{ or} \\ & \quad j = r+s-2 \text{ with } k = r+s-1, r+s, \text{ or (if } r = 2) \\ & \quad j = s \text{ with } k = s+2, \text{ and } k = s \text{ with } j = s+2 \\ 0 & \text{for all other values of } j \text{ and } k. \end{cases}$$

It should be noted that

$$\langle \alpha_s, \alpha_s \rangle = 0.$$

(m) For $r \geqslant 3$ and $s \geqslant 2$, with the index s' of (24.64) chosen to be $s+1$, the *Cartan matrix* of $D(r/s)$ is the $(r+s) \times (r+s)$ matrix

$$\mathbf{A} = \left[\begin{array}{c|c|c} \mathbf{A}(A_{s-1}) & \begin{matrix} 0 \\ \vdots \\ 0 \\ -1 \end{matrix} & \mathbf{0} \\ \hline 0 \cdots 0 \quad -1 & 0 & 1 \quad 0 \cdots 0 \\ \hline \mathbf{0} & \begin{matrix} -1 \\ 0 \\ \vdots \\ 0 \end{matrix} & \mathbf{A}(D_r) \end{array}\right],$$

where the sth row and column have been partitioned off for extra clarity, and where $\mathbf{A}(A_{s-1})$ and $\mathbf{A}(D_r)$ are the Cartan matrices of the simple complex Lie algebras A_{s-1} and D_r respectively (cf. Appendix F, Sections 1 and 4).

Similarly for $r \geqslant 3$ and $s = 1$ (with $s' = 1' = 2$) the Cartan matrix of $D(r/s)$ is the $(r+1) \times (r+1)$ matrix

$$\mathbf{A} = \left[\begin{array}{c|c} 0 & 1 \quad 0 \cdots 0 \\ \hline -1 & \\ 0 & \\ \vdots & \mathbf{A}(D_r) \\ 0 & \end{array}\right].$$

For $r = 2$ and $s \geqslant 2$ (with $s' = s + 1$) the Cartan matrix is the $(s+2) \times (s+2)$ matrix

$$\mathbf{A} = \left[\begin{array}{c|c|cc} & 0 & & \\ & \vdots & & \\ \mathbf{A}(A_{s-1}) & 0 & \multicolumn{2}{c}{\mathbf{0}} \\ & -1 & & \\ \hline 0 \cdots 0 \quad -1 & 0 & 1 & 1 \\ \hline \multirow{2}{*}{$\mathbf{0}$} & -1 & 2 & 0 \\ & -1 & 0 & 2 \end{array}\right],$$

and finally for $r = 2$ and $s = 1$ (with $s' = 1' = 2$) (i.e. for $D(2/1)$) the Cartan matrix is the 3×3 matrix

$$\mathbf{A} = \begin{bmatrix} 0 & 1 & 1 \\ -1 & 2 & 0 \\ -1 & 0 & 2 \end{bmatrix}. \tag{M.13}$$

(n) The *generalized Dynkin diagrams* corresponding to $D(r/s)$ for $r \geqslant 3$, $s \geqslant 1$ and for $r = 2$, $s \geqslant 1$ are given in Figures M.7 and M.8 respectively.

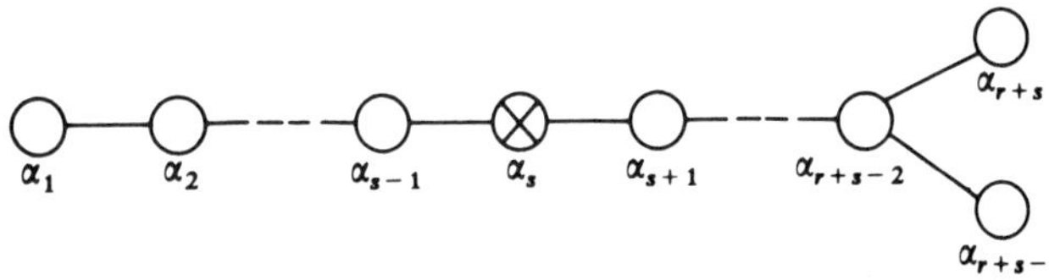

Figure M.7 Generalized Dynkin diagram for $D(r/s)$ (for $r \geqslant 3$ and $s \geqslant 1$) with the "distinguished" choice of simple roots.

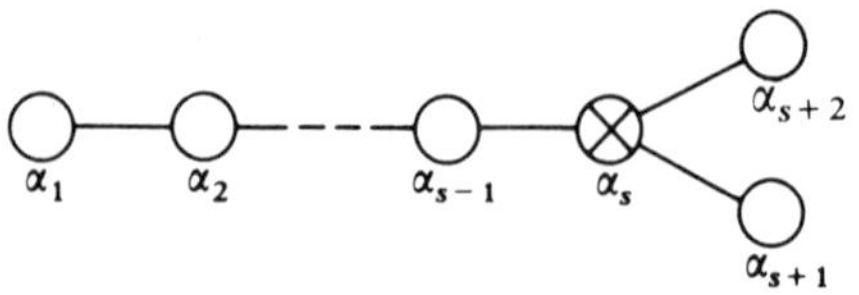

Figure M.8 Generalized Dynkin diagram for $D(2/s)$ (for $s \geqslant 1$) with the "distinguished" choice of simple roots.

(o) For *every odd root* α of $D(r/s)$ (for $r \geqslant 2$ and $s \geqslant 1$)

$$\langle \alpha, \alpha \rangle = 0.$$

(p) The *representation of* $\tilde{\mathscr{L}}_0$ *on* $\tilde{\mathscr{L}}_1$ is an *irreducible* representation of $\tilde{\mathscr{L}}_0$ that is the direct product of the $2s$-dimensional irreducible representation of $\tilde{\mathscr{L}}_0^{s1} (= C_s)$ whose highest weight is $\Lambda = \Lambda_1$ with the $2r$-dimensional irreducible representation of $\tilde{\mathscr{L}}_0^{s2} (= D_r)$ whose highest weight is $\Lambda = \Lambda_1$. The Dynkin index of the first of these irreducible representations is $\{2(s+1)\}^{-1}$, and since the direct product representation reduces to $2r$ copies of this irreducible representation on $\tilde{\mathscr{L}}_0^{s1}$, the Dynkin index γ_{D}^1 is $r/(s+1)$. Similarly $\gamma_{\mathrm{D}}^2 = s/(r-1)$.

Thus the coefficients of (24.21) are

$$1 - \gamma_{\mathrm{D}}^1 = -\frac{r-s-1}{s+1}$$

and

$$1 - \gamma_{\mathrm{D}}^2 = \frac{r-s-1}{r-1},$$

which are zero if $r = s+1$ (corresponding to the zero Killing form of this case), but are non-zero otherwise.

6 The basic type II classical simple complex Lie superalgebras $D(2/1;\alpha)$, with α a complex parameter taking all values other than 0, -1 and ∞

(a) The *even subalgebra* $\tilde{\mathscr{L}}_0$ of $\tilde{\mathscr{L}}_s = D(2/1;\alpha)$ (where α is a complex parameter) is $\tilde{\mathscr{L}}_0 = \mathscr{L}_0^{ss} = \tilde{\mathscr{L}}_0^{s1} \oplus \tilde{\mathscr{L}}_0^{s2} \oplus \tilde{\mathscr{L}}_0^{s3}$ with $\tilde{\mathscr{L}}_0^{sp} = A_1$ for $p = 1, 2, 3$. (No confusion should arise through having two different types of quantity denoted by α, namely this parameter and a general root of $\tilde{\mathscr{L}}_s$, since it will always be clear from the context which is intended in any particular situation.) As $\tilde{\mathscr{L}}_0 = A_1 \oplus A_1 \oplus A_1$, its dimension m is 9 and its *rank* l is 3.

Let a_j^p ($p = 1, 2, 3$; $j = 1, 2, 3$) be the basis elements of $\tilde{\mathscr{L}}_0^{sp}(= A_1)$ chosen so that

$$[a_j^p, a_{j'}^{p'}]_- = -\delta^{pp'} \sum_{j''=1}^{3} \varepsilon_{jj'j''} a_{j''}^p, \tag{M.14}$$

which implies that for each $p\,(= 1, 2, 3)$ the basis elements a_j^p ($j = 1, 2, 3$) satisfy the commutation relations (10.30)

(b) The *odd part* $\tilde{\mathscr{L}}_1$ of $\tilde{\mathscr{L}}_s = D(2/1; \alpha)$ (where α is a complex parameter) is defined to have dimension $n = 8$, the 8 basis elements Q_{rst} ($r, s, t = 1, 2$) being chosen so that

$$\left.\begin{aligned} [a_j^1, Q_{rst}]_- &= \sum_{r'=1}^{2} (\mathbf{a}_j)_{r'r} Q_{r'st}, \\ [a_j^2, Q_{rst}]_- &= \sum_{s'=1}^{2} (\mathbf{a}_j)_{s's} Q_{rs't}, \\ [a_j^3, Q_{rst}]_- &= \sum_{t'=1}^{2} (\mathbf{a}_j)_{t't} Q_{rst'} \end{aligned}\right\} \tag{M.15}$$

for all $j = 1, 2, 3$ and $r, s, t = 1, 2$. Here (as in (10.29))

$$\mathbf{a}_1 = \frac{1}{2}\begin{bmatrix} 0 & i \\ i & 0 \end{bmatrix}, \quad \mathbf{a}_2 = \frac{1}{2}\begin{bmatrix} 0 & 1 \\ -1 & 0 \end{bmatrix}, \quad \mathbf{a}_3 = \frac{1}{2}\begin{bmatrix} i & 0 \\ 0 & -i \end{bmatrix}. \tag{M.16}$$

As this representation of A_1 with $\boldsymbol{\Gamma}(\mathbf{a}_j) = \mathbf{a}_j$ (for $j = 1, 2, 3$) is equivalent to the 2-dimensional irreducible representation $\mathbf{D}^{1/2}$ of A_1 (cf. Chapter 12, Section 3), (M.15) imply that the representation of $\tilde{\mathscr{L}}_0$ on $\tilde{\mathscr{L}}_1$ is equivalent to the 8-dimensional irreducible representation $\mathbf{D}^{1/2} \otimes \mathbf{D}^{1/2} \otimes \mathbf{D}^{1/2}$ of $A_1 \oplus A_1 \oplus A_1$.

The remaining generalized Lie products of basis elements are given by

$$\begin{aligned} [Q_{rst}, Q_{r's't'}]_+ = {} & 2\kappa_1 C^2_{ss'} C^2_{tt'} \sum_{j=1}^{3} (\mathbf{C}^2 \mathbf{a}_j)_{rr'} a_j^1 \\ & + 2\kappa_2 C^2_{rr'} C^2_{tt'} \sum_{j=1}^{3} (\mathbf{C}^2 \mathbf{a}_j)_{ss'} a_j^2 \\ & + 2\kappa_3 C^2_{rr'} C^2_{ss'} \sum_{j=1}^{3} (\mathbf{C}^2 \mathbf{a}_j)_{tt'} a_j^3, \end{aligned} \tag{M.17}$$

where

$$\mathbf{C}^2 = \begin{bmatrix} 0 & 1 \\ -1 & 0 \end{bmatrix} \tag{M.18}$$

and κ_1, κ_2 and κ_3 are three complex numbers. It follows from (M.16) and

(M.18) that

$$\mathbf{C}^2\tilde{\mathbf{a}}_j = -\mathbf{a}_j\mathbf{C}^2 \tag{M.19}$$

for $j = 1, 2, 3$. As $\mathbf{C}^2$ is antisymmetric and $\mathbf{C}^2\mathbf{a}_j$ is symmetric, the right-hand side of (M.17) is symmetric with respect to the interchange of Q_{rst} with $Q_{r's't'}$, as is required by (21.8). It requires a fairly long argument (cf. Scheunert *et al.* 1976b) to show that the generalized Jacobi relations (21.15) are satisfied if and only if

$$\kappa_1 + \kappa_2 + \kappa_3 = 0. \tag{M.20}$$

(c) If Q_{rst} and $Q_{r's't'}$ are replaced by μQ_{rst} and $\mu Q_{r's't'}$ in (M.17), where μ is any complex number, the same form of generalized Lie product is obtained but with κ_1, κ_2 and κ_3 replaced by $\mu^2\kappa_1, \mu^2\kappa_2$ and $\mu^2\kappa_3$ respectively. This implies that the Lie superalgebras specified by $\{\kappa_1, \kappa_2, \kappa_3\}$ and $\{\mu^2\kappa_1, \mu^2\kappa_2, \mu^2\kappa_3\}$ are isomorphic for any $\mu \in \mathbb{C}$. It is natural to introduce a complex number α by the definition

$$\alpha = \kappa_2/\kappa_1, \tag{M.21}$$

so that α is left invariant in such a rescaling. With this definition, (M.20) implies that

$$\kappa_3/\kappa_1 = -1 - \alpha. \tag{M.22}$$

Consequently the Lie superalgebra $\tilde{\mathscr{L}}_s = D(2/1; \alpha)$ depends on the *one* complex number α alone.

It is obvious that if κ_1, κ_2 and κ_3 are *permuted* in (M.17), the effect is merely the same as *reordering* the simple Lie algebras $\tilde{\mathscr{L}}_0^{s1}$, $\tilde{\mathscr{L}}_0^{s2}$ and $\tilde{\mathscr{L}}_0^{s3}$ in $\tilde{\mathscr{L}}_0$, a process that leaves $\tilde{\mathscr{L}}_s$ unchanged. Thus a permutation of $\{\kappa_1, \kappa_2, \kappa_3\}$ leaves $\tilde{\mathscr{L}}_s$ invariant. However, the permutations $\{\kappa_1, \kappa_2, \kappa_3\} \to \{\kappa_2, \kappa_1, \kappa_3\}$ and $\{\kappa_1, \kappa_2, \kappa_3\} \to \{\kappa_1, \kappa_3, \kappa_2\}$ correspond to replacing α by α^{-1} and $-1-\alpha$ respectively. Thus for any $\alpha \in \mathbb{C}$ the Lie superalgebras specified by α, α^{-1} and $-1-\alpha$ are *isomorphic*.

(d) If $\kappa_1 = 0$ (but $\kappa_2 \neq 0$ and $\kappa_3 \neq 0$) then the set of basis elements a_j^2 $(j = 1, 2, 3)$, a_j^3 $(j = 1, 2, 3)$ and Q_{rst} $(r, s, t = 1, 2)$ form a proper graded invariant subalgebra of $\tilde{\mathscr{L}}_s$, so in this case $\tilde{\mathscr{L}}_s$ is *not* simple. The same result is true for the cases $\kappa_2 = 0$ $(\kappa_3 \neq 0, \kappa_1 \neq 0)$ and $\kappa_3 = 0$ $(\kappa_1 \neq 0, \kappa_2 \neq 0)$. Thus, by (M.21) and (M.22), *If* $\alpha = 0, -1$ *or* ∞ *then* $\tilde{\mathscr{L}}_s = D(2/1; \alpha)$ *is not simple.*

(e) If $\alpha = 1$ (or equivalently $\alpha = -2$ or $-\frac{1}{2}$) then $D(2/1; \alpha)$ is isomorphic to $D(2/1)$. (See subsection (o) below.)

(f) As the Dynkin index of the 2-dimensional irreducible representation $\mathbf{D}^{1/2}$

of A_1 is $\frac{1}{4}$, and the representation $\mathbf{D}^{1/2} \otimes \mathbf{D}^{1/2} \otimes \mathbf{D}^{1/2}$ of $A_1 \oplus A_1 \oplus A_1$ reduces to four copies of $\mathbf{D}^{1/2}$ on each $\tilde{\mathscr{L}}_0^{sp}$ $(p = 1, 2, 3)$, it follows that $\gamma_{\mathbf{D}}^p = 1$ for each $\tilde{\mathscr{L}}_0^{sp}$ $(p = 1, 2, 3)$. Theorem XIII of Chapter 25, Section 3(*a*) then implies that the *Killing form* of $\tilde{\mathscr{L}}_s = D(2/1;\alpha)$ is *identically zero.*

Let $B'(\ ,\ )$ be the bilinear form defined on the basis elements of $\tilde{\mathscr{L}}_s = D(2/1;\alpha)$ by

$$B'(Q_{rst}, Q_{r's't'}) = C^2_{rr'} C^2_{ss'} C^2_{tt'}$$

(for $r, s, t = 1, 2$),

$$B'(a_j^p, Q_{rst}) = 0$$

(for $j, p = 1, 2, 3$ and $r, s, t = 1, 2$) and

$$B'(a_j^p, a_{j'}^{p'}) = -\frac{1}{2\kappa_p}\delta^{pp'}\delta_{jj'}$$

(for $j, j', p, p' = 1, 2, 3$). Then it is easily shown that $B'(\ ,\ )$ is supersymmetric, consistent and invariant (provided, of course, that $\kappa_1 \neq 0$, $\kappa_2 \neq 0$ and $\kappa_3 \neq 0$, i.e. that $\alpha \neq 0, -1$ or ∞).

(g) A convenient basis of the *Cartan subalgebra* $\mathscr{H}_s$ of $\tilde{\mathscr{L}}_s = D(2/1;\alpha)$ (for $\alpha \neq 0, -1$ or ∞) is provided by

$$h^p = a_3^p$$

for $p = 1, 2, 3$, and a useful set of *linear functionals* on $\mathscr{H}_s$ may be defined by

$$\varepsilon^p(h^q) = i\delta^{pq}$$

(for $p, q = 1, 2, 3$).

(h) $\tilde{\mathscr{L}}_s = D(2/1;\alpha)$ (for $\alpha \neq 0, -1$ or ∞) has six *even roots*, which fall into two classes:

(1) $\alpha = \varepsilon^p$ for $p = 1, 2, 3$ (for which the corresponding root subspace $\tilde{\mathscr{L}}_{s\alpha}$ is one-dimensional with basis element $a_1^p + ia_2^p$);

(2) $\alpha = -\varepsilon^p$ for $p = 1, 2, 3$ (for which $\tilde{\mathscr{L}}_{s\alpha}$ is one-dimensional with basis element $a_1^p - ia_2^p$).

(i) $\tilde{\mathscr{L}}_s = D(2/1;\alpha)$ (for $\alpha \neq 0, -1$ or ∞) has eight *odd roots*:

$$\alpha = \tfrac{1}{2}(\pm\varepsilon^1 \pm \varepsilon^2 \pm \varepsilon^3).$$

(For each of these $\tilde{\mathscr{L}}_{s\alpha}$ is one-dimensional, Q_{rst} being the basis element for $-\frac{1}{2}\{(-1)^r\varepsilon^1 + (-1)^s\varepsilon^2 + (-1)^t\varepsilon^3\}$.)

(j) The *distinguished* set of *simple roots* of $\tilde{\mathscr{L}}_s = D(2/1;\alpha)$ (for $\alpha \neq 0, -1$ or ∞) may be taken to be

(1) simple even roots

$$\alpha_2 = -\varepsilon^1,$$
$$\alpha_3 = -\varepsilon^2;$$

(2) simple odd root

$$\alpha_1 = \tfrac{1}{2}(\varepsilon^1 + \varepsilon^2 + \varepsilon^3).$$

The number of simple roots is $L = l = 3$.

(k) In terms of these simple roots, the expressions for the *positive even* roots of $\tilde{\mathscr{L}}_s = D(2/1; \alpha)$ (for $\alpha \neq 0$, -1 or ∞) are

$$-\varepsilon^1 = \alpha_2, \quad -\varepsilon^2 = \alpha_3, \quad \varepsilon^3 = 2\alpha_1 + \alpha_2 + \alpha_3.$$

It should be noted that

$$\beta = \varepsilon^3 = 2\alpha_1 + \alpha_2 + \alpha_3$$

is the extension of a simple root of $\tilde{\mathscr{L}}_0^{ss}$ that is *not* a simple root of $\tilde{\mathscr{L}}_s$.

(l) In terms of these simple roots α_1, α_2 and α_3, the expressions for the *positive odd* roots of $\tilde{\mathscr{L}}_s = D(2/1; \alpha)$ (for $\alpha \neq 0$, -1 or ∞) are:

$$\tfrac{1}{2}(\varepsilon^1 + \varepsilon^2 + \varepsilon^3) = \alpha_1,$$
$$\tfrac{1}{2}(-\varepsilon^1 + \varepsilon^2 + \varepsilon^3) = \alpha_1 + \alpha_2,$$
$$\tfrac{1}{2}(\varepsilon^1 - \varepsilon^2 + \varepsilon^3) = \alpha_1 + \alpha_3,$$
$$\tfrac{1}{2}(-\varepsilon^1 - \varepsilon^2 + \varepsilon^3) = \alpha_1 + \alpha_2 + \alpha_3.$$

(m) For $p = 1, 2, 3$

$$h_{\varepsilon^p} = -2i\kappa_p h^p.$$

(n) The quantities $\langle \alpha_j, \alpha_k \rangle$ of $\tilde{\mathscr{L}}_s = D(2/1; \alpha)$ (for $\alpha \neq 0$, -1 or ∞) are given by

$$\begin{array}{lll} \langle \alpha_1, \alpha_1 \rangle = 0, & \langle \alpha_1, \alpha_2 \rangle = -\kappa_1, & \langle \alpha_1, \alpha_3 \rangle = -\kappa_2, \\ \langle \alpha_2, \alpha_1 \rangle = -\kappa_1, & \langle \alpha_2, \alpha_2 \rangle = 2\kappa_1, & \langle \alpha_2, \alpha_3 \rangle = 0, \\ \langle \alpha_3, \alpha_2 \rangle = 0, & \langle \alpha_3, \alpha_1 \rangle = -\kappa_2, & \langle \alpha_3, \alpha_3 \rangle = 2\kappa_2. \end{array}$$

It should be noted that these are all real only if κ_1 and κ_2 are both real. Moreover, although the introduction of a multiplicative factor into the form $B'(\ ,\ )$ of (f) above will result in all the expressions $\langle \alpha_j, \alpha_k \rangle$ being multiplied by a common factor, ratios such as $\langle \alpha_2, \alpha_3 \rangle / \langle \alpha_2, \alpha_2 \rangle = \kappa_2/\kappa_1 = \alpha$ are left unchanged. It follows that if α is *non-real* then the set $\langle \alpha_j, \alpha_k \rangle$ (for $j, k = 1, 2, 3$) *cannot* all be made real.

(o) The *Cartan matrix* of $\tilde{\mathscr{L}}_s = D(2/1;\alpha)$ (for $\alpha \neq 0, -1$ or ∞) is the 3×3 matrix

$$\mathbf{A} = \begin{bmatrix} 0 & 1 & \alpha \\ -1 & 2 & 0 \\ -1 & 0 & 2 \end{bmatrix}.$$

When $\alpha = 1$ this coincides with the Cartan matrix of $D(2/1)$ (cf. (M.13)), confirming that $D(2/1;1)$ is isomorphic to $D(2/1)$ (cf. subsection (e) above).

(p) For *every odd root* α of $\tilde{\mathscr{L}}_s = D(2/1;\alpha)$ (for $\alpha \neq 0, -1$ or ∞)

$$\langle \alpha, \alpha \rangle = 0.$$

(q) The *generalized Dynkin diagram* corresponding to $\tilde{\mathscr{L}}_s = D(2/1;\alpha)$ (for $\alpha \neq 0, -1$ or ∞) is given in Figure M.9.

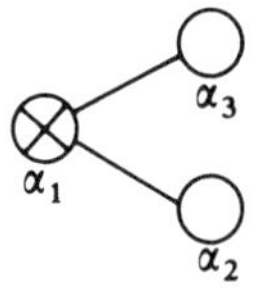

Figure M.9 Generalized Dynkin diagram for $D(2/1;\alpha)$ (for $\alpha \neq 0, -1$ or ∞) with the "distinguished" choice of simple roots.

7 The basic type II classical simple complex Lie superalgebra $F(4)$

(a) The *even subalgebra* $\tilde{\mathscr{L}}_0$ of $\tilde{\mathscr{L}}_s = F(4)$ is $\tilde{\mathscr{L}}_0 = \tilde{\mathscr{L}}_0^{ss} = \tilde{\mathscr{L}}_0^{s1} \oplus \tilde{\mathscr{L}}_0^{s2}$, with $\tilde{\mathscr{L}}_0^{s1} = A_1$ and $\tilde{\mathscr{L}}_0^{s2} = B_3$, so its dimension m is 24, and its *rank* l is 4.

Let A_j ($j = 1, 2, 3$) be the basis elements of $\tilde{\mathscr{L}}_0^{s1}$ ($= A_1$) chosen so that

$$[A_j, A_{j'}]_- = i \sum_{j''=1}^{3} \varepsilon_{jj'j''} A_{j''} \tag{M.23}$$

($j, j' = 1, 2, 3$) (as in (12.11)). Let M_{pq}, for $p, q = 1, 2, \ldots, 7$ and with $p < q$, be the basis elements of $\tilde{\mathscr{L}}_0^{s2}$ ($= B_3$). With the definition $M_{qp} = -M_{pq}$, they may be assumed (as in (G.13)) to satisfy the commutation relations

$$[M_{pq}, M_{p'q'}]_- = \delta_{qp'} M_{pq'} - \delta_{qq'} M_{pp'} + \delta_{pp'} M_{qq'} - \delta_{pq'} M_{qp'}$$

(for $p, q, p', q' = 1, 2, \ldots, 7$). Of course

$$[A_j, M_{pq}]_- = 0$$

(for $j = 1, 2, 3$ and $p, q = 1, 2, \ldots, 7$ (with $p \neq q$)).

(b) The *odd part* $\tilde{\mathscr{L}}_1$ of $\tilde{\mathscr{L}}_s = F(4)$ is defined to have dimension $n = 16$, the 16 basis elements Q_{kr} ($k = 1,2$ and $r = 1,2,\ldots,8$) being chosen so that

$$[A_j, Q_{kr}]_- = -\sum_{k'=1}^{2} (i\mathbf{a}_j)_{k'k} Q_{k'r} \tag{M.24}$$

(for $j = 1,2,3$; $k = 1,2$ and $r = 1,2,\ldots,8$), where $\mathbf{a}_1, \mathbf{a}_2$ and $\mathbf{a}_3$ are defined in (M.16), and

$$[M_{pq}, Q_{kr}]_- = \sum_{r'=1}^{8} (\mathbf{\Gamma}(M_{pq}))_{r'r} Q_{kr'} \tag{M.25}$$

(for $p, q = 1,2,\ldots,7$ (with $p < q$), $k = 1,2$ and $r = 1,2,\ldots,8$). Here $\mathbf{\Gamma}(M_{pq})$ is the 8-dimensional irreducible representation of B_3 that is defined in terms of the 8×8 matrices $\boldsymbol{\gamma}_p$ ($p = 1,2,\ldots,7$), which are assumed to satisfy the Clifford algebra

$$\boldsymbol{\gamma}_p\boldsymbol{\gamma}_q + \boldsymbol{\gamma}_q\boldsymbol{\gamma}_p = 2\delta_{pq}\mathbf{1}_8, \tag{M.26}$$

by

$$\mathbf{\Gamma}(M_{pq}) = -\tfrac{1}{2}\boldsymbol{\gamma}_p\boldsymbol{\gamma}_q \tag{M.27}$$

(for $p, q = 1,2,\ldots,7$ (with $p \neq q$)). A convenient choice is

$$\left.\begin{aligned}
\boldsymbol{\gamma}_1 &= \boldsymbol{\sigma}_1 \otimes \mathbf{1}_2 \otimes \mathbf{1}_2,\\
\boldsymbol{\gamma}_2 &= \boldsymbol{\sigma}_2 \otimes \mathbf{1}_2 \otimes \mathbf{1}_2,\\
\boldsymbol{\gamma}_3 &= \boldsymbol{\sigma}_3 \otimes \boldsymbol{\sigma}_1 \otimes \mathbf{1}_2,\\
\boldsymbol{\gamma}_4 &= \boldsymbol{\sigma}_3 \otimes \boldsymbol{\sigma}_2 \otimes \mathbf{1}_2,\\
\boldsymbol{\gamma}_5 &= \boldsymbol{\sigma}_3 \otimes \boldsymbol{\sigma}_3 \otimes \boldsymbol{\sigma}_1,\\
\boldsymbol{\gamma}_6 &= \boldsymbol{\sigma}_3 \otimes \boldsymbol{\sigma}_3 \otimes \boldsymbol{\sigma}_2,\\
\boldsymbol{\gamma}_7 &= \boldsymbol{\sigma}_3 \otimes \boldsymbol{\sigma}_3 \otimes \boldsymbol{\sigma}_3
\end{aligned}\right\} \tag{M.28}$$

(which is a reordering of the set given in Appendix L, Section 2). Here $\boldsymbol{\sigma}_1, \boldsymbol{\sigma}_2$ and $\boldsymbol{\sigma}_3$ are the Pauli spin matrices, which are related to the $\mathbf{a}_j$ of (M.24) by $\mathbf{a}_j = \frac{1}{2}i\boldsymbol{\sigma}_j$ ($j = 1,2,3$).

The remaining generalized Lie products of basis elements are given by

$$[Q_{kr}, Q_{k'r'}]_+ = 2i\sum_{j=1}^{3} C^8_{rr'}(\mathbf{C}^2\mathbf{a}_j)_{kk'} A_j + \frac{1}{3}\sum_{p,q=1(p\neq q)}^{7} C_{kk'}(\mathbf{C}^8\boldsymbol{\gamma}_p\boldsymbol{\gamma}_q)_{rr'} M_{pq}$$

where $\mathbf{C}^2$ is defined by (M.18) and $\mathbf{C}^8$ is the 8×8 matrix such that

$$\mathbf{C}^8\tilde{\boldsymbol{\gamma}}_p = -\boldsymbol{\gamma}_p\mathbf{C}^8$$

for $p = 1, 2, \ldots, 7$. For a proof that the generalized Jacobi relations (21.14) and (21.15) are satisfied see Scheunert *et al.* (1976b). For more information on the Clifford algebra matrices γ_p and the matrix $\mathbf{C}^8$ see Appendix L, Section 2. Other descriptions of $F(4)$ have been given by DeWitt and van Nieuwenhuizen (1982) and Sudbery (1983).

(c) It follows from (M.24) and (M.25) that *the representation of* $\tilde{\mathscr{L}}_0$ *on* $\tilde{\mathscr{L}}_1$ is the 16-dimensional *irreducible* representation of $\tilde{\mathscr{L}}_0$ that is the direct product of the 2-dimensional irreducible representation $\mathbf{D}^{1/2}$ of A_1 (cf. Chapter 12, Section 3) whose highest weight is $\Lambda = \Lambda_1$ with the 8-dimensional irreducible representation of B_3 whose highest weight is $\Lambda = \Lambda_3$. The Dynkin indices of these two latter irreducible representations are $\frac{1}{4}$ and $\frac{1}{5}$ respectively, so $\gamma_{\mathbf{D}}^1 = 8 \times \frac{1}{4} = 2$ and $\gamma_{\mathbf{D}}^2 = 2 \times \frac{1}{5} = \frac{2}{5}$. As neither has the value 1, Theorem XIII of Chapter 25, Section 3(*a*) implies that the Killing form of $\tilde{\mathscr{L}}_s = F(4)$ is *not* identically zero.

(d) A convenient basis of the *Cartan subalgebra* $\mathscr{H}_s$ of $\tilde{\mathscr{L}}_s = F(4)$ is provided by

$$h_1^1 = iA_3$$

(which provides a basis of the Cartan subalgebra $\mathscr{H}_s^1$ of $\tilde{\mathscr{L}}_0^{s1} = A_1$), and

$$h_j^2 = iM_{2j-1,2j}, \quad j = 1, 2, 3$$

(which provides a basis of the Cartan subalgebra $\mathscr{H}_0^2$ of $\tilde{\mathscr{L}}_0^{s2} = B_3$ (cf. (G.15)). It follows from (M.27) and (M.28) that

$$\left.\begin{aligned} \mathbf{\Gamma}(h_1^2) &= -\tfrac{1}{2}i\boldsymbol{\sigma}_3 \otimes \mathbf{1}_2 \otimes \mathbf{1}_2, \\ \mathbf{\Gamma}(h_2^2) &= -\tfrac{1}{2}i\mathbf{1}_2 \otimes \boldsymbol{\sigma}_3 \otimes \mathbf{1}_2, \\ \mathbf{\Gamma}(h_3^2) &= -\tfrac{1}{2}i\mathbf{1}_2 \otimes \mathbf{1}_2 \otimes \boldsymbol{\sigma}_3. \end{aligned}\right\} \tag{M.29}$$

Two useful sets ε_1^1 and $\varepsilon_1^2, \varepsilon_2^2, \varepsilon_3^2$ of *linear functionals* on $\mathscr{H}_s$ may be defined by

$$\varepsilon_1^1(h) = \begin{cases} i & \text{if } h = h_1^1, \\ 0 & \text{if } h \in \mathscr{H}_0^2, \end{cases}$$

and for $p, q = 1, 2, 3$

$$\varepsilon_p^2(h) = \begin{cases} i\delta_{jp} & \text{if } h = h_j^2,\ j = 1, 2, 3, \\ 0 & \text{if } h \in \mathscr{H}_0^1. \end{cases}$$

(e) $\tilde{\mathscr{L}}_s = F(4)$ has 20 *even roots*, which fall into 8 classes:

(1) $\alpha = \varepsilon_1^1$ (for which $\dim \tilde{\mathscr{L}}_{s\alpha} = 1$ with basis element $i(A_1 + iA_2)$);

(2) $\alpha = -\varepsilon_1^1$ (for which $\dim \tilde{\mathscr{L}}_{s\alpha} = 1$ with basis element $i(A_1 - iA_2)$);

(3) $\alpha = \varepsilon_p^2$ for $p = 1, 2, 3$ (for which dim $\tilde{\mathscr{L}}_{s\alpha} = 1$ with basis element $M_{2p,7} - iM_{2p-1,7}$);

(4) $\alpha = -\varepsilon_p^2$ for $p = 1, 2, 3$ (for which dim $\tilde{\mathscr{L}}_{s\alpha} = 1$ with basis element $M_{2p,7} + iM_{2p-1,7}$);

(5) $\alpha = \varepsilon_p^2 + \varepsilon_q^2$ for $p, q = 1, 2, 3$ (with $p < q$) (for which dim $\tilde{\mathscr{L}}_{s\alpha} = 1$ with basis element $-M_{2p,2q} + iM_{2p,2q-1} + iM_{2p-1,2q} + M_{2p-1,2q-1}$);

(6) $\alpha = -(\varepsilon_p^2 + \varepsilon_q^2)$ for $p, q = 1, 2, 3$ (with $p < q$) (for which dim $\tilde{\mathscr{L}}_{s\alpha} = 1$ with basis element $-M_{2p,2q} - iM_{2p,2q-1} - iM_{2p-1,2q} + M_{2p-1,2q-1}$);

(7) $\alpha = \varepsilon_p^2 - \varepsilon_q^2$ for $p, q = 1, 2, 3$ (with $p < q$) (for which dim $\tilde{\mathscr{L}}_{s\alpha} = 1$ with basis element $-M_{2p,2q} - iM_{2p,2q-1} + iM_{2p-1,2q} - M_{2p-1,2q-1}$);

(8) $\alpha = -(\varepsilon_p^2 - \varepsilon_q^2)$ for $p, q = 1, 2, 3$ (with $p < q$) (for which dim $\tilde{\mathscr{L}}_{s\alpha} = 1$ with basis element $-M_{2p,2q} + iM_{2p,2q-1} - iM_{2p-1,2q} - M_{2p-1,2q-1}$).

The first two of these are extensions of the roots of $\tilde{\mathscr{L}}_0^{s1} = A_1$, and the rest are extensions of the roots of $\tilde{\mathscr{L}}_0^{s2} = B_3$ (cf. Appendix G, Section 2).

(f) $\tilde{\mathscr{L}}_s = F(4)$ has 16 *odd roots*:

$$\alpha = \tfrac{1}{2}(\pm\varepsilon_1^1 \pm \varepsilon_1^2 \pm \varepsilon_2^2 \pm \varepsilon_3^2).$$

For each of these, $\tilde{\mathscr{L}}_{s\alpha}$ is one-dimensional, and with the index r of Q_{kr} being replaced by a set of three indices $\{r_1, r_2, r_3\}$ (which each take the values 1 or 2) in order to correspond to the direct-product structure in (M.29), $Q_{kr_1r_2r_3}$ is the basis element for

$$\tfrac{1}{2}\{(-1)^{k+1}\varepsilon_1^1 + (-1)^{r_1}\varepsilon_1^2 + (-1)^{r_2}\varepsilon_2^2 + (-1)^{r_3}\varepsilon_3^2\}.$$

(g) The *distinguished* set of *simple roots* of $\tilde{\mathscr{L}}_s = F(4)$ may be taken to be

(1) simple even roots

$$\alpha_2 = -\varepsilon_1^2,$$
$$\alpha_3 = \varepsilon_1^2 - \varepsilon_2^2,$$
$$\alpha_4 = \varepsilon_2^2 - \varepsilon_3^2;$$

(2) simple odd root

$$\alpha_1 = \tfrac{1}{2}(\varepsilon_1^1 + \varepsilon_1^2 + \varepsilon_2^2 + \varepsilon_3^2).$$

The number of simple roots is $L = l = 4$.

(h) In terms of these simple roots, the expressions for the *positive even* roots of

$\tilde{\mathscr{L}}_s = F(4)$ are

$$\alpha_2, \quad \alpha_3, \quad \alpha_4, \quad \alpha_2 + \alpha_3, \quad \alpha_3 + \alpha_4, \quad 2\alpha_2 + \alpha_3, \quad \alpha_2 + \alpha_3 + \alpha_4,$$
$$2\alpha_2 + \alpha_3 + \alpha_4, \quad 2\alpha_2 + 2\alpha_3 + \alpha_4, \quad 2\alpha_1 + 3\alpha_2 + 2\alpha_3 + \alpha_4.$$

It should be noted that

$$\beta = \varepsilon_1^1 = 2\alpha_1 + 3\alpha_2 + 2\alpha_3 + \alpha_4$$

is the extension of a simple root of $\tilde{\mathscr{L}}_0^{s1} = A_1$ that is *not* a simple root of $\tilde{\mathscr{L}}_s$.

(i) In terms of these simple roots $\alpha_1, \alpha_2, \alpha_3$ and α_4, the expressions for the *positive odd* roots of $\tilde{\mathscr{L}}_s = F(4)$ are

$$\alpha_1, \quad \alpha_1 + \alpha_2, \quad \alpha_1 + \alpha_2 + \alpha_3, \quad \alpha_1 + 2\alpha_2 + \alpha_3, \quad \alpha_1 + \alpha_2 + \alpha_3 + \alpha_4,$$
$$\alpha_1 + 2\alpha_2 + 2\alpha_3 + \alpha_4, \quad \alpha_1 + 2\alpha_2 + \alpha_3 + \alpha_4, \quad \alpha_1 + 3\alpha_2 + 2\alpha_3 + \alpha_4.$$

(j) The quantities $\langle \alpha_j, \alpha_k \rangle$ of $\tilde{\mathscr{L}}_s = F(4)$ (calculated using (24.21)) are given by

$$\langle \alpha_2, \alpha_2 \rangle = \tfrac{1}{6},$$
$$\langle \alpha_3, \alpha_3 \rangle = \langle \alpha_4, \alpha_4 \rangle = \tfrac{1}{3},$$
$$\langle \alpha_1, \alpha_2 \rangle = \langle \alpha_2, \alpha_1 \rangle = -\tfrac{1}{12},$$
$$\langle \alpha_2, \alpha_3 \rangle = \langle \alpha_3, \alpha_2 \rangle = \langle \alpha_3, \alpha_4 \rangle = \langle \alpha_4, \alpha_3 \rangle = -\tfrac{1}{6}$$

with $\langle \alpha_j, \alpha_k \rangle = 0$ for all other pairs. In particular,

$$\langle \alpha_1, \alpha_1 \rangle = 0.$$

(k) The *Cartan matrix of* $\tilde{\mathscr{L}}_s = F(4)$ is the 4×4 matrix

$$\mathbf{A} = \begin{bmatrix} 0 & 1 & 0 & 0 \\ -1 & 2 & -2 & 0 \\ 0 & -1 & 2 & -1 \\ 0 & 0 & -1 & 2 \end{bmatrix}.$$

(l) For *every odd root* α of $\tilde{\mathscr{L}}_s = F(4)$

$$\langle \alpha, \alpha \rangle = 0.$$

(m) The *generalized Dynkin diagram* corresponding to $\tilde{\mathscr{L}}_s = F(4)$ is given in Figure M.10.

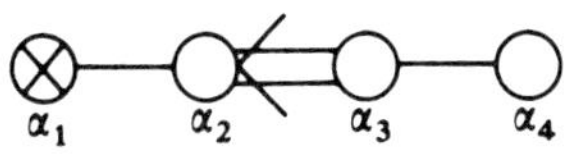

Figure M.10 Generalized Dynkin diagram for $F(4)$ with the "distinguished" choice of simple roots.

8 The basic type II classical simple complex Lie superalgebra $G(3)$

(a) The *even subalgebra* $\tilde{\mathscr{L}}_0$ of $\tilde{\mathscr{L}}_s = G(3)$ is $\tilde{\mathscr{L}}_0 = \tilde{\mathscr{L}}_0^{ss} = \mathscr{L}_0^{s1} \oplus \tilde{\mathscr{L}}_0^{s2}$, with $\tilde{\mathscr{L}}_0^{s1} = A_1$ and $\tilde{\mathscr{L}}_0^{s2} = G_2$, so its dimension m is 17 and its *rank* l is 3.

Let $A_j (j = 1, 2, 3)$ be the basis elements of $\tilde{\mathscr{L}}_0^{s1} (= A_1)$, their commutation relations being assumed to be

$$[A_j, A_{j'}]_- = i \sum_{j''=1}^{3} \varepsilon_{jj'j''} A_{j''}$$

$(j, j' = 1, 2, 3)$ (as in (12.11)). The basis elements of $\tilde{\mathscr{L}}_0^{s2} (= G_2)$ may be taken to be the 14 linearly independent elements of the set F_{pq} (for $p, q = 1, 2, \ldots, 7$), where $F_{qp} = -F_{pq}$ and

$$\sum_{p,q=1}^{7} \xi_{pqr} F_{pq} = 0$$

for $r = 1, 2, \ldots, 7$, where ξ_{pqr} are real numbers such that

(1) ξ_{pqr} is completely antisymmetric with respect to interchanges of any pair of the three indices p, q and r;

(2) $\xi_{pqr} = 1$ if $\{p, q, r\} = \{1, 2, 3\}$, $\{1, 4, 5\}$, $\{1, 7, 6\}$, $\{2, 4, 6\}$, $\{2, 5, 7\}$, $\{3, 4, 7\}$ or $\{3, 6, 5\}$;

(3) $\xi_{pqr} = 0$ if $\{p, q, r\}$ is neither one of the sets in (2) nor is a permutation of such a set.

Then

$$[F_{pq}, F_{p'q'}]_- = 3\delta_{pp'} F_{qq'} - 3\delta_{qp'} F_{pq'} + 3\delta_{qq'} F_{pp'} - 3\delta_{pq'} F_{qp'} - \sum_{p'',q''=1}^{7} \xi_{pqp''} \xi_{p'q'q''} F_{p''q''}$$

(for $p, q, p', q' = 1, 2, \ldots, 7$). Of course,

$$[A_j, F_{pq}]_- = 0$$

(for $j = 1, 2, 3$ and $p, q = 1, 2, \ldots, 7$).

(b) The *odd part* $\tilde{\mathscr{L}}_1$ of $\tilde{\mathscr{L}}_s = G(3)$ is defined to have dimension $n = 14$, the 14 basis elements Q_{kr} (for $k = 1, 2$ and $r = 1, 2, \ldots, 7$) being chosen so that

$$[A_j, Q_{kr}]_- = -\sum_{k'=1}^{2} (i\mathbf{a}_j)_{k'k} Q_{k'r} \tag{M.30}$$

(for $j = 1, 2, 3$; $k = 1, 2$ and $r = 1, 2, \ldots, 7$), where $\mathbf{a}_1, \mathbf{a}_2$ and $\mathbf{a}_3$ are defined in

(M.16), and

$$[F_{pq}, Q_{kr}]_- = 2\delta_{pr}Q_{kq} - 2\delta_{qr}Q_{kp} - \sum_{s=1}^{7} \eta_{pqrs}Q_{ks}, \qquad \text{(M.31)}$$

where

$$\eta_{pqrs} = \delta_{ps}\delta_{qr} - \delta_{pr}\delta_{qs} + \sum_{t=1}^{7} \xi_{pqt}\xi_{rst}.$$

The remaining generalized Lie products of basis elements are given by

$$[Q_{kr}, Q_{k'r'}]_+ = -4i\delta_{rr'} \sum_{j=1}^{3} (\mathbf{C}^2\mathbf{a}_j)_{kk'}A_j - \tfrac{1}{2}C^2_{kk'}F_{rr'}$$

(for $k, k' = 1, 2$ and $r, r' = 1, 2, \ldots, 7$), where $\mathbf{C}^2$ is given by (M.18). (See Scheunert *et al.* (1976b) for further details and proofs that the generalized Jacobi relations (21.14) and (21.15) are satisfied.) Other descriptions of $G(3)$ have been given by DeWitt and van Nieuwenhuizen (1982) and Sudbery (1983).

(c) It follows from (M.30) and (M.31) that *the representation of* $\tilde{\mathscr{L}}_0$ on $\tilde{\mathscr{L}}_1$ is the 14-dimensional *irreducible* representation of $\tilde{\mathscr{L}}_0$ that is the direct product of the 2-dimensional irreducible representation $\mathbf{D}^{1/2}$ of A_1 (cf. Chapter 12, Section 3) whose highest weight is $\Lambda = \Lambda_1$ with the 7-dimensional irreducible representation of G_2 whose highest weight is $\Lambda = \Lambda_2$. The Dynkin indices of these two latter irreducible representations are both $\frac{1}{4}$, so $\gamma^1_{\mathrm{D}} = 7 \times \frac{1}{4} = \frac{7}{4}$ and $\gamma^2_{\mathrm{D}} = 2 \times \frac{1}{4} = \frac{1}{2}$. As neight has the value 1, Theorem XIII of Chapter 25, Section 3(*a*) implies that the *Killing form* of $\tilde{\mathscr{L}}_s = G(3)$ is *not* identically zero.

(d) Taking the extension of the simple roots of $\tilde{\mathscr{L}}^{s1}_0 = A_1$ to be α^1_1 and the extensions of the simple roots of $\tilde{\mathscr{L}}^{s2}_0 = G_2$ to be α^2_1 and α^2_2, the 14 *even roots* of $G(3)$ are

$$\pm\alpha^1_1, \quad \pm\alpha^2_1, \quad \pm\alpha^2_2, \quad \pm(\alpha^2_1 + \alpha^2_2), \quad \pm(\alpha^2_1 + 2\alpha^2_2),$$
$$\pm(\alpha^2_1 + 3\alpha^2_2), \quad \pm(2\alpha^2_1 + 3\alpha^2_2)$$

(cf. Appendix F, Section 9). For each of these the corresponding root subspace $\tilde{\mathscr{L}}_{s\alpha}$ is one-dimensional

(e) In terms of α^1_1, α^2_1 and α^2_2, the 14 *odd roots* of $\tilde{\mathscr{L}}_s = G(3)$ are

$$\pm\tfrac{1}{2}\alpha^1_1, \quad \pm\tfrac{1}{2}\alpha^1_1 \pm (\alpha^2_1 + \alpha^2_2), \quad \pm\tfrac{1}{2}\alpha^1_1 \pm \alpha^2_2, \quad \pm\tfrac{1}{2}\alpha^1_1 \pm (\alpha^2_1 + 2\alpha^2_2).$$

For each of these the corresponding root subspace $\tilde{\mathscr{L}}_{s\alpha}$ is one-dimensional.

(f) The *distinguished* set of *simple roots* of $\tilde{\mathscr{L}}_s = G(3)$ may be taken to be

(1) simple even roots

$$\alpha_2 = \alpha_2^2,$$
$$\alpha_3 = -\alpha_1^2 - 3\alpha_2^2;$$

(2) simple odd root

$$\alpha_1 = \tfrac{1}{2}\alpha_1^1 + \alpha_1^2 + \alpha_2^2.$$

The number of simple roots is $L = l = 3$.

(g) In terms of these simple roots, the expressions for the *positive even* roots of $\tilde{\mathscr{L}}_s = G(3)$ are

$$\alpha_2, \quad \alpha_3, \quad \alpha_2+\alpha_3, \quad 2\alpha_2+\alpha_3, \quad 3\alpha_2+\alpha_3, \quad 3\alpha_2+2\alpha_3, \quad 2\alpha_1+4\alpha_2+2\alpha_3.$$

It should be noted that

$$\beta = \alpha_1^1 = 2\alpha_1 + 4\alpha_2 + 2\alpha_3$$

is the extension of a simple root of $\tilde{\mathscr{L}}_0^{s1} = A_1$ that is *not* a simple root of $\tilde{\mathscr{L}}_s$.

(h) In terms of these simple roots α_1, α_2 and α_3 the expressions for the *positive odd* roots of $\tilde{\mathscr{L}}_s = G(3)$ are

$$\alpha_1, \quad \alpha_1+\alpha_2, \quad \alpha_1+\alpha_2+\alpha_3, \quad \alpha_1+2\alpha_2+\alpha_3, \quad \alpha_1+3\alpha_2+\alpha_3,$$
$$\alpha_1+3\alpha_2+2\alpha_3, \quad \alpha_1+4\alpha_2+2\alpha_3.$$

(i) The quantities $\langle \alpha_j, \alpha_k \rangle$ of $\tilde{\mathscr{L}}_s = G(3)$ (calculated using (24.21)) are given by

$$\langle \alpha_2, \alpha_2 \rangle = \tfrac{1}{6}, \quad \langle \alpha_3, \alpha_3 \rangle = \tfrac{1}{2},$$
$$\langle \alpha_1, \alpha_2 \rangle = \langle \alpha_2, \alpha_1 \rangle = -\tfrac{1}{12},$$
$$\langle \alpha_2, \alpha_3 \rangle = \langle \alpha_3, \alpha_2 \rangle = -\tfrac{1}{4},$$

with $\langle \alpha_j, \alpha_k \rangle = 0$ for all other pairs. In particular,

$$\langle \alpha_1, \alpha_1 \rangle = 0.$$

(j) The *Cartan matrix* of $\tilde{\mathscr{L}}_s = G(3)$ is the 3×3 matrix

$$\mathbf{A} = \begin{bmatrix} 0 & 1 & 0 \\ -1 & 2 & -3 \\ 0 & -1 & 2 \end{bmatrix}.$$

(k) For the *odd roots* α of $\tilde{\mathscr{L}}_s = G(3)$

$$\langle \alpha, \alpha \rangle = \begin{cases} -\tfrac{1}{6} & \text{for } \alpha = \pm\tfrac{1}{2}\alpha_1^1 = \pm(\alpha_1+2\alpha_2+\alpha_3), \\ 0 & \text{for all other odd roots.} \end{cases}$$

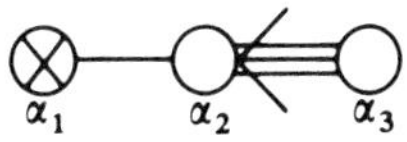

Figure M.11 Generalized Dynkin diagram for $G(3)$ with the "distinguished" choice of simple roots.

(l) The *generalized Dynkin diagram* corresponding to $\tilde{\mathscr{L}}_s = G(3)$ is given in Figure M.11.

9 The strange type I classical simple complex Lie superalgebras $P(r), r \geqslant 2$

(a) A realization of $\tilde{\mathscr{L}}_s = P(r)$ (for $r \geqslant 2$) is provided by the set of complex $(2r+2) \times (2r+2)$ matrices $\mathbf{M}$ with grading partitioning

$$\mathbf{M} = \begin{bmatrix} \mathbf{A} & \mathbf{B} \\ \mathbf{C} & -\mathbf{A} \end{bmatrix},$$

where $\mathbf{A}$, $\mathbf{B}$ and $\mathbf{C}$ are all complex $(r+1) \times (r+1)$ matrices, $\mathbf{A}$ being such that $\operatorname{tr} \mathbf{A} = 0$, $\mathbf{B}$ being symmetric and $\mathbf{C}$ being antisymmetric. The dimensions of $\tilde{\mathscr{L}}_0$ and $\tilde{\mathscr{L}}_1$ are given respectively by $m = r(r+2)$ and $n = (r+1)^2$. This Lie superalgebra $P(r)$ is sometimes referred to in the literature as $b(r)$.

(b) The *even subalgebra* $\tilde{\mathscr{L}}_0$ of $\tilde{\mathscr{L}}_s = P(r)$ (for $r \geqslant 2$) is $\tilde{\mathscr{L}}_0 = \tilde{\mathscr{L}}_0^{ss} = \tilde{\mathscr{L}}_0^{s1} = A_r$, this algebra being isomorphic to the set of complex $(r+1) \times (r+1)$ traceless matrices $\mathbf{A}$. The *rank of* $\tilde{\mathscr{L}}_s$ is given by $l = r$.

(c) A convenient basis of the *Cartan subalgebra* $\mathscr{H}_s$ of $\tilde{\mathscr{L}}_s = P(r)$ (for $r \geqslant 2$) is provided by

$$\mathbf{h}_j^1 = \mathbf{e}_{j,j} - \mathbf{e}_{j+1,j+1} - \mathbf{e}_{j+r+1,j+r+1} + \mathbf{e}_{j+r+2,j+r+2}, \quad j = 1, 2, \ldots, r.$$

(This provides a basis for the Cartan subalgebra $\mathscr{H}_0^1$ of $\tilde{\mathscr{L}}_0^{s1} = A_r$.)

(d) A useful set of *linear functionals* $\varepsilon_1^1, \varepsilon_2^1, \ldots, \varepsilon_{r+1}^1$ on $\mathscr{H}_s$ may be defined by

$$\varepsilon_p^1(\mathbf{h}_j^1) = \delta_{jp} - \delta_{j+1,p} \qquad \text{(M.32)}$$

(for $j = 1, 2, \ldots, r$ and $p = 1, 2, \ldots, r+1$).

(e) $\tilde{\mathscr{L}}_s = P(r)$ (for $r \geqslant 2$) has $r(r+1)$ *even roots*:

$$\alpha = \varepsilon_p^1 - \varepsilon_q^1 \quad \text{for } p, q = 1, 2, \ldots, r+1 \ (p \neq q)$$

(for which the corresponding root subspace $\tilde{\mathscr{L}}_{s\alpha}$ is one-dimensional and has basis $\mathbf{e}_{p,q} - \mathbf{e}_{q+r+1,p+r+1}$). These are extensions of roots of $\tilde{\mathscr{L}}_0^{s1} = A_r$.

(f) $\tilde{\mathscr{L}}_s = P(r)$ (for $r \geqslant 2$) has $(r+1)^2$ *odd roots*, which fall into two classes:

(1) $\alpha = \varepsilon_p^1 + \varepsilon_q^1$ for $p, q = 1, 2, \ldots, r+1$ $(p \leqslant q)$ (for which dim $\tilde{\mathscr{L}}_{s\alpha} = 1$ with basis element $\mathbf{e}_{p,q+r+1} + \mathbf{e}_{q,p+r+1}$);

(2) $\alpha = -(\varepsilon_p^1 + \varepsilon_q^1)$ for $p, q = 1, 2, \ldots, r+1$ $(p < q)$ (for which dim $\tilde{\mathscr{L}}_{s\alpha} = 1$ with basis element $\mathbf{e}_{p+r+1,q} - \mathbf{e}_{q+r+1,p}$).

It should be noted that for $\alpha = 2\varepsilon_p^1$ (for $p = 1, 2, \ldots, r+1$) it is *not* true that $-\alpha$ is a root. This demonstrates that $P(r)$ (for $r \geqslant 2$) must be a *strange* classical simple Lie superalgebra, since otherwise Theorem XVIII of Chapter 25, Section 3(*b*) would be violated.

(g) For $\tilde{\mathscr{L}}_s = P(r)$ (for $r \geqslant 2$) the *representation of* $\tilde{\mathscr{L}}_0$ *on* $\tilde{\mathscr{L}}_1$ is the direct sum of *two* irreducible representations of $\tilde{\mathscr{L}}_0 = A_r$ with dimensions $\frac{1}{2}(r+1)(r+2)$ and $\frac{1}{2}r(r+1)$. They have Dynkin indices $(r+3)/2(r+1)$ and $(r-1)/2(r+3)$, so $\gamma_D^1 = 1$, implying, by Theorem XIII of Chapter 25, Section 3(*a*), that the *Killing form* of $P(r)$ (for $r \geqslant 2$) is *zero* (which is as expected, since $P(r)$ (for $r \geqslant 2$) is *strange*).

(h) For $P(1)$ the basis elements of $\tilde{\mathscr{L}}_0$ together with $\mathbf{e}_{13}$ form the basis of a graded invariant subalgebra, so $P(1)$ is *not simple*.

(i) For $P(2)$ (M.32) implies that $\varepsilon_3^1 = -\varepsilon_1^1 - \varepsilon_2^1$, so the odd roots, when expressed in terms of ε_1^1 and ε_2^1 are

$$\pm(\varepsilon_1^1 + \varepsilon_2^1), \quad \pm\varepsilon_2^1, \quad \pm\varepsilon_1^1, \quad -2(\varepsilon_1^1 + \varepsilon_2^1), \quad 2\varepsilon_1^1, \quad 2\varepsilon_2^1.$$

That is, for the odd roots $\alpha = \varepsilon_1^1, \varepsilon_2^1$ and $-(\varepsilon_1^1 + \varepsilon_2^1)$, 2α is *also* an odd root.

(j) For $P(3)$ (M.32) implies that $\varepsilon_4^1 = -\varepsilon_1^1 - \varepsilon_2^1 - \varepsilon_3^1$, so the odd roots, when expressed in terms of $\varepsilon_1^1, \varepsilon_2^1$ and ε_3^1, are

$$\pm(\varepsilon_1^1 + \varepsilon_2^1), \quad \pm(\varepsilon_1^1 + \varepsilon_3^1), \quad \pm(\varepsilon_2^1 + \varepsilon_3^1), \quad 2\varepsilon_1^1, \quad 2\varepsilon_2^1,$$
$$2\varepsilon_3^1, \quad 2\varepsilon_4^1 = -2(\varepsilon_1^1 + \varepsilon_2^1 + \varepsilon_3^1),$$

and

$$\pm(\varepsilon_1^1 + \varepsilon_4^1) = \pm(-1)(\varepsilon_2^1 + \varepsilon_3^1),$$
$$\pm(\varepsilon_2^1 + \varepsilon_4^1) = \pm(-1)(\varepsilon_1^1 + \varepsilon_3^1),$$
$$\pm(\varepsilon_3^1 + \varepsilon_4^1) = \pm(-1)(\varepsilon_2^1 + \varepsilon_2^1).$$

Thus for $\alpha = \pm(\varepsilon_1^1 + \varepsilon_2^1)$, $\pm(\varepsilon_1^1 + \varepsilon_3^1)$ and $\pm(\varepsilon_2^1 + \varepsilon_3^1)$

$$\dim \tilde{\mathscr{L}}_{s\alpha} = 2.$$

(k) Although (M.32) implies that for all $r \geqslant 1$

$$\varepsilon^1_{r+1} = -\sum_{j=1}^{r} \varepsilon^1_j ,$$

no anomalous behaviour such as that described in subsections (i) and (j) occurs for $P(r)$ for $r \geqslant 4$.

10 The strange type II classical simple complex Lie superalgebras $Q(r)$, $r \geqslant 2$

(a) Consider the set $Q'(r)$ of $(2r+2) \times (2r+2)$ complex matrices $\mathbf{M}$ with grading partitioning

$$\mathbf{M} = \begin{bmatrix} \mathbf{A} & \mathbf{B} \\ \mathbf{B} & \mathbf{A} \end{bmatrix},$$

where $\mathbf{A}$ is an arbitrary $(r+1) \times (r+1)$ complex matrix and $\mathbf{B}$ is an $(r+1) \times (r+1)$ complex matrix such that $\operatorname{tr} \mathbf{B} = 0$. This set $Q'(r)$ contains a one-dimensional invariant subalgebra with basis element $\mathbf{1}_{2r+2}$.

$Q(r)$ (for $r \geqslant 2$) is *defined* to be the Lie superalgebra obtained from $Q'(r)$ by factoring out this one-dimensional invariant subalgebra. (See Chapter 25, Section 2 for a general treatment of this process.) This implies that if $\mathbf{M}$ and $\mathbf{N}$ are any two matrices of $Q'(r)$ such that

$$\mathbf{M} - \mathbf{N} = \kappa \mathbf{1}_{2r+2}, \tag{M.33}$$

where κ is some complex number, then $\mathbf{M}$ and $\mathbf{N}$ are to be regarded as corresponding to the same element of $Q(r)$. With this proviso, no confusion will arise if the elements of $Q(r)$ are denoted by the corresponding elements of $Q'(r)$. The dimensions of $\tilde{\mathscr{L}}_0$ and $\tilde{\mathscr{L}}_1$ respectively are given respectively by $m = n = r(r+2)$. This Lie superalgebra $Q(r)$ is sometimes referred to in the literature as $d(r)$.

(b) The *even subalgebra* $\tilde{\mathscr{L}}_0$ of $\tilde{\mathscr{L}}_s = Q(r)$ (for $r \geqslant 2$) is $\tilde{\mathscr{L}}_0 = \tilde{\mathscr{L}}_0^{ss} = \tilde{\mathscr{L}}_0^{s1} = A_r$, this algebra being isomorphic to the set of complex $(r+1) \times (r+1)$ matrices $\mathbf{A}$ factored by the one-dimensional subalgebra with basis element $\mathbf{1}_{r+1}$. The *rank* of $\tilde{\mathscr{L}}_s$ is given by $l = r$.

(c) A convenient basis of the *Cartan subalgebra* $\mathscr{H}_s$ of $\tilde{\mathscr{L}}_s = Q(r)$ (for $r \geqslant 2$) is provided by

$$\mathbf{h}^1_j = \mathbf{e}_{j,j} - \mathbf{e}_{j+1,j+1} + \mathbf{e}_{j+r+1,j+r+1} - \mathbf{e}_{j+r+2,j+r+2}, \quad j = 1, 2, \ldots, r.$$

(This provides a basis for the Cartan subalgebra $\mathscr{H}^1_0$ of $\tilde{\mathscr{L}}_0^{s1} = A_r$.)

(d) A useful set of *linear functionals* $\varepsilon_1^1, \varepsilon_2^1, \ldots, \varepsilon_{r+1}^1$ on $\mathscr{H}_s$ may be defined by

$$\varepsilon_p^1(\mathbf{h}_j^1) = \delta_{jp} - \delta_{j+1,p}$$

(for $j = 1, 2, \ldots, r$ and $p = 1, 2, \ldots, r+1$).

(e) $\tilde{\mathscr{L}}_s = Q(r)$ (for $r \geqslant 2$) has $r(r+1)$ *even roots*

$$\alpha = \varepsilon_p^1 - \varepsilon_q^1 \quad \text{for } p, q = 1, 2, \ldots, r+1 \ (p \neq q)$$

(for which the *even* root subspace $\tilde{\mathscr{L}}_{s\alpha} \cap \tilde{\mathscr{L}}_0$ is one-dimensional and has basis element $\mathbf{e}_{p,q} + \mathbf{e}_{q+r+1,p+r+1}$). These are extensions of roots of $\tilde{\mathscr{L}}_0^{s1} = A_r$.

(f) $\tilde{\mathscr{L}}_s = Q(r)$ (for $r \geqslant 2$) has $r(r+2)$ *odd roots*, which fall into two classes:

(1) $\alpha = \varepsilon_p^1 - \varepsilon_q^1$ for $p, q = 1, 2, \ldots, r+1$ $(p \neq q)$ (for which the *odd* root subspace $\tilde{\mathscr{L}}_{s\alpha} \cap \tilde{\mathscr{L}}_1$ is one-dimensional and has basis element $\mathbf{e}_{p,q+r+1} + \mathbf{e}_{p+r+1,q}$);

(2) $\alpha = 0$ (for which $\dim \tilde{\mathscr{L}}_{s\alpha} = r$, the basis elements being $\mathbf{e}_{p,p+r+1} - \mathbf{e}_{p+1,p+r+2} + \mathbf{e}_{p+r+1,p} - \mathbf{e}_{p+r+2,p+1}$ for $p = 1, 2, \ldots, r$).

(g) It should be noted that the roots $\alpha = \varepsilon_p^1 - \varepsilon_q^1$ (for $p, q = 1, 2, \ldots, r+1$, with $p \neq q$) are *both even and odd*, and that for these $\dim \tilde{\mathscr{L}}_{s\alpha} = 2$.

(h) For $\tilde{\mathscr{L}}_s = Q(r)$ (for $r \geqslant 2$) the *representation of* $\tilde{\mathscr{L}}_0$ *on* $\tilde{\mathscr{L}}_1$ is the adjoint representation of $\tilde{\mathscr{L}}_0 = A_r$, which is an *irreducible* representation of $\tilde{\mathscr{L}}_0 = A_r$ with dimension $r(r+2)$ and Dynkin index 1. Thus $\gamma_{\mathbf{D}}^1 = 1$, implying, by Theorem XIII of Chapter 25, Section 3(*a*), that the *Killing form* of $Q(r)$ (for $r \geqslant 2$) is *zero*. (See also subsection (i) below.)

(i) For $\tilde{\mathscr{L}}_s = Q(r)$ (for $r \geqslant 2$) *every* bilinear supersymmetric consistent invariant form is identically *zero*. To see this, suppose that $B'(\ ,\ )$ is such a form and suppose that $B'(\ ,\ )$ is not identically zero. Then, by Theorem XIX of Chapter 25, Section 3(*b*), if $a_\alpha \in \tilde{\mathscr{L}}_{s\alpha}$ and $a_{-\alpha} \in \tilde{\mathscr{L}}_{s(-\alpha)}$,

$$[a_\alpha, a_{-\alpha}] = B'(a_\alpha, a_{-\alpha}) h_\alpha. \tag{M.34}$$

Take $\alpha = \varepsilon_p^1 - \varepsilon_q^1$ for some p and q ($p, q = 1, 2, \ldots, r+1$, with $p \neq q$) and let $a_\alpha = \mathbf{e}_{p,q} + \mathbf{e}_{q+r+1,p+r+1}$ and $a_{-\alpha} = \mathbf{e}_{q,p+r+1} + \mathbf{e}_{q+r+1,p}$, so that $a_\alpha \in \tilde{\mathscr{L}}_0$ but $a_{-\alpha} \in \tilde{\mathscr{L}}_1$. The consistency condition (21.84) then implies that $B'(a_\alpha, a_{-\alpha}) = 0$, so that (M.34) gives $[a_\alpha, a_{-\alpha}] = 0$. However, direct calculation shows that

$$[a_\alpha, a_{-\alpha}] = \mathbf{e}_{p,p+r+1} - \mathbf{e}_{q+r+1,q},$$

yielding a contradiction. Thus the assumption that $B'(\ ,\ )$ can be chosen to be non-zero is false.

(j) For $Q(1)$ the basis elements of the odd part $\tilde{\mathscr{L}}_1$ are $\mathbf{e}_{14}+\mathbf{e}_{32}$, $\mathbf{e}_{41}+\mathbf{e}_{23}$ and $\mathbf{e}_{13}-\mathbf{e}_{24}+\mathbf{e}_{31}-\mathbf{e}_{42}$. Direct calculation (using (M.33)) shows that $[a, b]=0$ for all $a, b \in \tilde{\mathscr{L}}_1$, implying that $\tilde{\mathscr{L}}_1$ is a graded invariant subalgebra of $\tilde{\mathscr{L}}_s$. Thus $\tilde{\mathscr{L}}_s = Q(1)$ is *not simple.*

(k) For each $r \geqslant 2$ the algebra $Q(r)$ is isomorphic to an (f, d) algebra of Gell-Mann, Michel and Radicati (cf. Michel 1970). To establish this connection, let $\mathbf{a}_1$, $\mathbf{a}_2, \ldots, \mathbf{a}_p$ be a basis for the compact real Lie algebra su$(p+1)$, where $p=(r+1)^2-1$. These matrices also form a basis for the complex Lie algebra A_p, and (cf. (16.83))

$$[\mathbf{a}_j, \mathbf{a}_k]_+ = -\frac{1}{(p+1)^2}\delta_{jk}\mathbf{1}_{p+1} + \sum_{l=1}^{p} s_{jk}^l \mathbf{a}_l, \tag{M.35}$$

where the coefficients s_{jk}^l are real and symmetric with respect to the interchange of *any* pair of indices.

The even basis elements of $Q(r)$ (for $r \geqslant 2$) may be taken to be of the form

$$\mathbf{a}_j^0 = \begin{bmatrix} \mathbf{a}_j & \mathbf{0} \\ \mathbf{0} & \mathbf{a}_j \end{bmatrix}, \quad j = 1, 2, \ldots, p,$$

while the odd basis elements can be taken to be of the form

$$\mathbf{a}_j^1 = \begin{bmatrix} \mathbf{0} & \mathbf{a}_j \\ \mathbf{a}_j & \mathbf{0} \end{bmatrix}, \quad j = 1, 2, \ldots, p.$$

If c_{jk}^l are the structure constants of su$(p+1)$, i.e. if

$$[\mathbf{a}_j, \mathbf{a}_k]_- = \sum_{l=1}^{p} c_{jk}^l \mathbf{a}_l,$$

then it follows that

$$[\mathbf{a}_j^0, \mathbf{a}_k^0]_- = \sum_{l=1}^{p} c_{jk}^l \mathbf{a}_l^0,$$

$$[\mathbf{a}_j^0, \mathbf{a}_k^1]_- = \sum_{l=1}^{p} c_{jk}^l \mathbf{a}_l^1$$

and (by (M.33) and (M.35))

$$[\mathbf{a}_j^1, a_k^1]_+ = \sum_{l=1}^{p} s_{jk}^l \mathbf{a}_l^0.$$

The coefficients c_{jk}^l and s_{jk}^l are (up to a factor) just the coefficients f_{jkl} and d_{jkl} of Gell-Mann, Michel and Radicati (cf. Example V of Chapter 16, Section 6).

Appendix N

Properties of the Complex Affine Kac–Moody Algebras

1 The complex untwisted affine Kac–Moody algebra $A_1^{(1)}$

(a) The generalized Dynkin diagram for $A_1^{(1)}$ is given in Figure N.1.

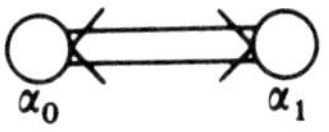

Figure N.1 Generalized Dynkin diagram for $A_1^{(1)}$.

(b) The generalized Cartan matrix for $A_1^{(1)}$ is

$$\mathbf{A} = \begin{bmatrix} 2 & -2 \\ -2 & 2 \end{bmatrix}.$$

(c) For $A_1^{(1)}$ the root δ is given by $\delta = \alpha_0 + \alpha_1$.

(d) For $\tilde{\mathscr{L}} = A_1^{(1)}$ the corresponding simple Lie algebra is $\tilde{\mathscr{L}}^0 = \tilde{\mathscr{L}}^s = A_1$, which has rank 1 and dimension 3. The highest root of A_1 is $\alpha_H = 2\Lambda_1 = \alpha_1$ (in the notation of Appendix F, Section 1), so that $\langle \alpha_H, \alpha_H \rangle^0 = \langle \alpha_H, \alpha_H \rangle^{(1)} = \frac{1}{2}$.

(e) For $\tilde{\mathscr{L}} = A_1^{(1)}$ the imaginary roots are such that $\dim \tilde{\mathscr{L}}_{j\delta}^{(1)} = 1$ (for all non-zero integers j).

(f) For $\tilde{\mathscr{L}} = A_1^{(1)}$, $\langle \alpha_j, \alpha_k \rangle^{(1)} = B_{jk}$ (for $j, k = 0, 1$), where $\mathbf{B} = \frac{1}{4}\mathbf{A}$.

(g) All real roots of $A_1^{(1)}$ are of the same length, i.e. $A_1^{(1)}$ is "simply laced".

(h) $A_1^{(1)}$ is self-dual.

2 The complex untwisted affine Kac–Moody algebras $A_l^{(1)}$, $l \geqslant 2$

(a) The generalized Dynkin diagram for $A_l^{(1)}$ (with $l \geqslant 2$) is given in Figure N.2.

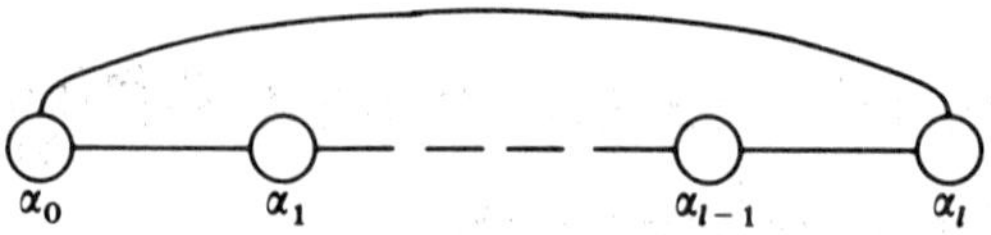

Figure N.2 Generalized Dynkin diagram for $A_l^{(1)}$ (for $l \geqslant 2$).

(b) The generalized Cartan matrix for $A_l^{(1)}$ (with $l \geqslant 2$) is the $(l+1) \times (l+1)$ matrix

$$\mathbf{A} = \begin{bmatrix} 2 & -1 & 0 & \cdots & 0 & 0 & -1 \\ -1 & 2 & -1 & \cdots & 0 & 0 & 0 \\ 0 & -1 & 2 & \cdots & 0 & 0 & 0 \\ \vdots & \vdots & \vdots & & \vdots & \vdots & \vdots \\ 0 & 0 & 0 & \cdots & 2 & -1 & 0 \\ 0 & 0 & 0 & \cdots & -1 & 2 & -1 \\ -1 & 0 & 0 & \cdots & 0 & -1 & 2 \end{bmatrix}.$$

In particular, for $l = 2$

$$\mathbf{A} = \begin{bmatrix} 2 & -1 & -1 \\ -1 & 2 & -1 \\ -1 & -1 & 2 \end{bmatrix},$$

and for $l = 3$

$$\mathbf{A} = \begin{bmatrix} 2 & -1 & 0 & -1 \\ -1 & 2 & -1 & 0 \\ 0 & -1 & 2 & -1 \\ -1 & 0 & -1 & 2 \end{bmatrix}.$$

(c) For $A_l^{(1)}$ (with $l \geqslant 2$) the root δ is given by

$$\delta = \alpha_0 + \alpha_1 + \cdots + \alpha_{l-1} + \alpha_l.$$

(d) For $\tilde{\mathscr{L}} = A_l^{(1)}$ (with $l \geqslant 2$) the corresponding simple Lie algebra is $\tilde{\mathscr{L}}^0 = \tilde{\mathscr{L}}^s = A_l$, which has rank l and dimension $(l+1)^2 - 1$. The highest root of A_l (with $l \geqslant 2$) is

$$\alpha_H = \Lambda_1 + \Lambda_l = \sum_{p=1}^{l} \alpha_p$$

(in the notation of Appendix F, Section 1), so that $\langle \alpha_{\mathrm{H}}, \alpha_{\mathrm{H}} \rangle^0 = \langle \alpha_{\mathrm{H}}, \alpha_{\mathrm{H}} \rangle^{(1)} = (l+1)^{-1}$.

(e) For $\tilde{\mathscr{L}} = A_l^{(1)}$ (with $l \geqslant 2$) the imaginary roots are such that $\dim \tilde{\mathscr{L}}_{j\delta}^{(1)} = l$ (for all non-zero integers j).

(f) For $\tilde{\mathscr{L}} = A_l^{(1)}$ (with $l \geqslant 2$) $\langle \alpha_j, \alpha_k \rangle^{(1)} = B_{jk}$ (for $j, k = 0, 1, \ldots, l$), where $\mathbf{B} = \{2(l+1)\}^{-1}\mathbf{A}$.

(g) All real roots of $A_l^{(1)}$ (with $l \geqslant 2$) are of the same length, i.e. $A_l^{(1)}$ (with $l \geqslant 2$) is "simply laced".

(h) $A_l^{(1)}$ (with $l \geqslant 2$) is self-dual.

3 The complex untwisted affine Kac–Moody algebras $B_l^{(1)}$, $l \geqslant 3$

(a) The generalized Dynkin diagram for $B_l^{(1)}$ (with $l \geqslant 3$) is given in Figure N.3.

Figure N.3 Generalized Dynkin diagram for $B_l^{(1)}$ (for $l \geqslant 3$).

(b) The generalized Cartan matrix for $B_l^{(1)}$ (with $l \geqslant 3$) is the $(l+1) \times (l+1)$ matrix

$$\mathbf{A} = \begin{bmatrix} 2 & 0 & -1 & 0 & \cdots & 0 & 0 & 0 \\ 0 & 2 & -1 & 0 & \cdots & 0 & 0 & 0 \\ -1 & -1 & 2 & -1 & \cdots & 0 & 0 & 0 \\ 0 & 0 & -1 & 2 & \cdots & 0 & 0 & 0 \\ \vdots & \vdots & \vdots & \vdots & & \vdots & \vdots & \vdots \\ 0 & 0 & 0 & 0 & \cdots & 2 & -1 & 0 \\ 0 & 0 & 0 & 0 & \cdots & -1 & 2 & -1 \\ 0 & 0 & 0 & 0 & \cdots & 0 & -2 & 2 \end{bmatrix}.$$

In particular, for $l = 3$

$$\mathbf{A} = \begin{bmatrix} 2 & 0 & -1 & 0 \\ 0 & 2 & -1 & 0 \\ -1 & -1 & 2 & -1 \\ 0 & 0 & -2 & 2 \end{bmatrix}.$$

(c) For $B_l^{(1)}$ (with $l \geqslant 3$) the root δ is given by

$$\delta = \alpha_0 + \alpha_1 + 2(\alpha_2 + \cdots + \alpha_l).$$

(d) For $B_l^{(1)}$ (with $l \geqslant 3$) the corresponding simple Lie algebra is $\tilde{\mathscr{L}}^0 = \tilde{\mathscr{L}}^s = B_l$, which has rank l and dimension $l(2l+1)$. The highest root of B_l (with $l \geqslant 3$) is

$$\alpha_H = \Lambda_2 = \alpha_1 + 2\sum_{p=2}^{l} \alpha_p$$

(in the notation of Appendix F, Section 2), which is a "long" root, so that $\langle \alpha_H, \alpha_H \rangle^0 = \langle \alpha_H, \alpha_H \rangle^{(1)} = (2l-1)^{-1}$.

(e) For $\tilde{\mathscr{L}} = B_l^{(1)}$ (with $l \geqslant 3$) the imaginary roots are such that $\dim \tilde{\mathscr{L}}_{j\delta}^{(1)} = l$ (for all non-zero integers j).

(f) For $\tilde{\mathscr{L}} = B_l^{(1)}$ (with $l \geqslant 3$) $\langle \alpha_j, \alpha_k \rangle^{(1)} = B_{jk}$ (for $j, k = 0, 1, \ldots, l$), where

$$\mathbf{B} = \frac{1}{2(2l-1)} \begin{bmatrix} 2 & 0 & -1 & 0 & \cdots & 0 & 0 & 0 \\ 0 & 2 & -1 & 0 & \cdots & 0 & 0 & 0 \\ -1 & -1 & 2 & -1 & \cdots & 0 & 0 & 0 \\ 0 & 0 & -1 & 2 & \cdots & 0 & 0 & 0 \\ \vdots & \vdots & \vdots & \vdots & & \vdots & \vdots & \vdots \\ 0 & 0 & 0 & 0 & \cdots & 2 & -1 & 0 \\ 0 & 0 & 0 & 0 & \cdots & -1 & 2 & -1 \\ 0 & 0 & 0 & 0 & \cdots & 0 & -1 & 1 \end{bmatrix}.$$

In particular, for $l = 3$

$$\mathbf{B} = \frac{1}{10} \begin{bmatrix} 2 & 0 & -1 & 0 \\ 0 & 2 & -1 & 0 \\ -1 & -1 & 2 & -1 \\ 0 & 0 & -1 & 1 \end{bmatrix}.$$

(g) For $B_l^{(1)}$ (with $l \geqslant 3$) $\alpha_0, \alpha_1, \ldots, \alpha_{l-1}$ are "long" roots, and α_l is a "short" root.

(h) The dual of $B_l^{(1)}$ (with $l \geqslant 3$) is $A_{2l-1}^{(2)}$.

4 The complex untwisted affine Kac–Moody algebras $C_l^{(1)}$, $l \geqslant 2$

(a) The generalized Dynkin diagram for $C_l^{(1)}$ (with $l \geqslant 2$) is given in Figure N.4.

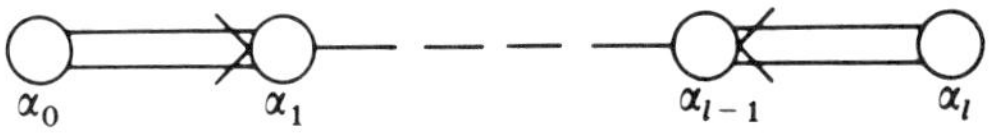

Figure N.4 Generalized Dynkin diagram for $C_l^{(1)}$ (for $l \geqslant 2$).

(b) The generalized Cartan matrix for $C_l^{(1)}$ (with $l \geqslant 2$) is the $(l+1) \times (l+1)$ matrix

$$\mathbf{A} = \begin{bmatrix} 2 & 0 & -1 & 0 & \cdots & 0 & 0 & 0 \\ -2 & 2 & -1 & 0 & \cdots & 0 & 0 & 0 \\ 0 & -1 & 2 & -1 & \cdots & 0 & 0 & 0 \\ 0 & 0 & -1 & 2 & \cdots & 0 & 0 & 0 \\ \vdots & \vdots & \vdots & \vdots & & \vdots & \vdots & \vdots \\ 0 & 0 & 0 & 0 & \cdots & 2 & -1 & 0 \\ 0 & 0 & 0 & 0 & \cdots & -1 & 2 & -2 \\ 0 & 0 & 0 & 0 & \cdots & 0 & -1 & 2 \end{bmatrix}.$$

In particular, for $l = 2$

$$\mathbf{A} = \begin{bmatrix} 2 & -1 & 0 \\ -2 & 2 & -2 \\ 0 & -1 & 2 \end{bmatrix},$$

and for $l = 3$

$$\mathbf{A} = \begin{bmatrix} 2 & -1 & 0 & 0 \\ -2 & 2 & -1 & 0 \\ 0 & -1 & 2 & -2 \\ 0 & 0 & -1 & 2 \end{bmatrix}.$$

(c) For $C_l^{(1)}$ (with $l \geqslant 2$) the root δ is given by

$$\delta = \alpha_0 + 2(\alpha_1 + \alpha_2 + \cdots + \alpha_{l-1}) + \alpha_l.$$

(d) For $C_l^{(1)}$ (with $l \geqslant 2$) the corresponding simple Lie algebra is $\tilde{\mathscr{L}}^0 = \tilde{\mathscr{L}}^s = C_l$, which has rank l and dimension $l(2l+1)$. The highest root of C_l (with $l \geqslant 2$) is

$$\alpha_H = 2\Lambda_1 = 2 \sum_{p=1}^{l-1} \alpha_p + \alpha_l$$

(in the notation of Appendix F, Section 3), which is a "long" root, so that $\langle \alpha_H, \alpha_H \rangle^0 = \langle \alpha_H, \alpha_H \rangle^{(1)} = (l+1)^{-1}$.

(e) For $C_l^{(1)}$ (with $l \geqslant 2$) the imaginary roots are such that $\dim \tilde{\mathscr{L}}_{j\delta}^{(1)} = l$ (for all non-zero integers j).

(f) For $C_l^{(1)}$ (with $l \geqslant 2$) $\langle \alpha_j, \alpha_k \rangle^{(1)} = B_{jk}$ (for $j, k = 0, 1, \ldots, l$), where

$$\mathbf{B} = \frac{1}{4(l+1)} \begin{bmatrix} 4 & -2 & 0 & 0 & \cdots & 0 & 0 & 0 \\ -2 & 2 & -1 & 0 & \cdots & 0 & 0 & 0 \\ 0 & -1 & 2 & -1 & \cdots & 0 & 0 & 0 \\ 0 & 0 & -1 & 2 & \cdots & 0 & 0 & 0 \\ \vdots & \vdots & \vdots & \vdots & & \vdots & \vdots & \vdots \\ 0 & 0 & 0 & 0 & \cdots & 2 & -1 & 0 \\ 0 & 0 & 0 & 0 & \cdots & -1 & 2 & -2 \\ 0 & 0 & 0 & 0 & \cdots & 0 & -2 & 4 \end{bmatrix}.$$

In particular, for $l = 2$

$$\mathbf{B} = \frac{1}{12} \begin{bmatrix} 4 & -2 & 0 \\ -2 & 2 & -2 \\ 0 & -2 & 4 \end{bmatrix},$$

and for $l = 3$

$$\mathbf{B} = \frac{1}{16} \begin{bmatrix} 4 & -2 & 0 & 0 \\ -2 & 2 & -1 & 0 \\ 0 & -1 & 2 & -2 \\ 0 & 0 & -2 & 4 \end{bmatrix}.$$

(g) For $C_l^{(1)}$ (with $l \geqslant 2$) α_0 and α_l are "long" roots, and $\alpha_1, \alpha_2, \ldots, \alpha_{l-1}$ are "short" roots.

(h) The dual of $C_l^{(1)}$ (with $l \geqslant 2$) is $D_{l+1}^{(2)}$.

5 The complex untwisted affine Kac–Moody algebras $D_l^{(1)}, l \geqslant 4$

(a) The generalized Dynkin diagram for $D_l^{(1)}$ (with $l \geqslant 4$) is given in Figure N.5.

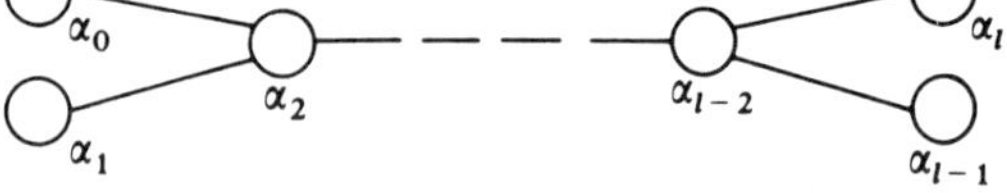

Figure N.5 Generalized Dynkin diagram for $D_l^{(1)}$ (for $l \geqslant 4$).

(b) The generalized Cartan matrix for $D_l^{(1)}$ (with $l \geqslant 4$) is the $(l+1) \times (l+1)$ matrix

$$\mathbf{A} = \begin{bmatrix} 2 & 0 & -1 & 0 & \cdots & 0 & 0 & 0 & 0 \\ 0 & 2 & -1 & 0 & \cdots & 0 & 0 & 0 & 0 \\ -1 & -1 & 2 & -1 & \cdots & 0 & 0 & 0 & 0 \\ 0 & 0 & -1 & 2 & \cdots & 0 & 0 & 0 & 0 \\ \vdots & \vdots & \vdots & \vdots & & \vdots & \vdots & \vdots & \vdots \\ 0 & 0 & 0 & 0 & \cdots & 2 & -1 & 0 & 0 \\ 0 & 0 & 0 & 0 & \cdots & -1 & 2 & -1 & -1 \\ 0 & 0 & 0 & 0 & \cdots & 0 & -1 & 2 & 0 \\ 0 & 0 & 0 & 0 & \cdots & 0 & -1 & 0 & 2 \end{bmatrix}$$

In particular, for $l = 4$

$$\mathbf{A} = \begin{bmatrix} 2 & 0 & -1 & 0 & 0 \\ 0 & 2 & -1 & 0 & 0 \\ -1 & -1 & 2 & -1 & -1 \\ 0 & 0 & -1 & 2 & 0 \\ 0 & 0 & -1 & 0 & 2 \end{bmatrix},$$

and for $l = 5$

$$\mathbf{A} = \begin{bmatrix} 2 & 0 & -1 & 0 & 0 & 0 \\ 0 & 2 & -1 & 0 & 0 & 0 \\ -1 & -1 & 2 & -1 & 0 & 0 \\ 0 & 0 & -1 & 2 & -1 & -1 \\ 0 & 0 & 0 & -1 & 2 & 0 \\ 0 & 0 & 0 & -1 & 0 & 2 \end{bmatrix}.$$

(c) For $D_l^{(1)}$ (with $l \geqslant 4$) the root δ is given by

$$\delta = \alpha_0 + \alpha_1 + 2(\alpha_2 + \cdots + \alpha_{l-2}) + \alpha_{l-1} + \alpha_1.$$

(d) For $D_l^{(1)}$ (with $l \geqslant 4$) the corresponding simple Lie algebra is $\tilde{\mathscr{L}}^0 = \tilde{\mathscr{L}}^s = D_l$, which has rank l and dimension $l(2l-1)$. The highest root of D_l (with $l \geqslant 4$) is

$$\alpha_{\mathrm{H}} = 2\Lambda_2 = \alpha_1 + 2\sum_{p=2}^{l-2} \alpha_p + \alpha_{l-1} + \alpha_l$$

(in the notation of Appendix F, Section 4), so that $\langle \alpha_{\mathrm{H}}, \alpha_{\mathrm{H}} \rangle^0 = \langle \alpha_{\mathrm{H}}, \alpha_{\mathrm{H}} \rangle^{(1)} = \{2(l-1)\}^{-1}$.

(e) For $D_l^{(1)}$ (with $l \geqslant 4$) the imaginary roots are such that $\dim \tilde{\mathscr{L}}_{j\delta}^{(1)} = l$ (for all non-zero integers j).

(f) For $D_l^{(1)}$ (with $l \geqslant 4$) $\langle \alpha_j, \alpha_k \rangle^{(1)} = B_{jk}$ (for $j, k = 0, 1, \ldots, l$), where $\mathbf{B} = \{4(l-1)\}^{-1}\mathbf{A}$.

(g) All real roots of $D_l^{(1)}$ (with $l \geqslant 4$) are of the same length, i.e. $D_l^{(1)}$ (with $l \geqslant 4$) is "simply laced".

(h) $D_l^{(1)}$ (with $l \geqslant 4$) is self-dual.

6 The complex untwisted affine Kac–Moody algebra $E_6^{(1)}$

(a) The generalized Dynkin diagram for $E_6^{(1)}$ is given in Figure N.6.

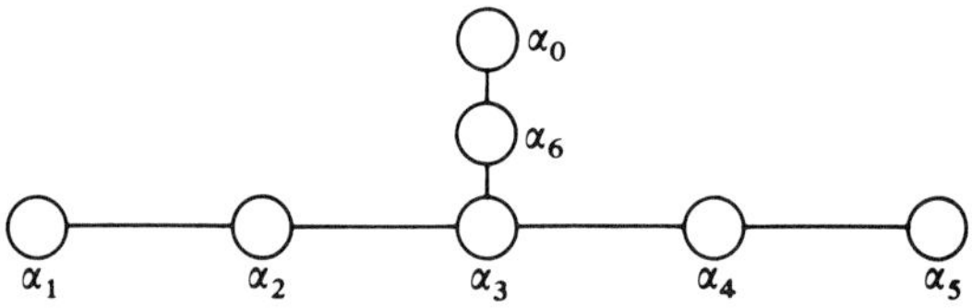

Figure N.6 Generalized Dynkin diagram for $E_6^{(1)}$.

(b) The generalized Cartan matrix for $E_6^{(1)}$ is

$$\mathbf{A} = \begin{bmatrix} 2 & 0 & 0 & 0 & 0 & 0 & -1 \\ 0 & 2 & -1 & 0 & 0 & 0 & 0 \\ 0 & -1 & 2 & -1 & 0 & 0 & 0 \\ 0 & 0 & -1 & 2 & -1 & 0 & -1 \\ 0 & 0 & 0 & -1 & 2 & -1 & 0 \\ 0 & 0 & 0 & 0 & -1 & 2 & 0 \\ -1 & 0 & 0 & -1 & 0 & 0 & 2 \end{bmatrix}.$$

(c) For $E_6^{(1)}$ the root δ is given by

$$\delta = \alpha_0 + \alpha_1 + 2\alpha_2 + 3\alpha_3 + 2\alpha_4 + \alpha_5 + 2\alpha_6.$$

(d) For $\tilde{\mathscr{L}} = E_6^{(1)}$ the corresponding simple Lie algebra is $\tilde{\mathscr{L}}^0 = \tilde{\mathscr{L}}^s = E_6$, which has rank 6 and dimension 78. The highest root of E_6 is

$$\alpha_{\mathrm{H}} = \Lambda_6 = \alpha_1 + 2\alpha_2 + 3\alpha_3 + 2\alpha_4 + \alpha_5 + 2\alpha_6$$

(in the notation of Appendix F, Section 5), so that $\langle \alpha_{\mathrm{H}}, \alpha_{\mathrm{H}} \rangle^0 = \langle \alpha_{\mathrm{H}}, \alpha_{\mathrm{H}} \rangle^{(1)} = \frac{1}{12}$.

(e) For $\tilde{\mathscr{L}} = E_6^{(1)}$ the imaginary roots are such that $\dim \tilde{\mathscr{L}}_{j\delta}^{(1)} = 6$ (for all non-zero integers j).

(f) For $\tilde{\mathscr{L}} = E_6^{(1)}$, $\langle \alpha_j, \alpha_k \rangle^{(1)} = B_{jk}$ (for $j, k = 0, 1, \ldots, 6$), where $\mathbf{B} = \frac{1}{24}\mathbf{A}$.

(g) All real roots of $E_6^{(1)}$ are of the same length, i.e. $E_6^{(1)}$ is "simply laced".

(h) $E_6^{(1)}$ is self-dual.

7 The complex untwisted affine Kac–Moody algebra $E_7^{(1)}$

(a) The generalized Dynkin diagram for $E_7^{(1)}$ is given in Figure N.7.

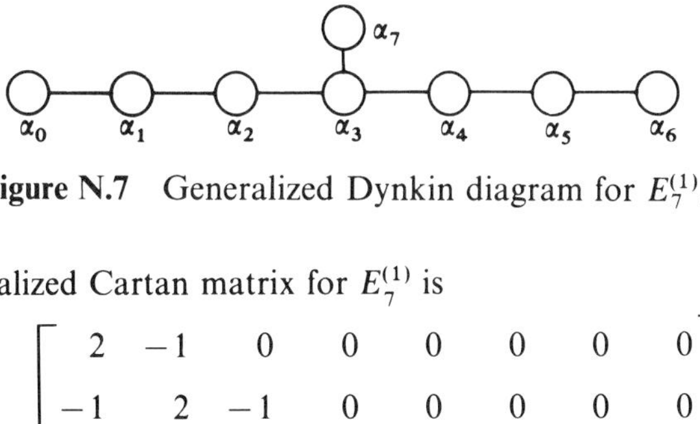

Figure N.7 Generalized Dynkin diagram for $E_7^{(1)}$.

(b) The generalized Cartan matrix for $E_7^{(1)}$ is

$$\mathbf{A} = \begin{bmatrix} 2 & -1 & 0 & 0 & 0 & 0 & 0 & 0 \\ -1 & 2 & -1 & 0 & 0 & 0 & 0 & 0 \\ 0 & -1 & 2 & -1 & 0 & 0 & 0 & 0 \\ 0 & 0 & -1 & 2 & -1 & 0 & 0 & -1 \\ 0 & 0 & 0 & -1 & 2 & -1 & 0 & 0 \\ 0 & 0 & 0 & 0 & -1 & 2 & -1 & 0 \\ 0 & 0 & 0 & 0 & 0 & -1 & 2 & 0 \\ 0 & 0 & 0 & -1 & 0 & 0 & 0 & 2 \end{bmatrix}.$$

(c) For $E_7^{(1)}$ the root δ is given by

$$\delta = \alpha_0 + 2\alpha_1 + 3\alpha_2 + 4\alpha_3 + 3\alpha_4 + 2\alpha_5 + \alpha_6 + 2\alpha_7.$$

(d) For $\tilde{\mathscr{L}} = E_7^{(1)}$ the corresponding simple Lie algebra is $\tilde{\mathscr{L}}^0 = \tilde{\mathscr{L}}^{\mathrm{s}} = E_7$, which has rank 7 and dimension 133. The highest root of E_7 is

$$\alpha_H = \Lambda_1 = 2\alpha_1 + 3\alpha_2 + 4\alpha_3 + 3\alpha_4 + 2\alpha_5 + \alpha_6 + 2\alpha_7$$

(in the notation of Appendix F, Section 6), so that $\langle \alpha_H, \alpha_H \rangle^0 = \langle \alpha_H, \alpha_H \rangle^{(1)} = \frac{1}{18}$.

(e) For $\tilde{\mathscr{L}} = E_7^{(1)}$ the imaginary roots are such that $\dim \tilde{\mathscr{L}}_{j\delta}^{(1)} = 7$ (for all non-zero integers j).

(f) For $\tilde{\mathscr{L}} = E_7^{(1)}$, $\langle \alpha_j, \alpha_k \rangle^{(1)} = B_{jk}$ (for $j, k = 0, 1, \ldots, 7$), where $\mathbf{B} = \frac{1}{36}\mathbf{A}$.

(g) All real roots of $E_7^{(1)}$ are of the same length, i.e. $E_7^{(1)}$ is "simply laced".

(h) $E_7^{(1)}$ is self-dual.

8 The complex untwisted affine Kac–Moody algebra $E_8^{(1)}$

(a) The generalized Dynkin diagram for $E_8^{(1)}$ is given in Figure N.8.

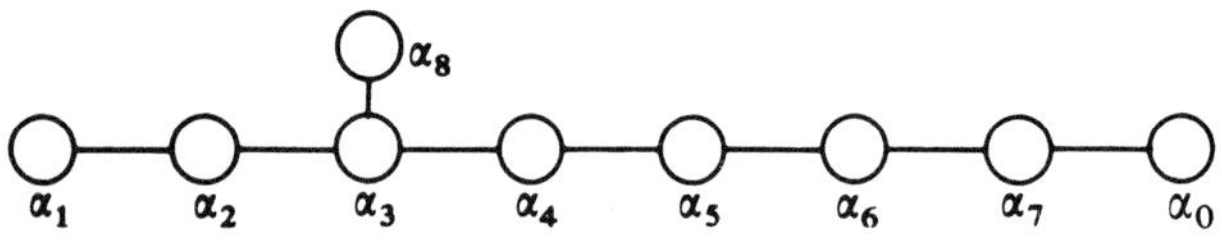

Figure N.8 Generalized Dynkin diagram for $E_8^{(1)}$.

(b) The generalized Cartan matrix for $E_8^{(1)}$ is

$$\mathbf{A} = \begin{bmatrix} 2 & 0 & 0 & 0 & 0 & 0 & 0 & -1 & 0 \\ 0 & 2 & -1 & 0 & 0 & 0 & 0 & 0 & 0 \\ 0 & -1 & 2 & -1 & 0 & 0 & 0 & 0 & 0 \\ 0 & 0 & -1 & 2 & -1 & 0 & 0 & 0 & -1 \\ 0 & 0 & 0 & -1 & 2 & -1 & 0 & 0 & 0 \\ 0 & 0 & 0 & 0 & -1 & 2 & -1 & 0 & 0 \\ 0 & 0 & 0 & 0 & 0 & -1 & 2 & -1 & 0 \\ -1 & 0 & 0 & 0 & 0 & 0 & -1 & 2 & 0 \\ 0 & 0 & 0 & -1 & 0 & 0 & 0 & 0 & 2 \end{bmatrix}.$$

(c) For $E_8^{(1)}$ the root δ is given by

$$\delta = \alpha_0 + 2\alpha_1 + 4\alpha_2 + 6\alpha_3 + 5\alpha_4 + 4\alpha_5 + 3\alpha_6 + 2\alpha_7 + 3\alpha_8.$$

(d) For $\tilde{\mathscr{L}} = E_8^{(1)}$ the corresponding simple Lie algebra is $\tilde{\mathscr{L}}^0 = \tilde{\mathscr{L}}^s = E_8$, which has rank 8 and dimension 248. The highest root of E_8 is

$$\alpha_H = \Lambda_7 = 2\alpha_1 + 4\alpha_2 + 6\alpha_3 + 5\alpha_4 + 4\alpha_5 + 3\alpha_6 + 2\alpha_7 + 3\alpha_8$$

(in the notation of Appendix F, Section 7), so that $\langle \alpha_H, \alpha_H \rangle^0 = \langle \alpha_H, \alpha_H \rangle^{(1)} = \frac{1}{30}$.

(e) For $\tilde{\mathscr{L}} = E_8^{(1)}$ the imaginary roots are such that $\dim \tilde{\mathscr{L}}_{j\delta}^{(1)} = 8$ (for all non-zero integers j).

(f) For $\tilde{\mathscr{L}} = E_8^{(1)}$, $\langle \alpha_j, \alpha_k \rangle^{(1)} = B_{jk}$ (for $j, k = 0, 1, \ldots, 8$), where $\mathbf{B} = \frac{1}{60}\mathbf{A}$.

(g) All real roots of $E_8^{(1)}$ are of the same length, i.e. $E_8^{(1)}$ is "simply laced".

(h) $E_8^{(1)}$ is self-dual.

9 The complex untwisted affine Kac–Moody algebra $F_4^{(1)}$

(a) The generalized Dynkin diagram for $F_4^{(1)}$ is given in Figure N.9.

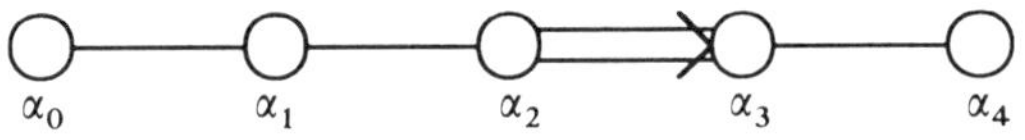

Figure N.9 Generalized Dynkin diagram for $F_4^{(1)}$.

(b) The generalized Cartan matrix for $F_4^{(1)}$ is

$$\mathbf{A} = \begin{bmatrix} 2 & -1 & 0 & 0 & 0 \\ -1 & 2 & -1 & 0 & 0 \\ 0 & -1 & 2 & -1 & 0 \\ 0 & 0 & -2 & 2 & -1 \\ 0 & 0 & 0 & -1 & 2 \end{bmatrix}.$$

(c) For $F_4^{(1)}$ the root δ is given by

$$\delta = \alpha_0 + 2\alpha_1 + 3\alpha_2 + 4\alpha_3 + 2\alpha_4.$$

(d) For $\tilde{\mathscr{L}} = F_4^{(1)}$ the corresponding simple Lie algebra is $\tilde{\mathscr{L}}^0 = \tilde{\mathscr{L}}^s = F_4$, which has rank 4 and dimension 52. The highest root of F_4 is

$$\alpha_H = \Lambda_1 = 2\alpha_1 + 3\alpha_2 + 4\alpha_3 + 2\alpha_4$$

(in the notation of Appendix F, Section 8), so that $\langle \alpha_H, \alpha_H \rangle^0 = \langle \alpha_H, \alpha_H \rangle^{(1)} = \frac{1}{9}$.

(e) For $\tilde{\mathscr{L}} = F_4^{(1)}$ the imaginary roots are such that $\dim \tilde{\mathscr{L}}_{j\delta}^{(1)} = 4$ (for all non-zero integers j).

(f) For $\tilde{\mathscr{L}} = F_4^{(1)}$, $\langle \alpha_j, \alpha_k \rangle^{(1)} = B_{jk}$ (for $j, k = 0, 1, \ldots, 4$), where

$$\mathbf{B} = \frac{1}{36} \begin{bmatrix} 4 & -2 & 0 & 0 & 0 \\ -2 & 4 & -2 & 0 & 0 \\ 0 & -2 & 4 & -2 & 0 \\ 0 & 0 & -2 & 2 & -1 \\ 0 & 0 & 0 & -1 & 2 \end{bmatrix}.$$

(g) For $\tilde{\mathscr{L}} = F_4^{(1)}$, α_0, α_1 and α_2 are "long" roots, and α_3 and α_4 are "short" roots.

(h) The dual of $F_4^{(1)}$ is $E_6^{(2)}$.

10 The complex untwisted affine Kac–Moody algebra $G_2^{(1)}$

(a) The generalized Dynkin diagram for $G_2^{(1)}$ is given in Figure N.10.

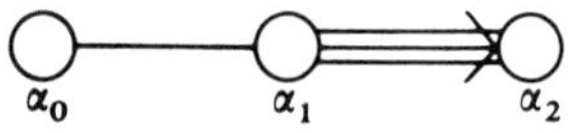

Figure N.10 Generalized Dynkin diagram for $G_2^{(1)}$.

(b) The generalized Cartan matrix for $G_2^{(1)}$ is

$$\mathbf{A} = \begin{bmatrix} 2 & -1 & 0 \\ -1 & 2 & -1 \\ 0 & -3 & 2 \end{bmatrix}.$$

(c) For $G_2^{(1)}$ the root δ is given by

$$\delta = \alpha_0 + 2\alpha_1 + 3\alpha_2.$$

(d) For $\tilde{\mathscr{L}} = G_2^{(1)}$ the corresponding simple Lie algebra is $\tilde{\mathscr{L}}^0 = \tilde{\mathscr{L}}^s = G_2$, which has rank 2 and dimension 14. The highest root of G_2 is

$$\alpha_H = \Lambda_1 = 2\alpha_1 + 3\alpha_2$$

(in the notation of Appendix F, Section 9), so that $\langle \alpha_H, \alpha_H \rangle^0 = \langle \alpha_H, \alpha_H \rangle^{(1)} = \frac{1}{4}$.

(e) For $\tilde{\mathscr{L}} = G_2^{(1)}$ the imaginary roots are such that $\dim \tilde{\mathscr{L}}_{j\delta}^{(1)} = 2$ (for all non-zero integers j).

(f) For $\tilde{\mathscr{L}} = G_2^{(1)}$, $\langle \alpha_j, \alpha_k \rangle^{(1)} = B_{jk}$ (for $j, k = 0, 1, 2$), where

$$\mathbf{B} = \frac{1}{24}\begin{bmatrix} 6 & -3 & 0 \\ -3 & 6 & -3 \\ 0 & -3 & 2 \end{bmatrix}.$$

(g) For $\tilde{\mathscr{L}} = G_2^{(1)}, \alpha_0$ and α_1 are "long" roots and α_2 is a "short" root.

(h) The dual of $G_2^{(1)}$ is $D_4^{(3)}$.

11 The complex twisted affine Kac–Moody algebra $A_2^{(2)}$

(a) The generalized Dynkin diagram for $A_2^{(2)}$ is given in Figure N.11.

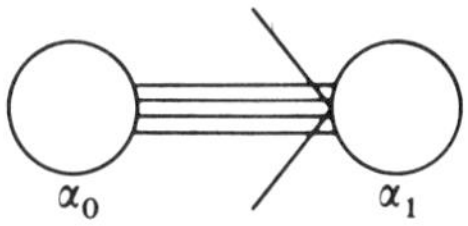

Figure N.11 Generalized Dynkin diagram for $A_2^{(2)}$.

(b) The generalized Cartan matrix for $A_2^{(2)}$ is

$$\mathbf{A} = \begin{bmatrix} 2 & -1 \\ -4 & 2 \end{bmatrix}.$$

(c) For $A_2^{(2)}$ the root δ is given by

$$\delta = \alpha_0 + 2\alpha_1.$$

(d) For $\tilde{\mathscr{L}}$ $(= \tilde{\mathscr{L}}^{(2)}) = A_2^{(2)}$ the corresponding simple Lie algebra $\tilde{\mathscr{L}}^0$ is A_2, which has rank 2 and dimension 8. The two-fold rotation τ of the roots of $\tilde{\mathscr{L}}^0$ is given (in the notation of Appendix F, Section 1) by

$$\tau(\alpha_1) = \alpha_2, \quad \tau(\alpha_2) = \alpha_1.$$

(e) For $\tilde{\mathscr{L}}$ $(= \tilde{\mathscr{L}}^{(2)}) = A_2^{(2)}$ the corresponding simple Lie algebra $\tilde{\mathscr{L}}_0^{0(2)}$ $(= \tilde{\mathscr{L}}^s)$ is A_1, which has rank 1 and dimension 3. Its Cartan subalgebra $\mathscr{H}^{0(2)}$ has basis element

$$h_{\alpha_1}^0 + h_{\alpha_2}^0.$$

Moreover, $\alpha_2(h^0) = \alpha_1(h^0)$ for all $h^0 \in \mathscr{H}^{0(2)}$, so the simple root system of $\tilde{\mathscr{L}}_0^{0(2)}$ is $\Pi^{(2)} = \{\alpha_1\}$.

(f) For $\tilde{\mathscr{L}}$ $(= \tilde{\mathscr{L}}^{(2)}) = A_2^{(2)}$ the representation Γ^1 of $\tilde{\mathscr{L}}_0^{0(2)}$ (cf. (26.145)) has highest weight $\lambda_H = 4\Lambda_1 = 2\alpha_1$ and is of dimension 5.

(g) For $\tilde{\mathscr{L}}$ $(= \tilde{\mathscr{L}}^{(2)}) = A_2^{(2)}$ the imaginary roots are such that $\dim \tilde{\mathscr{L}}_{j\delta}^{(2)} = 1$ (for all non-zero integers j).

(h) For $\tilde{\mathscr{L}}$ $(= \tilde{\mathscr{L}}^{(2)}) = A_2^{(2)}$, $\langle \alpha_j, \alpha_k \rangle^{(1)} = B_{jk}$ (for $j, k = 0, 1$), where

$$\mathbf{B} = \frac{1}{2}\begin{bmatrix} 4 & -2 \\ -2 & 1 \end{bmatrix}.$$

(i) For $\tilde{\mathscr{L}} = A_2^{(2)}$, α_0 is a "long" root and α_1 is a "short" root.

(j) The dual of $A_2^{(2)}$ is isomorphic to $A_2^{(2)}$, but involves a relabelling of the simple roots.

12 The complex twisted affine Kac–Moody algebras $A_{2l}^{(2)}, l \geqslant 2$

(a) The generalized Dynkin diagram for $A_{2l}^{(2)}$ (with $l \geqslant 2$) is given in Figure N.12.

Figure N.12 Generalized Dynkin diagram for $A_{2l}^{(2)}$ (for $l \geqslant 2$).

(b) The generalized Cartan matrix for $A_{2l}^{(2)}$ (with $l \geqslant 2$) is the $(l+1) \times (l+1)$ matrix

$$\mathbf{A} = \begin{bmatrix} 2 & -1 & 0 & 0 & \cdots & 0 & 0 & 0 \\ -2 & 2 & -1 & 0 & \cdots & 0 & 0 & 0 \\ 0 & -1 & 2 & -1 & \cdots & 0 & 0 & 0 \\ 0 & 0 & -1 & 2 & \cdots & 0 & 0 & 0 \\ \vdots & \vdots & \vdots & \vdots & & \vdots & \vdots & \vdots \\ 0 & 0 & 0 & 0 & \cdots & 2 & -1 & 0 \\ 0 & 0 & 0 & 0 & \cdots & -1 & 2 & -1 \\ 0 & 0 & 0 & 0 & \cdots & 0 & -2 & 2 \end{bmatrix}.$$

In particular, for $l = 2$

$$\mathbf{A} = \begin{bmatrix} 2 & -1 & 0 \\ -2 & 2 & -1 \\ 0 & -2 & 2 \end{bmatrix},$$

and for $l = 3$

$$\mathbf{A} = \begin{bmatrix} 2 & -1 & 0 & 0 \\ -2 & 2 & -1 & 0 \\ 0 & -1 & 2 & -1 \\ 0 & 0 & -2 & 2 \end{bmatrix}.$$

(c) For $A_{2l}^{(2)}$ (with $l \geqslant 2$) the root δ is given by

$$\delta = \alpha_0 + 2\alpha_1 + 2\alpha_2 + \cdots + 2\alpha_{l-1} + 2\alpha_l.$$

(d) For $\tilde{\mathscr{L}}$ $(= \tilde{\mathscr{L}}^{(2)}) = A_{2l}^{(2)}$ (with $l \geqslant 2$) the corresponding simple Lie algebra $\tilde{\mathscr{L}}^0$ is A_{2l}, which has rank $2l$ and dimension $4l(l+1)$. The two-fold rotation τ of the roots of $\tilde{\mathscr{L}}^0$ is given (in the notation of Appendix F, Section 1, with l replaced by $l^0 = 2l$) by

$$\tau(\alpha_k) = \alpha_{2l+1-k} \quad (\text{for } k = 1, 2, \ldots, l).$$

(e) For $\tilde{\mathscr{L}}$ $(= \tilde{\mathscr{L}}^{(2)}) = A_{2l}^{(2)}$ (with $l \geqslant 2$) the corresponding simple Lie algebra $\tilde{\mathscr{L}}_0^{0(2)}$ $(= \tilde{\mathscr{L}}^s)$ is B_l, which has rank l and dimension $l(2l+1)$. Its Cartan subalgebra $\mathscr{H}^{0(2)}$ has basis elements:

$$h_{\alpha_k}^0 + h_{\alpha_{2l+1-k}}^0 \quad (\text{for } k = 1, 2, \ldots, l)$$

Moreover, $\alpha_{2l+1-k}(h^0) = \alpha_k(h^0)$ for all $h^0 \in \mathscr{H}^{0(2)}$ and $k = 1, 2, \ldots, l$, so the simple root system of $\tilde{\mathscr{L}}_0^{0(2)}$ $(= B_l)$ is $\Pi^{(2)} = \{\alpha_1, \alpha_2, \ldots, \alpha_l\}$ (as in Appendix F, Section 2).

(f) For $A_{2l}^{(2)}$ (with $l \geqslant 2$) the representation Γ^1 of $\tilde{\mathscr{L}}_0^{0(2)}$ (cf. (26.145)) has highest weight

$$\lambda_{\mathrm{H}} = 2\Lambda_1 = 2\alpha_1 + 2\alpha_2 + \cdots + 2\alpha_{l-1} + 2\alpha_l$$

and is of dimension $l(2l+3)$.

(g) For $\tilde{\mathscr{L}}$ $(= \tilde{\mathscr{L}}^{(2)}) = A_{2l}^{(2)}$ (with $l \geqslant 2$) the imaginary roots are such that $\dim \tilde{\mathscr{L}}_{j\delta}^{(2)} = l$ (for all non-zero integers j).

(h) For $\tilde{\mathscr{L}}$ $(= \tilde{\mathscr{L}}^{(2)}) = A_{2l}^{(2)}$ (with $l \geqslant 2$) $\langle \alpha_j, \alpha_k \rangle^{(1)} = B_{jk}$ (for $j, k = 0, 1, \ldots, l$), where

$$\mathbf{B} = \frac{1}{2(2l-1)} \begin{bmatrix} 4 & -2 & 0 & \cdots & 0 & 0 & 0 \\ -2 & 2 & -1 & \cdots & 0 & 0 & 0 \\ 0 & -1 & 2 & \cdots & 0 & 0 & 0 \\ \vdots & \vdots & \vdots & & \vdots & \vdots & \vdots \\ 0 & 0 & 0 & \cdots & 2 & -1 & 0 \\ 0 & 0 & 0 & \cdots & -1 & 2 & -1 \\ 0 & 0 & 0 & \cdots & 0 & -1 & 1 \end{bmatrix}.$$

(i) $\tilde{\mathscr{L}}$ $(= \tilde{\mathscr{L}}^{(2)}) = A_{2l}^{(2)}$ (with $l \geqslant 2$) has real roots of *three* different lengths, α_0 being a "long" root, α_l a "short" root, and $\alpha_1, \alpha_2, \ldots, \alpha_{l-1}$ neither "long" nor "short".

(j) The dual of $A_{2l}^{(2)}$ is isomorphic to $A_{2l}^{(2)}$ (for $l \geqslant 2$), but involves a relabelling of the simple roots.

13 The complex twisted affine Kac–Moody algebras $A_{2l-1}^{(2)}, l \geqslant 3$

(a) The generalized Dynkin diagram for $A_{2l-1}^{(2)}$ (with $l \geqslant 3$) is given in Figure N.13.

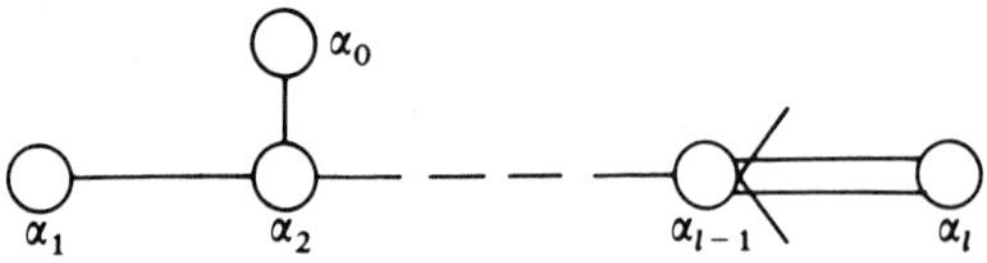

Figure N.13 Generalized Dynkin diagram for $A_{2l-1}^{(2)}$ (for $l \geqslant 3$).

(b) The generalized Cartan matrix for $A_{2l-1}^{(2)}$ (with $l \geqslant 3$) is the $(l+1) \times (l+1)$ matrix

$$\mathbf{A} = \begin{bmatrix} 2 & 0 & -1 & 0 & \cdots & 0 & 0 & 0 \\ 0 & 2 & -1 & 0 & \cdots & 0 & 0 & 0 \\ -1 & -1 & 2 & -1 & \cdots & 0 & 0 & 0 \\ 0 & 0 & -1 & 2 & \cdots & 0 & 0 & 0 \\ \vdots & \vdots & \vdots & \vdots & & \vdots & \vdots & \vdots \\ 0 & 0 & 0 & 0 & \cdots & 2 & -1 & 0 \\ 0 & 0 & 0 & 0 & \cdots & -1 & 2 & -2 \\ 0 & 0 & 0 & 0 & \cdots & 0 & -1 & 2 \end{bmatrix}.$$

In particular, for $l = 3$

$$\mathbf{A} = \begin{bmatrix} 2 & 0 & -1 & 0 \\ 0 & 2 & -1 & 0 \\ -1 & -1 & 2 & -2 \\ 0 & 0 & -1 & 2 \end{bmatrix}.$$

(c) For $A^{(2)}_{2l-1}$ (with $l \geqslant 3$) the root δ is given by

$$\delta = \alpha_0 + \alpha_1 + 2(\alpha_2 + \cdots + \alpha_{l-1}) + \alpha_l.$$

(d) For $\tilde{\mathscr{L}} \ (= \tilde{\mathscr{L}}^{(2)}) = A^{(2)}_{2l-1}$ (with $l \geqslant 3$) the corresponding simple Lie algebra $\tilde{\mathscr{L}}^0$ is A_{2l-1}, which has rank $2l-1$ and dimension $4l^2 - 1$. The two-fold rotation τ of the roots of $\tilde{\mathscr{L}}^0$ is given (in the notation of Appendix F, Section 1, with l replaced by $l^0 = 2l - 1$) by

$$\tau(\alpha_k) = \alpha_{2l-k} \quad (\text{for } k = 1, 2, \ldots, l-1),$$

and

$$\tau(\alpha_l) = \alpha_l.$$

(e) For $\tilde{\mathscr{L}} \ (= \tilde{\mathscr{L}}^{(2)}) = A^{(2)}_{2l-1}$ (with $l \geqslant 3$) the corresponding simple Lie algebra $\tilde{\mathscr{L}}^{0(2)}_0 \ (= \tilde{\mathscr{L}}^s)$ is C_l, which has rank l and dimension $l(2l+1)$. Its Cartan subalgebra $\mathscr{H}^{0(2)}$ has basis elements:

$$h^0_{\alpha_k} + h^0_{\alpha_{2l-k}} \quad (\text{for } k = 1, 2, \ldots, l-1),$$

together with $h^0_{\alpha_l}$.

Moreover, $\alpha_{2l-k}(h^0) = \alpha_k(h^0)$ for all $h^0 \in \mathscr{H}^{0(2)}$ and $k = 1, 2, \ldots, l-1$, so the simple root system of $\tilde{\mathscr{L}}^{0(2)}_0 \ (= C_l)$ is $\Pi^{(2)} = \{\alpha_1, \alpha_2, \ldots, \alpha_l\}$ (as in Appendix F, Section 3).

(f) For $A^{(2)}_{2l-1}$ (with $l \geqslant 3$) the representation $\boldsymbol{\Gamma}^1$ of $\tilde{\mathscr{L}}^{0(2)}_0$ (cf. (26.145)) has highest weight

$$\lambda_{\mathrm{H}} = \Lambda_2 = 2\alpha_1 + 2(\alpha_2 + \cdots + \alpha_{l-1}) + \alpha_l$$

and is of dimension $2l^2 - l - 1$.

(g) For $\tilde{\mathscr{L}} \ (= \tilde{\mathscr{L}}^{(2)}) = A^{(2)}_{2l-1}$ (with $l \geqslant 3$) the imaginary roots are such that

$$\dim \tilde{\mathscr{L}}^{(2)}_{j\delta} = \begin{cases} l, & \text{if } j = \pm 2, \pm 4, \ldots, \\ l-1 & \text{if } j = \pm 1, \pm 3, \ldots \end{cases}.$$

(h) For $\tilde{\mathscr{L}} \ (= \tilde{\mathscr{L}}^{(2)}) = A^{(2)}_{2l-1}$ (with $l \geqslant 3$) $\langle \alpha_j, \alpha_k \rangle^{(1)} = B_{jk}$ (for $j, k = 0, 1, \ldots, l$), where

$$\mathbf{B} = \frac{1}{4(l+1)} \begin{bmatrix} 2 & 0 & -1 & 0 & \cdots & 0 & 0 & 0 \\ 0 & 2 & -1 & 0 & \cdots & 0 & 0 & 0 \\ -1 & -1 & 2 & -1 & \cdots & 0 & 0 & 0 \\ 0 & 0 & -1 & 2 & \cdots & 0 & 0 & 0 \\ \vdots & \vdots & \vdots & \vdots & & \vdots & \vdots & \vdots \\ 0 & 0 & 0 & 0 & \cdots & 2 & -1 & 0 \\ 0 & 0 & 0 & 0 & \cdots & -1 & 2 & -2 \\ 0 & 0 & 0 & 0 & \cdots & 0 & -2 & 4 \end{bmatrix}.$$

(i) For $\tilde{\mathscr{L}}\ (= \tilde{\mathscr{L}}^{(2)}) = A^{(2)}_{2l-1}$ (with $l \geqslant 3$) $\alpha_0, \alpha_1, \ldots, \alpha_{l-1}$ are "short" roots and α_l is a "long" root.

(j) The dual of $A^{(2)}_{2l-1}$ is $B^{(1)}_l$ (for $l \geqslant 3$).

14 The complex twisted affine Kac–Moody algebras $D^{(2)}_{l+1}$, $l \geqslant 2$

(a) The generalized Dynkin diagram for $D^{(2)}_{l+1}$ (with $l \geqslant 2$) is given in Figure N.14.

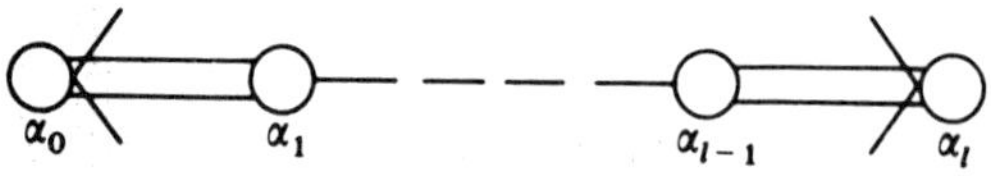

Figure N.14 Generalized Dynkin diagram for $D^{(2)}_{l+1}$ (for $l \geqslant 2$).

(b) The generalized Cartan matrix for $D^{(2)}_{l+1}$ (with $l \geqslant 2$) is the $(l+1) \times (l+1)$ matrix

$$\mathbf{A} = \begin{bmatrix} 2 & -2 & 0 & \cdots & 0 & 0 & 0 \\ -1 & 2 & -1 & \cdots & 0 & 0 & 0 \\ 0 & -1 & 2 & \cdots & 0 & 0 & 0 \\ \vdots & \vdots & \vdots & & \vdots & \vdots & \vdots \\ 0 & 0 & 0 & \cdots & 2 & -1 & 0 \\ 0 & 0 & 0 & \cdots & -1 & 2 & -1 \\ 0 & 0 & 0 & \cdots & 0 & -2 & 2 \end{bmatrix}.$$

In particular, for $l = 2$

$$\mathbf{A} = \begin{bmatrix} 2 & -2 & 0 \\ -1 & 2 & -1 \\ 0 & -2 & 2 \end{bmatrix},$$

and for $l = 3$

$$\mathbf{A} = \begin{bmatrix} 2 & -2 & 0 & 0 \\ -1 & 2 & -1 & 0 \\ 0 & -1 & 2 & -1 \\ 0 & 0 & -2 & 2 \end{bmatrix}.$$

(c) For $D_{l+1}^{(2)}$ (with $l \geqslant 2$) the root δ is given by

$$\delta = \alpha_0 + \alpha_1 + \alpha_2 + \cdots + \alpha_{l-1} + \alpha_l.$$

(d) For $\tilde{\mathscr{L}}\ (= \tilde{\mathscr{L}}^{(2)}) = D_{l+1}^{(2)}$ (with $l \geqslant 2$) the corresponding simple Lie algebra $\tilde{\mathscr{L}}^0$ is D_{l+1}, which has rank $l+1$ and dimension $(l+1)(2l+1)$. The two-fold rotation τ of the roots of $\tilde{\mathscr{L}}^0$ is given (in the notation of Appendix F, Section 4, with l replaced by $l^0 = l+1$) by

$$\tau(\alpha_k) = \alpha_k \quad (\text{for } k = 1, 2, \ldots, l-1)$$

and

$$\tau(\alpha_l) = \alpha_{l+1}, \quad \tau(\alpha_{l+1}) = \alpha_l.$$

(e) For $\tilde{\mathscr{L}}\ (= \tilde{\mathscr{L}}^{(2)}) = D_{l+1}^{(2)}$ (with $l \geqslant 2$) the corresponding simple Lie algebra $\tilde{\mathscr{L}}_0^{0(2)}\ (= \tilde{\mathscr{L}}^s)$ is B_l, which has rank l and dimension $l(2l+1)$. Its Cartan subalgebra $\mathscr{H}^{0(2)}$ has basis elements:

$$h_{\alpha_k}^0 \quad (\text{for } k = 1, 2, \ldots, l-1),$$

together with $h_{\alpha_l}^0 + h_{\alpha_{l+1}}^0$.

Moreover, $\alpha_{l+1}(h^0) = \alpha_l(h^0)$ for all $h^0 \in \mathscr{H}^{0(2)}$, so the simple root system of $\tilde{\mathscr{L}}_0^{0(2)}\ (= B_l)$ is $\Pi^{(2)} = \{\alpha_1, \alpha_2, \ldots, \alpha_l\}$ (as in Appendix F, Section 2).

(f) For $D_{l+1}^{(2)}$ (with $l \geqslant 2$) the representation $\boldsymbol{\Gamma}^1$ of $\tilde{\mathscr{L}}_0^{0(2)}$ (cf. (26.145)) has highest weight

$$\lambda_{\mathrm{H}} = \Lambda_1 = \alpha_1 + \alpha_2 + \cdots + \alpha_{l-1} + \alpha_l$$

and is of dimension $2l+1$.

(g) For $\tilde{\mathscr{L}}\ (= \tilde{\mathscr{L}}^{(2)}) = D_{l+1}^{(2)}$ (with $l \geqslant 2$) the imaginary roots are such that

$$\dim \tilde{\mathscr{L}}_{j\delta}^{(2)} = \begin{cases} l & \text{if } j = \pm 2, \pm 4, \ldots, \\ 1 & \text{if } j = \pm 1, \pm 3, \ldots \end{cases}.$$

(h) For $\tilde{\mathscr{L}}$ $(=\tilde{\mathscr{L}}^{(2)})=D_{l+1}^{(2)}$ (with $l\geqslant 2$) $\langle\alpha_j,\alpha_k\rangle^{(1)}=B_{jk}$ (for $j,k=0,1,\ldots,l$), where

$$\mathbf{B}=\frac{1}{4(2l-1)}\begin{bmatrix} 2 & -2 & 0 & \cdots & 0 & 0 & 0 \\ -2 & 4 & -2 & \cdots & 0 & 0 & 0 \\ 0 & -2 & 4 & \cdots & 0 & 0 & 0 \\ \vdots & \vdots & \vdots & & \vdots & \vdots & \vdots \\ 0 & 0 & 0 & \cdots & 4 & -2 & 0 \\ 0 & 0 & 0 & \cdots & -2 & 4 & -2 \\ 0 & 0 & 0 & \cdots & 0 & -2 & 2 \end{bmatrix}.$$

(i) For $\tilde{\mathscr{L}}$ $(=\tilde{\mathscr{L}}^{(2)})=D_{l+1}^{(2)}$ (with $l\geqslant 2$) α_0 and α_l are "short" roots and $\alpha_1,\alpha_2,\ldots,\alpha_{l-1}$ are "long" roots.

(j) The dual of $D_{l+1}^{(2)}$ is $C_l^{(1)}$ (for $l\geqslant 2$).

15 The complex twisted affine Kac–Moody algebra $E_6^{(2)}$

(a) The generalized Dynkin diagram for $E_6^{(2)}$ is given in Figure N.15.

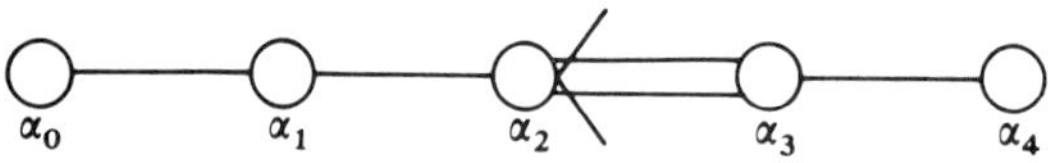

Figure N.15 Generalized Dynkin diagram for $E_6^{(2)}$.

(b) The generalized Cartan matrix for $E_6^{(2)}$ is

$$\mathbf{A}=\begin{bmatrix} 2 & -1 & 0 & 0 & 0 \\ -1 & 2 & -1 & 0 & 0 \\ 0 & -1 & 2 & -2 & 0 \\ 0 & 0 & -1 & 2 & -1 \\ 0 & 0 & 0 & -1 & 2 \end{bmatrix}.$$

(c) For $E_6^{(2)}$ the root δ is given by

$$\delta=\alpha_0+2\alpha_1+3\alpha_2+2\alpha_3+\alpha_4.$$

(d) For $\tilde{\mathscr{L}}$ $(=\tilde{\mathscr{L}}^{(2)})=E_6^{(2)}$ the corresponding simple Lie algebra $\tilde{\mathscr{L}}^0$ is E_6, which has rank 6 and dimension 78. The two-fold rotation τ of the roots of

$\tilde{\mathscr{L}}^0$ is given (in the notation of Appendix F, Section 5) by

$$\tau(\alpha_k) = \alpha_{6-k} \quad (\text{for } k = 1, 2, \ldots, 5)$$

and

$$\tau(\alpha_6) = \alpha_6.$$

(e) For $\tilde{\mathscr{L}}\,(= \tilde{\mathscr{L}}^{(2)}) = E_6^{(2)}$ the corresponding simple Lie algebra $\tilde{\mathscr{L}}_0^{0(2)}\,(= \tilde{\mathscr{L}}^s)$ is F_4, which has rank 4 and dimension 52. Its Cartan subalgebra $\mathscr{H}^{0(2)}$ has basis elements

$$h^0_{\alpha_1} + h^0_{\alpha_5}, \quad h^0_{\alpha_2} + h^0_{\alpha_4}, \quad h^0_{\alpha_3}, \quad h^0_{\alpha_6}.$$

Moreover, $\alpha_4(h^0) = \alpha_2(h^0)$ and $\alpha_5(h^0) = \alpha_1(h^0)$ for all $h^0 \in \mathscr{H}^{0(2)}$, so the simple root system of $\tilde{\mathscr{L}}_0^{0(2)}\,(= F_4)$ is $\Pi^{(2)} = \{\alpha_1, \alpha_2, \alpha_3, \alpha_6\}$. This corresponds to a *different* labelling from that given in Appendix F, Section 8. Denoting the simple roots of that section by α_1^F, α_2^F, α_3^F and α_4^F, the correspondence with the present labelling is $\alpha_1 \leftrightarrow \alpha_4^F$, $\alpha_2 \leftrightarrow \alpha_3^F$, $\alpha_3 \leftrightarrow \alpha_2^F$, and $\alpha_6 \leftrightarrow \alpha_1^F$.

(f) For $E_6^{(2)}$ the representation Γ^1 of $\tilde{\mathscr{L}}_0^{0(2)}$ (cf. (26.145)) has highest weight

$$\lambda_H = \Lambda_1 = \Lambda_4^F = 2\alpha_1 + 3\alpha_2 + 2\alpha_3 + \alpha_4$$

and is of dimension 26.

(g) For $\tilde{\mathscr{L}}\,(= \tilde{\mathscr{L}}^{(2)}) = E_6^{(2)}$ the imaginary roots are such that

$$\dim \tilde{\mathscr{L}}^{(2)}_{j\delta} = \begin{cases} 4 & \text{if } j = \pm 2, \pm 4, \ldots, \\ 2 & \text{if } j = \pm 1, \pm 3, \ldots \end{cases}.$$

(h) For $\tilde{\mathscr{L}}\,(= \tilde{\mathscr{L}}^{(2)}) = E_6^{(2)}$, $\langle \alpha_j, \alpha_k \rangle^{(1)} = B_{jk}$ (for $j, k = 0, 1, \ldots, 4$), where

$$\mathbf{B} = \frac{1}{36}\begin{bmatrix} 2 & -1 & 0 & 0 & 0 \\ -1 & 2 & -1 & 0 & 0 \\ 0 & -1 & 2 & -2 & 0 \\ 0 & 0 & -2 & 4 & -2 \\ 0 & 0 & 0 & -2 & 4 \end{bmatrix}.$$

(i) For $\tilde{\mathscr{L}}\,(= \tilde{\mathscr{L}}^{(2)}) = E_6^{(2)}$, α_0, α_1, and α_2 are "short" roots, and α_3 and α_4 are "long" roots.

(j) The dual of $E_6^{(2)}$ is $F_4^{(1)}$.

16 The complex twisted affine Kac–Moody algebra $D_4^{(3)}$

(a) The generalized Dynkin diagram for $D_4^{(3)}$ is given in Figure N.16.

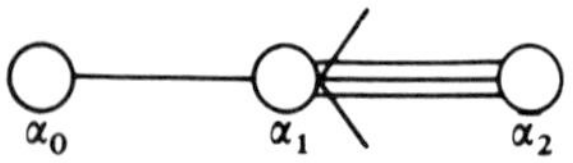

Figure N.16 Generalized Dynkin diagram for $D_4^{(3)}$.

(b) The generalized Cartan matrix for $D_4^{(3)}$ is

$$\mathbf{A} = \begin{bmatrix} 2 & -1 & 0 \\ -1 & 2 & -3 \\ 0 & -1 & 2 \end{bmatrix}.$$

(c) For $D_4^{(3)}$ the root δ is given by

$$\delta = \alpha_0 + 2\alpha_1 + \alpha_2.$$

(d) For $\tilde{\mathscr{L}}$ $(= \tilde{\mathscr{L}}^{(3)}) = D_4^{(3)}$ the corresponding simple Lie algebra $\tilde{\mathscr{L}}^0$ is D_4, which has rank 4 and dimension 28. The three-fold rotation τ of the roots of $\tilde{\mathscr{L}}^0$ is given (in the notation of Appendix F, Section 4) by

$$\tau(\alpha_1) = \alpha_3, \quad \tau(\alpha_2) = \alpha_2, \quad \tau(\alpha_3) = \alpha_4, \quad \tau(\alpha_4) = \alpha_1.$$

(e) For $\tilde{\mathscr{L}}$ $(= \tilde{\mathscr{L}}^{(3)}) = D_4^{(3)}$ the corresponding simple Lie algebra $\tilde{\mathscr{L}}_0^{0(3)}$ $(= \tilde{\mathscr{L}}^s)$ is G_2, which has rank 2 and dimension 14. Its Cartan subalgebra $\mathscr{H}^{0(3)}$ has basis elements:

$$h_{\alpha_1}^0 + h_{\alpha_3}^0 + h_{\alpha_4}^0, \quad h_{\alpha_2}^0.$$

Moreover, $\alpha_3(h^0) = \alpha_4(h^0) = \alpha_1(h^0)$ for all $h^0 \in \mathscr{H}^{0(3)}$, so the simple root system of $\tilde{\mathscr{L}}_0^{0(3)}$ $(= G_2)$ is $\Pi^{(3)} = \{\alpha_1, \alpha_2\}$. This corresponds to a *different* labelling from that given in Appendix F, Section 9. Denoting the simple roots of that section by α_1^F and α_2^F, the correspondence with the present labelling is $\alpha_1 \leftrightarrow \alpha_2^F$ and $\alpha_2 \leftrightarrow \alpha_1^F$.

(f) For $D_4^{(3)}$ the representations Γ^1 and Γ^2 of $\tilde{\mathscr{L}}_0^{0(3)}$ (cf. (26.145)) both have highest weight

$$\lambda_H = \Lambda_1 = \Lambda_2^F = 2\alpha_1 + \alpha_2$$

and are of dimension 7. Also $\Delta_1^{0(3)} = \Delta_2^{0(3)} = \{\pm\alpha_1, \pm(\alpha_1 + \alpha_2), \pm(2\alpha_1 + \alpha_2)\}$.

(g) For $\tilde{\mathscr{L}}$ $(= \tilde{\mathscr{L}}^{(3)}) = D_4^{(3)}$ the imaginary roots are such that

$$\dim \tilde{\mathscr{L}}_{j\delta}^{(3)} = \begin{cases} 2 & \text{if } j \bmod 3 = 0, j \neq 0, \\ 1 & \text{if } j \bmod 3 = 1 \text{ or } j \bmod 3 = 2. \end{cases}$$

(h) For $\tilde{\mathscr{L}}$ $(= \tilde{\mathscr{L}}^{(3)}) = D_4^{(3)}$, $\langle \alpha_j, \alpha_k \rangle^{(1)} = B_{jk}$ (for $j, k = 0, 1, \ldots, 4$), where

$$\mathbf{B} = \frac{1}{24}\begin{bmatrix} 2 & -1 & 0 \\ -1 & 2 & -3 \\ 0 & -3 & 6 \end{bmatrix}.$$

(i) For $\tilde{\mathscr{L}}$ $(= \tilde{\mathscr{L}}^{(3)}) = D_4^{(3)}$, α_0 and α_1 are "short" roots, and α_2 is a "long" root.

(j) The dual of $D_4^{(3)}$ is $G_2^{(1)}$.

Appendix O

Proofs of Certain Theorems on Kac–Moody and Virasoro Algebras

1 Proofs of Theorems I, III and IV of Chapter 27, Section 2

Theorem I Let λ be any weight of a standard irreducible representation of an untwisted affine Kac–Moody algebra $\tilde{\mathscr{L}}$. Then there exists a non-negative integer p (which depends on λ) such $\lambda - k\delta$ is in the δ-string of weights containing λ for *every* integer $k \geqslant -p$.

Proof Let p be such that $\lambda + p\delta$ is the highest weight of the form $\lambda - k\delta$. (The existence of p is ensured by the requirement that the representation has a highest weight.) Let ψ_1 be an element of the carrier space V corresponding to the weight $\lambda + p\delta$. Choose any $\beta \in \Delta^0$. If e^{β}_{δ} and $e^{-\beta}_{-\delta}$ are defined by (26.120) then, by (26.126),

$$[e^{\beta}_{\delta}, e^{-\beta}_{-\delta}]_- = \langle \beta, \beta \rangle^0 h_\delta, \tag{O.1}$$

and so, by (26.106) and (27.17),

$$[\Phi(e^{\beta}_{\delta}), \Phi(e^{-\beta}_{-\delta})]_- = \langle \beta, \beta \rangle^0 c_\Lambda 1. \tag{O.2}$$

It should be noted that $c_\Lambda > 0$ for a standard irreducible representation and $\langle \beta, \beta \rangle^0 > 0$. By the definition of ψ_1, $\Phi(e^{\beta}_{\delta})\psi_1 = 0$ (since otherwise $\Phi(e^{\beta}_{\delta})\psi_1$ would correspond to weight $\lambda + (p+1)\delta$). However, $\Phi(e^{-\beta}_{-\delta})\psi_1 \neq 0$, since otherwise the left-hand side of (O.2) acting on ψ_1 would be zero, but the right-hand side would not be so. Let $\psi_2 = \Phi(e^{-\beta}_{-\delta})\psi_1$, which corresponds to weight $\lambda + (p-1)\delta$. If $\Phi(e^{-\beta}_{-\delta})\psi_2 = 0$ then ψ_1 and ψ_2 would provide a basis for a 2-dimensional representation of the commutation relations (O.1). Taking the trace of both sides would give a contradiction, since the left-hand side would be zero (being a commutator) whereas the right hand side would be $2\langle\beta,\beta\rangle^0 c_\Lambda$, which is positive. A similar argument shows that $\Phi(e^{-\beta}_{-\delta})^j\psi_1 \neq 0$ for every positive integer value of j. Thus $\lambda+(p-j)\delta$ is a weight for all positive integer values of j.

Theorem III For a standard irreducible representation of an untwisted affine Kac–Moody algebra $\tilde{\mathscr{L}}$

(a) the highest weight Λ is a maximal dominant weight;

(b) at least one member of the set m_k $(k=0,1,\ldots,l)$ of (27.28) is non-zero;

(c) for every maximal weight $\lambda_{\max}$ there exists an element S of the Weyl group $\mathscr{W}$ of $\tilde{\mathscr{L}}$ and a unique dominant maximal weight $\lambda^+_{\max}$ such that $\lambda_{\max}=S\lambda^+_{\max}$; and

(d) the set of maximal dominant weights is *finite*.

Proof (a) This follows immediately from the definitions of maximal dominant weight and highest weight.

(b) Suppose that $m_j=0$ for $j=0,1,\ldots,l$. Then, with

$$\lambda_{\max}=\Lambda-\sum_{k=0}^{l} q_k\alpha_k$$

(cf. (27.8)), it follows that

$$0=\frac{2\langle\Lambda,\alpha_j\rangle}{\langle\alpha_j,\alpha_j\rangle}-\sum_{k=0}^{l} 2q_k\frac{\langle\alpha_k,\alpha_j\rangle}{\langle\alpha_j,\alpha_j\rangle}.$$

Thus, by (26.50) and (27.9), $\mathbf{Aq}=\mathbf{n}$, where $\mathbf{q}$ and $\mathbf{n}$ are the column vectors with entries q_j and n_j respectively. However, for a standard irreducible representation $\mathbf{n}\geqslant\mathbf{0}$ (cf. the discussion before (26.67)), so by the defining condition (A3) of an affine Kac–Moody algebra (see Chapter 26, Section 4), $\mathbf{Aq}=\mathbf{n}=\mathbf{0}$, i.e. $n_j=0$ for all $j=0,1,\ldots,l$, which is specifically excluded in the definition of a standard irreducible representation.

(c) It will first be proved that the set $\{S\lambda_{\max}\,|\,S\in\mathscr{W}\}$ contains a *dominant* maximal weight. For any weight

$$\lambda=\Lambda-\sum_{k=0}^{l} q_k\alpha_k$$

(cf. (27.8)) define

$$\mathrm{ht}(\Lambda-\lambda)=\sum_{j=0}^{l} q_j.$$

As $q_j \geqslant 0$ for all $j = 0, 1, \ldots, l$, and $q_j = 0$ for $j = 0, 1, \ldots, l$ only for the special case $\lambda = \Lambda$, it follows that $\mathrm{ht}(\Lambda - \lambda) \geqslant 0$. Let $\lambda'_{\max}$ be the member of the set $\{S\lambda_{\max} | S \in \mathscr{W}\}$ for which $\mathrm{ht}(\Lambda - S\lambda_{\max})$ is a minimum. Then

$$\mathrm{ht}(\Lambda - S_{\alpha_j}\lambda'_{\max}) \geqslant \mathrm{ht}(\Lambda - \lambda'_{\max})$$

for all $j = 0, 1, \ldots, l$. However, if

$$\lambda'_{\max} = \Lambda - \sum_{k=0}^{l} q'_k \alpha_k,$$

then

$$\mathrm{ht}(\Lambda - \lambda'_{\max}) = \sum_{k=0}^{l} q'_k,$$

and also

$$\mathrm{ht}(\Lambda - S_{\alpha_j}\lambda'_{\max}) = \sum_{k=0}^{l} q'_k + \frac{2\langle \lambda'_{\max}, \alpha_j \rangle}{\langle \alpha_j, \alpha_j \rangle}.$$

Thus $2\langle \lambda'_{\max}, \alpha_j \rangle / \langle \alpha_j, \alpha_j \rangle \geqslant 0$ for $j = 0, 1, \ldots, l$. As $\lambda'_{\max}$ is a maximal weight (by part (c) of Theorem II of Chapter 27, Section 2), $\lambda'_{\max}$ must be a *dominant* maximal weight.

It remains to be shown that $\lambda'_{\max}$ is *unique*. As

$$\mathrm{ht}(\Lambda - S_{\alpha_j}\lambda'_{\max}) = \mathrm{ht}(\Lambda - \lambda'_{\max}) + \frac{2\langle \lambda'_{\max}, \alpha_j \rangle}{\langle \alpha_j, \alpha_j \rangle},$$

it follows that

$$\mathrm{ht}(\Lambda - S_{\alpha_j}\lambda'_{\max}) > \mathrm{ht}(\Lambda - \lambda'_{\max})$$

unless

$$S_{\alpha_j}\lambda'_{\max} = \lambda'_{\max}.$$

The same result is true if S_α (with $\alpha = \alpha_j$) is replaced by $S_\beta S_\alpha$ (with $\alpha = \alpha_j$ and $\beta = \alpha_k$ but with $j \neq k$), since

$$S_{\alpha_k} S_{\alpha_j} \lambda'_{\max} = \lambda'_{\max} - \frac{2\langle \lambda'_{\max}, \alpha_j \rangle}{\langle \alpha_j, \alpha_j \rangle}\alpha_j - \frac{2\langle \lambda'_{\max}, \alpha_k \rangle}{\langle \alpha_k, \alpha_k \rangle}\alpha_k + \frac{2\langle \lambda'_{\max}, \alpha_j \rangle A_{kj}}{\langle \alpha_j, \alpha_j \rangle}\alpha_k$$

and since $A_{kj} \leqslant 0$. Continuing, one finds that $\mathrm{ht}(\Lambda - S\lambda'_{\max}) > \mathrm{ht}(\Lambda - \lambda'_{\max})$ for *every* $S \in \mathscr{W}$ unless $S\lambda'_{\max} = \lambda'_{\max}$. Thus $\lambda'_{\max}$ is unique.

(d) Suppose that

$$\lambda_{\max} = \Lambda - \sum_{k=0}^{l} q_k \alpha_k.$$

Then, by (26.50), (27.9), and (27.28),

$$m_j = n_j - \sum_{k=0}^{l} q_k A_{jk} \tag{O.3}$$

for $j=0,1,\ldots,l$. It has to be shown that there are only a *finite* number of sets $\{m_0, m_1, \ldots, m_l\}$ and $\{q_0, q_1, \ldots, q_l\}$ with *non-negative integer* entries that satisfy this condition (O.3) for fixed $\{n_0, n_1, \ldots, n_l\}$ *and* that correspond to a *maximal* weight. The essential ideas come out most clearly in the simplest example where $\tilde{\mathscr{L}} = A_1^{(1)}$, so that

$$\mathbf{A} = \begin{bmatrix} 2 & -2 \\ -2 & 2 \end{bmatrix}.$$

In this case adding the equations (O.3) with $j=0$ and $j=1$ gives

$$m_0 + m_1 = n_0 + n_1, \tag{O.4}$$

so there are only a finite number of possible values for the set $\{m_0, m_1\}$, namely

$$m_1 = (n_0 + n_1) - m_0, \quad \text{where } m_0 = 0, 1, \ldots, n_0 + n_1. \tag{O.5}$$

Thus (O.3) with $j=0$ gives

$$2q_0 - 2q_1 = n_0 - m_0, \tag{O.6}$$

which *only* has *integer* solutions for q_0 and q_1 if $n_0 - m_0$ is divisible by 2. When this is the case, writing $n_0 - m_0 = 2N$, (O.6) gives $q_0 = q_1 + N$ for $q_1 = 0, 1, 2, \ldots$. Denoting the corresponding weight by $\lambda_{\text{max}}(q_1)$, it is clear that $\lambda_{\text{max}}(q_1) = \lambda_{\text{max}}(0) - q_1\delta$, so only $\lambda_{\text{max}}(0)$ is a true *maximal* weight. This line of argument can clearly be extended to any untwisted affine Kac–Moody algebra $\tilde{\mathscr{L}}$, since there is always a condition analogous to (O.4) because $\det \mathbf{A} = 0$.

Theorem IV For the basic representation (with highest weight Λ_0) of a simply laced untwisted affine Kac–Moody algebra $A_l^{(1)}$ $(l \geqslant 1)$, $D_l^{(1)}$ $(l \geqslant 4)$, $E_6^{(1)}$, $E_7^{(1)}$ or $E_8^{(1)}$

(a) the *only* maximal dominant weight is Λ_0;

(b) the set of *maximal* weights is given by

$$\lambda_{\text{max}} = \Lambda_0 + \alpha - \frac{\langle \alpha, \alpha \rangle}{\langle \alpha_0, \alpha_0 \rangle}\delta, \tag{27.29}$$

where α runs over all members of the root lattice Q^0 of $\tilde{\mathscr{L}}^0$ $(= \tilde{\mathscr{L}}^s)$;

(c) the set of *all* weights is given by

$$\lambda = \Lambda_0 + \alpha - \left(j + \frac{\langle \alpha, \alpha \rangle}{\langle \alpha_0, \alpha_0 \rangle} \right)\delta, \tag{27.30}$$

where α runs over all members of the root lattice Q^0 of $\tilde{\mathscr{L}}^0$ $(= \tilde{\mathscr{L}}^s)$ and j takes all non-negative integer values.

Proof (a) The essential point is revealed by considering the proof of part (d) of Theorem III above for the present case $n_0 = 1$, $n_1 = 0$. There are only two sets given by (O.5), namely $\{0, 1\}$ and $\{1, 0\}$. Only for the latter set is $n_0 - m_0$ divisible by 2, and in this case $N = 0$. Consequently the maximal dominant weight corresponds to $q_0 = 0$ and $q_1 = 0$; that is, to $\lambda_{max} = \Lambda_0$. The argument for all other simply laced Kac–Moody algebras is similar.

(b) From part (a) of the present theorem, part (c) of Theorem III of Chapter 27, Section 2 and part (c) of Theorem II of the same section it follows that the set of maximal weights is just $\{S\Lambda_0 | S \in \mathscr{W}\}$. However, by Theorem III of Chapter 26, Section 5(*d*), every $S \in \mathscr{W}$ can be written in the form $S = T_\omega S^0$, where T_ω is a translation (cf. (26.181)) and S^0 is an element of $\mathscr{W}^0$, the Weyl group of $\tilde{\mathscr{L}}^0$. But $S_\alpha \Lambda_0 = \Lambda_0$ for $\alpha = \alpha_j$ with $j = 1, 2, \ldots, l$ (by (26.54) and (26.171)), so the set of maximal weights is just $\{T_\omega(\Lambda_0) | \omega \in \mathrm{Q}^{0\vee}\}$. As $\tilde{\mathscr{L}}^0$ is simply laced, every $\omega \in \mathrm{Q}^{0\vee}$ is the dual of some element of Q^0. Thus the set of maximal roots is just $\{T_{\alpha^\vee}(\Lambda_0) | \alpha \in \mathrm{Q}^0\}$. However, by (26.172), (26.179) and (26.191),

$$T_{\alpha^\vee}(\Lambda_0) = \Lambda_0 + \alpha - \frac{\langle \alpha, \alpha \rangle}{\langle \alpha_0, \alpha_0 \rangle} \delta,$$

thereby establishing (27.29).

(c) This follows immediately from part (b) of the present theorem and part (b) of Theorem II of Chapter 27, Section 2.

2 Proofs of Theorems I and II of Chapter 27, Section 4

Theorem I (a) The operators of the basic representation (with highest weight Λ_0) of a simply laced untwisted affine Kac–Moody algebra $\tilde{\mathscr{L}}$ act on the carrier space V defined by (27.70) and are given by

(i)
$$\Phi(C) = 1, \tag{27.87}$$

where 1 is the identity operator;

(ii)
$$\Phi(d) = -N_\mathrm{F} - N_\mathrm{L}, \tag{27.88}$$

where

$$N_\mathrm{F} = \sum_{j=1}^{\infty} \sum_{r=1}^{l} A_j^{r\dagger} A_j^r \tag{27.89}$$

and

$$N_\mathrm{L}(\psi_\mathrm{F} \otimes \mathrm{e}^\gamma) = \tfrac{1}{2}\hat{\gamma} \cdot \hat{\gamma}(\psi_\mathrm{F} \otimes \mathrm{e}^\gamma) \tag{27.90}$$

for all $\gamma \in Q^0$;

(iii) for $\alpha \in \Delta^0$

$$\Phi(t^0 \otimes H^0_\alpha)(\psi_F \otimes e^\gamma) = \hat{\boldsymbol{\alpha}} \cdot \hat{\boldsymbol{\gamma}}(\psi_F \otimes e^\gamma) \tag{27.91}$$

for all $\gamma \in Q^0$;

(iv) for $j = 1, 2, \ldots$ and $\alpha \in \Delta^0$

$$\Phi(t^j \otimes H^0_\alpha) = \hat{\boldsymbol{\alpha}} \cdot \mathbf{A}_j \tag{27.92}$$

and

$$\Phi(t^{-j} \otimes H^0_\alpha) = \hat{\boldsymbol{\alpha}} \cdot \mathbf{A}^\dagger_j; \tag{27.93}$$

(v) for $j = 0, \pm 1, \pm 2, \ldots$ and $\alpha \in \Delta^0$

$$\Phi(t^j \otimes E^0_\alpha) = U_j(\alpha) c_\alpha. \tag{27.94}$$

(b) The weights of the basic representation are given by

$$\lambda = \Lambda_0 + \gamma - (m + \tfrac{1}{2}\hat{\boldsymbol{\gamma}} \cdot \hat{\boldsymbol{\gamma}})\delta, \tag{27.95}$$

where γ runs over all members of the root lattice Q^0 of $\tilde{\mathscr{L}}^0$ and $m = 0, 1, 2, \ldots$ with every vector

$$\Psi(n^1_1, \ldots, n^r_j, \ldots) \otimes e^\gamma \tag{27.96}$$

such that

$$m = \sum_{j=1}^{\infty} \sum_{r=1}^{l} j n^r_j \tag{27.97}$$

being a weight vector $\psi(\lambda)$ with this weight as eigenvalue.

(c) For all $a \in \mathscr{L}_c$, the compact real form of $\tilde{\mathscr{L}}$, and for all $\psi, \phi \in V$

$$(\Phi(a)\psi, \phi) = -(\psi, \Phi(a)\phi), \tag{27.98}$$

so the basic representation is "unitary".

Proof (a) It has to be shown that $[\Phi(a), \Phi(b)]_- = \Phi([a, b]_-)$ for every pair of basis elements a and b of $\tilde{\mathscr{L}}$. The only cases that involve any complications are the following.

(1) For $j = 0, \pm 1, \pm 2, \ldots$ and $\alpha \in \Delta^0$, when acting on any state vector of the form $\psi_F \otimes e^\gamma$,

$$[\Phi(d), \Phi(t^j \otimes E^0_\alpha)]_- = \frac{1}{2\pi i} \oint z^{j-1} [-N_F - N_L, U(\alpha, z) c_\alpha]_- \, dz$$

(by (27.89), (27.94) and (27.86))

$$= -\frac{1}{2\pi i} \oint z^{j-1} \{[N_F, \exp\{i\hat{\boldsymbol{\alpha}} \cdot \mathbf{Q}_<(z)\}]_- M_\alpha(z) \exp\{i\hat{\boldsymbol{\alpha}} \cdot \mathbf{Q}_>(z)\}$$

$$+\exp\{i\hat{\boldsymbol{\alpha}}\cdot\mathbf{Q}_<(z)\}[N_{\rm L},M_\alpha(z)]_-\exp\{i\hat{\boldsymbol{\alpha}}\cdot\mathbf{Q}_>(z)\}$$

$$+\exp\{i\hat{\boldsymbol{\alpha}}\cdot\mathbf{Q}_<(z)\}M_\alpha(z)[N_{\rm F},\exp\{i\hat{\boldsymbol{\alpha}}\cdot\mathbf{Q}_>(z)\}]_-\}c_\alpha\,{\rm d}z$$

(by (27.81))

$$=-\frac{1}{2\pi i}\oint z^{j-1}\left\{\sum_{k=1}^{\infty}\hat{\boldsymbol{\alpha}}\cdot\mathbf{A}_k^\dagger z^k+\hat{\boldsymbol{\gamma}}\cdot\hat{\boldsymbol{\gamma}}+\tfrac{1}{2}\hat{\boldsymbol{\alpha}}\cdot\hat{\boldsymbol{\gamma}}+\sum_{k=1}^{\infty}\hat{\boldsymbol{\alpha}}\cdot\mathbf{A}_k z^{-k}\right\}U(\alpha,z)c_\alpha\,{\rm d}z$$

(by (27.82), (27.83), (27.71) and (27.84))

$$=-\frac{1}{2\pi i}\oint z^j\frac{{\rm d}U(\alpha,z)}{{\rm d}z}c_\alpha\,{\rm d}z$$

$$=\frac{j}{2\pi i}\oint z^{j-1}U(\alpha,z)c_\alpha\,{\rm d}z$$

$$=j\Phi(t^j\otimes E_\alpha^0)\quad\text{(by 27.94) and (27.86))}$$

$$=\Phi([d,t^j\otimes E_\alpha^0]_-)\quad\text{(by (27.64)).}$$

(2) For $k=1,2,\ldots;\ j=0,\pm 1,\pm 2,\ldots$ and $\alpha,\beta\in\Delta^0$

$$[\Phi(t^k\otimes H_\beta^0),\Phi(t^j\otimes E_\alpha^0)]_-=\frac{1}{2\pi i}\oint z^{j-1}[\hat{\boldsymbol{\beta}}\cdot\mathbf{A}_k,U(\alpha,z)c_\alpha]_-\,{\rm d}z$$

(by (27.92), (27.94), and (27.86))

$$=\frac{1}{2\pi i}\oint z^{j-1}\{[\hat{\boldsymbol{\beta}}\cdot\mathbf{A}_k,\exp\{i\hat{\boldsymbol{\alpha}}\cdot\mathbf{Q}_<(z)\}]_-M_\alpha(z)\exp\{i\hat{\boldsymbol{\alpha}}\cdot\mathbf{Q}_>(z)\}\,{\rm d}z$$

(by (27.81) and (27.71))

$$=\frac{1}{2\pi i}\oint z^{j-1}(\boldsymbol{\alpha}\cdot\hat{\boldsymbol{\beta}}z^k)U(\alpha,z)c_\alpha\,{\rm d}z$$

(by (26.264))

$$=\hat{\boldsymbol{\alpha}}\cdot\hat{\boldsymbol{\beta}}\Phi(t^{j+k}\otimes E_\alpha^0)\quad\text{(by (27.86))}$$

$$=\Phi([t^k\otimes H_\beta^0,t^j\otimes E_\alpha^0]_-)\quad\text{(by (27.61)).}$$

The corresponding result with $t^k\otimes H_\beta^0$ replaced by $t^{-k}\otimes H_\beta^0$ has a similar proof.

(3) For $j,k=0,\pm 1,\pm 2,\ldots;\alpha,\beta\in\Delta^0$ and $\gamma\in{\rm Q}^0$, by (27.94),

$$[\Phi(t^j\otimes E_\alpha^0),\Phi(t^k\otimes E_\beta^0)]_-(\psi_{\rm F}\otimes{\rm e}^\gamma)$$

$$=[U_j(\alpha)c_\alpha,U_k(\beta)c_\beta]_-(\psi_{\rm F}\otimes{\rm e}^\gamma)$$

$$=\{U_j(\alpha)U_k(\beta)\varepsilon(\alpha,\beta+\gamma)\varepsilon(\beta,\gamma)-U_k(\beta)U_j(\alpha)\varepsilon(\beta,\alpha+\gamma)\varepsilon(\alpha,\gamma)\}(\psi_{\rm F}\otimes{\rm e}^\gamma)$$

(by (27.80))

$$= \varepsilon(\alpha,\beta)\varepsilon(\alpha+\beta,\gamma)\{U_j(\alpha)U_k(\beta) - (-1)^{\eta(\alpha,\beta)}U_k(\beta)U_j(\alpha)\}(\psi_F \otimes e^{\gamma}) \tag{O.7}$$

(by (27.65) and (27.67)), where

$$\eta(\alpha,\beta) = \hat{\boldsymbol{\alpha}}\cdot\hat{\boldsymbol{\beta}}.$$

However, by (27.82), (27.83) and (27.71),

$$[Q^r_>(z), Q^s_<(\zeta)]_- = \sum_{j=1}^{\infty}\left(\frac{\zeta}{z}\right)^j \frac{\delta^{rs}}{j},$$

where the series converges only if $|\zeta/z| < 1$, so that

$$[Q^r_>(z), Q^s_<(\zeta)]_- = \delta^{rs}\ln\left(1 - \frac{\zeta}{z}\right) \tag{O.8}$$

for $|\zeta| < |z|$, but the left-hand side of (O.8) is not well defined for $|\zeta| \geqslant |z|$. From this it follows that

$$\exp\{i\hat{\boldsymbol{\alpha}}\cdot\mathbf{Q}_>(z)\}\exp\{i\hat{\boldsymbol{\beta}}\cdot\mathbf{Q}_<(\zeta)\}$$
$$= \left(1 - \frac{\zeta}{z}\right)^{\eta(\alpha,\beta)}\exp\{i\hat{\boldsymbol{\beta}}\cdot\mathbf{Q}_<(\zeta)\}\exp\{i\hat{\boldsymbol{\alpha}}\cdot\mathbf{Q}_>(z)\}, \tag{O.9}$$

provided that $|\zeta| < |z|$. Define the *normal product* $:U(\alpha,z)U(\beta,\zeta):$ by

$$:U(\alpha,z)U(\beta,\zeta):$$
$$= \exp\{i\hat{\boldsymbol{\alpha}}\cdot\mathbf{Q}_<(z)\}\exp\{i\hat{\boldsymbol{\beta}}\cdot\mathbf{Q}_<(\zeta)\}M_{\alpha,\beta}(z,\zeta)\exp\{i\hat{\boldsymbol{\alpha}}\cdot\mathbf{Q}_>(z)\}\exp\{i\hat{\boldsymbol{\beta}}\cdot\mathbf{Q}_>(\zeta)\}, \tag{O.10}$$

where

$$M_{\alpha,\beta}(z,\zeta)(\psi_F \otimes e^{\gamma}) = z^{\chi(\alpha,\gamma)}\zeta^{\chi(\beta,\gamma)}(\psi_F \otimes e^{\alpha+\beta+\gamma}) \tag{O.11a}$$

and

$$\chi(\alpha,\gamma) = \hat{\boldsymbol{\alpha}}\cdot\hat{\boldsymbol{\gamma}} + \tfrac{1}{2}\hat{\boldsymbol{\alpha}}\cdot\hat{\boldsymbol{\alpha}}. \tag{O.11b}$$

(In the "normal product" all creation operators $A_j^{r\dagger}$ appear to the *left* of all annihilation operators A_j^r.) As

$$M_\alpha(z)M_\beta(\zeta)(\psi_F \otimes e^{\gamma}) = z^{\eta(\alpha,\beta)}M_{\alpha,\beta}(z,\zeta)(\psi_F \otimes e^{\gamma})$$

for all $\gamma \in Q^0$, it follows from (O.9)–(O.11a, b) that

$$U(\alpha,z)U(\beta,\zeta) = (z-\zeta)^{\eta(\alpha,\beta)} :U(\alpha,z)U(\beta,\zeta): \tag{O.12}$$

provided that $|\zeta| < |z|$. Similarly

$$U(\beta,\zeta)U(\alpha,z) = (\zeta - z)^{\eta(\alpha,\beta)} :U(\beta,\zeta)U(\alpha,z): \tag{O.13}$$

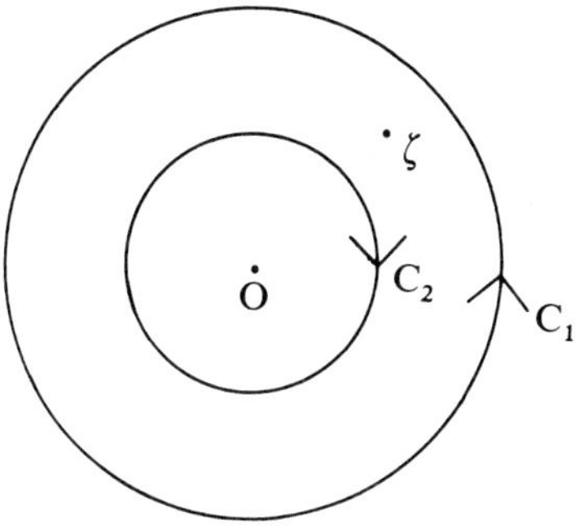

Figure O.1 The contours C_1 and C_2 in the complex z-plane.

for $|z| < |\zeta|$. However,

$$:U(\alpha, z)U(\beta, \zeta): = :U(\beta, \zeta)U(\alpha, z):,$$

so (O.13) can be rewritten as

$$(-1)^{\eta(\alpha,\beta)}U(\beta, \zeta)U(\alpha, z) = (z - \zeta)^{\eta(\alpha,\beta)} :U(\alpha, z)U(\beta, \zeta): \qquad \text{(O.14)}$$

for $|z| < |\zeta|$. Thus, by (27.86) and (O.12),

$$U_j(\alpha)U_k(\beta) = \left(\frac{1}{2\pi i}\right)^2 \oint_{C} d\zeta\, \zeta^{k-1} \oint_{C_1} dz\, z^{j-1}(z - \zeta)^{\eta(\alpha,\beta)} :U(\alpha, z)U(\beta, \zeta); \quad , \qquad \text{(O.15)}$$

where C is a contour enclosing $\zeta = 0$ in the complex ζ-plane, and C_1 is the contour in the z-plane on which $|z| > |\zeta|$ that is shown in Figure O.1. Similarly, by (27.86) and (O.14),

$$-(-1)^{\eta(\alpha,\beta)}U_k(\beta)U_j(\alpha)$$
$$= \left(\frac{1}{2\pi i}\right)^2 \oint_{C} d\zeta\, \zeta^{k-1} \oint_{C_2} dz\, z^{j-1}(z - \zeta)^{\eta(\alpha,\beta)} :U(\alpha, z)U(\beta, \zeta); \quad , \qquad \text{(O.16)}$$

where C_2 is the contour in the z-plane on which $|z| < |\zeta|$ that is indicated in Figure O.1. Let C_ζ be the contour in the z-plane obtained from C_1 and C_2 as shown in Figure O.2, the two segments joining C_1 and C_2 being arbitrarily close to each other. Then

$$U_j(\alpha)U_k(\beta) - (-1)^{\eta(\alpha,\beta)}U_k(\beta)U_j(\alpha)$$
$$= \left(\frac{1}{2\pi i}\right)^2 \oint_{C} d\zeta\, \zeta^{k-1} \oint_{C_\zeta} dz\, z^{j-1}(z - \zeta)^{\eta(\alpha,\beta)} :U(\alpha, z)U(\beta, \zeta): \quad . \qquad \text{(O.17)}$$

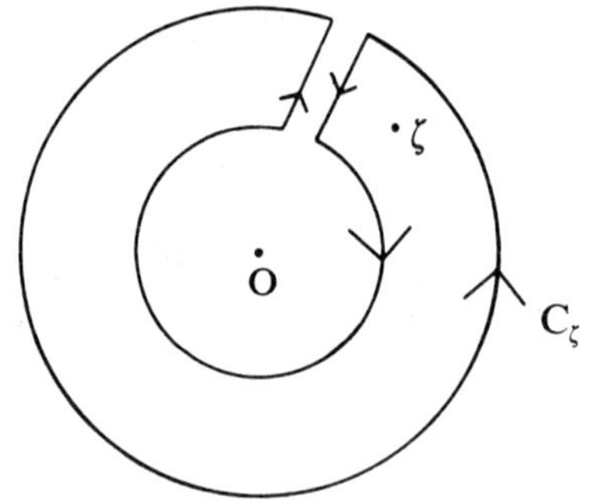

Figure O.2 The contour C_ζ in the complex z-plane.

The problem now reduces to the consideration of the integral

$$I = \frac{1}{2\pi i}\oint_{C_\zeta} dz\, z^{j-1}(z-\zeta)^{\eta(\alpha,\beta)} : U(\alpha, z)U(\beta, \zeta): \quad , \tag{O.18}$$

the essential feature of the contour C_ζ being that it encloses the point $z = \zeta$. The "normal product" $:U(\alpha, z)U(\beta, \zeta):$ is analytic at $z = \zeta$. It follows from (27.52)–(27.54) that if $\alpha, \beta \in \Delta^0$ then

$$\eta(\alpha,\beta) = \hat{\boldsymbol{\alpha}}\cdot\hat{\boldsymbol{\beta}} = \begin{cases} 0, 1 \text{ or } 2 & \text{if } \alpha+\beta \notin \Delta^0 \text{ and } \alpha+\beta \neq 0, \\ -1 & \text{if } \alpha+\beta \in \Delta^0, \\ -2 & \text{if } \alpha+\beta = 0, \end{cases}$$

so there are three cases to be considered.

(A) For $\eta(\alpha,\beta)(=\hat{\boldsymbol{\alpha}}\cdot\hat{\boldsymbol{\beta}}) \geqslant 0$ the integrand of (O.18) is analytic within C_ζ, so, by Cauchy's Integral Theorem, the value of the integral I is zero, and so in this case (O.17) gives

$$U_j(\alpha)U_k(\beta) - (-1)^{\eta(\alpha,\beta)}U_k(\beta)U_j(\alpha) = 0. \tag{O.19}$$

(B) For $\eta(\alpha,\beta)(=\hat{\boldsymbol{\alpha}}\cdot\hat{\boldsymbol{\beta}}) = -1$ the integral I is equal to the residue of the integrand at $z = \zeta$, which is $\zeta^{j-1}:U(\alpha,\zeta)U(\beta,\zeta):$. However, by (O.10) and (O.11), with $\eta(\alpha,\beta) = -1$,

$$:U(\alpha,\zeta)U(\beta,\zeta): = \zeta U(\alpha+\beta, \zeta),$$

so in this case

$$U_j(\alpha)U_k(\beta) - (-1)^{\eta(\alpha,\beta)}U_k(\beta)U_j(\alpha) = \frac{1}{2\pi i}\oint d\zeta\, \zeta^{j+k-1}U(\alpha+\beta,\zeta)$$
$$= U_{j+k}(\alpha+\beta). \tag{O.20}$$

(C) For $\eta(\alpha, \beta)(=\hat{\boldsymbol{\alpha}}\cdot\hat{\boldsymbol{\beta}}) = -2$, since the integrand of (O.18) has a pole of order 2 at $z=\zeta$, the residue at $z=\zeta$ is

$$\frac{\mathrm{d}}{\mathrm{d}z}(z^{j-1} :U(\alpha, z)U(-\alpha, \zeta):)$$

evaluated at $z=\zeta$. However, by (O.10) and (O.11),

$$\begin{aligned} z^{j-1} &:U(\alpha, z)U(-\alpha, \zeta): (\psi_{\mathrm{F}} \otimes \mathrm{e}^{\gamma}) \\ &= z^{j} \exp\{i\hat{\boldsymbol{\alpha}}\cdot(\mathbf{Q}_{<}(z) - \mathbf{Q}_{<}(\zeta)\} \\ &\quad \times \exp\{i\hat{\boldsymbol{\alpha}}\cdot(\mathbf{Q}_{>}(z) - \mathbf{Q}_{>}(\zeta)\} \left(\frac{z}{\zeta}\right)^{\eta(\alpha,\gamma)} \zeta(\psi_{\mathrm{F}} \otimes \mathrm{e}^{\gamma}) \end{aligned}$$

so

$$\begin{aligned} &\frac{\mathrm{d}}{\mathrm{d}z}(z^{j-1} :U(\alpha, z)U(-\alpha, \zeta):)|_{z=\zeta}(\psi_{\mathrm{F}} \otimes \mathrm{e}^{\gamma}) \\ &= \left[jz^{j-1} + z^{j}\left\{ \frac{\mathrm{d}}{\mathrm{d}z}(\exp\{i\hat{\boldsymbol{\alpha}}\cdot(\mathbf{Q}_{<}(z) - \mathbf{Q}_{<}(\zeta)\}) \right.\right. \\ &\quad \left.\left. + \frac{\mathrm{d}}{\mathrm{d}z}(\exp\{i\hat{\boldsymbol{\alpha}}\cdot(\mathbf{Q}_{>}(z) - \mathbf{Q}_{>}(\zeta)\}) + \frac{\mathrm{d}}{\mathrm{d}z}\left(\frac{z}{\zeta}\right)^{\eta(\alpha,\gamma)} \right\}\right]\zeta\Big|_{z=\zeta} (\psi_{\mathrm{F}} \otimes \mathrm{e}^{\gamma}) \\ &= \left\{ j\zeta^{j} + \sum_{n=1}^{\infty} \hat{\boldsymbol{\alpha}}\cdot(\mathbf{A}_n^{\dagger}\zeta^{j+n-1} + \mathbf{A}_n\zeta^{j-n-1}) + \hat{\boldsymbol{\alpha}}\cdot\hat{\boldsymbol{\gamma}}\zeta^{j} \right\}(\psi_{\mathrm{F}} \otimes \mathrm{e}^{\gamma}). \end{aligned}$$

Thus, by (O.17), in this case

$$\begin{aligned} &\{U_j(\alpha)U_k(\beta) - (-1)^{\eta(\alpha,\beta)}U_k(\beta)U_j(\alpha)\}(\psi_{\mathrm{F}} \otimes \mathrm{e}^{\gamma}) \\ &= \frac{1}{2\pi i}\oint\left(j\zeta^{j+k-1} + \sum_{n=1}^{\infty} \hat{\boldsymbol{\alpha}}\cdot\mathbf{A}_n^{\dagger}\zeta^{j+k+n-1} \right. \\ &\quad \left. + \sum_{n=1}^{\infty} \hat{\boldsymbol{\alpha}}\cdot\mathbf{A}_n\zeta^{j+k-n-1} + \hat{\boldsymbol{\alpha}}\cdot\hat{\boldsymbol{\gamma}}\zeta^{j+k-1} \right)\mathrm{d}\zeta\,(\psi_{\mathrm{F}} \otimes \mathrm{e}^{\gamma}) \\ &= \begin{cases} (j + \hat{\boldsymbol{\alpha}}\cdot\hat{\boldsymbol{\gamma}})(\psi_{\mathrm{F}} \otimes \mathrm{e}^{\gamma}) & \text{if } j+k=0, \\ \hat{\boldsymbol{\alpha}}\cdot\mathbf{A}_{j+k}^{\dagger}(\psi_{\mathrm{F}} \otimes \mathrm{e}^{\gamma}) & \text{if } j+k \geqslant 1, \\ \hat{\boldsymbol{\alpha}}\cdot\mathbf{A}_{-(j+k)}(\psi_{\mathrm{F}} \otimes \mathrm{e}^{\gamma}) & \text{if } j+k \leqslant -1, \end{cases} \\ &= \{\Phi(t^{j+k} \otimes H_{\alpha}^{0}) + j\Phi(C)\}(\psi_{\mathrm{F}} \otimes \mathrm{e}^{\gamma}) \qquad \text{(O.21)} \end{aligned}$$

(by (27.87) and (27.91)–(27.93)).

Hence, from (O.7) and (O.19)–(O.21),

$$[\Phi(t^j\otimes E^0_\alpha), \Phi(t^k\otimes E^0_\beta)]_-$$

$$= \begin{cases} 0 & \text{if } \alpha+\beta\notin\Delta^0 \text{ and } \alpha+\beta\neq 0, \\ \varepsilon(\alpha,\beta)\varepsilon(\alpha+\beta,\gamma)U_{j+k}(\alpha+\beta)(\psi_F\otimes e^\gamma) & \text{if } \alpha+\beta\in\Delta^0 \\ \varepsilon(\alpha,-\alpha)\varepsilon(0,\gamma)\{\Phi(t^{j+k}\otimes H^0_\alpha)+j\Phi(C)\}(\psi_F\otimes e^\gamma) & \text{if } \beta=-\alpha, \end{cases}$$

$$= \begin{cases} 0 & \text{if } \alpha+\beta\notin\Delta^0 \text{ and } \alpha+\beta\neq 0, \\ \varepsilon(\alpha,\beta)\Phi(t^{j+k}\otimes E^0_{\alpha+\beta})(\psi_F\otimes e^\gamma) & \text{if } \alpha+\beta\in\Delta^0, \\ \{\Phi(t^{j+k}\otimes H^0_\alpha)+j\Phi(C)\}(\psi_F\otimes e^\gamma) & \text{if } \beta=-\alpha, \end{cases}$$

(by (27.94), (27.80) and (27.67)–(27.69))

$$= \Phi([t^j\otimes E^0_\alpha, t^k\otimes E^0_\beta]_-) \quad \text{(by (27.62))}.$$

(b) By (27.74) and (27.75),

$$A^{r\dagger}_j A^r_j \Psi(n^1_1,\ldots,n^r_j,\ldots) = jn^r_j\Psi(n^1_1,\ldots,n^r_j,\ldots),$$

so, from (27.88)–(27.90), $\Psi(n^1_1,\ldots,n^r_j,\ldots)\otimes e^\gamma$ is an eigenvector of $\Phi(d)$ with eigenvalue

$$-\sum_{j=1}^{\infty}\sum_{r=1}^{l} jn^r_j - \tfrac{1}{2}\hat{\boldsymbol{\gamma}}\cdot\hat{\boldsymbol{\gamma}}.$$

However, from (26.169), (26.175), (26.70) and (26.83), for λ given by (27.95)

$$\lambda(d) = -m - \tfrac{1}{2}\hat{\boldsymbol{\gamma}}\cdot\hat{\boldsymbol{\gamma}},$$

so if (27.97) is satisfied then $\Psi(n^1_1,\ldots,n^r_j,\ldots)\otimes e^\gamma$ is an eigenvector of $\Phi(d)$ with eigenvalue $\lambda(d)$.

Similarly, by (27.91), $\Psi(n^1_1,\ldots,n^r_j,\ldots)\otimes e^\gamma$ is an eigenvector of $\Phi(t^0\otimes H^0_\alpha)$ with eigenvalue $\hat{\boldsymbol{\alpha}}\cdot\hat{\boldsymbol{\gamma}}$, this eigenvalue being equal to $\lambda(t^0\otimes H^0_\alpha)$ by (26.95), (26.96), (26.168) and (27.95).

Thus for all $h\in\mathscr{H}$

$$\Phi(h)(\Psi(n^1_1,\ldots,n^r_j,\ldots)\otimes e^\gamma) = \lambda(h)(\Psi(n^1_1,\ldots,n^r_j,\ldots)\otimes e^\gamma),$$

where λ is given by (27.95).

(c) The basis elements of $\mathscr{L}^0_c$ can be taken to be iH^0_α (for $\alpha=\alpha_k$ with $k=1,2,\ldots,l$) together with $E^0_\alpha - E^0_{-\alpha}$ and $i(E^0_\alpha + E^0_{-\alpha})$ (for $\alpha\in\Delta^0_+$). (By (27.40), the latter pairs are $\eta(e^0_\alpha + e^0_{-\alpha})$ and $i\eta(e^0_\alpha - e^0_{-\alpha})$, which are members of $\mathscr{L}^0_c$ because of Theorem IV of Chapter 14, Section 2.) Consequently, (26.194) implies that (27.98) is equivalent to the conditions

$$(\Phi(a)\psi,\phi) = (\psi,\Phi(a)\phi)$$

(for all $\psi, \phi = V$) for $a = d$, C and $t^0\otimes H^0_\alpha$, which are clearly satisfied, together with

$$(\Phi(t^j \otimes H^0_\alpha)\psi, \phi) = (\psi, \Phi(t^{-j} \otimes H^0_\alpha)\phi) \tag{O.22}$$

for all $j = 1, 2, \ldots$ and $\alpha \in \Delta^0$, and

$$(\Phi(t^j \otimes E^0_\alpha)\psi, \phi) = (\psi, \Phi(t^{-j} \otimes E^0_{-\alpha})\phi) \tag{O.23}$$

for all $j = 0, \pm 1, \pm 2, \ldots$ and $\alpha \in \Delta^0$. The validity of (O.22) follows immediately from (27.92) and (27.93), so it remains only to consider (O.23), which can be restated as

$$\Phi(t^j \otimes E^0_\alpha)^\dagger = \Phi(t^{-j} \otimes E^0_{-\alpha}). \tag{O.24}$$

However, by (26.287),

$$\Phi(t^j \otimes E^0_\alpha)^\dagger = c^\dagger_\alpha U_j(\alpha)^\dagger = U_j(\alpha)^\dagger c_{-\alpha}$$

(by (27.80)). Moreover, (27.81) implies that

$$U(\alpha, z)^\dagger = U(-\alpha, 1/z^*),$$

so, by (27.85),

$$U_j(\alpha)^\dagger = U_{-j}(-\alpha);$$

therefore (O.24) is indeed satisfied.

Theorem II The two sets of elements

(i) $|0\rangle \otimes \mathrm{e}^\gamma$, for every non-zero root γ of $\tilde{\mathscr{L}}^0$, and

(ii) $A^{s\dagger}_1 |0\rangle \otimes \mathrm{e}^0$, for $s = 1, 2, \ldots, l$,

together form a basis for the *adjoint* representation of $\tilde{\mathscr{L}}^0$ (the operators $\Phi(t^0 \otimes H^0_\alpha)$ and $\Phi(t^0 \otimes E^0_\alpha)$ being as defined in (27.91) and (27.94) respectively).

Proof (a) Consider first the operators $\Phi(t^0 \otimes H^0_\alpha)$. By (27.91),

$$\Phi(t^0 \otimes H^0_\alpha)(|0\rangle \otimes \mathrm{e}^\gamma) = \hat{\boldsymbol{\alpha}} \cdot \hat{\boldsymbol{\gamma}}(|0\rangle \otimes \mathrm{e}^\gamma)$$

and

$$\Phi(t^0 \otimes H^0_\alpha)(A^{s\dagger}_1 |0\rangle \otimes \mathrm{e}^0) = 0,$$

which are both consistent with the identifications (27.99) and (27.100), since (27.61) and (27.60) show that

$$[t^0 \otimes H^0_\alpha, t^0 \otimes E^0_\gamma]_- = \hat{\boldsymbol{\alpha}} \cdot \hat{\boldsymbol{\gamma}}(t^0 \otimes E^0_\gamma)$$

and

$$[t^0 H^0_\alpha, t^0 H^0_s]_- = 0.$$

(b) Consider now the operators $\Phi(t^0 \otimes E^0_\alpha)$, where α is a non-zero root of

$\tilde{\mathscr{L}}^0$. By (27.54),

$$\Phi(t^0 \otimes E^0_\alpha) = U_0(\alpha) c_\alpha,$$

and, since the operators c_α leave this subspace invariant and (27.86) implies that the only non-zero contributions to $U_0(\alpha)$ come from the z^0 terms of the power series expansion of $U(\alpha, z)$, it suffices to examine only these terms.

(A) Consider the subset $|0\rangle \otimes e^\gamma$ (for $\hat{\boldsymbol{\gamma}} \cdot \hat{\boldsymbol{\gamma}} = 2$) first. By (27.83), (27.84) and (O.11b)

$$M_\alpha(z) \exp\{i\hat{\boldsymbol{\alpha}} \cdot \mathbf{Q}_>(z)\}(|0\rangle \otimes e^\gamma) = z^{\chi(\alpha,\gamma)}(|0\rangle \otimes e^{\alpha+\gamma}).$$

There are three subcases to be examined.

(*i*) *$\alpha + \gamma$ is also a non-zero root of $\tilde{\mathscr{L}}^0$.* In this case $\hat{\boldsymbol{\alpha}} \cdot \hat{\boldsymbol{\gamma}} = -1$, so $\chi(\alpha, \gamma) = 0$ and hence

$$M_\alpha(z) \exp\{i\hat{\boldsymbol{\alpha}} \cdot \mathbf{Q}_>(z)\}(|0\rangle \otimes e^\gamma) = (|0\rangle \otimes e^{\alpha+\gamma}),$$

so the coefficient of z^0 in $U_0(\alpha)$ must come from the z^0 term in $\exp\{i\hat{\boldsymbol{\alpha}} \cdot \mathbf{Q}_<(z)\}(|0\rangle \otimes e^{\alpha+\gamma})$, which is merely $|0\rangle \otimes e^{\alpha+\gamma}$. Thus, by (27.80),

$$\Phi(t^0 \otimes E^0_\alpha)(|0\rangle \otimes e^\gamma) = c(\alpha, \gamma)(|0\rangle \otimes e^{\alpha+\gamma}),$$

which is consistent, under the identification (27.99), with

$$[t^0 \otimes E^0_\alpha, t^0 \otimes E^0_\gamma]_- = c(\alpha, \gamma)(t^0 \otimes E^0_{\alpha+\gamma})$$

(cf. (27.62)).

(*ii*) $\alpha + \gamma = 0$, *i.e.* $\gamma = -\alpha$. In this case

$$M_\alpha(z) \exp\{i\hat{\boldsymbol{\alpha}} \cdot \mathbf{Q}_>(z)\}(|0\rangle \otimes e^\gamma) = z^{-1}(|0\rangle \otimes e^0),$$

so the coefficient of z^0 in $U_0(\alpha)$ must come from the z^1 term in $\exp\{i\hat{\boldsymbol{\alpha}} \cdot \mathbf{Q}_<(z)\}(|0\rangle \otimes e^0)$, which is

$$\sum_{s=1}^{l} \hat{\boldsymbol{\alpha}}(H^0_s) A^{s\dagger}_1 |0\rangle \otimes e^0.$$

As $c_\alpha e^{-\alpha} = c(\alpha, -\alpha) e^{-\alpha} = e^{-\alpha}$ (by (27.80) and (27.69)), it follows that

$$\Phi(t^0 \otimes E^0_\alpha)(|0\rangle \otimes e^{-\alpha}) = \sum_{s=1}^{l} \hat{\boldsymbol{\alpha}}(H^0_s)(A^{s\dagger}_1 |0\rangle \otimes e^0),$$

which is consistent, under the identifications (27.99) and (27.100), with

$$[t^0 \otimes E^0_\alpha, t^0 \otimes E^0_{-\alpha}]_- = t^0 \otimes H^0_\alpha = t^0 \otimes \left(\sum_{s=1}^{l} \hat{\boldsymbol{\alpha}}(H^0_s) H^0_s \right)$$

(cf. (27.62)).

(*iii*) $\alpha + \gamma$ *is not a root of* $\mathscr{L}^0$. In this case

$$M_\alpha(z) \exp\{i\hat{\boldsymbol{\alpha}}\cdot\mathbf{Q}_>(z)\}(|0\rangle \otimes \mathrm{e}^\gamma) = z(|0\rangle \otimes \mathrm{e}^{\alpha+\gamma}),$$

and so the coefficient of z^0 in $U_0(\alpha)$ can only come from a z^{-1} term in $\exp\{i\hat{\boldsymbol{\alpha}}\cdot\mathbf{Q}_<(z)\}(|0\rangle \otimes \mathrm{e}^{\alpha+\gamma})$, but there are no such terms. Thus in this case

$$\Phi(t^0 \otimes E^0_\alpha)(|0\rangle \otimes \mathrm{e}^\gamma) = 0,$$

which is consistent, under the identification (27.99), with the commutation relation

$$[t^0 \otimes E^0_\alpha, t^0 \otimes E^0_\gamma]_- = 0.$$

(B) Now consider $\Phi(t^0 \otimes E^0_\alpha)$ acting on $A^{st\dagger}_1|0\rangle \otimes \mathrm{e}^0$. By (27.80) and (27.68),

$$c_\alpha(A^{st\dagger}_1|0\rangle \otimes \mathrm{e}^0) = A^{st\dagger}_1|0\rangle \otimes \mathrm{e}^0.$$

Then, by (27.71), (27.72) and (27.83),

$$Q^r_>(z)(A^{st\dagger}_1|0\rangle \otimes \mathrm{e}^0) = iz^{-1}\delta^{rs}(|0\rangle \otimes \mathrm{e}^0),$$

and so, by (27.50) and (27.8),

$$M_\alpha(z)\exp\{i\hat{\boldsymbol{\alpha}}\cdot\mathbf{Q}_>(z)\}(A^{st\dagger}_1|0\rangle \otimes \mathrm{e}^0) = z(A^{st\dagger}_1|0\rangle \otimes \mathrm{e}^\alpha) - \hat{\alpha}(H^0_s)(|0\rangle \otimes \mathrm{e}^\alpha).$$

However, $\exp\{i\hat{\boldsymbol{\alpha}}\cdot\mathbf{Q}_<(z)\}$ acting on $z(A^{st\dagger}_1|0\rangle \otimes \mathrm{e}^\alpha)$ produces no z^0 terms, and the z^0 term in $\exp\{i\hat{\boldsymbol{\alpha}}\cdot\mathbf{Q}_<(z)\}$ $(|0\rangle \otimes \mathrm{e}^\alpha)$ is merely $|0\rangle \otimes \mathrm{e}^\alpha$, so

$$\Phi(t^0 \otimes E^0_\alpha)(A^{st\dagger}_1|0\rangle \otimes \mathrm{e}^0) = -\hat{\alpha}(H^0_s)(|0\rangle \otimes \mathrm{e}^\alpha)$$

which is consistent, under the identifications (27.99) and (27.100), with

$$[t^0 \otimes E^0_\alpha, t^0 \otimes H^0_s]_- = -\hat{\alpha}(H^0_s)(t^0 \otimes E^0_\alpha).$$

3 Proof of Theorem I of Chapter 27, Section 5

Theorem I The set of operators defined by

$$\Phi(t^j \otimes a^0_m) = \frac{1}{2}\sum_{r,s=1}^{D_0} \Gamma(a^0_m)_{rs} \sum_p :B^r_{j-p} B^s_p: \tag{27.117}$$

for $j = 0, \pm 1, \pm 2, \ldots$ and $m = 1, 2, \ldots, n^0$, and by

$$\Phi(c) = \tfrac{1}{2}\gamma 1, \tag{27.118}$$

where γ is the Dynkin index of the representation $\boldsymbol{\Gamma}$ of $\tilde{\mathscr{L}}^0$ (cf. (16.12)) form a representation of the derived algebra $[\tilde{\mathscr{L}}, \tilde{\mathscr{L}}]_-$ of the untwisted affine Kac–

Moody algebra $\tilde{\mathscr{L}}$. (In (27.117) the index p runs over all integer values in the Ramond case and over all half-integer values in the Neveu–Schwarz case, the carrier space of the representation being the corresponding Fock space.)

Proof It is easily shown from (27.117) using (27.116) that

$$\begin{aligned}
&[\Phi(t^j \otimes a_m^0), \Phi(t^k \otimes a_n^0)]_- \\
&= \frac{1}{4} \sum_{r,s,r',s'=1}^{D_0} \Gamma(a_m^0)_{rs} \Gamma(a_n^0)_{r's'} \sum_{p,q} [B_{j-p}^r B_p^s, B_{k-q}^{r'} B_q^{s'}]_- \\
&= \frac{1}{2} \sum_{r,s=1}^{D_0} (\mathbf{\Gamma}(a_m^0)\mathbf{\Gamma}(a_n^0))_{rs} \sum_p B_{j-p}^r B_{k+p}^s \\
&\quad - \frac{1}{2} \sum_{r,s=1}^{D_0} (\mathbf{\Gamma}(a_n^0)\mathbf{\Gamma}(a_m^0))_{rs} \sum_p B_{j+k-p}^r B_p^s,
\end{aligned} \tag{O.25}$$

the antisymmetry of the matrices $\mathbf{\Gamma}(a_m^0)$ being used in obtaining the last line. However, (27.116) implies that for $j+k \neq 0$

$$B_{j-p}^r B_{k+p}^s = :B_{j-p}^r B_{k+p}^s:$$

and

$$B_{j+k-p}^r B_p^s = :B_{j+k-p}^r B_p^s: \quad ,$$

whereas for $j+k=0$

$$B_{j-p}^r B_{k+p}^s = \begin{cases} :B_{j-p}^r B_{-j+p}^s: & \text{if } p > j, \\ :B_{j-p}^r B_{-j+p}^s: + \delta^{rs} 1 & \text{if } p \leqslant j, \end{cases}$$

and

$$B_{j+k-p}^r B_p^s = \begin{cases} :B_{-p}^r B_p^s: & \text{if } p > 0, \\ :B_{-p}^r B_p^s: + \delta^{rs} 1 & \text{if } p \leqslant 0. \end{cases}$$

Thus from (O.25)

$$\begin{aligned}
&[\Phi(t^j \otimes a_m^0), \Phi(t^k \otimes a_n^0)]_- \\
&= \Phi(t^{j+k} \otimes [a_m^0, a_n^0]_-) + \tfrac{1}{2} j \operatorname{tr}\{\mathbf{\Gamma}(a_m^0)\mathbf{\Gamma}(a_n^0)\}\, \delta^{j+k,0} 1.
\end{aligned} \tag{O.26}$$

However, by (16.12),

$$\operatorname{tr}\{\mathbf{\Gamma}(a_m^0)\mathbf{\Gamma}(a_n^0)\} = \gamma B^0(a_m^0, a_n^0), \tag{O.27}$$

$B^0(\ ,\ )$ being the Killing form of $\tilde{\mathscr{L}}^0$ (and $\mathscr{L}_c^0$), so (O.26), (O.27), (27.188) and (26.89) imply that

$$[\Phi(t^j \otimes a_m^0), \Phi(t^k \otimes a_n^0)]_- = \Phi([t^j \otimes a_m^0, t^k \otimes a_n^0]_-)$$

for all $j, k = 0, \pm 1, \pm 2, \ldots$ and $m, n = 1, 2, \ldots, n^0$. Moreover, (27.118) and

(26.90) show that

$$[\Phi(t^j \otimes a_m^0), \Phi(c)]_- = 0 = \Phi([t^j \otimes a_m^0, c]_-)$$

so this set of operators form a representation of the derived algebra $[\tilde{\mathscr{L}}, \tilde{\mathscr{L}}]_-$.

4 Proofs of Theorems I and II of Chapter 28, Section 3

Theorem I With the operators L_J defined by (28.20), the following hold.

(a) The commutation relations with the A_k^s are

$$[L_J, A_k^s]_- = -kA_{J+k}^s \tag{28.23}$$

for $s = 1, 2, \ldots, D_0$ and for all integers k and J.

(b) The operators L_J provide a unitary irreducible highest weight representation of the Virasoro algebra (28.4) in which the carrier space is the bosonic Fock space V_F, and for which the central charge is given by

$$C_V = D_0. \tag{28.24}$$

The vacuum state $|0\rangle$ of V_F is the highest weight vector of the representation; that is,

$$L_J|0\rangle = 0 \quad \text{for } J > 0. \tag{28.25}$$

Moreover,

$$L_0|0\rangle = \left\{\frac{1}{2}\sum_{r=1}^{D_0} (\alpha^r)^2\right\}|0\rangle, \tag{28.26}$$

so that

$$h = \frac{1}{2}\sum_{r=1}^{D_0} (\alpha^r)^2. \tag{28.27}$$

Proof It follows from (28.14) and (28.19) that

$$:A_j^r A_k^r: = \begin{cases} A_j^r A_k^r & \text{if } j \leqslant k, \\ A_j^r A_k^r + k\delta_{j+k,0}1 & \text{if } j > k. \end{cases} \tag{O.28}$$

Thus, by (28.20) and (O.28),

$$[L_J, A_k^s]_- = \frac{1}{2}\sum_{j=-\infty}^{\infty}\sum_{r=1}^{D_0} [A_{J+j}^r A_{-j}^r, A_k^s]_- = -kA_{J+k}^s. \tag{O.29}$$

Hence for $J > 0$, by (O.28) and (O.29),

$$[L_J, L_K]_- = \left[L_J, \frac{1}{2} \sum_{k=-\infty}^{\infty} \sum_{s=1}^{D_0} A^s_{K+k} A^s_{-k} \right]_-$$

$$= \frac{1}{2} \sum_{k=-\infty}^{\infty} \sum_{s=1}^{D_0} \{ -(K+k) A^s_{J+K+k} A^s_{-k} + k A^s_{K+k} A^s_{J-k} \}$$

$$= \frac{1}{2} \sum_{k=-\infty}^{\infty} \sum_{s=1}^{D_0} \{ -(K+k) : A^s_{J+K+k} A^s_{-k} : + k : A^s_{K+k} A^s_{J-k} : \}$$

$$+ \frac{1}{2} \sum_{k=1}^{J} \sum_{s=1}^{D_0} k(J-k) \delta_{J+K,0} 1.$$

But

$$\sum_{k=1}^{J} k(J-k) = \tfrac{1}{6} J(J^2 - 1),$$

so, for $J > 0$,

$$[L_J, L_K]_- = (J-K) L_{J+K} + \tfrac{1}{12} J(J^2 - 1) D_0 \delta_{J+K,0} 1. \tag{O.30}$$

It is easily checked by a similar argument that (O.30) is also valid for $J \leqslant 0$. Equation (O.30) is of the form of (28.4) with C_V given by (28.6) and c_V by (28.24).

Equations (28.25) follow immediately from (28.21) and (28.22) on using (28.18), and (28.26) comes from (28.22) (with $J = 0$) on using (28.17) and (28.18).

Theorem II (a) The operators L_J defined by (28.34) (with κ given by (28.35)) satisfy the Virasoro algebra (28.4) and (28.5) provided that C_V has the form (28.6) and the eigenvalue c_V is given by

$$c_V = \frac{2c_\Lambda n^0}{2c_\Lambda + 1}. \tag{28.39}$$

(b) Moreover,

$$[L_J, \Phi(t^k \otimes a^0_s)]_- = -k \Phi(t^{J+k} \otimes a^0_s) \tag{28.40}$$

for all $J, K = 0, \pm 1, \pm 2, \ldots$ and $s = 1, 2, \ldots, n^0$, and

$$[L_J, \Phi(c)]_- = 0 \tag{28.41}$$

for $J = 0, \pm 1, \pm 2, \ldots$.

(c) The operator L_0 is related to $\Phi(d)$ by

$$L_0 = -\Phi(d) + \left\{ \Lambda(d) + \frac{C_2(\Lambda^0)}{\kappa} \right\} 1, \tag{28.42}$$

where $C_2(\Lambda^0)$ is the eigenvalue of the second-order Casmir operator of the corresponding irreducible representation of the complex simple Lie algebra $\tilde{\mathscr{L}}^0$ with highest weight $\Lambda^0 = \sum_{k=1}^{l} n_k \Lambda_k^0$, where Λ_k^0 are the fundamental weights of $\tilde{\mathscr{L}}^0$.

(d) If $\psi(\Lambda)$ is the highest weight vector of the Kac–Moody algebra representation then

$$L_J\psi(\Lambda) = 0 \quad \text{for } J > 0 \tag{28.43}$$

and

$$L_0\psi(\Lambda) = \frac{C_2(\Lambda^0)}{\kappa}\psi(\Lambda), \tag{28.44}$$

($C_2(\Lambda^0)$ being as in part (c)).

(e) With the definition (28.34),

$$L_J^\dagger = L_{-J} \tag{28.45}$$

for all $J = 0, \pm 1, \pm 2, \ldots$.

Proof For $J > 0$ the result (28.43) follows immediately from (28.31), (28.36) and (28.37). With $J = 0$ and $\psi = \psi(\Lambda)$, (28.31) and (28.37) imply that

$$L_0\psi(\Lambda) = -\frac{1}{\kappa}\sum_{r=1}^{n^0} \Phi(t^0 \otimes a_r^0)\Phi(t^0 \otimes a_r^0)\psi(\Lambda).$$

However, as was noted in Chapter 27, Section 2, when acting on $\psi(\Lambda)$ the operators $\Phi(t^0 \otimes a_r^0)$ act as operators of the irreducible representation of $\tilde{\mathscr{L}}^0$ with highest weight

$$\Lambda^0 = \sum_{k=1}^{l} n_k \Lambda_k^0,$$

so, by Theorem I of Chapter 16, Section 2,

$$L_0\psi(\Lambda) = \frac{C_2(\Lambda^0)}{\kappa}\psi(\Lambda),$$

where $C_2(\Lambda^0)$ is the eigenvalue of the second-order Casimir operator of this representation, thereby establishing (28.44). The result (28.41) is an immediate consequence of (28.33).

Turning to the proof of (28.40) (which is the most difficult part of the theorem), it follows from (26.90) and (28.38) that

$$\begin{aligned}&[:\Phi(t^{J+j} \otimes a_r^0)\Phi(t^{-j} \otimes a_r^0):, \Phi(t^k \otimes a_s^0)]_- \\ &\quad = [\Phi(t^{J+j} \otimes a_r^0)\Phi(t^{-j} \otimes a_r^0), \Phi(t^k \otimes a_s^0)]_-,\end{aligned}$$

and hence, by (28.30) and (28.34),

$$[L_J, \Phi(t^k \otimes a_s^0)]_- \psi = A\psi + B\psi \tag{O.31}$$

for all $\psi \in \mathrm{V}$, where

$$A\psi = -\frac{1}{\kappa}\sum_{j=-\infty}^{\infty}\sum_{r,p=1}^{n^0} c_{rs}^p\{\Phi(t^{J+j+k}\otimes a_p^0)\Phi(t^{-j}\otimes a_r^0) + \Phi(t^{J+j}\otimes a_r^0)\Phi(t^{k-j}\otimes a_p^0)\}\psi$$

and

$$B\psi = \frac{1}{\kappa}\sum_{j=-\infty}^{\infty}\sum_{r=1}^{n^0}\{(J+j)\delta_{J+j+k,0}\delta_{rs}\Phi(t^{-j}\otimes a_r^0) + (-j)\delta_{k-j,0}\delta_{rs}\Phi(t^{J+j}\otimes a_r^0)\}c_\Lambda\psi.$$

Clearly the expression for $B\psi$ simplifies to give

$$B\psi = -\frac{2kc_\Lambda}{\kappa}\Phi(t^{J+k}\otimes a_s^0)\psi, \tag{O.32}$$

so all the problems lie with $A\psi$, which, by virtue of the antisymmetry of the structure constants c_{rs}^p, can be rewritten as

$$A\psi = -\frac{1}{\kappa}\sum_{j=-\infty}^{\infty}\sum_{r,p=1}^{n^0} c_{rs}^p\{\Phi(t^{J+j+k}\otimes a_p^0)\Phi(t^{-j}\otimes a_r^0) - \Phi(t^{J+j}\otimes a_p^0)\Phi(t^{k-j}\otimes a_r^0)\}\psi. \tag{O.33}$$

An *incorrect* result can be obtained by writing the single infinite sum in (O.33) as two infinite sums, i.e. as

$$\sum_{j=-\infty}^{\infty}\Phi(t^{J+j+k}\otimes a_p^0)\Phi(t^{-j}\otimes a_r^0)\psi - \sum_{j=-\infty}^{\infty}\Phi(t^{J+j}\otimes a_p^0)\Phi(t^{k-j}\otimes a_r^0)\psi, \tag{O.34}$$

and then putting $-j' = k - j$ in the second sum, which then appears to be identical with the first, suggesting that the total sum is zero. A more careful analysis, which will be given shortly, shows that this is not true.

(The situation here is essentially the same as that occurring with the series of real numbers

$$\sum_{j=1}^{\infty}(a_{j+1} - a_j), \tag{O.35}$$

where it is assumed that $a_j \to a$ as $j \to \infty$ and that $a \neq 0$. The mth partial sum

s_m of (O.35) is defined by

$$s_m = \sum_{j=1}^{m} (a_{j+1} - a_j),$$

so that

$$s_m = a_{m+1} - a_1.$$

Then $s_m \rightarrow a - a_1$ as $m \rightarrow \infty$, and, since (O.35) is *defined* as the limit of s_m as $m \rightarrow \infty$, it follows immediately that

$$\sum_{j=1}^{\infty} (a_{j+1} - a_j) = a - a_1.$$

However, the analogue of the careless argument leading to (O.34) is that in which (O.35) is written as

$$\sum_{j=1}^{\infty} a_{j+1} - \sum_{j=1}^{\infty} a_j,$$

which is equal to

$$\sum_{j=2}^{\infty} a_j - \sum_{j=1}^{\infty} a_j,$$

and hence to $-a_1$, which does not agree with (O.35).)

The *correct* way of dealing with (O.33) is to introduce partial sums, and in particular to note that for any weight vector $\psi(\lambda)$

$$A\psi(\lambda) = \lim_{m\rightarrow\infty} \{A_m\psi(\lambda)\}, \tag{O.37}$$

where $A_m\psi(\lambda)$ is given by (O.33), but with the upper limit ∞ replaced by m. For simplicity it will be assumed that $k \geqslant 0$, but a similar argument can be given for $k < 0$. For any $\psi(\lambda)$ there exists a non-negative integer K (which depends on λ) such that $\Phi(t^j \otimes a_r^0)\psi(\lambda) = 0$ for all $j > K$ (and $r = 1, 2, \ldots, n^0$). It follows that

$$A_m\psi(\lambda) = -\frac{1}{K} \sum_{r,p=1}^{n^0} c_{rs}^p \left\{ \sum_{j=-K}^{m} \Phi(t^{J+j+k} \otimes a_p^0)\Phi(t^{-j} \otimes a_r^0) \right.$$
$$\left. - \sum_{j=k-K}^{m} \Phi(t^{J+j} \otimes a_p^0)\Phi(t^{k-j} \otimes a_r^0) \right\} \psi(\lambda),$$

and so

$$A_m\psi(\lambda) = -\frac{1}{K} \sum_{r,p=1}^{n^0} c_{rs}^p \left\{ \sum_{j=m-K+1}^{m} \Phi(t^{J+j+k} \otimes a_p^0)\Phi(t^{-j} \otimes a_r^0) \right\} \psi(\lambda).$$

By (28.30), this reduces to

$$A_m\psi(\lambda) = -\frac{1}{\kappa}\sum_{r,p=1}^{n^0} c^p_{rs} \sum_{j=m-K+1}^{m} \left\{ \sum_{q=1}^{n^0} c^q_{pr}\Phi(t^{J+k}\otimes a^0_q) - (J+j+k)\delta^{J+k,0}c_\Lambda\delta_{rp} \right.$$

$$\left. + \Phi(t^{-j}\otimes a^0_r)\Phi(t^{J+j+k}\otimes a^0_p) \right\}\psi(\lambda).$$

However, for sufficiently large m (i.e. for $J+m+1>K$) the last term is zero, and, since $c^p_{ps}=0$,

$$A_m\psi(\lambda) = -\frac{k}{\kappa}\sum_{p,q,r=1}^{n^0} c^p_{rs}c^q_{pr}\Phi(t^{J+k}\otimes a^0_q)\psi(\lambda).$$

As $\tilde{\mathscr{L}}^0$ is simple, (11.41) and the antisymmetry property of the c^p_{rs} give

$$\begin{aligned}\sum_{r,p=1}^{n^0} c^p_{rs}c^q_{pr} &= -\sum_{r,p=1}^{n^0}\{\mathbf{ad}(a^0_s)\}_{pr}\{\mathbf{ad}(a^0_q)\}_{rp} \\ &= -B^0(a^0_s, a^0_q) \quad \text{(by (13.1))} \\ &= \delta_{sq} \quad \text{(by (28.29))},\end{aligned}$$

so

$$A_m\psi(\lambda) = -\frac{k}{\kappa}\Phi(t^{J+k}\otimes a^0_s)\psi(\lambda).$$

Taking the limit as $m\to\infty$, by (O.37),

$$A\psi(\lambda) = -\frac{k}{\kappa}\Phi(t^{J+k}\otimes a^0_s)\psi(\lambda). \tag{O.38}$$

Thus, from (O.31), (O.32) and (O.38),

$$[L_J, \Phi(t^k\otimes a^0_s)]_- = -k\Phi(t^{J+k}\otimes a^0_s),$$

provided that

$$\kappa = 1 + 2c_\Lambda.$$

To prove that the operators L_J satisfy (28.4), it suffices to note that (28.40) and the Jacobi identity show that $[L_J, L_K]_- - (J-K)L_{J+K}$ commutes with $\Phi(t^k\otimes a^0_s)$ (for all $k=0,\pm 1,\pm 2,\ldots$ and $s=1,2,\ldots,n^0$), as well as with $\Phi(c)$. However, part (c) of Theorem VII of Chapter 27, Section 1 implies that the latter operators form an *irreducible* representation of the derived algebra $[\tilde{\mathscr{L}},\tilde{\mathscr{L}}]_-$. Hence, by Schur's Lemma,

$$[L_J, L_K]_- - (J-K)L_{J+K} = \gamma(J,K)1, \tag{O.39}$$

where $\gamma(J,K)$ is a constant that depends on J and K, which can be evaluated by considering the action of both sides of (O.39) on the highest state vector $\psi(\Lambda)$. It is obvious from (28.43) that if $J > 0$ and $K > 0$ then $\gamma(J, K) = 0$. A straightforward (but lengthy) piece of algebra demonstrates more generally that

$$\gamma(J, K) = \frac{2c_{\Lambda}n^0}{\kappa}\delta_{J+K,0}.$$

The argument leading to (28.42) is similar. Equations (26.91) and (28.40) show that $L_0 + \Phi(d)$ also commutes with all $\Phi(t^k \otimes a_s^0)$ (for all $k = 0, \pm 1, \pm 2, \ldots$ and $s = 1, 2, \ldots, n^0$) as well as with $\Phi(c)$, i.e. with all the operators of $[\tilde{\mathscr{L}}, \tilde{\mathscr{L}}]_-$, so Schur's Lemma implies that

$$L_0 + \Phi(d) = \mu 1, \tag{O.40}$$

where μ is some constant. Again μ can be found by acting with both sides of (O.40) on $\psi(\Lambda)$, when (28.44) shows that

$$\mu = \frac{C_2(\Lambda^0)}{\kappa} + \Lambda(d).$$

Finally, by Theorem V of Chapter 27, Section 2 and (26.194),

$$\Phi(t^j \otimes a_r^0)^\dagger = -\Phi(t^{-j} \otimes a_r^0)$$

for $j = 0, \pm 1, \pm 2, \ldots$ and $r = 1, 2, \ldots, n^0$, so (28.45) follows from (28.36) and (28.37).

References

Ademollo, M., Brink, L., D'Adda, A., D'Auria, R., Napolitano, E., Sciuto, S., Del Giudice, E., Di Vecchia, P., Ferrara, S., Gliozzi, F., Musto, R. and Pettorino, R. (1976a). *Phys. Lett.* **62B,** 105–110.

Ademollo, M., Brink, L., D'Adda, A., D'Auria, R., Napolitano, E., Sciuto, S., Del Giudice, E., Di Vecchia, P., Ferrara, S., Gliozzi, F., Musto, R. and Pettorino, R. (1976b). *Nucl. Phys.* **B111,** 77–110.

Ademollo, M., Brink, L., D'Adda, A., D'Auria, R., Napolitano, E., Sciuto, S., Del Giudice, E., Di Vecchia, P., Ferrara, S., Gliozzi, F., Musto, R. and Pettorino, R. (1976c). *Nucl. Phys.* **B114,** 297–316.

Affleck, I. (1987). In *Infinite Lie Algebras and Conformal Invariance in Condensed Matter and Particle Physics* (ed. K. Dietz and V. Rittenberg), pp. 1–16. World Scientific, Singapore.

Agrawala, V. K. (1979). *Hadronic J.* **2,** 830–839.

Agrawala, V. K. (1981). *Hadronic J.* **4,** 444–496.

Altschuler, D. (1986). *Mod. Phys. Lett.* **A1,** 557–564.

Alvarez, O., Windey, P. and Mangano, M. (1986). *Nucl. Phys.* **B277,** 317–331.

Alvarez-Gaumé, L. and Witten, E. (1983). *Nucl. Phys.* **B234,** 269–330.

Aneva, B. L., Mikhov, S. G. and Stoyanov, D. T. (1976). *Theor. Math. Phys.* **27,** 502–508 [*Teor. Mat. Fiz.* **27,** 307–316 (1976)].

Aneva, B. L., Mikhov, S. G. and Stoyanov, D. T. (1977). *Theor. Math. Phys.* **31,** 394–406 [*Teor. Mat. Fiz.* **31,** 177–189 (1976)].

Aneva, B. L., Mikhov, S. G. and Stoyanov, D. T. (1978). *Theor. Math. Phys.* **35,** 383–390 [*Teor. Mat. Fiz.* **35**, 163–170 (1976)].

Apikyan, S. A. (1987). *Mod. Phys. Lett.* **A2,** 317–329.

Aratyn, H. and Damgaard, P. H. (1986). *Phys. Lett.* **179B,** 51–56.

Baake, M. (1988). *J. Math. Phys.* **29,** 1753–1757.

Backhouse, N. B. (1977). *J. Math. Phys.* **18,** 239–244.

Backhouse, N. B. (1979). *J. Phys.* **A12,** 21–30.

Backhouse, N. B. (1980). *Ann. Isr. Phys. Soc.* **3,** 312–313.

Backhouse, N. B. (1982). *Physica* **114A,** 410–412.

Bagger, J., Nemeschansky, D. and Yankielowicz, S. (1988). *Phys. Rev. Lett.* **60,** 389–392.
Balantekin, A. B. (1982). *J. Math. Phys.* **23,** 486–489.
Balantekin, A. B. (1984). *J. Math. Phys.* **25,** 2028–2030.
Balantekin, A. B. and Bars, I. (1981a). *J. Math. Phys.* **22,** 1149–1162.
Balantekin, A. B. and Bars, I. (1981b). *J. Math. Phys.* **22,** 1810–1818.
Balantekin, A. B. and Bars, I. (1982). *J. Math. Phys.* **23,** 1239–1247.
Bando, M., Kuramoto, T., Maskawa, T. and Uehara, S. (1984). *Phys. Lett.* **138B,** 94–98.
Bars. I. (1984). In *Vertex Operators in Mathematics and Physics* (ed. J. Lepowsky, S. Mandelstam and I. M. Singer), pp. 373–392 (Mathematical Sciences Research Institute, Vol. 3). Springer-Verlag, New York.
Bars, I., Morel, B. and Ruegg, H. (1983). *J. Math. Phys.* **24,** 2253–2262.
Batchelor, M. (1980). *Trans. Amer. Math. Soc.* **258,** 257–270.
Batchelor, M. (1984). In *Mathematical Aspects of Superspace* (ed. C. J. S. Clarke, A. Rosenblum and H. J. Siefert), pp. 91–134. Reidel, Dordrecht.
Bedding, S. (1984). *Nucl. Phys.* **B236,** 368–380.
Bedding, S. (1985). *Phys. Lett.* **157B,** 183–185.
Bednar, M. and Sachl, V. F. (1979). *J. Math. Phys.* **19,** 1487–1492.
Bednar, M., McKellar, B. H. J. and Sachl, V. F. (1984). *J. Phys.* **A17,** 1579–1592.
Belavin, A. A., Polyakov, A. M. and Zamolodchikov, A. B. (1984a). *J. Stat. Phys.* **34,** 763–774.
Belavin, A. A., Polyakov, A. M. and Zamolodchikov, A. B. (1984b). *Nucl. Phys.* **B241,** 333–380.
Bellucci, S. (1987a). *Phys. Rev.* **D35,** 1296–1304.
Bellucci, S. (1987b). *Z. Phys.* **C33,** 551–560.
Bellucci, S. (1987c). *Z. Phys.* **C34,** 558.
Berele, A. and Regev, A. (1983). *Bull. Amer. Math. Soc.* **8,** 337–339.
Berezin, F. A. (1977). "Lie Superalgebras". Moscow Institute of Theoretical and Experimental Physics Report ITEP-66. [Reprinted in Berezin, F. A. and Kirillov, A. A. (1986). *Introduction to Superanalysis.* Reidel, Dordrecht.]
Berezin, F. A. (1979). *Sov. J. Nucl. Phys.* **29**, 857–866.
Berezin, F. A. and Leites, D. A. (1975). *Sov. Math. Dokl.* **16,** 1218–1222.
Bernard, D. and Thierry-Mieg, J. (1987). *Commun. Math. Phys.* **111,** 181–246.
Bernstein, I. N. and Leites, D. A. (1980). *C. R. Acad. Bulgare Sci.* **33,** 1049–1051.
Bershadsky, M. A., Knizhnik, V. G. and Teitelman, M. G. (1985). *Phys. Lett.* **151B,** 31–36.
Bhattacharya, G. and Chau, L. L. (1984). In *Group Theoretical Methods in Physics, Trieste, 1983* (ed. G. Denardo, G. Ghirardi and T. Weber), pp. 153–158 (Lecture Notes in Physics, Vol. 201). Springer-Verlag, Berlin.
Bincer, A. M. (1983). *J. Math. Phys.* **24,** 2546–2549.
Binegar, B. (1986). *Phys. Rev.* **D34,** 525–532.
Binétruy, P. (1988). In *19e Ecole d'Eté de Physique des Particules, 1987.* G. I. F., Paris.
Boucher, W., Friedan, D. and Kent, A. (1986). *Phys. Lett.* **172B,** 316–322.
Bourbaki, N. (1971). *Eléments de Mathématique*, Fascicule XXVI: *Groupes et Algèbres de Lie*, Chap. 1. Hermann, Paris. [English translation: *Elements of Mathematics: Lie Groups and Lie Algebras*, Chap. 1. Springer-Verlag, Berlin (1989).]
Bowcock, P. and Goddard, P. (1987). *Nucl. Phys.* **B285,** 651–670.
Boyer, C. and Gitler, S. (1980). In *Group Theoretical Methods in Physics, Cocoyoc, Mexico, 1980* (ed. K. B. Wolf), pp. 532–536 (Lecture Notes in Physics, Vol. 135). Springer-Verlag, Berlin.

Boyer, C. and Gitler, S. (1984). *Trans. Amer. Math. Soc.* **285**, 241–268.
Brauer, R. and Weyl, H. (1935). *Amer. J. Math.* **57**, 425–449.
Breitenlohner, P. and Freedman, D. Z. (1982). *Ann. Phys. (NY)* **144**, 249–281.
Breitenlohner, P. and Kabelschacht, A. (1979). *Nucl. Phys.* **B148**, 96–106.
Brink, L. (1985). In *Supersymmetry* (ed. K. Dietz, R. Flume, G. von Gehlen, and V. Rittenberg), pp. 89–134. Plenum Press, New York.
Brink, L. and Nielsen, H. B. (1973). *Phys. Lett.* **45B**, 332–336.
Brink, L. and Schwarz, J. H. (1977). *Nucl. Phys.* **B121**, 285–295.
Brink, L., Olive, D. I. and Scherk, J. (1973). *Nucl. Phys.* **B61**, 173–198.
Brink, L., Di Vecchia, P. and Howe, P.(1976). *Phys. Lett.* **65B**, 471–474.
Brower, R. (1972). *Phys. Rev.* **D6**, 1655–1662.
Brower, R. and Friedman, K. A. (1973). *Phys. Rev.* **D7**, 535–539.
Buchmüller, W., Peccei, R. D. and Yanagida, T. (1983). *Nucl. Phys.* **B227**, 503–546.
Burges, C. J. C., Freedman, D. Z., Davis, S. and Gibbons, G. W. (1986). *Ann. Phys (NY)* **167**, 285–316.
Burgess, C. P. (1985) *Nucl. Phys.* **B259**, 473–492.
Candelas, P., Horowitz, G. T., Strominger, A. and Witten, E. (1985). *Nucl. Phys.* **B258**, 46–74.
Chang, D. and Kumar, A. (1987). *Phys. Lett.* **193B**, 181–186.
Chari, V. and Pressley, A. (1988). *J. Algebra* **113**, 438–464.
Chen, J. Q. and Chen, X. G. (1983). *J. Phys.* **A16**, 3435–3456.
Chen, J. Q., Chen, X. G. and Gao, M. J. (1983a). *J. Phys.* **A16**, L47–L53.
Chen, J. Q., Chen, X. G. and Gao, M. J. (1983b). *J. Phys.* **A16**, 1361–1376.
Chen, J. Q., Gao, M. J. and Chen, X. G. (1984a). *J. Phys.* **A17**, 481–500.
Chen, J. Q., Gao, M. J. and Chen, X. G. (1984b). *J. Phys.* **A17**, 1941–1961.
Clavelli, L. (1986). In *Lewes String Theory Workshop* (ed. L. Clavelli and A. Halprin), pp. 1–20. World Scientific, Singapore.
Coleman, S. and Mandula, J. (1967). *Phys. Rev.* **159**, 1251–1256.
Copson, E. T. (1935). *Theory of Functions of a Complex Variable.* Oxford University Press.
Corrigan, E. F. (1986). *Phys. Lett.* **169B**, 259–263.
Corrigan, E. F. and Goddard, P. (1974). *Nucl. Phys.* **B68**, 189–202.
Cremmer, E. (1981). In *Superspace and Supergravity. Proceedings of the Nuffield Workshop, Cambridge, 1980* (ed. S. W. Hawking and M. Roček), pp. 267–282. Cambridge University Press.
Cremmer, E. and Scherk, J. (1976). *Nucl. Phys.* **B103**, 399–425.
Crispim-Romão, J., Ferber, A. and Freund, P. G. O. (1977). *Nucl. Phys.* **B126**, 429–435.
Cummins, C. J. and King, R. C. (1987a). *J. Phys.* **A20**, 3103–3120.
Cummins, C. J. and King, R. C. (1987b). *J. Phys.* **A20**, 3121–3133.
de Azcárraga, J. A., Lukierski, J. and Vindel, P. (1988). *Fortschr. Phys.* **36**, 453–478.
Delduc, F. and Gourdin, M. (1984). *J. Math. Phys.* **25**, 1651–1661.
Deser, S. and Zumino, B. (1976). *Phys. Lett.* **65B**, 369–373.
de Sitter, W. (1917). *Mon. Not. R. Astron. Soc.* **78**, 3–28.
DeWitt, B. (1984). *Supermanifolds.* Cambridge University Press.
DeWitt, B. S. and Van Nieuwenhuizen, P. (1982). *J. Math. Phys.* **23**, 1953–1963.
Di Vecchia, P., Del Guidice, E. and Fubini, S. (1972). *Ann. Phys. (NY)* **70**, 378–398.
Di Vecchia, P., Petersen, J. L. and Zheng, H. B. (1985a). *Phys. Lett.* **162B**, 327–332.
Di Vecchia, P., Knizhnik, V. G., Petersen, J. L. and Rossi, P. (1985b). *Nucl. Phys.* **B253**, 701–726.
Di Vecchia, P., Petersen, J. L. and Yu, M. (1986a). *Phys. Lett.* **172B**, 211–215.

Di Vecchia, P., Petersen, J. L., Yu, M. and Zheng, H. B. (1986b). *Phys. Lett.* **174B**, 280–284.
Dixon, L., Harvey, J. A., Vafa, C. and Witten, E. (1985). *Nucl. Phys.* **B261**, 678–686.
Dixon, L., Harvey, J. A., Vafa, C. and Witten, E. (1986). *Nucl. Phys.* **B274**, 285–314.
Dobrev, V. K. (1986a). *Lett. Math. Phys.* **11**, 225–234.
Dobrev, V. K. (1986b). In *Proceedings of the 13th International Conference of Differential Geometric Methods in Theoretical Physics* (ed. H. D. Doebner and T. D. Palev), pp. 348–370. World Scientific, Singapore.
Dobrev, V. K.. (1986c). "Characters of the irreducible highest weight modules over the Virasoro and super Virasoro algebras". ICTP Trieste Report IC/86/123.
Dobrev, V. K. (1986d). In *Topological and Geometrical Methods in Field Theory* (ed. J. Hietarinta), pp. 93–102. World Scientific, Singapore.
Dobrev, V. K. (1987a). *Phys. Lett.* **186B**, 43–51.
Dobrev, V. K. (1987b). *Commun. Math. Phys.* **111**, 75–80.
Dobrev, V. K. and Ganchev, A. Ch. (1988). *Mod. Phys. Lett.* **A3**, 127–138.
Dobrev, V. K. and Petkova, V. B. (1985). *Phys. Lett.* **162B**, 127–132.
Dobrev, V. K. and Petkova, V. B. (1986a). In *Conformal Groups and Related Structures: Physical Results and Mathematical Background* (ed. A. O. Barut and H. D. Doebner), pp. 291–299 (Lecture Notes in Physics, Vol. 261). Springer-Verlag, Berlin.
Dobrev, V. K. and Petkova, V. B. (1986b). In *Conformal Groups and Related Structures: Physical Results and Mathematical Background* (ed. A. O. Barut and H. D. Doebner), pp. 300–308 (Lecture Notes in Physics, Vol. 261). Springer-Verlag, Berlin.
Dolan, L. (1984a). *Phys. Rep.* **109**, 1–94.
Dolan, L. (1984b). In *Vertex Operators in Mathematics and Physics* (ed. J. Lepowsky, S. Mandelstam and I. M. Singer), pp. 353–372 (Mathematical Sciences Research Institute, Vol. 3). Springer-Verlag, New York.
Dolan, L. (1985). In *Applications of Group Theory in Physics and Mathematical Physics* (ed. M. Flato, P. Sally and G. Zuckerman), pp. 307–324 (Lectures in Applied Mathematics, Vol. 21). American Mathematical Society, Providence, Rhode Island.
Dolan, L. (1986). In *Lewes String Theory Workshop* (ed. L. Clavelli and A. Halprin), pp. 103–132. World Scientific, Singapore.
Dondi, P. H. and Jarvis, P. D. (1981). *J. Phys.* **A14**, 547–563.
Dondi, P. H. and Sohnius, M. F. (1974). *Nucl. Phys.* **B81**, 317–329.
Dragon, N., Ellwanger, U. and Schmidt, M. G. (1987). *Prog. Part. Nucl. Phys.* **18**, 1–92.
Eguchi, T. and Taormina, A. (1988a). *Phys. Lett.* **200B**, 315–322.
Eguchi, T. and Taormina, A. (1988b). *Phys. Lett.* **210B**, 125–132.
Eichenherr, H. (1985). *Phys. Lett.* **151B**, 26–30.
Ellis, J. (1986). In *Superstrings and Supergravity* (ed. A. T. Davies and D. G. Sutherland), pp. 399–518. Scottish Universities Summer School in Physics, Edinburgh.
Ellis, J., Gaillard, M. K., Maiani, L. and Zumino, B. (1980). In *Unification of the Fundamental Particle Interactions* (ed. S. Ferrara, J. Ellis and P. van Nieuwenhuizen), pp. 69–88. Plenum Press, New York.
Fairlie, D. B., Nuyts, J. and Zachos, C. K. (1988a). *Commun. Math. Phys.* **117**, 595–614.
Fairlie, D. B., Nuyts, J. and Zachos, C. K. (1988b). *Phys. Lett.* **202B**, 320–324.
Farmer, R. J. (1986a). *J. Phys.* **A19**, L97–L100.
Farmer, R. J. (1986b). *J. Phys.* **A19**, 321–328.
Farmer, R. J. and Jarvis, P. D. (1983). *J. Phys.* **A16**, 473–487.
Farmer, R. J. and Jarvis, P. D. (1984). *J. Phys.* **A17**, 2365–2387.
Fayet, P. (1975a). *Phys. Lett.* **58B**, 67–70.

Fayet, P. (1975b). *Nuovo Cim.* **31A**, 626–640.
Fayet, P. (1976). *Nucl. Phys.* **B113**, 135–155.
Fayet, P. (1979). *Nucl. Phys.* **B149**, 137–169.
Fayet, P. (1980). In *Unification of the Fundamental Particle Interactions* (ed. S. Ferrara, J. Ellis and P. van Nieuwenhuizen), pp. 587–620. Plenum Press, New York.
Fayet, P. (1984a). *Phys. Rep.* **105**, 21–51.
Fayet, P. (1984b). In *Supersymmetry and Supergravity '84* (ed. B. de Wit, P. Fayet and P. van Nieuwenhuizen), pp. 114–158. World Scientific, Singapore.
Fayet, P. and Ferrara, S. (1977). *Phys. Rep.* **32**, 249–334.
Fayet, P. and Iliopoulos, J. (1974). *Phys. Lett.* **51B**, 461–464.
Feigin, B. L. and Fuchs, D. B. (1982). *Funct. Anal. Appl.* **16**, 114–126 [*Funkt. Analis i ego Prilozh.* **16**, (2), 47–63 (1982)].
Feigin, B. L. and Fuchs, D. B. (1983). *Funct. Anal. Appl.* **17**, 241–242 [*Funkt. Analis i ego Prilozh.* **17**, (3), 91–92 (1983)].
Feingold, A. J. (1984). In *Vertex Operators in Mathematics and Physics* (ed. J. Lepowsky, S. Mandelstam and I. M. Singer), pp. 185–206. (Mathematical Sciences Research Institute, Vol. 3). Springer-Verlag, New York.
Feingold, A. J. and Frenkel, I. B. (1983). *Math. Ann.* **263**, 82–144.
Feingold, A. J. and Lepowsky, J. (1978). *Adv. Math.* **29**, 271–309.
Ferrara, S. (1974). *Nucl. Phys.* **B77**, 73–90.
Ferrara, S. (1984). *Phys. Rep.* **105**, 5–19.
Ferrara, S. (1985). In *Supersymmetry* (ed. K. Dietz, R. Flume, G. Von Gehlen and V. Rittenberg), pp. 245–340. Plenum Press, New York.
Ferrara, S. (ed.) (1987). *Supersymmetry*, 2 vols. World Scientific, Singapore.
Ferrara, S. and Savoy, C. A. (1982). In *Supergravity '81* (ed. S. Ferrara and J. G. Taylor), pp. 47–83. Cambridge University Press.
Ferrara, S. and Zumino, B. (1974). *Nucl. Phys.* **B79**, 413–421.
Ferrara, S., Wess, J. and Zumino, B. (1974). *Phys. Lett.* **51B**, 239–241.
Ferrara, S., Savoy, C. A. and Zumino, B. (1981). *Phys. Lett.* **100B**, 393–398.
Firth, R. J. and Jenkins, J. D. (1975). *Nucl. Phys.* **B85**, 525–534.
Flato, M. and Fronsdal, C. (1984). *Lett. Math. Phys.* **8**, 159–162.
Frampton, P. H. (1974). *Dual Resonance Models.* Benjamin/Cummings, New York.
Frampton, P. H. (1986). *Dual Resonance Models and Superstrings.* World Scientific, Singapore.
Freedman, D. Z. (1979). In *Recent Developments in Gravitation: Cargèse 1978* (ed. M. Levy and S. Deser), pp. 549–561. Plenum Press, New York.
Freedman, D. Z. and Nicolai, H. (1984). *Nucl. Phys.* **B237**, 342–366.
Frenkel, I. B. (1980). *Proc. Nat. Acad. Sci. USA* **77**, 6303–6306.
Frenkel, I. B. (1981). *J. Funct. Anal.* **44**, 259–327.
Frenkel, I. B. (1982). In *Lie Algebras and Related Topics, New Brunswick, NJ, 1981*, pp. 71–110 (Lecture Notes in Mathematics, Vol. 933). Springer-Verlag, Berlin.
Frenkel, I. B. (1985). In *Applications of Group Theory in Physics and Mathematical Physics* (ed. M. Flato, P. Sally and G. Zuckerman), pp. 325–354 (Lectures in Applied Mathematics, Vol. 21). American Mathematical Society, Providence, Rhode Island.
Frenkel, I. B. and Kac, V. G. (1980). *Invent. Math.* **62**, 23–66.
Frenkel, I. B., Lepowsky, J. and Meurman, A. (1984). *Proc. Nat. Acad. Sci. USA* **81**, 3256–3260.
Freund, P. G. O. (1976). *J. Math. Phys.* **17**, 424–426.
Freund, P. G. O. and Kaplansky, I. (1976). *J. Math. Phys.* **17**, 228–231.
Friedan, D., Qiu, Z. and Shenker, S. (1984a). *Phys. Rev. Lett.* **52**, 1575–1578.

Friedan, D., Qiu, Z. and Shenker, S. (1984b). In *Vertex Operators in Mathematics and Physics* (ed. J. Lepowsky, S. Mandelstam and I. M. Singer), pp. 419–450 (Mathematical Sciences Research Institute, Vol. 3). Springer-Verlag, New York.

Friedan, D., Qiu, Z. and Shenker, S. (1985). *Phys. Lett.* **151B**, 37–43.

Friedan, D., Qiu, Z. and Shenker, S. (1986). *Commun. Math. Phys.* **107**, 535–542.

Fubini, S. and Veneziano, G. (1970). *Nuovo Cim.* **67A**, 29–47.

Fubini, S., Gordan, D. and Veneziano, G. (1969). *Phys. Lett.* **29B**, 679–682.

Gabber, O. and Kac, V. G. (1981). *Bull. Amer. Math. Soc.* **5**, 185–189.

Galperin, A. S., Litov, L. B. and Soroka, V. A. (1983). *J. Phys.* **G9**, 133–138.

Galperin, A. S., Invanov, E., Kalitzin, S., Ogievetsky, E. and Sokatchev, E. (1984a). *Class. Quantum Grav.* **1**, 469–498.

Galperin, A. S., Ivanov, E., Kalitzin, S., Ogievetsky, E. and Sokatchev, E. (1984b). In *Supersymmetry and Supergravity '84* (ed. B. de Wit, P. Fayet and P. van Nieuwenhuizen), pp. 449–469. World Scientific, Singapore.

Galperin, A. S., Ivanov, E., Ogievetsky, E. and Sokatchev, E. (1985a). *Class. Quantum Grav.* **2**, 601–616.

Galperin, A. S., Ivanov, E., Ogievetsky, E. and Sokatchev, E. (1985b). *Class. Quantum Grav.* **2**, 617–630.

Gastmans, R., Sevrin, A., Troost, W. and Van Proeyen, A. (1987). *Int. J. Mod. Phys.* **A2**, 195–216.

Gervais, J.-L. (1970). *Nucl. Phys.* **B21**, 192–204.

Gliozzi, F., Scherk, J. and Olive, D. I. (1976). *Phys. Lett.* **65B**, 282–286.

Gliozzi, F., Scherk, J. and Olive, D. I. (1977). *Nucl. Phys.* **B122**, 253–290.

Goddard, P. (1987). In *Proceedings of the Workshop on "Super Field Theories"* (ed. H. C. Lee, V. Elias, G. Kunstatter, R. B. Mann and K. S. Viswanathan), pp. 233–264. Plenum Press, New York.

Goddard, P. and Olive, D. I. (1984). In *Vertex Operators in Mathematics and Physics* (ed. J. Lepowsky, S. Mandelstam and I. M. Singer), pp. 51–96. (Mathematical Sciences Research Institute, Vol. 3). Springer-Verlag, New York.

Goddard, P. and Olive, D. I. (1985). *Nucl. Phys.* **B257**, 226–252.

Goddard, P. and Olive, D. I. (1986). *Int. J. Mod. Phys.* **A1**, 303–414.

Goddard, P. and Olive, D. I. (eds) (1988). *Kac–Moody and Virasoro Algebras.* World Scientific, Singapore.

Goddard, P. and Thorn, C. (1972). *Phys. Lett.* **40B**, 235–238.

Goddard, P., Goldstone, J., Rebbi, C. and Thorn, C. B. (1973). *Nucl. Phys.* **B56**, 109–135.

Goddard, P., Kent, A. and Olive, D. I. (1985). *Phys. Lett.* **152B**, 88–92.

Goddard, P., Kent, A. and Olive, D. I. (1986a). *Commun. Math. Phys.* **103**, 105–120.

Goddard, P., Nahm, W., Olive, D. I. and Schwimmer, A. (1986b). *Commun. Math. Phys.* **107**, 179–212.

Goddard, P., Nahm, W., Olive, D. I., Ruegg, H. and Schwimmer, A. (1987a). *Commun. Math. Phys.* **112**, 385–408.

Goddard, P., Olive, D. I. and Waterson, G. (1987b). *Commun. Math. Phys.* **112**, 591–612.

Golitzin, G. (1988). *J. Algebra* **117**, 198–226.

Gomes, J. F. (1986). *Phys. Lett.* **171B**, 75–76.

Goodman, R. (1985). In *Infinite Dimensional Groups with Applications* (ed. V. G. Kac), pp. 125–136 (Mathematical Sciences Research Institute, Vol. 4). Springer-Verlag, New York.

Goodman, R. and Wallach, N. R. (1984a). *J. Reine Angew. Math.* **347**, 69–133.

Goodman, R. and Wallach, N. R. (1984b). *J. Reine Angew. Math.* **352**, 220.
Goto, T. (1971). *Prog. Theor. Phys.* **46**, 1560–1569.
Gourdin, M. (1986a). *J. Math. Phys.* **27**, 61–65.
Gourdin, M. (1986b). *J. Math. Phys.* **27**, 1980–1986.
Gourdin, M. (1987). *J. Math. Phys.* **28**, 2007–2017.
Grassmann, H. G. (1911). *Gessammelte Mathematische und Physikalische Werke.* B. G. Teuber, Leipzig.
Green, H. S. and Jarvis, P. D. (1983). *J. Math. Phys.* **24**, 1681–1687.
Green, M. B. (1983). *Surv. High Energy Phys.* **3**, 127–160.
Green, M. B. (1986). In *Supersymmetry, Supergravity, and Superstrings '86* (ed. B. de Wit, P. Fayet, and M. Grisaru), pp. 135–238. World Scientific, Singapore.
Green, M. B. and Schwarz, J. H. (1981). *Nucl. Phys.* **B181**, 502–530.
Green, M. B. and Schwarz, J. H. (1982a). *Phys. Lett.* **109B**, 444–448.
Green, M. B. and Schwarz, J. H. (1982b). *Nucl. Phys.* **B198**, 252–268.
Green, M. B. and Schwarz, J. H. (1982c). *Nucl. Phys.* **B198**, 441–460.
Green, M. B. and Schwarz, J. H. (1983). *Nucl. Phys.* **B218**, 43–88.
Green, M. B. and Schwarz, J. H. (1984a). *Phys. Lett.* **136B**, 367–370.
Green, M. B. and Schwarz, J. H. (1984b). *Nucl. Phys.* **B243**, 285–306.
Green, M. B. and Schwarz, J. H. (1984d). *Phys. Lett.* **149B**, 117–122.
Green, M. B. and Schwarz, J. H. (1985a). *Phys. Lett.* **151B**, 21–25.
Green, M. B. and Schwarz, J. H. (1985b). *Nucl. Phys.* **B255**, 93–114.
Green, M. B., Schwarz, J. H. and Brink, L. (1982). *Nucl. Phys.* **B198**, 474–492.
Green, M. B., Schwarz, J. H. and Witten, E. (1987a). *Superstring Theory*, Vol. 1: *Introduction.* Cambridge University Press.
Green, M. B., Schwarz, J. H. and Witten, E. (1987b). *Superstring Theory*, Vol. 2: *Loop Amplitudes, Anomalies and Phenomenology.* Cambridge University Press, Cambridge.
Gross, D. J., Harvey, J. A., Martinec, E. and Rohm, R. (1985a). *Phys. Rev. Lett.* **54**, 502–505.
Gross, D. J., Harvey, J. A., Martinec, E. and Rohm, R. (1985b). *Nucl. Phys.* **B256**, 253–284.
Gross, D. J., Harvey, J. A., Martinec, E. and Rohm, R. (1986). *Nucl. Phys.* **B267**, 75–124.
Grosser, D. (1975a). *Lett. Nuovo Cim.* **12**, 455–458.
Grosser, D. (1975b). *Lett. Nuovo Cim.* **12**, 649–652.
Grosser, D. (1976). *Nucl. Phys.* **B116**, 286–290.
Gruber, B., Santhanam, T. S. and Wilson, R. (1984). *J. Math. Phys.* **25**, 1253–1261.
Günaydin, M. and Marcus, N. (1985). *Class. Quantum Grav.* **2**, L19–L23.
Günaydin, M., van Nieuwenhuizen, P. and Warner, N. P. (1985). *Nucl. Phys.* **B255**, 63–92.
Günaydin, M., Sierra, G. and Townsend, P. K. (1986). *Nucl. Phys.* **B274**, 429–447.
Haag, R., Łopuszanski, J. T. and Sohnium, M. (1975). *Nucl. Phys.* **B88**, 257–274.
Haber, H. E. and Kane, G. L. (1985). *Phys. Rep.* **117**, 75–263.
Han, Q. Z. (1981). *Nuovo Cim.* **64A**, 391–405.
Han, Q. Z. and Sun, H. Z. (1983). *Commun. Theor. Phys.* **2**, 1137–1144.
Han, Q. Z., Song, X. C., Li, G. D. and Sun, H. Z. (1981). *Phys. Energ. Fortis. Phys. Nucl.* **5**, 546–553.
Heidereich, W. (1982). *Phys. Lett.* **110B**, 461–464.
Hlavaty, L. and Niederle, J. (1980). *Lett. Math. Phys.* **4**, 301–306.
Horsley, R. (1978). *Nucl. Phys.* **B138**, 493–516.

Howe, P. S., Stelle, K. S. and Townsend, P. K. (1983a). *Nucl. Phys.* **B214**, 519–531.
Howe, P. S., Sierra, G. and Townsend, P. K. (1983b). *Nucl. Phys.* **B221**, 331–348.
Hughes, J. W. B. (1981). *J. Math. Phys.* **22**, 245–250.
Hughes, J. W. B. and King, R. C. (1987). *J. Phys.* **A20**, L1047–L1052.
Hull, C. M. (1987). In *Proceedings of the Workshop on "Super Field Theories"* (ed. H. C. Lee, V. Elias, G. Kunstatter, R. B. Mann and K. S. Viswanathan), pp. 77–168. Plenum Press, New York.
Hurni, J. P. (1987). *J. Phys.* **A20**, 1–14.
Hurni, J. P. and Morel, B. (1982). *J. Math. Phys.* **23**, 2236–2243.
Hurni, J. P. and Morel, B. (1983). *J. Math. Phys.* **24**, 157–163.
Hurni, J. P., Morel, B., Ruegg, H., Sciarrino, A. and Sorba, A. (1984). In *Group Theoretical Methods in Physics, Trieste, 1983* (ed. G. Denardo, G. Ghirardi and T. Weber), pp. 53–56. (Lecture Notes in Physics, Vol. 201). Springer-Verlag, Berlin.
Ibáñez, L. E. (1987). In *Proceedings of the Workshop on "Super Field Theories"* (ed. H. C. Lee, V. Elias, G. Kunstatter, R. B. Mann and K. S. Viswanathan), pp. 293–320. Plenum Press, New York.
Iliopoulos, J. and Zumino, B. (1974). *Nucl. Phys.* **B76**, 310–332.
Ivanov, E. A. and Sorin, A. S. (1980). *Theor. Math. Phys.* **45**, 862–873 [*Teor. Mat. Fiz.* **45**, 30–45 (1980)].
Iwasaki, Y. and Kikkawa, K. (1973). *Phys. Rev.* **D8**, 440–449.
Jackiw, R. (1987). In *Proceedings of the Workshop on "Super Field Theories"* (ed. H. C. Lee, V. Elias, G. Kunstatter, R. B. Mann and K. S. Viswanathan), pp. 191–208. Plenum Press, New York.
Jacob, M. (ed.) (1974). *Dual Theory.* North Holland, Amsterdam.
Jadczyk, A. Z. and Pilch, K. A. (1981). *Commun. Math. Phys.* **78**, 373–390.
Jain, S., Mandal, G. and Wadia, S. R. (1987). *Phys. Rev.* **D36**, 3116–3142.
Jakobsen, H. P. and Kac, V. G. (1985). In *Non-Linear Equations in Classical and Quantum Field Theory, Meudon and Paris VI, 1983/84* (ed. N. Sanchez), pp. 1–20 (Lecture Notes in Physics, Vol. 226). Springer-Verlag, Berlin.
Jantzen, J. C. (1979). *Modulen mit linem Löchster Gewicht. III* (Lecture Notes in Mathematics, Vol. 750). Springer-Verlag, Berlin.
Jarvis, P. D. and Green, H. S. (1979). *J. Math. Phys.* **20**, 2115–2122.
Jarvis, P. D. and Murray, M. K. (1983). *J. Math. Phys.* **24**, 1705–1710.
Jevicki, A. (1988). *Int. J. Mod. Phys.* **A3**, 299–364.
Julia, B. (1985). In *Applications of Group Theory in Physics and Mathematical Physics* (ed. M. Flato, P. Sally and G. Zuckerman), 355–374 (Lectures in Applied Mathematics, Vol. 21). American Mathematical Society, Providence, Rhode Island.
Kac. V. G. (1968). *Math. USSR. Izv.* **2**, 1271–1311 [*Izv. Akad. Nauk SSSR Ser. Mat. Tom.* **32**, (6), 1923–1967 (1968)].
Kac, V. G. (1974). *Funct. Anal. Appl.* **8**, 68–70 [*Funkt. Analis i ego Prilozh.* **8**, (1), 77–78 (1974)].
Kac, V. G. (1975). *Funct. Anal. Appl.* **9**, 263–265 [*Funkt. Analis i ego Prilozh.* **9**, (3), 91–92 (1975)].
Kac, V. G. (1977a). *Adv. Math.* **26**, 8–96.
Kac, V. G. (1977b). *Commun. Math. Phys.* **53**, 31–64.
Kac, V. G. (1977c). *Commun. Algebra* **5**, 889–897.
Kac, V. G. (1978a). *Adv. Math.* **30**, 85–136.
Kac, V. G. (1978b). In *Differential Geometrical Methods in Mathematical Physics II* (ed. K. Bleuler, H. R. Petry and A. Reetz), pp. 597–626 (Lecture Notes in Mathematics, Vol. 676). Springer-Verlag, Berlin.

Kac, V. G. (1979). In *Group Theoretical Methods in Physics, Austin, 1978* (ed. W. Beiglböck, A. Böhm and E. Takasugi), pp. 441–445 (Lecture Notes in Physics, Vol. 94). Springer-Verlag, Berlin.

Kac, V. G. (1980). *Adv. Math.* **35**, 264–273.

Kac, V. G. (1985a). *Infinite Dimensional Lie Algebras.* Cambridge University Press.

Kac, V. G. (1985b). In *Infinite Dimensional Groups with Applications* (ed. V. G. Kac), pp. 167–216. (Mathematical Sciences Research Institute, Vol. 4). Springer-Verlag, New York.

Kac, V. G. and Kazhdan, D. A. (1979). *Adv. Math.* **34**, 97–108.

Kac, V. G. and Peterson, D. H. (1980). *Bull. Amer. Math. Soc.* **3**, 1057–1061.

Kac, V. G. and Peterson, D. H. (1981). *Proc. Nat. Acad. Sci. USA.* **78**, 3308–3312.

Kac, V. G. and Peterson, D. H. (1985). In *Symposium on Anomalies, Geometry, Topology* (ed. W. A. Bardeen and A. R. White), pp. 276–298. World Scientific, Singapore.

Kac, V. G. and Raina, A. (1988). *Bombay Lectures on Highest Weight Representations of Infinite-Dimensional Lie Algebras.* World Scientific, Singapore.

Kac, V. G. and Sanielevici, M. N. (1988). *Phys. Rev.* **D37**, 2231–2237.

Kac, V. G. and Todorov, I. T. (1985). *Commun. Math. Phys.* **102**, 337–347.

Kac, V. G. and Todorov, I. T. (1986). *Commun. Math. Phys.* **104**, 175.

Kac, V. G. and Wakimoto, M. (1986). In *Conformal Groups and Related Structures: Physical Results and Mathematical Background* (ed. A. O. Barut and H. D. Doebner), pp. 345–372 (Lecture Notes in Physics, Vol. 261). Springer-Verlag, Berlin.

Kac, V. G. and Wakimoto, M. (1988). *Adv. Math.* **70**, 156–236.

Kac, V. G., Kazhdan, D. A., Lepowsky, J. and Wilson, R. L. (1981). *Adv. Math.* **42**, 83–112.

Kaku, M. (1986). In *Lewes String Theory Workshop* (ed. L. Clavelli and A. Halprin), pp. 69–102. World Scientific, Singapore.

Kaku, M. (1987). *Int. J. Mod. Phys.* **A2**, 1–76.

Kaku, M. (1988). *Introduction to Superstrings.* Springer-Verlag, New York.

Kaku, M., Townsend, P. K. and van Nieuwenhuizen, P. (1977a). *Phys. Rev. Lett.* **39**, 1109–1112.

Kaku, M., Townsend, P. K. and van Nieuwenhuizen, P. (1977b). *Phys. Lett.* **69B**, 304–308.

Kane, G. L. (1985). In *Supersymmetry* (ed. K. Dietz, R. Flume, G. Von Gehlen, and V. Rittenberg), pp. 355–392. Plenum Press, New York.

Kaplansky, I. (1982). *Commun. Math. Phys.* **86**, 49–54.

Kaplansky, I. and Santharoubane, L. J. (1984). In *Vertex Operators in Mathematics and Physics* (ed. J. Lepowsky, S. Mandelstam, and I. M. Singer), pp. 217–232. (Mathematical Sciences Research Institute, Vol. 3). Springer-Verlag, New York.

Kass, S. N. and Patera, J. (1987). In *Proceedings of the Workshop on "Super Field Theories"* (ed. H. C. Lee, V. Elias, G. Kunstatter, R. B. Mann and K. S. Viswanathan), pp. 265–274. Plenum Press, New York.

Kato, M. and Matsuda, S. (1987a). *Phys. Lett.* **184B**, 184–190.

Kato, M. and Matsuda, S. (1987b). *Prog. Theor. Phys.* **78**, 158–165.

Kazama, Y. (1987). *Czech. J. Phys.* **37**, 312–328.

Keck, B. W. (1975). *J. Phys.* **A8**, 1819–1827.

Kent, A. (1985). *Rend. Circ. Mat. Palermo* (2) Suppl. **9**, 129–135.

Kent, A. and Riggs, H. (1987). *Phys. Lett.* **198B**, 491–496.

Ketov, S. V. (1988). *Fortschr. Phys.* **36**, 361–425.

Kim, J. (1984). *J. Math. Phys.* **25**, 2037–2051.

King, R. C. (1983). In *Group Theoretical Methods in Physics, Istanbul, 1982* (ed. M. Serdaroğlu and E. Inönü), pp. 41–47 (Lecture Notes in Physics, Vol. 180). Springer-Verlag, Berlin.
King, R. C. (1986). In *Group Theoretical Methods in Physics, Seoul, 1985* (ed. Y. M. Cho), pp. 199–204. World Scientific, Singapore.
Kiritsis, E. (1987). *Phys. Rev.* **D36**, 3048–3065.
Kiritsis, E. (1988a). *Int. J. Mod. Phys.* **A3**, 1871–1906.
Kiritsis, E. (1988b). *J. Phys.* **A21**, 297–306.
Kleeman, R. (1985). *J. Math. Phys.* **26**, 2405–2412.
Knizhnik, V. G. and Zamolodchikov, A. B. (1984). *Nucl. Phys.* **B247**, 83–103.
Kostant, B. (1977). In *Differential Geometrical Methods in Mathematical Physics* (ed. K. Bleuler and A. Reetz), pp. 177–306 (Lecture Notes in Mathematics, Vol. 570). Springer-Verlag, Berlin.
Kugo, T., Ojima, I. and Yanagida, T. (1984). *Phys. Lett.* **135B**, 402–408.
Kuperschmidt, B. A. (1985). *Phys. Lett.* **113A**, 117–120.
Leites, D. A. (1980a). *C.R. Acad. Bulgare Sci.* **33**, 1053–1055.
Leites, D. A. (1980b). *Funct. Anal. Appl.* **14**, 106–109 [*Funkt. Analis i ego Prilozh.* **14**, (2), 35–38 (1980)].
Leites, D. A. (1980c). *Russ. Math. Surv.* **35**, 1–64 [*Usp. Mat. Nauk* **35**, 3–57 (1980)].
Lemire, F. W. and Patera, J. (1982). *J. Math. Phys.* **23**, 1409–1414.
Lepowsky, J. (1985a). *Proc. Nat. Acad. Sci. USA.* **82**, 8295–8299.
Lepowsky, J. (1985b). In *Applications of Group Theory in Physics and Mathematical Physics* (ed. M. Flato, P. Sally and G. Zuckerman), pp. 375–398. (Lectures in Applied Mathematics, Vol. 21). American Mathematical Society, Providence, Rhode Island.
Lepowsky, J. and Primc, M. (1984a). In *Number Theory Seminar, 1982* (ed. D. V. Chudnowski, G. V. Chudnowski, H. Cohn and M. B. Nathanson), pp. 194–251 (Lecture Notes in Mathematics, Vol. 1052). Springer-Verlag, Berlin.
Lepowsky, J. and Primc, M. (1984b). In *Vertex Operators in Mathematics and Physics* (ed. J. Lepowsky, S. Mandelstam and I. M. Singer), pp. 143–162 (Mathematical Sciences Research Institute, Vol. 3). Springer-Verlag, New York.
Lepowsky, J. and Primc, M. (1985). *Contemp. Math.* **46**, 1–84.
Lepowsky, J. and Wilson, R. L. (1978). *Commun. Math. Phys.* **62**, 43–53.
Lepowsky, J. and Wilson, R. L. (1984). *Invent. Math.* **77**, 199–290.
Lepowsky, J., Mandelstam, S. and Singer, I. M. (eds) (1984).. *Vertex Operators in Mathematics and Physics* (Mathematical Sciences Research Institute, Vol. 3). Springer-Verlag, New York.
Lerche, W. (1983). *Nucl. Phys.* **B238**, 582–600.
Lerche, W. and Lust, D. (1984). *Nucl. Phys.* **B244**, 157–172.
Levstein, F. (1988). *J. Algebra* **114**, 489–518.
Li, G. H. and Sun, Z. Z. (1986). *J. Phys.* **A19**, 3487–3511.
Li, K. and Warner, N. P. (1988). *Phys. Lett.* **211B**, 101–106.
Llewellyn Smith, C. H. (1984). *Phys. Rep.* **105**, 53–70.
Łopuszanski, J. T. and Wolf, M. (1981). *Nucl. Phys.* **B184**, 133–179.
Lukierski, J. and Nowicki, A. (1982). *Fortschr. Phys.* **30**, 75–98.
Lukierski, J. and Nowicki, A. (1985). *Phys. Lett.* **151B**, 382–386.
Lynker, M. and Schwimmrigk, R. (1988). *Phys. Lett.* **207B**, 216–220.
MacDonald, I. G. (1981). In *Séminaire Bourbaki 1980/81, Exposés 561–578*, pp. 258–276 (Lecture Notes in Mathematics, Vol. 901). Springer-Verlag, Berlin.
Mack, G. and Salam, A. (1969). *Ann. Phys.* **53**, 174–202.

Mainland, G. B. and Tanaka, K. (1975). *Phys. Rev.* **D12**, 2394–2396.
Mandelstam, S. (1974). *Phys. Rep.* **13**, 259–353.
Mandelstam, S. (1984). In *Vertex Operators in Mathematics and Physics* (ed. J. Lepowsky, S. Mandelstam and I. M. Singer), pp. 15–36 (Mathematical Sciences Research Institute, Vol. 3). Springer-Verlag, New York.
Mandia, M. (1987). *Mem. Amer. Math. Soc.* No. 362, Vol. 65, pp. 1–146.
Mansouri, F. and Schaer, C. (1980). *Phys. Lett.* **101B**, 51–54.
Marcu, M. (1980a). *J. Math. Phys.* **21**, 1277–1283.
Marcu, M. (1980b). *J. Math. Phys.* **21**, 1284–1292.
Marcus, N. and Sagnotti, A. (1982). *Phys. Lett.* **119B**, 97–99.
Matsuo, Y. (1987). *Prog. Theor. Phys.* **77**, 793–797.
Mazouni, E. E. (1987). *C.R. Acad. Sci. Paris* **304**, 1–4.
McAvaty, C, and Niederle, J. (1980). In *Group Theoretical Methods in Physics, Cocoyoc, Mexico, 1980* (ed. K. B. Wolf), pp. 537–539 (Lecture Notes in Physics, Vol. 135). Springer-Verlag, Berlin.
Meurman, A. and Rocha-Caridi, A. (1986). *Commun. Math. Phys.* **107**, 263–294.
Meurman, A. and Santharoubane, L. J. (1988). *Commun. Algebra.* **16**, 27–36.
Mezinescu, L. (1977). *J. Math. Phys.* **18**, 453–455.
Michel, L. (1970). In *Group Representations in Mathematics and Physics.* (*Battelle Seattle 1969 Rencontres*) (ed. V. Bargmann), pp. 36–143. (Lecture Notes in Physics, Vol. 6). Springer-Verlag, Berlin.
Mickelsson, J. (1985). *J. Math. Phys.* **26**, 377–382.
Mickelsson, J. (1986). In *Conformal Groups and Related Structures: Physical Results and Mathematical Background* (ed. A. O. Barut and H. D. Doebner), pp. 372–378 (Lecture Notes in Physics, Vol. 261). Springer-Verlag, Berlin.
Mickelsson, J. (1987). *Czech. J. Phys.* **37**, 387–388.
Miki, K. (1988). *Mod. Phys. Lett.* **A3**, 191–200.
Misra, K. C. (1984). In *Vertex Operators in Mathematics and Physics* (ed. J. Lepowsky, S. Mandelstam and I. M. Singer), pp. 163–184 (Mathematical Sciences Research Institute, Vol. 3). Springer-Verlag, New York.
Mitra, B. (1987). *J. Phys.* **A20**, 4065–4074.
Mohapatra, R. N. (1986). *Unification and Supersymmetry.* Springer-Verlag, New York.
Molotkov, V. V., Petrova, S. G. and Stoyanov, D. T. (1976). *Theor. Math. Phys.* **26**, 125–131 [*Teor. Mat. Fiz.* **26**, 188–197 (1976)].
Moody, R. V. (1967). *Bull. Amer. Math. Soc.* **73**, 217–221.
Moody, R. V. (1968). *J. Algebra* **10**, 211–230.
Moody, R. V. (1969). *Canad. J. Math.* **21**, 1432–1454.
Moody, R. V. (1975). *Proc. Amer. Math. Soc.* **48**, 43–52.
Morel, B., Sciarrino, A. and Sorba, P. (1985). *J. Phys.* **A18**, 1597–1613.
Morel, B., Sciarrino, A. and Sorba, P. (1986). *Phys. Lett.* **166B**, 69–74.
Morel, B., Patera, J., Sharp, R. T. and Van der Jeugt, J. (1987). *J. Math. Phys.* **28**, 1673–1682.
Nahm, W. (1978). *Nucl. Phys.* **B135**, 149–166.
Nahm, W. and Scheunert, M. (1976). *J. Math. Phys.* **17**, 868–879.
Nahm, W., Rittenberg, V. and Scheunert, M. (1976). *Phys. Lett.* **61B**, 383–384.
Nam, S. (1986). *Phys. Lett.* **172B**, 323–327.
Nam, S. (1987). *Phys. Lett.* **187B**, 340–346.
Nambu, Y. (1970). In *Proceedings of the International Conference on Symmetries and Quark Models* (ed. R. Chand), pp. 269–278. Gordon and Breach, New York.
Nandi, S. (1975). *Nucl. Phys.* **B97**, 178–188.

Nanopoulos, D. V. (1984). *Phys. Rep.* **105**, 71–80.
Narain, K. S. (1986). *Phys. Lett.* **169B**, 41–46.
Narain, K. S., Sarmadi, M. H. and Witten, E. (1987). *Nucl. Phys.* **B279**, 369–379.
Nath, P. and Arnowitt, R. (1980). In *Unification of the Fundamental Particle Interactions* (ed. S. Ferrara, J. Ellis and P. van Nieuwenhuizen), pp. 411–434. Plenum Press, New York.
Nepomechie, R. I. (1986). *Phys. Rev.* **D34**, 1129–1135.
Neveu, A. and Schwarz, J. H. (1971a). *Nucl. Phys.* **B31**, 86–112.
Neveu, A. and Schwarz, J. H. (1971b). *Phys. Rev.* **D4**, 1109–1111.
Nicolai, H. (1984). In *Supersymmetry and Supergravity '84* (ed. B. de Wit, P. Fayet and P. van Nieuwenhuizen), pp. 368–399. World Scientific, Singapore.
Nicolai, H. (1987). *Prog. Part. Nucl. Phys.* **19**, 1–32.
Nicolai, H. and Sezgin, E. (1984). *Phys. Lett.* **143B**, 389–395.
Nilles, H. P. (1984). *Phys. Rep.* **110**, 1–162.
Nilles, H. P. (1986). In *Supersymmetry, Supergravity, and Superstrings '86* (ed. B. de Wit, P. Fayet and M. Grisaru), pp. 37–70. World Scientific, Singapore.
Nwachuku, C. O. and Rashid, M. A. (1985). *J. Math. Phys.* **26**, 1914–1920.
Ogievetsky, V. and Sokatchev, E. (1977). *J. Phys.* **A10**, 2021–2030.
Olive, D. I. (1987). In *Infinite Lie Algebras and Conformal Invariance in Condensed Matter and Particle Physics* (ed. K. Dietz and V. Rittenberg), pp. 51–58. World Scientific, Singapore.
O'Raifeartaigh, L. (1975a). *Phys. Lett.* **56B**, 41–44.
O'Raifeartaigh, L. (1975b). *Nucl. Phys.* **B96**, 331–352.
O'Raifeartaigh, L. (1975c). *Lecture Notes on Supersymmetry*. Communications of the Dublin Institute for Advanced Studies, Series A, No. 22.
Pais, A. and Rittenberg, V. (1975). *J. Math. Phys.* **16**, 2062–2073.
Pais, A. and Rittenberg, V. (1976). *J. Math. Phys.* **17**, 598.
Palev, T. D. (1978). *Int. J. Theor. Phys.* **17**, 985–992.
Palev, T. D. (1980). *J. Math. Phys.* **21**, 1293–1298.
Palev, T. D. (1981). *J. Math. Phys.* **22**, 2127–2132.
Palev, T. D. (1985). *J. Math. Phys.* **26**, 1640–1660.
Palev, T. D. (1986). *J. Math. Phys.* **27**, 1994–2001.
Palev, T. D. (1987a). *J. Math. Phys.* **28**, 272–291.
Palev, T. D. (1987b). *C.R. Acad. Bulgare Sci.* **40**, 33–36.
Palev, T. D. and Stoytchev, O. T. (1982). *C.R. Acad. Bulgare Sci.* **35**, 733–736.
Parker, M. (1980). *J. Math. Phys.* **21**, 689–697.
Parker, M. (1981). *Bull. Soc. Math. Belg.* **A33**, 129–139.
Patera, J. (1986). In *Group Theoretical Methods in Physics, Seoul, 1985* (ed. Y. M. Cho), pp. 289–292. World Scientific, Singapore.
Paton, J. and Chan, H. M. (1969). *Nucl. Phys.* **B10**, 516–520.
Pickup, C. and Taylor, J. G. (1981). *Nucl. Phys.* **B188**, 577–593.
Pilch, K., van Nieuwenhuizen, P. and Sohnius, M. F. (1985). *Commun. Math. Phys.* **98**, 105–117.
Polyakov, A. M. (1981a). *Phys. Lett.* **103B**, 207–210.
Polyakov, A. M. (1981b). *Phys. Lett.* **103B**, 211–213.
Pressley, A. and Segal, G. (1986). *Loop Groups*. Oxford University Press.
Qiu, Z. (1987a). *Phys. Lett.* **188B**, 207–213.
Qiu, Z. (1987b). *Phys. Lett.* **198B**, 497–502.
Raby, S. (1983). In *Supersymmetry and Supergravity '82* (ed. S. Ferrara, J. G. Taylor and P. van Nieuwenhuizen), pp. 304–334. World Scientific, Singapore.
Ramond, P. (1971). *Phys. Rev.* **D3**, 2415–2418.

Ramond, P. and Schwarz, J. H. (1976). *Phys. Lett.* **64B**, 75–77.
Rebbi, C. (1974). *Phys. Rep.* **12**, 1–73.
Rittenberg, V. (1978). In *Group Theoretical Methods in Physics, Tübingen, 1977* (ed. P. Kramer and A. Rieckers), pp. 3–21 (Lecture Notes in Physics, Vol. 79). Springer-Verlag, Berlin.
Rittenberg, V. (1986). In *Conformal Groups and Related Structures: Physical Results and Mathematical Background* (ed. A. O. Barut and H. D. Doebner), pp. 328–344 (Lecture Notes in Physics, Vol. 261). Springer-Verlag, Berlin.
Rittenberg, V. and Scheunert, M. (1982). *Commun. Math. Phys.* **83**, 1–9.
Rittenberg, V. and Schwimmer, A. (1987). *Phys. Lett.* **195B**, 135–138.
Rittenberg, V. and Sokatchev, E. (1981). *Nucl. Phys.* **B193**, 477–502.
Rocha-Caridi, A. (1984). In *Vertex Operators in Mathematics and Physics* (ed. J. Lepowsky, S. Mandelstam and I. M. Singer), pp. 451–474. (Mathematical Sciences Research Institute, Vol. 3). Springer-Verlag, New York.
Rocha-Caridi, A. (1987). In *Infinite Lie Algebras and Conformal Invariance in Condensed Matter and Particle Physics* (ed. K. Dietz and V. Rittenberg), pp. 59–80. World Scientific, Singapore.
Rocha-Caridi, A. and Wallach, N. R. (1983). *Invent. Math.* **72**, 57–75.
Rocha-Caridi, A. and Wallach, N. R. (1984). *Math. Z.* **185**, 1–21.
Rogers, A. (1980). *J. Math. Phys.* **21**, 1352–1365.
Rogers, A. (1981). *J. Math. Phys.* **22**, 939–945.
Rogers, A. (1984a). In *Mathematical Aspects of Superspace* (ed. C. J. S. Clarke, A. Rosenblum and H. J. Siefert), pp. 135–148. Reidel, Dordrecht.
Rogers, A. (1984b). In *Mathematical Aspects of Superspace* (ed. C. J. S. Clarke, A. Rosenblum and H. J. Siefert), pp. 149–160. Reidel, Dordrecht.
Rogers, A. (1985a). *J. Math. Phys.* **26**, 385–392.
Rogers, A. (1985b). *J. Math. Phys.* **26**, 2749–2753.
Rogers, A. (1986a). *Commun. Math. Phys.* **105**, 375–384.
Rogers, A. (1986b). *J. Math. Phys.* **27**, 710–717.
Rogers, A. (1987a). *Phys. Lett.* **193B**, 48–54.
Rogers, A. (1987b). *Commun. Math. Phys.* **113**, 353–368.
Ross, G. G. (1984). In *Supersymmetry and Supergravity '84* (ed. B. de Wit, P. Fayet and P. van Nieuwenhuizen), pp. 191–211. World Scientific, Singapore.
Rothstein, M. J. (1986). *Trans. Amer. Math. Soc.* **297**, 159–180.
Ruegg, H. (1983). In *Group Theoretical Methods in Physics, Istanbul, 1982* (ed. M. Serdaroğlu and E. İnönü), 265–268 (Lecture Notes in Physics, Vol. 180). Springer-Verlag, Berlin.
Saidi, E. H. (1988a). *Int. J. Mod. Phys.* **A3**, 861–874.
Saidi, E. H. (1988b). *J. Math. Phys.* **29**, 1949–1957.
Sakai, N. and Tanii, Y. (1984). *Phys. Lett.* **146B**, 38–40.
Salam, A. and Strathdee, J. (1974a). *Nucl. Phys.* **B76**, 477–482.
Salam, A. and Strathdee, J. (1974b). *Nucl. Phys.* **B80**, 499–505.
Salam, A. and Strathdee, J. (1974c). *Phys. Lett.* **49B**, 465–467.
Salam, A. and Strathdee, J. (1974d). *Phys. Lett.* **51B**, 353–355.
Salam, A. and Strathdee, J. (1975a). *Nucl. Phys.* **B84**, 127–131.
Salam, A. and Strathdee, J. (1975b). *Phys. Rev.* **D11**, 1521–1535.
Salam, A. and Strathdee, J. (1975c). *Nucl. Phys.* **B97**, 293–316.
Salam, A. and Strathdee, J. (1978). *Fortschr. Phys.* **26**, 57–142.
Salam, A. and Strathdee, J. (1982). *Ann. Phys. (NY)* **141**, 316–352.
Sanchez, O. A. V. (1988). *Trans. Amer. Math. Soc.* **307**, 597–614.
Sasaki, R. and Yamanaka, I. (1986). *Prog. Theor. Phys.* **75**, 706–726.

Savoy-Navarro, A. (1984). *Phys. Lett.* **105B**, 91–120.
Scherk, J. (1975). *Rev. Mod. Phys.* **47**, 123–164.
Scherk, J. (1979). In *Recent Developments in Gravitation: Cargèse 1978* (ed. M. Levy and S. Deser), pp. 479–517. Plenum Press, New York.
Scherk, J. and Schwarz, J. H. (1974a). *Nucl. Phys.* **B81**, 118–144.
Scherk, J. and Schwarz, J. H. (1974b). *Phys. Lett.* **52B**, 347–350.
Scherk, J. and Schwarz, J. H. (1975). *Phys. Lett.* **57B**, 463–466.
Scheunert, M. (1979a). *The Theory of Lie Superalgebras* (Lecture Notes in Mathematics, Vol. 716). Springer-Verlag, Berlin.
Scheunert, M. (1979b). *J. Math. Phys.* **20**, 712–720.
Scheunert, M. (1983a). *J. Math. Phys.* **24**, 2658–2670.
Scheunert, M. (1983b). *J. Math. Phys.* **24**, 2671–2680.
Scheunert, M. (1983c). *J. Math. Phys.* **24**, 2681–2688.
Scheunert, M. (1984). In *Differential Geometric Methods in Mathematical Physics, Jerusalem, 1982* (ed. S. Sternberg), pp. 115–124 (Mathematical Physics Studies, Vol. 6). Reidel, Dordrecht.
Scheunert, M. (1985). In *Supersymmetry* (ed. K. Dietz, R. Flume, G. Von Gehlen and V. Rittenberg), pp. 1–44. Plenum Press, New York.
Scheunert, M., Nahm, W. and Rittenberg, V. (1976a). *J. Math. Phys.* **17**, 1626–1639.
Scheunert, M., Nahm, W. and Rittenberg, V. (1976b). *J. Math. Phys.* **17**, 1640–1644.
Scheunert, M., Nahm, W. and Rittenberg, V. (1977a). *J. Math. Phys.* **18**, 146–154.
Scheunert, M., Nahm, W. and Rittenberg, V. (1977b). *J. Math. Phys.* **18**, 155–162.
Schwarz, J. H. (1972). *Nucl. Phys.* **B46**, 61–74.
Schwarz, J. H. (1973). *Phys. Rep.* **8**, 269–335.
Schwarz, J. H. (1981). *Nucl. Phys.* **B185**, 221–232.
Schwarz, J. H. (1982a). Caltech Report CALT-68-906.
Schwarz, J. H. (1982b). *Phys. Rep.* **89**, 223–322.
Schwarz, J. H. (1984a). *Comments Nucl. Part. Phys.* **13**, 103–115.
Schwarz, J. H. (1984b). In *Supersymmetry and Supergravity '84* (ed. B. de Wit, P. Fayet and P. van Nieuwenhuizen), pp. 426–448. World Scientific, Singapore.
Schwarz, J. H. (ed.) (1985a). *Superstrings: The First 15 Years of Superstring Theory*, 2 vols. World Scientific, Singapore.
Schwarz, J. H. (1985b). In *Applications of Group Theory in Physics and Mathematical Physics* (ed. M. Flato, P. Sally and G. Zuckerman), pp. 117–138 (Lectures in Applied Mathematics, Vol. 21). American Mathematical Society, Providence, Rhode Island.
Schwarz, J. H. (1986). In *Superstrings and Supergravity* (ed. A. T. Davies and D. G. Sutherland), pp. 301–360. Scottish Universities Summer School in Physics, Edinburgh.
Schwarz, J. H. (1987). *Int. J. Mod. Phys.* **A2**, 593–644.
Schweber, S. S. (1961). *An Introduction to Relativistic Quantum Field Theory*. Harper and Row, New York.
Schwimmer, A. and Seiberg, N. (1987). *Phys. Lett.* **184B**, 191–196.
Sciarrino, A. and Sorba, P. (1986). *J. Phys.* **A19**, 2241–2248.
Segal, G. (1981). *Commun. Math. Phys.* **80**, 301–342.
Serganova, V. V. (1984). *Funct. Anal. Appl.* **17**, (3), 200–207 [*Funkt. Analis i ego Prilozh.* **17**, (3), 46–54 (1983)].
Serganova, V. V. (1985). *Math. USSR. Izv.* **24**, 539–551 [*Izv. Akad. Nauk SSSR. Ser. Mat.* **48**, 585–598 (1984)].
Serre, J.-P. (1966). *Algèbres de Lie Semi-simple Complexes*. Benjamin, New York.
Serre, J.-P. (1973). *A Course in Arithmetic*. Springer-Verlag, New York.
Sevrin, A., Troost, W. and Van Proeyen, A. (1988). *Phys. Lett.* **208B**, 447–450.

Sharp, R. T., Van der Jeugt, J. and Hughes, J. W. B. (1985). *J. Math. Phys.* **26**, 901–912.
Shore, G. M. (1984). *Nucl. Phys.* **B248**, 123–140.
Siegel, W. (1988). *An Introduction to String Field Theory.* World Scientific, Singapore.
Sierra, G. and Townsend, P. K. (1983). *Proceedings of the XIXth Winter School and Workshop of Theoretical Physics, Karpacz, Poland, Feb. 1983.* pp. 396–430. World Scientific, Singapore.
Sohnius, M. F. (1978). *Nucl. Phys.* **B138**, 109–121.
Sohnius, M. F. (1983). In *Proceedings of the XIXth Winter School and Workshop of Theoretical Physics, Karpacz, Poland, Feb. 1983.* pp. 520–532. World Scientific, Singapore.
Sohnius, M. F. (1985). *Phys. Rep.* **128**, 39–204.
Sohnius, M. F. and West, P. C. (1981). *Phys. Lett.* **100B**, 245–250.
Sohnius, M. F., Stelle, K. S. and West, P. C. (1980). In *Unification of the Fundamental Particle Interactions* (ed. S. Ferrara, J. Ellis and P. van Nieuwenhuizen), pp. 187–244. Plenum Press, New York.
Sohnius, M. F., Stelle, K. S. and West, P. C. (1981). In *Superspace and Supergravity. Proceedings of the Nuffield Workshop, Cambridge, 1980* (ed. S. W. Hawking and M. Roček), pp. 283–329. Cambridge University Press.
Sokatchev, E. (1975). *Nucl. Phys.* **B99**, 96–108.
Song, X. C. and Han, Q. Z. (1980). In *Proceedings of the 1980 Guangzhou Conference on Theoretical Particle Physics*, pp. 1528–1542. Kexue Chubanshe, Peking.
Song, X. C., Han, Q. Z. and Sun, H. Z. (1980). *Kexue Tongboo* **25**, 201–207 [*Kexue Tongboa* **25**, (3), 105–207 (1980)].
Sorba, P. and Torresani, B. (1988). *Int. J. Mod. Phys.* **A3**, 1451–1474.
Stelle, K. S. (1985). In *Supersymmetry* (ed. K. Dietz, R. Flume, G. Von Gehlen and V. Rittenberg), pp. 535–576. Plenum Press, New York.
Strathdee, J. (1987). *Int. J. Mod. Phys.* **A2**, 273–300.
Strominger, A. and Witten, E. (1985). *Commun. Math. Phys.* **101**, 341–362.
Sudbery, A. (1983). *J. Math. Phys.* **24**, 1986–1988.
Sugawara, H. (1968). *Phys. Rev.* **170**, 1659–1662.
Sun, H. Z. and Han, Q. Z. (1981). *Scientia Sinica* **24**, 914–923.
Sun, H. Z. Yu, Y. Q. and Han, Q. Z. (1980). In *Proceedings of the 1980 Guangzhou Conference on Theoretical Particle Physics*, pp. 1543–1556. Kexue Chubanshe, Peking.
Synge, J. L. (1960). *Relativity: The General Theory.* North-Holland, Amsterdam.
Taylor, J. G. (1980). *Nucl. Phys.* **B169**, 484–500.
Taylor, J. G. (1982). In *Supergravity '81* (ed. S. Ferrara and J. G. Taylor), pp. 17–46. Cambridge University Press.
Thierry-Mieg, J. (1983). "Table of irreducible representations of the basic classical Lie superalgebras osp(m/n), su(m/n), ssu(n/n), $D(2/1, \alpha)$, $G(3)$, $F(4)$". Lawrence Berkeley Laboratory Report LBL 19353.
Thierry-Mieg, J. (1984a). *Phys. Lett.* **138B**, 393–396.
Thierry-Mieg, J. (1984b). In *Group Theoretical Methods in Physics, Trieste, 1983* (ed. G. Denardo, G. Ghirardi and T. Weber), pp. 94–98 (Lecture Notes in Physics, Vol. 201). Springer-Verlag, Berlin.
Thorn, C. B. (1984a). *Nucl. Phys.* **B248**, 551–569.
Thorn, C. B. (1984b). In *Vertex Operators in Mathematics and Physics* (ed. J. Lepowsky, S. Mandelstam and I. M. Singer), pp. 411–418 (Mathematical Sciences Research Institute, Vol. 3). Springer-Verlag, New York.
Todorov, I. T. (1985). *Phys. Lett.* **153B**, 77–81.
Tsohantjis, I. and Cornwell, J. F. (1989). To be published.

Van der Jeugt, J. (1987a). *J. Phys.* **A20**, 809–824.
Van der Jeugt, J. (1987b). *Lett. Math. Phys.* **14**, 285–292.
Van der Waerden, B. L. (1974). *Group Theory and Quantum Mechanics.* Springer-Verlag, Berlin and New York.
Van Holten, J. W. and Van Proeyen, A. (1982). *J. Phys.* **A15**, 3763–3783.
Van Proeyen, A. (1987). In *Proceedings of the Workshop on "Super Field Theories"* (ed. H. C. Lee, V. Elias, G. Kunstatter, R. B. Mann and K. S. Viswanathan), pp. 547–556. Plenum Press, New York.
Vasiliev, M. A. (1987). *Fortschr. Phys.* **35**, 741–770.
Vasiliev, M. A. (1988). *Nucl. Phys.* **B307**, 319–347.
Virasoro, M. A. (1970). *Phys. Rev.* **D1**, 2933–2936.
Volkas, R. R. and Joshi, G. C. (1988). *Phys. Rep.* **159**, 303–386.
Volkov, D. V. (1987). *Ukrain. Fiz. Zh.* **32**, 1782–1794.
Volkov, D. V. and Akulov, V. P. (1973). *Phys. Lett.* **46B**, 109–110.
Wakimoto, M. (1986). *Commun. Math. Phys.* **104**, 605–609.
Wallach, N. R. (1987). *Math. Z.* **196**, 303–314.
Wang, X. D. (1988). *Kexue Tongbao* **33**, 1152–1155.
Waterson, G. (1986). *Phys. Lett.* **171B**, 77–80.
Weinberg, S. (1964a). *Phys Rev.* **133**, B1318–B1332.
Weinberg, S. (1964b). *Phys. Rev.* **133**, B882–B896.
Weinberg, S. (1969). *Phys. Rev.* **181**, 1893–1899.
Wess, J. and Bagger, J. (1983). *Supersymmetry and Supergravity.* Princeton University Press.
Wess, J. and Zumino, B. (1974a). *Nucl. Phys.* **B70**, 39–50.
Wess, J. and Zumino, B. (1974b). *Phys. Lett.* **49B**, 52–54.
Wess, J. and Zumino, B. (1974c). *Nucl. Phys.* **B78**, 1–13.
West, P. C. (1986a). In *Superstrings and Supergravity* (ed. A. T. Davies and D. G. Sutherland), pp. 125–208. Scottish Universities Summer School in Physics, Edinburgh.
West, P. C. (1986b). In *Superstrings and Supergravity* (ed. A. T. Davies and D. G. Sutherland), pp. 301–360. Scottish Universities Summer School in Physics, Edinburgh.
West, P. C. (1986c). *Introduction to Supersymmetry and Supergravity.* World Scientific, Singapore.
Weyl, H. (1925). *Math. Z.* **23**, 271–309.
Weyl, H. (1926a). *Math. Z.* **24**, 328–376.
Weyl, H. (1926b). *Math. Z.* **24**, 377–395.
Wigner, E. P. (1937). *Ann. Math.* **40**, 149–204.
Williams, D. and Cornwell, J. F. (1984). *J. Math. Phys.* **25**, 2922–2932.
Windey, P. (1986). *Commun. Math. Phys.* **105**, 511–518.
Witten, E. (1985). *Nucl. Phys.* **B258**, 75–110.
Witten, E. (1986). *Nucl. Phys.* **B268**, 79–112.
Witten, E. and Olive, D. I. (1978). *Phys. Lett.* **78B**, 97–101.
Wybourne, B. G. (1984). *J. Phys.* **A17**, 1573–1578.
Yamron, J. P. and Siegel, W. (1986). *Nucl. Phys.* **B263**, 70–92.
Yang, M. and Wybourne, B. G. (1986). *J. Phys.* **A19**, 2003–2017.
Yang, S. K. (1988). *Phys. Lett.* **209B**, 242–246.
Yu, M. (1987a). *Phys. Lett.* **196B**, 345–348.
Yu, M. (1987b). *Nucl. Phys.* **B294**, 890–904.

Zamolodchikov, A. B. and Fatteev, V. A. (1986). *Zh. Exp. Teor. Fiz.* **90**, 1553–1566.
Zeng, G. J. (1987). *J. Phys.* **A20**, 5425–5434.
Zhang, Y. Z. (1987). *J. Phys.* **A20**, 3099–3102.
Zumino, B. (1977). *Nucl. Phys.* **B127**, 189–201.
Zwirner, F. (1988). *Int. J. Mod. Phys.* **A3**, 49–161.

Subject Index